Environmental Microbiology

Environmental Microbiology

RG BUCKLEY

CBSPD

CBS Publishers & Distributors Pvt Ltd

New Delhi • Bengaluru • Chennai • Kochi • Kolkata • Lucknow • Mumbai

Hyderabad • Jharkhand • Nagpur • Patna • Pune • Uttarakhand

Environmental Microbiology

ISBN: 978-81-239-2833-3

First Edition: 2016
Reprint: 2019, 2023

Published by **Satish Kumar Jain** and produced by **Varun Jain** for

CBS Publishers & Distributors Pvt Ltd

4819/XI Prahlad Street, 24 Ansari Road, Daryaganj, New Delhi 110 002, India.
Ph: 011-23289259, 23266861, 23266867   Website: www.cbspd.com
Fax: 011-23243014   e-mail: delhi@cbspd.com;

Corporate Office: 204 FIE, Industrial Area, Patparganj, Delhi 110 092
Ph: 011-4934 4934   Fax: 011-4934 4935   e-mail: publishing@cbspd.com;
                                                  publicity@cbspd.com

Branches

- **Bengaluru:** Seema House 2975, 17th Cross, KR Road, Banasankari 2nd Stage, Bengaluru 560 070, Karnataka, India
  Ph: +91-80-26771678/79   Fax: +91-80-26771680   e-mail: bangalore@cbspd.com
- **Chennai:** 7, Subbaraya Street, Shenoy Nagar, Chennai 600 030, Tamil Nadu, India
  Ph: +91-44-26680620, 26681266   Fax: +91-44-42032115   e-mail: chennai@cbspd.com
- **Kochi:** 42/1325, 1326, Power House Road, Opp KSEB, Power House, Ernakulum Kochi 682 018, Kerala, India
  Ph: +91-484-4059061-65,67   Fax: +91-484-4059065   e-mail: kochi@cbspd.com
- **Kolkata:** 147, Hind Ceramics Compound, 1st Floor, Nilgunj Road, Belghoria, Kolkata-700056, West Bengal, India
  Ph: +033-25633055, 033-25633056   e-mail: kolkata@cbspd.com
- **Lucknow:** Basement, Khushnuma Complex, 7 Meerabai Marg (Behind Jawahar Bhawan),Lucknow-226001, UP, India
  Ph: +0522-4000032   e-mail: tiwari.lucknow@cbspd.com
- **Mumbai:** PWD Shed, Gala no 25/26, Ramchandra Bhatt Marg, Next to JJ Hospital Gate no. 2, Opp. Union Bank of India, Noorbaug, Mumbai-400009, Maharashtra, India
  Ph: 022-66661880/89   e-mail: mumbai@cbspd.com

Representatives

- Hyderabad   0-9885175004   • Jharkhand   0-9811541605   • Nagpur   0-9421945513
- Patna   0-9334159340   • Pune   0-9923910676   • Uttarakhand   0-9716462459

Printed at Glorious Printer, Dilshad Garden, Delhi, India

Preface

Microbiology is the study of living organisms of microscopic size, which include bacteria, fungi, virus, protozoa and algae. Micro-organism and their activities are increasingly central to many of the concerns of society. The problems of global environment, the need to recycle natural resources, the discovery of recombinant DNA and the resulting high technology of genetic engineering have placed microbiology in the limelight.

In industrial microbiology, micro-organisms are used to make products such as antibiotics, vaccines, steroid, alcohols, organic solvents, vitamins, amino acids, and enzymes and hormones. Micro-organisms leach valuable minerals from low-grade ores. Microbial ecologists are employing micro-organisms in bioremediation and resource development, waste-water treatment, etc. Thus, micro-organisms play a significant role in the production of biofertilisers, biopesticides, biofuels, etc.

Environmental microbiology is blessed, or possibly cursed, with a multitude of terms that often are contradictory, sometimes misused, and frequently misunderstood even by the specialists. The ability to use language properly is an art and helps us to understand one another. But semantic sensitivity is frequently not the hallmark of the experimentalist, and this lack of sensitivity often is the basis for confusion and disagreement.

Environmental microbiology has a much wider scope. It is more than the study of pathogens and bioremediation. There are many different areas of this field. These include the discovery and identification of new microbes and their products that may have practical application for protection of environment, human health, and various commercial applications.

Thus different areas and microbial issues of the modern environmental microbiology are: aero-microbiology, biogeochemistry, agriculture and soil microbiology, water quality, waste-water treatment, biotechnology, resources production, resource recovery.

Since environmental micro-organism affect so many aspects of life and are easily transported between environments, the field of environmental microbiology interfaces with a number of subspecialities, including soil, aquatic and aeromicrobiology, as well as bioremediation, water quality, occupational health, and infection control, food safety and industrial microbiology. The subject matter of reference cum textbook on Environmental microbiology is accommodated in 38 chapters under nine sections.

Section I deals with general considerations. Chapter 1 deals with fundamentals of microbiology. The kinds of micro-organisms encountered in the biosphere are as varied and diverse as the environment present on the earth. Various organisms such as bacteria, eucarya, archaea, viruses, algae, fungi and protozoa are briefly discussed. Chapter 2 is devoted to neighbourhoods and community involvement. The chapter explains that, about the most important thing which must be understood about micro-organism in their environment is that no microbe exists by itself.

Section II concentrates on the general methodology of environmental microbiology. Chapter 3 focuses on general methodology: a review and introduces the readers that no microbe is an island and explains the important things as to how micro-organisms interact among themselves and with their surroundings. Chapter 4 is devoted to prokaryotic diversity: form, ecophysiology and habitat. This chapter not only discusses the diversity of microbes and their functions in the more usual environmental sense, but also includes a focus on microbes associated, either extracellularly on intracellularly, with macrobes, including animals, insects and plants.

Section III focuses on water microbiology in public health. Chapter 5 concentrates on water microbiology. Waterborne transmission is a highly effective means of spreading infections agents to a large portion of the population. Considering this chapter 6 is devoted to waterborne transmission of infectious agents. Chapter 7 explains the detection of micro-organisms in environmental freshwaters and drinking waters. Chapter 8 focuses on detection of protozoan parasites in source and finished drinking waters. Protozoa are unicellular bound genetic material or nuclei and other assorted cellular organelles. Chapter 9 is devoted to detection of bacteria, viruses and parasitic protozoa in shellfish. This chapter discusses those shellfish that are edible bivalve molluscs of the class Pelecypoda and include the species commonly referred to as oysters, mussels, clams and cockles. Chapter 10 concentrates on control of micro-organisms in source water and drinking water. Pathogenic micro-organisms frequently can be found in untreated water and, as a result, can cause infections which may lead to illness and even to death among those people who ingest that water. To reduce these risks, drinking water often is treated on a community basis. Considering this chapter 11 deals with estimating the risk of infectious disease associated with pathogens in drinking water. Chapter 12 is devoted to methods of enumeration and activities of micro-organisms. Various methods discussed are cultural, physiological and immunological.

Section IV focuses on waste-water treatment. A great many different groups of viruses are found in environmental water which represent exogenous contaminants, whose hosts are nonaquatic plants and animals. Keeping this in mind chapter 13 deals with detection of viruses in environmental water, sewage and sewage sludges. Chapter 14 concentrates on detection of bacterial pathogens in waste-water and sludge. The number and variety of bacteria present in waste-water and associated solids are legion. Their sources are the excrement of humans and animals, and other waste material find their way into sewage, and the microbial flora in the source water. Various traditional and molecular biological methods for the detection and enumeration of bacterial pathogens are discussed.

Section V is devoted to aquatic and extreme environments. Cultivation of communities and consortia provides an understanding of the community level mechanisms needed for specific environmental processes to occur. Aquatic microbiology in broad terms is the study of micro-organisms and microbial communities in water environment. Chapter 15 deals with basic concepts of aquatic and extreme environment. Primary production is the production of organic compounds from atmospheric or aquatic carbon dioxide, principally through the processes of photosynthesis with chemosynthesis being much less important. Considering this chapter 16 focuses on primary productivity and producers. Chapter 17 concentrates on bacterial secondary productivity. Chapter 18 discusses community structure: Bacterial and Archaea. Community structure is generally considered to be information related to the types of organisms present in the environment and the relative proportions of those types. Chapter 19 deals with protistan community structure. Protistan assemblages of aquatic ecosystems have become the focus of a concentrated research effort in aquatic ecology. Various methods of assessing protistan community structure and molecular approaches are discussed. Chapter 20 focuses on phagotrophy in aquatic microbial food webs. Determination of predator-prey interactions and pathways of energy and elemental flows within microbial food web, in vital to understanding the structure and functioning of ecosystems. Chapter 21 deals with biogeochemical cycles. Various geochemical cycles such as carbon cycle, nitrogen cycle, phosphorus cycle, and sulphur cycles are discussed in detail.

Section VI focuses on soil, subsurface and landfill. The microbiology of landfills plays an important role as this waste option is the most popular to date and likely to remain so for the immediate future. Chapter 22 concentrates on groundwater chemistry and subsurface microbiology. Chapter 23 explains the microbial ecology of groundwater systems. The chapter discusses the scope of subsurface microbial ecology, microbial diversity and niches in aquifers systems, and population interactions. The study of microbial processes and how they effect groundwater geochemistry begins with consideration of the abundance and distribution of bacteria. Considering this chapter 24 is devoted to abundance and distribution of bacteria in the subsurface. Chapter 25 focuses on microbiological sampling of subsurface environments.

Subsurface environments, whether they are shallow local flow systems or deep regional flow systems, share a common characteristics. Chapter 26 deals with microbial acclimation to groundwater contamination. The chapter discusses the microbial response to environmental changes, mechanism of acclimation and factors affecting microbial acclimation.

Section VII is devoted to aeromicrobiology which in its wider sense is the study of all forms microbial life in air. To an environmental microbiologist, aeromicrobiology is the study of various aspects of intramural (indoor) and extramural (outdoor) aerobiology in relation with the airborne transmission of environmentally relevant micro-organisms including viruses, bacteria, fungi, yeasts and protozoans. Besides these microbes, other biological materials like pollens, insect debris, animal danders (chief source of allergic disorders) are also found in air. Chapter 27 focuses on aeromicrobiology: a review. Chapter 28 deals with bioaerosol control and biosafety in laboratory. The chapter discusses the various mechanisms used to control bioaerosol such as ventilation, filtration, UV treatment, biocidal agents and physical isolation. Chapter 29 and 30 concentrates on beneficial and pathogenic microbes in agriculture and airborne *Mycobacterium* spp.

Section VIII focuses on biotransformation and biodegradation. Biodegradation is the process of breakdown of generally complex, organic pollutants (contaminants) to smaller, simpler products by the activities of micro-organisms. Chapter 31 is devoted to biodegradation: a review. Chapter 32 concentrates on biodegradation of phenol. Chapter 33 deals with biodegradation and bioremediation of halogenated organic compounds. Chapter 34 focuses on microbial fertilisers. Chapter 35 deals with microbial production of biosurfactants.

Section IX is devoted to special topics. Chapter 36 deals with environmentally friendly products and processes of the 21st century. Chapter 37 focuses on microbial production of energy from industrial waste. Chapter 38 focuses on patents and secret processes.

This reference textbook '*Environmental Microbiology*' is designed to fulfil the requirements of undergraduates and postgraduates in the disciplines of environmental microbiology, microbial ecology, biotechnology, biochemical engineering. The book will also useful for scientists, researchers or professionals interested in exploring the role of micro-organisms in public health and waste-water engineering.

Glossary and index have been provided at the end for quick reference. Diagrams, figures and tables supplement the text. All the topics have been covered into a cogent and lucid style to help the reader grasp the information quickly and easily.

RG Buckley

Contents at a Glance

Preface   v

SECTION I

General Considerations   **1–41**

1. Fundamentals of Microbiology   3–22
2. No Microbe is an Island   23–41

SECTION II

General Methodology   **43-66**

3. General Methodology: A Review   45–50
4. Prokaryotic Diversity: Form Ecophysiology and Habitat   51–66

SECTION III

Water Microbiology in Public Health   **67–202**

5. Water Microbiology   69–71
6. Waterborne Transmission of Infectious Agents   72–94
7. Detection of Micro-organisms in Environmental Freshwaters and
   Drinking Waters   95–111
8. Detection of Protozoan Parasites in Source and Finished Drinking Waters   112–130
9. Detection of Bacteria, Viruses and Parasitic Protozoa in Shellfish   131–144
10. Control of Micro-organisms in Source Water and Drinking Water   145–167
11. Estimating the Risk of Infectious Disease Associated with Pathogens in
    Drinking Water   168–178
12. Methods of Enumeration and Activities of Micro-organism   179–202

SECTION IV

Waste-water Treatment   **203-232**

13. Detection of Viruses in Environmental Water, Sewage and Sewage Sludges   205–219
14. Detection of Bacterial Pathogens in Waste-water and Sludge   220–232

SECTION V

Aquatic and Extreme Environment   *233–361*

15. Basic Concepts of Aquatic and Extreme Environment   235–260
16. Primary Productivity and Producers   261–278
17. Bacterial Secondary Productivity   279–288
18. Community Structure: Bacteria and Archaea   289–297
19. Protistan Community Structure   298–311
20. Phagotrophy in Aquatic Microbial Food Webs   312–320
21. Biogeochemical Cycling   321–361

SECTION VI

Soil, Subsurface and Landfill   *363–434*

22. Groundwater Chemistry and Subsurface Microbiology   365–370
23. Microbial Ecology of Groundwater Systems   371–392
24. Abundance and Distribution of Bacteria in the Subsurface   393–409
25. Microbiological Sampling of Subsurface Environments   410–421
26. Microbial Acclimation to Groundwater Contamination   422–434

SECTION VII

Aeromicrobiology   *435–481*

27. Aeromicrobiology: A Review   437–455
28. Bioaerosol Control and Biosafety in Laboratory   456–464
29. Beneficial and Pathogenic Microbes in Agriculture   465–474
30. Airborne *Mycobacterium* Spp.   475–481

SECTION VIII

Biotransformation and Biodegradation   *483–547*

31. Biodegradation: A Review   485–490
32. Biodegradation of Phenol   491–502
33. Biodegradation and Bioremediation of Halogenated Organic Compounds   503–510
34. Microbial Fertilisers   511–534
35. Microbial Production of Biosurfactants   535–547

SECTION IX

Special Topics   *549–575*

36. Environmental Friendly Products and Processes for the 21st Century   551–559
37. Microbial Production of Energy from Industrial Waste   560–566
38. Patents and Secret Processes   567–575

   Glossary   577–596
   References   597
   Index   599–605

Contents

Preface   v
Contents at a Glance   ix

SECTION I

General Considerations   *1–41*

1. Fundamentals of Microbiology   **3–22**

Introduction   3
Bacteria   3
    Bacteria and its Growth   6
    Bacterial Respiration   8
    Reproduction and Growth of Bacteria   8
    Factors Affecting Bacteria   9
    Germicides and Bacteriostatic Agents   10
    Antimetabolite   10
    Bacteria Identification   11
    Bacteriology   11
    Multiple Fermentation Tube Method   12
    Most Probable Number (MPN)   12
    Membrane Filter Method   13
Eucarya   13
Archaea   15
Viruses   16
    Viral Ecology   17
Algae   18
Fungi   19
Protozoa   19
    Characteristics   19
Environmental Microbiology   20

2. No Microbe is an Island   **23–41**

Introduction   23
Neighbourhood, Cliques and Cliches   24
Understanding the Nature of Community Life   27
    Involvement and Homeostasis are Evolutionary Consequences   28
    Being Kind and Helpful   29
    Using and Abusing   29
    Hunters and the Hunted   31
Why They are Where They are When They are There   33
    Boundaries and Barriers   33
    Staying at Home versus Venturing to New Places   34

Travelling by Air, Sea or Land 35
Prepared for the Climate and What about the Location? 36
Nutrients in the Food? 38
Settling in with Relatives (In-Laws and Outlaws) 39
Understanding Ecology 41

SECTION II

General Methodology 43–66

3. General Methodology: A Review 45–50

Introduction 45
Challenges and New Tools 45
  Microscopy 46
  Pure Culture Approaches 46
  Community-Level Physiological Profile (CLPP) 48
  Problem with Interpretation of Biochemical Mass 48
  Nucleic Acid Sequence Analysis 48
  Microbiological, Chemical and Molecular Tools 49
  Statistical Techniques 49

4. Prokaryotic Diversity: Form Ecophysiology and Habitat 51–66

Introduction 51
Recognition of Prokaryote Diversity: Background 52
Limitations in Comprehension of Prokaryote Diversity 52
Factors Affecting Limitations in Our Understanding 53
Morphologies and Sizes of Prokaryotic Cells 55
Physiological Diversity Among the Prokaryotes 57
Diversity: Insights from Macromolecular Analyses 62
Aspects of Behavioural Diversity Among Prokaryotes 64
Diversity of Habitats Exploited for Growth of Prokaryotes 65
  Enlarging Perspectives 66

SECTION III

Water Microbiology in Public Health 67–202

5. Water Microbiology 69–71
Introduction 69
Human Diseases Associated with Water 69
Routes by which Human Acquire Waterborne Infections 70

6. Waterborne Transmission of Infectious Agents 72–94

Introduction 72
Classification of Water-related Diseases 77
Study of Waterborne Diseases 80
  Epidemic and Endemic Waterborne Disease 80

Microbial Agents Associated with Waterborne Disease 81
  Enteric and Aquatic Bacteria 81
  Enteric Viruses 83
  Enteric Protozoa 84
Emerging Waterborne Pathogens 85
  Recently Recognised and Suspected Waterborne Pathogens 85
  Recently Recognised Waterborne Pathogens in Developing Countries 90
  Waterborne Disease of Unknown Etiology 91
Prevention and Control of Waterborne Disease 92
  Measurements of Microbiological Water Quality 92
  Multiple-barrier Approach 93
  Risk Assessment Approaches 94
  Recommendations and Research Needs 94

7. Detection of Micro-organisms in Environmental Freshwaters and Drinking Waters **95–111**

Introduction 95
Commonly Used Indicators 96
  Total Coliforms 96
  Thermotolerant Coliforms (Fecal Coliforms) 96
  Escherichia Coli 97
  Fecal Streptococci and Enterococci 97
  Staphylococci 98
  Bacteriophages 98
Other Commonly Used Indicators 99
  Sulphite-reducing Clostridia 99
  Pseudomonas spp. 99
  Mycobacteria 99
  H_2S Producers 100
  Aeromonas spp. 100
  HPC 101
  Sterols and Other Chemical Indicators of Sewage Contamination 101
  FC/FS Ratio 101
Significance of Indicators to Public Health 102
  Current Regulations Regarding Drinking and Surface Waters 102
General Methods 103
  Sample Collection, Transport and Storage 103
  MPN (Multiple-Tube Technique) 103
  MF 105
  Presence-Absence (P-A) 107
  Enzyme-Specific Tests 108
  Bacteriophages 108
  Rapid Tests 109
  Recovery of Injured Bacteria . 110
  Molecular Methods 110

Surface Freshwaters		111
Drinking Waters		111

8. Detection of Protozoan Parasites in Source and Finished Drinking Waters — **112–130**

Introduction	112
Biology and Ecology	113
Detection and Quantitation	114
Free-Living Protozoa	114
Parasitic Protozoa	115
Emerging Protozoans of Potential Health Significance in Water	128

9. Detection of Bacteria, Viruses and Parasitic Protozoa in Shellfish — **131–144**

Introduction	131
Etiology of Mollusc-Associated Outbreaks of Illness	131
Pathogenic Bacteria Transmitted by Bivalve Molluscs	132
Enteric Viruses Transmitted by Bivalve Molluscs	132
Enteropathogenic Protozoa	134
Methods for the Detection of Pathogens in Shellfish	134
General Steps in the Isolation and Detection of Pathogens from Shellfish	134
Conventional Methods for the Detection of Bacteria in Shellfish	136
Conventional Methods for the Detection of Enteric Viruses in Shellfish	136
Conventional Methods for the Detection of Human Protozoan Parasites in Shellfish	137
Molecular Methods for the Detection of Pathogens in Shellfish	137
Barriers to the Application of Molecular Methods for Detection of Pathogens in Shellfish	143
Research Needs for the Detection of Pathogens in Shellfish	144

10. Control of Micro-organisms in Source Water and Drinking Water — **145–167**

Introduction	145
Choosing Water Sources and Watershed Protection	145
Surface Water Sources	145
Groundwater	146
Watershed Protection	146
Treatment and Disinfection	148
Methods of Disinfection Commonly Used in Water Treatment	149
Kinetics of Disinfection	150
Chemical Methods	150
Chlorine	150
Control of Gas Chlorination	152
Sodium Hypochlorite	154
Disinfection with Chlorine Dioxide	157
Generation of Chlorine Dioxide	157
Sodium Chlorite	158
Advanced Oxidation Process	159
Disinfection with Ozone	159
Physical Methods	162
Disinfection with Ultraviolet Light	162

11. Estimating the Risk of Infectious Disease Associated with Pathogens in Drinking Water   **168–178**

Introduction   168
Nature of the Hazards Associated with Pathogenic Micro-organisms in Aquatic
    Environment   169
Routes of Pathogen Acquisition   170
Defining the Main Variables Incorporated in Risk Estimates for the Level of Infectious
    Disease Associated with Ingestion of Water   170
    Defining the Exposure Level   170
    Defining the Probability of Infection   170
    Defining the Probability of Illness   173
    Defining the Probability of Death   173
Calculating the Risk Estimation   175
    Format of the Equations   175
    Validation of the Risk Estimation Technique   176
    Caveat Emptor Principle   177

12. Methods of Enumeration and Activities of Micro-organism   **179–202**

Introduction   179
Cultural Methods   179
    Cultural Methods for Enumeration of Bacteria   179
    Cultural Media for Bacteria   182
    Cultural Methods for Fungi   185
    Cultural Methods for Algae and Cyanobacteria   185
    Cell Culture-Based Detection Methods for Viruses   186
Physiological Methods   187
    Microbial Activity Measurements in Pure Culture   187
    Microbial Activity Measurements in Environmental Samples   188
Immunological Methods   192
    Antibody   193
    Immunoassays   194
Nucleic Acid-Based Methods of Analysis   199
    Nucleic Acid-Based Methods   200

SECTION IV

Waste-water Treatment   *203–232*

13. Detection of Viruses in Environmental Water, Sewage and Sewage Sludges   **205–219**

Introduction   205
Mechanisms Involved in Viral Concentration Methods   206
    Passive Adsorption   206
    Directed Adsorption   206
    Ultrafiltration   206
    Direct Physico-chemical Flocculation and Phase Separation   207
    Affinity Chromatography   207

Directed Adsorption, Elution and Reconcentration of Viruses   207
  Sampling Apparatuses and Processing Equipment   209
  Cartridge Filtration Apparatus   209
  Sterilisation Requirements   211
  Techniques for Field Sterilisation of Sampling Equipment   211
Detection of Viruses   212
  Viral Concentration Techniques   213
Viral Assay Techniques   219

14. Detection of Bacterial Pathogens in Waste-water and Sludge   **220–232**

Introduction   220
Traditional Methods for the Detection and Enumeration of Bacterial Pathogens 221
  Sampling Requirements, Transportation and Preservation   221
  Enrichment and Isolation   222
  Salmonella Detection   222
  Salmonella Isolation Media   223
  Other Bacterial Pathogens   226
Molecular Biological Methods   227
  Polymerase Chain Reaction (PCR)   227
  Fluorescent Antibodies (FA) Techniques   231
  Enzyme-Linked Immunosorbent Assays (ELISA)   231

SECTION V

Aquatic and Extreme Environment   *233–361*

15. Basic Concepts of Aquatic and Extreme Environment   **235–260**

Introduction   235
Microbial Habitats in the Aquatic Environment   235
  Planktonic Environment   235
  Microbial Mats   239
  Biofilms   240
  Aquatic Environments   243
  Brackish Water   249
  Marine Water   250
  Subterranean Water   255
Environmental Determinants that Govern Extreme Environments   255
  Air-Water Interface   255
  High Temperature   255
  High Solute   256
  Low pH   257
  High Pressure   257
  No Nutrient   258
Aquatic Microbes in the News   258
  Giant Marine Bacteria   258
  Aquatic Microbes: Food for the Future   259

16. Primary Productivity and Producers   261–278

Introduction   261
Process of Primary Production and Relevant Micro-organisms   261
Ecological and Biogeochemical Significance   262
Methods for Measuring Primary Production in Aquatic Habitats   264
    CO_2 Uptake by Gas Analysis   264
    Uptake of ^{14}C-Labelled CO_2 265

17. Bacterial Secondary Productivity   279–288

Introduction   279
Thymidine (TdR) Incorporation into DNA   281
    Rationale, Advantages and Disadvantages   281
    Procedures for Estimating BBP from Incorporation Rates of TdR   284
Leucine Incorporation into Protein   285
    Rationale, Advantages and Disadvantages   285
Experiments to Determine Empirical Conversion Factors for the TdR and Leu Methods   286
Adenine Incorporation into DNA and RNA   286
    Rationale, Advantages and Disadvantages   286
Other Methods   287

18. Community Structure: Bacteria and Archaea   289–297

Introduction   289
Microscopy   289
Culturing   290
Immunological Approaches   290
Lipid Analysis   291
Low-Molecular-Weight RNA Profiles   291
DNA-DNA Hybridisation   291
Other DNA-Based Approaches   292
16S rRNA-Based Approaches   292
Probes   296
Probe-Microautoradiography Combination   297

19. Protistan Community Structure   298–311

Introduction   298
Assessing Protistan Community Structure   299
    Protistan Species Diversity   299
    Protistan Abundance and Biomass   301
    Describing Protistan Community Structure   301
Temporal and Spatial Changes in Community Structure   309
    Freshwater versus Marine Ecosystems   309
    Benthic versus Pelagic Ecosystems   309
    Depth and Seasonal Distributions   310
Molecular Approaches to Studying Protistan Community Structure   311

20. Phagotrophy in Aquatic Microbial Food Webs   312–320

Introduction   312
Trophic Interactions Among Aquatic Microbes   312

Bacterivory	313
Manipulation Methods	314
Prey Addition Methods	316
Digestive Enzyme Activity Approach	317
Protistan Herbivory	318
Other Trophic Pathways in Microbial Food Webs	320

21. Biogeochemical Cycling — **321–361**

Introduction	321
Biogeochemical Cycles	321
Gaia Hypothesis	322
Carbon Cycle	324
Carbon Reservoirs	324
Carbon Fixation and Energy Flow	325
Carbon Respiration	327
Microbial Nitrogen Cycling	328
N_2 Fixation	328
Nitrification	332
Dissimilatory Nitrate Reduction and Denitrification	335
Other Aspects	338
Sulphur Cycling	339
Sample Collection and Handling	339
Sulphate Reduction	340
Sulphur Oxidation	349
Inhibitors	352
Analyses	353
Phosphorus Cycling	353
Measurement of Phosphorus Uptake and Incorporation by Using ^{32}P and Differential Filtration	354
Alkaline Phosphate as an Indicator of Phosphorus Limitation	358

SECTION VI

Soil, Subsurface and Landfill — *363–434*

22. Groundwater Chemistry and Subsurface Microbiology — **365–370**

Introduction	365
Subsurface Microbiology and the Geosciences	366
Subsurface Microbiology and Microbial Ecology	366
Subsurface Microbiology and Contaminant Biodegradation	368

23. Microbial Ecology of Groundwater Systems — **371–392**

Introduction	371
Scope of Subsurface Microbial Ecology	372
Methods in Subsurface Microbial Ecology	376
Culture Methods	377
Direct Observation	377
Phospholipid Fatty Acid Analysis	378

Activity Measurements in Microcosms   378
Geochemical Methods   381
Molecular Microbial Ecology   381
Microbial Diversity and Niches in Aquifer Systems   383
Measurement of Diversity   383
Niches and Sources of Microbial Diversity   384
Stress and Microbial Diversity   384
Population Interactions   385
Neutralism   385
Commensalism   386
Synergism and Symbiosis   386
Competition   387
Antagonism, Parasitism and Predation   388
r and K Strategies in Microbial Ecosystems   389
r and K Strategies in the Aquifer Environment   390

24. Abundance and Distribution of Bacteria in the Subsurface   393–409

Introduction   393
Classification of Subsurface Environments   393
Unsaturated Zone   395
Unsaturated Zone as a Microbial Habitat   397
Biomass Measurements in Soil Microbiology   398
Distribution of Bacteria in the Unsaturated Zone   401
Local Flow Systems   402
Local Flow Systems as a Microbial Habitat   402
Distribution of Bacteria in Local Flow Systems   404
Intermediate Flow Systems   404
Intermediate Flow Systems as a Microbial Habitat   405
Distribution of Bacteria in Intermediate Flow Systems   405
Microbial Processes in Confining Beds   405
Regional Flow Systems   407
Early Observations from Petroleum Reservoirs   407
Distribution of Bacteria in Regional Flow Systems   407

25. Microbiological Sampling of Subsurface Environments   410–421

Introduction   410
Sampling the Unsaturated Zone   411
Hand Augering   411
Air Drilling and Coring   412
Sampling Local Flow Systems   412
Split-Spoon Sampling   412
Push-Tube (Shelby Tube) Sampling Methods   413
Direct Push Sampling   413
Aseptic Technique with Split-Spoon, Shelby Tube and Direct Push Sampling   413
Sampling Intermediate and Regional Systems   415
Mud-Rotary Drilling   415
Drilling Fluids   415
Mud-Rotary Coring   417

Drilling-fluid Contamination of Cored Sediments   418
    Down-hole Saturation Contamination   418
    Core Seepage Contamination   419
    Core-Fracture Contamination   419
    Evaluating Drilling Fluid Contamination   419
Sampling Groundwater for Micro-organisms   419

26. Microbial Acclimation to Groundwater Contamination   **422–434**

Introduction   422
Microbial Response to Environmental Changes   422
Mechanisms of Acclimation   424
    Induction   424
    Catabolite Repression   426
    Genetic Mutations   427
    Acclimation to Available Electron Acceptors   427
    Production of Proteins in Response to Chemical Stresses   427
Factors Affecting Microbial Acclimation   428
    Rates of Acclimation   428
    Concentration Effects   428
    Cross-Acclimation of Xenobiotic Compounds   429
    Chemical Structure of Xenobiotics   429
Acclimation to Xenobiotics in Groundwater Systems   429
    Acclimation Response in a Contaminated Aquifer   429
    Acclimation Response in Pristine Aquifer Sediments 430
    Acclimation of Eukaryotic Micro-organisms   430
    Acclimation in Bioremediation Technology   430
Acclimation to Metal Toxicity   431
    Metal Detoxification Mechanisms   431
    Plasmid-Encoded Metal Resistance Mechanisms   432
    Acclimation to Mercury Toxicity   433

SECTION VII

Aeromicrobiology   *435–481*

27. Aeromicrobiology: A Review   **437–455**

Introduction   437
Important Airborne Pathogens   437
Important Airborne Toxins   440
Nature of Bioaerosols   441
Atmosphere   442
Aeromicrobiological Pathway   442
    Launching   442
    Transport   443
    Deposition   444

Mathematical Modelling 446
  Point Source Modelling 446
  Area Source Modelling 446
  Indoor Air Modelling 446
Sampling Devices for the Collection of Bioaerosols 447
  Impingement 447
  Impaction 448
  Centrifugation 449
  Filtration and Deposition 450
Microbial Survival in the Air 451
  Relative Humidity 451
  Temperature 452
  Radiation 452
  Oxygen, OAF and Ions 452
Extramural Aeromicrobiology 453
  Agriculture 453
  Waste Disposal 454
  Germ Warfare 455

28. Bioaerosol Control and Biosafety in Laboratory **456–464**

Introduction 456
Method of Controlling Bioaerosols 456
  Ventilation 456
  Filtration 456
  Biocidal Control 457
  Isolation 457
Biosafety in the Laboratory 458
  Biological Safety Cabinets 458
  Biosafety Laboratories 463

29. Beneficial and Pathogenic Microbes in Agriculture **465–474**

Introduction 465
Soil–Plant–Micro-organism System 465
Rhizosphere Environment 465
Organic Compounds Released by Plants 467
  Exudates, Secretions, and Lysates 468
  Root Border Cells and Mucigel 468
  Factors Affecting the Release of Compounds 469
Rhizosphere Populations 469
  Microflora 469
  Microfauna 470
Beneficial Root-Microbial Interactions 470
  Biological Dinitrogen Fixation 470
Pathogenic Microbes in Agriculture 471
  Plant Disease Caused by Fungi 471
  Diseases Caused by Bacteria 472
  Plant Diseases Caused by Viruses 472
  Soil Biological Control of Plant Diseases 473

30. **Airborne *Mycobacterium* Spp.** 475–481

Introduction 475
Bacteriology, Epidemiology and Pathogenesis of Mycobacterial Infections 475
  M. tuberculosis Complex 475
  Mycobacteria other than *M. tuberculosis* 477
Sampling Considerations 478
Sampling and Analytical Methods 478
  Sampling for Airborne Mycobacteria 478
  Analytical Methods 478

SECTION VIII

Biotransformation and Biodegradation 483–547

31. **Biodegradation: A Review** 485–490

Introduction 485
Complete Biodegradation 485
  Partial Biodegradation 485
Environmental Factors Affecting Biodegradation 487
  Oxygen 488
  Organic Matter Content 488
  Nitrogen 489
  Other Environmental Factors 490

32. **Biodegradation of Phenol** 491–502

Introduction 491
Materials and Methods 492
  Micro-organisms 492
  Nitrogen Concentration 492
  Sensitivity to Phenol 492
  Adaptation of the Test Fungi to Phenol 492
  Age of Cultures 493
  Analytical Methods 493
  Phenol Biodegradation 493
Results and Discussion 493
  Sensitivity to Phenol 493
  Effect of Adaptation 494
  Effect of Nitrogen Concentration on Phenol Degradation 497
  Effect of Culture Age 498
  Effect of Glucose Concentration 498
  Effect of Carbon Source 501
  Relationship between Growth and Degradation of Phenol 501

33. **Biodegradation and Bioremediation of Halogenated Organic Compounds** 503–510

Introduction 503
Microbial Degradation of Halogenated Organic Compounds 504
  Chlorinated Ethenes 504
  Chlorinated Benzenes 506

Polychlorinated Biphenyls 507
Organochlorine Insecticides 508
Chlorinated Herbicides 508
Chlorinated Phenols 509

34. Microbial Fertilisers **511–534**

Introduction 511
Nitrogen: Preparation and Usage 515
Manufacture of *Rhizobium* Inoculants 517
Soil + Farmyard Manure-based Cultures 520
Quality Control of Legume Inoculants 520
Methods of Application of Nitragin (*Rhizobium* Inoculants) 523
Problems in Legume Establishment 525
Azotobacterin 528
Preparation and Usage 528
Field Experiments (Usage) 529
Blue-Green Algae 531
Preparation of Blue-green Algae Inoculants 531
Method of Application of Blue-green Algae 531
Nature and Production of Extra-cellular Substances 532
Role of Blue-green Algae in Rice Fields 533
Phosphobacterin 533
Preparation of Inoculants of Phosphate Solubilising Bacteria 533
Field Experiments (Usage) 533
Media Used for Rhizobium Culture 534

35. Microbial Production of Biosurfactants **535–547**

Introduction 535
Types of Biosurfactants 536
Glycolipids 536
Phospholipids 537
Lipopeptides and Lipoproteins 537
Polymeric 537
Experimental Procedure 537
Production of Biosurfactants 537
Rapid Method for Monitoring Maximum Biosurfactant Production Obtained by Acetone
Precipitation 541
Isolation of the Biosurfactant 541
Estimation of Biomass 541
Chemical Analysis 541
Protein Estimation 541
Discussion and Conclusions 545

SECTION IX

Special Topics *549–575*

36. Environmental Friendly Products and Processes for the 21st Century **551–559**

Introduction 551

Environmental Friendly Products (Green Products)   553
   New Bioflocculant Produced by *Rhodococcus erythropolis*   553
   New Super Polysaccharide Bioabsorbent from *Alcaligenes latus*   555
Environmental Friendly Process (Green Process)   559

37. Microbial Production of Energy from Industrial Waste   **560–566**

Introduction   560
Anaerobic Digestion of Complex Substrates   560
Performance of Anaerobic Digesters   561
Conventional Process Configurations   562
   Limitations of Conventional Digestion and Process Improvement Approaches 563
   Novel Bioreactor Designs   563
   Two Phase Digestion—An Innovative Fermentation Mode   566
   Digestion of Preheated Feed and Use of Biostimulants   566
   Biomethanation of Industrial Wastes   566

38. Patents and Secret Processes   **567–575**

Introduction   567
Composition of a Patent   569
Subject Matter and Characteristics of a Patent   569
Who is the Inventer: Who Owns an Invention   571
Protection of the Rights of the Inventer: Infringement   573
Cost of a Patent   573
Possible Changes in United States Patent Law   574
Patents in Other Countries   574

Glossary   577–596

References   597

Index   599–605

SECTION I

General Considerations

1. Fundamentals of Microbiology   3

2. No Microbe is an Island   23

Chapter 1

Fundamentals of Microbiology

INTRODUCTION

The kinds of micro-organisms encountered in the biosphere are as varied and diverse as the kinds of environments present on earth. If one considers the diversity of potential habitats in the biosphere (surface sediments in freshwater or saltwater bodies, subsurface sediments in either aerobic or anaerobic environments, the water column of deep or shallow water bodies, hot hydrothermal waters, frozen sediments in the Arctic and Antarctic plains, extreme pressure in deep ocean waters, and the bodies of higher plants and animals), it's little wonder that micro-organisms display such astonishing diversity.

In spite of this diversity, there are only three primary lineages, referred to as domains, to which all life on earth (not just micro-organisms) belong. These domains are the Archaea the Bacteria, and the Eucarya. If we include viruses, which many microbiologists consider to be noncellular life-forms, there are just four basic kinds of micro-organisms to consider when studying groundwater systems.

The bacteria are easily distinguished from eucarya on the basis of cellular architecture. The bacteria, also called procaryotes, are characterised by the lack of a true nucleus (*pro* meaning early or primitive, and karyo meaning nucleus) and includes the bacteria and the cyanobacteria. Cyanobacteria were formerly called 'blue-green algae', but are now recognised as members of the Domain Bacteria. The eucaryotic cell has a true nucleus (*eu* meaning true) and includes algae, fungi, and protozoa.

The archaea, which include the methane-producing 'bacteria', are morphologically indistinguishable from bacteria and were once considered to be procaryotic micro-organisms. Studies in the last 30 years, however, have shown that archaea are not bacteria and that they represent a previously unrecognised domain. Archaea are restricted to anaerobic environments, such as organic-rich sediments and the intestines of higher animals, or hypersaline environments. Geologically, the archaea are important because they inhabit virtually all subsurface environments, are an important source of commercial methane, and greatly impact the chemistry of groundwater systems.

Viruses are distinct from other types of micro-organisms in that they are obligate parasites. That is, they do not have the capability to live and reproduce without having a host cell to provide energy. Viruses are important in subsurface microbiology primarily because groundwater may transport viruses to wells or other drinking water supplies and thus spread infectious diseases. It is almost certainly true that viruses use subsurface bacteria as hosts and therefore are probably present wherever bacteria are present.

BACTERIA

Bacterial micro-organisms are characterised by their distinctive and relatively simple cellular structure. Figure 1.1 shows the kinds of structures that are typically observed in bacteria.

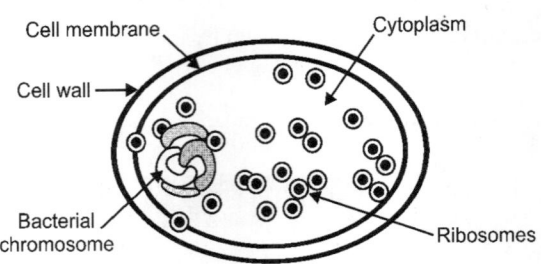

Fig. 1.1. Cellular structure of bacteria.

The bacterial chromosome consists of a single molecule of DNA that, in spite of many loops and twists, is arranged into a closed circle. This closed circle arrangement is found only in procaryotes and the archaebacteria. The DNA is otherwise identical to that of other organisms and is characterised by its double helix structure. The DNA carries genetic information needed for the cell to carry out metabolism and growth, as well as carrying the information needed for replication. Each bacterial chromosome consists of anywhere between 2000 and 10,000 units of heredity called genes. Genes are segments of the DNA strand that code for a particular protein or polypeptide. Most of the time, procaryotic cells have just one copy of their chromosomal DNA and are, therefore, referred to as being haploid. Just before cell division, however, two or more copies of the bacterial chromosome may be present. These masses of DNA are sometimes visible under the microscope and are referred to as nucleoids.

Bacteria may also contain smaller circles of DNA, distinct from chromosomal DNA, that are termed plasmids. Plasmids are not involved in cell replication but are nevertheless very important. Plasmids code for enzymes or other proteins that have specific functions in helping the micro-organism deal with its environment. For example, plasmids often code for proteins that detoxify or otherwise neutralise antibiotics. Plasmids may also code for proteins that aid in the decomposition of particular organic compounds, enabling the bacteria to use those compounds as an energy source. Much research has gone into identifying plasmids that code for the decomposition of toxic chemicals, because such capability could increase the effectiveness of bioremediation strategies.

Ribosomes are small, dark structures that are embedded in the cytoplasm of the cell. Ribosomes are protein assembly structures and provide a surface upon which amino acids can be brought together and assembled in the proper sequence. Ribosomes in procaryotes are made from two subunits. The smaller subunit is called the 30S subunit and the larger one is called the 50S subunit. The S stands for 'Svedberg' units, a measure of the rate of sedimentation in an ultracentrifuge and hence a measure of molecular size. Because the 50S subunit settles faster than the smaller 30S subunit, it is proportionally more massive. Each subunit consists of an RNA molecule, which on its own may be either 23S, 16S or 5S and associated proteins that aid in assembling amino acids.

The cytoplasm of procaryotes always contains individual ribosomes. Often, however, ribosomes are arranged into complexes that are called polysomes. These polysomes work simultaneously to assemble different polypeptides which may constitute parts of one protein. Sometimes polysomes are embedded into the cell membrane.

The cell membrane, also termed the cytoplasmic membrane, acts as the boundary between the interior of the cell and the outside environment. As such, the cell membrane has numerous functions that regulate the chemical environment inside and outside the cell. Figure 1.2 shows a schematic diagram of the cell membrane. It consists of approximately 60 per cent protein and 40 per cent phospholipids. The phospholipids are arranged into a bilayer, in which the hydrophobic (non-water-soluble) portions point outward. This arrangement helps the cell to regulate its water balance. Embedded in the phospholipids are proteins that exhibit a variety of arrangements. The purpose of cell membrane proteins is to regulate the transport of chemicals into and out of the cell. Some proteins, for example, act as 'ports' for bringing simple sugars into the cell, where they can be utilised for energy. Other proteins act as 'switches', transferring electrons in the cell's electron transport system.

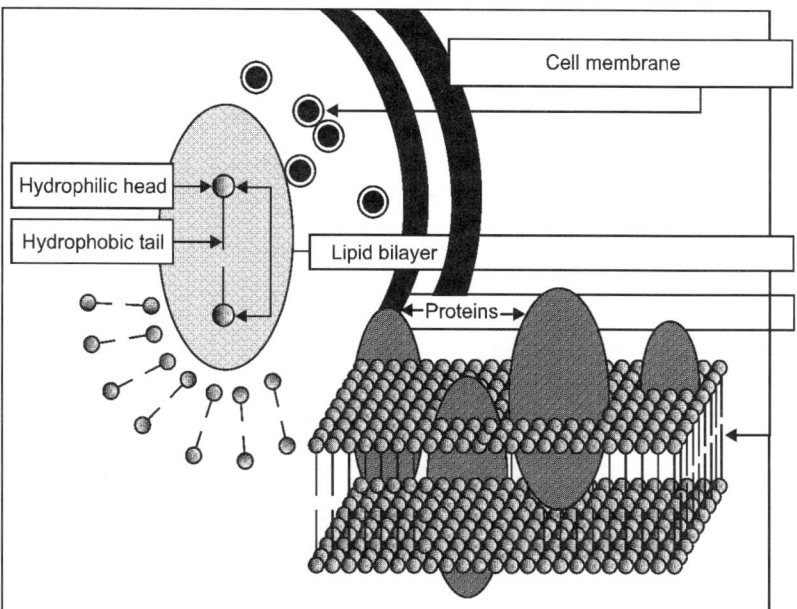

Fig. 1.2. Schematic diagram of the cell membrane, showing the arrangement of proteins and phospholipids.

The cell membrane is semipermeable and allows some substances, such as water, to cross in response to concentration gradients. This type of transport occurs spontaneously and does not require the cell to expend energy. Other substances are actively carried across the cell membrane by specialised proteins. This type of transport is termed facilitated transport and requires the cell to expend energy.

The cell membrane is enclosed within the cell wall (Fig. 1.1), which gives the cell rigidity and helps to protect it against osmotic stress. A substance called peptidoglycan, which is unique to procaryotic micro-organisms, provides much of the cell wall's structural strength. Peptidoglycan is a three-dimensional polymer of sugars and amino acids that are cross-linked with short peptide bridges.

There are two major types of cell walls (Fig. 1.3). The gram-positive cell wall consists of an inner membrane with a relatively thick layer of peptidoglycan covering it. There are also varying amounts of teichoic acids, polymers of sugar alcohols and phosphates, present in gram-positive cell walls. This thick peptidoglycan layer has the characteristic that it retains the crystal violet pigment in Gram's stain, even when washed with ethyl alcohol—hence the term 'gram-positive'.

The gram-negative cell wall (Fig. 1.3) has a layer of phospholipids and lipoproteins outside a thinner peptidoglycan layer. The gram-negative cell wall does not retain Gram's stain when washed with ethyl alcohol. In the gram-negative cell wall, there is a space between the cell membrane and the peptidoglycan layer, termed the periplasmic space. The periplasmic space is absent from gram-positive organisms and reflects a basic difference in how substrate-degrading enzymes are utilised by the two types of micro-organisms.

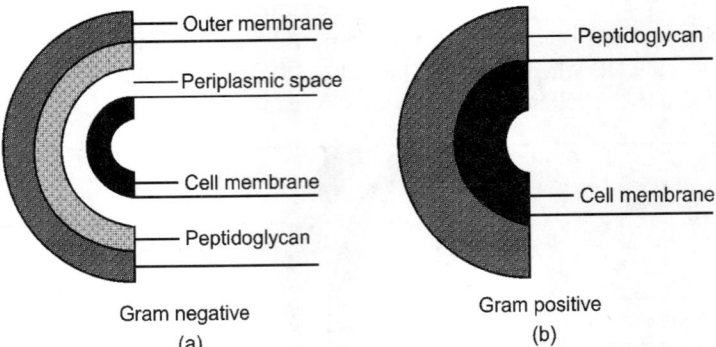

Fig. 1.3. Diagram showing the configuration of: (a) the gram-negative and (b) the gram-positive cell wall.

Bacteria often have external coatings on their cell walls. These coatings are called the glycocalyx and consist of either polysaccharides or proteins. If the coating is hard and dense, it is often referred to as a capsule. If the coating is soft and pliable, it is often termed a slime layer. Cells growing on culture media often produce a thick glycocalyx, and this is what gives individual bacterial colonies a smooth appearance. Cells that do not produce abundant glycocalyx often produce colonies that appear rough.

One important function of the glycocalyx is that it facilitates the attachment of bacteria to surfaces in the environment. For example, *Streptococcus mutans*, a bacterium that lives in the mouths of human beings, produces a thick glycocalyx slime designed to allow it to stick to its host's teeth (thus avoiding being swallowed and killed). *S. mutans* produces this polysaccharide glycocalyx from sucrose (refined sugar), so that eating a candy bar is a boon for slime production. The 'cottony' feeling in your mouth that becomes noticeable 10 or 15 minutes after eating candy comes from the slime produced from sucrose and used to coat your teeth. This slime, also called plaque, contributes greatly to tooth decay.

In the same way that *S. mutans* uses a glycocalyx to cling to teeth, bacteria in sediments use these coatings to cling to mineral surfaces. This ability helps bacteria gain access to nutrients associated with sediment particles and provides a stable environment for subsequent reproduction.

Bacteria and its Growth

Morphology

Bacteria of various species show differences as shown in Fig. 1.4. Three general morphological categories into which all bacteria fall are cocci, bacilli and spirilla with modifications of these forms.

Cocci are round cells, sometimes slightly flattened where they are adjacent to one another. They exist in pairs, as diplococci, in chains as streptococci, in groups of four as tetrads and in packets of eight, as sarcinae. Bacilli are rod-shaped. The length of the cell varies, even in a single species, under the influence of age or environmental conditions. They too occur singly or in chains. Bacteria of the coliform group, normally used as an index of sewage pollution in water, are bacilli. Spirilla are curved. The length of the cell and the number of convolutions varies with the organism. Some of them are shaped somewhat like a corkscrew. Most of them are rigid and motile. A special group of spirilla known as spirochetes are not rigid but flexible and are long and slender. Two well-known pathogens are spirochetes, *Treponema pallium*, which causes syphilis and *Leptospira*, which causes leptospirosis.

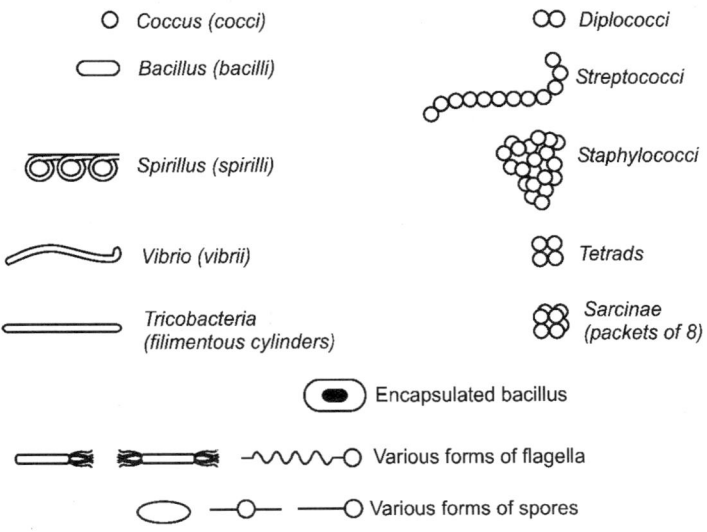

Fig. 1.4. Physical forms of bacteria and cocci.

Metabolism

Metabolism, meaning life processes, includes nutrition, respiration and reproduction. All of these processes are complex even for bodies as small as bacterial cells.

Enzymes of bacteria

A bacterial cell must obtain its food from the environment external to the cell wall. Since this wall is permeable only to substances in true solution, solid food material must be made soluble before it becomes available for cell nutrition. Enzymes (exo-enzymes), chemical catalysts manufactured by the cell, accomplish the liquefaction process. Enzymes are not altered in the process and can function as long as food is available. Once the soluble food has passed through the cell wall, other enzymes (endo-enzymes) reduce the chemically complex substances to simple compounds which the cell can utilise in its metabolic processes.

Autotrophic and heterotrophic bacteria

A distinction can be made between species of bacteria on the basis of the kind of food they require. Those which consume simple inorganic chemicals are called autotrophic and those which require chemically complex organic food are called heterotrophic. Nitrifying bacteria which inhabit the soil and subsist on ammonia, nitrite and oxygen are autotrophic. Saprophytic bacteria, which feed on dead animal and plant tissue, are heterotrophic.

All living matter is made up of proteins, carbohydrates and fats, together with inorganic (mineral) constituents. Proteins, carbohydrates and fats all contain carbon, hydrogen and oxygen. Proteins and especially the animal proteins, also contain nitrogen, phosphorus and sulphur. Saprophytes, heterotrophic bacteria, feed on dead organic matter and reduce these chemically complex materials to simpler substances. Nitrogen-bearing material, protein, is converted progressively by saprophytes to many compounds, each of which is less complex than the original. Among the simplest of these is ammonia, an end-product derived from the nitrogen fraction. At this point the function of the saprophytes is complete and autotrophic bacteria take over.

Autotrophic bacteria called nitrosomonas are capable of using ammonia as food and through their metabolism, convert the ammonia to nitrite. These bacteria, like the saprophytes, are found in natural environment, principally soil and to accomplish their purpose they require oxygen, because the conversion of ammonia (NH_3) to nitrite (NO_2^-) is an oxidative process.

Another group of autotrophic bacteria called nitrobacter, also found in natural environment, converts nitrite (NO_2^-) to nitrate (NO_3^-), which is an oxidative reaction requiring available oxygen.

Therefore, there are three groups of bacteria each performing a very important part in the conversion of dead organic material to stable chemical compounds, a process which is essential. In order to complete the description of the various transformations which nitrogen undergoes, the role played by living plants and animals must be mentioned. Nitrate (NO_3^-) and ammonia are food materials for plants and are incorporated into the protein fraction of plant structure. When plants decay, the cycle of nitrogen transformation through the action of bacteria begins again. If plants are consumed by animals, the nitrogen in the plant is incorporated into animal protein or evacuated as waste material of the animal body and then the nitrogen cycle is carried on as described through the action of bacteria.

Culture media

The study of laboratory cultures of bacteria for purposes of identification or research is a highly specialised branch of bacteriology. Time and effort is expended in developing culture media suitable for growth of specific bacterial species. Usually organic food supplemented with mineral salts is required for successful growth. In addition to food, various other substances are necessary components of a culture medium. Included are inorganic salts such as phosphates and sulphates, amino acids, minute amounts of iron and copper and vitamins.

Bacterial Respiration

Various species of bacteria may differ with respect to their respiration of oxygen. Bacteria, like all other plants and animals, require oxygen either in the gaseous state or combined with other elements in a chemical compound. The bacteria which require gaseous oxygen in their environment are called aerobes, while those which require the complete absence of gaseous oxygen are called anaerobes. There are many species which thrive better in the absence of oxygen but which can tolerate it as well. These are described as facultative anaerobes.

Reproduction and Growth of Bacteria

Bacteria have no sexual faculties and reproduction takes place whenever environmental conditions are favourable and is accomplished through a process called fission. Bacterial cells become constricted, usually near the centre. As the cell ages, the constriction progresses with time until eventually the cell completely divides to produce two separate cells.

In favourable environmental conditions, an average bacterial cell divides approximately every 20–30 minutes. Within a few hours, if all cells survive, a progression of this type would produce many millions of descendants from a single cell as shown in Fig. 1.5. There are limiting factors of growth and survival in bacterial existence just as there are in human existence and an ordinary bacterial culture 24 hours old contains about 20 million individual cells per millilitre.

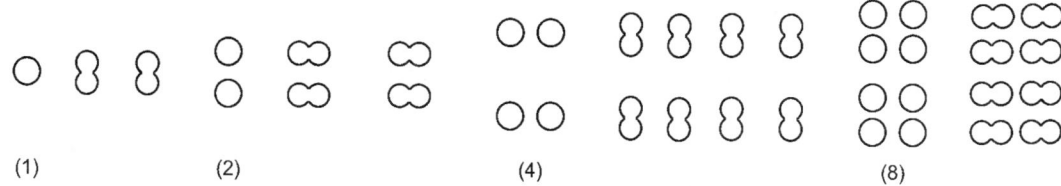

Fig. 1.5. Reproduction of bacteria.

In a closed system where the food supply is satisfactory but limited and the by-products of cell metabolism are retained, the rate of growth gradually decreases to zero. Thereafter, the death rate gradually increases until a maximum is reached and the climax is attained with the death of all cells. A curve showing the relationship of bacterial number to time under such conditions is shown in Fig. 1.6.

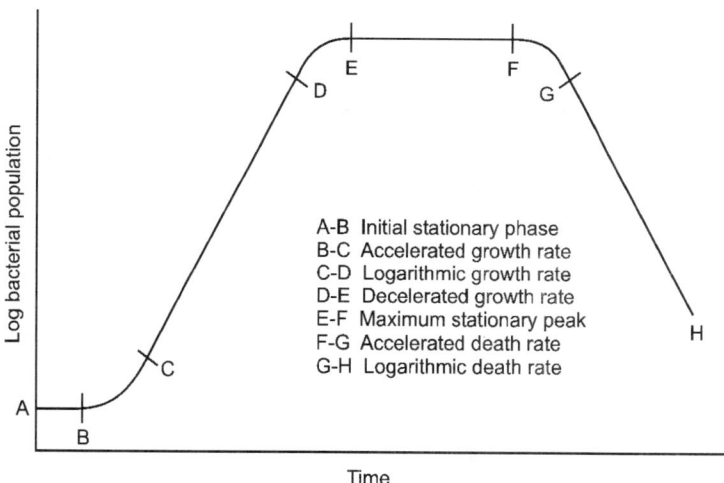

A-B Initial stationary phase
B-C Accelerated growth rate
C-D Logarithmic growth rate
D-E Decelerated growth rate
E-F Maximum stationary peak
F-G Accelerated death rate
G-H Logarithmic death rate

Fig. 1.6. Growth and death of bacteria in a closed cultural system.

Factors Affecting Bacteria

Other conditions that affect the growth and survival of bacteria are light, heat, drying, bacterial agents, bacteriostatic agents and antimetabolite of various kinds.

Heat

Heat requirements for optimum growth of bacteria vary with the species. Many bacteria thrive best at temperatures near that of the human body or slightly below (32°–37°C). These are mesophilic or moderate heat-loving bacteria. Those which grow best at low temperatures, slightly above the freezing point of water, are called cryophilic or cold-loving bacteria and those which require higher temperatures for growth, that is, temperatures between 40°C and 65°C, are described as heat-loving or thermophilic. Bacteria are not destroyed by low temperatures, but most reproduce very slowly or not at all under such conditions. When transferred from the frozen state to a more suitable environment, they immediately carry on a normal life cycle. Extreme heat, on the other hand, destroys all bacterial species, although those which are in the spore stage will withstand much more heat than those which are not. Moist heat, steam, is more effective in the destruction of bacteria than dry heat. A temperature above which no cells of a bacterial species can survive is considered to be the thermal death point of the species.

Light

Certain wavelengths of light are very destructive to bacterial cells. These lie in the region of the spectrum known as the ultra-violet. Light rays below 2900 angstrom units (290 millimicrons wavelength) are especially destructive. A necessary condition, however, is for the light ray to strike the cell directly. Ultraviolet light may be used to destroy bacteria in water, provided the water, is sufficiently free of suspended matter to permit the light ray to strike the bacterial cells. Water itself has some absorptive effect on ultra-violet light and therefore even in clear water, the distance between the bacteria cells and the light source must be relatively short.

Drying

Bacteria cannot reproduce without moisture. Drying of food materials is a method for preservation from decomposition by bacteria. Spore-forming bacteria may survive in a dry environment but they cannot function normally. If moisture becomes available, bacterial spores will vegetate and resume a normal life cycle.

Osmosis

This is a physical phenomenon which depends upon the relative concentration of soluble substances, usually salts, inside and outside bacterial cell walls. If the concentrations are in equilibrium, the cell is said to be in an isotonic environment which is favourable. If the concentration of electrolytes outside the wall is greater than that within, there is a tendency for water to pass out of the cell to restore equilibrium. This may result in destruction of the cell through the effect of shrinking or plasmolysis. If the concentration of soluble substances outside the cell walls is lower than that within, water will tend to pass through the cell wall from the outside and cause the cell to swell and perhaps burst and is called plasmoptysis.

A common practice for preserving of meat and vegetables is pickling the product in a strong salt solution. Under such conditions, bacteria present in the food are destroyed by plasmolysis. There is no practical application of this principle in water treatment. However, the length of time intestinal bacteria can survive in water is dependent in some degree on the effects of osmosis on the bacterial cells.

Germicides and Bacteriostatic Agents

Germicides are substances which destroy a bacterial cell on contact and a bacteriostatic agent is a substance which prevents the cell from reproducing itself. This latter situation indirectly brings about the destruction of the bacterial culture since without reproduction there can be no continuation of life. The best known germicide in water treatment is chlorine.

Antimetabolite

Antimetabolites are substances which destroy or alter metabolic agents or growth factors essential to the normal life processes of a bacteria cell. Such action results in the destruction of bacteria since without normal growth and reproduction, life itself ceases. Some of the antibiotic drugs used in treating disease are antimetabolites.

Bacteria Identification

Bacteria are identified through a systematic application of procedures which are designed to:

1. Secure a culture of bacteria, that is, a very large number of living bacterial cells in or on a medium which provides adequate food for them.
2. Successively sub-culture or secure a separation of the individual species from each other.
3. Determine the cultural characteristics of each species.
4. Determine the morphological characteristics of each species by examination of stained preparations of bacterial cells.

There are many kinds of bacterial cultures, that is, growths in solutions or liquid suspensions of nutrient materials, growths in suspensions of the nutrients in jelly-like substances such as agar or gelatine, growths on the surface of animal or vegetable tissue and growths on the flesh or in the bloodstream of animals. There are many materials used as nutrients, ranging from simple inorganic salts to organic carbohydrates, such as sugars, to relatively simple protein substances, such as egg albumin and to the extremely complex proteins of animal tissue. Sub-culturing is the process of securing growth of the bacteria in colonies and fishing a colony, for re-growth on fresh medium. In fishing, a sterile wire is gently touched to a surface of the colony, thus removing a few cells which cling to the tip of the wire. These cells are then immediately transferred to a fresh lot of sterile medium in another vessel. Incubation of the transplants then produces new colonies. A successive number of such transfers eventually produces a pure culture, that is, a culture of a single species.

Small amounts of pure cultures are inoculated into specialised media to determine growth characteristics. These media often consist of carbohydrates of various kinds differing in chemical structure and complexity. Growth is usually indicated by production of a cloudy effect. The number of suspended cells increases and acid and/or gas are produced as the carbohydrates are decomposed by the bacteria. A wide variety of stains and staining techniques are available to enable the bacteriologist to observe size, shape, spore formation, presence of flagella, etc., under the microscope.

Bacteriology

Even distilled water contains sufficient nutrients to support bacterial growth. The only way sterile water can be obtained is by treating it with chemicals, such as chlorine, to destroy the bacteria, by heating it or, under special circumstances, irradiating it with ultra-violet light. All natural water, whether from surface or ground sources or from precipitation, is contaminated to some degree with bacteria. These are mostly saprophytes, which feed on organic matter present in the water though contact with earth and vegetation, although some are autotrophic bacteria, which feed on simple inorganic salts and dissolved gases. Those of greatest interest in water treatment are in the water as a result of its pollution with sewage. These bacteria may include organisms pathogenic for human. There are diseases of bacterial origin transmittable from human to human through sewage contaminated water. A group of bacteria whose normal habitat is the large intestine of human and animals is known as the coliform group of bacteria. In general, these bacteria (comprising more than 30 individual species) conform to the requirements of an ideal index of sewage pollution of water, which are:

1. Present when sewage is present.
2. Absent when sewage is absent.
3. Survives longer in water than any of the pathogenic species.
4. Easily isolated and identified.

The coliform group includes all of the aerobic and facultative anaerobic, Gram negative, non-spore-forming, rod-shaped bacteria which ferment lactose with gas formation within 48 hours at 35°C. Some of the terminology in this definition requires interpretation. Two bacteriological methods are available for estimating the degree of sewage pollution of a water sample: the multiple tube fermentation method and the membrane filter method. The multiple tube fermentation method is standard and applicable to all types of water. The membrane filter method is standard for water free from turbidity and algae and filter-clogging materials after adequate parallel testing has demonstrated that it yields, for a particular water, information comparable to that of the multiple tube method.

Multiple Fermentation Tube Method

The objective in this method is to demonstrate the presence or absence of bacteria which will ferment the sugar lactose added to a nutrient liquid medium and produce gas within a specified period of 48 hours. A necessary piece of equipment is the fermentation tube. This is a small culture tube inverted within a larger tube and completely filled with the medium during sterilisation.

The medium is protected from chance bacterial contamination by a plug of cotton at the mouth of the larger tube. When the medium is inoculated with a portion of the water sample and the fermentation tube is placed in an incubator, bacteria multiply in numbers as evidenced by increasing turbidity of the medium with passage of time. If lactose-fermenting bacteria are present, gas will collect in the inner tube. Any amount of gas produced within the specified incubation period is presumptive evidence of the presence of coliform bacteria in the water sample. Evidence of the presence of coliform bacteria is necessary since not all lactose fermenters belong to the coliform group. Confirmation is obtained by transferring a drop of the bacterial culture to another fermentation tube containing a different medium — one which permits the growth of coliform bacteria only. Production of gas in this medium establishes the presence of the sewage organisms in the original sample. It is pertinent to note that one cannot deposit the original inoculum of water into the confirmation medium directly because the bacteria being sought are not in their natural environment, the intestinal tract and for that reason are unable to thrive in the rigorous confirmatory medium without preliminary increase in their numbers and viability.

Most Probable Number (MPN)

It is important not only to detect pollution of water but also to estimate its degree, specifically the number of coliform bacteria per unit volume. In a test involving the use of a medium in which the original number of bacteria introduced is greatly increased before their presence is evident, a precise count of bacteria cells is impossible. A suitable alternative is to inoculate several fermentation tubes with a series of dilutions of the sample, to observe which produce gas and calculate the number of bacteria using a mathematical formula based upon the laws of probability. A suitable combination of fermentation tube implants for a tap sample from a drinking water supply is 5 tubes with 10 ml portions, 1 tube with a 1 ml portion and 1 tube with 0.1 ml portion. The MPN results for the most commonly found combinations of positive and negative tubes in this series are as follows:

10 ml portion	1 ml portion	1/10 ml portion	MPN
– – – – –	–	–	less than 2.2
+ – – – –	–	–	2.2
+ + – – –	–	–	5.0
+ + + – –	–	–	8.8

(Contd ...)

10 ml portion	1 ml portion	1/10 ml portion	MPN
+ + + + –	–	–	15
+ + + + +	–	–	38
+ + + + +	+	–	240
+ + + + +	+	+	2400 or more

The most probable number (MPN) is not an exact measure of coliform bacteria in the sample, but an estimate.

Membrane Filter Method

Membrane filters are very thin films of cellulose manufactured in a way to produce a porous structure. Water passes freely through the membrane with slight suction, but particles, even those as small as bacteria, are retained on the surface. Starting with the sterile membrane placed on a sterilised funnel-shaped holder mounted on a suction-type receiving flask, a sample of known volume is filtered with the aid of suction applied to the receiving flask. The membrane filter is then removed with sterilised forceps and placed in a sterile culture dish on a sterile absorbent pad and saturated with special nutrient medium. Colonies develop after incubation for 18–22 hours at 35°C and these may be counted with the aid of a 10-power stereomicroscope. Only those which have a dark purplish-green colour with a metallic sheen are considered members of the coliform group. Results are reported as colonies per 100 ml of sample.

The membrane filter method yields results roughly comparable to the fermentation tube method when the sample is free from suspended matter. The more suspended matter present in the sample, the greater the discrepancy between the two methods. The technique is more easily performed than the fermentation tube method. Results are obtained more quickly and with special equipment, the test may be made in the field.

EUCARYA

Algae, fungi, and protozoa have cells that are much different from bacterial cells and are classified as eucaryotes. In contrast to the relatively simple structures present in bacterial cells, the eucaryotic cell is extremely complex. The eucaryote is characterised by the presence of a distinct nucleus (Fig. 1.7). Many, but not all, eucaryotes have a cell wall, and all eucaryotes have a cell membrane. Also present are large folded structures, called the endoplasmic reticulum, oval membrane disks called Golgi bodies, and organelles called mitochondria, lysosomes, and chloroplasts (Fig. 1.7).

The cell wall in eucaryotes performs many of the same functions as that in bacteria, such as protecting against osmotic shock and giving tensile strength to the cell. The cell walls of eucaryotes are formed from polysaccharides, but peptidoglycan is absent. The cell membrane in eucaryotes is similar to that of procaryotes, the main differences being in the composition of the lipids. The eucaryotic cell membrane carries out facilitated and active transport, as in procaryotic cells. Many of the metabolic functions carried out in procaryotic cell membranes, however, have been taken over by specialised organelles in the eucaryote.

In the eucaryotic cell, energy production occurs in the mitochondria, photosynthetic activity is carried out by chloroplasts, and digestive enzymes are stored in lysosomes. All of these functions are carried out in the cell membranes of bacteria. Protein assembly in eucaryotes is carried out by ribosomes, as in procaryotes. However, eucaryotic subunits are larger (40S and 60S) than in procaryotes. Furthermore, eucaryotic ribosomes are often arranged along the surface of a folded membrane, the endoplasmic reticulum. This structure aids in the synthesis of very complex proteins that require numerous steps during assembly. Further processing of enzymes is performed by Golgi bodies separate from the endoplasmic reticulum. Algae are photosynthetic eucaryotes and have the distinction of producing most of the world's oxygen. As such, algae contain photosynthetic chloroplasts. Algae have cell walls composed of cellulose, pectin, or silica, and may live singly or in colonies. There are six groups of algae recognised: green algae, euglenids, diatoms, dinoflagellates, brown algae, and red algae.

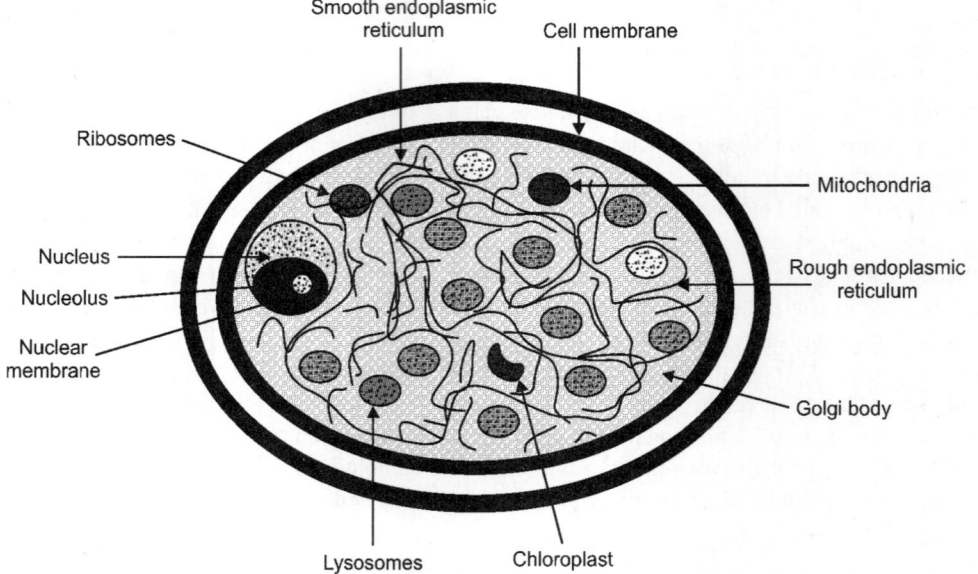

Fig. 1.7. Structures present in eucaryotic cells.

Unlike the photosynthetic algae, the fungi are heterotrophic; that is, they are decomposers and obtain energy and nutrients from pre-existing organic carbon sources. The fungi have developed mechanisms for degrading almost any kind of organic carbon compound found in nature. Most fungi are obligate aerobes, meaning that they must have oxygen in order to respire. However, some fungi, notably the yeasts, are capable of fermentation.

The special function of fungi in the environment is to recycle the remains of plant and animal debris. In this way, fungi often compete directly with heterotrophic bacteria for available resources. The greater complexity of eucaryotic structure, which allows the synthesis of many degrading enzymes lacking in the procaryotes, often gives fungi a competitive advantage over bacteria. Fungi, for example, are particularly adept at degrading lignins in decaying plants, a class of compounds that bacteria attack, but less efficiently. Also, the metabolic flexibility of fungi allow them to live in particularly stressful environments, such as hypersaline lakes. This resistance to osmotic stress is the main reason that moulds, which are a type of fungi, are able to grow on preserved foods, such as jams, that are resistant to attack by bacteria.

The protozoa are single-celled heterotrophic eucaryotes that are characterised by extremely complex cell structure. The most familiar protozoa to geoscientists are foraminifera, radiolaria, and dinoflagellates, all of which are prominently represented in the fossil record. Some protozoans are predatory and are able to attack and ingest other micro-organisms. Many predatory protozoans are specifically adapted to feeding on bacteria, and this grazing is often a limiting factor in bacterial populations in both surface and groundwater systems. Most protozoa, however, belong to floating communities called zooplankton.

ARCHAEA

The distinction between bacterial and eucaryotic micro-organisms was first defined in terms of the internal structures that were visible with a microscope (Figs. 1.1 and 1.7). On this basis, as well as on the basis of differences between the biochemistry of their enzyme systems, biologists constructed a tree in which all cells could be classified under either the procaryotic or eucaryotic 'stems'.

There is also a very 'small' one that is 5S and is only 120 nucleotides long. Intermediate between these two is the 16S rRNA that is about 1540 nucleodides long. The 23S was too long to be conveniently sequenced in the early 1970s whereas the 5S did not contain enough nucleotides to give much information. Thus, Woese's group settled on the 16S as the rRNA unit of choice and began sequencing the 16S rRNA of many different types of procaryotes and eucaryotes. There are three kinds of rRNA in bacteria. The 'large' ribosomal unit has a Svedberg unit value of 23S and is approximately 2900 nucleotide.

As the sequencing proceeded, a surprise emerged. As expected, eucaryotes were most closely related to eucaryotes. Among the bacteria, however, it appeared as if the methane-producing bacteria were distinct from other types of bacteria. In fact, the methanogens were no more closely related to the other bacteria than they were to eucaryotes. The implications of this finding were straightforward. There were not just two but three stems in the tree (Fig. 1.8). Woese named these three stems the eubacteria (true bacteria), the eucaryotes (true nucleus), and the archaebacteria (ancient bacteria). This nomenclature reflected Woese's belief that archaebacteria evolved very early in earth's history. More recently, however, Woese has suggested that using the word bacteria in archaebacteria implies is misleading, and proposed the term 'Archaea' instead.

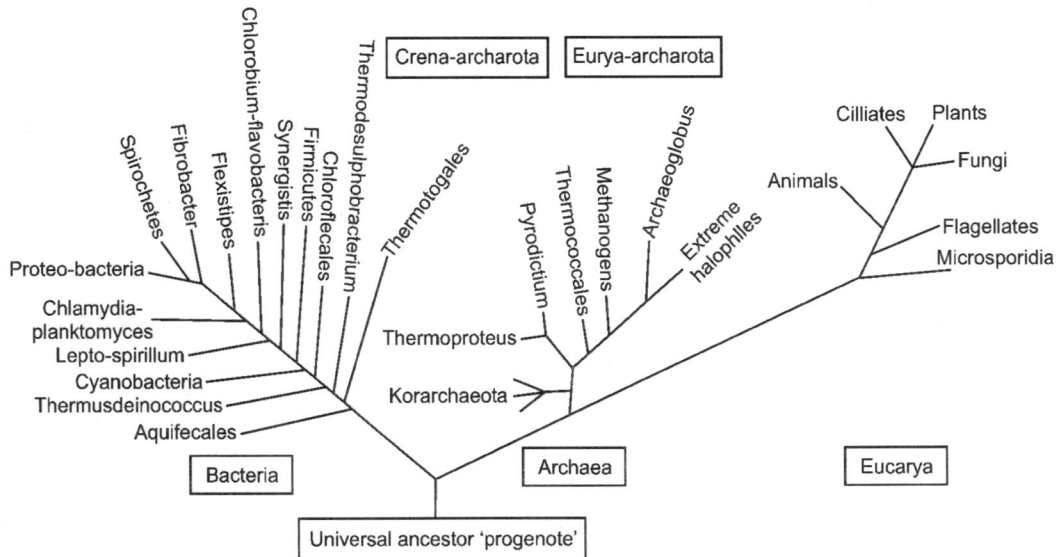

The universal phylenogenetic tree of life

Fig. 1.8. Diagram showing the relatedness of the three domains.

In addition to their distinctive 16S rRNA sequences, archaea exhibit other differences from the procaryotes. The lipids of both bacteria and eucaryotes consist mainly of two straight-chain fatty acids bound at one end to a glycerol molecule by an ester linkage (–CO–O–). In archaea, on the other hand, the glycerol and the acid chains have an ether (–O–) link. In addition, the cell walls of bacteria contain peptidoglycan, whereas the cell walls of archaea contain pseudopeptidoglycan or just protein.

The archaea include three distinct kingdoms. Kingdom Crenarchaeota consists mainly of thermophilic (heat-loving) organisms. Kindom Euryarchaeota includes the methanogens (microbes that produce methane from carbon dioxide and hydrogen), extreme halophiles (microbes that live in concentrated salt brines), and thermoacidophiles (microbes that live under extreme conditions of heat and low pH). Interestingly, a third kingdom, Kingdom Korarchaeota, has been proposed based solely on the 16S rRNA sequences of uncultured micro-organisms found in terrestrial hot springs. Of these types, the methanogens most commonly affect the geochemistry of groundwater. Early in the earth's history, methanogens could have existed almost anywhere. Today they are restricted to environments where oxygen has been excluded and where hydrogen and carbon dioxide are available. Because subsurface environments are often anoxic, they are extensively colonised by methanogens.

VIRUSES

Viruses are very small infectious agents. They are too small to be seen with a light microscope, and their presence was discovered indirectly.

Viruses are not cells, for example, because they cannot reproduce independently and because they have no independent metabolism. One of the classic (and unresolved) debates in microbiology is whether or not viruses are living organisms. When it was first shown that 'inanimate' crystals of the tobacco mosaic virus could cause the disease, many people argued that viruses were merely toxic chemicals. On the other hand, these particular toxic chemicals appeared to have the ability to reproduce themselves; was this not a clear characteristic of life? Not necessarily, came the rejoinder. Many crystalline substances 'grow' from supersaturated solutions. Was this not 'reproduction'? This debate has never been settled to everyone's satisfaction, and most people have come to the conclusion that there is not much to be gained by arguing about it. It does point out, however, some of the unique features of viruses.

At the simplest level, a virus is simply a genome, genetic information stored on DNA or RNA, wrapped in an exterior coat of protein. In the tobacco mosaic virus (Fig. 1.9a), a single strand of RNA is surrounded by protein subunits (called capsomeres) that link together to form the protein coat (called a capsid). Viral particles are called virions and are characterised by a regular geometrical arrangement of the capsomeres. Common shapes of virions are spheres, cylinders, wedges, or prisms. Some virions are covered by an envelope and are termed enveloped; others have no envelope and are termed naked.

Viruses that are parasitic to bacteria are called bacteriophages and have a somewhat more complicated structure (Fig. 1.9b). The head of the virus contains the nucleic acid genome (in this case, DNA) that is set upon the contractile tail sheath. At the base of the tail sheath is the tail pin and the tail fibres. These structures enable bacteriophages, such as the T4 bacteriophage, to attack a bacterium. First, the tail fibres find and attach to a target protein on the cell wall. Second, the cell wall is pierced by the tail pin. Finally, the tail sheath contracts and injects DNA into the bacterium. The DNA promptly takes over the cell's metabolism and redirects it to manufacture more virus particles.

It is convenient to classify viruses according to the nucleic acid that makes up its genome. The DNA viruses include the herpesvirus, which causes herpes simplex types I and II, and adenoviruses, which cause symptoms of the common cold. The RNA viruses include enteroviruses, which cause polio and gastrointestinal illness; rhinoviruses, which also cause symptoms of the common cold; and the HIV virus, which causes the deadly disease AIDS in humans.

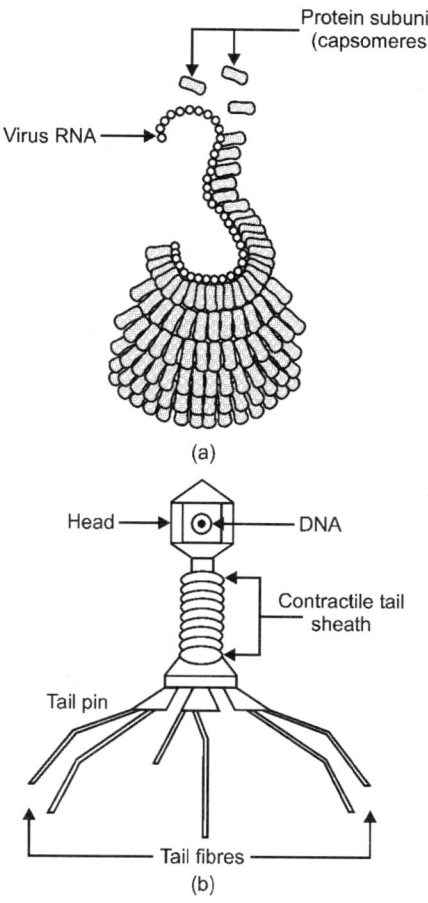

Fig. 1.9. Structures present in (a) tobacco mosaic virus, and (b) a typical bacteriophage.

Viral Ecology

Given the history of how viruses came to be studied by microbiologists, it is entirely understandable that most of what is known about viruses deals with how they infect humans, animals, and plants. Very little attention has been given to the role that viruses play in the ecology of natural systems.

In the late 1980s, however, microbiologists made the surprising discovery that a millilitre of ordinary seawater often contained as many as 10^8 viral particles. Further studies showed that viruses were intimately involved in the ecology of photosynthetic cyanobacteria that form the base of marine ecosystems. Specifically, viruses infect cyanobacteria, hijack their cellular metabolism into reproducing more viral particles, and kill them. For cyanobacteria, life is a race to reproduce before they are infected and killed by the viruses. For viruses, on the other hand, 'life' consists of trying to locate a cyanobacteria, attach to specific proteins on the bacterial cell wall, and injecting their DNA. If, after a few hours, a viron cannot locate and infect a host, it loses its ability to attach to bacterial cells and, in effect, 'dies'.

The ecological advantage for a viron infecting a cyanobacteria is obvious—it allows the viron to reproduce. However, the ecological advantage for the cyanobacteria is less obvious. Studies have shown that cyanobacteria introduced into virus-free seawater do not thrive, as one might expect. Rather, their rate of growth slows down and stops. The reason for this puzzling behaviour, it turns out, is due to the cycling of nutrients. In natural seawater, the death of individual cyanobacteria due to viral infection is a principal source of essential nutrients such as nitrogen and phosphorus for growing cells. Without that source of nutrients, the cyanobacteria can't grow. Thus, while viral infection is a disaster for individual cyanobacteria, it actually benefits the health of the overall cyanobacterial population. This is a classic example of a host-parasite relationship that has benefits for both species.

Other than the studies of seawater cited above, very little is known about the ecology of virus-host interactions in terrestrial environments or in groundwater systems. Nevertheless, by analogy to the ecology of seawater, it may be anticipated that such relationships exist, and that they may be integral parts of the ecology of these systems.

ALGAE

Algae are organisms usually found in surface water exposed to sunlight and during the summer months frequently become so numerous as to produce blooms which appear to completely cover whole sections of ponds or lakes. Algal cells contain chlorophyll, the green colouring matter of plants and under the influence of sunlight, chlorophyll enables the organism to combine water and carbon dioxide to form complex chemicals and produce oxygen as a by-product. The process is photosynthesis. Production of oxygen is beneficial in that it promotes oxidation of organic debris, supports life and favours the existence of aerobic saprophytic bacteria which decompose organic matter without the liberation of foul-smelling gases. Chemical reactions involving changes in pH and hardness of the water are also brought about through the catalytic action of chlorophyll during daylight hours. Hardness of the water decreases and pH rises during the day in water reservoirs having prolific growths of algae. Unfortunately, the process reverses when darkness sets in with the result that a water plant operator may find the chemical characteristics of the water he is treating to be quite different in the daytime than at night.

Algae are the chief cause of tastes and odours in a water supply. The tastes and odours produced by these organisms are characteristic and the presence of an offending species can often be detected by an alert plant operator by the taste and odour of the water even before it becomes obvious. Daily microscopic examination of the raw water, especially during the summer months, should be routine practice in all surface water plants to detect an increase in algae. There are at least six classes of algae common to water supplies. Some of them, particularly those of the cyanophyceae or blue-green class, are toxic.

Fungi are members of the plant kingdom but, unlike algae, they have no chlorophyll in their cellular structure. Water forms of fungi include phycomycetes, similar to algae and ascomycetes, commonly called yeasts and moulds. Some classifications place schizomycetes, of which class bacteria are members, under fungi.

Schizomycetes is the classification name for bacteria, which were discussed previously. There are several orders of bacteria which are found in water and whose cells are large enough to be segregated by sand filtration and thus be observed in a microscopic examination. These are often referred to as higher bacteria. Included in this group are Chlamydobacteriales, Actinomycetales, Thiobacteriales and Beggiatoales. These are usually indicative of polluted water. Chlamydobacteriales are also called iron bacteria because iron is often found deposited in the thick gelatinous sheath of the organism.

Some of them move about freely in the whole body of water while others are found most frequently in bottom mud. One or more varieties of protozoa are sure to be found in samples of fresh water since they are very numerous. They are sometimes responsible for disagreeable tastes in water.

Amoeba and Paramecium are two other protozoan species. These are common water forms. *Entamoeba hystolytica* is a species of amoeba responsible for amebic dysentery in human. It may be transmitted from one individual to another through polluted water.

Rotifera, Crustacea, Bryozoa and Porifera or microbial sponges, are found in freshwater, sometimes in great numbers. They are too small to be seen without a microscope. When their population is great, clogging is a problem. Larger forms, including roundworms (Nematoda), Hydra, Nais, bloodworms (Chironomus), etc., frequently are troublesome in water supplies causing taste, odours, clogged filters and perhaps even more seriously, consumer anger because some are large enough to be seen. Insects and insects larvae are a similar source of friction. In unfiltered supplies, these problems are inevitable but occasionally they can occur in filtered water.

FUNGI

The fungi are heterotrophic organisms possessing a chitinous cell wall. The majority of species grow as multicellular filaments called hyphae forming a mycelium; some fungal species also grow as single cells. Sexual and asexual reproduction of the fungi is commonly via spores, often produced on specialised structures or in fruiting bodies. Some species have lost the ability to form specialised reproductive structures, and propagate solely by vegetative growth. Yeasts, moulds, and mushrooms are examples of fungi. The fungi are a monophyletic group that is phylogenetically clearly distinct from the morphologically similar slime moulds (myxomycetes) and water moulds (oomycetes). The fungi are more closely related to animals than plants, yet the discipline of biology devoted to the study of fungi, known as mycology, often falls under a branch of botany.

Occurring worldwide, most fungi are largely invisible to the naked eye, living for the most part in soil, dead matter, and as symbionts of plants, animals, or other fungi. They perform an essential role in all ecosystems in decomposing organic matter and are indispensable in nutrient cycling and exchange. Some fungi become noticeable when fruiting, either as mushrooms or moulds. Many fungal species have long been used as a direct source of food, such as mushrooms and truffles and in fermentation of various food products, such as wine, beer, and soya sauce. More recently, fungi are being used as sources for antibiotics used in medicine and various enzymes, such as cellulases, pectinases, and proteases, important for industrial use or as active ingredients of detergents. Many fungi produce bioactive compounds called mycotoxins, such as alkaloids and polyketides that are toxic to animals including humans. Some fungi are used recreationally or in traditional ceremonies as a source of psychotropic compounds. Several species of the fungi are significant pathogens of humans and other animals, and losses due to diseases of crops (e.g. rice blast disease) or food spoilage caused by fungi can have a large impact on human food supply and local economies.

PROTOZOA

Protozoa are unicellular eukaryotes, (singular protozoan). While there is no exact definition of the term, most scientists use protozoan to refer to a unicellular heterotrophic protist, like amoebas and ciliates. The term algae is used for the photosynthetic micro-organisms. However, the distinction between the two is often vague. For example the alga Dinobryon has chloroplasts for photosynthesis but it can also feed on organic matter and is motile.

Characteristics

Protozoa usually range from 10–50 μm but can grow up to 1 mm, and are easily seen under a microscope. Protozoa exist throughout aqueous environments and soil, occupying a range of trophic levels. As predators, they prey upon unicellular or filamentous algae, bacteria, and microfungi. Protozoa play a role as both herbivores and consumers in the decomposer link of the food chain. Protozoa also play a vital role in controlling bacteria populations and biomass. Protozoa may absorb food via their cell membranes, some, e.g. amoebas, surround food and engulf it, and yet others have openings or 'mouth pores' into which they sweep food. All protozoa digest their food in stomach-like compartments called vacuoles.

As components of the micro- and meiofauna, protozoa are an important food source for microinvertebrates. Thus, the ecological role of protozoa in the transfer of bacterial and algal production to successive trophic levels is important. Protozoa such as the malaria parasites (*Plasmodium* spp.), trypanosomes and leishmania are also important as parasites and symbionts of multicellular animals.

Some protozoa have life stages alternating between proliferative stages (e.g. trophozoites) and dormant cysts. As cysts, protozoa can survive harsh conditions, such as exposure to extreme temperatures and harmful chemicals, or long periods without access to nutrients, water, or oxygen for a period of time. Being a cyst enables parasitic species to survive outside of the host, and allows their transmission from one host to another. When protozoa are in the form of trophozoites (Greek, *tropho*= to nourish), they actively feed and grow. The process by which the protozoa takes its cyst form is called encystation, while the process of transforming back into trophozoite is called excystation.

Protozoa can reproduce by binary fission or multiple fission. Some protozoa reproduce sexually, some asexually, while some use a combination, (e.g. Coccidia). An individual protozoan is hermaphroditic.

ENVIRONMENTAL MICROBIOLOGY

Environmental pollution control microbiology is currently an exciting and challenging area of science and engineering. Environmental pollution control microbiology is concerned with solving broad spectrum of environmental pollution problems that affect people around the world from a microbiological point of view. On one hand, environmental pollution control microbiology is concerned with protecting people from pathogenic micro-organisms; and on the other hand, it is concerned with the application of microbiology to solve a wide range of environmental pollution problems. Emphasis is placed on the environment rather than on individuals. Both science and engineering are involved in environmental pollution control. Scientists determine the environmental problems and evaluate different solutions. Engineers utilise the information supplied by scientists to design the most efficient systems to solve the environmental problems. By working together as a team, scientists and engineers are meeting the challenges of environmental pollution control in our modern world.

The concepts of environmental pollution control microbiology have their roots in conventional microbiology with its concerns for pathogenic micro-organisms and public health. The roots are even deeper in chemistry, which forms the basis of all reactions between the chemical compounds that make up our world and in biochemistry, which focuses on the chemistry of biological systems. The deepest roots are in civil engineering, which provides the basis for all the other areas of engineering contributing to the design and construction of pollution control facilities. Chemical engineering is one of the newest areas of engineering to become involved in solving environmental problems. It joins mechanical engineering and electrical engineering in contributing their special expertise. Environmental Pollution Control Microbiology was developed to show how microbiology, chemistry and engineering are combined to provide real solutions to environmental pollution problems.

The very breadth of microbiology poses a challenge to anyone with a major interest in pollution control microbiology. It is not possible for everyone to be an expert in all areas of microbiology. Yet, it is important to have a good understanding of the different areas to provide communication with the technical specialists in the different areas of microbiology. There is nothing wrong in developing a special interest in one area of microbiology, as long as you keep your focus on the total picture and not on the details. It is important to realise that while pure cultures are critical in the study of microbiology, it is equally important to understand the competition that exists between the different organisms. The fastest growing pure culture organism may not be the predominant organism in the real environment. Laboratory conditions often fail to stimulate the natural environment and allow entirely different growth patterns to occur. The keys to pollution control microbiology are in examining the different organisms in their natural environment and in understanding the how and why of their growth characteristics in that environment.

Chemistry has a very special place in environmental pollution control microbiology, the same as in the other branches of microbiology. Chemistry is a basic science concerned with the composition of all matter. Chemistry helps to explain how materials react with each other.

Environmental chemistry helps make the bridge between environmental pollution control microbiology and all of the major areas of chemistry. As with microbiology, it is not necessary to be a specialist in all areas of chemistry; but one must have knowledge of the major areas of chemistry that affect microbiological systems in the environment.

Engineering is the third cornerstone in the foundation for environmental pollution control microbiology. Engineering is concerned with the various structures and systems employed in solving environmental pollution problems. Because of the diversity of environmental problems and the large number of different solutions, there are many different types of engineers involved.

The success of activated sludge is a primary example of pollution control microbiology at its best. While waste-water treatment removed contaminants from the water, large quantities of sludge remained to be returned to the environment. Since time began, waste-water sludges were always returned back to the environment as fertiliser. Unfortunately, pathogenic micro-organisms were transmitted through the waste-water sludge to some crops used for human consumption. Research found that anaerobic treatment could reduce the survival of pathogenic micro-organisms in the waste-water sludges; but it was Karl Imhoff's development of a practical design to treat waste-water sludges that resulted in large-scale treatment systems for waste-water sludges. Anaerobic treatment of waste-water sludges developed slowly over the years until researchers became interested in a more detailed examination of the anaerobic micro-organisms responsible for sludge digestion. This is one of the more interesting areas of environmental pollution control microbiology that is still evolving.

Many environmental pollution problems and solutions deal with micro-organisms. In order to solve the microbial pollution problems, environmental engineers require a background in environmental pollution control microbiology. By understanding how the microbes live and grow in the environment, the environmental engineer can develop the best possible solution to biological waste problems. After construction of the desired treatment facilities, the environmental engineer is in position to see that they are properly operated to produce the desired results at the least cost. Engineering solutions to environmental problems are still evolving. There are no set patterns to solve all environmental problems. Each problem has its own unique challenges that require careful evaluation of all the different technical skills. One thing is certain; engineers will be involved in all the solutions.

Soil is a complex mixture of different chemical particles. As water moves through the void spaces around the particles, it carries micro-organisms and chemical nutrients for their metabolism. The tiny soil particles filter out the large micro-organisms and furnish surfaces on which the micro-organisms grow. As the water moves deeper into the soil, it forms groundwater. It is not surprising that the groundwater has the best microbiological characteristics for domestic waster supplies. The deep soil is simply not a suitable environment for micro-organisms. On the other hand, surface water provide a real challenge for treatment when used for domestic water supplies. The water treatment processes making up municipal systems focus on removing micro-organisms and other contaminants in the water. The concepts of each treatment unit will be examined from a microbiological point of view. Once the public has used the water, it is collected in sewerage systems and carried to the waste-water treatment plants. Microbial action has already begun when the waste-water enters the public sewerage system and continues throughout the treatment process.

Environmental pollution control microbiology will stimulate students in environmental studies and in environmental engineering to recognise the role that environmental microbiology plays in solving environmental problems. Although most of the emphasis will always be placed on the pathogenic micro-organisms, most of the micro-organisms in the environment are non-pathogenic. Learning to use mixtures of micro-organisms to control the major environmental systems has become a major challenge for environmental engineers. Environmental pollution control microbiology should help operators of municipal waste-water and industrial waste treatment plants gain a better understanding of how their biological treatment unit's work and what should be done to obtain maximum treatment on a daily basis. Design engineers should gain a better understanding of how microbes provide the desired treatment and what the limitations are for good design. Even regulatory personnel can obtain a better understanding of the limits of their regulations and what concentration of contaminants can be allowed in the environment. Lastly, it is hoped that some members of the general public will gain a better appreciation of the environment in which they exist and what can be done to maintain the quality of that environment for everyone to enjoy.

Thus, environmental microbiology is the study of micro-organisms which exist in natural or artificial environment. The evolution of life on this planet would have depended on water, the molecule whose bipolar nature induces the three-dimensional configuration of our biomolecules. Also, both the presence and metabolic activities of micro-organisms found in surface water has been continuously examined.

Chapter 2

No Microbe is an Island

INTRODUCTION

The most important thing which must be understood about micro-organisms in their environments is that no microbe exists by itself. Since in the environment no microbe is an island existing independently. There are only a few, extremely rare instances when a single species exists naturally in pure culture. The interactions and interdependence between members of different species are a consequence of the evolution of those species into their occupation of specific niches. The second most important thing to recognise is that, as we look around nature's neighbourhoods and observe the types of interactions that exist within the communities of macrobial organisms which live there, all of those types of interactions likewise exist in the microbial world. These interactions include such things as peaceful coexistences, competitions, hunting and hiding, and the use of chemical attractants, repellents, and toxicants. In fact, these approaches to surviving in community life would have existed within the microbial realm long before the first macro-organism evolved.

The microbes form the basis of our biological heritage as humans and indeed are the basis upon which all macro-organisms, including vascular plants and animals, evolved. Microbes form the understructure which supports what we perceive as being the macrobial realm. The fact that this assemblage of microbes and macrobes has evolved together binds it together. No single species selected out of this assemblage could survive for very long on its own in the wild without the biological activity provided by those species which occupy the connecting niches. This is a guiding concept in species ecology.

We also know that the life which exists on this planet evolved by using the hydrogen bonding bipolar nature of water molecules in their liquid state to serve as an organising force. Through this force, biomolecules are arranged into their functional configurations by virtue of their having evolved hydrophilic and hydrophobic zones. Thus, the life on this planet requires water. The pervasiveness of this biology is such that it seems that on this planet active indigenous microbial life can be found in any environmental site where the laws of physics can allow water to exist in its liquid state. The additional fact that many types of organisms have evolved mechanisms which allow them to survive periods of freezing or even desiccation means that much of this planet is available for supporting life. Microbes would have formed the initial basis of the food chains for all of this planet, serving as its primary producers. Vascular plants have taken over the role of primary production in those land surface areas of the planet where the air temperature is above the freezing point of water for at least part of the year. Vascular plants also have taken over the primary production role in many of the shallow aquatic areas where sunlight reaches in abundance. However, deep inside the earth the microbes still hold sway, and phytoplankton in the oceans are assumed to serve as the earth's primary source of fixed carbon. Microbes also perform the primary production role in such intriguing places as the communities of sea ice endofaunal organisms, where microfauna and meiofauna (small size category of macro-organisms) live in the brine channels which exist within the Arctic and Antarctic pack ice, and in the unconsolidated layers of platelet ice which underlie the pack ice. Microbes apparently live within, and due to the absence of sunlight must serve as the primary producers for, the water of lake. At the other end of the production line, microbes are the key to the recycling of available nutrients, including the important aspect of cellulose recycling within the gut of termites and the recycling which occurs in deep sea sediments.

Microbial communities function as interacting, coevolved assemblage. Of course, as we first begin to look at microbial communities and the interactions among microbial populations, we immediately are faced with three significant problems. First, an enormous number of different micro-organisms exist in a wide array of highly dissimilar environments and communities. These micro-organisms include archaea and bacteria, fungi, algae, protozoa, and viruses. Second is the enormously complex task of understanding and profiling the interactions among microbial populations. Such interactions not only are numerous and dissimilar but often are poorly characterised or totally undefined. Furthermore, there is difficulty in understanding the interactive role and major impact of invertebrate and vertebrate animals and vascular plant populations on micro-organisms, a subject which is not within the usual scope of interest of microbiologists. Although the more complicated reality is that microbial populations often are intimately linked with coevolved invertebrates that frequently consume micro-organisms, vascular plants that provide readily available organic nutrients, and vertebrate animals that may consume all of the above. At times it does seem that evolution works its way down strange pathways. Third, there exists a woeful lack of agreement about the various definitions that are appropriate for characterising microbial populations, communities, environments, and interactions.

This chapter is not designed to be encompassing. The topic of microbial communities and interactions would easily fill an entire monograph. Rather, the text will serve as a prelude, presenting an introduction to some of the principal issues of environmental microbiology.

NEIGHBOURHOOD, CLIQUES AND CLICHES

Environmental microbiology is blessed, or possibly cursed, with a multitude of terms that often are contradictory, sometimes misused, and frequently misunderstood even by the specialists. The ability to use language properly is an art and helps us to understand one another. But semantic sensitivity is frequently not the hallmark of the experimentalist, and this lack of sensitivity often is the basis for confusion and disagreement. Some of the more critical terms are defined in this section of the chapter, totally or partially, with the definitions used.

Defining the ecology of a species is a complex issue, but the issue can be divided into different ecological aspects. The first aspect can be consideration of the terms boundary and barrier. The descriptions provided here will be brief, but the concepts of boundaries and barriers will be taken up a bit more extensively later in the chapter. The word boundary, as used here, represents an abstract concept which describes the environmental limits beyond which the physiological capabilities of a species will not allow its members to function. This boundary is established by the physical needs and biochemical capacities of the members of that particular species. The members of a species are obliged to live within their species' boundary. This concept of a boundary is not something marked out as if a territory on a map; instead, it is a statement of the functional requirements and limitations of that species. Often, the boundaries of different species overlap spatially and biochemically. It is within these areas of overlap that biological interactions between the members of those species can occur. If the boundaries of two species do not overlap, then biologically the members of those two species will not interact with one another.

Territories, or physical locations, which meet the requirements for being encompassed by the functional boundary of a species represent potentially habitable sites for the members of that species. Those territories may be either contiguous or separated by barriers. Barriers are obstacles and can be classified by their nature as either physical, chemical, or biological (Table 2.1).

Table 2.1. Examples of barriers applying to different species.

Physical

    Thermal

    Acoustic (usually ultrasonic)

    Pressure (barometric, hydrostatic, osmotic)

    Radiation (electronic, neutronic, photonic, protonic)

    Impaction (includes gravitational settling)

    Adhesion (adsorption, electrostatic, van der Waals)

    Filtration (size exclusion)

    Geographic features

    Atmospheric factors (include meteorological aspects such as humidity, precipitation, and prevailing winds)

Chemical

    Ionic (includes pH and salinity)

    Surfactant

    Oxidant

    Alkylant

    Desiccant

    Denaturant

Biological

    Immunological (includes specific as well as non-specific)

    Naturally induced (intrinsic response)

      Naturally transferred (lacteal, transovarian, transplacental, etc.)

    Artificially induced (includes cytokine injection and vaccination)

      Artificially transferred (includes injection with antiserum and tissue transfers such as transfusion and grafting)

    Biomolecular resistance (not immune related)

      Lack of receptor molecules

      Molecular attack mechanisms (includes nucleotide-based restrictions)

      Antibiotic compounds (metabolic inhibitors, either intrinsic or artificially supplied)

    Competitive (other species in ecological competition), including predation or parasitism against the subject species by members of some other species or, if the subject species is an infectious agent, competition against a vector which the subject species might need in order to achieve transmission between host organisms

The next ecological aspect, which is to consider that the ecology of a species has two components: its niche and its habitat. Each of these components can be defined both in a potential perspective and in a more limited operational perspective. The major difference between these components is that, whereas a niche is biologically defined, a habitat is physically defined. The word habitat describes only the place where a species lives. The word niche describes how this species fits into that place, representing its interactions and accommodations with respect to the abiotic and biotic conditions found there. Thus, for any particular species, the description of its niche consists largely of the interactions between the members of that particular species and other species which occupy connecting niches. However, the definition of niche also includes changes in the environment effected by the species' biological activity. Among ecologists, the niche of a species represents what that species does; that is, its function in its natural habitat. The potential niche of an organism, expressed in terms of its total, genetically provided capacity for biological interactions, may be a marvellously broad thing and enables potential evolution of that species. In nature, however, competition and other interactions with and against different species narrow this potential niche to a smaller, more functionally restricted operational niche. The physically defined term habitat brings us to the point of identifying territories, places to which we could point and sometimes even identify on a map. A habitat is a site, generally having some physical uniformity, which possesses those characteristics that appear to be of ecological importance for supporting residence by the members of a species. The potential habitat of a species may be very large but operationally may be more restricted due to deterring interactions such as competition against or predation by the members of other species. Some species end up being restricted to perhaps only a single and frequently unique habitat, whereas other species are cosmopolitan and successfully reside in a wide variety of seemingly dissimilar habitats. An individual habitat may be a specific layer in the depth profile of a characterisable soil type, the intestinal tract of members of a particular animal species, the outer surfaces of a plant root, or some zone within lake bottom sediments. Frequently, however, owing to the small size of micro-organisms, the concern of the microbial ecologist is the microhabitat, inasmuch as individual propagules, filaments, or cells often are restricted to a site that is no larger than several millimetres or sometimes even micrometres. Thus, an organism may be found to exist only within a specific depth range in the water column of a particular lake and interact only with the other species found in that vicinity. This may be the case even though the boundary of that species might be broad enough that the organism could exist at other depths and in different bodies of water. Yet that one place then represents the operational habitat of that species, even if its potential habitat could be much larger. The interactions of this species with other organisms existing within that small vicinity are a defining part of this species' operational niche, even if its potential niche could be far larger. The reasons for restriction to that operational habitat and operational niche may be the existence of barriers, which may include a biological inability to successfully compete against, or may result from predation by, the members of other species found at other depths within that water column or found within other bodies of water.

Within a particular site, we will find what we call a community. Perhaps we could consider the term neighbourhood to define the site itself, and community then could define those which live therein. The community comprises all members of all species that occupy a particular site. These members of the community generally represent different populations. A population of a particular organism is an assemblage of individual organisms having common characteristics. In microbiology, the term population variously may be used to represent all individuals of a single species, subspecies, variety, race, or other subspecies designation. Within a population, there may be a large number of individual cells, or a population may be composed of a multicellular filament, such as is found among the fungi and algae. An appropriate term for an individual microbe belonging to a population is propagule. However, the reader should be aware that many specialists use the term community to encompass only a particular category of organisms. For this reason, some would describe an environment as having separate communities of bacteria, fungi, algae, plants, or animals. The community of a soil in general is somewhat different from that of the subsoil or underlying aquifer, and the community in the water column is not the same as that inhabiting the underlying sediment. A term synonymous with community but not widely used in English is biocoenosis. Although clear distinctions between communities and populations thus exist, the words are frequently and mistakenly used as synonyms.

Consortium is a term used more commonly for bacteria in associations than for assemblages of other organisms, but this concept does not have to be limited to just bacteria. It is a useful word when it refers to a collection of organisms that have some functional association with one another and when we talk about the properties which describe particular communities. For example, one species may provide growth factors for a second or serve to eliminate inhibitory compounds that affect a second species. To some degree, a consortium may represent a commensal or protocooperative interaction whose basis has not yet been well established. Unfortunately, however, mixtures of bacteria that have yet to be separated into pure cultures are frequently designated consortia; in these instances, the mixture may not represent a functional association but, rather, reflects the lack of isolation of individual components of the mixture.

UNDERSTANDING THE NATURE OF COMMUNITY LIFE

The degree of species diversity within communities can vary enormously. Communities typically may have several, many, or innumerable species. Although microbiologists are prone to work with pure cultures, communities having only one species are rare in nature, except in those monospecific communities associated with disease processes or in environments so extreme that few species can survive or proliferate. Clearly, therefore, at least in nature there must be some benefit gained by one species from coexistence with other species. Although the benefits are unquestionable, the basis for those benefits and the mechanisms associated with the interactions in natural communities are rarely understood and rarely studied.

We know that in order to become a member of a community, the propagules of a species not only must reach the environment where that community exists, but also must successfully cope with the detrimental interactions occurring there, including competition with the species which already exist in that site. Some sites seem not only readily accessible to many different types of micro-organisms, but also supportive towards the survival and proliferation of those new arrivals. These receptive sites thus often end up with a high species diversity. Soils, surface waters, sediments, and other environments such as sites rich in decaying organic materials support a multitude of species of bacteria, fungi, protozoa, and sometimes algae. On the other hand, sites with high salinity, extreme pH, low nutrient levels, or high light intensity often are characterised by low species diversity, which may be reflective of a low receptivity.

An old truism states that micro-organisms are potentially everywhere. That truism is little more than a platitude, and as a statement it takes on a degree of ecological inaccuracy when it is used as the basis for assuming that species or genera that are widely disseminated are also widely established. Potentially cosmopolitan species frequently are able to grow, in culture at least, very rapidly in the absence of other organisms so that a few propagules or a small biomass can multiply to reach a high cell density or large biomass in a matter of hours or days. Such largely unchecked proliferation occurs in nature, however, and effective checks exist to prevent the unbridled multiplication of widely disseminated propagules. The operation of these various checks is the basis for the appropriate second half of the truism, namely, that the environment selects. Hence, the full and ecologically appropriate statement is: micro-organisms are potentially everywhere, but the environment selects.

The impact of that selection is evidenced by the characteristic communities of individual habitats. Many of the bacteria and protozoa present in surface waters are thus not characteristic of soils, and the types of heterotrophs (organisms which derive their operating energy from organic compounds found in their surrounding environment, as contrasted with autotrophic organisms, which derive their operating energy either from photosynthesis [photoautotrophic] or chemosynthesis [chemoautotrophic]) found in soils are often quite different from those that would be found on the surfaces of leaves. The forces of selection are often nonbiological, and an area that has a low pH, is exposed to high light intensity, has no available oxygen, or contains low concentrations of readily available carbon sources supports a community that is able to cope with these restraints or limitations. Such abiotic (not biological in origin) factors are often reasonably easy to demonstrate. However, more difficult to establish and scientifically more interesting are the biotic (of biological origin) stresses that are operating in environments in which major abiotic stresses do not determine community composition.

Involvement and Homeostasis are Evolutionary Consequences

The indigenous populations that make up the community are responsible for the biotic balance that is maintained. They regulate the population densities or biomass of the individual component species of the community, and they act to prevent the establishment of invading species. These various biotic interactions determine the stability of established microbial communities, and they reflect what is designated homeostasis. From a biological perspective, homeostasis is the numerical and biochemical equilibrium resulting from the activities of the different species which comprise the community. From an ecological standpoint, homeostasis is the capacity of a community to maintain its stability and integrity in an environment subject to abiotic and biotic modifications. These homeostatic mechanisms are constantly operating, and although they often modified somewhat, they generally are not altered appreciably unless an enormous perturbation occurs. Characteristically, it is abiotic perturbations that may upset the community. Only in rare instances is a biotic influence, such as the mere introduction of propagules of a non-indigenous micro-organism, sufficient to alter the community to an appreciable extent. Indeed, in many instances, the new arrival is not even allowed to become established.

Ignoring the potency, as well as the limitations, of homeostatic mechanisms has led to a number of viewpoints that run counter to observations in nature. For example, individuals who want to introduce bacteria or fungi into soils, subsoils, or aquifers for bioremediation assume that the introduced species will become established and bring about the destruction of a contaminant of concern. In doing so, it is assumed that just because humans judge that the introduced micro-organism has a beneficial trait (in this case the capacity to grow on and thus destroy an organic pollutant), that organism indeed will become established and perform what is desired of it. While the humans performing the experiment may think that some characteristic of the introduced microbe is important from their own perspective, the microbial community may perceive things otherwise. The capacity of the microbe to use a particular substrate for growth, although of paramount importance in culture, simply may not be sufficient for ecological success: the organism also must be able to cope with the various factors associated with homeostasis. An attribute that is a necessary requirement for growth is not *a priori* an attribute sufficient for establishment, and both necessary and sufficient traits must be present. An introduced organism must be able to compete effectively for limited resources other than the specific organic compound which laboratory studies may have proven can serve as its carbon source. The introduced organism also must be able to cope with the stresses associated with predation and parasitism, which are of great importance in many environments. In contrast with the vendors of micro-organisms, who always hope for success from their introductions, are other individuals fearfully concerned that the environmental introduction of genetically engineered micro-organisms could result in successful establishment. Many of the latter individuals must optimistically presume that all introductions will fail and that modest changes in the genotype of an existing organism will not result in its establishment. Natural events often prove humans to have been wrong in their assumptions. It is necessary to consider not only the strengths but also the limitations of homeostatic mechanisms. While the operation of homeostatic mechanisms does eliminate most introductions, these mechanisms are not omnipotent, and some introduced organisms do succeed in becoming established. Borrowing an analogy from the world of macro-organisms, farmers know that sowing a new variety of plant, one that has major beneficial attributes, is not sufficient to obtain high yields of that variety. The introduced crop species frequently is unable to cope with competition by weeds, parasitism by insects and plant pathogens, poor soil structure, and other stresses. Few introduced species become established, regardless of whether the introduction has the capacity to do good or is potentially injurious. Yet in some instances, an alien species does become successfully established, as evidenced by the major upsets that have occurred because of invasion by plants and animals. Attempts have been made to predict the capacity of an introduced organism to become established, and studies have been directed towards establishing the traits that might be used to predict the outcome of an introduction. To date, the outcome of that research is modest. Our inability to predict the success or failure of introductions is a reflection of incomplete knowledge of the various components of homeostasis.

Being Kind and Helpful

To the general public, micro-organisms are frequently considered to be solely harmful. This is evident in the use of the word germ (commonly referring to a micro-organism thought to cause disease but perhaps also definable as a microbe unwelcome in the place where it has been found). Yet altruism is a widespread attribute among all major categories of micro-organisms. Admittedly, it is true that competition, parasitism, and predation are of great importance and that disease-producing micro-organisms are widespread. Nevertheless, the good that many heterotrophic and autotrophic populations, including the natural and often protective microfauna associated with plants and animals, do for their neighbours should not be overlooked.

Using and Abusing

Some apparent types of beneficence are commonly categorised, from the ecological viewpoint at least, by the terms commensalism, protocooperation, and symbiosis. Although a clear distinction among those three terms, in fact a continuum exists so that the range of each of these types of beneficial interaction merges into the next. Indeed, it is likely that there is a continuous evolution so that organisms which at one time were commensals evolve to exist in protocooperative relationships and those that are involved in protocooperation will, with time, evolve into highly dependent and mutually beneficial symbioses.

In a commensal association, one species benefits a second but the first gets no apparent advantage in return. The second is deemed the commensal. Such interactions are evident between heterotrophic bacteria as a commensal and algae excreting photosynthetically fixed carbon and apparently between populations in a biodegradative process. The mechanisms underlying these commensal associations are numerous, but only a few have been characterised. In one such type of commensalism, one population converts a compound which metabolically is unavailable to a second population into a product that can be used as a nutrient source by the second species. The first gets no benefit from the association, but the second is provided with something essential for its replication. Sometimes the second population is auxotrophic, meaning that its members cannot grow in the absence of required compounds which are termed its growth factors. An organism which excretes a growth factor may allow for the development of this auxotroph as a commensal. Studies of water and soils indicate that frequently as many as three-fourths of the indigenous species are able to excrete one or more vitamins or amino acids which may be capable of serving as growth factors. Another type of commensalism is evident in environments containing organic or inorganic inhibitors, and a species that destroys an organic inhibitor or somehow detoxifies an inorganic ion will serve an altruistic role for its sensitive neighbour.

Because terrestrial and aquatic communities contain a high percentage of micro-organisms that are auxotrophic, commensalism must be biologically favourable in those communities. An organism that relies on its neighbours for the synthesis of a carbon source, growth factors, or detoxifying enzymes does not need to expend energy of its own to either make or detoxify those substances. Thus, although the commensal is dependent upon the helpful neighbours found in that environment, energetically the commensal has a competitive advantage, all other factors being equal, over some other organism that rejects dependence by choosing to expend energy to make all of its vital compounds or detoxify all harmful substances. In turn, a chain or network of commensal reliances can evolve. If, energetically, evolution thus benefits a commensal, it would also likely favour the development of a second species that is not fully self-reliant. Mutually, those two species might energetically gain from some cooperation, with each providing some substance needed by the other. Energetically, the second species would be more competitive against otherwise similar populations that instead develop their own full armament of enzymes and physiological mechanisms. In this way, two populations that rely on one another will have, with time, a selective advantage over partially or fully independent populations for as long as those two species coexist. This is the basis for protocooperation, an association in which each of the two interactants needs and benefits the other.

This two-member association, when functioning as a unit, has greater fitness than do the two species functioning independently. The identities of the associates involved in a protocooperative association are not fixed, in the sense that it is neither obligatory nor specific that only these two species can fulfil those roles for one another. Well-studied protocooperative relationships include those in which there is interspecies H_2 transfer and the fermentation of polysaccharides. Synergism refers to a process in which two species cooperate in causing a change that otherwise neither species could perform alone, or else their accomplishing that change independently would occur more slowly than when they are together. The synergants could, however, live in that environment independently of one another. Synergism often seems to result from protocooperation. The two interactants may function together because one provides something to the other, such as a carbon or energy source, a growth factor O_2, or detoxifying enzymes, that serves as the basis for the benefit.

In a simple protocooperation, only a single benefit needs to be conferred from one population upon the second. However, if a simple protocooperation gives the interactants additional fitness, so too might greater degrees of integration. Each population may, thus, evolve to contribute more than one requisite to the second population. When this occurs, the association becomes somewhat less flexible, and the identities of the two interactants become somewhat more restricted because each must provide the full complement of benefits required by the second. This seems to be the basis for the tight interactions that characterise symbiotic, or mutualistic, associations.

Symbioses, those instances when different organisms live together as one, are evident among many microbial groups and in many different environments. In a symbiotic relationship, neither organism could naturally live independently in that particular environment. Some symbioses involve two species of micro-organisms. Others involve one microbial species and either a plant or animal species. In the orchid symbiosis, by contrast, there are three species, the orchid, a fungus, and a tree; the first two are symbionts, but the fungus is a parasite of the tree. Among the intermicrobial symbioses, a few have been the subject of considerable scrutiny, and sometimes the contributions of the symbionts to their partners have been well-defined. For example, the lichen, an association of an alga with a fungus, has long been of interest to botanists, although to few microbiologists, and the nature of the symbiosis is becoming ever more clear. Endosymbiotic associations, in which one of the organisms lives within the body of the other, represent an even greater commitment to symbiosis. An example of an endosymbiotic relationship is bacteria residing within protozoa. Several of the symbioses involving plants have been intensively investigated and remain the subject of active research in many countries. That research has been driven in part by the benefits to human society of such relationships as the fungus-root symbiosis termed a mycorrhiza and the associations between legumes and members of the genera Rhizobium and Bradyrhizobium. In both instances, the micro-organisms are heterotrophs, although now photosynthetic rhizobia are known to occur, and the micro-organism must, thus, get carbon from the plant. In turn, the root nodule bacterium provides nitrogen in a form usable to the higher plant. The precise contribution of the mycorrhizal fungus to the plant is still the subject of considerable controversy. Seemingly, the simple provision of carbon by the plant to the micro-organism, and of nitrogen by the rhizobia to the legume, does not adequately explain the basis for a symbiosis. Further understanding might come from the knowledge that nonlegumes also provide carbon through their root excretions and that many bacteria are able to fix nitrogen yet do not enter into symbiotic relationships.

Hunters and the Hunted

Many types of micro-organisms feed upon other micro-organisms. Often, the same microbes which did the feeding then serve as prey and hosts for other micro-organisms and macro-organisms in elaborate food chains. There is a vast body of literature on laboratory studies of some of these general kinds of relationships, but far less is known about the ecological role played in nature by those species that parasitise or prey upon micro-organisms. This shortfall in knowledge surely affects us, because it is almost certain that ultimately some of the predator-prey relationships must affect either human health or farm production. At least in part, this shortfall in terms of understanding how microbial predator-prey relationships affect natural systems probably exists because it is easier to carry out laboratory studies with only two organisms living in culture media than to establish the very significant but far more complex interactions which naturally occur in highly heterogeneous microbial communities. Sadly, however, our understanding of the importance of specific predator-prey systems also may be biased because all too often it depends on the interest of the researcher and the ease of working with some organisms as compared to others. Thus, the more visible and stepwise simpler topics and trophic chains tend to be better examined. The complex and more difficult but potentially far more important topics then get ignored. In turn, the relative abundance of literature may lead the non-ecologist to conclude that a particular group of parasites or predators is indeed very important in nature and that all other groups are unimportant. The truth, then, as to which players and roles are the most significant is not yet revealed.

Among the microbial predators are protozoa, myxobacteria, acrasiomycetes, and a number of chlorophyll-containing flagellates that are often but not always classified as algae. Of these various groups, there is incontrovertible evidence for the importance only of protozoa in determining the structure or function of microbial communities. Waste-waters, many surface waters, and the rhizosphere of actively growing plants often contain large numbers of protozoa, and a high percentage of the individual protozoa are present in a trophic stage that is associated with active feeding on bacteria. A single protozoan cell, upon division, consumes 10^3 to 10^4 bacteria.

Moreover, by use of eukaryotic inhibitors, it has been shown that inhibition of protozoan growth is associated with the maintenance of large numbers of potential prey individuals and often the existence within a community of populations that otherwise would have been suppressed. Thus, eliminating protozoa can alter homeostasis.

Nevertheless, despite their abundance and activity, protozoa generally do not eliminate their prey. It may appear that a predator which is able to markedly reduce the size of a large population is somehow unable to effectively destroy a small population. Instead, however, it seems the case that the protozoa and most other predators are prudent. The elimination of prey by an obligate predator would result in the subsequent elimination of the predator species itself. This prudence in feeding appears to be density dependent rather than due to altruism and often has been interpreted in terms of the balance between energy needed for hunting and energy gained by the hunt. Thus, when a predator population is getting more energy by feeding on the prey, as would occur at high prey densities, than it needs to hunt for those prey, it will continue to graze. On the other hand, at low prey densities, the predator must use considerable energy to move to the few surviving prey cells or to bring those survivors to itself for consumption, and it, thus, will not continue to feed on that species.

The options then available to the predatory protozoan would be to begin feeding upon another species, die of starvation, or encyst and hope for a happier future when its prey species would again be more abundant. This is not to say that the predator cannot eliminate a single prey species, which it will do if the density of alternative bacterial prey is sufficient to sustain feeding. Thus, the effects of protozoa may be viewed in terms of reducing the size of the bacterial community and also, under circumstances when the bacterial community is large, possibly eliminating individual species. Still, soils contain higher bacterial densities, at least by total counts, than are predicted on the basis of the capacity of the indigenous protozoa to reduce the bacterial densities. For example, the total count of bacteria in soil may be in the vicinity of $10^9/cm^3$, but in solution the protozoa may reduce the bacterial density to approximately $10^6/cm^3$. Therefore, it is likely that there is a refuge in soil that possibly results from the inability of protozoa to penetrate pores large enough for bacteria but too small for the predator or from the inability of protozoa to feed on bacteria that are adsorbed to particulate surfaces. Adsorption onto particulates would effectively convert the bacteria into objects too large to be engulfed by the protozoa. It also is possible that susceptibility to grazing by protozoa is checked by other factors, perhaps including evolutionary pressure favouring potential prey organisms which can effectively conceal themselves by changing their 'signature' of secreted compounds.

Less defined and still subject to extensive discussion are the roles of virus and *Bdellovibrio* species in nature. The viruses are acellular; bdellovibrios are cellular and in appearance are much like small bacteria. The bacteria, algae, fungi, protozoa, and archaea, plus all species of plants and animals, are affected by viruses, which are obligate intracellular parasites. Bdellovibrios can replicate in very complex defined media, but presumably in nature they exist as obligate intracellular parasites of bacteria. The ecology of their host, or prey, is integral in the ecology of both virus and *Bdellovibrio* species. Indeed, as intracellular parasites, the viruses and bdellovibrios rely on a host organism for their development and even their very existence in nature.

WHY THEY ARE WHERE THEY ARE WHEN THEY ARE THERE

'What are the major global pollutants?' This would be an interesting question to put to an audience of intelligent lay individuals. Most of the answers could be predicted. Typically, similar answers might be obtained if one addressed the same question to many ecologists, microbiologists, and other scientists. The answers probably would include air pollutants, ozone, pesticides, polychlorinated biphenyls, heavy metals, etc. However, the available morbidity and mortality data give answers that are quite different and perhaps unexpected. Indeed, on the basis of global morbidity and mortality data, the major pollutants are micro-organisms, and they are the chief environmental stressors which deleteriously affect human health. Pneumonic infections such as tuberculosis and influenza, diarrheal illnesses such as cholera, typhoid fever, and bacterial and amoebic dysenteries, and circulatory infestations such as malaria are among the major causes of human death and suffering. This may not seem to be true for many individuals living in developed countries, who are healthy and well-nourished and for whom good sanitation and wealthier lifestyles can afford the luxury of pre-occupation with the psychological issues associated with chemical pollutants. However, the threat from microbial contaminants cannot be ignored without perilous consequences. Understanding the ecology of these microbial contaminants provides us with clues that can help us to minimise the peril.

Boundaries and Barriers

One of the concepts which is important in consideration of the ecology of the members of a species is that in nature each species has a boundary. As indicated earlier in this chapter, boundaries are operationally defined, and the physiological limitations of a species will not allow its members to actively function in the conditions found beyond that species' boundary. For example, we know that almost no species of fish can survive for very long when out of water. Thus, the interface between the atmosphere and the surface of the water represents a boundary for those species of fish. Equally, most terrestrial species cannot survive unaided for very long when submerged in water, so the water's surface likewise serves as a boundary aspect for those terrestrial species. The ecological boundaries of many amphibians and reptiles cross the water's surface, attested to by the fact that those organisms have both aquatic and terrestrial capabilities. The concept of a boundary includes a complicated mixture of factors, sometimes very highly species specific. However, we often can identify both biotic and abiotic components for some of those factors; i.e. oxygen requirements (biotic) may restrict the altitude or depth (abiotic) at which the members of a species can function. A species cannot move its boundary except through evolution. For most species, this evolution must be biological in nature. Humans, however, can move their boundary by cultural evolution. For example, although we lack effective body insulation by fur, the development of clothing and weather-resistant shelters allows us to live even in polar environments, places whose physical conditions normally would make them beyond humans' boundary. Development of the aqualung, an early portable, self-contained underwater breathing apparatus (scuba), and modern re-breathing technology represents a cultural evolution which allows humans to exist underwater at least temporarily. Within the confines of its boundary, a species will find at least one habitat in which its members successfully can reside (otherwise, it would become extinct). While a particular species cannot cross and survive beyond its own boundary, its habitat may be sufficiently diverse that, while moving within that habitat, the members of that species may cross the boundaries of many other species. This overlap of boundaries allows biological interactions between the members of different species.

The parts of a species' habitat may be contiguous or may be separated by barriers (Table 2.1). These barriers can be physically defined and are determinable by physical measurements. They also may be tangible. For example, a species' oxygen requirements may turn geographical features into barriers. Barriers are not fixed with regard to time and space. Rather, they can appear, disappear, and move with time (e.g. as mountain ranges rise and fall, glaciers advance and retreat, continents shift toward either more polar or more equatorial zones, and competing species evolve or become extinct). Species tend to evolve survival mechanisms which allow their members to successfully move as barriers move and even evolve mechanisms which allow them to cross those barriers. Indeed, some species have evolved to a situation in which cyclical migrations between parts of their habitat are necessary for continuance of the species. For example, seasonal migrations occur among many species of bats, birds, butterflies, caribou, salmon, and whales. Even maple trees migrate within the northern hemisphere, albeit on a longer time scale, advancing and retreating with the glacial cycle. In the cases of monarch butterflies and maple trees, no individual member of the species completes the full migration. The capability of a species' members to migrate also facilitates colonisation of new areas. Cyclical migrations could be viewed as highly evolved forms of dispersal.

Staying at Home versus Venturing to New Places

What does the importance of micro-organisms as major pollutants have to do with dispersal? Dispersal of the host population can reduce the general incidence of disease. For example, sometimes wealthier humans can flee to other, presumably more healthful areas. In those preferred destinations, costly medical capabilities such as prophylaxis and chemotherapy have some impact upon either reducing the incidence of or ameliorating the misery resulting from microbially induced diseases. Sadly, these luxurious options do not exist in remote or poor areas of developing countries. For people in much of the, world, the major means of controlling or preventing these infectious diseases is by interfering with microbial dispersal. Dispersal, thus, not only is important for populations of humans, animals, and, plants as they try to survive the onslaught by pathogenic, microbes but also is critical for viruses and other types of infectious or otherwise dependent micro-organisms in their own efforts to maintain their existence by keeping up with their dispersed hosts. Dispersal also is a key factor for free-living archaea, bacteria, fungi, protozoa, and algae, which must seek new home territories in order to avoid competition against their parental populations for available space and nutrients.

Some of the scientific interest in microbial dispersal comes from basic research. There have been major breakthroughs in the use of genetic and biochemical techniques to understand chemotaxis, which is, movement in response to a chemical stimulus. In addition, a number of very useful mathematical models have been developed in aerobiology and for use in predicting microbial dissemination through aquifers and soils. Interest in microbial dispersal also comes from concern with bacteria and viruses that cause diseases of importance to humans and with fungi that, as a consequence of their aerial movement, are contributors to decline in food and feed production. The disposal on land of agricultural and urban wastes containing pathogens has also resulted in considerable research and monitoring of microbial movement.

Potential new habitats regularly become available for microbial colonisation. For example, habitats appear as seedlings emerge from seeds and with the extension of growing root or shoot tissues, during wounding which creates access to the interior tissues of multicellular beings (including damage to fruits), with the first appearance of a newborn infant, and at the site of bruised tissues. All of these habitats contain sterile sites which could be inhabited by a variety of micro-organisms that might find the sites through successful dispersion. Similarly, possibilities occur when nutrients arrive at new locations in water bodies because of vertical and horizontal mixing of water, and new environments are created both by soil erosion and by the building of dams and ponds. The first half of the earlier statement that micro-organisms are everywhere, or at least potentially everywhere, is clearly a platitude when one considers that the issue is not whether a micro-organism will appear but whether members of a particular species will have reached a site in which they can grow. However, the organism also must have evolved the capability of staying at that site long enough to grow. Staying around to grow often involves adhesion to the existing surfaces and the formation of complex biofilms. Humans are just another part of the biofilm which has evolved and spread itself over and into the nooks and crannies of this planet. Finally, after adhering, the organism would have to successful compete for nutrients and defend itself against any organisms already found in that site.

Sometimes microbial communities completely consume the supply of limiting nutritional elements or other limiting resources available in the specific sites where they reside. If there is no further input or regeneration of a limiting nutrient or resource, then the species that make up those communities will die away at those sites. Should this happen in all of the ecosystems within which some particular rare species resides, and if the members of those populations have not evolved adequate survival mechanisms, extinction will occur.

Avoiding extinction, thus, can require that a species have evolved a means of escape either in time (via a persistence mechanism) or in space (via a dispersal mechanism). Escape in time can occur by the establishment of starvation existence, wherein an individual shuts down unneeded metabolic activity and hopes for the arrival of nutrients before it dies from starvation. Starvation modes are common among aquatic bacteria species. Escape in time also can occur by the combined development of metabolic inactivity and protection of an environmentally resistant structure. This combined approach is represented by the spores formed by fungi and by some species of bacteria, the cysts and oocysts formed by many types of protozoa, and the nucleocapsid structure formed by viruses. Escape in space dictates dispersal, leaving in search of a more favourable environment.

Travelling by Air, Sea or Land

Micro-organisms have developed many means for dispersal, and each species must have one or more methods for accomplishing its migration. A few microbial groups have evolved specialised structures to launch their cells into the air or send them swirling into the water. These organisms depend upon those vehicles to deliver their progeny to a new home. There are parasitic species which possess mechanisms which cause changes in the behaviour of their animal or plant hosts that result in dispersal of the parasite. Still other micro-organisms utilise very efficient vectors, such as mosquitoes and biting flies, which almost unerringly deliver the progeny microbes to the correct address. Some means of dispersal have a higher risk of failure than others, and species with more efficient dispersal mechanisms do not require as great a level of production and shedding of new propagules because each of their released propagules will have a greater likelihood of success at encountering a hospitable site.

Active dispersal, in which the physiology of the micro-organism controls its transport (automobility), has been the subject of considerable research, probably because it is more comforting to the microbiologist to have his or her pet control its own fate. Active dispersal, either by motility or by growth of filaments, may be somewhat random rather than resulting from a specifically directed movement (taxis) or growth in a specific direction (tropism). Still, the motility of protozoa and the growth of either bacterial chains or the filaments of fungi in soil and algae in water are sufficient to result in those species successfully encountering and colonising new environments. As a consequence, considerable research has been done on the effects of chemical stimuli. In the short term, active microbial motility generally is restricted to limited distances because of the energy requirement for movement either toward or away from a chemical stimulus. In the long term, the result can be global dispersals and migrations.

In contrast with active dispersal, in which the extent of spread is limited, passive dispersal can result in relatively quick dissemination of an organism to locations metres, kilometres, or hundreds of kilometres away from its original reservoir. Aerial dispersal is frequently of considerable importance to fungi. For organisms dispersed through the atmosphere to be successful, they must produce enormous numbers of propagules to survive the random chance that one or several might alight in an environment in which that species can stay around and grow. A propagule that is transported through the air must have mechanisms to overcome three major hazards: radiation, desiccation, and extremes of temperature. It is not clear in all instances what physiological adaptations are responsible for successful resistance against such environmental hazards. However, the presence of thick walls and dark pigments in many fungi transported through the air and of carotenoid pigments in many aerially dispersed bacteria suggests that these features are important adaptations for this mode of transport. The extent of migration of some of these organisms is truly impressive. For example, in a single year *Helminthosporium maydis*, a pathogen of corn, may spread over areas of thousands of square kilometres, doing enormous financial damage to the intensively planted corn crops of North America. Similarly, spores of the fungal genus *Hemileia* apparently have spread from Angola to Brazil via trade winds which cross the Atlantic Ocean, a distance of more than 1000 km.

Passive dispersal in water and soils has also been the subject of considerable inquiry, in part because of public health problems. For example, algae associated with red tides move for some distances and then suddenly create blooms (population explosions visible to the eye because of the pigmentation of these organisms), often only to be decimated by viruses. Still, the toxic products from those blooms can bioaccumulate in the food chain and may have major consequences for aquatic fauna and people who consume seafood animals from the area of the algal bloom. The interest in passive dispersal through soil is often a result of concern with the vertical migration of bacteria or viruses that cause human disease and the entry of those micro-organisms into aquifers underlying soils into which the micro-organisms were inadvertently or deliberately introduced. However, appreciable research on the vertical movement important to microbial colonisation of roots has also been conducted.

Bacteria, viruses, or fungi that are transmitted by living vectors typically have efficient dispersal mechanisms and require fewer propagules for the species to be maintained.

Prepared for the Climate and What about the Location?

Climate? Even geography? Yes, climate is an important consideration for microbes, and there is a microbial geography. Restricted distributions on a macroscale as well as on a microscale characterise all groups of organisms, and the microbes are no exception. The literature dealing with the geography of microbial groups is often unknown to laboratory scientists, but an investigation of microbial communities in natural environments quickly shows marked and sometimes extreme localisation of microbial groups. For example, geographical distributions are evident among aquatic and terrestrial algae; free-living and pathogenic fungi; protozoa in marine water, freshwater, or soil; and bacteria in countless habitats. Many genera, and often even species, seem cosmopolitan, but their widespread distribution does not mask the restricted nature of their occurrence within particular regions, sites, or microenvironments.

A key element for any particular species of alga, archaeon, bacterium, fungus, or protozoan to successfully exist in an environment is the ability to endure all of the abiotic stresses characteristic of that environment. These stresses include essentially the same list of things which can serve as either physical or chemical barriers: factors such as unsuitable pH or temperature (what is just fine for one species may be too high or too low for another species), occasional drying or freezing in some environments, intense solar radiation, high pressures deep in the ocean, or salinity in certain terrestrial or aquatic ecosystems. The combination of these factors often produces amazing outcomes. Some algae can be found broadly distributed over the surfaces of lakes and oceans but exist in patches at each site. Some of these patches are no more than a few centimetres across, whereas other patches extend over areas of more than 300 km^2. Well-known to phycologists are the limited distributions of diatoms. Some diatom species are present only in subtropical or tropical waters, whereas others characteristically are found only in the Arctic or Antarctic and even there often are stratified by depth. Other algae have snowfields as their habitats. Bacteria such as *Beijerinckia* species similarly have a restricted distribution, being commonly but not solely present in soils of the tropics or subtropics. Often, the biogeography of a micro-organism that is transmitted by a living vector or that is an obligate parasite is determined largely or exclusively by the biogeography of its vector or host. This is true of both the protozoan genera *Plasmodium* and *Trypanosoma*, for example, and the explanations for their distributions, thus, are quite simple. This also is true for vector-borne viruses, which can exist only within the same geographical areas where both their host and the vectoring species can be found. In the case of insect-vectored protozoans and likewise insect-vectored viruses, success requires a community interaction of the residential predatory parasitic species (remember that despite the terminology often used by protozoologists, both the protozoans and the viruses are parasitic) which lives inside the vector, a predatory but non-residential parasitic vectoring species (the mosquito), and the free-living host species which simultaneously serves as prey for both of the predators. Not quite as simple to explain are rare distribution patterns of pathogenic micro-organisms that do not have a known vector and whose distribution does not seem obligately associated with a particular known host organism. A notable example is *Coccidioides immitis*, which has a unique distribution in the Western Hemisphere. This organism causes disease in humans, but although that host is largely global, the fungus is restricted to certain localised geographical and climatic sites. This fungus is characteristically found in the soils of certain semi-arid regions that typically are exposed to high temperatures, receive little rainfall, and frequently have high salinities. Although *C. immitis* is transmitted by wind, wind movement alone cannot account for its biogeography. Even if the fungus is cold intolerant, that possible sensitivity would not explain why it is not present in warm, humid areas to which it may be carried by the wind. Having failed to find a climatic reason for this geographical restriction of the habitat of *C. immitis*, we may in time find that the answer lies in an examination of biotic factors. It is possible that this narrow operational habitat of *C. immitis* is related to the need for a supporting community composition and interactive community structure, which then determine the sites where the organism is successful and, thus, is likely to be found. Conversely, then, an unfavourable community composition or structure may prohibit the organism from residing in other potential habitats.

Natural geographic and climatic restrictions are called zonations. Zonation is evident even at a micro-environmental level. Marked horizontal and vertical differences in either the distributions or the occurrences of algae, bacteria, and fungi are evident in waters, sediments, and soils. Such microscale zonations almost certainly reflect biologically important differences in the physical and chemical characteristics of the environment. However, in only a few cases have the causes of this highly localised micro-environmental distribution been established by either physical or chemical analyses. Among the factors either known or postulated to be important in microscale biogeography are nutrient concentration and type, temperature, pH, oxygen, grazing by zooplankton, mechanical barriers, and inhibitory substances (some of which may be biotic, while others certainly are abiotic).

The fitness traits of the initial colonists in environments that are largely free of micro-organisms or that have been drastically disturbed are often reasonably simple to predict. Such environments are not unknown, and they exist, for example, on the previously uncolonised surfaces of roots growing through soil, on plant materials, that become bruised, and in waters that receive sudden influxes of organic or inorganic materials. The fitness traits associated with successful colonisation, and with an organism's hoped establishment, frequently represent the basis for a capacity to use the organic nutrients present at the new site. For example, with root-dwelling microbes, those nutrients may be either compounds excreted by the emerging root segments or constituents of the tissues that become bruised and, thus, accessible for microbial utilisation. Frequently, the initial colonist pre-empts the site so that a propagule that arrives later but has the same enzymatic capacity will be unable to multiply. In these instances of pre-emptive colonisation, the key fitness trait is frequently associated with dispersal (getting there first).

Finding common traits among organisms can indicate a similar ancestry, albeit horizontal gene transfer can scramble phenotypic traits and complicate our efforts at understanding how phenotypic traits and physiology relate to ancestry. Still, the very fact that species can be distinguished is evidence that each has some uniqueness with respect to its combination of biochemical, physiological, and morphological traits. Some of these distinguishing traits help explain the presence of a species in one environment but its absence in another and the relative abundance and activity of the various inhabitants which exist in a given environment. This combination of a species' traits provides the basis for natural selection, accounts for the geography of a species, and explains the role (i.e. the niche) of a species in its environment. The sum of the traits of all species present within an ecosystem represents an identifying signature for that ecosystem.

Nutrients in the Food?

An obvious need that must be satisfied for an organism to become established is the presence of all nutrients that it requires. Often, in aquatic environments, phosphorus is a limiting nutrient for photoautotrophic organisms. Many natural terrestrial ecosystems contain most and sometimes all inorganic nutrients in concentrations sufficient to maintain a reasonably large community. Heterotrophic organisms often seem more finicky than autotrophs. Thus, when considering the needs for heterotrophic organisms and any single specific site, we often find that overall the supply of energy resources is the limiting factor and that the kinds of carbon sources available will affect the selection of organisms which can maintain a presence in that site. In most environments containing readily or slowly available organic molecules, a variety of dissimilar propagules able to use those carbon sources arrive, yet only a few become established. In this instance, as in so many cases in environmental microbiology when causation is being sought, the presence of a suitable energy source is a necessary but not sufficient requirement for establishment.

How can we understand what sufficiency requires? We gain this knowledge by understanding the basis for natural selection, or selection in nature. From an environmental viewpoint, the basis upon which selection operates is the set of fitness traits that underlies ecological success and sometimes the achievement of community dominance by the members of a species. These traits are the specific biochemical, physiological, or morphological characteristics that determine a species' boundary, barriers, habitat, and niche. Nature tries many different approaches to solving a particular problem. Evolution is based upon the fact that those solutions which work are retained and hopefully will again prove adequate at some future point. The issue for scientists is to identify those attributes of organisms that are necessary (the organism could not do without them) and sufficient (with those, the organism will have a chance at competitive success) to enable the organism to survive and occasionally multiply in particular environments. After we first show the importance of tolerance to the abiotic factors that are detrimental to one or another group of organisms and secondly understand the need for a supply of inorganic nutrients and a carbon or energy source, then what? The methodologies associated with enrichment or selective culture techniques are ideal for examining natural selection under artificial conditions. But we must understand the associated limitations of these methodologies and of artificial conditions. The technical limitations focus on the fact that the use of such methods to examine the community of organisms present in an environmental sample typically results in finding only the one organism that grows fastest under those artificial conditions. In contrast, the competitive conditions in nature are such that many inhabitants of natural communities grow slowly. Even more confounding to our efforts is the reality that most of the organisms existing in nature do not appear to grow at all under conditions imposed by laboratory media. The reason why they do not grow is simply that those organisms evolved to grow in nature and have only recently been introduced to that brand or composition of culture medium. Hopefully, newer biochemical tools will help us in sorting out many of the remaining answers, allowing us to achieve success in areas where present cultural techniques are not sufficient tools.

Settling in with Relatives (In-Laws and Outlaws)

Abundant literature on the biochemistry, physiology, and morphology of micro-organisms has been derived from studies of pure (axenic) cultures of organisms conducted under artificial conditions, usually in liquid media, and in the absence of any other species. However, it is far from certain which of those biochemical, physiological, or morphological properties either truly are of ecological significance or facilitate the establishment of a particular microbe in a particular site. Many facile extrapolations have been made from *in vitro* to *in vivo* conditions, but rarely have these extrapolations been verified as being ecologically relevant. We can surmise that if an organism identified in culture is unable to survive and grow when exposed to artificially simulated environmental stresses, it is unlikely to be an inhabitant of an environment in which those stresses occur naturally. Some environmental stresses are easy to establish and duplicate in the laboratory, and it is simple to show by process of elimination which of these may be important in determining the absence of a specific organism in a specific habitat. Conversely, understanding the presence of a particular organism in a particular place and time can be very difficult. Many species are transported to environments in which they are able to tolerate all the abiotic stresses, but they still do not become established. What lies behind this failure of establishment? We must surmise that it stems from the homeostatic mechanisms operating in the community in which the new arrival alights and that these mechanisms are extremely effective at eliminating many of the arrivals. Yet the exact reasons for eliminating or allowing the continued presence of the new potential inhabitants are rarely understood.

We do know that with time, the initial colonising species present at a site can become displaced as more recently arrived organisms become abundant within the community. These displacements represent natural successions. The initially dominant species often seem to become of less significance to the community as a succession proceeds, and some of the initially dominant species may be eliminated totally. Those displaced species succeed overall by having dispersed propagules to colonise other, more recently available sites before their parental population loses its foothold on the present site. Yet, despite these successional eliminations of specific species, often the overall trend of succession with time is towards an increasing species diversity. As succession proceeds, the identities of the fitness traits associated with allowing any single species to maintain a place within the community become increasingly less certain. However, we do know that among the factors that either contribute to or determine selection during succession are: the availability of nutrients that are synthesised by the temporally preceding species, the alterations in concentrations of inorganic nutrients and the formation of toxic products by those earlier arrivals, successful competition for limiting resources (especially the supply of organic carbon), and the inevitable appearance of organisms that parasitise either the pioneering species or the subsequent colonists. The last factor tells us that if a species chooses to remain in the community, it must successfully dodge or repel parasitic organisms such as viruses and grazing organisms such as protozoa and invertebrates.

The displacement of organisms which occurs during the process of colonisation and succession ultimately leads to the climax community. This is the assemblage of organisms most characteristic of any habitat. Often the communities of chief concern to environmental microbiologists are the climax communities, because they tend to be the most prevalent. The climax community tends to reproduce itself and may remain similar in composition with the passage of time. The organisms making up the climax community interact in a variety of ways. The relative numbers of each of the component species may fluctuate and may even go through cycles. However, as a general rule, the component populations of a climax community are not eliminated. At this stage, the nature of the interactions between species is difficult to unravel, and the specific fitness traits that underlie any particular organism's position in the community may seem impossible to sort out. In some instances, a community is dominated by a species that has pre-emptively colonised the site, and its role is associated with its presence at the site before other organisms arrived. In other instances, however, a variety of organisms endowed with appropriate physiological capacities have reached the site, and they, thus, may end up competing over some factor that they all need but which is in limited supply. Undoubtedly, competition is one of the major interactions in climax communities. As micro-organisms usually grow readily, when growth freely is allowed, one or more factors in the environment must become limiting, and these limiting factors serve as the basis for competition. For communities dominated by heterotrophic bacteria and fungi, the limiting factor is frequently the supply of available carbon. Many environments, such as soils and sediments, contain large amounts of organic matter, but much of that organic matter is not in a readily available (biologically utilisable) form. Thus, the heterotrophic organisms that are successful are often those that are able to make use of the less readily available organic materials. In communities containing chemoautotrophs, the limiting factor is frequently the supply of the inorganic compound or ion that serves as their energy source. In surface waters, the limiting factor for the photoautotrophic algae or cyanobacteria is often the concentration of either phosphorus or nitrogen.

By way of summation, what determines the outcome of competition and obtaining a place in the community? It is tempting to suggest that the successful competitor is the organism that grows most rapidly. However, the fact that many of the dominant organisms in natural environments do not grow quickly suggests that it is imprudent to extrapolate to natural environments from growth rates obtained by studies performed with pure cultures under non-competitive circumstances. Clearly, more is involved than simply growth rate *per se* because of other stresses which exist in nature. Likewise, we must consider the need for an organism to be transported to sites where there is an available supply of limiting nutrients and the need to avoid predation after it has arrived. Sometimes, dominance does result simply from being the first arrival, but even those first arrivals may lose dominance and become displaced during community successions. Perhaps it is simply 'peacefully fitting in' which describes the recipe for success. Alternatively, it may be subterfuge. Both thievery and enslavement of other species may be involved. The success of new arrivals and their inclusion in the community also may depend in part upon the biochemical weaponry which those species bring with them, since microbes seem to have invented the concepts of laying siege and attack.

UNDERSTANDING ECOLOGY

Perhaps the key aspect of understanding ecology is development of a comprehension that the individual species are connected, i.e. that the ecology of one species coordinates with those of other species. This coordination usually includes both macrobes and microbes. All of these species then function together as a community. An ecologist has been euphemistically defined as an individual whose feet are firmly planted in midair, possibly because of the enormous time which must be spent pondering it all. Perhaps, like Buddhist monks, we seek some 'true understanding of nature' as we regard both interrelations among the activities of different species and their interactions with the physical environment which surrounds all of them. Much of ecology including environmental microbiology, is concerned with gaining insight through the process of basic science. Indeed, basic science can help us to find the keys which then allow us to open some of the seemingly locked doors beyond which lies a better understanding of nature. However, it is also clear that environmental microbiology is partly an applied science. It has much to offer to our knowledge regarding the maintenance and restoration of environmental quality, prevention of the transmission of diseases of animals and plants, and approaches to improving human health.

The guardianship of environmental quality relies upon understanding of the role of micro-organisms in preventing pollution, destroying noxious organic materials before their concentration becomes objectionable, or destroying toxic chemicals before they have an impact on humans, animals, or plants. Indeed, using information which they have gained through their studies, environmental microbiologists now are engaged in bioremediation technologies, designing ways to enhance the capacity of microbes to bring about the destruction of pollutants.

The epidemiology of communicable diseases is, to a significant degree, an extension of evaluations of microbial dispersal. The spread of viruses, bacteria, fungi, and protozoa, thus, has a tremendous impact on the protection of plant, animal, and human health. Plant pathologists have long recognised the significance of information on the ecology not only of the disease-producing fungi, bacteria, and viruses but also of other micro-organisms, many competitive and some even protective, that reside in the same habitats. Many plant diseases are not effectively controlled by chemical agents or by sanitation procedures, and it is the activity of those other, non-pathogenic members of microbial communities that must be used to form the basis for effective control of particular diseases. Much of environmental microbiology research is interdisciplinary. For example, aerobiologists and soil microbiologists frequently interact with plant pathologists. Preventive animal husbandry and human medicine have acquired valuable information from environmental microbiology. Notably, in the field of medical treatment, the development of antibiotics and other forms of chemotherapy has provided us with some powerful tools, many of which have been based upon knowledge of upsets and restorations in microbial communities.

Thus, knowledge of microbial community structure and community function has a key role in improving our lives. Furthermore, this knowledge will aid us in understanding ways of maintaining the environment and its microbial communities. As John Donne correctly stated, 'No man is an *Iland*'. Carrying his understanding a step further, humans and all macro-organisms are connected to the microbial community and have evolved from it, and our future depends upon that community.

SECTION II

General Methodology

3. General Methodology: A Review   45

4. Prokaryotic Diversity: Form Ecophysiology and Habitat   51

Chapter 3

General Methodology: A Review

INTRODUCTION

Microbial ecology is the study of how micro-organisms interact among themselves and with their surroundings. Understanding the structure and function of a microbial community requires more than recognition of interactions between microbial species. It requires quantitative information on the numbers, kinds, and activities of micro-organisms present within that ecosystem. While it has been possible to obtain an accurate assessment of the microbial biomass within most ecosystems, an accurate estimate of microbial diversity is less straightforward. The number of microbial species in an ecosystem can be potentially very large, with as many as 4000 different genotypes represented within a gram of soil; therefore, determining the number of each kind of micro-organism is a daunting task. Furthermore, many of these micro-organisms have been identified only as nucleic acid sequences obtained from environmental samples, and no representatives are yet available in culture, thus making it impossible to assign specific physiological functions to these organisms; however, it has been possible to assign general physiological attributes to organisms affiliated with some phylogenetically defined groups (e.g. methanogens, sulphate reducers, and nitrifiers).

Finally, the species concept for bacteria is still very much in flux, as it is for other organisms. Thus, different investigators may use different metrics of diversity, depending on the specific goals of the study. Coupled with these problems is the realisation that microbial activity occurs on very small scales (micrometres to millimetres), which are much smaller than most sampling scales (which are often on the order of centimetres).

CHALLENGES AND NEW TOOLS

These challenges have led to a revolution in the development of new tools to study microbial ecology. New methods that permit the investigator to detect, identify, and quantify micro-organisms directly in the environment, to assess the microenvironments in which these micro-organisms reside, and to determine the activity and physiological state of micro-organisms in their natural habitats have been developed. Additionally, methods to culture entire microbial communities permit the study of important community-level considerations, such as the rules that govern the development of community functions, e.g. nutrient cycling and homeostasis. These methods provide a format to study higher-level processes in a controlled and reproducible manner. Coupled with the development of these new technologies is an every increasing ability to culture unusual micro-organisms and to track the evolutionary heritage of a micro-organism.

Most of the sections in this chapter deal with ways to quantify the numbers and kinds of micro-organisms within an ecosystem. As mentioned above, such information is fundamental to any ecological investigation. Given the complexity of microbial communities, it is essential to use an approach or combination of approaches that provides the information needed to answer the particular questions under study. Rather than inventory all approaches, this section emphasises the key conceptual and analytical principles of each method so that the investigator will know which approach is best for the question being addressed.

Microscopy

Microscopy has been one of the primary approaches used by microbiologists for hundreds of years. The development of new hardware and the advent of software for digital image acquisition and analysis have put microscopy at the forefront of microbial ecology research. By using analytical image analysis, it is now possible to assess the microenvironment, the 'activity' state of an organism, and that organism's taxonomic status.

The development of fluorescent probes makes it possible to obtain quantitative information on the taxonomic and phylogenetic affiliation and the physiological state of individual cells in complex associations of organisms as they occur in their natural environment.

An important advantage of scanning confocal laser microscopy is that it provides a format for nondestructive, *in situ* optical sectioning of biological material without fixation, thereby avoiding many of the artifacts frequently encountered with light and electron microscopy.

Pure Culture Approaches

Pure culture approaches have traditionally played an important role in many microbiological investigations, providing a relatively inexpensive assessment of the numbers, kinds, and metabolic activities of micro-organisms. Also, pure culture methodology is still the definitive approach to demonstrate that an activity is microbially mediated.

A cursory review of the literature quickly reveals that a great number and variety of media have been used to cultivate prokaryotic and eukaryotic micro-organisms, which makes it appear that culturing is an empirical art.

Although there is some art, it is an art guided by an understanding of the basic concepts of microbial nutrition and the factors that control microbial growth.

Cultivation of microeukaryotes

Cultivation of microeukaryotes requires many of the approaches developed for the cultivation of prokaryotes. The cultivation of algae most closely mirrors that of bacteria, since many of the protocols were developed to culture blue-green algae, which we now know are prokaryotes. The cultivation of protozoa is less straightforward. One must first consider whether the protozoan uses dissolved or particulate organic carbon.

For the latter, the choice of the appropriate particulate food or prey organism is important, since protozoa are often selective feeders. Continued refinements of procedures have led to an increasing number of microeukaryotes that are available in laboratory culture as monoprotist, monoaxenic, or axenic cultures. These procedures permit a more thorough understanding of the physiology, evolution, and taxonomy of microeukaryotes than could have been obtained by studying morphological characteristics alone.

Viruses

The viruses that are most often encountered in environmental samples include bacteriophages and animal and human enteric viruses. The presence of the latter in potable water sources is a major public health concern. Thus, it is important to be able to detect low levels of infectious units of these disease-causing viruses. Since viruses cannot be propagated in artificial media, the enumeration of an infectious animal virus requires its propagation in a living host cell.

The development of cell culture and concentration methodologies has made it possible to detect very low levels of environmentally important animal and human viruses. Cell cultures have replaced the need to use laboratory animals, and the optimisation of cell culture methodologies combined with the development of concentration or extraction procedures allows very low numbers of infectious units to be detected in environmental samples.

Pure culture

Except in certain diseases, pure cultures of micro-organisms rarely exist in nature. Indeed, the activities of free-living micro-organisms are most effective only when organisms are present in association with other groups of organisms, usually as structured communities encased in self-produced polysaccharide matrices called biofilms. Laboratory culture of these organisms in association with one another is essential in order to understand the community-level mechanisms involved in specific processes and the effects of perturbations on these associations. Such microbial assemblages can be maintained in laboratory systems by controlling the substrate(s) and its concentration and rate of input, often using systems that provide for both temporal and spatial variations of these parameters. These biofilms have many characteristics of natural communities.

They arise spontaneously, remain relatively constant over time, and exhibit homeostatic behaviour. Also, distinct assemblages of micro-organisms with clearly differentiated transition zones (ecotones) form in response to nutrient gradients.

Systems to study biofilms and other microbial communities should become important tools to test fundamental ecological questions regarding community structure and function and provide a realistic model to test hypotheses in many areas of environmental microbiology, including bioremediation and waste-water treatment.

Culture-dependent methods

Culture-dependent methods to quantify the numbers and kinds of micro-organisms in a sample have a strong bias in that these methods detect only those organisms that grow under the selected set of conditions. It is generally accepted that rarely are more than 1 per cent of the bacteria resident in a sample recovered in culture. For many environments, the physiological status of an organism or the lack of understanding of its nutritional requirements may prevent its successful detection.

An effective and quantitative way to measure microbial biomass *in situ* without the need to culture or to distinguish individual cells in an environmental sample is to measure the amount of certain cellular components, as noted below. If the cellular component is universally distributed, has a short residence time after cell death, and is expressed at relatively constant levels throughout the growth cycle, then this component can be used as a measure of biomass. A number of components could potentially be used for biomass estimation. However, because all intact cells contain polar lipids (primarily phospholipids) and because organisms without membranes are not viable, phospholipid analysis has become an important method for determining microbial biomass.

Community-Level Physiological Profile (CLPP)

One problem with the interpretation of biochemical biomass data is that the units of measurement are different from those traditionally used by microbiologists, i.e. micromoles of a component rather than the number of cells per unit amount of sample. However, conversion factors provide a mechanism to relate these different measures of biomass.

In addition, phospholipid fatty acid (PLFA) analysis can provide insight into community composition as well as physiological status. Total community PLFA patterns accurately mirror the shifts in community composition and provide an approach to correlate community composition to specific metabolic properties. However, PLFA analysis may not provide a definitive interpretation of shifts in specific microbial groups because of overlapping lipid compositions among different groups of micro-organisms unless the species of interest contains unique lipid components. The combination of signature lipid biomarkers with gene probing should expand the specificity and scope of community compositional analyses.

Problem with Interpretation of Biochemical Mass

Community-level physiological profile (CLPP) analysis (often referred to as the biolog method) is a rapid and relatively inexpensive means of gathering large amounts of information about whole communities of micro-organisms to differentiate among them. The technique is used to examine community responses by inoculating whole environmental samples directly into plates containing wells with different substrates. The patterns of substrate utilisation that emerge from this analysis represent the profiles of the whole community in so far as they are not biased by inferring community properties based on a selected number of isolates.

These profiles do not represent the activities of every member of the community but only of those that grow rapidly under the conditions imposed by the analyses. This method's cell growth requirement means that the CLPP information cannot be used to infer ecological relevance. At best, the results may indicate potential function; at least there may be cells present that are capable of using the substrate. CLPPs have provided groupings of communities similar to those obtained by DNA- and PLFA-based methods, suggesting that the CLPP approach responds to differences in community structure in a manner similar to that of other established techniques. Keeping in mind the constraints discussed above, CLPP analysis does provide large amounts of information that can be used to discriminate between microbial communities.

Nucleic Acid Sequence Analysis

There is no doubt that the development of methods based on nucleic acid sequence analysis has revolutionised environmental microbiology studies. For example, molecular approaches based on 16S rRNA sequence analysis allow the direct measurement of the abundance, diversity, and phylogeny of micro-organisms in almost any environment. By using nucleic acid probes, taxonomic or phylogenetic status of individual cells in complex microbial assemblages can be determined at the scale of individual bacteria (e.g. micrometers). Although the molecular approaches provide the framework for explicitly describing the diversity of the microbial community in terms of the number of unique phylotypes or genotypes which are present, further study is needed to determine how this information relates to more traditional taxonomic ranks such as species.

Also, the proportional recovery of specific sequences cannot be directly equated with traditional measurements of species abundance. Given these differences in resolution, molecular techniques such as DNA hybridisation, per cent G+C content, DNA reassociation and restriction fragment length polymorphism do provide a number of direct methods to discriminate between microbial communities and to quantify specific populations.

The application of PCR to microbial ecology has eliminated the need for large amounts of DNA required for some of the methods mentioned above and is especially useful in comparing ribosomal DNA (rDNA) sequences. Some DNA-based approaches compare individual rDNA sequences to a database of previously encountered sequences in order to assess diversity. Other approaches (terminal restriction fragment length polymorphism and denaturing gradient gel electrophoresis) examine the diversity of electrophoretic patterns of DNA fragments termed 'fingerprints' derived from the manipulation of the rDNA sequence. However, these other approaches also provide phylogenetic information about the populations present.

Microbiological, Chemical and Molecular Tools

The microbiological, chemical, and molecular tools described above are very useful and can be used to provide great insight into the working of micro-organisms in their natural environments. However, these approaches often necessitate the destruction of the sample, and there is often a significant delay before the results are obtained. New technologies such as bioreporter genes can, in certain instances, provide information on the workings of individual cells or populations within a community in real time with continuous monitoring.

Bioreporter gene technology allows the expression of almost any gene to be detected by providing an easily detectable gene product when the gene is transcribed and translated. The application of bioluminescent bioreporters is often limited because oxygen, the aldehyde substrate (which may not be available to the cell), and an active enzyme (which is heat labile) are needed for the reaction. The use of green fluorescent protein (GFP) as a bioreporter has greater applicability than bioluminescence approaches because GFP is heat stable, does not need an additional reactant once the chromophore is formed, and requires oxygen only for the correct folding of the protein. Modifications to GFP to make it susceptible to degradation by cellular housekeeping proteases should allow GFP to be used as bioreporter for real-time gene expression.

Novel detection technologies such as surface plasmon resonance and resonant mirror biosensors have been enhanced by the development of more sensitive detection methods and instruments such as charge-coupled devices to detect a greater range of compounds in many different environmental settings. These technologies now provide a format to characterise the physical and chemical conditions that control and reflect the activity of biofilms and other multispecies assemblages on the scale of individual cells.

Thus, the management and administrative procedures needed to ensure the quality of the data generated by research activities. The reliability of scientific information is undergoing increasing scrutiny in the public sector. This is particularly true in environmental microbiology, where this information is often used for risk assessment and liability decisions.

Thus, we must all be aware of the procedures needed to ensure the continuous reliability of the data and the quality of the measurement process.

Statistical Techniques

A brief survey has been provided of the statistical techniques that can be used for sample design and analysis of environmental and experimental data and some examples of how to use these approaches.

These methods summarises exploratory approaches (both numerical and graphical) used to measure central tendency and variability of the data. Hypothesis testing and analysis of variance to compare the mean and variance of two or more groups of samples described. Regression analysis can be used to examine the relationships between sets of variables and is often used to examine the dependence of microbial populations or activity on environmental parameters. Multivariate statistics provides several methods that can be used to interpret data sets with large numbers of variables and to partition samples into similar groups. The latter is a task that is quite common to taxonomy but is also applicable to many environmental studies. The heterogeneous distribution of micro-organisms and microbial activities in natural environments has hindered the ability to predict accurately the *in situ* rates of biodegradation as well as other microbial processes. The use of geostatistics, in particular variogram analysis, to examine the spatial distribution of micro-organisms has been discussed.

Chapter 4

Prokaryotic Diversity: Form Ecophysiology and Habitat

INTRODUCTION

The prokaryotes are a group of organisms that lack a cell nucleus (= karyon), or any other membrane-bound organelles. They differ from the eukaryotes, which have a cell nucleus. Most are unicellular, but a few prokaryotes such as myxobacteria have multicellular stages in their life cycles.

Although the meaning of the word 'diversity' is ordinarily quite clear, this is not always so when it is used in connection with living organisms. The connotations of the word with regard to large organisms (macrobes) are usually different than when it is applied to smaller ones (microbes). One obvious difference is that with the former group, it is ordinarily organisms with distinctive anatomical and morphological features visible to the eye that are the objects of the habitat census, and it is these traits that lead to enumeration or identification of different genera and species present. The sustained presence of an organism is taken as evidence of its significance — the ability to thrive — in the habitat. Usually there is little regard to, or indeed knowledge of, the organism's precise functions in that habitat. By contrast, establishing the diversity of micro-organisms present by relying on morphological features alone is not really possible; instead, specific physiological capabilities also need to be deciphered in order to establish the different types of microbes present. Invariably, this has required the isolation of pure cultures and determination of the cell traits. The specific biochemical and physiological activities of microbial cells are of profound significance for the habitat. The high surface-to-volume ratio of these very small cells endows them with high biochemical and physiological activities, and this in turn marks them as having major impacts on chemical aspects of their habitats. This impact of nutrient consumption and production formation as a consequence of microbial growth is such that the organisms are properly regarded as determinants of their habitat environments and not merely, as is the case for so many macrobes, organisms that either survive or do not as they respond to environmental changes.

Thus, diversity in the context of the macrobe is more an index of the suitability or degree of habitat health for existing, or desired, macrobiotas, while in the microbial context the diversity revealed by a census leads to establishment of the bases for and maintenance of the chemical as well as many physical features of the habitat. This chapter not only discusses the diversity of microbes and their functions in the more usual environmental sense (association with soils and waters, for example) but also includes a focus on microbes associated, either extracellularly or intracellularly, with macrobes, including animals, insects, and plants. This seems appropriate for several reasons. One is that the alimentary tract of many macrobes has long been considered a tube within the organism, with openings at each end to the exterior environment in which the macrobe exists; invariably the lumen, wall, and other anatomical aspects of such a tube are colonised by microbes. The factors that affect initial microbial colonisation and subsequent survival are, at least in part, those of the environment exterior to the macrobe. A second reason is that phenomena such as the bacterial invasion of plant root cells that results in formation of the oft-described 'nodule' associated with symbiotic assimilation of dinitrogen and the bacterial invasion that results in formation of the light-emitting organs of some marine organisms are reflections of the existence and persistence of microbes both in the macrobes and the environment in which the latter live. For similar reasons, other significant and often not yet well-understood interactions of microbes and macrobes are included in this discussion of the diversity of prokaryotes and of the habitats in which they are present.

RECOGNITION OF PROKARYOTE DIVERSITY: BACKGROUND

Although it has been more than three centuries since what we now call prokaryotes were first seen, it has been only slightly more than a century that the organisms termed bacteria, in particular, began to be studied extensively. As many have pointed out in general terms, especially Stanier in an environmental microbiological context, our perceptions that microbes as a group possess extremely diverse characteristics and live and function in quite diverse habitats (many of which seem 'extreme' to those acquainted only with macrobiotas) had their origins in the extraordinary, novel, and were among the major contributors to the breadth and depth of our comprehension of prokaryote diversity, the crucial functional roles these microbes play in the biosphere, and, accordingly, the recognition that in the absence of these diverse microbial activities other forms of life on this planet would promptly cease to exist. Even though our present understanding of prokaryotic diversity, in terms of form, function, and habitats occupied, is enormous in comparison with that enjoyed some 50 years ago, we are nonetheless regularly made aware of the real limits of that understanding. The enormous success of enrichment culture and related isolation approaches and considerations, along with application of new analytical techniques, has enabled us to see (literally, in some instances) that we understand very little about so many of the presumptive prokaryotes present in nearly every habitat we explore. The 'humbling experience' that resulted from the study of natural populations with then-state-of-the-art light microscopes has become ever more common and daunting as newer approaches to light, electron, and other forms of microscopy have evolved and as tools of a molecular biological nature (e.g. analysis of DNA and RNA sequences and use of RNA-targeted oligonucleotide probes) have been brought to bear in the examination of mixed microbial populations of many habitats.

LIMITATIONS IN COMPREHENSION OF PROKARYOTE DIVERSITY

The past several decades have indeed been heady times in what might be termed the 'golden age of environmental microbiology' (by analogy to the label applied to the period of about a century past in which there were stellar accomplishments establishing specific microbes as causes of specific diseases). It is a rare issue of a major journal that does not contain a report of the cultivation and characterisation of an organism that had been refractory to study in the laboratory or the presence of one with apparently novel features detected in a habitat that had seemingly already been well explored or a report showing that the microbial content of some novel habitat is worthy of inspection. Even so, it is not possible at this time to describe completely either the enormous diversity in features of the many prokaryotes currently recognised, the wide range of habitats in which they are known to occur, or the microbial composition of the populations present in those habitats. There is no little irony in the fact that as we learn more and more about ever more members of the Bacteria and Archaea, we come to realise that we comprehend less and less, both qualitatively and quantitatively, of the diverse features of different prokaryotes and of environmental factors affecting, and reflecting, their growth and persistence. Twenty years ago, some might have surmised that perhaps as much as 40 per cent of the prokaryotic world was recognised and understood; today it is probably overly optimistic to suggest that the figure is more than 5 per cent, if even that. Such has been the progress in the more than 300 years since van Leeuwenhoek not only discovered bacteria but described some diversity in morphology (e.g. cocci and rods), function (e.g. motile versus nonmotile cells), and habitat (e.g. a pepper-grain infusion, tooth and gum surfaces, and sputum). Louis Pasteur, when he observed the 'vibrion butyrique' some 200 years later, laid the foundations of the biochemical, physiological, and genetic diversity that continues to be studied and enlarged today.

Just what we consider, or how we define, 'species' in the prokaryotic context has been perplexing for decades. Recent, thoughtful essays point out how the rather narrow concept we now (mostly) employ actually diminishes the true dimensions of prokaryotic and eukaryotic microbial diversity. These considerations, taken together with those of horizontal (or lateral) gene transfer, undoubtedly will assume considerable import in the coming millennium as recognition of the enormous functional significance of prokaryotic life continues to enlarge.

In this assessment of prokaryotic diversity and the range of habitats in which these organisms are known to multiply (or at least survive), the phrase 'as known at present', even though it is not regularly stated, needs to be kept uppermost in mind. This limitation, then, constitutes a challenge to those interested in enlarging our understanding of microbes' contribution to life on the planet. In the space available for it, this chapter is intended to be not so much a vade mecum as a descriptive outline of the Bacteria, the Archaea, and some new chemical boundaries of their habitats.

FACTORS AFFECTING LIMITATIONS IN OUR UNDERSTANDING

Many different factors are responsible for the fact that we are still unable to describe the prokaryotic world in its entirety. As in any of the sciences, it is quite often major advances in the development and application of new technologies that permit significant increases in our comprehension. In addition, a regular impediment is the limited ability of humans to apply new findings in an integrative way to the microhabitat dimensions and features that are so difficult to describe adequately. Among the factors which continue to be hurdles that we must overcome are those noted below. Only in the most rare instances does the size or shape of a prokaryote alone, invariably deduced by microscopic examination, permit one to identify it and equate its properties with those of an organism already described. This limitation is no less serious if one encounters a novel or 'new' organism. The quite considerable range in sizes and shapes sometimes makes it difficult to determine whether an object is in fact an organism or is instead some inanimate material. In addition, many cells are sufficiently small so as to be at or below the limits of resolution of the light microscope and hence not visible or visible only with special optics (e.g. dark-field microscopy, as for some spirochetes). Still another important consideration is that for a single cell or even a few cells to be detected in a typical field of view by light microscopy, on the order of 10^6 cells per ml needs to be present. Clearly, it would be easy to overlook the presence of a less abundant organism. It is a fact that no single laboratory growth environment (e.g. nutrients or other chemical and physical factors) permits the multiplication and subsequent isolation and culture of even a small subset of the Bacteria or Archaea that we know at present. In addition, in many instances cells are attached to or localised on surfaces or form biofilms from which it is difficult to remove these cells without damaging them. Accordingly, for many habitats it is not readily possible either to enumerate the prokaryote population present or to isolate and then describe thoroughly the types present. Pure cultures, as such, are rarely found in habitats other than the laboratory. Even though we continue to explore the metabolic and physiologic properties of pure (as well as some simple, mixed) cultures in controlled laboratory environments and to use these insights to make reasoned suggestions about functional attributes that might occur in natural habitats, we obviously lack such information about the capabilities of those organisms not amenable to cultivation. Of no little importance are the many demonstrations of the interactions of different prokaryotes with each other, or with other biotas, with the result that the behaviours of members of mixed populations (mixed cultures) are seen to differ from those predicted on the basis of study of component pure cultures. These are but a few reasons why a particular organism's presence, in and of itself, in a habitat does not necessarily provide evidence concerning the precise *in situ* activities of that organism and its contributions to habitat functions.

Another complication is that many bacteria exhibit an enormous degree of metabolic flexibility (e.g. nutrients consumed, metabolic end products, and mode of energy conservation); these traits are often regulated by numerous and complex environmental factors that are difficult to ascertain with accuracy for the microscale environments occupied. In contrast, not all prokaryotes possess such a broad spectrum of traits; instead, many possess a specialised metabolism or physiology. Detection of significant numbers of these types of organisms in a habitat may allow a more confident prediction of their roles in the habitat than is possible for the more metabolically flexible ones.

Highly significant contributions to our present understanding of the relationships between particular microbes and cause-effect relationships in many habitats have resulted from use of the enrichment culture approach.

Many who have carefully employed this approach to determine: (i) whether organisms with already known attributes were present in samples from a habitat being examined; and (ii) whether a novel transformation of compound 'C' to compound 'D', or some other alteration of a habitat feature, could in fact be mediated by a microbe have recognised the need to quantify the numbers of the particular organisms present as an important aspect of establishing causal relationships. Such quantification often involves use of most-probable-number studies employing particular enrichment culture media and environmental conditions as well as determination of biotransformation rates under, for example, conditions closely approximating those of the habitat in question.

Even though application of relatively newly described 'molecular probe' technologies is argued to obviate the need for bringing organisms into culture in order to describe the microbial composition of the habitat studied, we need to recognise some caveats associated with this methodology. One is that, in general, the probes are designed on the basis of our knowledge of molecular signatures of well-described and understood organisms, that is, organisms whose properties have been discerned from the study of pure cultures. As long as the habitats being examined using these probe methodologies contain no heretofore novel, unrecognised organisms with, accordingly, previously undescribed molecular signatures, the phrase 'there is no longer a need to culture' is to a degree correct. On the other hand, because of the inherent metabolic and/or physiologic flexibility, noted above, of many prokaryotes, even if probe studies reveal the presence of (for example) sulphate-reducing bacteria in a habitat, this information alone does not establish whether it is sulphate, sulphite, thiosulphate, some other sulphur-containing electron acceptor, or even nitrate that is being anaerobically respired by the population or whether it is lactate or some other electron donor that is serving as the energy source. Still, the potential and utility of the molecular probe approaches, when used in combination with those of classic enrichment culture for habitat analysis, are very likely to lead to greater comprehension, as demonstrated by an elegant study of the ecology of methanotrophs, a physiological group with a notably limited nutritional spectrum for growth. But perhaps the most significant factors responsible for the limitations in our understanding of bacterial diversity relate, first and foremost, to the relatively small numbers of analysts who have devoted themselves to these tasks and, second, to the fact that often their fascinating revelations led, and continue to lead, many of them to a type of 'reductionism', namely, the exploration and elucidation of the properties of the organism(s) newly isolated.

Because the elucidation of novel properties of such isolates was and continues to be in itself both stimulating and rewarding, individuals with interests in surveys and elucidation of microbial diversity have had their attentions refocused. There is probably no better explanation for the enormous numbers of newly described prokaryotes possessing newly recognised biochemical, morphological, and physiological properties other than that there has been a real increase over the past three or so decades in the numbers of investigators who have pursued this type of scholarship. Even so, the total number of practitioners of this art and science is not large, and their goals, significant as they are, often have not been widely appreciated or understood, nor has the research been well funded; regrettably, 'a fishing expedition' or 'there is no hypothesis' is a term too often seen in reviews of requests for funding of promising explorations. Many thoughtful students of prokaryotic diversity have long recognised that only a fraction of the presumed prokaryotes seen in habitat samples or appearing in enrichment cultures have ever been brought into pure culture (or into stable mixed culture) so that traits could be elucidated and the organisms could be identified. Seemingly insignificant modifications of the culture medium or incubation conditions often resulted pronounced changes in the populations that developed. Even when similar modifications were made in isolation media in attempts to further cultivate such organisms, the efforts were often unsuccessful. Clearly, the organisms' nutritional needs (in the broad sense of these words) remained unmet. Possible explanations are legion.

An oft-noted situation is that the populations developing in an enrichment medium with or without agar present as a solidifying agent (an example of the accumulative versus the separative enrichment culture approach) are recognisably different. Many investigators note that some organisms that do appear in accumulative (liquid) enrichments are unable to grow in an otherwise presumably identical agar-solidified medium. This discordance between what can be seen and what can be cultivated has been brought into renewed view, literally speaking, as a result of current analytical approaches based on extraction of nucleic acids from populations collected from native habitats to permit amplification and cloning of genes for rRNA. In many instances, rRNA with sequences unlike those already characterised from cultivated organisms is found, thereby indicating the presence of a novel, undescribed organism(s) in the habitat sampled. Thus, many factors that had long been recognised by those seeking to enlarge the understanding of different prokaryotes present in different habitats have now become more widely appreciated and led to new emphases and approaches in pursuit of long-sought goals.

MORPHOLOGIES AND SIZES OF PROKARYOTIC CELLS

We no longer consider prokaryotic cells to have shapes once described simply as spheres, rods, vibrioids, or spirals, for in addition we recognise cells that are essentially square, others that are nearly rectangular, some that approximate a triangle, and others that approximate a star. Many of these different morphological types are characterised as well by the presence of constrictions, protuberances, lobes, or other geometrically irregular aspects of their surfaces. Cells may exist as single entities or as units forming chains, clumps, or filaments. Table 4.1 provides a selected list of the diverse morphological traits of representative organisms, although significant aspects of flagellar architecture differences in the Bacteria and Archaea are not enumerated here, for example.

Table 4.1. Examples of diversity in morphology of selected prokaryotes.

Morphology or other cell anatomical feature	Representative genera
Coccus	
Single	*Acidianus, Megasphaera*
In chains	*Lactococcus, Streptococcus*
In other groupings	*Pediococcus, Sarcina*
Coccus-rod, lobed	*Sulfolobus*
Coccus ↔ rod[a]	*Arthrobacter*
Circular (or nearly so)	*Cyclobacterium*
Cyst (or microcyst) formed[a]	*Azotobacter, Sporocytophaga*
Endospore formed[a]	*Bacillus, Clostridium, Acetonema, Sporosarcina, Thermoactinomyces*
Exospore formed[a]	*Methylocystis*
Irregular	*Nocardia, Mycobacterium, Streptomyces*[a]
Myxospore formed[a]	*Myxococcus, Stigmatella*
Rectangle	*Methanopyrus*
Rod	
Long, single	*Bacillus*
Short, single	*Pseudomonas*
Often in chains	*Bacillus, Lactobacillus*
Spiral, small	*Bdellovibrio, Desulfovibrio, Rhodospirillum, Methanospirillum*
Sheathed	*Thermotoga, Sphaerotilus*
Square	Not yet named
Stalked[a]	*Asticcacaulis, Caulobacter*
Triangle	*Pyrodictium*
Vibrioid	*Bdellovibrio, Marinomonas, Vibrio*
Wall-less	*Mycoplasma, Thermoplasma*

[a]Cell undergoes morphogenesis.

This morphological diversity is a contributing factor to the difficulty sometimes encountered in distinguishing cells from inanimate material. Of additional significance is the fact that the morphology of many prokaryotes has an inherent degree of inconstancy, that is, that the morphology of a given organism can undergo change depending upon the extracellular environment or the stage of growth, for example. In addition, cells subjected to drying for purposes of staining, for instance, sometimes appear discernibly different from cells in the living, unstained state. The range of sizes for different prokaryotic cells is quite large. There is an overlap in the size range of pro- and eukaryotic cells. One unicellular organism may be barely sufficiently large to be resolved by and thus seen in a light microscope, while others, conversely, can be just large enough to be seen with the unaided human eye. The recent identification of one 'protist' as, instead, yet another large prokaryote generates wonder about just what upper size limits for prokaryotes will be ascertained and for how long such a consideration will be regarded as fixed. At the other end of the size spectrum, just how small a micro-organism can be, or have been, has been the subject of detailed contention. Table 4.2 provides a selected listing of the cell sizes currently recorded for prokaryotes.

Table 4.2. Diversity in prokaryote cell size.

Size (μm)[a]	Representative organism(s)
<0.2	Some 'picoplankton'
0.3–0.5	*Veillonella*
2–3	*Megasphaera*
5–25	*Thiovulum*
0.6–1.2 × 2.5–5.8	*Bacillus*
1.1–1.5 × 2–6	*Escherichia*
5–6 × 8–12	*Chromatium*
2.5–4 × 40–100	*Thiospirillum*
1–100 × 500–200	*Beggiatoa*
80 × 600	*Epulopiscium*
100–750	*Thiomargarita*

[a]Approximate diameter (× length where appropriate).

PHYSIOLOGICAL DIVERSITY AMONG THE PROKARYOTES

In the biological world, there are no known parallels to the abilities of the Bacteria to utilise an enormous array of energy sources to support growth. Presumably this represents the consequences of mutation and adaptation that occurred during the lengthy evolutionary history of these organisms on the planet. The recognition that beneath the widely disparate nutritional and environmental needs for the growth of different bacteria there was an underlying unity in their physiological attributes was a major conceptual contribution that had a marked practical influence on the development, nature, and extent of our understanding of the significance of prokaryotic diversity.

This encompassing view was that energy was conserved and made available for life processes as a result of cellularly mediated, coupled oxidation-reduction reactions and that as a group the Bacteria exploited in this way nearly every conceivable source of energy available. Then novel, this concept of an extraordinary versatility, along with the wide range of environments in which different prokaryotes were shown to grow, led to the notion that prokaryotes are unique in their ability to inhabit and thrive in environmental 'extremes' (i.e. those not conducive to growth of macrobes). One outgrowth of these considerations was the recognition that every naturally occurring organic compound is subject to attack (usually meaning utilisation as a growth-supporting nutrient) by one or another bacterium and mineralised as a result of participation in the biogeochemical cycles. We have come to realise that not all compounds synthesised by animals and plants are subject to biodegradation at identical rates (the slow attack on lignocellulosic material is one example) and that the initial degradative steps of some biosynthesised materials are carried out not by Bacteria or Archaea but instead by other microbes. The initial contributions of Beijerinck and Winogradsky that established the utilisation of inorganic ions and molecules for energy conservation coupled to growth have since been extended in terms of both the scope of the Bacteria and Archaea involved and the range of inorganic entities able to serve in this capacity. It is customary to categorise and contrast the several ways prokaryotes employ the coupled oxidation-reduction reactions listed below:

1. Aerobic respiration: Molecular oxygen serves as the oxidant in a redox reaction and appears in reduced form as water, one end product of this metabolism.
2. Anaerobic respiration: In environmental conditions where molecular oxygen is absent or in limited supply, an inorganic ion such as nitrate, sulphate, or carbonate serves as the terminal

oxidant and becomes reduced to dinitrogen (or ammonia), sulphide, or methane, respectively. It is now recognised that a variety of other ions such as oxidised (ferric) iron or organic molecules such as fumarate, trimethylamine oxide, dimethyl sulphoxide, or the sulphonic acids can also serve as terminal oxidants for a variety of anaerobically respiring prokaryotes.

3. Fermentation: An organic compound, most often a metabolic intermediate that results from oxidation of the organic compound serving as the energy source, serves as the terminal oxidant, and accordingly, a more reduced organic molecule is a metabolic end product(s).

4. Phototrophy: Radiant energy is absorbed by chlorophyll-, bacteriochlorophyll-, or 'accessory pigment' containing pigment complexes, resulting in an excitation of electrons present in the complex and leading to an oxidation and charge separation. When water is employed as the ultimate reductant for the sequential reactions, the processes are characterised by evolution of molecular oxygen ('oxygenic photosynthesis'); when either inorganic or organic compounds replace water, molecular oxygen is not formed, and the processes are termed, instead, 'anoxygenic photosynthesis'.

Table 4.3 provides examples of some additional aspects of these processes.

Table 4.3. Some nutritional aspects of physiological diversity.

Electron donor(s) utilised (examples)	*Electron acceptor, reduced end product(s)*[a]
In aerobic respiration	Molecular oxygen, *water*
Organic molecules: carbohydrates, amino acids, purines, pyrimidines, lipids, fatty acids, alcohols, hydrocarbons (both aliphatic and aromatic), sulphonic and aromatic acids	
Inorganic molecules or ions: carbon monoxide, molecular hydrogen, metallic sulphides, ammonium, nitrite, ferrous and manganous salts, elemental sulphur	
In anaerobic respiration	
Organic molecules: much as given above; possible exceptions are some sulphonic and aromatic acids, gaseous hydrocarbons	Nitrate, nitrite; *nitrite, nitrogen gas, 'ammonium'*
Inorganic molecules or ions: much as given above	Sulphate, sulphite, elemental sulphur; *sulphite, sulphide* Fumarate, succinate; dimethyl sulphoxide, *dimethyl sulphide*; ferric salts, *ferrous salts*; trimethylamine oxide, *trimethylamine*
But methanogens typically utilise primarily hydrogen gas, formate, or acetate	Carbonate, *methane*
In fermentation	Organic molecules, protons; *alcohols, fatty acids, ketones, hydrogen gas*
Organic molecules: carbohydrates, purines, pyrimidines	
In phototrophy	Carbonate, *cellular components*
Organic molecules: alcohols, fatty acids, organic acids (e.g. malate, succinate, benzoates)	
Inorganic compounds or ions: hydrogen gas, ferric, sulphide, elemental sulphur, thiosulphate	
But, for cyanobacteria, water	

[a] Reduced end products are listed in italics.

We may summarise, then, one aspect of the prokaryotes physiological diversity by noting that some can inhabit and multiply only in habitats that are in regular contact with the earth's atmosphere and the molecular oxygen it contains, employing aerobic respiration (i.e. are strictly or obligatorily aerobic). Other prokaryotes function only in the absence of air, employing either fermentation or anaerobic respiration, and thus pursue other modes of energy acquisition and conservation; we term such organisms strictly or obligatorily anaerobic (but might also describe them as strictly fermentative or as living strictly by anaerobic respiration).

Another group may possess, for example, the ability either to live by aerobic or anaerobic respiration; the adjective 'facultative' is added to describe such an organism's respiration.

The same term is used in describing the ability of an organism to live either by aerobic respiration or by fermentation (e.g. facultatively fermentative) and in the description of a bacterium able to live either by anoxygenic phototrophy or by aerobic respiration.

We recognise that evolution has resulted in a continuum of traits rather than a set of neatly packaged ones. The inadequacy of words such as 'facultative' to describe the physiologies of a bacterium able to live by either fermentation, aerobic respiration, or anoxygenic phototrophy, for example, becomes readily apparent.

Inadequacy of different sorts, such as not knowing that a nutritional lifestyle based on acetogenesis remained to be discovered in a distinct morphological prokaryotic group (spirochetes), that phosphite oxidation could be coupled to dissimilatory sulphate reduction by anaerobes, or that N_2-dependent growth was a trait of metal-metabolising bacteria, serves as one large constraint on understanding the roles of prokaryotes in biosphere functions. This is true, as well, for N_2 assimilation in marine microbiotas.

It is also customary to categorise physiological traits in yet other ways. One such way refers to the source(s) of carbon assimilated for biosynthesis: 'autotrophy' describes the ability of an organism to utilise carbon dioxide as the principal source of carbon (save perhaps the need for vitamins or an amino acids), while 'heterotrophy' or 'organotrophy' describes the use of carbon atoms of organic molecules as the principal carbon source.

When categorisation is focused on the source of energy to be conserved, an organism doing so at the expense of reduced inorganic ions, or molecules such as hydrogen gas, is regarded as a 'lithotroph'. The term 'organotroph' describes an organism utilising organic, molecules; note the use of the identical word to describe both the carbon and the energy sources.

Table 4.4 lists examples of descriptive terms often used in categorising what are indeed different lifestyles of prokaryotes.

Prokaryote physiological diversity is not, however, restricted only to relationships to molecular oxygen or to the ability to utilise radiant energy to capture energy. Optimal growth under conditions of low pH (acidophiles) or high pH (alkaliphiles) is characteristic of many different prokaryotes. Similar diversity exists in temperature optima for growth: cold-loving organisms (psychrophiles) contrast with those that are unable to grow at temperatures less than ca. 80°C, some of which (hyperthermophiles) have been shown to grow at temperatures in excess of 110°C. Among other traits in which prokaryotes show remarkable variation from organism to organism are the ability to tolerate or the need for inorganic salts (as for the strictly 'halophilic' subgroup of the *Archaea*) and the ability of some organisms (oligotrophs) to grow only in environments with low nutrient concentrations, while others (copiotrophs) either tolerate or require much higher nutrient levels. Yet another trait noted for certain organisms is a requirement for pressure in excess of that on the earth's surface (barophiles). Table 4.5 summarises ranges of selected traits.

Table 4.4. Some terms used in relation to bacterial growth.

Acidophiles: organisms with growth optima at ca. pH 1–5

Alkaliphiles: organisms with growth optima at ca. pH > 8

Anaerobes: organisms that are unable to use (i.e. consume) molecular oxygen

  Obligate: those that cannot grow in the presence of molecular oxygen

  Oxyduric: those not killed by (i.e. tolerant of) molecular oxygen

  Oxylabile: those killed by the presence of molecular oxygen

  Aerotolerant: those able to grow in the presence of molecular oxygen even though they do not use it

Aerobes: organisms that use molecular oxygen in redox reactions coupled to energy conservation

  Obligate: those that are unable to grow in the absence of molecular oxygen

  Facultative (or euryoxic): those that are able to grow in the absence of molecular oxygen

Microaerophiles: organisms that require molecular oxygen for growth, but can tolerate its presence only when present at low levels (often ca. 10 per cent of atmospheric levels)

Mixotrophs: organisms that utilise both autotrophic and heterotrophic means, usually simultaneously, of conserving energy and assimilating nutrients for growth

Phototrophs: organisms that use radiant energy (light) as a source of energy for growth

  Obligate: those that are unable to grow in the absence of light

  Facultative: those able to grow by gaining (conserving) energy in the absence of light

Symbiosis: two or more dissimilar organisms that interact and live together

Syntrophy: the relationship between proton-reducing organisms and other organisms that consume hydrogen gas

  Obligate: the relationship when the proton-reducing organism is unable to grow in the absence of the hydrogen-consuming one

Table 4.5. Environmental extremes in which prokaryotes are thought to multiply.

Characteristic	Value
Pressure	1 to ca. 1000 atm[a]
Temperature	−1.2 to 110–113°C
Depth	2000 metres (terrestrial subsurface) to ca. 10,000 metres (ocean floor)
Salt concentration	Up to ca. 4–5 M
Acidity or alkalinity	pH of < 1 to 11–12
Available water	a_w (water activity) as low as 0.6

[a] 1 atm = 101.29 kPa.

As noted above, the scope of nutritional diversity among the prokaryotes is not only impressive but of important consequence in affecting the ability of these microbes to colonise and to thrive in nearly every imaginable habitat. Some of the bacteria, are able to oxidise one-carbon compounds (those containing no carbon-to-carbon bonds) and to assimilate the oxidised moiety for synthesis of the small and macromolecules characteristic of nearly any prokaryotic (or, for that matter, eukaryotic) cell. While many of these organisms are also able to oxidise and assimilate methanol carbon, they are generally incapable of assimilating the carbon of other simple or complex organic molecules as the sole carbon and energy source for growth. By contrast with these 'methanotrophs', another group, the 'methylotrophs', differs in two ways. Methylotrophs cannot utilise methane for growth but can grow at the expense of methanol, formate, and methylated amines; more strikingly, however, they can grow at the expense of acetate, other organic acids, amino acids, etc. Most of the well-studied methanotrophs are quite specialised in terms of nutrients utilisable for growth, while the methylotrophs are much less so.

Another example of a nutritional specialisation is shown by at least one *Bacillus* species which is able to grow readily with urate, a purine, or compounds such as allantoin and allantoate, both of which are intermediates in the degradative pathway for urate oxidation and assimilation. However, in contrast to the scores of other *Bacillus* species, *Bacillus fastidious* is unable to grow at the expense of a range of sugars, polysaccharides, amino acids, simple proteins, etc. In similar fashion, some *Cytophaga* species, unable to grow with amino acids serving as carbon and energy sources when first isolated from a natural habitat, are strictly dependent upon polymers such as cellulose or chitin as a source of carbon and energy for growth. Other cytophagas, by contrast, are unable to utilise cellulose or chitin for growth, but they are able to grow on complex media containing peptides, amino acids, yeast extract, or simple sugars. These specificities contrast to those of some pseudomonads and their relatives which are able to use any of scores of organic compounds, including simple organic acids, the range of amino acids, and benzenoid and polycyclic compounds, to support cell multiplication.

Such nutritional specificity occurs not only in regard to utilisation of organic compounds by these organotrophs. Autotrophs — organisms able to synthesise the majority, if not all, of the carbon skeletons of their cellular compounds by reducing carbon dioxide to the 'level' of carbohydrate (representative of the overall C:H:O ratio of cells) — vary in the reductant they employ for this purpose. Some are able to oxidise hydrogen gas or may use a compound such as hydrogen sulphide from which to capture energy for growth, or they may be able to grow as organotrophs by use of organic compounds. However, not all chemolithotrophs are so versatile; some 'sulphur-oxidising' chemolithotrophs use any of several reduced sulphur compounds to acquire energy, while others display much more selectivity and specificity.

For another group of autotrophs (e.g. *Nitrosomonas* spp.), ammonia serves as an energy source, while nitrite does not, yet *Nitrobacter* spp. oxidise nitrite to form nitrate but are unable to gain energy from ammonia. Reduced forms of iron and manganese are suitable electron donors in energy-conserving processes for yet other groups of autotrophs.

At least two other aspects of physiological diversity are significant for the discussions that follow. One of these deals with the fact that although a particular organic molecule may not function as a sole source of carbon and energy for growth of a particular organism, the compound may be metabolisable (i.e. it may be either oxidised or reduced in whole or in part, and all or part of the molecule may be assimilated into cellular components) as long as the organism is utilising a different molecule as a carbon and energy source. This phenomenon, first termed 'co-oxidation' when it was described in reference to oxidation of ethane and other gaseous hydrocarbons by a methanotroph, has been shown to be of more general significance and distribution in prokaryotes and is now considered an important aspect of many bioremediation processes. A second trait, which has become of more widespread interest and significance since it was initially recognised, is syntrophy. In anoxic environments, the degradation of reduced organic compounds (e.g. simple alcohols, fatty acids, and certain aromatic acids), resulting in the accumulation of acetate, carbon dioxide, hydrogen gas, and formate (among other end products), is energetically unfavourable.

On the other hand, if the concentrations of these products are kept sufficiently low, the overall energetics for growth become somewhat more favourable. Methanogens and sulphate-reducing bacteria (sulphidogens), commonly present in anoxic habitats, function to consume hydrogen gas and thus lower its partial pressure in the microhabitat. Thus, such cocultures function to permit the biodegradation of compounds that would otherwise be refractory to attack by pure cultures. The term 'interspecies hydrogen transfer' has been used to describe such interactions as well.

DIVERSITY: INSIGHTS FROM MACROMOLECULAR ANALYSES

Because it appears that prokaryotes have been present on this planet for perhaps four-fifths of its age, there has been ample time for repeated mutations, their accumulation and natural selection, and an introduction and accumulation of widely varied traits in these organisms. For many years, the lack of a substantial traditional fossil record impeded critical considerations of evolutionary change in the microbial world. It was not until Zuckerkandl and Pauling's proposal of macromolecular traits as indicators of evolutionary changes that we could ponder events of the past. Data from both protein and rRNA sequences are now used widely for this purpose and for making inferences about phylogenetic relationships. The evolution of prokaryote diversity has become a subject of intense interest and significance, regularly revisited and reviewed; an international symposium on the topic is notable for its scope.

To many, the implications of sequence information suggest strongly that life on the planet should be classified in three major categories: archaea, bacteria, and eucarya. A widely accepted phylogenetic tree (Fig. 4.1) is based on the inferences that the groups termed archaea and eucarya diverged from ancestors of the bacteria, first as a single lineage and only later diverging and becoming separately recognisable entities. The timing of this divergence, and particularly the estimates for the divergence point of the eukaryotes and prokaryotes, has been considered often and in different ways, but most recently and carefully by Doolittle and colleagues.

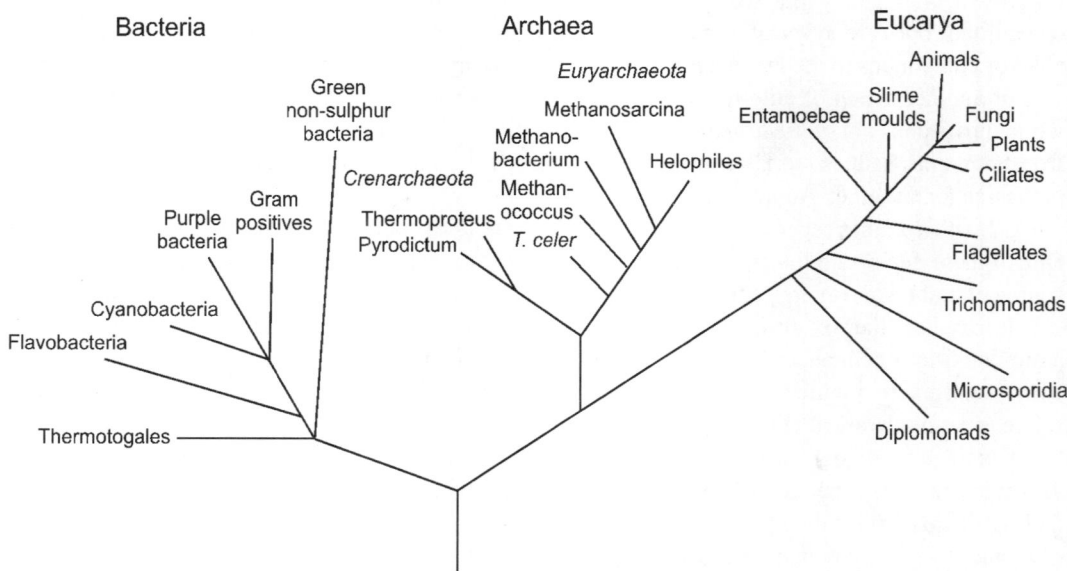

Fig. 4.1. A phylogenetic tree based on evaluation of 16S rRNA sequences. The three major lineages of life (archaea, bacteria, and eucarya) as shown. *T. celer, Thermococcus celer.*

The potential for lateral gene transfer in natural populations remains a matter of concern in using sequence information for phylogenetic considerations. As the number of genomic sequences increases, many are surprised by the chimeras that appear in the archaeal and eukaryotic genomes that have been examined. The effect of this recognition of many examples of trees that are noncongruent is to allow the possibility that horizontal (or lateral) gene transfer has occurred more frequently in the course of evolution than has often been considered.

The basis for 'type III' secretion systems evident in different bacteria that are involved in certain parasitic or mutualistic microbe-macrobe interactions (in both animals and plants) represents but one example of such new insights. Accordingly, the oft-accepted consensus of cellular evolution as outlined in the following paragraphs seems likely to be subject to considerable revision as the details of genomic sequences become clearer.

In the line leading to the bacteria, an initial bifurcation leads to thermophilic cells now represented by genera such as Aquifex, Fervidobacterium, and Thermotoga. It is a matter of curiosity and concern that Aquifex, regularly placed at the bottom of the tree, is nonetheless known as an aerobic, respiring chemolithotroph; this represents an anomaly, as the planet presumably was devoid of molecular oxygen at the time of this evolutionary event.

Many but not all practitioners consider that the green nonsulphur phototrophs were among the next groups to diverge. Perhaps at about the same time a group that possessed cell walls devoid of peptidoglycan emerged: the *Planctomycetales*, including genera such as *Gemmata, Isosphaera, Pirellula*, and *Planctomyces*.

The remaining well-studied and recognised bacterial groupings designated the flavobacteria, cyanobacteria, gram-positive bacteria (of which there appear to be at least two major subgroups), and purple bacteria or proteobacteria (a group itself considered to warrant live subdivisions: alpha, beta, gamma, delta, and epsilon) appear to have emerged later as a single radiation which then underwent divergence.

Even before this refined evidence about likely events in bacterial evolution, it was deemed likely that thermophilic anaerobes were likely candidates as the most ancient of living beings. Arguments over whether autotrophs preceded heterotrophs or *vice versa* abounded but were based more on conjecture than on anything approaching persuasive evidence.

Among the candidates for the earliest of organisms were the methanogenic bacteria, prokaryotes which are now considered not to be bacteria *per se*, but instead members of the archaea, appearing later on the evolutionary scene.

A current view of the archaea is that they are as unlike the bacteria as they are unlike the eucarya. Their possession of unusual ether-linked lipids and the lack of peptidoglycan in their cell walls are among their distinguishing characteristics. The group is now divided into the sub-categories of methanogens, thermo- and hyperthermophiles, and strict halophiles. Some limits of our understanding of the archaea are demonstrated by the diversity recognised after the surprising discovery of new types of these organisms in cold, marine environments.

Although the eucarya are usually cited as characteristically containing a membrane-bound nucleus, they are also usually considered to possess subcellular organelles termed chloroplasts or mitochondria. It may be significant for evolutionary considerations that in the eucarya, three groups at the base of the tree — the diplomonads, the microsporidians, and the trichomonads — are devoid of mitochondria.

One of the many surprises arising from such approaches to the study of phylogeny, and of special interest for considerations of prokaryotic diversity, has been the demonstration that some biochemical and physiological traits appear in quite distantly related organisms. For example, sulphate reduction occurs both in bacteria and in archaea; the same is true for autotrophic assimilation of carbon dioxide as well as the trait of hyperthermophily.

ASPECTS OF BEHAVIOURAL DIVERSITY AMONG PROKARYOTES

With samples of mixed microbial populations are obtained from either natural habitats or laboratory cultures, or when pure cultures are examined using any one of several types of light microscopy, we quickly note that nor all organisms have identical aspects of behaviour. Some organisms swim through liquids with ease and rapidity, others regularly do so more slowly; some dart about seemingly randomly as if in a frenzy, while others move for longer distances and appear to be swimming smoothly; in some, no swimming motility is ever discerned. We now recognise that, in nearly every instance, such motility reflects the presence and function of either a single flagellum or multiple flagella and the arrangement of these on the cell surface(s). Many of the bacteria termed spirochetes are seen to swim much better through viscous liquids than they do in typical water samples or laboratory media; this may be a reflection of the position of their flagella (axial filaments, or endoflagella) between the cell wall and an external sheath of the cell. This location clearly does not permit the flagellum to rotate in the extracellular milieu and propel the cell in a manner similar to that of, for example, the well-studied enteric bacteria. Usually the behaviour of swimming flagellated bacteria has been described as alternating 'runs' and 'tumbles'. The observations made with the enteric bacteria may be of limited application, however. *Azospirillum* spp., pseudomonads, and rhizobia are among organisms observed not only to behave differently but to have different intracellular regulatory mechanisms involved.

A group of bacteria with rather diverse physiological properties move in a different manner, not by swimming through liquids but rather by migrating over insoluble substrates which provide solid surfaces (e.g. agar, cellulose, chitin, and other cells) on which to move. This motility, for which locomotor organelles have as yet not been visualised and for which, accordingly, the mechanism for mobility remains unknown, has often been termed 'gliding'. The implication of type IV pili in one form of myxobacterial motility and in the motility termed 'twitching' (characteristic of some acinetobacters and pseudomonads) is a welcome development in visualisation, although the mechanistic basis (e.g. the motor) for these functions of pili remain unclear; an earlier explanation, invoked and then discarded, for the gliding motility of some cyanobacteria has recently been reassessed and supported; polysaccharide extrusion through a 'wall' pore complex seems plausible for certain cyanobacteria. Cells can be seen to move (glide) on a solid surface, and the colony can spread a considerable distance (often more than a centimetre) beyond the point of cell deposition (inoculation). Such colonies usually are characterised as much thinner (very little vertical height) than those typical of enteric bacteria or pseudomonads, for example, growing on an identical nutrient-poor medium. Nutrient concentration usually exerts a significant effect on the ability of cells to glide and hence to form spreading colonies; the ability of a cell to glide is ordinarily not expressed on a nutrient-rich medium such as nutrient broth, and the gliding ability of an organism is thus not noted when such a medium is used for growth. Among the best studied of the bacteria known to possess gliding ability are those of genera such as *Beggiatoa, Cytophaga Flexibacter*, and *Myxococcus*; this is far from an inclusive list.

The ability of a prokaryote to swim through liquids in the absence of flagella is known for at least one cyanobacterium (a *Synechucoccus* sp.); no locomotor organelles have yet been discerned, although careful and extensive studies employing a variety of sensitive analytic techniques have given little credence to several plausible explanations. This organism is not known to glide.

It has already been suggested that cells are able to alter their physiological phenotypic properties, within limits imposed by their genotype, as indicated by the fact that some bacteria can live by either phototrophy or aerobic respiration, others can live by fermentation or aerobic respiration, and still others can live by aerobic or anaerobic respiration. Such environmental effects are myriad but also occur with respect to anatomical features and motility phenomena. One such example is the motility of some flagellated bacteria on solid surfaces; this behaviour is termed 'swarming'. Certain *Serratia*, *Proteus*, and *Vibrio* species, among others, synthesise additional, specialised flagella as a result of contact with a solid surface, and it is these additional flagella that enable the organism to spread. Surface features, among them the concentration of agar used in solidifying the medium, have significant effects on events in the cell's interior that in turn affect synthesis of these appendages for motility.

The ability of cells to respond to concentration gradients of attractants (often nutrients) or repellents (chemotaxis) by employing temporal sensing mechanisms has been extensively studied as one category of behavioural responses. Although phototrophic eukaryotic cells have long been known to exhibit a response (phototaxis) to light gradients, no unambiguous instance of this phenomenon has yet been established in prokaryotic phototrophs. It is important to note that studies of the *in situ* behaviour of prokaryotic populations have established phenomena such as diel migration which undoubtedly reflect both chemo- and phototactic (and possibly other types of) behaviour of members of these communities. Such behaviour is, then, not an artifact of pure culture study. Another sort of response is reflected in the ability of some cells to undergo either cellular or colonial morphogenesis in connection with entrance into a dormant or nongrowing (resting) state. The formation of the remarkably resistant endospore in the cell interior of the organisms grouped in genera such as *Acetonema*, *Bacillus*, *Clostridium*, *Desulphotomaculum*, *Sporomusa*, *Sporosarcina*, and *Thermoactinomyces*, for example, generally is regarded as a response to an encounter with nutrient-poor conditions. That seems to be true as well for myxospores and microcysts formed by myxobacteria and cytophagas, in which the entire cell shortens, thickens, and becomes spherical (or nearly so). This is probably the explanation for exospore formation by some methanotrophs. Other examples of changes in cell morphology resulting in specialised functions are the formation of heterocysts in some cyanobacteria and the formation of akinetes in others and stalk formation in prosthecate bacteria, as typified by *Caulobacter* spp.

These morphogenetic (behavioural) responses clearly make it more difficult to distinguish a potential cell from debris in samples taken from the environment, as noted above. Application of specific molecular probes is likely to be of enormous aid in identifying organisms of these types *in situ* and in assessing their growth (versus dormancy) state in natural habitats.

DIVERSITY OF HABITATS EXPLOITED FOR GROWTH OF PROKARYOTES

Why is it that the prokaryotes have been known to occur in so many different habitats and that our comprehension of 'new places' in which they are found is increasing? Certainly, as may be obvious from the degree of physiological diversity outlined above, prokaryotes have the potential to exploit (to multiply in) habitats judged 'extreme' in comparison with those that support the existence of animals and plants.

Yet another factor must be reflected in the ability of these metabolically diverse and very active cells to enter a state of inactivity, or dormancy, and be able to survive for periods of time far longer than the generation times deduced for them in studies of laboratory cultures. Although dormancy is a trait often attributed to cysts or endospores, for example, it is well recognised (even though it is more difficult to study inactive, dormant cells than it is to examine traits of exponentially growing ones) that dormancy is a significant phenomenon in the persistence of an enormous variety of cells.

If we accept, as we should, the dictum that prokaryotes are found virtually everywhere (although perhaps not in the totally cosmopolitan scope of the past, it follows that the entirety of habitats and the microbes therein cannot be enumerated, let alone described, in a limited space. Nonetheless, given the dynamic and exciting state of the study of microbial diversity, selected recent descriptions, reflecting either novel revelations or reconsiderations of earlier findings, are noted very briefly below.

Enlarging Perspectives

As ever-increasing numbers of quite different habitats are examined by use of macromolecular sequence and related analytical approaches, it becomes ever more evident that prokaryotic diversity has been regularly underestimated by classic isolation and cultivation approaches. It is not only seemingly extreme habitats such as hot springs where this was demonstrated early, but it is true for habitats at the other extreme — quite cold — as well as for those that are quite anoxic and the deep ocean, where we encounter indications of once unexpected population diversity and microbial impact.

Nonetheless, it is far from correct to believe that direct observational approaches to habitat analysis are playing only small roles in expanding our comprehension of the impact of prokaryotes in different habitats. The presence of purple sulphur bacteria that are seen in the form of macroscopically visible 'berries' in salt marsh ponds and the dense accumulation (as plates) of another phototrophic bacterium in a meromictic saline lake both remind us of our limited understanding of the roles such prokaryotes play in these and other habitats. Another example of a transformation long wondered about but only recently demonstrated is that of the oxidation of ammonium under anaerobic respiratory (denitrifying) conditions; even more exciting are the indication that the prokaryote involved is a previously unknown member of the *Planctomycetales* and the prospect that new aspects of bacterial physiology may exist in this poorly studied group. It is worth noting that continued microscopic examination played an important part in establishing cause and effect for this microbe, which has not yet been brought into pure culture. Establishing a microbial role in filamentous sulphur formation involved, again microscopic analysis as an important ingredient in unravelling this phenomenon.

The unaided eye, too, remains a significant part of the exploration of the microbial world. Although dense mats of members of the genus *Thioploca* had been shown to oxidise sulphide by reducing nitrate near the coasts of Chile and Peru, these organisms were present in amounts far smaller than expected in a similar situation in Namibian shelf sediments; instead, giant (up to 750 μm in diameter) cells of a previously unknown prokaryote given the name *Thiomargarita namibiensis* were readily visible and extensively studied. Our perception of just how large prokaryote dimensions might be was thereby challenged. A reminder that so many habitats on the planet have a very limited molecular oxygen content is made evident by the emerging significance of bacterial respiration of, for example, arsenic and selenium and the evidence that hydrocarbons such as hexadecane can be metabolised via methanogenesis. Can the long-sought evidence for anaerobic oxidation of the gaseous hydrocarbon methane be far away?

Studies of *Buchnera* spp., the prokaryotic intracellular symbionts of aphids, are but one reminder that not all organisms of environmental significance are free-living but rather often exist in mutualistic states. The phenomenon of bacterial luminescence, so long associated solely with the marine habitat, is now associated with terrestrial biotas as well. Study of one squid-vibrio interaction makes clear that the microbe-macrobe interaction that may result in a mutualistic (as opposed to parasitic) outcome has effects on the 'normal' development of both the microbe and macrobe.

In short, minihabitats are abundant and significant. These range from the halophilic habitats near salt-excreting plants to the internal workings of the magnetic bacteria, whose magnetism is a trait distributed far more widely than at first imagined. When does 'mini' become 'macro'? The answer may be one of scale, but it is clear that we can control growth of a single organism in a mixed population of activated sludge. It seems that it is only the lack of ingenuity, or of trying, that limits our understanding microbial populations and their impact on our habitats.

SECTION III

Water Microbiology in Public Health

5. **Water Microbiology**   **69**

6. **Waterborne Transmission of Infectious Agents**   **72**

7. **Detection of Micro-organisms in Environmental Freshwaters
   and Drinking Waters**   **95**

8. **Detection of Protozoan Parasites in Source and Finished
   Drinking Waters**   **112**

9. **Detection of Bacteria, Viruses, and Parasitic Protozoa
   in Shellfish**   **131**

10. **Control of Micro-organisms in Source Water
    and Drinking Water**   **145**

11. **Estimating the Risk of Infectious Disease Associated
    with Pathogens in Drinking Water**   **168**

12. **Methods of Enumeration and Activities of Micro-organism**   **179**

Chapter 5

Water Microbiology

INTRODUCTION

One of the most important aspects of water microbiology, from a human perspective, is the fact that we acquire numerous diseases from micro-organisms found in water. Some of these diseases represent intoxications resulting from the ingestion of microbially produced toxins. One category of intoxications is linked to drinking water which contains toxins produced by cyanobacteria such as *Anabaena* and *Microcystis*. A second category of intoxications is associated with dinoflagellates such as *Gambierdiscus*, *Gonyaulax*, and *Ptychodiscus*; these are aquatic micro-organisms which produce neurotoxins that can become biologically concentrated in the tissues of reef fish and shellfish. Humans acquire this second category of toxins by ingesting the affected animals.

HUMAN DISEASES ASSOCIATED WITH WATER

The majority of human diseases associated with water are, however, infectious in nature. The magnitude of human morbidity and mortality associated with these infectious diseases has led to the development of epidemiological surveillance studies. The associated pathogens include numerous bacteria viruses and protozoa. These water-related infectious hazards can be characterised according to various schemes. One such approach divides the hazards into four categories based upon the source of the involved pathogen and the route by which human recipients come in contact with that pathogen. Those categories could be defined as follows:

1. Infections that are waterborne: These are recognised as resulting from physical contact with microbially contaminated water. Infection may occur during bathing or recreational or occupational aquatic activities or after ingesting contaminated water, ice that has been made from contaminated water, or food items that have come into contact with either contaminated water or contaminated ice.

2. Infections caused by aquatic organisms: These result from pathogenic micro-organisms or worms (by definition macro-organisms) which naturally spend at least a part of their normal life cycle either living directly in the water or else within intermediate vertebrate or invertebrate hosts that reside in aquatic environments. These diseases are acquired either by ingestion of or immersion in the water or by inadvertent ingestion of infested intermediate host animals.

3. Infections which have water-related insect vectors: The associated pathogens are acquired not from water but rather as a consequence of humans being bitten by invertebrate vectors (usually mosquitoes) whose life cycle depends upon access to water either on the land surface or in

uncovered containers. Thus, the incidence of the resulting diseases may increase during periods of land surface flooding and during monsoon seasons.

4. Infections associated with inadequate hygiene: These are caused by pathogens which are not acquired from water but rather are acquired by either contact with or ingestion of microbially contaminated material (generally fecally contaminated). The resulting disease hazards could be ameliorated by using clean water for sanitation purposes, including bodily cleansing and the washing of plates and drinking glasses.

ROUTES BY WHICH HUMAN ACQUIRE WATERBORNE INFECTIONS

This section of the manual of environmental microbiology address various issues related to the routes by which humans acquire waterborne infections. These routes of infection are diagrammed in Fig. 5.1.

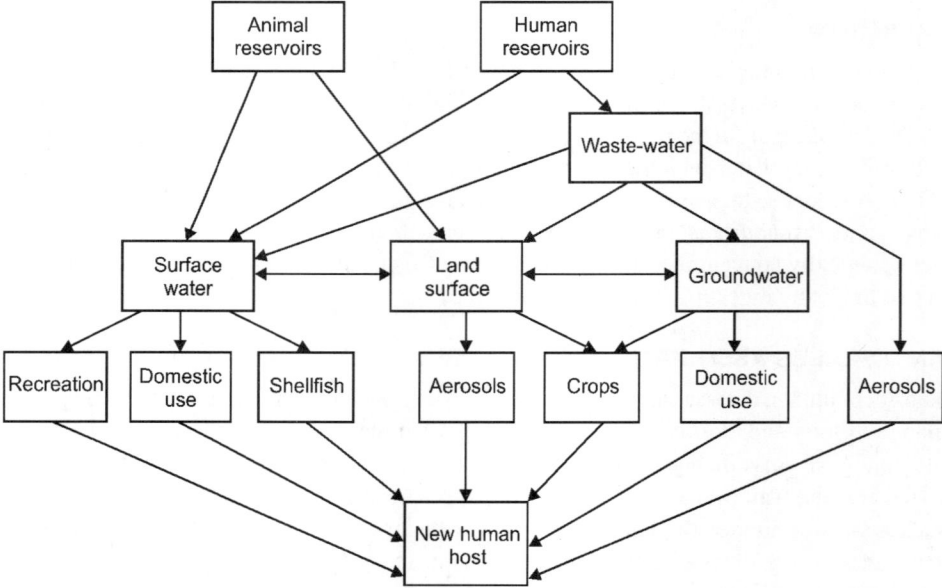

Fig. 5.1. Water-related environmental routes by which infectious agents are transmitted to susceptible individuals. The animals symbolically represented are cows as a source of the micro-organism which causes cryptosporidiosis and rodents as a source of those micro-organisms which cause giardiasis and campylosis.

The reservoirs for pathogenic micro-organisms found in environmental waters can be humans, animals, or the environment itself as summarised in Table 5.1. However, it commonly is presumed that many of those micro-organisms which infect humans and are found in our aquatic resources originate from human sources. This human-related contamination can occur during either defecation in water or recreational activities conducted in water. In addition, domestic waste-water seems to be of particular importance as a contributor of the pathogenic contaminants found in aquatic environments; the attendant public health concerns have resulted in the development of methods for studying and reducing the levels of pathogens in waste-water. Waste-water treatment efforts may help to reduce the incidence of problems resulting from the discharge of waste-water into environmental surface water, including illness among swimmers, contamination of drinking water, and disease that can occur when bivalve molluscan shellfish harvested from contaminated waters are subsequently consumed by humans. Waste-water treatment will reduce aquifer contamination, which can result either indirectly following the percolation of surface-applied waste-water into the subsurface or directly during subsurface injection of waste-water. The treatment of waste-water also is intended to reduce the contamination of crops that may occur when waste-water is eventually discharged onto land surfaces.

Table 5.1. Examples of infectious disease hazards associated with water.

Source of micro-organisms	Disease	Causative microbial genus (genera)
Human reservoirs	Cholera	Vibrio
	Encephalitis	Enterovirus
	Entamoebiasis	Entamoeba
	Gastroenteritis	Astrovirus, Calicivirus, Coronavirus, Rotavirus
	Hepatitis	Calicivirus, Hepatovirus
	Meningitis	Enterovirus
	Paratyphoid	Salmonella
	Shigellosis	Shigella
	Tularemia	Francisella
	Typhoid fever	Salmonella
Animal reservoirs	Campylobacteriosis	Compylobacter
	Cryptosporidiosis	Cryptosporidium
	Enteric fever	Salmonella
	Giardiasis	Giardia
	Leptospirosis	Leptospira
	Tularemia	Francisella
Environmental reservoirs	Encephalitis	Naegleria
	Cholera	Vibrio
	Legionellosis	Legionella
	Wound infections	Vibrio

Waterborne Transmission of Infectious Agents

INTRODUCTION

Waterborne transmission is a highly effective means of spreading infectious agents to a large portion of the population. Large quantities of enteric organisms can be introduced into the aquatic environment by contamination of the water by feces of infected persons (or animals) that is discharged into sewers or unprotected waterways. In contrast to person-to-person transmission, where a mobile infected person has more opportunity to transmit disease, infectious agents even from bedridden infected persons can play a role in waterborne disease transmission because pathogens in soiled bedding and clothing may be released into water during washing. When such contaminated water mixes with unprotected and/or inadequately treated drinking water, large numbers of susceptible hosts can be exposed and become infected. Waterborne disease outbreaks infecting as few as 10 or as many as hundreds of thousands of people have been reported, and Ewald has observed that pathogens that are most frequently waterborne tend to be the most virulent enteric pathogens.

Numerous infectious agents have been transmitted by ingestion of, contact with, or inhalation of water (Tables 6.1 and 6.2). Disease outcomes associated with waterborne infections include mild to life threatening gastroenteritis, hepatitis, skin infections, wound infections, conjunctivitis, respiratory infections, and generalised infections. Most waterborne infections are acute; however, some have chronic sequelae such as hemolytic-uremic syndrome associated with *Escherichia coli* O157:H7; myocarditis and diabetes associated with coxsackieviruses; peptic and duodenal ulcers and stomach cancer associated with *Helicobacter pylori*; reactive arthritis associated with *Yersinia*, *Shigella*, and *Salmonella*; pericarditis, orchitis, and splenic or liver abscesses due to *Salmonella enterica* serovar Typhi; and Guillain-Barré syndrome associated with *Campylobacter*. Some waterborne micro-organisms are frank pathogens, some are opportunistic pathogens, and some are toxigenic. The total number of potentially pathogenic micro-organisms is unknown but may be thousands. New infectious agents of disease continue to be recognised. There are five critical elements in the transmission of infectious agents through water: (i) the source of the infectious agents, (ii) specific water-related modes of transmission, (iii) attributes of the organism that allow it to survive and possibly multiply and to move into and within the aquatic environment, (iv) the infectious dose and virulence factors of the organism, and (v) host susceptibility factors.

Some agents of waterborne diseases are indigenous aquatic organisms, such as *Legionella* species, *Vibrio* species, *Aeromonas hydrophila*, and *Pseudomonas aeruginosa*. Control of these infections may depend on controlling exposure to water containing such organisms or, when possible, treating the water to remove or inactivate the infectious agents. Most microbial waterborne pathogens of concern originate in the enteric tract of humans or animals and enter the aquatic environment via fecal contamination.

Table 6.1. Illnesses acquired by ingestion of water.

Agent	Source	Incubation period	Clinical syndrome	Duration
Viruses				
Astrovirus	Human feces[a]	1–4 days	Acute gastroenteritis	2–3 days; occasionally 1–14 days
Calicivirus (Norwalk virus, Snow Mountain agent, and other related viruses)	Human feces[a]	1–3 days	Acute gastroenteritis with predominant nausea and vomiting	1–3 days
Enteroviruses (polioviruses, coxsackie-viruses, echoviruses)	Human feces	3–14 days (usually 5–10 days)	Febrile illness, respiratory illness, meningitis, herpangina, pleurodynia, conjunctivitis, myocardiopathy, diarrhea, paralytic disease, encephalitis, ataxia, diabetes	Variable
Hepatitis A virus	Human feces	15–50 days (usually 25–30 days)	Fever, malaise, jaundice, abdominal pain, anorexia, nausea	1–2 weeks to several months
Hepatitis E virus	Human feces	15–65 days (usually 35–40 days)	Fever, malaise, jaundice, abdominal pain, anorexia, nausea	1–2 weeks to several months
Rotavirus A	Human feces[a]	1–3 days	Acute gastroenteritis with predominant nausea and vomiting	5–7 days
Rotavirus B	Human feces[a]	2–3 days	Acute gastroenteritis	3–7 days
Bacteria				
Aeromonas hydrophila	Freshwater	Unknown	Watery diarrhea	Avg. 42 days
Campylobacter jejuni	Human and animal feces	3–5 days (1–7 days)	Acute gastroenteritis, possible bloody and mucoid feces, possible Guillain-Barre syndrome	1–4 years occasionally >10 days

(Contd ...)

Agent	Source	Incubation period	Clinical syndrome	Duration
Enterohemorrhagic *Escherichia coli* O157:H7	Human and animal feces	3–8 days	Watery, then grossly bloody diarrhea; vomiting; possible hemolytic-uremic syndrome	1–12 days, (usually 7–10 days)
Enteroinvasive *Escherichia coli*	Human feces	2–3 days	Possible dysentery with fever	1–2 weeks
Enteropathogenic *Escherichia coli*	Human feces	2–6 days	Watery to profuse watery diarrhea	1–3 weeks
Enterotoxigenic *Escherichia coli*	Human feces	12–72 hr	Watery to profuse watery diarrhea	3–5 days
Plesiomonas shigelloides	Fresh surface water, fish, crustaceans, wild and domestic animals	1-2 days	Bloody and mucoid diarrhea, abdominal pain, nausea, vomiting	Avg. 11 days
Salmonellae	Human and animal feces	8–48 hr	Loose, watery, occasionally bloody diarrhea; possible reactive arthritis	3– 5 days
Salmonella enterica serovar Typhi	Human feces and urine	7–28 days (avg. 14 days)	Fever, malaise, headache, cough, nausea, vomiting, abdominal pain; possible pericarditis, orchitis, and splenic or liver abscesses	Weeks to months
Shigellae	Human feces	1–7 days	Possible dysentery with fever, possible reactive arthritis	4–7 days
Vibrio cholerae O1	Human feces	9–72 hr	Profuse, watery diarrhea; vomiting; rapid dehydration	3–4 days
Vibrio cholerae non-O1	Human feces	1–5 days	Watery diarrhea	3–4 days
Yersinia enterocolitica	Animal feces and urine	2–7 days	Abdominal pain; mucoid, occasionally bloody diarrhea; fever; possible reactive arthritis	1–21 days (avg. 9 days)

(Contd ...)

Agent	Source	Incubation period	Clinical syndrome	Duration
Parasites				
Balantidium coli	Human and animal feces	Unknown	Abdominal pain, occasional mucoid or bloody diarrhea	Unknown
Cryptosporidium parvum	Human and animal feces	1–2 weeks	Profuse, watery diarrhea	4–21 days
Entamoeba histolytica	Human feces	2–4 weeks	Abdominal pain, occasional mucoid or bloody diarrhea	Weeks to months
Giardia lambia	Human and animal feces	5–25 days	Abdominal pain; bloating; flatulence; loose, pale, greasy stools	1–2 weeks to months and years
Algae				
Cyanobacteria (*Anabaena* spp., *Aphanizomenon* spp., *Microcystis* spp.)	Cyanobacterial blooms in water	Few hours	Toxin poisoning (blistering of mouth, gastroenteritis, pneumonia)	Variable
Helminths				
Dracunculus medinensis (guinea worm)	Larvae discharged from worms protruding from skin of infected person	8–14 months (usually 12 months)	Blister, localised arthritis of joints adjacent to site of infection	Months

[a] Animal strains of these viruses are believed to be not pathogenic for humans.

The concentration of these pathogens in a community water supply will depend in part on the number of infected persons and/or animals in the community and the opportunities for feces from these individuals to enter the water supply.

Control of these diseases rests on sanitation measures and waste-water treatment to prevent the introduction of feces containing these organisms into drinking water supplies or recreational waters and on adequate water treatment to remove or inactivate these organisms in drinking water.

Several water-related modes of transmission of infectious agents are discussed in this chapter. Many enteric pathogens transmitted via ingestion of feces-contaminated water can also be transmitted from person to person by contact with fecally contaminated hands or fomites or by consumption of fecally contaminated food. In endemic situations with poor sanitation and hygiene, the attributable risk due to water may be difficult to determine because of the risks from many other routes of transmission. Different classes of organisms have specific attributes, such as size and charge, which determine their movement and survival in the aquatic environment and their susceptibility to various water and waste-water treatment processes. These are discussed in more detail for the different groups of micro-organisms in the following section. Knowledge of these attributes can aid in the design of effective barriers or control strategies.

Table 6.2. Illnesses acquired by recreational contact with water[a].

Agent	Source	Incubation period	Clinical syndrome	Duration
Viruses				
Adenovirus (serotypes 3, 7, 1, 4, and 14)	Humans	4-12 days	Conjunctivitis, pharyngitis, fever	7–15 days
Bacteria				
Aeromonas hydrophila	Freshwater and brackish water	8–48 hr	Wound infections	Weeks to months
Legionellae	Freshwater, soil	Legionnaires' disease: 2–14 days (usually 5–6 days)	Legionnaires' disease: pneumonia with anorexia, malaise, myalgia and	Legionnaires' disease: variable (usually weeks to months)
		Pontiac fever: 5–66 hr (usually 24-48 hr)	headache, rapid fever and chills, cough, chest pain, abdominal pain, and diarrhea	Pontiac fever: 2–7 days
			Pontiac fever: fever, chills, myalgia, headache	
Leptospira spp.	Urine from infected domestic and wild animals	2–20 days (usually 7–12 days)	Leptospirosis (headache, chills, fever, myalgia, nausea, neck or joint pain)	Few days to 3 weeks
Mycobacterium spp. (*M. marinum, M. balnei, M. platy, M. kansasii, M. szulgai*)	Marine or brackish waters, freshwater	2–4 weeks	Lesions of skin or subcutaneous tissues	Months
Pseudomonas spp.	Water	Unknown	Dermatitis, ear infections, conjunctivitis	Unknown
Vibrio spp. (*V. alginolyticus, V. parahaemolyticus, V. vulnificus, V. mimicus*)	Marine water	*V. vulnificus:* 24 hr *V. parahemolyticus:* 4–48 hr	*V. vulnificus:* acute gastroenteritis, wound infections. septicemia *V. parahemolyticus:* acute gastroenteritis, wound infections, ear infections	*V. vulnificus:* septicemia (fatal in 2–4 days) *V. parahemolyticus:* usually 3 days
Other				
Cyanobacteria (*Anabaena, Aphanizomen,* and *Microcystis* species)	Cyanobacterial blooms in marine water or freshwater	Few hours	Dermatitis	

(Contd ...)

Agent	Source	Incubation period	Clinical syndrome	Duration
Naegleria fowleri	Freshwater in warm climates, soil, decaying vegetation	3–7 days	Meningoencephalitis, headache, anorexia, fever, nausea and vomiting; usually fatal	10 days
Acanthamoeba species	Water		Subcutaneous abscesses, conjunctivitis	8 days to several months
Schistosoma species	Feces and urine of infected animals and birds	Few minutes to hours	Dermatitis, prickly sensation, itching	Years

[a] Agents acquired through ingestion of water are not included in this table.

The infectious dose (ID) is the number of organisms required to cause infection and varies considerably by type of organism. In general, enteric viruses and protozoa have low infectious doses, typically between 1 and 50 tissue culture infectious units, plaque-forming units, cysts, or oocysts. Larger doses of bacterial pathogens tend to be required to cause infection. The observed median infectious dose (ID_{50}, defined as the number of organisms resulting in a 50 per cent infection rate) for enteric bacteria ranges from 10^2 to 10^8. Because such data are derived from studies of healthy, adult human volunteers under controlled conditions (vehicle of administration and gastric acidity), care must be taken when extrapolating these findings to other populations, such as malnourished children in developing countries or human immunodeficiency virus (HIV)-infected persons. For example, studies of *Vibrio cholerae* indicate that 10^6 organisms caused no illness when ingested with water by fasting volunteers. However, the same inoculum ingested with food or sodium bicarbonate caused illness in 90 to 100 per cent of the exposed volunteers because these vehicles decreased the protective gastric acidity. The virulence of micro-organisms varies by type and strain and by the route of infection. Microbial virulence factors include the ability of the organism to attach to and penetrate host cells; spread and multiply within the host; cause damage to host cells via lytic infection, production of toxins, or activities which may cause initiation of an injurious host response (such as loss of blood clotting); evade or resist host defence mechanisms (ranging from stomach acidity to various components of the immune system); and resist antimicrobial therapy.

Infection and development of clinical symptoms depend on a number of specific and non-specific host factors, such as age, immune status, gastric acidity, nutritional status, vitamin A deficiency, and possibly genetic predisposition. The majority of poliovirus and hepatitis A virus infections in young children are asymptomatic. For enteric viruses like group A rotavirus and astrovirus, symptomatic infections are common among children younger than 2 years of age. *E. coli* O157:H7 infections are more severe in young children and the elderly because of the weaker immune systems of these groups of individuals and are more likely to result in hemolytic-uremic syndrome.

CLASSIFICATION OF WATER-RELATED DISEASES

The classification of water-related diseases by Bradley provides a valuable framework for understanding the relationship between infectious-disease transmission and water. This classification system facilitates the planning of effective prevention and control measures for a variety of water-related diseases, depending on the type of agent and transmission route involved. Bradley described four main categories of water-related infections: 'water-borne infections', 'water-washed infections', 'water-based infections', and 'infections with water-related insect vectors'.

The 'water-borne infections' described by Bradley are those classically recognised as waterborne disease, such as typhoid and cholera, where an enteric micro-organism enters the water source through fecal contamination and transmission occurs by ingestion of contaminated water. Transmission by this route depends on: (i) the amount of fecal contamination in the water, the concentration of pathogens in the contaminated water (determined by the number of infected persons in the community), and the survival of the pathogenic organism in water, (ii) the infectious dose of the organism, and (iii) individual ingestion (exposure) of the contaminated water. Control of these infections is generally achieved through improvement of microbiological water quality, either through water treatment and/or through source protection.

The 'water-washed infections' described by Bradley are diseases due to poor personal and/or domestic hygiene. These diseases are not due to the presence of infectious agents in water but, rather, to the lack of readily accessible water. This limits washing of hands and utensils and thus permits the transmission of infectious agents, such as *Shigella* species, by fecally contaminated hands and utensils. Transmission is again related to the presence of feces from an infected individual, the infectious dose, the amount of fecal contamination on the hands or surface, and the survival of the organism on surfaces, lack of water for bathing also facilitates the spread of diseases that affect the eyes and skin, such as trachoma, conjunctivitis, and scabies. Control of these diseases is achieved through provision of greater quantities of water, closer and easier access to water, and education to improve personal and domestic hygiene.

The 'water-based infections' described by Bradley are worm infections in which the pathogen must spend part of its life cycle in the aquatic environment. This category is further subdivided into diseases acquired by ingestion of water and diseases acquired by contact with water. The prototype infections in this category are dracunculiasis, which is due to ingestion of water contaminated with guinea worms (*Dracunculus medinensis*), and schistosomiasis, which is transmitted by contact with water contaminated with species of the trematode genus *Schistosoma*. The original source of the guinea worm is larvae discharged from the female worm, which lies in a vesicle usually on the lower leg or foot of an infected human. The larvae are discharged when the vesicle is immersed in water and are then ingested by a copepod (genus *Cyclops*), where they develop into the infective stage. Humans become infected when they ingest water containing the copepods. The eggs of schistosome worms enter the aquatic environment is the urine or feces of an infected human. The eggs hatch in the water to produce miracidia, which infect snails, develop into the infective stage, and are shed by the snails into the water over a period of months. Humans become infected when the free-swimming infective larvae penetrate the skin during water contact. Although schistosomiasis is typically considered a tropical disease, there are reports of schistosomal dermatitis (swimmer's itch) in the United States that appear to be associated with *Schistosoma* species. Control of dracunculiasis and schistosomiasis is achieved through protection of the water source and the user by limiting skin contact with water and by eradicating the intermediate hosts.

The types of water contact diseases most frequently encountered in industrialised countries are those associated with recreational exposure to contaminated marine water, freshwater lakes, ponds, creeks, or rivers and occasionally treated water in swimming pools, wave pools, hot tubs, and whirlpools. While many recreational water-related outbreaks are associated with enteric organisms and ingestion of water contaminated with feces, some diseases of the ears, eyes, and skin are associated with actual water contact and some systemic illnesses are associated with penetration of a pathogen through an open wound or abrasion. Reported recreational water-related outbreaks have involved *Giardia*, *Cryptosporidium*, *Shigella sonnei*, and *E. coli* O157:H7, which presumably entered the gastrointestinal tract via ingestion. Other recreational water-related outbreaks have involved ingestion of, contact with, or inhalation of indigenous aquatic organisms such as *Naegleria*, *Pseudomonas*, *Legionella*, several Vibrio species, and several *Mycobocterium* species. Epidemiological studies indicate that *Staphylococcus aureus* skin and ear infections are often associated with recreational use of water and that the source of these organisms may be other bathers or the water. *Vibrio vulnificus* can cause serious wound infections when a skin injury occurs in marine water or from contact of pre-existing wounds with marine water. Cyanobacterial toxins have been associated with contact irritation experienced by some individuals after bathing in marine or freshwaters. An additional cause of recreational water-associated infections is *Leptospira*; these species are neither enteric organisms nor aquatic organisms, but they enter water via the urine of infected domestic and wild animals.

The 'infections with water-related insect vectors' described by Bradley are those transmitted by insects that breed in water, such as mosquito vectors of malaria, or insects that bite near water, like the tsetse fly vectors of sleeping sickness. Control of these infections is achieved through the application of pesticides, destruction of insect breeding grounds, and construction of piped water supplies.

Two additional modes of transmission that are water related are inhalation of water aerosols and consumption of raw or undercooked contaminated shellfish or fish. The major pathogens associated with aerosol transmission are *Legionella* species, especially *L. pneumophila*, the etiologic agent of legionnaires' disease and Pontiac fever. *Legionella* species are ubiquitous in water and soil and are capable of prolonged survival and reproduction in the aquatic environment. Growth within free-living amoebae appears to enhance their survival and provide protection from routine disinfection. Outbreaks of legionellosis have been associated with aerosols from cooling towers and evaporative condensers of large buildings or with hot- and cold-water systems in hospitals, hotels, and other institutions. *Legionella* can proliferate in hot-water tanks maintained at 30 to 54°C, and exposure occurs from aerosols created from showerheads. Control of these infections is achieved through minimising exposure to contaminated aerosols and routine cleaning and disinfection of water systems with adequate doses of chlorine or ozone.

The potential for aerosol transmission of *Mycobacterium avium* and other non-tuberculous mycobacteria and the risk to the immunocompromised population continues to be a concern. Like *Legionella*, these organisms are frequently isolated in environmental and treated water systems and are able to colonise and propagate within water distribution systems. Wendt reported the isolation of non-tuberculous mycobacteria from aerosol samples near the James River in Virginia. The isolates (mostly *M. intracellulare*) were biochemically similar to those recovered from human clinical specimens, suggesting that airborne mycobacteria derived from freshwater might be a significant source of infection.

Bivalve molluscan shellfish serve as vehicles of enteric disease transmission because of their ability to concentrate enteric organisms from fecally contaminated water in their tissue. Numerous outbreaks have been attributed to the consumption of raw or undercooked oysters, clams, and mussels. Many pathogens, including hepatitis A and E viruses, human caliciviruses (HuCV) (Norwalk-like viruses [NLV]), pathogenic *E. coli*, *S. enterica* serovar Typhi, and species of *Shigella*, *Vibrio*, *Plesiomonas*, and *Aeromonas*, have been implicated in shellfish-borne disease. Shellfish and some species of fish may also serve as vehicles for algal toxins. Toxic species of *Gonyaulax* and *Gymnodinium* are concentrated by filter-feeding mollusks and can cause paralytic shellfish poisoning among shellfish consumers. Reef-feeding fish can concentrate toxic dinoflagellates of the genus *Gambierdiscus*, which cause ciguatera seafood poisoning in consumers.

STUDY OF WATERBORNE DISEASES

Epidemic Waterborne Disease

Most of the information on the risk factors and etiologic agents of waterborne disease comes from investigations of waterborne disease outbreaks by state and local health departments and the surveillance programme maintained by the Centres for Disease Control and Prevention and the Environmental Protection Agency (EPA). It is difficult to recognise waterborne disease outbreaks because many infections due to waterborne agents are relatively mild and self-limited and so the infected persons may not seek medical care. For those who do receive medical care, many waterborne infections may not be diagnosed or reported to public health authorities by a health care provider or laboratory. The outbreak surveillance system is based on voluntary reporting by state health departments and clearly represents only a fraction of the true incidence of waterborne disease outbreaks. Data on waterborne outbreaks in the United States have been summarised by Craun and others. These data indicate that since the early 1980s, protozoa have become the major etiologic agents associated with waterborne disease outbreaks in the United States. Recently recognised etiologic agents, such as *E. coli* O157:H7 and *Cryptosporidium*, are being reported more frequently and in new settings, such as outbreaks associated with recreational water use.

The majority of waterborne outbreaks are classified by this surveillance system as acute gastrointestinal illness of unknown etiology (AGI). Stool examinations by hospital laboratories typically include culture for *Salmonella*, *Shigella*, and *Campylobacter*. In addition, many laboratories will test for group A rotavirus in specimens from young children, *Giardia*, and, more recently, *Cryptosporidium* at the request of a physician. The clinical symptoms suggest that many of the AGI outbreaks may be due to viral agents, such as HuCV. Recent advances in the detection of these agents in clinical and environmental samples have led to an increased understanding of their importance in epidemic gastroenteritis. In addition, there are a number of newly recognised etiologic agents for which there is some evidence of an association with waterborne disease, such as *Cyclospara cayetanensis*, cyanobacteria, *H. pylori*, *Mycobacterium* species, microsporidia, and *Aeromonas*. These agents are discussed below (see 'Emerging Waterborne Pathogens').

Endemic Waterborne Disease

The primary disease caused by infectious agents in drinking water is acute gastrointestinal illness. In industrialised countries there have been few studies of endemic gastrointestinal disease associated with the consumption of drinking water. Two randomised intervention trials were conducted in Canada to examine the risk of gastrointestinal illness associated with the consumption of conventionally treated municipal drinking water that met current microbiological standards. In the first study, 606 households were recruited and 299 of these households were supplied with reverse-osmosis filters that provided additional in-home water treatment. Gastrointestinal symptoms were recorded in family health diaries. Water samples from the surface water source, treatment plant, distribution system, and study households were analysed for several indicator bacteria and culturable viruses. Over a 15-month period, a 35 per cent higher rate of gastrointestinal symptoms was observed in the 307 study households drinking municipal tap water without in-home treatment than in the 299 study households supplied with reverse-osmosis filters. The symptoms and serologic evidence suggested that much of this increased illness might have been due to low levels of enteric viruses in the municipal water supply, which originated from a river contaminated by human sewage. Using a similar design, a second intervention study was conducted in the same community with 1400 families randomly allocated to four groups of 350. One group consumed conventionally treated water that met current North American drinking water standards. The second group consumed tap water from a continuously purged tap. The third group consumed tap water that was bottled at the treatment plant. The fourth group consumed purified bottled water (tap water treated by reverse osmosis or spring water).

The health of the families was monitored for a 16-month period. The groups consuming tap water and continuously purged tap water experienced 14 and 19 per cent more illness, respectively, than did the families consuming purified bottled water. Higher rates of illness were observed in children 2 to 4 years of age. Smith concluded that 14 to 40 per cent of the reported gastrointestinal illnesses were attributable to tap water that met current water standards and that contamination in the water distribution system was partly responsible for these illnesses. A similar study recently completed in Melbourne, Australia, with 600 families found no difference in self-reported gastroenteritis rates between the 300 families with a water treatment device (1-μm absolute filter cartridge and UV treatment chamber) and the 300 families with a sham device (www.waterquality.crc.org.au/wqsweb.htm).

A longitudinal study of French alpine villages that used untreated groundwater for their drinking water supplies observed a weak relationship between rates of acute gastrointestinal disease and the presence of fecal streptococcus indicator bacteria in the public water system over a 15-month study period. Illness data were collected through active surveillance of physicians, pharmacists, and school-teachers. Weekly water samples were collected from frequently used taps in the distribution system of each village and were analysed for several bacterial indicator organisms.

Ecological studies have attempted to find a relationship between endemic rates of hepatitis A infection reported in 75 counties and municipal source water quality and/or level of water treatment for all water supplies in the counties. However, no statistically significant associations were observed. A time series study attempted to link turbidity in the Philadelphia water supply with visits to hospital emergency rooms by patients experiencing gastrointestinal symptoms. However, a number of methodological problems with this investigation make the study results questionable. Two case-control studies have found the consumption of unfiltered surface water or shallow well water to be a risk factor for endemic giardiasis.

MICROBIAL AGENTS ASSOCIATED WITH WATERBORNE DISEASE

The commonly recognised waterborne pathogens include several groups of enteric and aquatic bacteria, enteric viruses, and three enteric protozoa (Table 6.1).

Enteric and Aquatic Bacteria

In the 19th century, *V. cholerae* and *S. enterica* serovar Typhi were the first waterborne pathogens to be recognised, and they have been responsible for tremendous morbidity and mortality worldwide. *Vibrio* species are aquatic bacteria that are well adapted to both the estuarine environment and the intestinal tract. Waterborne enteric bacteria include both human-associated and zoonotic species. *Campylobacter* and *Salmonella* species are found in the intestinal tracts of numerous domestic and wild animals. Therefore, contamination of water by animal feces also poses a human health risk. For enteric bacteria, such as *S. enterica* serovar Typhi and *Shigella*, infections are generally limited to humans. The ID of enteric bacteria depends on several host factors, including gastric acidity, and on the vehicle of transmission. Volunteer studies have shown that when bacteria are ingested with milk, the ID_{50} is approximately 10^2 CFU for *Shigella* species, approximately 10^7 CFU for *S. enterica* serovar Typhi, and approximately 10^8 CFU for enterotoxigenic *E. coli*. Diagnosis of these infections is typically made by culturing the micro-organisms from clinical specimens, detecting bacterial antigens or antibodies by enzyme immunoassays, or by detecting specific genes using molecular biology methods such as nucleic acid probe hybridisation.

In the developing world, classic waterborne bacterial infections due to *V. cholerae* and *S. enterica* serovar Typhi continue to be a problem. In the United States, recent drinking, water-related outbreaks due to bacterial agents have been due predominantly to *Shigella* species (15 outbreaks) followed by *Campylobacter* (7 outbreaks). The majority of recreational water-related outbreaks due to bacterial agents have been dermatitis infections caused by *Pseudomonas* (57 outbreaks) followed by shigellosis (20 outbreaks) and *E. coli* O157:H7 infection (8 outbreaks).

Data on the occurrence of pathogenic bacteria in water have been summarised by Emde. However, many of the data on microbial occurrence in raw water and treated water are believed to be obsolete because they were gathered using insensitive analytical methods with selective recovery media that underestimate the bacterial levels by as much as 100- to 1000-fold. The isolation of bacterial pathogens from water is difficult because of their low concentrations and stressed conditions, interference from competing aquatic micro-organisms, and the fastidious nutrient requirements of some pathogens. Enrichment procedures are often used to promote the growth of the organisms before they are transferred to selective media for identification. However, this limits the ability to quantify the pathogens in the sample.

The persistence of enteric bacteria in the aquatic environment depends on the species and on a variety of environmental factors (temperature, pH, sunlight, predation and/or competition by indigenous aquatic micro-organisms, dissolved organics, attachment to particulates, association with vectors such as amoebae, protozoa, or copepods, and presence of salts and other solutes), which have been reviewed. The classic study by McFeters demonstrated that the time needed for a 50 per cent reduction in the population of several enteric bacteria in well waters at 20°C ranged from 2.4 hours for some *Salmonella* species to 26.8 hours for some *Shigella* species. More recently, it was recognised that some enteric bacteria have the ability to enter a dormant state, referred to as viable but non-culturable, in which they can survive for long periods, and can still be infectious at high doses. Difficulty in recovering stressed bacteria from water samples may lead to false-negative results when evaluating microbial water quality. The public health significance of enteric bacteria in this state is unknown. Some enteric bacteria have the ability to multiply in the aquatic environment under favourable nutrient and temperature conditions.

The majority of bacterial pathogens are removed or inactivated by standard water treatment practices. The removal capabilities of specific processes have been reviewed. Most of the recent outbreaks associated with bacterial pathogens in drinking water have been due to consumption of untreated groundwater. Enteric bacteria are relatively susceptible to disinfection.

The CT (product of disinfectant concentration in milligrams per litre and contact time in minutes) values for 99 per cent inactivation by free chlorine (pH 6 to 7) vary from 0.034 to 0.05 for enteric bacteria like *E. coli* at 50°C to 15 and 28 for aquatic bacteria such as *L. pneumophila* at 20°C and *M. fortuitum*, respectively.

Enteric Viruses

The enteric viruses are more recently recognised waterborne pathogens. Human rotaviruses, Norwalk virus (NV) (and other HuCV), and human astroviruses were first described in the early to mid-1970s. Depending on the virus, these infections may be diagnosed by commercially available enzyme immunoassays for antigen or antibodies (group A rotavirus, enteric adenovirus types 40 and 41, and hepatitis A virus), electron microscopy, tissue culture (enterovirus, adenovirus, astrovirus, and rotavirus), or molecular methods such as PCR and probe hybridisation (HuCV, astrovirus, rotavirus, adenovirus, enterovirus, and hepatitis E virus). Unlike the enteric bacteria, concern about waterborne transmission of enteric viruses is generally limited to the strains that have humans as their natural reservoir. Although there are animal strains of many of these viruses, animal-to-human transmission is believed to be uncommon. The infectious dose of these agents is low, typically in the range of 1 to 10 infectious units. Human volunteer studies with group A human rotavirus estimated the median infectious dose to be between 5 and 6 focus-forming units.

Enteric viruses tend to be more persistent in the aquatic environment than most enteric bacteria; however, their survival depends on numerous physical, chemical, and microbial characteristics of the water as well as the virus type. The estimated time for about a 50 per cent reduction of hepatitis A virus in groundwater at 25°C is approximately 14 days. Their prolonged survival and small size enable viruses to move greater distances in soil and water. Unlike aquatic or enteric bacteria, it is not possible for enteric viruses to multiply in the environment. Cultivable enteric viruses have been detected in surface waters, groundwaters, and treated drinking waters in concentrations ranging from 647 PFU per litre to 1 PFU per 1000 litres. Virus recovery from water samples is relatively poor, and some important enteric viruses (HuCV) cannot be cultured *in vitro*.

However, the development of new molecular amplification techniques and nucleic acid hybridisation has improved the sensitivity of virus detection in water. A recent groundwater survey of sites in 35 states used reverse transcription-PCR amplification and detected enterovirus RNA in approximately 16 per cent of 427 wells, rotavirus RNA in 15 per cent of 425 wells, and hepatitis A virus RNA in 7 per cent of 429 wells.

Inactivation and/or removal of enteric viruses by water treatment processes had been reviewed and varies by virus type and treatment conditions. Reported CT values for 99 per cent inactivation of poliovirus type 1 and rotavirus at 5°C with free chlorine (pH 6 to 7) are 1.1 to 2.5 and 0.01 to 0.05, respectively. Experiments by Sobsey demonstrated that the CT value for 99.99 per cent inactivation of hepatitis A virus (pH 6, 5°C) was 2.3 for dispersed virus and 29 for cell-associated virus. A human volunteer study that examined the infectivity of NV in water treated with different doses of chlorine indicated that the virus remained infectious after a 3.75 mg/litre dose of chlorine and a contact time of 30 minutes. However, more recent studies by Shin indicate that NV is no more resistant to disinfection than are other enteric viruses based on reverse transcription-PCR detection of viral RNA.

Inadequate diagnostic technology has limited the detection of many enteric viruses in both clinical and environmental samples. Consequently, enteric viruses have not been frequently identified as the etiologic agents of waterborne disease outbreaks.

Many outbreaks currently reported as AGI are thought to be due to viral agents because of the epidemiological and clinical features of the outbreak. Recent advances in the diagnosis and detection of enteric viruses through the application of molecular techniques should increase the recognition of waterborne disease outbreaks due to viral agents.

Enteric Protozoa

Since 1981, enteric protozoa have become the leading cause of waterborne disease outbreaks where an etiologic agent was determined. The enteric protozoa are relatively recently recognised waterborne pathogens. Human cryptosporidiosis was first described in 1976, and the first reported waterborne outbreak occurred in 1984. There is evidence that *Cryptosporidium* is the third most common enteric pathogen worldwide. Diagnosis is typically made by immunofluorescence microscopy or enzyme immunoassay; however, cryptosporidiosis is probably greatly underdiagnosed. A prospective surveillance study in one state indicated that only 0.5 to 1 per cent of all fecal samples submitted were examined for *Cryptosporidium* and that 3.7 to 8.5 per cent of stool samples examined contained oocysts. Although both trophozoites and cysts/oocysts are shed in feces the cysts are the infective form and, like the enteric viruses, the infectious dose for these agents is low. A human challenge study with *Cryptosporidium parvum* determined that the median infectious dose is 132 oocysts. Water contamination by both human and animal feces is an important mode of transmission for *Giardia* and *Cryptosporidium*. Humans are the only host for *Entamoeba histolytica*.

The thick-walled protozoan cysts and oocysts are environmentally resistant, but there are few data on their survival in water. Their size (*Giardia* cysts are 7 to 14 µm in diameter, and *Cryptosporidium* oocysts are 4 to 6 µm) tends to limit extensive migration through soil; however, groundwater has become contaminated by protozoan cysts during flooding. Methods for the recovery and detection of *Giardia* and *Cryptosporidium* cysts or oocysts in water are cumbersome and relatively inefficient and do not distinguish between viable and nonviable cysts or oocysts. Surveys of raw and treated water supplies indicate that the occurrence of *Cryptosporidium* oocysts is widespread. Two surveys of almost 300 surface water supplies in the United States revealed that 55 to 77 per cent of surface water samples contained *Cryptosporidium* oocysts. Water analysis at 66 US and Canadian surface water treatment plants demonstrated that up to 27 per cent of treated drinking water samples contained low levels of *Cryptosporidium* oocysts. However, current methods may underestimate the occurrence of these organisms in water due to low efficiency of recovery or may overestimate the risk to human health due to an inability to assess the viability of the cysts and oocysts in a way that relates to their potential for human infectivity. Studies by LeChevallier and Norton suggest that approximately 90 per cent of cysts and 66 per cent of oocysts detected in raw water may not be viable. Like enteric viruses, it is not possible for enteric protozoa to multiply in the environment.

The cysts and oocysts of enteric protozoa are relatively resistant to chlorine disinfection. CT values for *G. lamblia* cysts for 99 per cent inactivation at pH 6 to 7 by free chlorine range from 47 to 150. The CT value for 99 per cent inactivation of *C. parvum* by free chlorine (pH 7) at 25°C is estimated to be 7200. Therefore, removal of these organisms by water treatment processes depends on effective coagulation/flocculation, sedimentation, and filtration. A recent case study of three surface water treatment facilities that used coagulation/flocculation, sedimentation, and filtration observed reductions of 2.24 to 2.78 $\log_{10}$ units for *Giardia* and 2.3 to 2.45 $\log_{10}$ units for *Cryptosporidium*.

As diagnostic methods for *Giardia* and *Cryptosporidium* improved, the recognition of their role in waterborne disease outbreaks increased dramatically. From 1986 through 1996, 21 per cent of the 172 outbreaks associated with drinking water were attributed to parasitic agents. Numerous recreational water-related outbreaks have also been associated with *Giardia* and *Cryptosporidium*. In 1993, *Cryptosporidium* was implicated in the largest recorded waterborne disease outbreak in United States, where it was estimated that over 4,00,000 people were infected. The majority of drinking water outbreaks attributed to enteric protozoa have been associated with surface water supplies that were unfiltered or for which flocculation and filtration processes were inadequate. *Cryptosporidium* outbreaks tend to have high attack rates, and fecal contamination from cattle is usually suspected to be the source of the oocysts. There have been no reported waterborne outbreaks of *Entamoeba histolytica* infection in the United States since 1971.

EMERGING WATERBORNE PATHOGENS

Emerging infections diseases have been defined as those whose incidence in humans has increased within the past two decades or threatens to increase in the near future. There are a number of newly recognised infectious agents that have recently been associated with outbreaks of waterborne disease or appear to have the potential for waterborne transmission. Emerging waterborne pathogens are receiving increased recognition by water utilities and regulators.

Recently Recognised and Suspected Waterborne Pathogens

Enteric viruses

HuCV comprise two genera in the family *Caliciviridae* and include NV and other previously named 'small round structured viruses' such as the Snow Mountain agent, Southampton virus, Hawaii virus, Mexico virus, Toronto virus, Lordsdale virus, Sapporo virus, and Desert Shield virus. These viruses are the leading cause of epidemic viral gastroenteritis in the United States and acquire their names from the locations where outbreaks have occurred. Numerous HuCV outbreaks linked to drinking water, recreational water, ice, shellfish, various food items, and environmental contamination have been documented. Until recently, insufficient diagnostic technology limited the study of the role of these viruses in both epidemic and endemic gastroenteritis. The cloning and sequencing of NV in 1990 led to rapid advances in the development of sensitive new molecular diagnostic methods and increased recognition of the importance of HuCV in waterborne and food-borne disease.

Seroprevalence data suggest that epidemic and endemic HuCV infections are common worldwide. Early studies in the United States reported that 50 to 79 per cent of adults have NV antibodies by the fifth decade of life. More recent studies have reported high HuCV seroprevalence in England, Kuwait, Brazil, Sweden, Japan, Mexico, Finland, Norway, Kenya, Chile, Venezuela, and South Africa. These high rates of antibody prevalence indicate that much of the population is frequently exposed to HuCV and may experience recurrent HuCV-associated gastroenteritis because antibodies do not appear to confer protection from illness.

Several other recently recognised enteric viruses are know to be transmitted by drinking water or recreational water, contaminated shellfish, and cold foods and from person to person. Human astroviruses, first described in 1975 by Appleton and Higgins are RNA viruses characterised as small round structured viruses based on their appearance by electron microscopy. They have been associated with acute gastroenteritis, mainly in children and the elderly, and have a worldwide distribution. The predominant clinical features are vomiting and diarrhea. The incubation period is between 1 to 3 days, and symptoms last 1 to 4 days. Outbreaks typically occur in institutional settings with children or the elderly. Astrovirus is generally diagnosed by electron microscopy. However, immunofluorescence, enzyme immunoassay, nucleic acid probe hybridisation, and PCR techniques have recently been developed to detect astroviruses in fecal specimens. There are no reports of the detection of these viruses in water. Evidence of waterborne transmission is provided by epidemiologic analyses of outbreaks.

Enteric adenoviruses (serotypes 40 to 41, also known as subgenus F) are DNA viruses that are associated with about 5 to 12 per cent of pediatric diarrhea and a median duration of infection of 8.6 and

12.2 days, respectively. Diagnosis is made by a commercially available enzyme immunoassay or by PCR. There is no evidence of waterborne transmission of enteric adenoviruses. However, nonenteric adenoviruses (serotypes 3, 7, 1, 4, and 14) have been associated with recreational water-related outbreaks of pharyngoconjunctival fever and conjunctivitis. Adenoviruses have been isolated from a sewage outlet at a lake and from a swimming pool.

Coronaviruses are pleomorphic, enveloped RNA viruses that are well-established causes of diarrhea in animals. They were first observed in feces of persons with gastroenteritis by electron microscopy in 1975, but since then they have also been frequently detected in the feces of healthy persons. Hence, there continues to be doubt about their etiologic role in human diarrhea. Except for one strain, these viruses have not been propagated *in vitro* and their diagnosis depends on observation of characteristic morphological features by electron microscopy. Epidemiologic evidence suggests that fecal-oral transmission and personal hygiene may be key factors in transmission and several studies have noted that the highest prevalence rates were among populations with low socio-economic status and poor personal hygiene. One study in Lesotho reported prevalence rates in children of 30 to 68 per cent but found no difference between prevalence rates in a village with an 'improved water supply' and a village with a 'traditional contaminated water supply'. These findings suggest that coronavirus infections may be related to water quantity and hygiene rather than to water quality. There have been no reported waterborne outbreaks of coronavirus infection.

Enteric and aquatic bacteria

Enterohemorrhagic *E. coli* O157:H7 is a pathogenic strain of *E. coli* that produces two potent toxins. This organism causes bloody diarrhea, and 2 to 7 per cent of infections result in hemolytic-uremic syndrome, in which the red blood cells are destroyed and the kidneys fail. In many parts of the United States and Canada, *E. coli* O157:H7 is the second or third most commonly isolated enteric bacterial pathogen. Children and the elderly are most susceptible to hemolytic-uremic syndrome. This severe disease was first recognised in 1982 and has the highest mortality rate of all waterborne diseases in the United States. The infectious dose of *E. coli* O157:H7 is believed to be low, like *Shigella*, and the incubation period is 12 to 60 hours. The reservoir appears to be healthy cattle, and transmission often occurs by ingestion of undercooked beef or raw milk. Person-to-person transmission is important among families and in child care centres. Diagnosis is made by culture followed by a commercially available latex agglutination test for *E. coli* O157:H7 antigen.

Aeromonas species are commonly found in water and soil. Reported densities in water range from 10^2–10^3 per ml in river water to 1–100 per litre in groundwater. There are reports of *Aeromonas* spp. in chlorinated waters, and evidence suggests that chlorination has only a temporary effect on *Aeromonas* and does not prevent its regrowth in distribution systems. High population densities appear to be related to fecal pollution and temperature, and aeromonads proliferate in domestic and industrial waste-waters. There is some evidence to suggest that a high proportion of environmental isolates may produce enterotoxins, and several reports have suggested an association between gastroenteritis and *Aeromonas* in drinking water. A 2-year clinical study in Iowa concluded that three strains of *Aeromonas* were capable of causing diarrhea and that consumption of untreated water was a risk factor for *Aeromonas* infection. Studies in London reported a correlation between water and fecal isolates of *A. sobria*. A case-control study in the United States of *A. hydrophila* gastroenteritis reported that drinking untreated water was a significant risk factor for infection. A high summer prevalence of nosocomial *A. hydrophila* infection was linked to high counts in hospital water storage tanks in France. A study in the Netherlands found that patients infected with cytotoxigenic strains of *Aeromonas* were more likely to have had recreational water contact during the week prior to onset of illness. However, other studies in the Netherlands and London found little similarity between aeromonads isolated from diarrheal feces and those found in drinking water. Concern about the possible health effects from these organisms in the Netherlands has led to the establishment of drinking water guidelines of <20 CFU/100 ml for drinking water leaving the treatment plant and <200 CFU/100 ml for drinking water in the distribution system.

A wide range of 'atypical' (non-tuberculosis) mycobacteria occur in the environment. Generally, these organisms cause illness mainly in immunosuppressed populations. Although opportunistic mycobacteria can infect almost any site in the body, they are most commonly associated with pulmonary disease, cervical lymphadenopathy, and localised skin and soft tissue infections. Cutaneous lesions have been associated with recreational water exposure to *M. marinum* (also called *M. balnei*). *M. avium* and *M. intracellulare* (together referred to as *M. avium* complex [MAC]) have been detected in soil, waste-water, surface water, groundwater, and drinking water. Reported concentrations of MAC in water distribution systems range from 0.08 to 45,000 CFU/100 ml. The incidence of MAC infection, which typically causes pulmonary and disseminated disease, in HIV-infected patients has dramatically increased, causing further speculation about the possible role of water in the transmission of this agent. A recent study that compared clinical mycobacterial isolates with environmental isolates from hospital hotwater systems suggested that exposure to these water sources was the transmission route for several groups of AIDS patients. Other species (*M. kansasii* and *M. xenopi*) have been isolated from domestic and hospital water supplies; however, the health significance of this finding is uncertain. MAC organisms can grow over a wide temperature range (15° to 45°C), can persist in water distribution systems for years, and are resistant to chlorination. Little is known about the susceptibility of these organisms to other disinfectants.

H. pylori, formerly referred to as *Campylobacter pylori*, were first described in 1982 by Marshall and Warren. This gram-negative, microaerophilic, spiral-shaped bacterium resides in the mucus layer lining of the stomach and upper gastrointestinal tract and is associated with indigestion and abdominal pain. Current evidence suggests that it is responsible for >90 per cent of duodenal ulcers and >80 per cent of gastric ulcers (www.cdc.gov/ncidod/dbmd/md/htm). Chronic infection is associated with gastric cancer, which is the most common cancer in developing countries, where more than half of the population is infected in early childhood (www.cdc.gov/ncidod/dbmd/md/htm). *H. pylori* infections occur throughout the world, and the prevalence of infection increases with age. Individuals in developing countries are more likely to become infected earlier in life, and by adulthood, infection rates range from approximately 40 per cent in developed countries to 80 per cent in developing countries. Effective antibiotic treatment can successfully eradicate the infection in most cases. If untreated, infection becomes chronic and probably persists for life.

Fecal-oral transmission of *H. pylori* infection has been suggested by several studies that implicated crowding, socioeconomic status, and consumption of raw, sewage-contaminated vegetables as risk factors for infection. Studies in Peru and Ethiopia have identified the type of water supply as a risk factor for infection and found that the water source appeared to be a more important risk factor than socioeconomic status. Other studies in Peru have argued that food prepared under unhygienic conditions is the major transmission route of *H. pylori* in developing countries.

A seroprevalence survey of 245 healthy children in Arkansas did not find a relation between *H. pylori* seropositivity and type of water supply (municipal or well). However, it is likely that differences in the degree of fecal contamination in the Peruvian and Ethiopian water sources compared to those in Arkansas may explain these conflicting reports.

Because it is a fastidious organism, *H. pylori* have been difficult to culture from environmental sources. Laboratory studies demonstrated that *H. pylori* survives in freshwater and sterile, distilled water (7°C) for 10 and 14 days, respectively. These studies also found evidence that *H. pylori* can survive for prolonged periods as viable, non-culturable coccoid bodies. Klein reported difficulty in detecting *H. pylori* from water samples because the culture plates were overgrown before the 3 to 4 days required to isolate *H. pylori*. Enroth and Engstrand reported the development of PCR methods to detect *H. pylori* in water. Subsequently, PCR was used to detect *Helicobacter* DNA in private wells, municipal tap water, and waste-water in Sweden and drinking water samples in Peru. Another study used monoclonal anti-*H. pylori* antibody and microscopy to detect actively respiring *H. pylori* in surface and shallow groundwater in Pennsylvania and Ohio and found that there was no significant correlation between the detection of either total coliforms or *E. coli* and *H. pylori* in water. Disinfection studies have shown that three strains of *H. pylori* were inactivated by more than four orders of magnitude after exposure to 0.5 mg of free chlorine per litre for 80 seconds. These findings suggest that *H. pylori* should be readily controlled by disinfection practices normally used for drinking water treatment.

Protozoa, fungi and algae

Cyanobacteria (blue-green algae) occur naturally in freshwater and brackish water worldwide. Although these are not infectious agents, some species produce toxins during algal blooms that are triggered by nutrient enrichment from natural waters, agricultural fertiliser runoff, or domestic or industrial effluents. Approximately 46 species of cyanobacteria have been associated with toxic effects in vertebrates. The most common freshwater toxic forms are *Microcystis* spp., *Planktothrix* (syn. *Oscillatoria*) *rubescens*, *Planktothrix* (syn. *Oscillatoria*) *agardhii*, *Anabaena* spp., *Aphanizomenon* spp., some *Oscillataria* spp., *Cylindrospermopsis raciborskii*, *Synechococcus* spp., *Gloeotrichia* spp., *Lyngbya* spp., *Nostoc* spp., *Schizothrix* spp., and *Synechocystis* spp. Toxic blooms have been reported in many parts of the United States, Canada, Europe, South Africa, Asia, Australia, and New Zealand. The conditions that influence the toxicity of a bloom are not known. However, temperature is believed to be one key factor. Surveys in several countries indicate that 75 per cent of bacterial blooms contain toxins.

Cyanobacterial toxins are of three main types: lipopolysaccharide endotoxins, hepatotoxins, and neurotoxins. Hepatotoxins appear to be more common than neurotoxins. Acute health effects in humans include gastroenteritis, liver damage, nervous system damage, pneumonia, sore throat, earache, and contact irritation of skin and eyes. The potential chronic health effects of long-term exposure to cyanobacterial toxins in drinking water are unknown. It has been suggested that the high rates of liver cancer in parts of China may be linked to cyanobacterial hepatotoxins in drinking water. A 1976 outbreak of intestinal illness in Pennsylvania was associated with a cyanobacterial bloom in a municipal water supply and affected 62 per cent of the population. In addition, there are several reports of adverse health effects related to recreational contact with water in the United States.

Control of cyanobacteria is problematic because several studies indicate that the toxins can remain potent for days after the organisms have been destroyed by copper sulphate or chlorination. Activated carbon and ozone are both effective for removing these toxins. Lime treatment is also an effective control method because it can precipitate phosphorus and sediment the cyanobacterial cells. Based on toxicity data from mouse bioassays, the Engineering and Water Supply Department of South Australia developed interim guidelines for acceptable numbers of cyanobacteria in water supplies. Recently the World Health Organisation suggested a provisional drinking water guideline for microcystin-LR of 1 μg/litre. New methods of chemical analysis, high-performance liquid chromatography, and commercially available immunoassays and enzyme assays are now available for detection of microcystins and saxitoxins. However, further research is needed on the acute and chronic toxicity of cyanobacterial toxins and better methods of monitoring the types and concentrations of cyanobacterial toxins in natural and treated water supplies are needed.

Cyclospora cayetanensis, formerly called cyanobacterium-like bodies or big *Cryptosporidium*, was identified as a new protozoan pathogen of humans in 1993. A member of the family Eimeriidae, it was first described by Ashford in 1979 as present in human stool specimens in Papua New Guinea. In 1986, Soave reported the first detection in the United States in stools from patients who had lived or travelled in developing countries. In the United States, *Cyclospora* is frequently found in AIDS patients with prolonged diarrhea. Under light microscopy, it appears as a refractile sphere, 8 to 10 μm in diameter. It has been associated with prolonged, self-limited watery diarrhea with an average duration of 40 days. This organism occurs in tropical climates worldwide and has been identified as the cause of diarrheal disease outbreaks in North, Central, and South America, Caribbean Countries, Southeast Asia, and Eastern Europe.

Evidence for waterborne transmission comes from a case-control study of a diarrhea outbreak among foreign residents in Nepal, where consumption of untreated water was identified as a risk factor. A case-control study of *Cyclospora* infection in Guatemala also found that the consumption of untreated water, river water, or spring water was significantly associated with risk of infection. The Nepal and Guatemala studies and two prospective cohort studies of Peruvian children indicate a seasonal pattern of infection, with peak incidence occurring in the warm summer months. In the United States, an outbreak of *Cyclospora* diarrhea occurred among the house staff of a Chicago hospital in the summer of 1990. Contaminated open-air, rooftop water storage tanks that supplied the house staff dormitory were believed to be the source of the outbreak. However, *Cyclospora* species could not be detected in water samples from the tanks. Another waterborne outbreak of cyclosporiasis was linked to the consumption of water from water coolers on a golf course in New York in June 1995. Two case reports have described *Cyclospora* infections in immunocompetent hosts. In one case the host developed prolonged diarrhea after exposure to sewage-contaminated water in his home. Sewage from a neighbouring dairy backed up into the basement of the patient's home, and *Cyclospora* was detected in sewage effluent samples taken from the pipe that served the dairy. In the other case, *Cyclospora* infection was associated with consumption of well water in New Hampshire after heavy snow melts had caused local flooding. *Cyclospora* oocysts have been detected by microscopy and PCR in sewage lagoons near areas of endemic cyclosporiasis in Lima.

'Microsporidia' is a general term that describes a large group of primitive, obligate intracellular protozoa commonly found in animals. The first human infection was reported in 1985. Currently, species of six genera, *Nosema*, *Encephalitozoon*, *Pleistophora*, *Enterocytozoon*, *Vittaforma*, and *Trachipleistophora*, and a general grouping, 'microsporidium' (for unclassified microsporidia), are known to cause human infections. With the exception of *Enterocytozoon bieneusi*, there seems to be little host specificity because human infections with nonmammalian genera have been described and because these organisms can infect a variety of tissues and organs (intestinal tract, eyes, liver, kidneys, heart, and central nervous system). *E. bieneusi* and *Encephalitozoon intestinalis* (formerly referred to as *Septata intestinalis*) primarily parasitise the intestine and can cause persistent diarrhea. The vast majority of reported cases of microsporidial infections have occurred among persons infected with HIV. A study of 5692 stool specimens collected from 1454 people from 1993 to 1996 were submitted to a reference laboratory in Lyon, France, for assessment of microsporidia. Intestinal microsporidia were identified in

338 persons: 261 persons infected with HIV (0.1 per cent prevalence), 16 transplant recipients, and 61 others. Epidemiological and laboratory studies indicate that *E. bieneusi* is an important cause of chronic diarrhea in AIDS patients, and there is no effective treatment against this organism. Studies of seasonal variation in enteric microsporidiosis among HIV-infected persons in New Orleans and southern California did not find any significant seasonal trends in the prevalence of microsporidiosis that could be evidence of waterborne transmission.

The role of microsporidia in waterborne disease is not known. Transmission of microsporidia is believed to be primarily fecal-oral. Infection is acquired through ingestion of small (1- to 2-μm-diameter) spores with thick walls, which make them enviromentally resistant and allow them to remain infective for up to 4 months in the environment. One purported waterborne outbreak of intestinal microsporidiosis has been reported. The monthly incidence of intestinal microsporidiosis detected at reference laboratory in Lyon, France, was monitored from May 1993 to December 1996. The prevalence in HIV-infected patients rose from a baseline level of about 0.1 per cent to a sharp peak of 1 per cent in July and August 1995. Analyses of risk factors associated with microsporidiosis during the period before the peak did not indicate any relationship between microsporidiosis and the water distribution subsystem. However, during the period of peak prevalence, the diagnosis of microsporidiosis was strongly associated with living in an area served by the 'medium elevation water distribution system'. The water supply in this area is river water 'naturally filtered by alluvia', chlorinated at two sites, and distributed to three independent subsystems with different elevations. Standard water quality measures did not indicate fecal contamination, and increased prevalence of *Cryptosporidium parvum* or *Giardia intestinalis* was not observed during the study period.

Methods to recover and detect microsporidian spores by immunofluorescence assays and PCR are still under development. The first document report of microsporidia in water was the repeated microscopic detection in 1987 of microsporidian spores in ditch water samples collected in Florida during a 1-year period. Spore concentrations of up to 3000 per ml of water were observed, and two of the five genera identified in the samples were genera that include human pathogens (*Nosema* and *Pleistophora*). More recently, Dowd reported PCR amplification of microsporidium DNA in 7 of 14 water concentrates and detected *Encephalitozoon intestinalis* in tertiary sewage effluent, surface water and groundwater; *Enterocytozoon bieneusi* in surface water; and *Vittaforma corneae* in tertiary effluent.

Recently Recognised Waterborne Pathogens in Developing Countries

Hepatitis E virus

Hepatitis E virus (HEV) is the only known agent of enterically transmitted non-A, non-B infectious hepatitis, although some evidence suggests that there may be others. A single serotype of virus that has single-stranded, positive-sense RNA with three open reading frames has been described. Because of its physico-chemical properties, HEV was provisionally classified in the family *Caliciviridae*. New data indicate that the organisation of the HEV genome is substantially different from that of other caliciviruses, and it is likely to be classified as a separate family. The recent development of serologic tests for HEV antibody based on recombinant DNA technology has enabled further study of the epidemiology of this virus, which is clinically and epidemiologically similar to hepatitis A virus. However, unlike hepatitis A virus, the majority of cases in epidemics occur among young adults (15 to 40 years of age), and there is a high case-fatality rate (up to 30 per cent) among pregnant women. Large outbreaks, involving thousands of cases, have been reported in developing areas in Africa, Asia, and Mexico and have been linked to fecally contaminated water and inadequate chlorination. In 1991, the largest documented waterborne

HEV outbreak affected an estimated 79,000 persons in Kanpur, India, and was associated with contaminated surface water. HEV RNA has been detected in samples from sewage treatment plants in India by using reverse transcriptase PCR and nucleic acid probe hybridisation. Currently, there are no data on the inactivation of this virus in water by disinfection and there are no reports of HEV detection in water. HEV has been detected in swine manure, and it has been suggested that pigs may provide a nonhuman reservoir.

Group B rotavirus

Group B rotavirus was first reported in connection with a waterborne outbreak in China in 1984 and differs from the group A rotavirus strains that commonly cause pediatric diarrhea. Group B rotavirus infections occur more frequently in adults than children, are associated with severe cholera-like illness, and have been found mainly in China. By electron microscopy, the virus is morphologically similar to group A rotavirus, but it is antigenically distinct. Diagnosis is made by electron microscopy and genome, electrophoresis or by enzyme immunoassay. Seroprevalence surveys for group B rotavirus antibody have reported rates of up to 41 per cent positive in China; 5 to 18 per cent positive in Thailand, Burma, and Hong Kong; 1 to 12.5 per cent positive in the United Kingdom, United States, and Canada; and 10 per cent positive in Kenya. More than 1 million cases were reported in China in 1982 to 1983, and many outbreaks involving tens of thousands of cases were attributed to fecally contaminated water. However, it is puzzling why so few cases of group B rotavirus infection have been identified since the 1980s. Group B rotaviruses have not yet been detected in water.

Vibrio cholerae O139

V. cholerae O139 is a new toxigenic strain of epidemic *V. cholerae* that has been called the Bengal strain. This is the first non-O1 *V. cholerae* strain ever reported to be associated with epidemic cholera, and it is causing a new pandemic in Asia, which began in India in October 1992 and is currently spreading into the Middle East. The clinical illness is indistinguishable from cholera caused by *V. cholerae* O1. However, this new epidemic is affecting persons of all ages in areas where most of the population, except for young children, have acquired immunity to *V. cholerae* O1. The Bengal strain is also significant in the history of cholera because recent evidence suggests that this strain resulted from genetic recombination and horizontal gene transfer, which allowed the acquisition of unique DNA. As with other strains of *V. cholerae*, transmission occurs through fecally contaminated water and food.

Waterborne Disease of Unknown Etiology

Waterborne disease surveillance in the United States indicates that no etiologic agent can be determined for more than 50 per cent of reported waterborne disease outbreaks. Part of this may be due to lack of appropriate clinical and/or environmental samples or to limitations of the diagnostic techniques for many of the newly recognised pathogens. However, there are some outbreaks that, despite thorough investigation, could not be attributed to any known etiologic agent, which suggests the existence of unrecognised agents of waterborne disease.

'Brainerd diarrhea' was first described in an outbreak with 122 cases that occurred in Brainerd, Minn., in 1983. The illness was characterised by chronic diarrhea with acute onset, marked urgency, lack of systemic symptoms, and failure to respond to antimicrobial therapy. The distinctive feature of the illness is the average duration of 12 to 18 months. In the Brainerd outbreak, transmission was linked to raw milk that had passed through hoses rinsed with poor-quality water. A subsequent outbreak in

Illinois in 1987 was clearly associated with the consumption of untreated well water that had sporadic problems with fecal coliform counts. Several other possible outbreaks have been suspected in Texas in 1985 and five other states, as well as in 58 passengers on a cruise ship who participated in group tours to the Galapagos Islands in 1992. In the Galapagos cruise ship outbreak, illness began 11 days (median) after the passengers boarded the ship and lasted 7 to 42 months. Epidemiological evidence linked the outbreak to the water supply on the ship, which was inadequately treated. Despite intense microbiological analyses of several outbreaks, no etiologic agent for this syndrome has been identified.

PREVENTION AND CONTROL OF WATERBORNE DISEASE

Historical surveillance data and epidemiologic-microbiologic studies of water and health have documented the relationship between improvements in water quality and water treatment practices and a reduction in morbidity and mortality associated with waterborne diseases. Reviewing studies of water quality, water quantity, and hygiene, Esrey concluded that improvements in microbiological water quality resulted in a 16 per cent median reduction of diarrhea morbidity (range, 0 to 90 per cent) and that improvements in both water quality and availability resulted in a 37 per cent median reduction of diarrhea morbidity (range, 0 to 82 per cent). Prevention and control of waterborne diseases requires accurate and rapid methods to measure microbiological water quality and to identify and evaluate risk factors for waterborne disease.

Measurements of Microbiological Water Quality

Detection of infectious agents in water

Many waterborne pathogens are difficult to detect and/or quantify in water, and methods to detect most of the newly recognised agents in environmental samples have still to be developed. The concentrations of enteric organisms are much lower in water than in clinical specimens, and so their detection in water starts with a concentration process such as filtration. This is followed by a process to recover the pathogen from the filter and then by an enrichment or amplification process by either culture or molecular biology methods. Typically the recovery efficiencies of these procedures are low, making it difficult to estimate the original concentration of the infectious agent in the water. Also, some methods, such as those used to detect *Giardia* cysts and *Cryptosporidium* oocysts, do not give an indication of the viability or infectivity of the organisms. Furthermore, many of these laboratory techniques are limited to specialised research or reference laboratories and are not done on a routine basis.

In most investigations of waterborne disease outbreaks, water is identified as the vehicle of transmission by epidemiologic evidence rather than by the detection of the infectious agent in water samples. For example, of the 32 waterborne infectious disease outbreaks reported in the United States between 1991 and 1992, the etiologic agent was detected in the water in only 4: *Shigella* (in one outbreak), *Giardia* (in two), and *Cryptosporidium* (in one). This may be because the contamination of the water supply was temporary and the infectious agent died off or was flushed out of the water system before the outbreak was recognised and appropriate water samples were collected. The longer the incubation period of the infectious agent, the longer it may be before the outbreak is recognised and water is suspected as the vehicle. The size and timing of the contamination event, the lag time until recognition of the outbreak, the survival characteristics and transport patterns of the agent, and the sensitivity and efficiency of the laboratory methods all affect the likelihood that an infectious agent will be detected in water.

Microbial indicator organisms

EPA and World Health Organisation standards for microbiological water quality are expressed in terms of total coliforms and fecal coliforms. These are groups of bacteria excreted by healthy humans and animals that serve as indicators of fecal contamination. Laboratory tests for total and fecal coliforms in water are much easier to perform than tests to detect pathogenic micro-organisms in water. Because these indicator organisms are excreted in large numbers by all individuals, the concentrations of these indicators are likely to be higher and a more constant fraction of the community fecal waste 'pool'. By contrast, specific pathogens are excreted only by infected individuals and their numbers in the community fecal waste pool depend on the excretion level of the particular pathogen and on the number of infected individuals in the community.

Ideally, microbial indicators should provide a measure of health risk associated with the ingestion of or contact with water. Total and fecal coliforms have many limitations as predictors of the risk of waterborne disease. Because of their shorter survival times in water and their greater susceptibility to water treatment processes, these indicator organisms tend to be poor models for enteric protozoa and viruses. Outbreaks of waterborne disease, especially protozoal outbreaks such as the Milwaukee epidemic, have been associated with water that met total and fecal coliform standards. Moreover, there are nonfecal sources for these indicator organisms and, in contrast to most enteric pathogens, total and fecal coliforms may multiply in aquatic environments that provide sufficient nutrients and optimal temperatures. Such characteristics may result in false-positive reports of water contamination. The performance of total and fecal coliforms as reliable indicators of tropical water quality is especially problematic and has been investigated and reviewed by Hazen. This is of particular concern in tropical developing countries, where there are a greater number of untreated, contaminated water sources and high morbidity and mortality associated with waterborne disease.

Many alternative indicator organisms have been investigated, and their advantages and limitations have been reviewed. *E. coli* and enterococci were included in the recent revision of the EPA recreational water standards. *Clostridium perfringens* and male-specific coliphage have been proposed as potential indicators of drinking water quality that may better model the survival and disinfection resistance of enteric protozoa and viruses. Epidemiologic-microbiologic studies have demonstrated that *E. coli* and enterococci are better indicators of tropical drinking water quality than are fecal coli forms. There is increasing recognition that no single organism can serve as an adequate indicator for all types of water and all routes of exposure.

Multiple-barrier Approach

In developed countries, waterborne disease prevention and control is based on a multiple-barrier approach that involves source water protection, water treatment, and distribution system management and protection. Waterborne disease outbreaks in the United States usually involve: (i) source contamination and the breakdown of one or more of the treatment barriers (disinfection or filtration), (ii) contamination of the distribution system, and (iii) the use of untreated water. The multiple-barrier approach emphasises the identification of all available barriers, assessment of the degree of vulnerability of each barrier to the passage of pathogens, recognition and anticipation of conditions under which pathogen risk increases, and maintenance of barriers at high levels of effectiveness. For waterborne diseases associated with aquatic micro-organisms, approaches to prevention and control may depend on treatment technology or on the control of exposure to untreated waters.

Risk Assessment Approaches

Risk assessment approaches have been useful in systematically identifying, analysing, quantifying, and characterising the risk of specific waterborne illnesses. These models are based on field data for the occurrence of specific micro-organisms in raw and treated water supplies, experimental data on removal or inactivation by various water treatment processes, and experimental dose-response data. However, the shape of the dose-response curve, especially in the low-dose region representative of waterborne exposure, is ill defined. Furthermore, it is difficult to model variations in microbial virulence factors and host-specific characteristics, such as age and immune status, that may affect individual exposure and susceptibility to infection and disease. Finally, for infectious agents with multiple transmission routes, it may be difficult to determine the attributable risk associated with waterborne transmission compared to other routes of transmission, especially in areas where waterborne diseases are endemic.

Recommendations and Research Needs

Although the incidence of waterborne diseases has been greatly reduced in areas with effective water treatment, control of waterborne transmission of infectious agents continues to be an important challenge for public health research. Methods to detect many of these agents in clinical and environmental samples must be developed or improved. There is a critical need for rapid methods to detect waterborne pathogens, both those typically associated with waterborne disease and other microbial pathogens that may be inadvertently or deliberately added to a water supply. New technologies such as DNA microarray chips, immunomagnetic separation and solid-phase laser-scanning cytometry, and automated cell counting of immunofluorescence-tagged cells (advanced analytical technologies rapid bacteria detector RBD-2000) may have potential for rapid, automated water quality analyses. There is also a need to develop simple methods to detect fecal contamination in water that can be readily used in developing countries with limited laboratory facilities and resources.

Information on the occurrence and persistence of enteric and aquatic pathogens in various types of water supplies should be updated, and the factors that contribute to virulence and waterborne transmission must be better characterised. For many waterborne pathogens, we need to improve our understanding of the effectiveness of various water treatment processes and disinfectants in removing and/or inactivating these micro-organisms. Finally, better surveillance of waterborne diseases would lead to earlier recognition and investigation of epidemic and endemic waterborne disease. Infections with *Giardia*, *Cryptosporidium*, and *E. coli* O157:H7 are now reportable diseases in most states. Routine stool examination for *Cryptosporidium* and other pathogens would greatly further our understanding of the disease burden associated with various enteric pathogens, but this requires a shift in the current practice of stool examination solely for individual treatment purposes to stool examination for the benefit of public health practice. Increased awareness among health professionals of the symptoms associated with more recently described waterborne illnesses, such as *Cyclospora* or cyanobacterial poisoning, may lead to greater reporting of these conditions and provide opportunities to study their waterborne transmission. Because the health risks posed by many waterborne infectious agents are still unknown, water utilities and public health professionals need to work together to maintain vigilance for any indication of waterborne transmission of disease in their communities.

Chapter 7

Detection of Micro-organisms in Environmental Freshwaters and Drinking Waters

INTRODUCTION

The presence of certain micro-organisms in water is used as an indication of possible contamination and an index of water quality. Heavy reliance has been placed on the presence of the coliform group of bacteria to determine the safety of potable water, while coliform bacteria and other organisms are used to signal conditions that could lead to adverse health consequences in recreational and shellfish harvesting water. Although this practice is far from perfect and there is considerable variety in the ways that different indicator micro-organisms occur in various geographical areas and situations, public health concerns have generally been well served by this approach.

The presence of indicator organisms will likely continue to be a valuable criterion water quality if more attention is given to the development and use of optimal methods for the recovery of these micro-organisms; however, data also exist indicating that indicator micro-organisms have limited utility under certain circumstances.

Gastroenteritis is the most common affliction associated with waterborne pathogens. Although for most of the population in developed countries gastroenteritis is usually minor and simply means hours of discomfort, in developing countries up to 10 million people die every year as a direct result of the consumption of contaminated water. However, the continuing increase in the proportion of immunologically compromised and elderly people in developed countries has focused more attention on possible waterborne-disease outbreaks in these countries as well. The seriousness of this reality is compounded by the fact that determining the microbiological quality of waters is more complex than previously thought.

The presence of enteric pathogens in drinking and recreational waters is of great concern. As a result of the danger to public health due to the presence of these pathogens, it is important to determine the microbiological safety of these waters. The ideal manner for doing this would be to analyse the waters for the presence of specific pathogens of concern. However, hundreds of different micro-organisms have been shown to be involved in waterborne-disease outbreaks; thus, it is impractical to look for every pathogen potentially present in water samples. Culture methods are usually used for bacteria and cell culture techniques for the detection of viruses, while microscopic methods are used for protozoa. However, target bacteria may not grow in culture media since they are frequently injured by exposure to environmental stressors such as disinfectants used during water treatment. Additionally, several enteric viruses cannot be cultured in the laboratory, and methods for the detection of protozoan pathogens are notoriously inefficient.

As a result, analyses for the presence of waterborne pathogens become complex and do not ensure complete safety of water for the consumer. Thus, members of the following groups of indicator micro-organisms are used to determine the biological safety of the waters.

COMMONLY USED INDICATORS

Total Coliforms

The coliform bacteria are aerobic and facultatively anaerobic, gram-negative, non-spore-forming, rod-shaped bacteria that ferment lactose with gas and acid production in 24 to 48 hours at 35°C. Coliform bacteria belong to the family *Enterobacteriaceae* and include *Escherichia coli*, as well as various members of the genera *Enterobacter*, *Klebsiella*, and *Citrobacter*. The above definition of the coliform bacteria has classically been translated into specific biochemical reactions or the appearance of characteristic colonies on commonly used media. However, the development of enzyme-specific media has allowed the application of cytochrome oxidase (negative) and β-galactosidase (positive) tests as additional criteria for the presence of the coliform group.

Classically, these bacteria have been used as indicators of possible fecal contamination or water pollution from sewage, and thus they are of sanitary significance. Although coliform bacteria can originate from the intestinal tracts of homeothermic animals, other bacteria numerically dominate that type of microbial community. In addition, experience has demonstrated that members of the coliform group can originate from nonenteric environments such as wastes from the wood industry and surfaces of redwood water tanks, biofilms within drinking-water distribution systems, and epilithic algal-mat communities in pristine streams. The persistence of these bacteria in aquatic systems is comparable to that of some of the waterborne bacterial pathogens, although they are much less persistent than enteric viruses and protozoa. In some aquatic environments, the survival rate of total coliforms is lower than that of pathogenic micro-organisms. Likewise, bacteria belonging to the coliform group are somewhat like many of the waterborne bacterial pathogens with respect to disinfection susceptibility but unlike the more persistent viruses and protozoa. These factors illustrate why experience and discretion are often vital in the correct interpretation of coliform data obtained from some aquatic environments.

The ability of some coliforms to grow in natural waters, the lack of correlation between the numbers of coliforms and those of pathogenic micro-organisms, and the detection of atypical strains are serious shortcomings for their general use as indicators of water quality.

The presence of bacteria that compose what has traditionally been termed the total-coliform group provides the basis of the primary standards for potable water in North America and indeed most of the world. The designation 'total-coliform group' is usually interpreted as the totality of bacteria that conform to the classical, nontaxonomic coliform definition, although the inclusion of the term 'total' has often led to the false assumption that all viable coliform bacteria are detected by a given method specified for the analysis of coliform bacteria.

Thermotolerant Coliforms (Fecal Coliforms)

The subset of the more comprehensive coliform or total-coliform group that is more definitive as an indicator of homeothermic fecal contamination consists of what are termed the fecal coliforms. However, 'thermotolerant coliforms' is a more scientifically accurate term for this group. These bacteria conform to all of the criteria used to define total coliforms plus the requirement that they grow and ferment lactose with the production of gas and acid at 44.5 ± 0.2°C. Dufour suggested the redefinition of the fecal coliform group with the addition of several biochemical characteristics, such as their complete inability to use citrate as the only carbon source, the presence of tryptophanase activity and lack of urease activity, the negative Voges-Proskauer test, and the positive methyl red test. Bacteria in this coliform subgroup have been found to have an excellent positive correlation with fecal contamination from warm-blooded animals. The physiological basis of the elevated-temperature phenotype in the fecal coliforms has been described as a thermotolerant adaptation of proteins to and their stability at the temperatures found in the enteric tracts of animals, which are both constant and higher than temperatures in most aquatic and terrestrial environments. However, some thermotolerant coliform bacteria that conform to this definition also belong to the genus *Klebsiella*, and have been isolated from environmental samples in the apparent absence of fecal pollution.

Such observations have been made in water receiving high levels of carbohydrate-rich industrial effluent and in water that has contact with plant material. Similarly, in the last 20 years other members of the thermotolerant coliform group, including *E. coli*, have been detected in some pristine areas of the world and associated with regrowth events in potable-water distribution systems. Thus, caution needs to be exercised when deciding whether the presence of these indicator micro-organisms does indeed represent fecal contamination and thus a threat to public health.

Escherichia Coli

Among thermotolerant coliforms, *E. coli* deserves further discussion. This bacterium has been shown to be a more specific indicator for the presence of fecal contamination than other members of the thermotolerant-coliform group of bacteria. In addition, *E. coli* conforms to taxonomic as well as functional identification criteria and is enzymatically distinguished by the lack of urease and presence of β-glucuronidase, both of which form the basis for recently developed differential methods discussed below. One of the disadvantages associated with this organism as an indicator is that it has been consistently found in pristine tropical rain forest aquatic and plant systems as well as soils and so may not be a reliable signal of fecal contamination in those environments. Additionally, *E. coli* has low survival capabilities in aquatic environments in temperate environments. *E. coli* has been effectively used as a water quality indicator for some time in Europe and has recently been incorporated into US drinking water regulations as a specific indicator of fecal contamination.

Fecal Streptococci and Enterococci

Fecal streptococci and enterococci, which are gram-positive bacteria, have received widespread acceptance as useful indicators of microbiological water quality, because (i) they show a high and close relationship with health hazards, mainly gastrointestinal symptoms, associated with bathing in aquatic environments, (ii) they are not as ubiquitous as coliforms, (iii) they are always present in feces of warm-blooded animals, (iv) they are unable to multiply in sewage-contaminated waters, and (v) their die-off is less rapid than that of coliforms in water, and persistence patterns are similar to those of potential waterborne pathogenic bacteria.

Fecal streptococci comprise species of different, sanitary significance and survival characteristics; in addition, the numbers of the species of this group are not the same in animal and human feces. Therefore, a clearer definition of fecal streptococci is necessary to establish a specific standard methodology of enumeration. The taxonomy of this group has been subject to extensive revision. According to Borrego and Figueras, the following species of the *Enterococcus* and *Streptococcus* genera may be included in the fecal streptococcal group: *Enterococcus faecalis, Enterococcus faecium,*

Enterococcus durans, Enterococcus hirae, Enterococcus avium, Enterococcus gallinarum, Enterococcus cecorum, Streptococcus bovis, Streptococcus equinus, Streptococcus alactolyticus, Streptococcus intestinalis, Streptococcus hyointestinalis, and *Streptococcus acidominimus.*

The enterococci listed above include all of the species described as members of the genus *Enterococcus* that also fulfil Sherman's criteria: growth at 10°C and 45°C, resistance to temperature of 60°C for 30 minutes, growth at pH 9.6 and at 6.5 per cent NaCl, and the ability to reduce 0.1 per cent methylene blue. These micro-organisms are more closely related to the presence of human feces than are other fecal streptococci.

Other members of the fecal streptococcal group such as *S. bovis* and *S. equinus* are somewhat more characteristic of fecal contamination from specific animals, although these indicator species do not survive in water as well as do many of the other classical indicator bacteria and some pathogens.

Staphylococci

Several epidemiological studies show high incidences of nondiarrheal diseases, such as eye, ear, mucus, and skin infections, which are associated with swimming. Many swimmers acquire staphylococcal skin infections after swimming in recreational waters, and thus, there is also a need for an indicator of water quality which addresses skin infections rather than gastrointestinal infections.

Evidence to support the use of coagulase-positive staphylococci, particularly *Staphylococcus aureus*, as indicators of water quality includes the following: (i) this micro-organism is stable and shows a great resistance to environmental conditions, especially in marine and chlorinated swimming pool waters, and (ii) its concentration in water has been shown to represent the load of micro-organisms being shed by swimmers.

Bacteriophages

The use of bacteriophages as indicators is not a new concept, since they have been previously proposed as fecal and enteric viral indicators. The term bacteriophage encompasses a wide group of viruses, and the bacteriophage groups proposed as indicators include somatic coliphages, F-specific RNA bacterio- phages, and phages of *Bacteroides fragilis.*

Somatic coliphages are specific viruses of *E. coli,* and they have been the most studied group. They are commonly used as indicators of fecal and/or sewage pollution, on the basis of their direct correlation with the presence of enteric viruses in marine waters, freshwater ecosystems, and sewage effluents. Recently, Borrego reviewed the indicator potential of somatic coliphages. He concluded that, considering the differences in origin and ecology between enteric viruses and somatic coliphages, it is doubtful that this phage group could successfully be used in all situations as viral indicators; however, somatic coliphages are good indicators of fecal pollution and of the microbiological quality of the water.

Several authors have proposed other bacteriophage groups as alternate indicators. Primrose proposed the use of F (male)-specific RNA bacteriophages (FRNA phages) as fecal pollution indicators, on the basis of their inability to replicate in the water ecosystem, since pili are not synthesised below 30°C. Havelaar and Hogeboom developed a host bacterium specific for the detection of this bacteriophage group, and proposed them as model viruses in water hygiene, because: (i) the group comprises viruses similar in size, shape, and genetic make-up to human enteric viruses, which are responsible for most waterborne diseases, (ii) it represents viruses which are more stable than human enteroviruses in environmental waters and more resistant to disinfection, and (iii) the concentrations found in environmental waters correlate with sewage contamination.

The low incidence of the FRNA phages in feces—human feces in particular—and their high concentration in sewage suggest that FRNA phages should multiply in sewerage system. Hence, the presence of FRNA phages in water is primarily an index of sewage pollution rather than fecal pollution. FRNA phages have been proposed as indicators of the environmental behaviour of enteric viruses (e.g. Norwalk virus) during water and waste-water treatment rather than as general indicators of fecal pollution.

Bacteroides fragilis is a strict anaerobe found in high concentrations in the human intestinal tract, and it dies rapidly when discharged into environmental waters. A phage of the strain HSP 40 of *B. fragilis* has been proposed as a specific index of human fecal pollution of waters, because (i) phages against this strain are human specific and are not isolated from the feces of other homeothermic animals, (ii) *B. fragilis* HSP 40 phages are consistently isolated from sewage, fecally polluted waters, and their sediments but not from unpolluted samples, (iii) the levels of phages are related to the pollution degree, (iv) *B. fragilis* phages always outnumber human enteric viruses, and (v) in model experiments, no replication of these phages has been observed under simulated environmental conditions. The low occurrence of these phages in waters with low and moderate levels of fecal pollution and the complex methodology for their recovery are the main drawbacks for the general use of these viruses as an indicator group.

OTHER COMMONLY USED INDICATORS

Sulphite-reducing Clostridia

The use of the sulphite-reducing members of the genus *Clostridium* (*Clostridium perfringens/Clostridium welchii*) as indicators of fecal pollution was originally proposed in the late 1800s. The presence of these micro-organisms in the feces of all warm-blooded animals is the basis for this practice, although they are considered ubiquitous in aquatic sediments, and the spore form explains their persistence. As a result of the longevity of the spores, these bacteria can be considered indicators of remote fecal pollution and of the disinfection processes of water.

Pseudomonas spp.

Members of the genus *Pseudomonas* are possibly the micro-organisms most often isolated from bodies of water. However, contrary to the previously discussed indicators, their presence does not necessarily indicate a possible risk to public health. Some species have been linked to infections associated with exposure to recreational waters and thus have been, proposed as indicators of recreational water quality. *Pseudomonas aeruginosa* was found to be more resistant than acid-fast bacteria during ozonation processes, which demonstrates its resistance to chemical disinfection and thus its usefulness in the analysis of recreational waters such as swimming pools, which receive chemical disinfection. This micro-organism has also been isolated from drinking waters.

Mycobacteria

The mycobacteria belong to a group of micro-organisms considered to be emerging pathogens of increasing importance. Their importance as etiological agents of waterborne disease is still not completely understood. Although some members of the genus Mycobacterium have been known and studied for a long time as the causative agents of tuberculosis (*Mycobacterium tuberculosis, Mycobacterium africanum,* and *Mycobacterium bovis*), the nontuberculous or atypical mycobacteria have gained notoriety as a result of an increased number of AIDS patients infected with these bacteria. About 50 species of atypical mycobacteria have been identified to date. There is no evidence of person-to-person transmission for any of these atypical species, and they are known to be widely distributed in soil and water environments. The two species most commonly associated with human disease are *Mycobacterium intracellulare* and *Mycobacterium avium*, which together form the group known as the *M. avium* complex (MAC). The MAC are slow growers, which makes it difficult to isolate them in the laboratory. In fact several decontamination techniques have been proposed in an attempt to eliminate the nontarget micro-organisms in environmental samples, but these methods may also affect the MAC. Neumann conducted a comparison of 12 methods for the isolation of MAC from different water samples and concluded that the method used should depend on the type of water being analysed.

H$_2$S Producers

Although most of the members of the total- and fecal-coliform groups are not H$_2$S producers, Manja developed a simple method which has been tested in developing countries and rural areas of North America. Although no correlation was observed between the presence of coliforms and H$_2$S producers, other studies have shown that the presence of the H$_2$S producers does correlate with other criteria of microbiological water quality. Thus, this test may be a good alternative for rural areas, as well as remote areas of the world, since it does not necessitate incubations sophisticated equipment, or trained personnel proposed the use of beehives as a means of keeping the temperature constant during incubation.

Aeromonas spp.

The taxonomy of the genus *Aeromonas* remained unclear for many years, although molecular approaches have established 14 species and a number of unnamed hybridisation groups (HGs) or genomospecies. However, a number of problems are associated with several described Aeromonas species: (i) the existence of mesophilic and psychrophilic strains, (ii) the limited number of strains that have been analysed for taxonomic purposes, and (iii) the limited sources of strains to be studied in defining a species. In addition, other confounding problems involve strains that previously fell within the *Aeromonas hydrophila* complex (HGs 1 through 3). This group has been found to be much more complex than originally thought, including organisms that share phenotypic features but do not belong to the same genomospecies. These organisms could in essence represent new species, but from a practical and medical standpoint, they should probably be referred to as genomovars or nomenspecies.

Several species of the genus *Aeromonas* (*A. hydrophila*, *Aeromonas caviae*, *Aeromonas veronii*. bt. sobria, *A. veronii* bt. *veronii*, *Aeromonas jandaei*, and *Aeromonas schubertii*) have been associated with different human diseases such as gastroenteritis and several nonenteric diseases. Most *Aeromonas*-associated infections seem to be related to water exposure of open wounds as well as the consumption of contaminated water or food. Underlying illnesses (e.g. diabetes, cirrhosis, an immunosuppressive state) play a major role in the acquisition and outcome of the diseases produced by *Aeromonas* spp.

Mesophilic members of the genus *Aeromonas* have been detected and isolated from brackish, fresh, and estuarine waters; sewage, treated effluents; and chlorinated and nonchlorinated water supplies throughout the world. In recent years, several studies have been carried out to determine the occurrence and levels of mesophilic aeromonads in both polluted and unpolluted water systems, as well as to determine their human health significance. Aeromonads have been isolated from surface freshwater with temperatures of 4° to 45°C, with the highest densities at 35°C, but these bacteria have not been isolated at temperatures above 45°C. No relationship with pH has been observed, and aeromonads have been isolated under pH conditions ranging from 5.2 to 9.8, although laboratory studies have indicated that *A. hydrophila* was unable to grow below pH 4 or above pH 10. *Aeromonas* spp. can be readily isolated from drinking-water distribution systems, where they appear to survive well when associated with pipe biofilms. Within water distribution systems aeromonads are able to proliferate at low temperatures (4°C). A highly significant correlation has been found between thermotolerant aeromonads and the trophic state of freshwater in lakes in the United States. However, other authors have been unable to predict the trophic status of several lakes in relation to these micro-organisms. In coastal or estuarine waters, halotolerant aeromonads are associated with sewage discharges.

Aeromonads are not considered to be normal inhabitants of the human gastrointestinal tract; thus, their presence in raw sewage and in treated effluents at concentrations equivalent to those of thermotolerant coliforms indicates that they are able to multiply in these waters. Several outbreaks of *Aerromonas* associated disease linked to either drinking water or recreational water contact have been reported and associated with gastrointestinal and extra-intestinal infections; however, the evidence is tenuous at best.

HPC

The heterotrophic plate count (HPC) method is used to take into consideration the enumeration of aerobic bacteria capable of growing in some commonly used media such as R_2A, HPC agar, or plate count agar. The level of bacteria determined by HPC indicates the overall microbiological status of the system and not necessarily the possibility of risk to public health, although such bacteria have been recently implicated as potential pathogens in drinking water.

Sterols and Other Chemical Indicators of Sewage Contamination

Alternative analytical approaches continue to be sought in lieu of classically used micro-organisms as indicators of fecal pollution, because of inadequacies associated with the failure of cultivation-based methods to detect indicator bacteria as well as concerns regarding the validity of the detection. The use of specific saturated sterols such as 5,β-cholestan-3,β-ol (coprostanol) has been investigated as a molecular signature of fecal contamination in water. Although not frequently used, this approach has been useful as a marker of fecal pollution in sediments because the compound is specific to the feces of higher animals including humans, it is biodegradable, and there is a relationship between its concentration and the degree of fecal pollution.

Caffeine and urobilin have also been proposed as indicators of the presence of sewage, since the only source of these compounds is domestic sewage. However, since these are chemical methods and are beyond the scope of this chapter, they are not discussed further.

FC/FS Ratio

Although both fecal (thermotolerant) coliforms and fecal streptococci are numerous in the feces of humans and animals, the ratio of fecal coliforms to fecal streptococci (FC/FS ratio) was proposed as an indicator of the origin of the of contamination. A ratio of greater than 4 is characteristic of human fecal contamination, whereas a ratio of less than 0.7 suggests animal waste. Although the index has been used successfully, its use is complicated by the differential die-off kinetics of the two bacterial groups as well as differences among individual species within those categories. Specifically, studies show that the fecal streptococci have a greater survival than fecal coliforms in water and that certain enterococci such as *Streptococcus bovis* and *S. equinus* die-off much more rapidly than the other fecal streptococci. Geldreich and Kenner suggested that the ratio would be valid only for less than 24 hours following the discharge of feces into the water being tested. The ratio is also influenced by the efficiency of methods and media used to detect fecal streptococci. Hence, the FC/FS ratio is not generally recommended as a totally unambiguous means of differentiating human from animal fecal pollution, although it has value in some instances if care is used in interpreting the results.

SIGNIFICANCE OF INDICATORS TO PUBLIC HEALTH

Analysis for the presence of indicators is a short-cut attempt to determine the microbiological quality and public health safety of waters. Very few water bodies are completely free of indicator micro-organisms. High concentrations of total coliforms can be found associated with plant material; thus, their presence in surface waters is not necessarily a cause for concern. Under some circumstances, even the presence of thermotolerant coliforms (e.g. *Klebsiella* spp. or *E. coli*) can be expected in surface waters. In these instances, other indicators, such as those discussed above, should be used to determine the possible threat to public health. However, the difference between untreated and treated waters in terms of the microbiota should be emphasised. The detection of thermotolerant coliforms in treated drinking waters should be a cause for concern, since current drinking-water treatment processes successfully eliminate indicator micro-organisms.

Current Regulations Regarding Drinking and Surface Waters

The standards of the US environmental protection agency and the world health organisation, as well as the European economic community, allow for a maximum number of micro-organisms that can be present in a given volume of drinking water. Table 7.1 gives some of the US and international standards for drinking and bathing waters. These standards are believed to provide optimal public health protection. In the United States, the maximum contaminant levels (MCLs) are the maximum permissible concentrations of these organisms and are legally enforceable. In Canada, the maximum acceptable concentration are nonenforceable guidelines.

Table 7.1. Bacteriological drinking-water and recreational freshwater standards or guidelines.

Source of standard	Maximum no. of indicated organisms permitted per 100 ml of water type:					
	Total coliforms[a]		Thermotolerant coliforms		Enterococci	Turbidity
	Drinking	Recreational[c]	Drinking	Recreational[c]	(recreational)	(NTU[b])
World health organisation	1–10	–	0	–	–	<1–5
Canada	<10	–	0	200[d]	35[e]	<1–5
European economic community	0	<10,000[f]	–	–	–	0–4
United States	0	200[g]	–	<2000[f]	–	1 (monthly)

[a] In systems analysing <40 samples per month, the maximum contaminant level specifies that no more than one sample per month may be total coliform positive. In systems analysing >40 samples per month, the maximum contaminant level specifies that no more than 5 per cent of the monthly samples may be total coliform positive.

[b] NTU, nephelometric turbidity units.

[c] Recreational refers to primary-contact (swimming) waters.

[d] Geometric mean of at least five samples when experience has shown that greater than 90 per cent of the thermotolerant coliforms are *E. coli*.

[e] Geometric mean of at least five samples taken during a period not to exceed 30 days.

[f] Compulsory limit. If exceeded in more than 20 per cent of samples with at least 14 days of sampling then bathing is prohibited.

[g] This is a US Environmental Protection Agency criterion. Since no uniform national standards exist, it may vary from state to state.

GENERAL METHODS

Sample Collection, Transport and Storage

It is just as important to use appropriate care in obtaining a representative sample as it is to analyse the sample correctly. The sample should be collected in sterile glass or polypropylene bottles or bags. Whenever it is suspected that a disinfectant is present, the sample should be amended with a solution of sodium thiosulphate ($Na_2S_2O_3$), which inactivates any residual halogen compounds present in the sample. An $Na_2S_2O_3$ concentration of 18 mg/litre will neutralise up to 5 mg of free (residual) chlorine per litre.

Some environmental samples may also contain high concentrations of zinc and copper. In such cases, it is recommended that the sampling containers be amended with EDTA (Na_2EDTA) at a concentration of 372 mg/litre. The presence of this chelating agent will reduce metal toxicity and is especially important if more than 4 hours elapses between collection and analysis of the sample.

Both compounds ($Na_2S_2O_3$ and Na_2EDTA) can be added together or separately to the sample bottle before sterilisation by autoclaving.

Additionally, the amount of elapsed time allowed between sample collection and analysis should not exceed 24 hours. The samples should be kept in a refrigerated container (or icebox) at a temperature below 10°C during transport and storage. The sample will not necessarily reach this temperature. If sample analysis within the 24-hour time limit is not possible, *in situ* analysis using portable equipment should be considered.

Water samples can be analysed for total coliform bacteria by a range of techniques, including the most probable number (MPN) using multiple-tube fermentation test, presence-absence (P-A), and membrane filtration (MF), as discussed below. Because of this variety of analytical approaches, a number of media are commonly used.

MPN (Multiple-Tube Technique)

MPN analysis is a statistical method based on the random (poisson) dispersion of micro-organisms in a given sample. The results are expressed in terms of the MPN of micro-organisms detected per volume of sample. Classically, this assay has been performed as a multiple-tube fermentation test. Although it is rather time consuming (to perform the presumptive, confirmed, and sometimes completed phases, a process which takes several days), most laboratory technicians around the world are fully trained in this technique and prefer it to other methods of water analysis. The MPN technique is also recommended for high-turbidity waters.

Classically, lauryl tryptose broth is used with the MPN method; with this technique, a small inverted tube is included to facilitate the detection of gas formation, and 0.01 gram of bromcresol purple per litre (final concentration) is used to determine acid production. The concentration of this medium is prepared so that the addition of water sample volumes of 10, 20, or 100 ml to the liquid medium will not reduce the ingredient concentration below that of the specified medium. Typically, five identical (20 ml) aliquots of drinking water are analysed as a set. Each tube is vigorously mixed with the medium about 25 times, taking care not to introduce air bubbles into the inverted tubes. These samples are then incubated at 35°C and examined for growth and for gas and acid production after 24 and 48 hours. The production of gas or acidic growth after 48 hours constitutes a positive presumptive reaction. Confirmation of presumptive reactions is done by inoculating the positive aliquots into brilliant green lactose bile broth tubes containing an inverted tube, incubating the samples for 24 and 48 hours at 35°C, and scoring tubes with gas as representing confirmed samples that contain coliform bacteria. Bacterial density and the 95 per cent confidence limits can be estimated with the use of MPN tables for the volumes and number of aliquots used.

The MPN method can also be used for the direct detection of coliforms and *E. coli* by using enzyme-specific tests as mentioned below. The characteristic colour and/or fluorescent end points specified by the manufacturer should be used to determine the presence or level of the target bacteria. The US Environmental Protection Agency has indicated that confirmation is not needed with the commercially available 4-methylumbelliferyl-β-D-glucuronidebased media. For the examination of surface waters, inoculate a series of medium tubes with a 10-fold dilutions of the sample. Using five- or three-tube series, inoculate each series with one dilution. If low concentrations of indicator bacteria are present; inoculate 100 ml of water into flasks containing 100 ml of the appropriate medium at double strength. If high concentrations of bacteria are suspected, a 10-fold dilution series should be performed. A sterile solution such as phosphate buffer (pH 7.2) or peptone water (0.1 per cent final solution, pH 6.8) should be used as diluent. Inoculate media as indicated in Table 7.2 for the presumptive test. Subsequently, all positive tubes should be subjected to the confirmation test by inoculation into a second medium as indicated in Table 7.3.

Table 7.2. Analysis of drinking water or surface water (including treated waste-water) by the multiple-tube (MPN) technique.

Type of analysis	Medium	Vol. (ml) inoculated into five- or three-tube series	Positive reaction
Total coliforms (incubated at 35°C ± 0.5°C)	Lactose broth or lauryl tryptose broth	10, 1, 0.1	Gas and/or acid
Thermotolerant coliforms			
Incubate at 44.5°C ± 0.2°C	EC medium	10, 1, 0.1	Gas and/or acid
Incubate at 35°C ± 0.5°C for 3 hours and transfer to 44.5°C ± 0.2°C for 21 hours	A-1[a]	10, 1, 0.1	Gas
Enterococci (incubates at 35°C ± 0.5°C)	Azide dextrose broth	10, 1, 0.1	Growth

[a] Although A-1 medium is recommended by the American Public Health Association only for analysis of marine waters and treated waste-water, several laboratories have tested it for analysis of surface and drinking waters with excellent results. Additionally, A-1 medium does not require a confirmation test.

Table 7.3. Confirmation procedures for the MPN technique.

Type of analysis	Medium	Positive reaction
Total coliforms (incubate at 35°C ± 0.5°C)	Brilliant green lactose bile broth and lauryl tryptose broth	Gas and/or acid
Thermotolerant coliforms (incubate at 44.5°C ± 0.2°C)	EC medium	Gas and/or acid
Enterococci (incubate at 35°C ± 0.5°C)	Pfizer selective Enterococcus medium[a]	Growth
		Brownish colonies (with halos)[b]

[a] No longer commercially available and therefore must be prepared separately.

[b] Brownish colonies confirm the presence of fecal streptococci. Colonies are transferred to brain heart infusion broth containing 6.5 per cent NaCl. Growth in the latter medium confirms the presence of enterococci.

MF

MF is possibly the method most widely used to detect coliform bacteria in water in North America and Europe. It is a simple test and lends itself well to the *in situ* analysis of samples as a result of the portability of the necessary equipment. The technique is based on the entrapment of the bacterial cells by a membrane filter (pore size, 0.45 μm). After the water is filtered, the membrane is placed on an appropriate medium and incubated.

Discrete colonies with typical appearance are counted after 24 to 48 hours. This technique is more precise than the multiple-tube MPN technique. The greatest limitation of the MF test is that it is useful only for low-turbidity waters and for waters that have low concentrations of nontarget (i.e. background) micro-organisms.

Samples of water that possess high turbidity clog the filter, and high concentrations of nontarget micro-organisms mask the presence of the target colonies. Tables 7.4 and 7.5 outline the basic MF method, including confirmation procedures.

Table 7.4. MF procedures for analysis of drinking water or surface water (including treated waste-water).

Type of analysis	Medium	Vol. (ml) filtered (depending on type of sample)[a]	Positive reaction
Total coliforms (incubated at 35°C ± 0.5°C)	LES Endo agar or *m*-Endo medium	100, 10, 1.0	Colonies with metallic green sheen
Thermotolerant coliforms (incubate at 44.5°C ± 0.2°C)	*m*-FC	100, 10, 1.0	Pale to deep-blue colonies
Fecal streptococci (incubate at 35°C ± 0.15°C)	*m-Enterococcus*	100, 10, 1.0	Pink to deep-red colonies
Enterococci (incubate at 41°C for 24 to 48 hours, then transfer membrane to EIA medium and incubate for 20 minutes at 41°C)	*m*-E	100, 10, 1.0	Pink to red colonies with a black or reddish brown precipitate on the underside after transfer of the membrane to EIAmedium

[a] When treated waste-water or water affected by sewage is being sampled, several dilutions may be analysed, and the 100- and 10-ml volumes may be skipped altogether, with higher dilutions used instead.

Table 7.5. Confirmation procedures for the MF technique.

Type of analysis	Medium	Positive reaction
Total coliforms (incubate at 35°C ± 0.5°C)	Brilliant green lactose bile broth and lauryl tryptose broth	Gas and/or acid
Thermotolerant coliforms (incubate at 44.5°C ± 0.2°C)	EC medium	Gas and/or acid
Fecal streptococci[a]	Bile esculin (incubate at 35°C ± 0.5°C)	Growth
	BHI–6% NaCl (incubate at 35°C ± 0.5°C)	Growth
	BHI (incubate at 45°C ± 0.5°C)	Growth

[a] Growth in brain heart infusion (BHI) broth at 45°C ± 0.5°C and bile esculin confirms the presence of fecal streptococci. Growth in bile esculin and BHI broth containing 6.5 per cent NaCl confirms the presence of enterococci.

Total coliforms

The MF analysis of water for total coliform bacteria can involve the use of LES Endo agar (Difco), *m*-Endo broth (Difco), or *m*-Coliform broth (BBL) as well as other media.

If liquid medium is used, ca. 2 ml of the broth can be added to certified absorbent pads, available from commercial source, in small sterile petri dishes, or 1.5 per cent agar-agar is added to the broth to make it semisolid. Characteristic colonies appear pink to dark red with a unique metallic green sheen (viewed at a magnification of ×10 to ×15 if necessary). Tergitol agar has been used to detect coliform bacteria as a means to follow the movement of the plume resulting from a marine sewage outfall. Only plates containing between 20 and 80 typical colonies should be counted. Confirmation can be done as described earlier by the selection of both representative typical and atypical colonies. The population density of the target bacteria in the original sample, usually described in CFU/100 ml, can be calculated from the volume filtered and dilutions used, if any. The percentage of the typical colonies that are confirmed as positive can be used to determine the verified coliform density.

Thermotolerant coliforms (fecal coliforms)

Fecal coliform medium (*m*-FC) is used to quantify thermotolerant coliforms in water samples when the MF technique is used. A solution of 1 per cent rosolic acid (dissolved in 0.02 N NaOH) can be added to the medium. However, the use of rosolic acid has been questioned by some European researchers in terms of its possible toxicity to organisms present in water already stressed because of chlorine or ozone disinfection. *m*-FC can be used as an agar-based medium (1.5 per cent agar added to the broth), or broth can be added to certified absorbent pads. Petri dishes containing filters are incubated at 44.5°C ± 0.2°C. Incubation with this degree of accuracy can best be achieved by using incubators or water baths for this specific purpose. Petri dishes can be incubated submerged with weights after they have been placed inside two tightly sealed plastic bags. Typical thermotolerant coliform colonies will appear in various shades of blue, although atypical *E. coli* may be pale yellow whereas nonthermotolerant coliform colonies are gray to cream coloured. Again, magnification of ×10 to ×15 can be used to observe colonies, and the desired range is 20 to 60 colonies per plate.

Typical colonies sometimes lose the characteristic appearance (green sheen and bluish colour for total and thermotolerant coliforms, respectively) after 24 hours; thus, reading plates at 24 hours is highly recommended.

Enterococci and fecal streptococci

Enterococcus medium (*m*-E) is used for the detection of enterococci in freshwater and marine waters. Although KF medium was used in the past, it has been found to be susceptible to false-positive results and is no longer recommended. The *m*-E agar is prepared by heating the following basal ingredients and amounts per litre: 10 grams of peptone, 15 grams of NaCl, 30 grams of yeast extract, 1 gram of esculin, 0.05 gram of cycloheximide, 0.15 gram of NaN_3, and 15 grams of agar. The mixture is then cooled to <46°C. Then 0.25 gram of nalidixic acid per litre is mixed in 5 ml of reagent-grade water per litre, a few drops of 0.1 N NaOH is added to dissolve the antibiotic, and the whole volume is added to the basal medium. Finally, 0.15 gram of 2,3,5-triphenyl tetrazolium chloride per litre is added and dissolved, and the entire mixture is dispensed into petri dishes. After the sample is processed the membrane filter is placed onto the solidified medium, incubated at 41°C, and observed after 24 and 48 hours. After incubation, the membrane is transferred very carefully to a petri dish containing EIA medium (1 gram of esculin, 0.5 gram of ferric citrate, and 15 grams of agar per litre of water [dissolve and sterilise by autoclaving]) and incubated at 41°C for 20 minutes. All pink to red colonies developing a black or reddish brown precipitate on the underside of the filter should be counted. For the detection of fecal streptococci in freshwaters and marine waters, *m-Enterococcus* (containing per litre of water, 20 grams of tryptone, 5 grams of yeast extract, 2 grams of glucose, 4 grams of K_2HPO_4, 0.4 gram of NaN_3, 0.1 gram of 2,3,5-triphenyl tetrazolium chloride, and 10 grams of agar) is used. The sample is processed, and the petri dishes are incubated at 35°C for 24 to 48 hours. Colonies should be counted after 24 and 48 hours, as some of the pinpoint colonies are not readily visible after 24 hours. Audicana designed and tested a modification of kanamycin-esculin-azide (KEA) agar, named oxolinic acid-aesculin-azide (OAA) agar, to improve the selectivity in the enumeration of fecal streptococci from water samples by MF. OAA agar showed higher specificity, selectivity, and recovery efficiencies than those obtained by using *m*-E and KF agars. In addition, no confirmation of typical colonies was needed when OAA agar was used, which significantly shortens the time of sample processing (only 24 hours) and increases the accuracy of the method.

Aeromonas

MF is used conjointly with selective and differential recovery media for the enumeration of aeromonads in water samples. Although an optimal and universal simple recovery medium is not available at present, *Aeromonas* medium (*m*-A), DNase-toluidine blue-ampicillin, *m*ADA (O/129), *m*SA (O/129), ampicillin blood (ASBA), ampicillin-dextrin, starch-glutamate ampicillin-penicillin, glutamate starch penicillin (GSP), prilampicillin-dextrin-ethanol (PADE), and starch ampicillin (SAA) can be used. The use of media containing blood agar also allows for the detection of enterotoxigenic *Aeromonas* strains. Lye and Dufour designed an *in situ* cytotoxic activity assay of gram-negative bacteria isolated from water. The procedure consists of the transfer of the membrane filter to the surface of strain Y-1 mouse adrenal cells overlaid with 1 per cent agar.

After 15 minutes, the filter is removed, and then Y-1 cells are incubated for an additional 24 hours. The release of putative cytotoxic and cytotonic products from the bacterial colonies is recognised by zones of cellular lysis and injury of Y-1 cells. Molecular techniques have also been applied for the rapid detection and identification of aeromonads from several sources.

Presence-Absence (P-A)

The microbiological monitoring of drinking water has historically relied on MPN and MF approaches to estimate population densities of indicator organisms. Although those methods have been useful, disadvantages such as space and time requirements prompted the idea of simply testing for the presence or absence of indicator bacteria in a standardised volume (i.e. 100 ml) of water, using a liquid medium resembling many employed in the MPN approach. This approach is easy to follow and very useful in small potable water systems where microbiological problems are more frequent, and it permits analysis of greater numbers of samples in larger systems to gain a more comprehensive microbiological evaluation of their distribution network. Therefore, the P-A concept of indicator organism occurrence has replaced the specified monthly average for compliance with US regulations in the current total coliform rule. Although this approach can yield valuable information on the prevalence of microbiological problems within a system, a weakness is that it fails to provide data on the magnitude of such occurrences. In addition, the common connotation of 'absence' can be misleading in the case of injured bacteria that are frequently present in treated drinking-water systems and fail to produce a positive test on established media, as discussed in more detail below.

The P-A analysis of drinking water for total coliforms entails the addition of 100 ml of sample to 50 ml of triple-strength P-A broth (single strength is as follows [per litre]: 13 grams of lactose broth, 17.5 grams of lauryl tryptose broth, and 0.0085 gram of bromcresol purple) in 250-ml bottles. Bottles containing aliquots of the water sample to be tested are incubated, and resulting end points are determined as described above. A distinct yellow colour results from the fermentation of lactose, and gas formation can be detected as bubbles with gentle shaking. Confirmation of such presumptive results can be done as described above and in Table 7.3.

Enzyme-Specific Tests

New criteria have been added to the traditional definition of coliform bacteria, presented earlier and based on classical microbiological characteristics and phenotypes. With this new approach, the presence of characteristic enzymatic activities permits the differentiation of the coliform group of bacteria and *E. coli* in the determination of microbiological water quality. The definition of a coliform or a fecal coliform bacterium basically relies on the activity of a single enzyme (β-galactosidase). The new enzymatic definition of total-coliform bacteria is based on the presence of β-galactosidase, and that of *E. coli* is based on the enzymatic action of β-glucuronidase. Although the common use of media that are based on these specific enzymatic activities to identify indicator bacteria in water has gained wide acceptance only in the past few years. In addition, a verification method using cytochrome oxidase and β-galactosidase has proved to be superior to the more traditional technique for confirming the presence of total coliforms in water. The end point characteristically incorporated into media designed for the enzymatic detection of coliforms and *E. coli* in water is the development of a specific colour or fluorescence. The activity of β-galactosidase results in the hydrolysis of substrates such as *ortho*-nitrophenyl-β-D-galactoside, which is colourless, to a coloured product, thereby indicating the presence of coliform bacteria. *E. coli* is characterised by the production of a fluorescent end product following the hydrolysis of 4-methylumbelliferyl-β-D-glucuronide by the action of the enzyme β-glucuronidase. A wide range of other chromogenic and fluorogenic substrates acted upon by β-glucuronidase that have been incorporated into various media are also available. Exhaustive comparative testing of at least one commercially available enzyme-based medium indicated that it performed as well as accepted media in the detection of both total coliforms and *E. coli* from drinking water and *E. coli* following chlorine injury, using mixed natural bacterial suspensions. However, chlorine-mediated injury results in the somewhat delayed development of both enzymatic end points in two of the commercially available media because of the extended lag phase that is characteristic of stressed bacteria.

Currently, a number of different media based on this principle have been developed for use in the MF, MPN, and P-A techniques. Commercially available media include Colisure, Colilert, *m*-ColiBlue, ColiComplete, and MicroSure. Similar media to detect coliforms and/or *E. coli* in water have also been described in the literature; representative examples of the new enzyme-specific media are Ml agar, *m*X, *m*-LGA, and mLGA. All of these media can be used in a P-A or MPN analysis, depending on the needs of the analyst.

Bacteriophages

The enumeration of bacteriophages is one of the simplest methods for the microbiological analysis of water. This technique utilises a host strain of *E. coli* since viruses (phages) are obligate intracellular parasites and thus need a metabolically active host in which to replicate. Several host strains can be used; each has its limitations. One of the most widely used hosts is *E. coli* C. This host strain allows for the replication of somatic phages (those phages having their receptor sites located on the bacterial cell wall). However, *E. coli* C3000 is a strain which has sex pili and thus allows for the replication of male-specific coliphages (those phages having the receptor sites on the sex pilus) as well as somatic phages. A third host is a genetically manipulated *Salmonella* strain that has the ability to produce sex pili but does not contain the receptor sites for coliphages on its cell wall, allowing only male-specific (F+ or F-specific) coliphages to replicate.

For the analysis of surface and drinking waters, a single-layer, direct plaque assay as outlined by Grabow and Coubrough can be used. In addition, the method allows for the analysis of 100-ml volumes, which allows for a numerical comparison with bacterial indicators. The medium consists of the following components per litre of water: 14 grams of meat extract, 4 grams of yeast extract, 4 grams of NaCl, 12 grams of peptone, 1 gram of sodium carbonate, 1 gram of magnesium chloride, and 12 grams of agar; a 13 per cent (wt/vol) $CaCl_2$ solution is prepared separately and autoclaved. The medium is autoclaved in 100-ml volumes and kept liquid at ca. 48°C, and 1 ml of the $CaCl_2$ solution is added. Five millilitres of an overnight culture of the appropriate host is added to the medium, and finally a 100-ml volume of the sample is mixed carefully and thoroughly (taking care not to create air bubbles) and poured into large petri dishes. The plates are incubated at 35°C for 6 to 24 hours and read for the presence of small clear areas which represent viral PFU. Care should be taken to include negative controls (i.e. plates that contain only the medium and the bacterial host). The total number of viral plaques is counted and expressed as PFU per 100 ml.

Rapid Tests

The need for tests to quickly determine the possible existence of water contamination of public health significance is long-standing. This need exists because the time required to perform the current tests is greater than the mean residence time for water within many potable distribution networks. Significant factors associated with that dilemma in drinking water and other relatively high-quality aquatic environments and systems include the low densities of ambient bacteria plus the need for at least 18 hours of bacterial growth before an observable end point is obtained with the classical techniques. The constraints associated with other systems containing higher concentrations of bacteria are clearly less demanding. However, for most applications, the rapid methods available should be viewed as research tools since at present they either lack sufficient sensitivity or are unsuitable for use in routine monitoring laboratories. However, a variety of rapid methods are in the developmental stage, and some appear very promising. Those seeking such a method need to critically evaluate which of the available methods are compatible with their specific needs. A variety of analytical approaches have been proposed for the rapid detection of bacteria in water and waste-water, although most are limited by sensitivity with respect to analysis of water of good microbiological quality.

The following is a list of some of the more attractive rapid methodological options and a brief statement about potential applications and noteworthy limitations of each.

1. The direct total microbial count or acridine orange direct count is often used to determine the total bacterial population in an aquatic system. This analytical approach requires a fluorescence microscope with appropriate optical filters and is of limited value in determining bacterial viability unless known organisms are used under defined conditions. The use of 5-cyano-2,3-ditolyl tetrazolium chloride (a fluorescent compound) and a contrasting counterstain allows determination of the total bacterial population as well as the fraction respiring in water and the assessment of biofilm disinfection. This method provides little information of value in bacterial

identification and is tedious, although image analysis technology has proven useful in making the technique less problematic and allows statistical concerns to be addressed more easily.

2. A rapid test that detects thermotolerant coliforms after 7 hours of incubation is available. This method has value in the examination of surface waters as well as unchlorinated sewage and might serve as an emergency test for detection of sewage or fecal contamination in potable water. However, the presence of injured bacteria in such systems might represent an explanation for false-negative results because of the prolonged lag that is characteristic of stressed bacteria in environmental samples.

3. ATP detection assays have been useful in the determination of bacterial population density. This approach is rather insensitive, requiring the presence of at least 1000 cells, and does not identify organisms, but it has been used in ecological and waste-water studies.

4. Radiometry using labelled substrates can be both rapid and sensitive in the detection of organisms that metabolise the labelled nutrient to CO_2. This method does not discriminate bacteria phylogenetically, although it provides information of functional significance. Despite the potential of this method, very few reports document its application in environmental analysis.

5. Fluorescently labelled antibodies have been used for the detection and identification of bacteria in environmental studies for some time. This method also requires a fluorescence microscope and is tedious unless image analysis is used, and it is constrained by the specificity of the antibodies used and does not indicate viability or physiological activity. However, this approach can be used in a variety of applications, including tracking a specific organism in the environment or detecting enteric bacteria. The lack of sensitivity of this method prohibits its use for the direct detection of pathogens in waters. Large volumes of water need to be concentrated before this technique becomes useful for pathogen detection.

6. A hybrid method that incorporates the use of fluorescent antibodies to identify bacteria along with 5-cyano-2,3-ditolyl tetrazolium chloride to determine cellular respiratory activity has recently been introduced. Although this method requires the use of a fluorescence microscope and can be tedious, it is unique in that it allows the rapid, simultaneous detection of specific bacteria and the discrimination of respiratory activity at the cellular level.

Recovery of Injured Bacteria

Indicator bacteria become injured in water and waste-water following sublethal exposure to a wide variety of chemical and physical environmental stressors, including disinfectants, metals, and UV irradiation. Such bacteria are unable to form colonies on most selective media, and between 10 and 90 per cent of the coliform bacteria in treated drinking water may be injured. As a consequence, injured cells are undetected in water, leading to an underestimation of a contamination event, and injured bacteria may also pass undetected into finished drinking water distribution networks, where they may eventually recuperate and colonise the system. A medium, m-T7, was developed to detect injured bacteria in drinking water and is commercially available. Although the use of media and methods designed to detect injured bacteria does not always result in the detection of greater numbers of indicator bacteria in all systems, it provides a more complete view of the water quality plus guidance in the diagnosis of problems within water distribution systems experiencing unexplained occurrences of excessive indicator bacteria.

Molecular Methods

PCR provides an option for the detection of micro-organisms in aquatic environments as well as sewage and sludge. Caution is needed in the interpretation of PCR results since this test allows for the amplification and detection of DNA sequences without regard for bacterial viability. In addition, materials in natural samples can interfere with the required reactions.

Phylogenetic identification of specific bacteria or groups of organisms can be achieved without cultivation through the use of fluorescently labelled oligonucleotide probes for rRNA sequences. This approach has gained wide application in ecological studies and has been used to detect an opportunistic pathogen (*Vibrio vulnificus*) in waters after concentration on membrane filters.

SURFACE FRESHWATERS

It should be kept in mind that the microbiology of untreated waters is drastically different from that of treated drinking waters. Thus, strong emphasis should be placed on the use of the appropriate indicators. In the case of recreational waters (excluding swimming pools and spas, which receive fecal contamination from bathers), it should be determined whether the recreational areas are subjected to fecal pollution. In any case the presence of bathers in the water will result in fecal deposition, and thus fecal contamination is likely. Thus, the use of the appropriate indicators to determine risk of enteric or other types of disease is of extreme importance. Risk can be determined most accurately if the area is known thoroughly. Indicators such as total coliforms may not be of much value in analyses of recreational waters, in spite of possible statistical correlations. *E. coli* and coliphages are much more reliable indicators of the presence of fecal pollution (and thus possible risk), unless the recreational waters are part of a tropical rain forest. In the latter case, the use of coliphages may be more advisable.

DRINKING WATERS

Traditional treatment is based on both physical and chemical barriers. Thus, indicator micro-organisms should not be present in finished waters. The presence of any indicator (total or thermotolerant coliforms, coliphages, or *E. coli*) in the finished waters suggests that one of the barriers is not functioning. The distribution network may be susceptible to contamination, and, therefore, the presence of thermotolerant coliforms in treated drinking water should be a cause for concern. However, coliforms have been shown to be able to grow in potable-water distribution systems as a result of the bacterial biofilms on pipe surfaces. The analyst should be aware of the different types of indicators, and proper use of these tools should he emphasised. To sum up, in this chapter we have reviewed those methods currently in use for the microbiological analysis of drinking waters and surface freshwaters, as well as those that are now being developed. The practice of using groups of bacteria or specific organisms such as *E. coli* as indicators of the possible presence of fecal contamination or pathogenic micro-organisms is and has been extremely useful to protect, public health. However, it should be kept in mind that indicators are only tools and have limitations. Numerous misunderstandings exist regarding the use of indicators. For example, analysis of levels of coliforms and thermotolerant coliforms is considered by many to be a simplistic method for determining and guaranteeing water quality. Such a dogmatic approach to water quality is counterproductive and may place public health in jeopardy.

There are exciting new developments in the area of rapid methods as well as the direct detection of pathogenic micro-organisms. The reader is exhorted to become familiar with the advantages and disadvantages of these new advances. However, there is no better way of approaching water quality management and analysis than being familiar with the watershed or distribution system under investigation. This knowledge and the correct and timely application of suitable analytical methods allow for early detection of anomalous conditions within the system. The resulting data can then be used to better manage the system and thus protect the public health against not only enteric but also other types of diseases.

Chapter 8

Detection of Protozoan Parasites in Source and Finished Drinking Waters

INTRODUCTION

Protozoa are unicellular micro-organisms which, unlike bacteria and viruses, possess membrane-bound genetic material or nuclei and other assorted cellular organelles. They exhibit various forms of locomotion and reproduction that have been used to categorise them into broad groups; however, establishment of the taxonomy of the protozoa is a dynamic, evolving process about which there is little agreement.

Free-living protozoa are ubiquitous in natural waters and moist soils and fall into three broad groups: amoebae, flagellates, and ciliates. They are often abundant in both surface water and groundwater supplies as part of the normal aquatic community. Free-living protozoa are of little concern to water plant operators, since they cause no treatment problems or health impacts. Identification of free-living protozoa to the genus and species levels requires a great deal of skill and training. Keys, line drawings, and pictures in *An Illustrated Guide to the Protozoa* aid in identifying unknown organisms at least to ordinal and familial levels. Included as representatives of free-living protozoa found in source waters are *Hartmanella* spp., *Acanthamoeba* spp., *Echinamoeba* spp., *Naegleria fowleri*, *Euglena* spp., *Tetrahymena* spp., and *Paramecium* spp.

Parasitic protozoa, which live in or on another organism like either a plant or an animal, may or may not be of concern to the water treatment plant operator, depending on the organism on which they live. This chapter deals primarily with human enteric protozoa and more specifically *Giardia* and *Cryptosporidium*. Within the human small and large intestines a number of protozoa including amoebae, flagellates, coccidians, and ciliates can be found. Many of these organisms are not recognised as pathogenic and consequently are dismissed as commensals. *Entamoeba coli* and *Trichomonas hominis* are recognised as examples of commensal organisms which live in another organism without harming it. Parasitic protozoa, on the other hand, are known to be harmful when living in or on another organism. Included as representatives of enteric protozoa are *Entamoeba histolytica*, *Giardia lamblia*, *Cryptosporidium parvum*, *Cyclospora* spp., *Toxoplasma gondii*, *Enterocytozoon bieneusi*, *Isospora belli*, *Isospora hominis*, and *Balantidium coli*.

Amoebae belong to the phylum Sarcomastigophora and generally possess a vesicular nucleus and pseudopods, which are retractable cytoplasmic protrusions used both for locomotion and for feeding. Thecate amoebae are distinguished from athecate or naked forms, the other main subgrouping, by a closely fitting envelope or shell secreted by the trophozoite. Many of the athecate amoebae are able to withstand adverse environmental conditions by transforming from a vegetative trophozoite form to a dormant, resistant cyst form. Cysts of certain amoebae have been isolated from the air in association with dust particles. *Hartmanella* spp., *Acanthamoeba castellanii*, *Echinamoeba* spp., and *N. fowleri* are athecate examples which can produce cysts. *Hartmanella* spp., *A. castellanii*, and *Echinamoeba* spp. are known to harbour and amplify *Legionella* spp., *Acanthamoeba* spp., and *N. fowleri* are normally free-living forms, but under the right conditions, they become opportunistic pathogens which are responsible for primary amoebomeningoencephalitis in humans. Furthermore, *Acanthamoeba* spp. also are well-documented causes of corneal keratitis in contact lens wearers.

Flagellates also belong to the phylum Sarcomastigophora and have vesicular nuclei but move by hair- or whip-like cylindrical organelles that are approximately 0.25 μm wide and that are known as flagella. Flagellar length and number in this group are variable. Members of this group may also possess chloroplasts, thecal plates, basal bodies, and collars around the flagellum. Some genera exhibit multicellular colonial forms. *Euglena* spp. are examples of free-living flagellates which contain chloroplasts. *Giardia* spp. are examples of parasitic flagellates.

Ciliates, which belong to the phylum Ciliophora, are distinguished from the other protozoan groups by their unique nuclei. They have both a large macronucleus which regulates cellular metabolism and a small micronucleus which is involved in genetics and sexual recombination. Many ciliates are phagotrophic, which means that they ingest nutrients through a mouth or cytostome. These organisms are covered with cilia which are used both in feeding and in movement. Cilia are organelles similar in structure to flagella. However, they differ from flagella in being generally shorter in length and in being interconnected through their intracytoplasmic basal structure. As a result of the interconnection, ciliary movement can be coordinated. *Tetrahymena* spp. and *Paramecium* spp. are examples of free-living ciliates, while *B. coli* is an example of a parasitic form.

The phylum Apicomplexa is sometimes referred to as the phylum Sporozoa. This group is exclusively parasitic and is responsible for both human and veterinary diseases. Malaria and coccidian diseases are well-known examples from humans and animals, respectively. The Apicomplexa are characterised by possession of an anterior apical complex. In addition, they have complicated life cycles. Unlike the phyla mentioned above, they possess no obvious locomotor organelles. Most members of this group have a vesicular nucleus, a Golgi apparatus, and mitochondria. Moreover, there is usually a sexual phase to the life cycle in this group.

BIOLOGY AND ECOLOGY

Free-living forms are found in air, water, and soil. Large numbers of bactivorous ciliates and amoebae can be found in sewage and other organic wastes. Furthermore, they may be found on the surface, bottom, banks, and slopes of most bodies of freshwater and marine water. Free-living forms may withstand temperatures ranging from freezing up to 28°C. Certain ecological stimuli may cause amoebae to encyst or to differentiate a flagellum.

Most enteric protozoans have two stages in their life cycles. The trophozoite is an actively feeding, growing, and reproducing stage. Stimuli in the host's intestinal tract induce most of the enteric protozoans to produce a resistant, dormant transmission form, which is referred to as either a cyst, an oocyst, or a spore. Whether commensal or parasitic, these protozoans have simple, direct life cycles, and are transmitted as fecal contaminants of food and/or water. Furthermore, person-to-person transmission is known to occur. Trophozoites generally do not survive outside their host unless they are propagated in a specialised culture medium. On the other hand, cysts, oocysts, and spores are different in that they are known to survive for long periods outside the host, especially under cool, moist conditions. Ambient temperatures of about 25°C or higher and desiccation are known to rapidly reduce the time that a cyst, an oocyst, or a spore can survive outside the host. Unlike clinical specimens, only cysts, oocysts, and spores of parasitic forms are of concern in environmental samples. Most troubling for the water treatment industry is the fact that many of these forms are more resistant to chemical disinfection than enteric bacteria and viruses.

Enteric parasitic protozoa are known to produce gastrointestinal distress including diarrhoea, flatulence, cramps, anorexia, and weight loss. However, they can produce a continuum of pathologies ranging from no symptoms to extremes of illness requiring hospitalisation.

Presently, *Giardia* and *Cryptosporidium* are of great concern to the water treatment industry because they are known to be the etiologic agents responsible for a number of episodes of waterborne gastroenteritis. Their significance is also increased, as they have a number of mammalian reservoir hosts that they can use besides humans. This results in an amplification in the numbers of cysts and/or oocysts, which are potential challenges to water impoundments and treatment plants.

DETECTION AND QUANTITATION

Whether looking for free-living or parasitic forms, no one collection method suffices for all protozoa. Free-living forms are usually present at a sufficient density, if they are present at all, so that they can be collected by a variety of grab-sample techniques from either sediment or the water above the sediment. Parasitic forms, on the other hand, most likely are suspended in the water and are usually very sparse. Consequently, they must be concentrated from large volumes of water.

Free-Living Protozoa

After collection, field samples of free-living forms may sometimes be observed directly under the microscope or they may be cultured. Each free-living form has optimal cultural requirements which cannot be covered in detail here. Parasitic forms, like *Giardia* and *Cryptosporidium*, can be cultured by very exacting aseptic procedures which do not lend themselves to routine use with field samples. Whatever the organism of interest, the investigator must pay close attention to the pH, the nutrients present, the toxic substances present (like hydrogen sulphide), temperature, osmotic effects, and other organisms in the sample.

Initial identification of free-living protozoa is done with taxonomic keys; however, a basic knowledge of protozoan morphology is required to use them. Fundamental keys have been developed but now are somewhat out of date taxonomically. A more recent, taxonomically accepted key has been published but requires considerable knowledge of protozoa on the part of the user. In many cases the only way to identify some free-living protozoa to the genus and species levels is to collaborate with experts in the field.

Detection and quantitation of protozoa usually require microscopic techniques including epifluorescence, bright-field, phase-contrast, and differential interference contrast (DIC) microscopies. Moreover, some protozoa are identified on the basis of exacting staining techniques. Very large forms may require observation with a dissecting microscope. Many free-living forms are very motile and must be slowed down to be seen at all. Substances like Protoslo 10 per cent methylcellulose, 1 per cent aqueous solution of a polyether, and 3 per cent polyacrylamide hydrazide have been used to increase the viscosity of the medium and slow protozoan movement. Whenever a microscope slide is to be observed for a protracted period, the coverslip must be sealed to the slide with either nail polish, Vaspar, or petroleum jelly to prevent evaporation. Air bubbles under a sealed coverslip can provide a limited source of oxygen for a living preparation. For photomicrography of motile living material, a flash attachment for the microscope is essential.

Free-living amoebae, like *Acanthamoeba* spp., *Hartmanella* spp., *Naegleria* spp., and others, are usually collected by the grab-sample technique. The volumes of the samples are generally small, ranging from 1 to 4 litres of fluid or 10 to 100 ml of sediment. Fluid samples are concentrated by centrifugation at 250 × gram for 5 to 8 minutes when necessary, while sediment samples are used directly. The fluid concentrates and sediments are then cultured on non-nutrient agar lawns of *Escherichia coli* which the amoebae use as a food source. Detection, purification, and cloning of the free-living amoebae are done by isolating organisms from plaques, places where the amoebae have consumed all the bacteria. Identification is based upon cyst and trophozoite morphologies. Collaboration with an expert in the field is required in many instances, and antibodies and genetic primers and probes can be obtained from the expert to confirm preliminary identifications.

Parasitic Protozoa

Detection of *G. lamblia* and *C. parvum* in source and finished waters has been and continues to be of great interest. *G. lamblia* is one of the most frequently reported parasitic waterborne pathogens. Furthermore, over the past 20 years *C. parvum* has also been responsible for numerous waterborne outbreaks of gastroenteritis. The method used at that time to detect these protozoan pathogens is known as the ICR method. It is an immunofluorescence detection procedure performed after concentration of large volumes of water. Initially, concentration of *Giardia* cysts and *Cryptosporidium* oocysts from water is done by retention on a nominal-porosity yarnwound filter. Retained particulates are eluted from the filter and reconcentrated by centrifugation. The pelleted *Giardia* cysts and *Cryptosporidium* oocysts are separated to some extent from other particulate debris by flotation on a Percoll-sucrose solution with a specific gravity of 1.1. A portion of the water layer— Percoll-sucrose interface is then placed on a membrane filter to form a monolayer, indirectly stained with fluorescent antibody, and examined with an epifluorescence microscope. Cysts and oocysts are classified by specific criteria (immunofluorescence, size, shape, and internal morphological characteristics), and the results are reported in terms of the presumptive and confirmed numbers of cysts and oocysts per 100 litres. Presumptive cysts and oocysts are defined as objects having the right size and shape as well as fluorescence characteristics. Confirmed cysts and oocysts are those in which internal morphological characteristics like nuclei, axonemes, and median bodies can be demonstrated in Giardia cysts and sporozoites can be seen in oocysts by DIC microscopy.

Because rates of recovery of *Giardia* cysts and especially *Cryptosporidium* oocysts in seeded samples were so low by the ICR method described above, a great deal of effort has gone into the development of an improved method, which is known as *Method* 1623; *Cryptosporidium* and *Giardia in Water by Filtration/IMS/FA* (method 1623). The first version of this method is known as method 1622: *Cryptosporidium* in water by filtration/IMS/FA (method 1622), which detected only *Cryptosporidium* oocysts, because the reagents necessary for detection of *Giardia* cysts were not commercially available. Key changes made during the development of method 1623 included the use of an absolute-porosity filter to concentrate the sample, reduction of the sample size from 100 to 10 litres, selective purification of the parasites from other particulates by immunomagnetic separation (IMS), and counterstaining with 4′,6-diamidino-2-phenylindole (DAPI) to elucidate nuclei, in addition to traditional staining with fluorescein isothiocyanate (FITC)-labelled antibodies and use of rigorous quality assurance. Rates of recovery for *Cryptosporidium* oocysts during validation studies of method 1623 were about 38 per cent, in contrast to the 11 per cent for the ICR method.

Quality control

Laboratories that use method 1623 are expected to operate a formal quality assurance (QA) programme. The minimum requirements of this programme consist of an initial demonstration of laboratory capability through performance of the initial performance and recovery (IPR) test, analysis of spiked samples to evaluate and document data quality, and analysis of standards and blanks as tests of continued performance. Periodic ongoing precision and recovery (OPR) and matrix spike (MS) samples must be spiked and processed within 24 hours of spiking with flow cytometer-counted *Giardia* cysts and *Cryptosporidium* oocysts.

Data that will allow an independent reviewer to validate and trace the processing and analysis steps leading to the final result are required. At a minimum the following information is required in the final report: (i) sample numbers and other identifiers, (ii) source of spiking suspensions, as well as lot number and date received, (iii) spike enumeration date, (iv) sample spiking dates and times, (v) volume filtered, (vi) filtration and concentration dates and times, (vii) initial pellet volume and resuspended pellet volume, (viii) staining completion dates and time, (ix) staining control results, (x) all required examination and confirmation information, (xi) examination and confirmation dates and times, (xii) lot numbers of elution, IMS, and staining reagents, (xiii) copies of bench sheets (a record on paper of information which accompanies the field sample), logbooks, and other recordings of raw data, and (xiv) any links to other related data outputs.

Before using method 1623 for routine analysis, analysts in the laboratory need to demonstrate control over the analytical process and generate acceptable precision and recovery data. To accomplish this task, each laboratory analyst needs to perform the IPR test. For the IPR test, use about 100 flow cytometry-counted cysts and oocysts to spike each 10-litre Cubitainer of reagent water; four replicates are required. Then, filter, elute, concentrate, purify by IMS, stain, and examine microscopically each of the four 10-litre reagent water samples. By using the results of the four analyses, compute the average per cent recovery (X) and the relative standard deviation of the recovery (s_r) for *Cryptosporidium*. Compare s_r and X with the corresponding limits for initial precision and recovery in Tables 8.1 and 8.2. However, if s_r and X meet the acceptance criteria, system performance is acceptable and analysis of blanks and field samples may begin. If s_r or X falls outside the range for recovery, system performance is unacceptable. In this event, correct the problem and repeat the test with additional samples of spiked reagent water.

Table 8.1. Quality control acceptance criteria for *Cryptosporidium*.

Performance test	Acceptance criteria
Initial precision and recovery	
Mean recovery (%)	21–100
Precision (maximum relative SD)	40
Ongoing precision and recovery (%)	19–100
MS and MS duplicate (for method modifications)	
Mean recovery (%)	13–111
Precision (maximum relative % difference)	61

The laboratory should analyse one MS sample to determine the effect of the matrix on the method's oocyst and cyst recovery when field samples of a particular matrix type are first received from a source for which the laboratory has never before analysed samples. The MS sample analysis is performed with an additional (second or duplicate) field sample. If the laboratory routinely analyses field samples from one or more sources, then analysis of 1 MS sample should be performed per 20 field samples. For example, when a laboratory receives the first field sample from a given site, the laboratory must obtain a second aliquot of this field sample to be used for the MS sample. When the laboratory receives the 21st field sample from this site, a separate aliquot of this 21st field sample must be collected, spiked, processed, and analysed.

Table 8.2. Quality control acceptance criteria for *Giardia.*

Performance test	Acceptance criteria
Initial precision and recovery	
Mean recovery (%)	17–100
Precision (maximum relative SD)	41
Ongoing precision and recovery (%)	16–100
MS and MS duplicate (for method modifications)	
Mean recovery (%)	15–118
Precision (maximum relative % difference)	30

In addition, analysis of method blanks is required to demonstrate that reagents and processing are not introducing contamination. The blank (negative control sample) is an unspiked 10-litres sample of reagent water. It is processed prior to analysis of the IPR test and the OPR test and prior to analysis of any of the field samples for any given week to demonstrate that the reagents and processing are not introducing contamination. The laboratory should, on an ongoing basis, demonstrate through analysis of the OPR (positive control) samples that the analysis system is within acceptable control limits. At least once a week spike about 100 flow cytometer-counted cysts and oocysts into 10 litres of reagent water, filter, elute, concentrate, separate (purify), stain, and examine the sample to verify all performance criteria. The laboratory must analyse 1 OPR sample for every 20 field samples, if more than 20 field samples are analysed in a week. Adjustment and/or recalibration of the analytical system shall be performed until all performance criteria are met. Only after all performance criteria are met may field samples be analysed. For each organism, compute the per cent recovery by the following equation:

$$P = 100 \times \frac{N}{T}$$

where, P is per cent recovery, N is the number of oocysts or cysts detected, and T is the number of oocysts or cysts spiked. Compare the per cent recovery with the limits for ongoing precision and recovery in Tables 8.1 and 8.2. If the recovery meets the acceptance criteria, system performance is acceptable and analysis of blanks and field samples may proceed. If, however, the recovery falls outside of the range given, then system performance is unacceptable. In this event, there may be a problem with the microscope or with the filtration or separation system. To resolve this, reanalyse the OPR sample and recollect and reanalyse field samples. All samples must be associated with an OPR sample that passes the acceptance critetia.

The laboratory should analyse 1 method blank and 1 OPR sample each week during which samples are analysed if 20 or fewer field samples are analysed during this period. The laboratory should analyse 1 laboratory blank and 1 OPR sample for every 20 field samples if more than 20 field samples are analysed in a week. If *Cryptosporidium* oocysts, *Giardia* cysts, or any potentially interfering organism or material is found in the blank, analysis of additional field samples is halted until the source of contamination eliminated and a blank shows no evidence of contamination. Any field sample in a batch associated with a contaminated blank that shows the presence of one or more oocysts or cysts is assumed to be contaminated and should be recollected, if possible. Any method blank in which oocysts or cysts are not detected is assumed to be uncontaminated and may be reported.

The laboratory should add results that pass the method acceptance criteria to initial and previous ongoing data and update a quality control (QC) chart to form a graphic representation of continued laboratory performance. The laboratory should develop a statement of laboratory accuracy (for reagent water, raw surface water, etc.) by calculating the average per cent recovery (X) and s_r. Express the accuracy as a recovery interval from $X - 2s_r$ to $X + 2s_r$. For example, if X equals 95 per cent and s_r equals 25 per cent, the accuracy is 45 to 145 per cent. If available, the laboratory should periodically analyse an external QC sample, such as material used for a performance evaluation or standard reference material. The laboratory also should periodically participate in interlaboratory comparison studies that use this protozoan recovery method. The specifications contained in method 1623 can be met if the analytical system is under control. The standards used for IPR and OPR tests should be identical, so that the most precise results will be obtained.

Internal and external calibration records of instruments and equipment must be kept on file in the laboratory's QA logbook.

Sampling

Field samples are collected in disposable plastic 10-litre Cubitainers, which are shipped to the laboratory for processing. Sample containers should be labelled immediately with the name of the site, the date and time of sampling, and the name of the person taking the sample. If residual disinfectant is present in the water, add 100 ml of 2 per cent (wt/volume) aqueous thiosulphate to the Cubitainers before addition of the 10 litre sample. Samples should be shipped to the laboratory on the day that they are collected and must arrive within 24 hours of sample collection. Store the Cubitainers with field samples at 4° to 8°C between the time of collection and the time of shipment to the laboratory and upon receipt at the laboratory until the sample is ready for filtration. Do not allow the sample to freeze. Although it is desirable to ship field samples on wet or chemical ice, it is not a requirement.

This procedure requires a minimum sample volume of 10 litres. Unless the sample is known to contain *Giardia*, *Cryptosporidium*, or some other infectious agent (e.g. during an outbreak), samples should be shipped as noninfectious material and should not be marked as infectious. When field samples are known to contain infectious agents or are suspected of containing infectious agents, it is recommended that the field samples be filtered in the field and that the filter be shipped to the laboratory to avoid violating DOT regulations.

Holding times

Once the field sample arrives in the laboratory, various holding times apply. Periodic OPR and MS samples must be spiked and processed within 24 hours of spiking with flow cytometer-counted *Giardia* cysts and *Cryptosporidium* oocysts. Laboratory filtration, elution, and concentration of a field sample must be completed within 72 hours of sample collection. The concentrate must be stored at 4° to 8°C if one is not proceeding immediately to IMS. IMS and sample staining must be completed within 24 hours of sample concentration. Stained slides must be stored at 4° to 8°C in the dark. Although immuno-fluorescence assay (FA), DAPI staining, and DIC microscopy examination and confirmation should be performed immediately after staining is complete, laboratories have up to 72 hours from time of completion of sample staining to complete the examination and confirmation of results for the samples. However, if fading or diffusion of the DAPI-stained sample is noticed, the laboratory must reduce this holding time and/or adjust the concentration of the DAPI staining solution so that fading or diffusion does not occur. Never allow freezing of any of these samples or reagents.

Sample filtration and elution

At present the Pall Gelman (Ann Arbor, Mich.) Envirochek capsule filter, the IDEXX Filtamax foam filter, and the Whatman Cryptest filter have been validated and found to be acceptable for concentration of 10 litres water samples by method 1623. Each of these filters has an upper limit on the volume of water that it can efficiently filter. For the sake of brevity, only the Gelman Envirochek capsule filter will be considered here. The manufacturer's directions should be followed if either of the other two filter options is used.

First assemble the sample concentration components, consisting of tubing, a pump, a flow controller, a capsule filter, and a sample container, as illustrated in Fig. 8.1. To prevent cross contamination between samples, use the tubing between the sample container and the Gelman Envirochek filter capsule only once. Connect the sampling system, minus the capsule, to a carboy filled with reagent water. Turn on the pump and adjust the flow rate to 2.0 litres/minute, allowing 2 to 10 litres of reagent water to flush the system. When the flow rate has been adjusted, turn off the pump and install the capsule filter in the line, securing the inlet and outlet ends with the appropriate clamps and fittings.

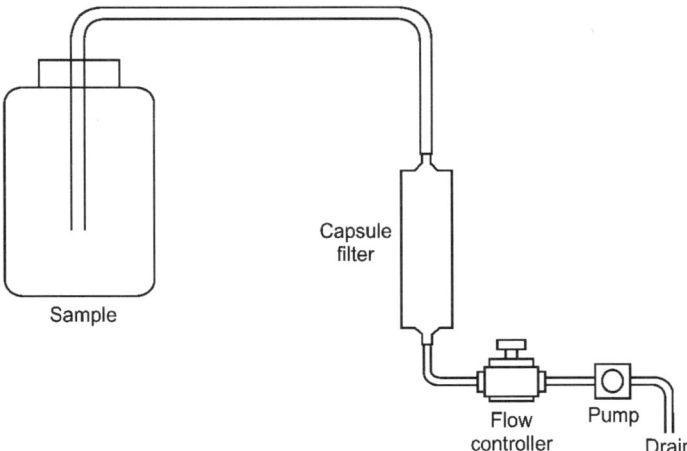

Fig. 8.1. Sampling apparatus for raw water.

Record the sample number, sample turbidity, and the name of the analyst filtering the sample on a bench sheet and on the capsule filter. Connect the sampling system to the field carboy of sample water, or transfer the sample water to the laboratory carboy used to regulate the flow rate as described above. It should be noted that if the field sample is transferred to a laboratory carboy, the laboratory carboy must be cleaned and disinfected before it is used with another field sample. Place the drain end of the sampling system tubing into an empty graduated container that has a capacity of 10 to 15 litres and that is calibrated at 9.0, 9.5, 10.0, 10.5, and 11.0 litres. This container is used to determine the sample volume filtered. Alternately, a totalising meter can be placed on the drain end of this apparatus to determine the sample volume filtered.

Allow the carboy discharge tube and filter capsule to fill with sample water. Vent residual air by using the bleed valve-vent port built into the filter capsule. Turn on the pump to start the flow of water through the filter, and verify that the flow rate is 2 litres/min. After the sample has passed through the filter, turn off the pump. Allow the pressure to decay until the flow stops. Disconnect the inlet end of the capsule filter assembly while maintaining the level of the inlet fitting above the level of the outlet fitting to prevent backwashing and the loss of oocysts and cysts from the filter. Restart the pump just long enough to allow as much water to drain from the filter capsule as possible. On the basis of the water level in the graduated container or the reading on the totalising meter, record the volume filtered on a bench sheet and the capsule filter label to the nearest quarter litre. Discard the contents of the graduated container. Loosen the outlet fitting and then close the inlet and outlet fittings with the caps provided with the capsule filter.

Assemble the laboratory shaker (shaker [model 3589]) equipped with side arms [model 3587–4]; Lab-line, Melrose Park, Ill.) with the clamps aligned vertically so that the filters will be aligned horizontally. Prepare sufficient elution buffer, which contains laureth-12, Tris solution, EDTA, and antifoam A, so that all samples to be eluted that day can be processed with the same batch of buffer. To prepare elution buffer, add 10 ml of 10 per cent (wt/volume) aqueous laureth-12 (PPG Industries, Gurnee, Ill.) to a 1000 ml graduated cylinder. Also add 10 ml of 1 M Tris (pH 7.4), 2.0 ml of 0.5 M disodium EDTA (pH 8.0), and 150 µl of antifoam A (Sigma-Aldrich, St. Louis, Mo.) to the graduated cylinder and dilute to 1000 ml with reagent water. Elutionn may require up to 275 ml of buffer per sample. Designate at least one 250 ml conical centrifuge tube, which will be used to receive and process the filter eluate, for each sample and label it with the sample number.

Record the name of the analyst performing the elution on the bench sheet. By using a ring stand or other means, clamp each filter capsule to be eluted in a vertical position with the inlet end up. Remove the inlet cap and allow the liquid level to stabilise. Pour elution buffer through the inlet fitting. Enough elution buffer must be added to just cover the pleated white membrane of the filter. Replace the inlet cap and transfer the filter capsule to a clamp on the shaker. Adjust the filter capsule in the clamp on the laboratory shaker so that the bleed valve is positioned at the top of the vertical axis (at the 12 o'clock position). Turn on the shaker, set the speed to 80 per cent of maximum (approximately 600 rpm), and agitate the capsule for 10 minutes. Remove the filter capsule from the shaker, remove the inlet cap, and pour the contents of the capsule into the 250 ml conical centrifuge tube. Again, clamp the capsule vertically with the inlet end up and again add enough elution buffer through the inlet fitting to just cover the pleated filter membrane. Replace the inlet cap, transfer the capsule to the shaker clamp, and adjust the capsule in the clamp so that this time the bleed valve is positioned on a horizontal axis (3 or 9 o'clock position). Turn on the shaker and agitate the capsule for an additional 5 minutes. Add the contents of the capsule to the 250 ml centrifuge tube.

Centrifuge the 250 ml centrifuge tube containing the capsule filter eluate at 1100 × g for 15 minutes with a swinging-bucket rotor. Allow the centrifuge to coast to a stop. Record the initial pellet volume (volume of solids) and the date and time that the concentration was completed on a bench sheet. What follows assumes that the entire pellet will be analysed. With a Pasteur pipette, carefully aspirate off the supernatant down to 1 ml above the pellet. If the sample is reagent water (e.g. an IPR or OPR sample), extra care must be taken to avoid the aspiration of oocysts and cysts during this step. If the packed pellet volume is less than or equal to 0.5 ml, add reagent water to the centrifuge tube to bring the total volume to 10 ml. However, if the packed pellet volume is greater than 0.5 ml, use the following formula to determine the total volume required in the centrifuge tube:

$$\text{Total volume (in millilitres) required} = \frac{\text{pellet volume}}{0.5 \text{ ml}} \times 10 \text{ ml}$$

For example, if the packed pellet volume is 0.8 ml, the total volume required is 16 ml. Add reagent water to the centrifuge tube to bring the total volume to the level calculated as described above. Vortex the tube for 10 to 15 seconds to resuspend the pellet. Record this resuspended volume on a bench sheet. Add additional reagent water to the centrifuge tube to bring the volume to an amount evenly divisible by 10 and then vortex the tube for an additional 10 to 15 seconds. For example, if the resuspended volume measured is 16 ml, add 4 ml of reagent water to bring the volume to 20 ml. Record this final resuspended volume on a bench sheet. Process the sample as described above as multiple, independent, 10 ml subsamples from here onward, including the preparation and examination of separate slides for each aliquot.

The IMS procedure is based on the Dynal Combo IMS kit and should be performed on a bench top with all materials at room temperature, ranging from 15 to 25°C. Prepare a 1 × dilution of SL-buffer-A from the 10 × SL-buffer-A (clear, colourless solution) supplied in the kit. Use reagent water as the diluent. A volume of 1.5 ml of 1 × SL-buffer-A will be required per sample or subsample with which the Dynal IMS procedure is performed. For every 1 ml of 1 × SL-buffer-A required, take 100 μl of 10 × SL-buffer-A and adjust the volume to 1 ml with the reagent water.

To a flat-sided Leighton sample tube (60 by 10 mm), add 1 ml of the 10 × SL-buffer-A (as supplied, not the diluted 1 × SL-buffer-A). Add 1 ml of 10 × SL-buffer-B (a magenta solution supplied with the kit) to the Leighton sample tube containing 10 × SL-buffer-A. Quantitatively transfer 10 ml of the water sample concentrate to the Leighton sample tube containing the 2 SL-buffers. Label the tube(s) with the sample number. Vortex the Dynabeads *Crypto*-Combo vial from the IMS kit for approximately 10 seconds to suspend the beads. Ensure that the beads are fully resuspended by inverting the tube and making sure that there is no residual pellet at the bottom. Add 100 μl of the resuspended Dynabeads *Crypto*-Combo beads to each Leighton sample tube containing the water sample concentrate and SL-buffers. Vortex the Dynabeads *Giardia*-Combo vial from the IMS kit for approximately 10 seconds to suspend the beads. Ensure that the beads are fully resuspended by inverting the tube and making sure that there is no residual pellet at the bottom. Add 100 μl of the resuspended Dynabeads *Giardia*-Combo beads to each Leighton sample tube containing the water sample concentrate, Dynabeads *Crypto*-Combo beads, and SL-buffers. Attach the sample tube to a Dynal rotating mixer and rotate at approximately 18 rpm for 1 hour at room temperature.

After the 1 hour incubation period, remove the sample tube from the rotating mixer and place the tube in the magnetic particle concentrator (MPC-l; Dynal, Inc.), with the flat side of the Leighton tube toward the magnet. Without removing the sample tube from the MPC-l, place the magnet side of the MPC-1 downward, so that the tube is horizontal and the flat side of the tube is facing down. Gently rock the sample tube by hand end to end through approximately 90°, tilting the cap end and base end of the tube up and down in turn. Continue the tilting action for 2 minutes with approximately one tilt per second. Ensure that the tilting action is continued throughout this period to prevent binding of low-mass, magnetic, or magnetisable material. If the sample in the MPC-1 is allowed to stand motionless for more than 10 seconds, repeat the tilting oscillation before continuing. Return the MPC-1 to the upright position, with the sample tube vertical and the Leighton tube cap on top. Immediately remove the cap and pour off all of the supernatant from the tube held in the MPC-1 into a suitable container. Do not shake the tube and do not remove the tube from MPC-1 until all the supernatant has been poured off. Now remove the sample tube from the MPC-1 and resuspend the sample by using 1 ml of 1 × SL buffer-A (prepared as described above from the 10 × SL buffer-A stock). Mix very gently to resuspend all material in the Leighton tube. Do not vortex. Quantitatively transfer all the liquid from the Leighton tube to a labelled, 1.5 ml microcentrifuge tube. Ensure that all of the liquid and beads are transferred.

Place the microcentrifuge tube into the second magnetic particle concentrator (MPC-M); Dynal, Inc.) with its magnetic strip in place. Without removing the microcentrifuge tube from the MPC-M, gently rock and roll the tube through 180° by hand. Continue for approximately 1 minute with approximately one 180° roll and rock per second. At the end of this step, the beads should produce a distinct brown dot at the back of the tube. Immediately aspirate the supernatant from the microcentrifuge tube and cap held in the MPC-M. If more than one sample is being processed, conduct three 90° rock-and-roll actions before removing the supernatant from each tube. Take care not to disturb the material attached to the wall of the tube adjacent to the magnet. *Do not shake the tube. Do not remove the tube from the MPC-M while conducting these steps.*

Remove the magnetic strip from the MPC-M. Add 50 µl of 0.1 N HCl to each microcentrifuge tube and then vigorously vortex the tube for 10 to 15 seconds. Place the microcentrifuge tube(s) back in the MPC-M without the magnetic strip in place and allow the unit to stand in a vertical position for at least 10 minutes at room temperature. Then, again remove each microcentrifuge tube from the MPC-M and vortex it vigorously for 5 to 10 seconds. Ensure that all of the sample is at the base of the tube and again place the microcentrifuge tube back in the MPC-M. Replace the magnetic strip in the MPC-M and allow the tube to stand undisturbed for approximately 10 seconds. Label a Dynal spot-on well slide for sample screening. Add 5 µl of 1.0 N NaOH to a slide sample well. Without removing the microcentrifuge tube from the MPC-M, transfer all of the sample from the microcentrifuge tube in the MPC-M to the sample well with the NaOH. Do not disturb the beads at the back wall of the tube. Ensure that all of the fluid is transferred. This completes the first magnetic bead dissociation.

Air dry the sample(s) onto the well slide(s). If a second magnetic bead dissociation is required (this may enhance the recovery of oocysts and cysts in some cases), do not discard the beads or microcentrifuge tube after the sample is transferred to a well slide. Again add 50 µl of 0.1 N HCl to the microcentrifuge tube and then vortex the tube vigorously for 10 to 15 seconds. Continue the procedure as described in the previous paragraph through the transfer of the second dissociation product to an unused well on an appropriately labelled well slide and air dry.

Positive and negative staining controls must be prepared for each batch of slides being stained. For the positive control, pipette 10 µl of positive antigen or 200 to 400 intact *Cryptosporidium* oocysts and 200 to 400 intact *Giardia* cysts to the centre of a well and spread them over the well area with a pipette tip. For the negative control, pipette 50 µl of phosphate-buffered saline (PBS; the formulation is given below) into the centre of a well and spread it over the well area with a pipette tip. Air dry the control slides. Apply 1 drop of a *Giardia* and *Cryptosporidium* combination FITC-labelled monoclonal antibody from a kit (e.g., Aquaglo) to each well. If specified in the antibody kit, apply 1 drop of counterstain to each well. Spread the fluid over the entire well with an applicator stick, if necessary. Do not allow the stick to scratch the treated surface of the slide. Use a different applicator stick for each well. Place the slides in a humid chamber in the dark and incubate them at room temperature for approximately 30 minutes. The humid chamber consists of a tightly sealed plastic container containing damp paper towels on top of which the slides are placed. Apply 1 drop of 1× wash buffer to each well. Tilt each slide on a clean paper towel with the long edge down. By using a clean Pasteur pipette, gently aspirate the excess detection reagent and counterstain from below each well. Avoid disturbing the samples. Do not allow the slides to dry. Apply 50 µl of working DAPI staining solution to each well. Working DAPI stain is prepared daily just before use by diluting 10 µl of a stock solution (2 mg of DAPI/ml of absolute methanol; store in the dark at 4 to 8°C) to 50 ml with PBS (add 1.07 gram of Na_2HPO_4, 0.39 gram of $NaH_2PO_4 \cdot 2H_2O$, and 8.5 gram of NaCl to 800 ml of reagent water; dissolve the salts and adjust the volume to 1 litre with reagent water; and adjust the pH to 7.2 with NaOH or HCl, as required). Allow the DAPI stain to stand at room temperature for approximately 1 minute. Then apply one drop of 1× wash buffer or PBS to each well. Tilt each slide on a clean paper towel with the long edge down. Gently aspirate the excess DAPI staining solution from below each well with a clean Pasteur pipette. Avoid disturbing the sample. Do not allow the slides to dry. Add 1 drop of mounting medium (dissolve 2 grams of 1,4-diazabicyclo[2.2.2]octane [DABCO] in 95 ml of warm glycerol-PBS [60 per cent glycerol, 40 per cent PBS]; adjust the volume to 100 ml with additional glycerol-PBS after the DABCO dissolves) to each well and apply a coverslip. Use a Kimwipe to remove excess mounting fluid from the edges of the coverslip. Seal the coverslip to the slide with clear nail polish. Record the date and time that the staining was completed on the bench sheet. If slides will not be read immediately, store them in a humid chamber in the dark at 4 to 8°C until they are ready for examination.

Microscopic examination

This section assumes use of a microscope, capable of epifluorescence and DIC or Hoffman modulation optics, with stage and ocular micrometers and ×20 (numerical aperture, 0.6) to ×100 (numerical aperture, 1.3) oil objectives. This section also assumes that the ocular micrometer has been calibrated. The epifluorescence portion of the microscope should be equipped with appropriate excitation and bandpass filters for examination of FITC-labelled specimens (exciter filter, 450 to 490 nm; dichroic beam-splitting mirror, 510 nm; barrier or suppression filter, 515 to 520 nm), and DAPI-labelled specimens (exciter filter, 340 to 380 nm; dichroic beam-splitting mirror, 400 nm; barrier or suppression filter, 420 nm). The slides are read by scanning each well in a systematic fashion. Either an up-and-down or a side to-side scanning pattern may be used (Fig. 8.2). Before examination of the sample wells, ensure that the positive and negative staining controls reacted appropriately. Make sure that the positive staining control contains oocysts and cysts within the numerical range expected and at the appropriate fluorescence for both FA and DAPI staining and that the negative staining control does not contain any oocysts or cysts. Use epifluorescence to scan the entire well for each sample at not less than ×200 total magnification for apple green fluorescence of oocyst and cyst shapes.

When ovoid or spherical objects that are 4 to 6 µm in diameter and that fluoresce a brilliant apple green with brightly highlighted edges are observed, switch the microscope to the UV filter block for DAPI staining and then to DIC microscopy. By using the UV filter block for DAPI staining the object may exhibit one of the following characteristics: (i) up to four distinct, sky-blue nuclei, (ii) intense blue internal staining, or (iii) light blue internal staining (no distinct nuclei) and a green rim. The first two conditions are recorded as DAPI positive, while the third condition is recorded as DAPI negative. By DIC microscopy, look for external or internal morphological characteristics atypical of *Cryptosporidium* oocysts (e.g. spikes, stalks, appendages, pores, one or two large nuclei filling the cell, red fluorescing chloroplasts, crystals, and spores). If atypical structures are not observed, then categorise each object that fluoresces apple green as (i) an empty *Cryptosporidium* oocyst-like object; (ii) a *Cryptosporidium* oocyst-like object with an amorphous structure; or (iii) a *Cryptosporidium* oocyst with internal structure (one to four sporozoites per oocyst). Record the shape and measurements to the nearest 0.5 μm at × 1000 total magnification for each such object. Although they are not a defining characteristic, surface oocyst folds may be observed in some specimens.

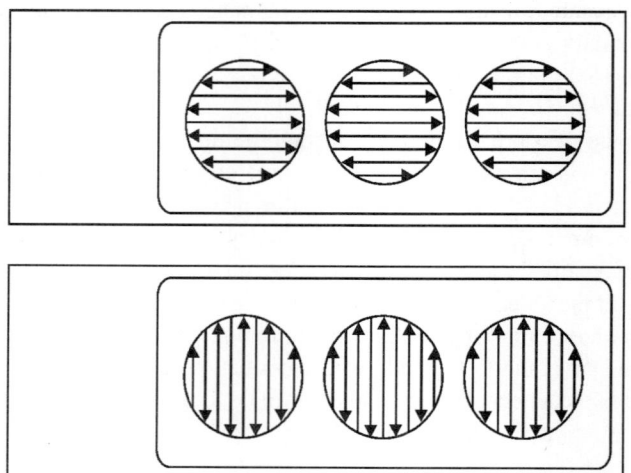

Fig. 8.2. Methods for scanning a well slide.

When round to oval objects (8 to 18 μm long by 5 to 15 μm wide) that fluoresce a brilliant apple green are observed, switch the microscope to the UV filter block for DAPI staining and then to DIC microscopy. By using the UV filter block for DAPI staining, the object may exhibit one or more of the following characteristics: (i) two to four sky-blue nuclei, (ii) intense blue internal staining, or (iii) light blue internal staining (no distinct nuclei) and a green rim.

The first two conditions are recorded as DAPI positive, while the third condition is recorded as DAPI negative. By DIC microscopy, look for external or internal morphological characteristics atypical of Giardia cysts (e.g. spikes, stalks, appendages, pores, one or two large nuclei filling the cell, red fluorescing chloroplasts, crystals, and spores). If these atypical structures are not observed, then identify such objects that are of the aforementioned size and shape and that fluoresce an apple green as presumptive *Giardia* cysts. Record the shape and measurements (to the nearest 0.5 μm at × 1000 magnification) for each such object as part of the presumptive count. If two or more internal morphological structures are observed at this point, record this as a confirmed *Giardia* cyst as well. Record presumptive and confirmed results for *Cryptosporidium* oocysts on a *Cryptosporidium* report form and presumptive and confirmed results for *Giardia* cysts on a *Giardia* report form.

Limitations

Overall limitations associated with this immunofluorescence detection method include the following: (i) the level of organism recovery is low. The recovery efficiency of the method is influenced by water quality and in particular by the water matrix being sampled; (ii) by this procedure about a half a day is needed to run one sample, and the procedure requires specialised, expensive equipment; (iii) non-specificity of the monoclonal antibodies used in the test kits allows a number of cross-reactions and potential false-positive results for both *Giardia* and *Cryptosporidium*. For example, algal cells of the same size and shape as *Giardia* cysts and *Cryptosporidium* oocysts react with the monoclonal antibodies used in this procedure. Furthermore, the *Cryptosporidium* and *Giardia* monoclonal antibodies are not species specific; (iv) the method does not determine the viability, much less the infectivity, of the organism detected; (v) the procedure does not identify the host of origin. Consequently, it is not known whether the particular isolate is a significant threat to humans; (vi) large amounts of algae and debris, which can physically obscure the observation of the cysts and oocysts on the microscope slides, are sometimes recovered, from water samples during the procedure; (vii) many algae and other animate and inanimate objects have pigments that autofluoresce at the excitation wavelengths for FITC and other common fluorochromes. This autofluoresce interferes with cyst and oocyst detection and makes reading of slides more time-consuming and tedious; and (viii) this method is dependent on the skill, experience, and training of the microscopist. After detecting objects of the right size shape, and fluorescence characteristic, the microscopist must demonstrate internal morphological characteristics by DIC or Hoffman modulation microscopy. Unfortunately, the microscopic portion of the analysis can be quite subjective even among experts. Moreover, this type of microscopic work is intense, demanding, and tiring. When the microscopist becomes too tired, the reliability of the results becomes questionable.

While these methods have successfully been used on some occasions to confirm epidemiological evidence that a waterborne outbreak has occurred and have been used in surveys of raw and finished waters, they do not lend themselves to use for easy, routine monitoring. In an effort to address the limitations of the monitoring methods, the US Environmental Protection Agency convened the protozoa, virus and coliphage monitoring workshop of experts in Cincinnati, Ohio, in August 1993. Additional workshops on the protozoan method occurred in Cocoa Beach, Fla., in January 1994 and in Cincinnati, Ohio, in May 1995. Out of these workshops came the recognition that a negative count and a low detection limit do not ensure pathogen-free water. Smith provided the following recommendations: (i) it was recommended that the results obtained by this method be interpreted with extreme caution, (ii) a training course of 3 to 4 days, duration is required before a laboratory can begin the laboratory approval process, (iii) microscopic work must be supervised by a senior analyst. Moreover, microscopic work by a single analyst should not exceed 4 hr/day nor more than 5 consecutive days/week. Intermittent rest periods during the 4 hr/day are encouraged.

Alternate approaches

Recognition of the limitations of the FA for *Giardia–Cryptosporidium* has stimulated research using other approaches to detection and identification including fluorescence-activated cell sorting (FACS), enzyme-linked immunosorbent assay (ELISA), confocal microscopy, PCR, and fluorescence *in situ* hybridisation (FISH). Presently, with the exception of ELISA, none of these approaches has been validated critically with environmental and finished water samples.

A number of clinical ELISA kits are commercially available and have been evaluated for their ability to detect spiked *Giardia* cysts and *Cryptosporidium* oocysts in concentrated environmental water samples. The detection limit reported for ELISA is 10 cysts or oocysts. Like the antibody-based FA for these organisms, detection by ELISA suffers from cross-reactions with some algae and is adversely affected by turbidity. However, because the ELISA technique can be done more rapidly than the FA procedure and does not require a skilled microscopist to read the result, it has potential for use for the screening of water samples to determine which ones might require more rigorous testing by a more exacting method.

Confocal microscopy in conjunction with FISH has been used to detect *Giardia* cysts and determine the *Giardia* species detected. Oligomeric probes to *Giardia* 16S rRNA, which were conjugated to either carboxymethylindocyanine dyes (Cy3 or Cy5) or fluorescein, were used to distinguish between

G. lamblia. Giardia muris, and *Giardia ardeae* cysts. When the fluorochrome FITC was used in conjunction with a detection monoclonal antibody and the ribosomal DNA probes, it was possible to detect cysts of *G. lamblia* in fecal samples as well as environmental samples taken from a sewage lagoon. Observation of the preparations was done with a krypton-argon laser-equipped confocal microscope which allowed the simultaneous visualisation of all the fluorochromes. The FISH assay preparations can also be evaluated with a standard epifluorescence microscope. Recently, a FISH assay has also been developed for *Cryptosporidium.* Once oocysts are isolated, they are treated with 50 per cent ethanol and are incubated at 80°C for 10 minutes to permeabilise them. Subsequently, the oocysts are incubated with the FISH probe at 48°C for 30 minutes. Used in conjunction with an FITC-labelled monoclonal antibody, the FISH assay was able to confirm the identities of the oocysts down to the species level. The major advance by use of this approach is that the rate of occurrence of false-positive results is reduced, and the need for DIC optics is eliminated. The drawback, however, is the great expense of and skill needed to operate either an epifluorescence microscope or a confocal microscope. If the investigator can forgo the simultaneous observation of two or more fluorochrome labels, then all that is needed is a conventional epifluorescence microscope equipped with the requisite number of cubes for the various fluorochrome labels. If a procedure that exploits FISH in conjunction with fluorescent detection antibodies could be developed for use in a FACS, then the microscope time required for examination of water samples would be greatly reduced, if not eliminaled.

A FACS is a very elaborate particle counter with a laser. The FACS, besides being able to sort particles, is able to sense the fluorescence of labelled particles or cells as well as determine their size. This instrumentation has been used to detect *Giardia* and *Cryptosporidium* in water samples in the United Kingdom and is being used in Australia with environmental samples. In combination with calcium carbonate flocculation recoveries of *Giardia* cysts and *Cryptosporidium* oocysts are reported to range from 92 to 104 per cent in spiked samples of waste-water, reservoir water, and river water. In some instances the FACS is capable of looking at much more of the sample than the FA procedure can. Because fluorescent antibodies are used in the procedure with the FACS and cross-reactions can occur with other objects, it is recommended that sorted fluorescent objects of the right size and shape be confirmed on an epifluorescence microscope equipped with DIC optics. Microscopic observation is much less fatiguing after processing with the FACS, as most of the contaminating debris and background fluorescence are eliminated by the FACS. This approach allows most of the sample to be sorted in 5 to 15 minutes. Moreover, since a greater percentage of the sample is observed in the procedure with the FACS than in the FA procedure, the probability of finding the organisms of interest is greatly increased. The major disadvantages of this technique are the skill required of the FACS operator and/or microscopist, the high cost of the instrument, and the quality and reliability of the antibodies.

Solid-phase cytometry is a variation of fluorescence-activated cell sorting or flow cytometry. However, unlike flow cytometry, in which particles in a fluid pass by a laser, in solid-phase cytometry the particles are immobilised on either a membrane or a glass slide which is scanned with a movable laser beam. Image analysis software used in conjunction with this instrumentation discriminates between the various fluorescent particles found during the scanning of the solid support and keeps a coordinate record of all the potential positive images. Using the microscope attached to this instrument, all the potential positive coordinates can be examined to confirm or deny the detected object. This technology has been used to detect *Cryptosporidium* oocysts in water. Solid-phase cytometric analysis was substituted for traditional microscopic fluorescent analysis in a variation of method 1622 in one study. When 100 oocysts were seeded at the beginning of the solid-phase cytometric variation of method 1622 and the traditional microscopic version of method 1622, average recovery rates were estimated to be 73 per cent for solid-phase cytometry and 49 per cent for the traditional microscopic version of method 1622. Another study suggested that the recovery efficiency of solid-phase cytometry ranged from 82.3 to 86.3 per cent. An advantage of this technique is that it can be performed much faster than the conventional epifluorescence microscopic assay. Disadvantages include the expense of the instrumentation, the drying and curing of the support membranes before analysis can be completed for some samples, and the inability of the software to deal with samples containing large numbers of organisms.

The detection of *Giardia* cysts and *Cryptosporidium* oocysts by amplification of specific regions of DNA by PCR has been reported. A single *Giardia* cyst was detected by using DNA coding for giardin as the target for PCR. Moreover, the total mRNA level increased significantly after induction of excystation in living *Giardia* cysts, making this the first successful molecular biology-based means of viability determination reported for this organism. Discrimination between *Giardia* species which are pathogenic and those which are not pathogenic for humans was also possible by PCR. Primers and probes for detection of *C. parvum* oocysts have been reported as well. Theory suggests that this procedure is just what is needed for the detection of organisms in environmental samples that are usually low in protozoan density. However, analysis of environmental samples by this technique can be compromised by the presence of inhibitors like humic acids. Consequently, PCR works well only with highly purified samples. To overcome this problem, it has been suggested that water sample concentrates should be purified by chromatography through Sephadex G-100 and Chelex-100 spin columns. To date, successful application of this purification procedure has been reported only for enteric viruses. Attempts to use column purification and sample dilution in conjunction with PCR detection of *Giardia* have not been very successful, and it has been concluded that other inhibitors of PCR besides organic compounds like humic acids are present. Recent purification of water samples by IMS has helped overcome the inhibitor problem in the detection of both *Giardia* cysts and *Cryptosporidium* oocysts. Other problems with PCR are that it is not quantitative and is susceptible to false-positive reactions from organisms that are long dead and deteriorated. In addition, Champliaud and other studied the specificities of eight different previously published *C. parvum* sequences targeting 18S rRNA. Their results showed that PCR amplification failed to differentiate *C. parvum* from *Cryptosporidium meleagridis* when the eight DNA sequences were tried. This work indicates that great caution should be used in interpreting PCR results.

A great deal of emphasis has been placed on determining viability with various fluorochromes like DAPI, propidium iodide, fluorescein diacetate, and SYTO-9 and SYTO-59. Although a correlation between fluorescence intensity and the ability of *C. parvum* oocysts to infect neonatal CD-1 mice in the study with SYTO-9 and SYTO-59 was found, the fluorescence profiles varied significantly between batches of oocysts. Another study compared *C. parvum* viability following ozone treatment of the oocysts by both the neonatal mouse CD-1 animal infectivity assay and *in vitro* fluorochrome assays like the SYTO-59 assay. In this investigation, which was conducted in laboratories both in the United States and in the United Kingdom, *in vitro* fluorochrome assays of ozone-exposed oocysts showed a marginal reduction of oocyst viability. In contrast, the neonatal CD-1 mouse assay demonstrated significant inactivation of the oocysts, and the results failed to correlate with those of any of the *in vitro* assays. At best, fluorochrome viability testing appears to be unreliable and variable between laboratories. The search for a good oocyst viability assay has now turned to coupling IMS purification of the water sample with cell culture followed either by fluorescence assay or PCR analysis of developmental stages. Besides not being validated for robustness and precision in multiple laboratories at this time, questions regarding the sensitivity and specificity of this approach have yet to be clarified.

Microscopic particulate analysis (MPA) is a method that has evolved from the immunofluorescence microscopic methods used to detect *Giardia* and *Cryptosporidium*. Initially, this method was developed to determine if groundwater was under the influence of surface water. In this protocol 500 gal (1893 litres) of water is sampled with a depth fibre filter over an 8 to 24 hours period. The particulates caught by the filter fibres are extracted with particle-free water containing Tween either by hand washing or with a stomacher blender. The particulates are concentrated by centrifugation. Samples with particulate volumes of less than 20 μl/100 gal (20 μl/379 litres) are observed directly, while samples with particulate volumes of greater than 20 μl/100 gal are floated on Percoll-sucrose (specific gravity, 1.15). Unlike the parent immunofluorescence procedure, the particulates are not stained with monoclonal antibodies. Instead, staining with Lugol's iodine may or may not be done, depending upon whether bright-field microscopy is being used. Observation of the particulates is done by either bright-field, phase-contrast, or DIC microscopy. All micro-organisms in the 1 to 400 μm size range are counted and measured with a calibrated ocular micrometer. These micro-organisms are referred to as bioindicators in this method and include diatoms, algae, *Giardia* cysts, coccidian oocysts, plant debris, pollen, rotifers, crustaceans, amoebae, nematodes, insect parts, and insect larvae. The bioindicators are identified taxonomically at least to the phylum level and, it is hoped, to the class level. In addition, inanimate particulates are tallied. Identification of *Giardia* cysts, coccidian oocysts, and/or enteric helminths are indicative of groundwater under the influence of surface water. The occurrence of pigment-bearing diatoms and algae also is indicative of groundwater under the influence of surface water. While this list of bioindicators is long, there is little agreement among individuals using this method as to what level of risk they represent. To overcome this problem, this consensus MPA protocol categorises each bioindicator into relative risk factor groups. On the basis of each bioindicator's density and risk factor category, the water is mathematically rated as to the level of risk for being under the influence of surface water. Recently, the MPA method has been adapted to evaluate water filtration plant efficiency. This is done by looking at pre- and postfiltered water samples. In experienced hands this method has great value. It, however, requires much greater expertise than the FA method from which it evolved. Besides extensive experience, analysts need strong backgrounds in limnology, parasitology, phycology, and invertebrate zoology. Even though this method is very laborious, it still provides a more detailed analysis of the water than is possible from more automated approaches like particle counting.

EMERGING PROTOZOANS OF POTENTIAL HEALTH SIGNIFICANCE IN WATER

Of all the cases of waterborne gastroenteritis, only about 50 per cent are attributable to a specific micro-organism, toxin, or chemical. Twenty-five years ago *Giardia* was thought to be an innocuous commensal and *Cryptosporidium* was basically an unknown. With vigilant epidemiology and the ability to monitor these protozoans, their significance has come to be appreciated. Like the emergent *Cryptosporidium* concerns of 20 years ago, there are now clinical reports of gastroenteritis caused by microsporidians, *Cyclospora*, *Isospora* spp., and *Blastocystis hominis*, in both immunocompetent and immunocompromised individuals. Since these organisms are enteric, are these emerging protozoans of potential health significance in water? At present, other than clinical detection approaches, there are no protocols for detection of them in environmental samples. This, however, does not preclude adaptation of existent techniques for *Giardia* and *Cryptosporidium* to look for them in water.

Microsporidians is a nontaxonomic term that describes obligate intracellular parasites belonging to the class Microspora of the protozoan phylum microspora. They are ubiquitous parasites that infect a variety of vertebrate and invertebrate hosts. From an anthropocentric point of view, insects of commercial significance like honeybees and silkworms are affected by this group. Moreover, moving up the phylogenetic tree, snails and commercial fish such as salmon, flounder, and monkfish are prone to microsporidian diseases. Only with the advent of the AIDS epidemic was this group of pathogens recognised as a cause of human disease. Six genera of microspondians have been recovered from humans: *Encephalitozoon*, *Nosema*, *Pleistophora*, *Enterocytozoon*, *Septata*, and *Microsporidium*, a genus for all forms as yet unclassified. Although rare, microsporidian infections are now being reported from immunocompetent people. Presently, classification is done on the basis of spore size (1.5 to 5 μm), nuclear configuration, the number of polar tube coils within the spore and developing forms, and the host cell-parasite relationship. The life cycle of these parasites is thought to be direct by either ingestion, inhalation, or inoculation. Two enteric forms, *Enterocytozoon bieneusi* and *Encephalitozoon intestinalis*, may be transmitted by the water route. Furthermore, *Encephalitozoon cuniculi* and *Encephalitozoon hellem* have been recovered from the central nervous system and the eye, respectively, but can disseminate to the urinary tract, from which spores can be released to the environment. Consequently, *Encephalitozoon* spp. are likely candidates for waterborne transmission. Since microsporidians have been detected in clinical specimens with indirect fluorescent poly- and monoclonal antibodies, this approach could be adapted for detection of microsporidians in water samples in a fashion similar to the FA used for *Giardia* and *Cryptosporidium*. A FISH probe useful for the detection of *E. hellem* in water samples has just been reported, however, use of FISH will require further development.

Cyclospora sp., which is an apicomplexan, is being incriminated as a cause of prolonged diarrhoea in the Third World and more recently in developed countries. The oocysts are 8 to 10 μm in diameter and have the acid-fast staining characteristic, like *Cryptosporidium* and *Isospora*. However, the oocysts of *Cyclospora* have the unique characteristic of autofluorescing when viewed by UV microscopy.

Three species of *Isospora* are reponed to be parasitic in humans. These apicomplexans are *Isospora belli*, *Isospora, hominis*, and *Isospora natalensis*. Confusion exists regarding the status of *I. belli* and *I. hominis*, as some investigators believe that they are the same organism. There is even a report that suggests that *I. hominis* is really a species of Sarcocystis. *I. natalensis* is rare, occurs only in South Africa. Unlike *C. parvum* oocysts, *I. belli* oocysts are not fully sporulated at the time of passing from the host. When sporulation, which requires about 48 hours at room temperature, is complete, the oocyst has two sporocysts, each containing four crescent-shaped sporozoites. Overall, *I. belli* oocysts are elongated oval structures, measuring 20 to 30 μm in length by 10 to 19 μm in width, with both ends being somewhat narrow. Oocysts of *I. belli* can be detected on the basis of their shape and ability to be acid-fast stained. Like *Cryptosporidium* oocysts and *Giardia* cysts, *I. belli* oocysts can be viewed by phase-contrast and Nomarski DIC microscopy. Although unproven, dogs have been suspected of being a reservoir host for *I. belli*.

I. hominis oocysts, which are ovoid and 25 to 33 μm in length, are fully sporulated and contain two sporocysts before being passed from the host. At the time of passage the oocyst has broken open and ripe sporocysts containing four sporozoites are passed in the feces. The sporocysts measure about 14 μm in length. Visualisation can be by acid-fast staining and bright-field microscopy or by either phase-contrast or Nomarski DIC microscopy with unstained material.

B. hominis is a yeast-like organism of questionable taxonomic status. Depending upon the source or expert consulted, *B. hominis* may or may not be classified as a pathogen. Generally, those classifying it as a pathogen do so because drug treatment eliminates this organism as well as the mild gastroenteritis that is present, when no other etiologic agent can be detected. Often confused with amoebic cysts,

B. hominis is of various sizes and has a large vacuole surrounded by peripheral granules, in addition to a single nucleus. Visualisation can be by trichrome staining in conjunction with bright-field microscopy.

To sumup protozoans are eukaryotic organisms which can live either a free-living or a parasitic existence. Some free-living forms, under the right conditions, can become opportunistic parasites. Enteric pathogenic protozoans, like *Giardia* and *Cryptosporidium*, which are now known to be transmitted by water, have been responsible for numerous waterborne outbreaks of gastroenteritis. The primary means for detection, since density levels in water are low, involves processing of a large volume of water by filtration, extraction of the particulates from the filter, concentration of the organisms from the particulates, and assays for the pathogens. The most widely used method for detection of protozoans has been the immunofluorescence assay. While method 1623 has improved upon the utility of the FAs for *Giardia* and *Cryptosporidium*, the procedure is still labourintensive and highly dependent on the skill of the microscopist. Even with the improvements to date, this technique is known to have a number of deficiencies including false positive relations, an inability to determine the viabilities of detected organisms, and a low average per cent recovery of cysts and oocysts.

Various attempts have been made to improve the immunofluorescence detection method. Rather than sampling by filtration and buoyant density centrifugation to purify the organisms, carbonate flocculation is reported to improve recoveries. PCR, cell culture, fluorescence-activated cell sorting, and solid-phase cytometry are being evaluated as alternate test procedures. As each of these approaches is relatively new and much more research is needed, it remains to be seen whether they will be equal to or better than the current fluorescence assay procedure.

Chapter 9

Detection of Bacteria, Viruses and Parasitic Protozoa in Shellfish

INTRODUCTION

Bacterial and viral disease outbreaks have been associated with all major types of edible bivalve molluscs, and human enteric pathogens have been isolated from shellfish obtained from both opened and closed harvesting areas. This chapter contains information that pertains to those shellfish that are edible bivalve molluscs of the class Pelecypoda and include the species commonly referred to as oysters, mussels, clams, and cockles. Since most of these organisms are filter feeders, with the Gastropoda group as a notable exception, they use siphoning organelles and mucous membranes to sieve suspended food particles from the aquatic environment as a source of food. If their surrounding water is contaminated by bacteria, viruses, or parasitic protozoa, these mucous membranes may entrap the pathogens and transfer them to the digestive tract. Since these molluscan shellfish are usually consumed whole and raw, they may act as passive carriers of human pathogens. Although the definition of shellfish can include both crustaceans and molluscs, in this chapter we will use the word 'shellfish' in reference to only bivalve molluscs.

The National Shellfish Sanitation Programme was developed in 1925 to target the prevention of shellfish-associated disease caused primarily by enteric bacteria. In the United States, this programme has established bacteriological standards for shellfish and their harvesting waters based on the fecal coliform index. For approved harvesting waters, the most probable number (MPN) of coliforms must not exceed 70 per 100 ml, with no more than 10 per cent of the samples exceeding 230 per 100 ml. Shellfish meat may contain no more than 230 fecal coliforms per 100 grams. While these standards have proven very effective in preventing outbreaks of disease due to enteric bacterial pathogens, there is little or no relationship between the levels of coliform indicator bacteria in water and shellfish and the presence of human enteric viruses or naturally occurring marine *Vibrio* species. Consequently, shellfish-associated outbreaks and cases of disease due to both human enteric viruses and *Vibrio* species continue to occur in the United States. Human enteric viruses appear to be the major cause of shellfish-associated disease, while the *Vibrio* species cause rarer but frequently more severe disease syndromes. The relative importance of parasitic protozoa in shellfish-associated enteric disease is currently unknown. Microbiological standards for enteric virus, *Vibrio*, or protozoan contamination in shellfish have not been established, primarily due to the lack of accessible detection methodology.

ETIOLOGY OF MOLLUSC-ASSOCIATED OUTBREAKS OF ILLNESS

Wallace retrospectively examined food-borne disease outbreaks in New York State, finding that contaminated seafood accounted for 19 per cent of all reported outbreaks and 10 per cent of all illnesses, with shellfish the most frequently implicated vehicle. Infectious diseases associated with molluscan shellfish consumption have been reviewed by Rippey.

Pathogenic Bacteria Transmitted by Bivalve Molluscs

Two general groups of pathogenic bacteria may be transmitted by shellfish. The first group is termed indigenous bacterial pathogens because these organisms are native to the marine environment, consisting predominantly of members of the family *Vibrionaceae*, including the genera *Vibrio*, *Aeromonas*, and *Plesiomonas*. The presence of these organisms is unrelated to fecal pollution and therefore cannot be monitored by using fecal coliform indices. The second group, referred to as nonindigenous bacteria pathogens, are not natural marine inhabitants, and their presence in shellfish arises either from direct fecal contamination by human or animal reservoirs or from poor general sanitation during harvesting, processing, or preparation of the food animals. Because most of the agents implicated in shellfish-associated disease are not reportable, scientists debate over the exact role that bacteria play in the prevalence of disease. While some have estimated that bacteria account for only about 2 per cent of shellfish-associated outbreaks of illness, both the incidence and the public awareness of shellfish-associated human *Vibrio* infections have increased during the last decade.

Enteric bacteria

Outbreaks of human disease associated with the presence of *Salmonella* and *Shigella* species in shellfish is well documented, even in the recent literature. When these agents are present, it is usually due to fecal contamination of harvesting sites but occasionally due to cross-contamination by human carriers, particularly food handlers. Although not an enteric bacterium, the primary source of *Staphylococcus aureus* in shellfish is also food handlers, and shellfish can readily support the growth of this organism when contaminated and subjected to temperature abuse (storage at an improper temperature). The illness associated with this species is one of intoxication rather than an infection.

Vibrios

The vibrios are part of the normal estuarine microbiota, have excellent survival capabilities in both marine and estuarine environments, and may be accumulated by shellfish during feeding.

Vibrio vulnificus has been of greater concern in the United States in recent years, since epidemiological evidence indicates that enteric exposure to *V. vulnificus* can result in a syndrome characterised by gastrointestinal disease followed by primary septicemia, with mortality rates approaching 50 per cent. Individuals with underlying liver dysfunction or circulatory problems, particularly related to diabetes, or those who are immunocompromised are especially at risk. A self-limiting gastrointestinal disease syndrome without subsequent septic disease can also be seen in healthy individuals consuming contaminated bivalve molluscs.

Enteric Viruses Transmitted by Bivalve Molluscs

Enteric viruses are parasites of animals and humans which are believed to be transmitted primarily by the fecal-oral route. They are excreted in large numbers (10^6 to 10^{10} virus particles/g) in the feces of infected individuals and can almost always be detected in domestic sewage effluents. Since mammalian viruses are extremely specific for both species and tissue, the source of human enteric virus contamination of shellfish harvesting beds is always human fecal pollution rather than general animal waste. Enteric viruses may enter the ocean or estuaries directly through the discharge of domestic sewage, sewage-contaminated rivers and streams, ocean disposal of domestic sewage sludge, and boat wastes. Indirect sources of contamination include land application and burial of sludge with subsequent runoff via rivers and perhaps groundwater to estuaries and the coastline. Enteric viruses are more resistant than bacterial pathogens to common sewage treatment processes, including chlorination. In laboratory studies, enteric viruses have been reponed to survive from 2 to 130 days in seawater, survival periods that far exceed those reported for coliform bacteria in similar environment. For these reasons and others, the fecal coliform index is inadequate for monitoring the presence of viral contamination in shellfish.

There are over 110 known enteric viruses that are excreted in human feces and ultimately find their way into domestic sewage. These viruses cause a wide variety of illnesses, such as hepatitis, fever, diarrhea, gastroenteritis, paralysis, meningitis, myocarditis, and others. The list of known enteric viruses has grown rapidly over the last two decades as better methods for their detection have become available. Regardless of the large number of enteric viruses, only hepatitis A virus (HAV), hepatitis E virus (HEV), the human caliciviruses (such as Norwalk-like viruses [NLVs] and other small round-structured viruses [SRSVs]), astroviruses, and small round (featureless) viruses (SRVs) have been epidemiologically linked in various degrees to shellfish-associated viral disease. Nonetheless, sporadic cases of illness that are not detected as outbreaks may potentially be attributable to shellfish consumption.

HAV

Infectious hepatitis, or hepatitis A, is perhaps the most serious viral illness transmitted by the ingestion of contaminated shellfish. HAV is currently classified as a member of the genus *Hepatovirus* of the family Picornaviridae and has many basic characteristics in common with the enterovirus group, i.e. single-stranded RNA genome, 28 nm diameter, and fecal-oral transmission. HAV causes a higher incidence of symptomatic infections than do enteroviruses, up to 95 per cent during outbreaks. Common symptoms of HAV infection are dark urine, nausea, vomiting, malaise, fever, chills, and jaundice. Fulminant hepatitis A infection, although rare, has a very high mortality rate. The high rate of HAV asymptomatic infections in children may contribute greatly to the nonepidemic prevalence of the disease in persons of other ages. Shellfish-associated hepatitis A is of greater concern in countries where there is widespread consumption of raw bivalve molluscs.

Human caliciviruses

Recent epidemiological estimates suggest that the human gastrointestinal viruses may be a more significant cause of food-borne disease than previously believed, perhaps being responsible for over 60 per cent of all cases. Generally, these viruses have a characteristic morphology of 32 cup-like depressions superimposed on an icosahedral symmetry and are 28 to 34 nm in diameter. Some human strains more commonly affect children, causing vomiting and diarrhoea, whereas other strains affect all age groups with more flulike symptoms (i.e. fever, malaise, and nausea). Attack rates are generally high. The Norwalk agent was the first recognised human gastroenteritis virus of clinical importance and is considered a prototype strain. This virus was discovered in 1972 in Norwalk, Ohio, by immune electron microscopy of an infectious stool filtrate. The first documented outbreak of shellfish-associated Norwalk virus gastroenteritis involved over 2000 persons throughout Australia. Norwalk virus has been reported consistently since 1980. More recent US outbreaks have been linked to the importation of depurated English clams, shellfish harvested from approved waters and grossly fecally contaminated waters, malfunctioning boat sewage disposal systems, and ill shellfish harvesters who routinely discharged their sewage overboard.

Since the discovery of the Norwalk virus, its infectivity and pathogenicity have been studied in adult human volunteers and its genome and that of related viruses have been sequenced. The genome size, buoyant density, and genomic organisation of these viruses are similar to those of other viruses in the family Caliciviridae, prompting investigators to classify these Norwalk-like SRSVs as belonging to this virus family. None of these human enteric viruses have been cultivated *in vitro*, and there exists no practical animal model for their propagation.

Other enteric viruses

A group of SRVs have been reported to be the cause of numerous outbreaks of shellfish-associated gastroenteritis. These viruses do not appear to be serologically related to the Norwalk virus or HAV. Diagnosis of SRVs is done by electron microscopy, and information on the characteristics of the viruses is virtually nonexistent. Astroviruses, given their name because of the characteristic five or six-pointed starlike form seen under electron microscopy have also been associated with bivalve mollusc consumption. HEV infection is transmitted mainly by sewage-contaminated water and person-to-person contact. The etiological agent of this disease, previously believed to be a calicivirus, has now been removed from the Caliciviridae family and placed in an 'unclassified' status. The disease caused by HEV can be more severe than that caused by HAV, with a high incidence of cholestasis and significant mortality (>15 per cent) in pregnant women during their second or third trimester. HEV is endemic in the Middle East, Africa, and India but not in the United States. The role of shellfish in the transmission of this disease is suspected but currently unconfirmed epidemiologically.

Enteropathogenic Protozoa

Protozoan parasites such as *giardia* and *cryptosporidium* species may be present in shellfish-growing waters as a result of contamination with animal farm runoff or as a result of treated and untreated sewage input.

METHODS FOR THE DETECTION OF PATHOGENS IN SHELLFISH

Many types of methodology have been used for the detection of pathogenic bacteria, viruses, and parasitic protozoa from bivalve molluscs. In this chapter, we present general processing methodology, including conventional detection approaches such as microscopic or cultural methods, as well as the newer molecular approaches that are increasingly applied to the detection of these pathogens in the bivalve mollusc food matrix.

General Steps in the Isolation and Detection of Pathogens from Shellfish

While the methods for the recovery and detection of micro-organisms in shellfish differ for bacteria, viruses, and parasitic protozoa, they all follow five or six general steps from sample collection to final identification (Table 9.1). In this chapter, only the steps which strongly impact the assay method are discussed. Sample collection and shucking as well as meat homogenisation are common to all analyses and are straightforward techniques, provided that they are done aseptically. A major difference between methods designed for virus and parasitic protozoa detection, compared to bacterial detection methods, is the need to purify and concentrate the pathogens away from the food matrix prior to testing. For viruses and parasitic protozoa, the approach used for pathogen extraction and concentration will not only affect the efficiency of recovery of targeted pathogens but will also influence the sample-associated toxicity or inhibition of the assay. In the case of bacterial pathogens, cultural enrichment steps are used to increase target pathogen numbers Although these methods may be considered reasonably standard, they can be complicated by high levels of background microflora or fastidious organisms, among which can be those capable of entering into a viable but nonculturable state. For subsequent molecular detection, nucleic acid extraction and concentration approaches also will affect detection accuracy. In all cases, assay inhibitors must be removed before application of molecular methods.

Table 9.1. General steps in the isolation of pathogens from shellfish.

Sample collection and shucking
Meat homogenisation
Virus extraction, cultural enrichment of bacteria or nucleic acid extraction and concentration
Removal of assay inhibitors
Assay by conventional or molecular methods

Various methods have been developed for the recovery of enteric viruses from bivalve molluscan shellfish and other seafood. Two general schemes have proven particularly successful, designated extraction-concentration and adsorption-elution-concentration. The goal in both cases is to separate viruses from shellfish meats, provide a low-volume aqueous solution that is free of cytotoxic material, and recover most of the viruses present in the shellfish sample. Both schemes employ conditions favouring the separation viruses from shellfish tissues, primarily through the use of filtration, precipitation, polyelectrolyte flocculation, and solvent extraction. However, the adsorption-elution-precipitation methods have received more common use in recent years. To this end, a modified adsorption-elution-precipitation procedure is described here. Briefly, 50 grams of oyster meats is diluted sevenfold with cold distilled water, the mixture is vigorously blended, and the viruses are adsorbed to the meat by reducing the conductivity to <2000 ppm and the pH to 5.0. After centrifugation at $1700 \times$ grams and 4°C for 15 minutes, the solid adsorbed viruses are eluted by resuspension in 350 ml of 0.05 M glycine– 0.14 M NaCl with subsequent pH adjustment to 7.5. Shellfish solids are removed by centrifugation at $1700 \times$ gram and 4°C for 15 minute, and viruses in the supernatant are concentrated by acid precipitation at pH 4.5. The resulting floc is sedimented by centrifugation at $1700 \times$ gram and 4°C for 15 minutes and resuspended in 10 ml of 0.1 M disodium phosphate buffer (pH 9.3 to 9.5), and then the pH is adjusted to 7.0 to 7.5. Cytotoxic components of the sample are precipitated by addition of Cat-Floc T to a final concentration of 0.1 per cent and removal of the precipitate by centrifugation at $3000 \times$ gram and 4°C for 20 minutes. The recovered supernatants are supplemented with gentamicin and kanamycin to final concentrations of 50 and 250 μg/ml, respectively, and the mixtures are incubated at room temperature for 2 hours. The final concentrates can be stored at –80°C until assay or further treatment. Various modifications have included further clarification of concentrates by ultrafiltration or freeze-thaw followed by centrifugation. The adsorption-elution-precipitation method is easily adapted to the different species of shellfish and has even been utilised for freshwater clams.

It is recommended that parasitic protozoa, like viruses, be concentrated from shellfish tissues prior to the application of detection methods. A method was developed by Toro that makes use of alkali for the digestion of the tissues, leaving the pathogens intact for their subsequent detection by use of immunofluorescence. The digestion of the tissue makes it possible to detect the intact oocysts and cysts while disposing of the tissue that would mask their presence, presumably due to matrix-associated inhibitors, or that would complicate their detection by microscopic visualisation. The approach involves the preparation of a shellfish homogenate in a waring blender (approximately 100 grams of shucked tissues) in the presence of 1 volume of 10 per cent (wt/v) KOH at 37°C for 30 minutes. This procedure will completely dissolve the tissues of some shellfish, depending on the type. The resulting solution can be centrifuged at $5000 \times$ grams for 10 minutes, the supernatant is discarded, and the pathogen-containing pellet is washed once by being mixed with PBS (0.01 M phosphate-buffered saline, pH 7.2) and then recentrifuged. After the wash step, the pellet is resuspended in a small volume of PBS, and detection methods can be applied. If the tissue is not completely dissolved, the homogenate is centrifuged (5000 × grams, 10 minutes) and the pellet is resuspended in a detergent solution (0.5 per cent sodium dodecyl sulphate [SDS], 0.5 per cent Tween 80 [wt/v]). The resulting mixture is then sonicated for 10 minutes in an ultrasonic bath at full power and centrifuged (1300 × grams). The supernatant is then poured into a fresh tube and recentrifuged (5000 × grams 10 minutes), and the pellet is resuspended in a small volume of PBS prior to the application of detection methods.

Conventional Methods for the Detection of Bacteria in Shellfish

Examination of food products for bacterial pathogens differs from the examination of clinical samples because of the potentially lower numbers of contaminants and the frequently poor physiological state of the organisms found in the food items. Conventional methods for the detection of bacteria in foods rely on the ability of the organism to multiply in specific selective enrichment media, followed by defined growth, morphology, or other distinguishing features on selective plating media. Therefore, detection of pathogens is dependent on both the availability of media specific for their isolation and the ability of the organism to grow. Consequently, conventional cultural methods most likely will fail to detect bacteria when they are in the viable but nonculturable state. Nonetheless, successful methods have been developed for the detection of culturable forms of *V. cholerae*, *V. parahaemolyticus*, and *V. vulnificus* in shellfish. The general approach involves preparation of a shellfish homogenate by 1:10 dilution in PBS (*V. vulnificus*), PBS supplemented with 2 to 3 per cent NaCl (*V. parahaemolyticus*), or alkaline peptone water (APW) (0.01 per cent peptone–0.01 per cent NaCl [wt/v], pH 8.5) (*V. cholerae*). APW dilutions of the resulting homogenate are made for enrichment and enumeration purposes. For *V. cholerae*, enrichment is performed by incubation of the homogenate and APW dilutions at 42°C for 6 to 24 hours; for *V. parahaemolyticus* and *V. vulnificus*, enrichment is done in a three-tube, multiple-dilution MPN format at 35° to 37°C for 16 to 18 or 12 to 16 hours, respectively. Selective plating is done with thiosulphate-citrate-bile salts-sucrose (TCBS) agar and/or modified cellobiose-polymyxin B-colistin (mCPC) agar. Typical colony colour, morphology and improvements in medium formulation are periodically reported. Preliminary biochemical tests for the identification of *V. cholerae* include the oxidase test, triple-sugar iron agar, Klinger's iron agar, arginine glucose slant, 1 per cent tryptone broth supplemented with 1 to 3 per cent NaCl, gelatine and gelatine salt agars, and Hugh Leifson glucose broth. For *V. vulnificus* and *V. parahaemolyticus*, biochemical confirmation is done by use of the oxidase test, motility test, arginine glucose slant, triple-sugar iron agar, O/129 disk assay, and the ONPG (*o*-nitrophenyl-β-D-galactopyranoside) test. Other biochemical tests may be appropriate and are commercially available. Standard serological tests are available for *V. cholerae* and *V. parahaemolyticus*, and various forms of immunological and molecular (DNA probe) detection technologies are in developmental phases.

Conventional Methods for the Detection of Enteric Viruses in Shellfish

The shellfish extracts obtained after extraction-concentration or adsorption-elution-concentration processes may be assayed for viruses by conventional mammalian cell culture techniques. There are two basic procedures for detection and quantification of viruses in cell culture, i.e. quantal methods (50 per cent tissue culture infective dose or MPN) and enumerative methods (plaque assay). Quantal methods rely on the development of widespread cell monolayer deterioration (cytopathic effects [CPE]), while enumerative methods detect localised cell monolayer damage under an agar overlay (a plaque).

Selection of the cell culture type depends almost entirely upon the viruses to be detected. Unfortunately, there is no single universal cell culture system that will detect all or even a majority of the more than 100 different human enteric viruses that can potentially contaminate shellfish. Several convenient and sensitive cell lines are commonly used for the detection of human enteroviruses, including the BGMK (buffalo green monkey kidney-derived), MA-104 (rhesus monkey kidney-derived), RD (human rhabdomyosarcoma-derived), and AGMK (primary African monkey kidney-derived) cell lines. Most studies on the incidence of enteric viral contamination in shellfish have been limited to the enterovirus family because of the ease of their detection in mammalian cell culture. Likewise, because of its ease of detection, Richards has proposed the use of poliovirus, a model human enterovirus, as an alternative indicator of virus contamination of shellfish and their harvesting waters.

However, it must be recognised that the epidemiologically important shellfish-borne enteric viruses do not replicate (NLVs, SRVs) or replicate poorly (HAV, astroviruses) in all currently used mammalian cell cultures. While the FRhK-4 (fetal rhesus monkey kidney-derived) cell line has useful in the propagation of a laboratory-adapted strain of HAV, its use in the detection of wild-type HAV is of limited value. At present, there are no mammalian cell culture lines available for the detection of NLVs, SLVs, SRVs, or astroviruses. Furthermore, lengthy assay time and high cost seriously limit the value of this detection methodology.

Conventional Methods for the Detection of Human Protozoan Parasites in Shellfish

Giardia and *Cryptosporidium* species cannot be grown in the laboratory in the same manner as bacteria; thus, other indirect methods need to be used for their detection. Traditional methods for the detection of parasitic agents from water sample concentrates have been based on immunofluorescent staining of filtered sample concentrates. In the few instances when these techniques have been applied to shellfish matrices, the samples have been prepared for immunofluorescence either by PBS washing of dissected gill tissue or else by using the alkaline digestion method of Toro.

Molecular Methods for the Detection of Pathogens in Shellfish

The advent of recombinant DNA technology has made it possible to use a segment of cloned DNA or RNA as the complementary sequence for the detection of viral genomes in clinical samples, infected cells, and, recently, environmental and food samples. Of particular value have been nucleic acid amplification techniques such as the PCR, since this provides the most feasible alternative for achieving the sensitivity limits needed for routine assay of viral contamination in environmental and shellfish samples. The method is readily adapted to the detection of RNA viruses by reverse transcribing the RNA before PCR amplification (RT-PCR). Nucleic acid amplification also offers the possibility of reducing overall detection time for culturable bacterial pathogens by replacing lengthy cultural enrichment steps with more rapid nucleic acid enrichment. Although there are a number of advantages to these molecular techniques (Table 9.2), it must be recognised that they are not yet ideal (Table 9.3). However, considerable progress has been made in adapting PCR and RT-PCR methods to the detection of enteric pathogens in shellfish.

Table 9.2. Advantages of molecular techniques for pathogen detection in shellfish.

More rapid detection of fastidious organisms (i.e. those which do not grow readily under laboratory conditions)

Detection of nonculturable pathogens

Simultaneous detection of multiple pathogens

Greater selectivity for the detection of groups of organisms or specific pathogens within a group

Table 9.3. Disadvantages of molecular techniques for pathogen detection in shellfish.

Lack ability to distinguish between viable and nonviable organisms

Environmental samples often need extensive preparation and cleaning prior to detection

Results are mostly qualitative

Quantitative techniques, when available, are usually very target specific

PCR methods for the detection of vibrio species in shellfish

V. vulnificus is an organism which can enter a viable but nonculturable state and is thus a prime candidate for detection methods which do not rely on conventional culture. Unfortunately, while studies have reported detection of pathogenic vibrios isolated from marine environments, only a limited number of studies have addressed molecular detection of the organisms in shellfish matrices. Hill used guanidinium isothiocyanate followed by chloroform extraction and ethanol precipitation to isolate DNA sequence of *V. vulnificus* from oyster homogenates seeded at the 100 CFU/g level. Koch used a short-term (6- to 8-hour) enrichment period to detect 10 CFU of seeded *V. cholerae* per gram of oyster meat from crude shellfish homogenates. Another study sought to detect a variety of bacterial pathogens (*Salmonella enterica* serovar Typhimurium, *V. vulnificus*, *V. cholerae*, and *V. parahaemolyticus*) in shellfish matrices by using a multiplex, nested PCR approach. In most of these studies, prior cultrual enrichment for up to 24 hours was required to achieve more desirable detection limits, but this may not necessarily be a drawback since it can save painstaking pathogen extraction procedures and provides some indication that the organisms may be viable. The need to confirm the identity of PCR amplification products by DNA hybridisation (oligoprobing) methods cannot be overemphasised, since shellfish matrices are notorious for promoting nonspecific DNA amplification which can result in false-positive results if not confirmed. Unfortunately, no methods specifically targeting pathogens in the viable but nonculturable state, or forgoing cultural enrichment procedures, have been reported at this time.

When further strain discrimination is desired, molecular typing methods can be used. For instance, pulsed-field gel electrophoresis (PFGE), random amplified polymorphic DNA (RAPD) analysis, restriction fragment length polymorphism (RFLP), and repetitive intergenic consensus sequence (ERIC) PCR have all been applied for further discrimination of *V. parahaemolyticus* strains. In the case of *V. vulnificus*, molecular typing methods have included PFGE, RAPD analysis, and ribotyping.

PCR methods for the detection of enteric viruses in shellfish

Shellfish meat represents a chemically complex matrix, containing many compounds which can interfere with the performance of PCR methodology. Three general approaches have been undertaken to reduce the level of interfering compounds and thereby facilitate the application of PCR to the detection of viral pathogens in shellfish. These approaches can be grouped into two categories. In the first category, termed direct nucleic acid extraction, total sample RNAs (including viral nucleic acids) are isolated and purified from the shellfish meat before application of RT-PCR. In the second category, termed virion concentration, the viruses are concentrated from the food matrix prior to RNA extraction and subsequent nucleic acid amplification using RT-PCR. Under this category, two virus concentration approaches have been used. One such approach is to capture the intact virion by using magnetic beads coated with antibodies specific to one or more epitopes of the virion capsid protein(s). The second approach involves a more general concentration and purification of intact virus particles from the complex matrix with simultaneous reduction in sample volume and removal of PCR inhibitors. Although the first category of approach, direct nucleic acid extraction, is usually a simpler procedure, destruction of the integrity of the virions during this process exposes naked viral RNA to potential degradation and destroys viral infectivity, thus precluding direct comparison with cell culture infectivity. A representative virus concentration and RT-PCR detection scheme is provided in Fig. 9.1.

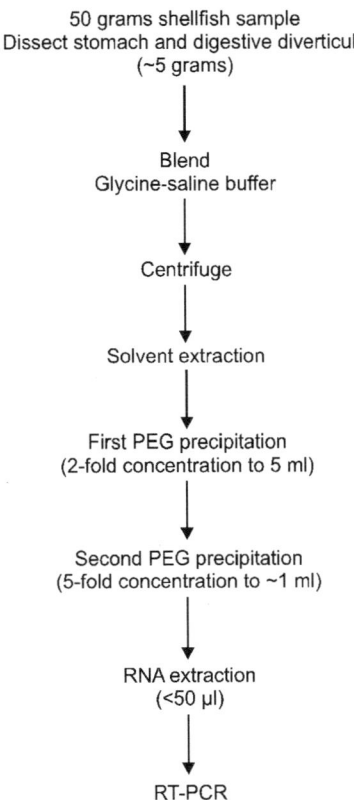

Fig. 9.1. Representative virus concentration and purification method for shellfish.

The most popular method for detection of viruses in environmental and shellfish samples has involved direct nucleic acid extraction. A frequently used method for the preparation of nucleic acid for hybridisation and RT-PCR involves extraction of RNA directly from shellfish extracts or environmental concentrates by SOS-proteinase K procedures, phenol-chloroform extraction, and ethanol precipitation. Further removal of inhibitors has been accomplished by treatment of crude nucleic acid extracts with the cationic detergent cetyltrimethylammonim bromide (CTAB). Briefly, nucleic acids from oyster extracts are obtained by initial digestion with proteinase K (50 μg/ml) in 10 mM Tris HCl (pH 7.5) 5 mM EDTA–0.5 per cent SDS at 56°C for 30 minutes. The digest is extracted twice with an equal volume of phenol-water-chloroform (68:18:14), and the aqueous phase is precipitated in ethanol. The resulting pellet is suspended in water, to which CTAB and NaCl are added to final concentrations of 1.4 per cent and 0.11 M, respectively. The mixture is incubated at room temperature for 30 minutes, and the nucleic acids are pelleted by centrifugation at 12100 × gram for 30 minutes. The pellet is resuspended in 1 M NaCl and precipitated in ethanol. The precipitated nucleic acids are suspended in 100 μl of water and used in RT-PCR in 1 to 20 μl aliquots. This method has been recently refined by use of dissected shellfish tissues (stomach and digestive diverticula) in place of whole shellfish and by the addition of an internal RNA standard to simultaneously evaluate the presence of RT-PCR inhibitors. The method has been tested collaboratively with shellfish tissues artificially contaminated with the Norwalk virus and reportedly has been used to detect virus in oysters implicated in a viral gastroenteritis outbreak.

The second category for virus detection from shellfish involves virion concentration from the food matrix prior to RNA extraction and subsequent nucleic acid amplification using RT-PCR. Early work in this area focused on developing methods to concentrate intact virions, with removal of enzymatic inhibitors, from oyster samples processed for virus isolation by use of adsorption-elution-precipitation methods. In this method, RT-PCR inhibitors were initially removed by two consecutive extractions using equal volumes of Freon. After centrifugation at $1200 \times$ gram and $5°C$ for 20 minutes, the resulting supernatant was adjusted to pH 7.3 to 7.4 and 0.3 M NaCl, supplemented with 4 to 12 per cent polyethylene glycol (PEG) 6000, and incubated overnight at $4°C$ to precipitate viruses. The precipitated viruses were recovered by centrifugation at $6000 \times$ gram and $4°C$ for 20 minutes and resuspended in one-seventh the original sample volume, using 50 mM Tris-0.2 per cent Tween 20, pH 9.0. The resuspended precipitates were held at room temperature for 30 minutes to aid virus elution and then subjected to centrifugation for 15 minutes at $8200 \times$ gram and room temperature to remove extraneous particulates. The resulting supernatant, with a volume of 1.5 to 2.0 ml, was a 10-fold concentrate of the initial extract volume, and these supernatants were termed PEG eluants. Viruses in 1 ml volumes of the PEG eluants were then precipitated for 15 minutes by addition of an equal volume of Pro-Cipitate. The solid phase (precipitate) was recovered by centrifugation for 15 minutes at room temperature and $13,800 \times$ gram. To elute adsorbed viruses from a precipitate, 4 ml of 50 mM Tris–0.2 per cent Tween 20 (pH 9.0) was added, and the mixture was gently rotated for 1 hour at room temperature and then centrifuged at $6000 \times$ gram and $15°C$ for 20 minutes to remove excess Pro-Cipitate. Viruses in the resulting supernatant were further concentrated by a secondary PEG precipitation using 5 to 10 per cent PEG and 0.3 M NaCl. After 2 hours at $4°C$ and gentle rotation, the solutions were centrifuged at $6000 \times$ gram and $4°C$ for 20 minutes. This virus-containing PEG precipitate was resuspended in 100 to 200 μl of 50 mM Tris–0.2 per cent Tween 20, pH 8.0, with 10 μl aliquots, representing 5 to 10 per cent of the total sample volume used in RT-PCRs. With this approach, processed samples could be readily assayed by both RT-PCR and cell culture infectivity, and the method has been successfully applied to the detection of enteroviruses in naturally contaminated field oyster specimens. More recent work has demonstrated that use of an increased Tris-Tween volume during elution of PEG pellets, sequential PEG precipitations, and/or the addition of other virus purification compounds, such as Viraffinity (LigoChem) may also aid in the removal of shellfish-associated PCR inhibitors while yielding acceptable overall sample concentration and virus recovery.

The above method has been somewhat simplified in recent years. One alternative has used PEG precipitation and Freon extraction to initially purify the viruses from shellfish samples as large as 50 grams, followed by extraction of total nucleic acids from 1 to 8 gram weight equivalents by use of guanidinium isothiocyanate and glass powder. The guanidinium isothiocyanate plays dual roles as a sample lysis buffer that also protects naked RNA from enzymatic degradation. This RNA can be further cleaned by being bound to glass powder, washed with ethanol and acetone, eluted in Tris buffer, and reprecipitated again in ethanol. Other investigators have reported similar guanidinium-based methods with additional RNA purification steps using silica gel or spin columns. These methods appear to result in significant removal of shellfish-related inhibitors with sample tolerances exceeding 5 grams equivalent weight. While precise detection limits were not documented, researchers have been able to detect human enteroviruses and NLVs in shellfish samples harvested from both pristine and fecally impacted sites and also from shellfish epidemiologically implicated in disease outbreaks. Similar methods have been reported for the detection of HAV in shellfish matrices.

An alternative, and perhaps more specific, approach in the general category of virion concentration has focused on initial virus purification by use of recognised methods such as ultrafiltration or polyelectrolyte flocculation followed by further concentration using antibody capture methods. Jansen and others were perhaps the first to report such an approach, when they detected HAV in stool specimens by an antibody capture-RT-PCR method. Desenclos and others implicated HAV in oyster outbreak specimens by immunocapture of virus, heat release of RNA, and subsequent RT-PCR, a method that has been further refined by Deng and Lopez-Sabater. These investigators initially purified HAV from 20 gram samples of artificially contaminated oysters and clams by elution in 100 ml of buffer, addition of 2 ml of a 1 per cent solution of Cat-Floc T, and sequential filtration through a GF/F filter and a 0.2-μm-pore-size membrane filter. Alternatively, viruses in the first filtrate were concentrated 10-fold (to 10 ml) by ultrafiltration through a PM 30 membrane followed by filtration through a 0.2 μm-pore-size membrane filter. The final sample concentrates were assayed for virus by antibody capture-RT-PCR. Basically, 0.5 ml polypropylene microcentrifuge tubes were precoated with diluted anti-HAV immunoglobulin G as previously described. Sample concentrate was added to each tube, and the tubes were incubated overnight at 4°C.

The tubes were washed six times with a Tris-KCl-Mg$_2$Cl buffer and then used directly in RT-PCR after heat release of the viral nucleic acid. In a later study, paramagnetic beads were coated with biotinylated human anti-HAV immunoglobulin G and the antibody-virus complex was gathered by use of a magnetic particle separator. This use of magnetic immunocapture resulted in a 500-fold sample concentration with removal of shellfish-related RT-PCR inhibitors. By use of this method, a detection limit of 10 PFU of HAV in 20 grams of oyster meat was reported for artificially contaminated product. A similar method, using magnetic beads coated with human antiserum, has been used for the detection of NLVs in environmental samples although applications to shellfish have not been reported.

None of the approaches under the two categories described above is ideal. For instance, although the RNA resulting from direct nucleic acid extraction is of relatively high purity, disadvantages to this method include incomplete recovery and/or potential degradation of RNA during the extraction procedure, incomplete removal of enzymatic inhibitors, and difficulty in making direct correlations with infectivity due to loss of virion integrity. The virion concentration approaches offer the advantage of significant sample volume reductions with recovery of infectious viruses, enabling direct comparison of detection by nucleic acid amplification and cell culture infectivity assay when available. In the case of antibody capture methods, a positive RT-PCR demonstrates the presence of antigen-associated viral RNA, which may be more highly associated with infectious virus than detection of free viral nucleic acid. Furthermore, previous studies have reported good correlation between detection by viral infectivity and RT-PCR when virion concentration approaches are used. However, these methods frequently require significant sample manipulations, many of which may result in substantial virus loss, which in turn may reduce overall detection limits. A further disadvantage of antibody capture may be the specificity of the approach, since only a single virus type can be detected in a single sample matrix. In reality, most recent advances combine sample concentration with effective nucleic acid extraction to further facilitate removal of RT-PCR-inhibitory compounds. In all cases, subsequent DNA oligoprobe hybridisation is necessary to confirm the identity of the RT-PCR amplification products. Representative studies in which these types of methods have been used to detect viral contamination in field, market, and outbreak-associated shellfish samples are summarised in Table 9.4.

Table 9.4. Recent studies reporting detection of human enteric viruses in field, market or outbreak-implicated shellfish.

Agent(s)	Food	Sample type(s)	Methods	Comments
NLV	Oysters (United Kingdom)	Field samples Outbreak samples Market samples (food only)	RT-PCR (single and nested) Southern hybridisation Sequencing	5/31 field samples positive 4/4 outbreak samples positive Nested RT-PCR increases detection sensitivity
NLV	Oysters (Japan)	Outbreak sample (clinical and food)	RT-PCR (nested) Cloning Sequencing	Coexistence of 2 different NLV genogroups in single oyster specimen
NLV	Oysters (US)	Outbreak samples (food only)	RT-PCR Southern hybridisation Sequencing	2/3 recalled outbreak specimens positive (GII)[a]
NLV Enterovirus Astrovirus Rotavirus	Oysters and mussels (France)	Field samples (food only)	RT-PCR Dot blot hybridisation Sequencing	Contamination rates for oysters were 17–27 per cent and for mussels were 13–52 per cent
Rotavirus	Oysters (France)	Market samples (food only)	RT-PCR (nested) Restriction digestion	Representative market oysters, mussels, and shrimp tested for HAV, human enteroviruses, NLVs, and rotavirus

Several investigators have combined traditional cell culture and molecular techniques by amplifying viruses from environmental samples in mammalian cell culture prior to RNA extraction of cell monolayers and subsequent nucleic acid amplification. This is termed a combined cell culture-PCR (or RT-PCR) procedure. For instance, Chung examined 31 oyster specimens harvested from fecally impacted waters and found 12 of 31 samples positive for human enteroviruses as determined by CPE and 6 of 31 positive by RT-PCR and oligoprobing of RNA extracts from CPE-negative cultures. In that study, HAV was also detected in RNA extracts of two CPE-positive samples by RT-PCR and oligoprobing. This approach accomplishes the dual purpose of increasing the number of copies of target nucleic acid and of incorporating an infectivity assay, but often its utility is limited by cytotoxicity associated with shellfish extracts, long testing times, high cost, and, most importantly, the absence of susceptible cell lines for some of the more epidemiologically important viruses such as the NLVs.

PCR methods for the detection of human protozoan parasites in shellfish

In general, considerably less effort has been put into the detection of parasitic protozoa from shellfish. Theoretically, the separation of cysts and oocysts from shellfish tissues by enzymatic digestion methods such as those described above would facilitate the direct isolation of nucleic acids from the final samples, taking into account both sample volume constraints and effective removal of inhibitors. Subjection of final sample concentrates to repeated cycles of washing in sterile water followed by sequential freeze-thaw facilitates nucleic acid extraction and removal of matrix inhibitors from shellfish samples. Only one published study has effectively detected *Cryptosporidium parvum* in commercial shellfish by using molecular methods, and this study used pooled haemolymph and gill washings that were extracted for DNA and subjected to nested PCR targeting amplification of 18S rRNA gene sequences. Species- and genotype-specific confirmation was done by RFLP and DNA sequence analysis. Indeed, species- and strain-specific typing methods for *Cryptosporidium species* are currently based on polymorphisms in the small-subunit RNA gene.

BARRIERS TO THE APPLICATION OF MOLECULAR METHODS FOR DETECTION OF PATHOGENS IN SHELLFISH

Although prototype molecular methods for the detection of human pathogens in shellfish have been reported, there are still several barriers to overcome before these methods are applicable to routine monitoring (Table 9.5). In order to obtain sample representation and detection sensitivity adequate for the low infectious dose of most enteric pathogens, in particular the viruses, large samples of shellfish will need to be processed. While some of the viral detection methods reported in the literature begin with large (50 grams) sample sizes, most effectively involve extraction from very small volumes of sample aliquots, which limits the sensitivity of the assay procedure from the very beginning. The approaches then applied to concentrate and purify the pathogens from the homogenate not only need to be reasonably efficient but also need to produce a concentrate that is low in both volume and enzymatic inhibitors. To complicate matters further, the prototype extraction and detection procedures differ for bacteria, viruses, and parasitic protozoa, as well as for different shellfish species. It appears that shellfish have extremely high levels of enzymatic inhibitors, with complex carbohydrates (particularly glycogen) being the major culprit. These compounds have remained recalcitrant to almost all removal processes, applied. In addition, the type of enzymatic inhibitors may differ with the shellfish species, potentially restricting the use of some methods to a single species of shellfish. The type and concentration of enzymatic inhibitors may also differ with harvesting season and geographic location.

Table 9.5. Barriers to the application of molecular techniques to the detection of pathogens in shellfish.

Large sample volumes of shellfish are required to obtain the necessary level of detection sensitivity

Extraction procedures differ for each pathogen group

High levels and a diversity of enzymatic inhibitors exist in shellfish meats

Enzymatic inhibitors may differ with shellfish species and harvesting location

Concentrations of inhibitors may vary with seasons

RESEARCH NEEDS FOR THE DETECTION OF PATHOGENS IN SHELLFISH

Research is needed in order to develop and refine the prototype protocols into collaboratively tested methods which could be used routinely and expeditiously to evaluate the microbiological safety of shellfish products (Table 9.6). In general, future, research needs for the routine application of PCR methods to the detection of microbiological contamination in shellfish include development of the following: (i) simple, rapid, and cost-effective pathogen extraction procedures, (ii) simple and reliable methods for the removal of enzymatic inhibitors, (iii) methods which are not restricted by shellfish species or harvesting location or season, and (iv) quantitative approaches for assessing the relative levels of contamination. Furthermore, additional experimentation is needed to establish the relationship between detection by molecular amplification techniques and the presence of infective virus particles or parasitic protozoan cysts and oocysts in shellfish and environmental samples. With sufficient developmental effort, effective methods for the rapid detection of human pathogenic micro-organisms in shellfish at naturally occurring levels of contamination should be available in future years.

Table 9.6. Future research needs to facilitate pathogen detection in shellfish.

Simple and rapid pathogen (or nucleic acid) extraction methods

Simple and reliable methods for the removal of enzymatic inhibitors

Standardised methods whose applicability is broader, i.e. not specific to just a single shellfish species or harvesting location

Quantitative methods for more readily assessing relative levels of contamination

Chapter 10

Control of Micro-organisms in Source Water and Drinking Water

INTRODUCTION

A more effective approach to ensuring potable water is a multibarrier approach. The multiple barriers should include source protection to prevent the entry of pollutants and pathogens, proper treatment and disinfection to reduce the numbers of pathogens and remove other pollutants, and proper distribution maintenance to ensure that treated water does not become recontaminated before it reaches the consumer.

Traditionally, the preparation of potable surface water has been done by a disinfection process. Chlorine and chlorine-based compounds have been around the longest and are probably the best known. Other methods for preparing potable water, such as ozone disinfection, reverse osmosis, and distillation, have been used intermittently in the United States and more widely in other countries. Recently, UV light has gained popularity for disinfecting ground and surface water.

In the last decade, however, there has been a growing change in the philosophy governing how we care for our source waters. Disinfection alone is a myopic approach and is no longer considered adequate. When drinking water sources become contaminated, the cost to industry and public health is high because of the need for additional treatment processes. There is a growing recognition that effective drinking water system management includes addressing the quality and protection of water sources. The 1996 safe drinking water act amendments go beyond monitoring and treating contaminated water to emphasising pollution prevention. They direct state drinking water agencies to examine the sources of drinking water as an additional layer of protection.

CHOOSING WATER SOURCES AND WATERSHED PROTECTION

Selection of a water treatment process is a complicated task that ultimately must result in delivery of potable water which is acceptable to the consumer at a reasonable cost. Factors which must be considered include the water supply, source quality, desired finished water quality, reliability of processing equipment, operational requirements and personnel capabilities, flexibility in dealing with changing water quality and equipment malfunctions, available space for construction of treatment facilities, waste, disposal constraints and capital and operating costs (including chemical availability).

Surface Water Sources

One problem associated with surface water is its ability to change rapidly as stresses are placed on the surrounding environment. Treatment processes for surface water sources must be able to respond to changes in the water quality. Runoff from rain can rapidly increase turbidity, thus decreasing the efficiency of existing disinfection practices. Spills and runoff of surface contaminants, such as fertilisers and insecticides, can inhibit treatment plant operations. Additional problems can arise from biological contamination, such as algal blooms, insects, and animal intrusion. Once a surface water source is chosen, consideration must be given to natural processes in relation to the placement of intake pipes. Spring and fall turnover can turn pristine waters with low nutrient loads into turbid waters with high concentrations of organics. During summer months, surface waters may stratify into distinct layers so that warmer water stays near the surface and cooler water is trapped below with little intermixing, resulting in anoxic conditions and solubilisation of iron and manganese as well as the production of sulphur compounds from anaerobic bacterial processes. Treatment plants must be able to respond with additional processes, such as increased sedimentation or flocculation times and perhaps the use of ion-exchange resins to remove manganese ions, as well as additional disinfection in order to maintain potable water which is acceptable to the consumer and free of risk.

There are several advantages to using surface waters as a source, if they can be exploited. The location and elevation of the water sources may offer the advantage of gravity flow to the treatment facility. Visibility of the water supply allows daily evaluation. Water levels and sources of contamination can be less costly to identify than those of groundwater. Finally, surface waters that are properly managed, which can be an enormous task, can also be used for recreational purposes.

Groundwater

Groundwater supplies are derived primarily from wells. Water from shallow or 'hand-dug' wells is not considered groundwater because it is usually under the influence of surface water via runoff and infiltration and hence can have many of the characteristics of surface water. True groundwater, in contrast, is relatively constant from season to season and is usually not affected by changes in the surface environment. This does not mean that variation in water quality from well to well does not exist. Depending on hydrological influences, groundwater quality can vary greatly between wells. In general, groundwater is more expensive than surface water, in large part because of the pumping requirements, but groundwater usually requires less treatment than most surface water with respect to microbial pathogens, organic matter, and turbidity. The microbial pathogens of concern in groundwater are the enteric viruses. Viruses, unlike bacteria and protozoan cysts, are not always retained by the soil matrix and, if environmental conditions (i.e. soil type and pH, rain events, and cation concentration) allow, can migrate to underground aquifers. Additionally, viruses which are complexed with particulate matter and kept in a cold environment have been documented to migrate great distances from the source of contamination and to survive for long periods outside the human host.

Watershed Protection

All too often, the burden of controlling pathogens in drinking water has been placed on treatment plant operations. A more comprehensive approach is the use of multiple barriers. One of the first barriers in place should be able to control the entrance of microbes into source waters via watershed protection. The watershed protection approach (WPA) of the Environmental Protection Agency (EPA) is a departure from what has been previously done and the way in which central, state, and local agencies have viewed or regulated natural resources. At the national level, the Safe Drinking Water Act and the Clean Water Act both directly regulate the introduction of contaminants into the nation's surface waters and groundwater. At the state and local levels, regulatory programmes can vary but are usually designed to protect source waters by restricting and/or regulating certain activities which could degrade their quality.

Upon closer examination, managing watersheds can be divided into nine subcategories: management units, management cycles, stakeholder involvement, strategic monitoring, assessment, prioritisation and targeting, development of management strategies, basin or watershed plans, and implementation. Management units are areas that are delineated by the state and contain many watersheds. Management cycles pertain to monitoring the management units. Each unit is examined cyclically so that all of a state's management units are examined in a set period, typically 5 years. Stakeholder involvement allows for more individuals and groups to become involved in the outcome of one or more watersheds. The theory behind more involvement is that the greater the number of individuals who have a concern about water quality, the more likely it is that there will be a concerted effort to maintain that water quality.

Strategic monitoring allows the identification of a problem and a determination of its extent. While increased monitoring helps to ensure watershed quality, monitoring of one watershed should not come at the price of another. Hence, monitoring should be done on a rotating basis and is contingent upon the number of watersheds and the available funding. Data analysis or assessment of a watershed help to identify problems and sources of contamination. Prioritisation and targeting allow watersheds to be ranked according to resource value and degree of impairment. The next two subcategories, development of management strategies and watershed plans, deal with the goals and assessment of such goals for the watershed. The goals for a watershed can include how the watershed will be used while still maintaining its quality. It is in this part of the plan that educating the public is helpful. The last phase is implementation. Here, the plans that have been developed and brought forward are put into place.

In designing an organisational and management plan, certain physical attributes of a wastershed, including its characteristics, the size and configuration of the reservoirs, and the flexibility of the system, must be taken into consideration.

Additionally, watersheds need to develop partnerships with local agriculture to limit or restrict the introduction of farm animal waste into the watershed. This approach to watershed management addresses both the introduction of pathogens into the watershed and the individual farmer's operational and financial circumstances.

Microbial monitoring remains one cornerstone of any watershed quality assessment. However, there are several aspects of a monitoring programme which need to be considered before the programme can make a substantial contribution. The programme design should include frequencies for sampling analysis, protocols, and the types of analysis needed. Detection of pathogens alone is not an adequate monitoring tool. Pathogens usually occur in low concentrations, and the analytical techniques available for their detection are often not very reliable or sensitive. This is not to underestimate the importance of pathogen monitoring but simply to emphasise that there still does not exist one 'magic bullet', and monitoring must combine the evaluation of several different organisms. Data collected by any monitoring programme must be interpreted carefully. While bacteria have long been used as the sole indicator of microbiological water quality, enough evidence has been established over the years to indicate that bacteria, while able to predict the presence of other bacterial contaminants or the introduction of fecal material, are not adequate predictors for the presence of viruses or protozoa. To accurately assess the presence of these organisms, they should be evaluated separately. Also, since any source water is a dynamic body and because pathogens usually occur in low numbers, a monitoring schedule covering at least 1 year, and preferably more, should be adhered to, and results must be evaluated at the end of that period. The monitoring programme must be designed to take into account the changing properties of the source water. For instance, if the water is used for recreational activities during the summer months, the frequency of monitoring should reflect the increased usage. The same may hold true for dry- versus wet-weather events. Discharges and runoff may occur only during heavy rains, and the pathogens that they bring may not be detected if monitoring is done only during the dry period.

Looking at all the parameters of watershed protection, breaking it down into many different parts, and ultimately trying to minimise or eliminate source water contamination completely are more beneficial and cost-effective than water treatment alone.

Watershed protection does not begin and end with surface waters alone. Groundwater accounts for more than 97 per cent of available freshwater supplies. On average, groundwater supplies approximately 40 per cent of the stream flow in the country. Groundwater and surface water are integral and related components of the hydrologic cycle, yet these resources are often managed by different programmes in central, state, and local agencies, where protection of one resource may have been achieved at the expense of another. Additionally, in many cases, drinking water wells which are constructed in shallow alluvial aquifers are influenced by surface water. Some of these wells actually draw from the hyporearic zone (that area of the river and sediments where there is constant interchange of water).

Groundwater presents different challenges than surface waters. Determining the source of groundwater contamination can be quite difficult. Due to retention and transport times, groundwater contamination is more characterised by non-point source contamination. Monitoring wells can be expensive, and sampling underground aquifers over a long period of time is labour-intensive.

TREATMENT AND DISINFECTION

The threat of microbiological contaminants in drinking water is eliminated by three complementary strategies: (i) preventing their access to the water source, (ii) employing water treatment to reduce their concentration in the water, and (iii) maximising the integrity of the distribution system for finished water. In recent years, greater emphasis has been directed toward providing effective water treatment to reduce microbiological contaminants. Today, there is increasing emphasis on employing both source protection and treatment to ensure that safe water is produced and on improving distribution system integrity to ensure that contamination does not occur during transport from the treatment plant to the consumer's tap. In the water treatment process, reducing microbiological contaminants is accomplished by two basic strategies, removing them from the water or inactivating them. Inactivated micro-organisms, although still present in the water, are no longer able to cause disease in the consumer. Processes that use inactivation as their strategy are traditionally referred to as disinfection.

In water works practice, the term disinfection is used to refer to two activities: (i) primary disinfection—the inactivation of micro-organisms in the water, and (ii) secondary disinfection—maintaining a disinfectant residual in the treated water distribution system (residual maintenance). The characteristics that make a disinfectant, therefore, for each of these purposes are not the same.

New treatment processes have also come to the fore that show promise for the removal or inactivation of chlorine-resistant organisms and others as well. Membranes are capable of removing pathogens much more effectively than traditional physical treatment processes like coagulation and granular media filtration. In fact, the removals that have been demonstrated using membranes are on the same order as the magnitude of inactivation of bacteria customarily achieved by chlorine. Disinfection with ultraviolet light (UV) is also effective for inactivating *Giardia* and *Cryptosporidium*. While chlorine remains the dominant drinking water disinfectant and disinfection (inactivation) remains the cornerstone of water treatment, this situation may change in the future.

METHODS OF DISINFECTION COMMONLY USED IN WATER TREATMENT

Five disinfection agents are commonly used in drinking water treatment today: (i) free chlorine, (ii) combined chlorine (chlorine combined with ammonia), (iii) chlorine dioxide, (iv) ozone, and (v) UV light. The first four agents are chemical oxidants whereas UV light involves the use of electromagnetic radiation. Of the five, by far the most common is free chlorine. Combined chlorine is also common, but its use is often limited to residual maintenance. Ozone is the strongest of the four oxidants and its use is becoming increasingly common, in part because of its stronger disinfecting properties and in part because it controls taste and odour compounds, specifically geosmin and methyl isoborneol. Information on each of these common disinfectants is summarised in Table 10.1.

Table 10.1. Characteristics of five most common disinfectants.

	Disinfectant				
Issue	*Free chlorine*	*Combined chlorine*	*Chlorine dioxide*	*Ozone*	*Ultraviolet light*
Effectiveness in disinfection					
Bacteria	Excellent	Good	Excellent	Excellent	Good
Viruses	Excellent	Fair	Excellent	Excellent	Fair
Protozoa	Fair to poor	Poor	Good	Good	Excellent
Endospores	Good to poor	Poor	Fair	Excellent	Fair
Frequency of use as primary disinfectant	Most common	Common	Occasional	Common	Emerging use
Regulatory limit on residuals	4 mg/l	4 mg/l	0.8 mg/l	–	–
Formation of chemical by-products					
Regulated by-products	Forms 4 THMs[a] and 5 HAAs[b]	Trace of THMs and HAAs	Chlorite	Bromate	None
By-products that may be regulated in future	Several	Cyanogen halides, NDMA	Chlorate	Biodegradable organic carbon	None known
Typical application					
Dose, mg/l (kg/ml)	1–6	2–6	0.2–1.5	1–5	20–100 mJ/cm^2
Dose, lb/mg	8–50	17–50	2–13	8–42	–
Chemical source	Delivered: as liquid gas in tank cars, 1 tonne and 68 kg (150 lb) cylinders or as liquid bleach. Onsite generation from salt and water using	Same sources for chlorine. Ammonia is delivered as aqua ammonia solution, liquid gas in cylinders or solid	ClO_2 is manufactured with an on-site generator from chlorine and chlorite. Same sources for chlorine. Chlorite as powder or stabilised liquid	Manufactured on-site using a corona discharge in dry air or pure oxygen. Oxygen is usually delivered as a liquid. Oxygen is also manufactured	Uses low-pressure or low-pressure high-intensity UV (254 nm) or medium-pressure UV (several wavelengths) lamps in the contactor itself

(Contd ...)

Issue	Disinfectant				
	Free chlorine	Combined chlorine	Chlorine dioxide	Ozone	Ultraviolet light
Bacteria	Excellent electrolysis. Calcium hypochlorite powder is used for very small applications	Good ammonium sulphate. Chlorine and ammonia are mixed in treatment process	Excellent solution	Excellent on-site in some large plants	Good
Typical contactor	In the past was added at beginning of plant and residual carried through. Increasingly engineered contactors are used	In the past was added at beginning of plant and residual carried through. Increasingly engineered contactors are used	In the past was added at beginning of plant and residual carried through. Increasingly engineered contactors are used	Has always been added in specially engineered contactors. These contactors are using more compartments and other techniques are being experimented with	Lamps are placed in gravity channels or in specially manufactured UV reactors. Because the contact time is so short, reactors must be tested for short circuiting

[a] THMs = trihalomethanes.
[b] HAAs = haloacetic acids.

All five disinfectants are best applied as a separate unit process. The three forms of chlorine are most often used in baffled, serpentine contact chambers or long pipelines when these are available. Both types of reactors can be designed to be highly efficient, closely approaching ideal plug flow. Ozone is generally introduced in bubble chambers where short circuiting can be a bigger problem. As a result, designs of ozone reactors often employ multiple chambers in series.

KINETICS OF DISINFECTION

For chemical disinfectants like chlorine, combined chlorine, chlorine dioxide and ozone, the specific mechanisms of micro-organism inactivation and the reaction kinetics are not well understood. The action of each chemical disinfecting agent depends on the properties of each micro-organism, the disinfectant and the water. As will be shown later, the reaction rates that have been observed can vary by as much as six orders of magnitude from one organism to the next, even for one reactant. Even for disinfection reactions where the reaction mechanism is well understood, for example, UV light, reaction rates vary by one and one-half orders of magnitude. Nevertheless, there is one simple kinetic model that is widely used and there is enough commonality in the behaviour of all these reactions to allow the development of some phenomenological laws that are useful in modelling the course of all of these reactions. As these disinfection processes are physio-chemical processes, they are also subject to the rules of analysis.

CHEMICAL METHODS

Chlorine

Chlorination has been the major line of defence against waterborne disease outbreaks in the United States and more widely in other countries. Chlorination is probably the oldest and most widely used form of drinking water disinfection. It has several advantages which make it appealing and have bolstered its popularity. Chlorine is relatively inexpensive and easily obtained, and it provides residual protection in distribution systems. Chlorine is a strong oxidising agent and can be used to modify the chemical character of water. In water, chlorine gas hydrolyses to form hypochlorous acid (HOCl). The hypochlorous acid undergoes further ionisation to form hypochlorite ions (OCl⁻).

Equilibrium concentrations of HOCl and OCl⁻ depend on the pH of the water. Alkaline pH shifts the equilibrium to the right, causing the formation of higher concentrations of HOCl. Both HOCl and OCl⁻ are commonly referred to as free chlorine. Upon chlorination of water, a portion of the chlorine reacts with compounds in the water, such as organics. The difference between the chlorine concentration applied to the water and that of the free chlorine is known as the chlorine demand (i.e. the concentration of chlorine added minus the concentration of free chlorine resulting) and is one of the physical attributes used to define the level of water treatment needed.

One of chlorine's greatest attributes is ease of use. While chlorine gas requires special handling, there are two forms of chlorine which are readily available and easy to use. One form is calcium hypochlorite $[Ca(OCl)_2]$ which is the predominant dry form and when dissolved in water contains approximately 70 per cent available chlorine. Sodium hypochlorite (NaOCl) is available in liquid form at concentrations between 5 and 15 per cent.

When chlorine reacts with ammonia in water, the resulting reactions can form chloramines: monochloramine, dichloramine, or trichloramine. The specific reaction products that are formed depend on the temperature and pH of the water, the reaction time, and the initial chlorine-to-ammonia ratio. In general, below pH 4.4, trichloramine is produced; above pH 8.5, monochloramine usually exists alone; and between pH 4.5 and 8.5, both mono-and dichloramine exist. Chloramines are thought to be less effective as disinfectants than hypochlorite.

The ability of hypochlorite to destroy microbial life is predominantly due to the ability of HOCl to oxidise proteins and other structures found on bacteria, viruses, and protozoa. Since HOCl is neutral and has a relatively low molecular weight, it is able to penetrate the organism with relative ease and hence has a high germicidal activity. The microbicidal activity of chlorine is greatly reduced at high pH, probably because at an alkaline pH, the predominant species of chlorine is OCl⁻. The negative charge of this molecule may prevent it from penetrating or coming in close contact with surface proteins as a result of electrostatic repulsive forces that exist between the chlorine and the carboxyl ends of proteins. Organisms that are relatively large and contain a 'waxy' outer coat, such as some protozoan cysts, may be more resistant to the oxidising potential of chlorine or require much longer contact times and hence not be adequately controlled by CT values of chlorine which have been established for the inactivation of bacteria (CT values are defined by the concentration of the disinfectant and the contact time). Leahy showed that 2.8 mg of chlorine per litre achieves a 99 per cent inactivation of *Giardia muris* cysts in 16 minutes of contact time at pH 7 and 25°C. Studies by Korich, demonstrated that 80 mg of chlorine per litre requires more than 90 minutes of contact time for a 90 per cent inactivation of *Cryptosporidium parvum* oocysts. All of these studies were done at the bench level in static systems. Work by Tilton has demonstrated that there is very little correlation between the CT developed with a static bench scale system and that developed with a dynamic pilot plant system: the latter usually requires a much greater CT than is indicated by static bench scale systems.

Advances in the microbiological examination of water have shown that hypochlorite is also less effective at inactivating viruses than at killing total and fecal coliforms, the historical indicator organisms. This may be in part because viruses do not rely on any metabolic (enzymatic) functions for survival in the environment. Harakeh reported that of six viruses tested, coxsackievirus B5 was the most resistant, with 99.99 per cent inactivation achieved at a dose of 18 mg of free chlorine per litre after 5 minutes of contact, whereas simian rotavirus was the most sensitive, with a dose of 5 mg of chlorine per litre required to obtain 99.99 per cent inactivation.

While chlorine has been widely used over the decades, it is not without its drawbacks. Since the early 1970s, it has been known that undesirable disinfection by-products (DBPs) can be produced when free chlorine interacts with selected organic precursors such as humic substances. This interaction produces a group of single-carbon compounds with halogen suustitutions referred to as total trihalomethanes (THMs). Because of the potential carcinogenicity of the THMs, chloroform in particular, the EPA has established a maximum contaminant level of 80 μg/litre for total THMs. Approaches to control THM formation are the removal of organic precursors that react with the chlorine, removal of THMs from finished water, and the use of alternative disinfectants.

Control of Gas Chlorination

Four methods have traditionally been used for controlling the feed rate of chlorine gas when it is being used for residual control in drinking water systems. Each is given in Fig. 10.1: (i) manual control, (ii) feed-forward control, (iii) residual feedback control, and (iv) compound loop control. Through the middle of the twentieth century manual control was the most common. A great deal of attention was required by the operator to ensure that a suitable residual was reliably provided, especially when the flow rate through the plant was adjusted. By the mid-1950s flow measurement and chlorine metering techniques improved to the point where feed-forward control began to appear. This important advance allowed automatic adjustment for flow but still required the operator to adjust for any changes in the water quality (chlorine demand) or any drifts in monitoring and feed rates.

By the mid-1960s direct residual control began to appear as well. In principle, the feedback method of control is more robust than feed-forward control, but residual measurement did not approach suitable levels of reliability and precision for two more decades. As a result, compound loop control evolved as a compromise. With this method, changes in flow could be accommodated via the flow signal and an additional control increment could be added via residual control for minor water quality changes. Properly maintained, compound loop control was the first system to provide reliable, continuous residual control. During the last decade of the twentieth century, computerised supervisory control systems had evolved to the point where these same inputs (flow and residual) could be combined with other measurements as well to provide improved reliability. None of these control systems, however, is an adequate substitute for vigilant attention from the operator.

Residual control system

With all of these alternatives, the sequence of events in the residual control system must be carefully designed and controlled. All the elements shown and labelled in the diagram as 'compound loop control' in Fig. 10.1 must be considered in designing the system and envisioning its method of control. The instructions of the supervisory control and data acquisition (SCADA) system must be conceived with a full understanding of the range of timing that can occur in all the other elements of the system.

The time required for the chlorinator to completely respond to an instruction from the SCADA must be considered. Normally this time is not too significant. The time required for the change in feed rate established by the chlorinator to be recognised at the eductor must be considered. This time to change feed rates is not normally too long either, but it can be too long when the chlorinator is located a long way from the injector and when the chlorine feed rate is very low. White suggests that this time be estimated by assuming that the change in pressure will be accommodated in a wave that travels about three times as fast as the gas flow in the line.

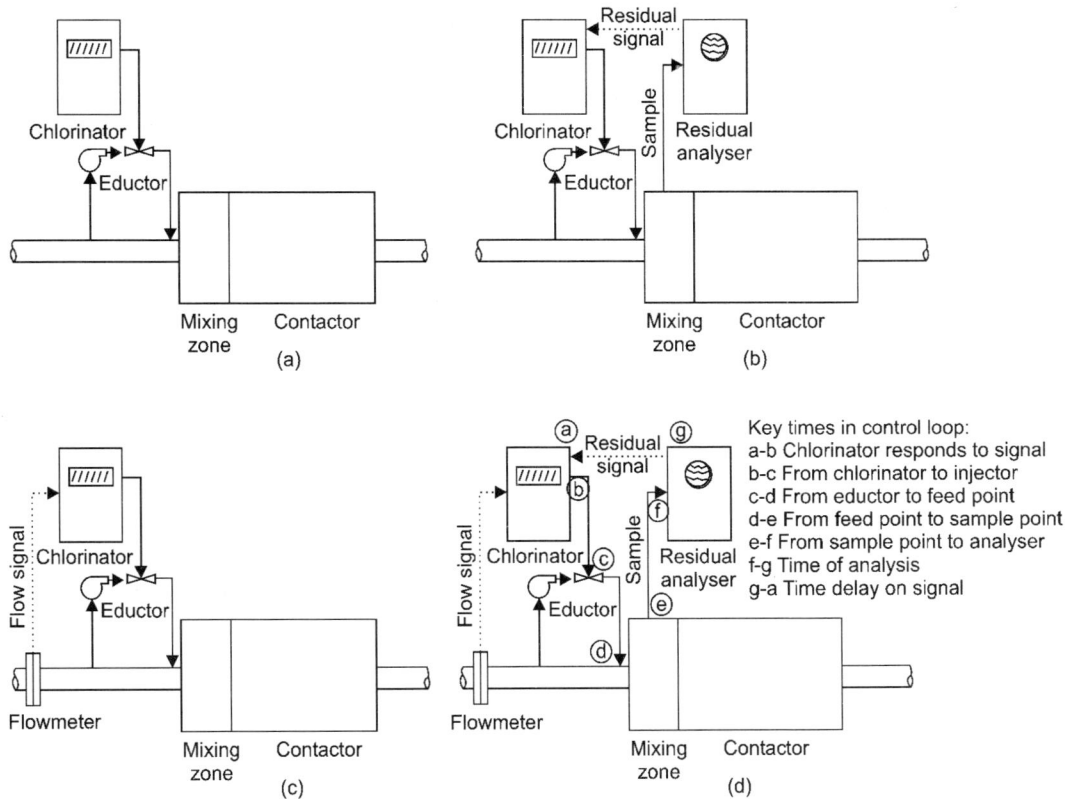

Fig. 10.1. Control of chlorine gas feed rate: (a) manual control; (b) feedback or residual control; (c) feed forward control; and (d) compound loop control.

Next, the time for the water in the chlorine water line to travel from the eductor to the application point must be considered. This time is a function of the distance between the eductor and the application point and the flow rate (velocity) in the chlorine water line. Again, designs with long distances between the eductor and the application point can cause trouble. Ideally the chlorine is stored near the application point so that both the time in the vacuum line and the time in the chlorine line are minimised. When nearby storage is not possible, it is usually best to lengthen the vacuum line, not the chlorine water line, as a signal ordinarily travels much faster down a vacuum line.

Sampling point

Another important constraint is the time between the application point and the sampling point. There is an inherent design conflict where these two times are concerned. Putting these points too close together can result in poor blending before the treated water reaches the sampling point. When this happens, the control system constantly searches for control but can never quite find it. Putting them too far apart can result in control-cycling problems of another kind. To avoid control problems, the residual for sample control must be taken after mixing is satisfactory. Depending on the method of chlorine introduction and the criteria used for mixing, the distance downstream to accomplish satisfactory blending is between 40 and 200 times the hydraulic radius of the water conduit. Because the design velocities in these conduits are usually constant, this problem becomes more aggravated with larger applications because of the larger conduit diameters that are used.

The time required for the sample to travel from the sample point to the analyser is often not considered either. Sample travel time can be a significant complication if the designer decides to locate the analyser in some central location far from the sampling site. The time for the analyser to assay the sample (normally between 15 and 20 seconds) can also be important in some applications. In designing such a control system, it is important to analyse all these times and the sequence in which they operate at both high and low-flow conditions, both early and late in the life of the design, to ensure that problems do not occur after the installation is complete.

Sodium Hypochlorite

When chlorine was first used for disinfection, it was often used in the form of hypochlorite. Calcium hypochlorite $Ca(OCl)_2$ is still used, particularly in small utilities, but liquid bleach or sodium hypochlorite (NaOCl), which came into use near the beginning of the Great Depression in the late 1920s, is the most widely used form of hypochlorite today. It is widely used not only in disinfection of water but also for a myriad of other household and industrial uses.

High-test calcium hypochlorite takes up less storage space than does liquid bleach, but liquid bleach is an easier chemical to handle and feed. Whereas chlorine gas is prepared by an electrolytic process that breaks sodium chloride solution into chlorine gas and sodium hydroxide, ironically, sodium hypochlorite is generally prepared by mixing sodium hydroxide and chlorine gas together:

$$2NaOH + Cl_2 \longrightarrow NaOCl + NaCl + H_2O \qquad ... (10.1)$$

On a weight basis, 1.128 kg of NaOH reacts with 1 kg of chlorine to produce 1.05 kg of NaOCl and 0.83 kg of NaCl. The process is complicated by the fact that the reaction generates a significant amount of heat. It is common practice to add an excess of NaOH because, as will be shown, hypochlorite is more stable at higher pH values. As a result, the density of one hypochlorite solution is not necessarily the same as another, even if both are of the same strength (per cent Cl_2). This density difference occurs because the final density depends on the amount of excess NaOH added during manufacture. Liquid bleach is usually delivered with a pH between 11 and 13. Hypochlorite can also be manufactured via onsite generation and this process is receiving more attention.

Stability of hypochlorite

Under some conditions, the strength of hypochlorite can significantly decline in just a few days. In fact, stability is one of the major issues that must be addressed in both designing and operating a hypochlorite facility. A utility should not undertake using hypochlorite unless it is prepared to dedicate time and energy to a regular programme of monitoring and controlling its decay. Of considerable significance is the fact that, when hypochlorite does decay, chlorate ion is one of the principal by-products of the reaction. The stability of hypochlorite is affected by the strength of the solution, the temperature at which it is stored, the pH and the contamination of heavy metals, which can catalyse its decay. Light is also a problem. As a general rule, the following actions accelerate the rate of decay:

1. Storing the hypochlorite at higher concentrations.
2. Storing the hypochlorite at lower pH.
3. Storing the hypochlorite at higher temperatures.

4. Storing the hypochlorite where it is exposed to sunlight.
5. The presence of certain heavy metals, notably copper and nickel.

Under these basic conditions the decomposition of hypochlorite ion to chlorate ion follows a disproportionation reaction, which exhibits second-order reaction kinetics and the following overall stoichiometry:

$$OCl^- + OCl^- \longrightarrow ClO_2^- + Cl^- \qquad \qquad ... (10.2)$$

$$OCl^- + ClO_2^- \longrightarrow ClO_3^- + Cl^- \qquad \qquad ... (10.3)$$

$$3OCl^- \longrightarrow ClO_3^- + 2Cl^- \qquad \qquad ... (10.4)$$

The second reaction, as given by Eq. 10.3, is the faster of the two. As a result, the first reaction is the rate-limiting step in the consumption of hypochlorite ion. Bleach also decays via a slow reaction that forms oxygen and an acid forming reaction that also forms chlorate ion. These reactions are shown below:

$$OCl^- + OCl^- \longrightarrow O_2 + 2Cl^- \qquad \qquad ... (10.5)$$

$$2HOCl + OCl^- \longrightarrow ClO_3^- + 2H^+ + 2Cl^- \qquad \qquad ... (10.6)$$

The pH at which a sodium hypochlorite solution is stored has important impacts on its rate of decay. Temperature is also an important consideration. Like most reactions, the decay of hypochlorite accelerates as the temperature increases. Gordon examined this question by comparing the rate of decay of three commercial bleaches, all at approximately the same temperature. In this work it was demonstrated that temperature is important and also that the rate of decay of most modern good-quality commercial bleaches is comparable.

It was discovered that hypochlorite solutions containing significant concentrations of chlorate ion were responsible for introducing chlorate ions into drinking water.

Feeding sodium hypochlorite

Experience with materials for the construction of large hypochlorite tanks has not been uniformly good. Early projects in Chicago had unsatisfactory experience with filament-wound fiberglass tanks and with underground concrete tanks with fiberglass lining. These tanks were replaced with hand lay-up fabricated fiberglass tanks using a vinyl resin binder and with plastic, continuous-weld, full-weight carbon steel tanks lined with a fiberglass reinforced polyester material at a thickness of 0.9 mm (35 mils). The latter gave acceptable performance. Properly fabricated fiberglass tanks or steel tanks with a rubber or polyvinyl chloride (PVC) lining give satisfactory service as well.

Hypochlorite is an extremely aggressive chemical and no equipment used to store or feed it can be expected to last indefinitely. Some particularly robust diaphragm and solenoid-metering pumps have been successfully used and this is the approach that is found in most plants. In very large plants (>380 ML/d or 100 mgd), White recommends metering the chemical by gravity from the storage tank through a Teflon-lined magnetic flowmeter and rate-modulating valve to the point of application.

In general, precautions should also be taken for the potential for precipitation of calcium carbonate whenever the hypochlorite is mixed with carrying water or at the application point with the water being treated. The high specific gravity of hypochlorite solution must be overcome to accomplish effective mixing at the point of application. This can be accomplished by using a diffuser and carrying water (be cautious about the potential formation of $CaCO_3$) or by the use of a pumped jet mixer like that often used for coagulants. Mixing can also be accomplished by introducing the hypochlorite at a point of significant turbulence.

Ammonia

Ammonia can be supplied for water treatment applications in three forms: as a pure liquid (anhydrous ammonia), dissolved in water (aqueous ammonia) or as a dry ammonium salt, usually ammonium sulphate. Ammonia is not an expensive chemical, but the relative cost of these alternative forms of ammonia varies from one location and one application to another.

Storage and feeding of anhydrous ammonia

At normal temperatures and pressures anhydrous ammonia (>99 per cent NH_3) is a gas. However, it can be easily liquefied and is commonly stored and transported in liquid form in pressurised containers of the same size and same design as those used for chlorine (they are usually of different colour). At atmospheric pressure, liquid anhydrous ammonia has a density of 680 kg/m^3 (42.6 lb/ft^3 or 5.7 lb/gal), approximately two-thirds that of water. Anhydrous ammonia can be fed by two methods: direct feed and solution feed. In direct gas feed, the ammonia gas is directly introduced into the water to be treated. Unless the plant is very small, this method often suffers from poor distribution at the application point because of the low-flow rate of ammonia gas. The solution feed method is analogous to the technology used to feed chlorine, except the vapour pressure of ammonia is higher. Precipitation of basic salts like $CaCO_3$ is often a problem in the vicinity of the application point.

Direct gas feed

Aqua ammonia can sometimes be fed directly to the process stream using a metering pump. Suitable metering pumps are commercially available. Progressive cavity pumps have also been successfully used. The storage tank is a permanent on-site facility and should have enough storage for at least 10 days of maximum usage. The tank should have a liquid-level monitor to allow monitoring of the inventory in the tank. The flow metering pump should be located in the proximity of the tank and below its hydraulic grade to minimise chances of ammonia vapourisation in the piping. If necessary, the metering pumps can be sheltered in a building.

Storage and feeding of ammonium sulphate

The most common salt of ammonia used in water treatment is ammonium sulphate, $(NH_4)_2SO_4$. This form of ammonia has the advantage that it does not raise the pH as much as the others do. As a result, it is easier to combine it with dilution water to obtain proper mixing. Mixing can be an important consideration when adding ammonia to water containing free chlorine to arrest the formation of DBPs.

Mixing

Adding chlorine to water that already has ammonia in it can result in a lot of undesirable reactions while mixing takes place. To prevent free ammonia and thus minimise nitrification, it is common for water systems today to add ammonia at a total dose that is at the peak of the breakpoint curve, namely about 1 mole of ammonia per mole of chlorine.

Managing combined chlorine (chloramine) residuals

Maintaining a combined chlorine residual involves some considerations that are not important when a free-chlorine residual is used. Chloramines have the advantage that their odour threshold is lower, that they are more effective in controlling growths on the pipe surfaces and that they are generally much more stable. It should be noted that combined chlorine residuals are subject to destruction by biological nitrification, especially if temperatures are warm and if ammonia is used in excess. Also there is recent evidence that the use of combined chlorine can result in the formation of low levels of 1-nitrosodimethylamine (NDMA), a suspected human carcinogen. Some of the conditions that aggravate NDMA formation, namely a high chlorine-to-ammonia ratio, are the same things that discourage nitrification.

Disinfection with Chlorine Dioxide

When the regulation of the by-products of chlorination began, along with ozone, chlorine dioxide was a fairly high profile disinfection alternative. Chlorine dioxide is widely used in continental Europe, particularly Germany, Switzerland and France and produces almost no identifiable organic by-products, except a few aldehydes and ketones, produced at low levels. Chlorine dioxide was known to produce two inorganic by-products, chlorite and chlorate ion. As a result, most applications of chlorine dioxide were on low-TOC water that did not require a high dose to overcome oxidant demand. Late in the 1980s, concern about the toxicity of chlorite ion and chlorine dioxide itself reached a peak. Also based on field experience, it was found that the use of chlorine dioxide was sometimes responsible for a very undesirable 'cat urine' odour. As a precautionary measure, the State of California banned the use of chlorine dioxide for the disinfection of drinking water and several other states followed.

Eventually, when the disinfectant by-product rule was promulgated, an MCL of 0.8 mg/l was set for chlorite ion and a maximum disinfectant residual limit (MDRL) of l mg/l was set for chlorine dioxide. No MCL was placed on chlorate ion, but utilities were encouraged to be cautious about the production of chlorate and again as a precautionary measure, the State of California has set an action level of 0.8 mg/l. Methods for reducing the concentration of chlorite ion downstream of the use of chlorine dioxide have been demonstrated and it has been established that the cat urine odour only occurs when chlorite ion is exposed to a free-chlorine residual.

As a result, it appears that chlorine dioxide may indeed play an important role in DBP control, particularly in systems using combined chlorine for residual maintenance and looking for a small boost in primary disinfection.

Generation of Chlorine Dioxide

The principal reactions that occur in most chlorine dioxide generators have been known for a long time. In industry, large-scale chlorine dioxide generators use chlorate as a feedstock, but for potable water applications chlorine dioxide is usually generated using a 25 per cent sodium chlorite solution. Although a sodium chlorite feedstock is a common starting point, a number of different approaches are used to convert the chlorite to chlorine dioxide. These include reactions with gaseous chlorine (Cl_2), aqueous chlorine (HOCl) or acid (usually hydrochloric acid, HCl). The reactions are:

$$2NaClO_2 + Cl_2(g) \longrightarrow 2ClO_2(g) + 2NaCl \qquad \ldots (10.7)$$

$$2NaClO_2 + HOCl \longrightarrow 2ClO_2(g) + NaCl + NaOH \qquad \ldots (10.8)$$

$$5NaClO_2 + 4HCl \longrightarrow 4ClO_2(g) + 5NaCl + 2H_2O \qquad \ldots (10.9)$$

The stoichiometry of Eq. 10.7 requires 0.5 kg of chlorine and 1.34 kg of sodium chlorite to produce 1 kg of chlorine dioxide. Several of the alternative approaches used for the generation of chlorine dioxide are summarised in Table 10.2. The differences between Eqs. 10.7, 10.8 and 10.9 help to explain how generators can differ even though the same feedstock chemicals are used and why some should be pH controlled and others are less dependent on pH. In most generators, more than one reaction may be taking place. Chlorine dioxide generators are relatively simple mixing chambers. The reactors are frequently filled with media (Teflon chips, ceramic or Raschig rings) to generate hydraulic turbulence for mixing. A sample petcock valve on the discharge side of the generator is desirable to allow for monitoring of the generation process.

Table 10.2. Chlorine dioxide generation alternatives.

Generator type	Main reactions, reactants, by-products, key reactions, and chemistry notes	Special attributes
Acid-chlorite: (direct acid system)	$5NaClO_2 + 4HCl \rightarrow 4ClO_2(g) + 5NaCl + 2H_2O$ Low pH ClO_3^- also possible Slow reaction rates	Chemical feed pump interlocks required; production limit ~10–15 kg/d (25–30 lb/d); maximum yield is ~80 per cent of stoichiometric yield
Aqueous chlorine-chlorite: (Cl_2 gas ejectors with chemical pumps for liquids or booster pump for ejector water)	$Cl_2 + H_2O \rightarrow HOCl + HCl$ $HOCl + 2NaClO_2 \rightarrow ClO_2(g) + NaCl + NaOH$ Low pH ClO_3^- also possible Relatively slow reaction rates Excess Cl_2 or acid to neutralise NaOH	Production rates limited to ~450 kg/d (1000 lb/d); high conversion but yield only 80–92 per cent; more corrosive effluent due to low pH (~2.8–3.5); three chemical systems pump HCl, hypochlorite, chlorite and dilution water to reaction chamber
Recycled aqueous chlorine-chlorite: (saturated Cl_2 solution via a recycling loop prior to mixing with chlorite solution)	$2HOCl + 2NaClO_2 \rightarrow 2ClO_2 + Cl_2 + 2NaOH$ Excess Cl_2 or HCl needed due to NaOH formed Concentration of ~3g/l required for maximum efficiency	Production rate limited to ~450 kg/d (1000 lb/d); yield of 92–98 per cent with ~10 per cent excess Cl_2 reported; highly corrosive to pumps; drawdown; calibration needed; maturation tank required after mixing
Gaseous chlorine-chlorite: (gaseous Cl_2 and 25 per cent solution of sodium chlorite; pulled by ejector into the reaction column)	$Cl_2(g) + 2NaClO_2 \rightarrow 2ClO_2(g) + 2NaCl$ Neutral pH Rapid reaction Potential scaling in reactor under vacuum due to hardness of feedstock	Production rates 2300–55000 kg/d (5000–1,20,000 lb/d); ejector-based, with no pumps; motive water is dilution water; near-neutral pH effluent; no excess Cl_2; turndown rated at 5–10X with yield of 95–99 per cent; less than 2 per cent excess Cl_2; highly calibrated flowmeters with minimum line pressure ~275 kPa (40 psig) needed

Sodium Chlorite

Sodium chlorite is used as a solution, normally with a concentration of approximately 25 per cent sodium chlorite or less. It is commercially available as a 38 or 25 per cent solution. The major safety concern for solutions of sodium chlorite is the unintentional and uncontrollable release of high levels of chlorine dioxide. Levels that approach an explosive mix can some times occur if the sodium chlorite is exposed to acid.

Another concern to be addressed with sodium chlorite is crystallisation. Like most salts, sodium chlorite solutions are prone to crystallisation at low temperatures and/or higher concentrations. When crystallisation occurs, it may obstruct flow in pipelines, valves and other equipment. Sodium chlorite is not stable as a powder. If dried, it is a fire hazard and can ignite when in contact with combustible materials. A sodium chlorite explosion may occur if too much water and inappropriate fire-fighting techniques are used to quench such a fire. Burning sodium chlorite will quickly generate enough heat to turn water to steam. At high temperatures, the breakdown products of sodium chlorite include oxygen. As a result, highly trained firefighters are required to extinguish closed containers or dry material that has been ignited. Stratification in holding tanks for sodium chlorite solutions may also occur and when it does, will adversely influence the chlorine dioxide yield in the generator. As stratification develops, the sodium chlorite solution being fed gradually changes from low to high density as the generator operates. In stratified tanks, excess chlorite will be fed to the generator as the bottom of the tank will have denser material and this material will have more chlorite than required. Similarly, the bulk tank will later yield chlorite that is too dilute. Although infrequent, such stratification is not readily apparent and may likely remain unnoticed by an operator unless the generator performance is evaluated frequently. Operators should be aware of the possibility of stratification and crystallisation during delivery conditions.

ADVANCED OXIDATION PROCESS

Disinfection with Ozone

The word ozone finds its origins in the Greek word ozein, which means 'to smell'. Ozone (O_3) is an allotrope of oxygen with three oxygen atoms. Ozone in air has a pungent odour that is noticeable to most persons at levels above 0.1 ppm$_v$. Because ozone is a strong oxidant, extended exposure to ozone-containing air is harmful.

Chemistry of ozone

At high concentrations (>23 per cent) ozone is unstable (explosive) and under ambient conditions it undergoes rapid decay. Therefore, unlike chlorine gas, it cannot be stored inside pressurised vessels and transported to the water treatment plant. Once dissolved in water, ozone begins a process of decay that results in the formation of the hydroxyl radical. Where contaminants and microbes are concerned, ozone reacts in two ways: (i) by direct oxidation, and (ii) through the action of hydroxyl radicals ($HO\cdot$) generated during its decomposition. The consensus is that action of ozone as a disinfectant is primarily dependent on its direct reactions; hence it is the residual of the ozone itself that is important.

Ozone decay and ozone demand

The overall rate of ozone decay in water is generally consistent with first order kinetics, although, like chlorine, it can be modelled successfully through the use of a parallel first-order decay model. Although simple reactions serve as good phenomenological models for ozone decay, it is unlikely that they correctly characterise the actual mechanism of decay. From work done in this area, it appears more likely that ozone decay consists of a large number of nth-order reactions operating in parallel that, in sum, appear to be simple first order. An introduction to ozone decay based on the models developed by Staehelin and Hoigne is provided in Fig. 10.2. The cyclic nature of the ozone decay process in pure water illustrated in Fig. 10.2a. The process must be initiated by a reaction between ozone and the hydroxide ion to form superoxide radicals (O_2^-) and peroxide ions (HO_2^-), a slow process. As a result, decay is accelerated at higher pH. Once completed the decay reactions enter a cyclic process represented in the figure in the shape of a circle. The cyclic reactions are promoted by ozone itself—hence the decay of ozone. If the concentration of ozone is increased, the cycle is accelerated.

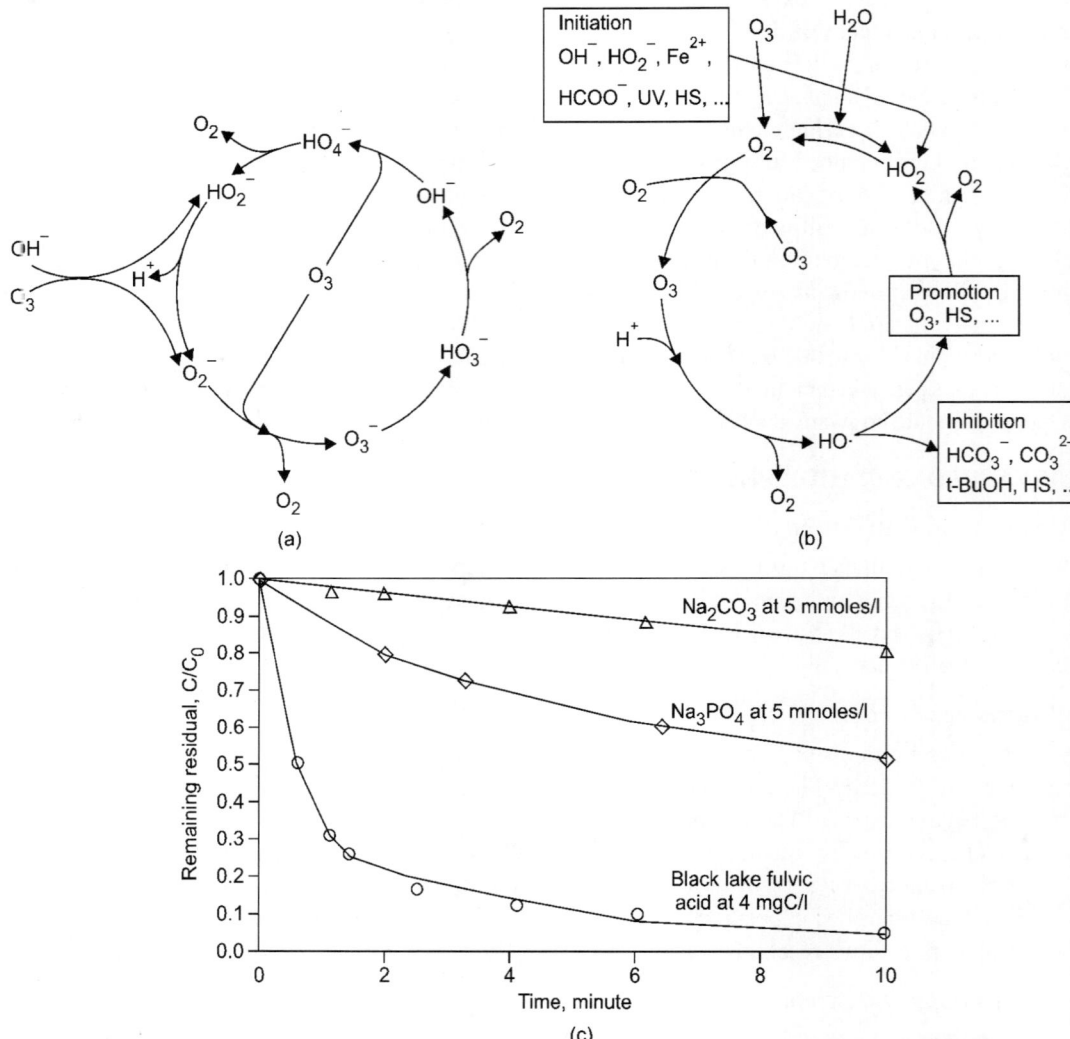

Fig. 10.2. Understanding ozone reaction pathways and decay of residual ozone in natural water: (a) influence of initiators, promotors and inhibitors; (b) the ozone decay wheel—reaction pathways in pure water; and (c) effect of fulvic acid and carbonate on ozone decay—all tests conducted at 20°C with GAC filtered, deionised tap water adjusted to pH 7 and C_0 ~ 8 mg/l.

In natural water, there are other 'initiators' besides hydroxide ion, as shown in Fig. 10.2b. Prominent among them are the ferrous ion and hydrogen peroxide. In natural water certain natural organic materials have also been shown to promote the cycle, accelerating decay. Finally, the continuation of the cycle depends on the action of the hydroxyl radical on the ozone residual. As a result, scavengers that react with the hydroxyl radical, removing it from the process, also slow the rate of decay. The carbonate and bicarbonate ions are important examples of such inhibitors. The data of Reckhow and co-workers, are shown in Fig. 10.2c to illustrate the action of fulvic acids as initiators and promoters and carbonate and bicarbonate ions as HO · traps or inhibitors. The factors that influence the stability of ozone residuals are summarised in Table 10.3.

Table 10.3. Factors that influence stability of aqueous ozone residuals.

Increases stability	Reduces stability
Low pH	High pH
High alkalinity	Low alkalinity
Low TOC	High TOC
Low temperature	High temperature

Conceptual design

Ozonation in a water treatment plant requires the addition of two components in the process treatment train: (i) a mass transfer device for dissolving the ozone into the water, and (ii) a contact chamber in which the disinfection reaction takes place. For several decades, the most common approach to ozonation has been to combine these two components by introducing the ozone into the water in large, deep basins using porous diffusers. Both mass transfer and the disinfection reaction then take place in these basins. Most often these contact basins are simulated at a pilot scale using tall bubble columns. The conceptual design of any ozonation system requires a means for estimating mass transfer of ozone into the water, an understanding of the kinetics of ozone decay, an understanding of the disinfection kinetics and a means for estimating dispersion in the reactor. All of these components have been applied successfully to model the disinfection of *C. parvum* in bubble columns at pilot scale. Nevertheless, significant challenges remain in addressing all these issues in the variety of approaches that are considered in full-scale design.

Generation of ozone

Ozone can be generated by photochemical, electrolytic and radiochemical methods, but the corona discharge method is the most commonly used in water treatment. In this method, oxygen is passed through an electric field that is generated by applying a high voltage potential across two electrodes separated by a dielectric material (Fig. 10.3). The dielectric material prevents arcing and spreads the electric field across the entire surface of the electrode.

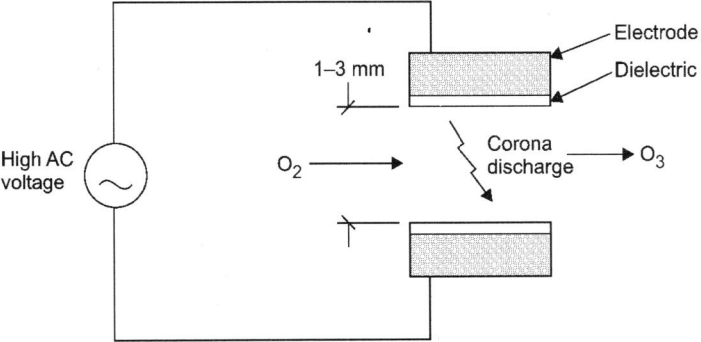

Fig. 10.3. Ozone generation by corona discharge.

As the oxygen molecules pass through the electric field, they are broken down to highly reactive oxygen singlets (O ·), which then react with other oxygen molecules to form ozone. The thickness of the gap through which the oxygen-rich gas stream passes is 1 to 3 mm wide. Because most of the energy used in ozone generation is lost as heat, cooling of the ozone generator is necessary to avoid overheating and subsequent decomposition of the ozone generated. Cooling is normally accomplished by passing a continuous stream of cooling water next to the ground electrode. Some of the key design factors that influence ozone generator performance are summarised in Table 10.4.

Table 10.4. Influence of increasing four key design factors on generator performance.

Design factor	Effect on ozone production
Frequency of applied current	Increase ozone production
Voltage of applied current	Increase ozone production
Gap between generator electrodes	Decrease ozone production
Dielectric constant of dielectric separating electrodes	Decrease ozone production

The equation below, while not intended to be quantitative, provides a general idea of the significance of a number of the variables of importance to the design of a corona discharge ozone generator:

$$Q_{O_3} \ \alpha \ \left(f \frac{V^2 A}{d\varepsilon} \right) Q_{O_2} \qquad \qquad \text{... (10.10)}$$

where,

Q_{O_3} = ozone generation, MT^{-1}
f = frequency of applied emf
V = emf across electrodes, V
A = surface area of electrodes, l^2
d = distance between electrodes, l
ε = dielectric constant

Oxygen source

Ozone can be generated directly from the oxygen in air or from pure oxygen. Pure oxygen is generated on-site from ambient air at larger plants or provided through the use of liquid oxygen (commonly referred to as LOX), which is generated off-site and transported to the plant. The most suitable method for providing oxygen for ozone generation in a particular plant depends on economic factors, the principal ones being the scale of the facility and the availability of industrial sources of liquid oxygen.

PHYSICAL METHODS

Disinfection with Ultraviolet Light

All of the disinfectants discussed previously in this chapter are oxidising chemicals. Disinfection can also be accomplished by other means, heat and electromagnetic radiation among them. Heat is used to disinfect or pasteurise, beverages and even to disinfect water through boiling. Electromagnetic radiation, specifically gamma radiation and UV radiation, is also used for disinfection: gamma radiation in the case of food products and UV radiation in the case of air, water and some medical surfaces. Of these, only UV radiation has so far found a place in the routine disinfection of drinking water. To date UV has had wide application to drinking water disinfection. Nevertheless its application has been rather erratic and it has not become an accepted standard.

In the United States, far more UV facilities have been installed to disinfect waste-water. The popularity of UV in waste-water disinfection stems largely from the fact that it does not contribute to effluent toxicity as does chlorine. Recent developments, however, have given the application of UV to drinking water a great deal more momentum. Interest in UV began to increase when it was found that medium-pressure UV, a relatively new, multiwavelength UV technology, was unusually effective in inactivating *C. parvum*. Flaws in earlier testing methodologies for the inactivation of protozoan cysts and oocysts were also identified in that study.

Later it became apparent that, contrary to earlier findings, low-pressure UV was also effective in inactivating *C. parvum*. It has been demonstrated that these technologies are effective in inactivating *G. lamblia* as well. Controlling these two pathogens has been particularly difficult using traditional chlorination technology.

Ultraviolet light

Ultraviolet light is the name used to describe electromagnetic radiation having a wavelength between 100 and 400 nm. As illustrated in Fig. 10.4, electromagnetic radiation of slightly shorter wavelength has been named X-rays and electromagnetic radiation of slightly longer wavelength, visible to the human eye, is referred to as visible light.

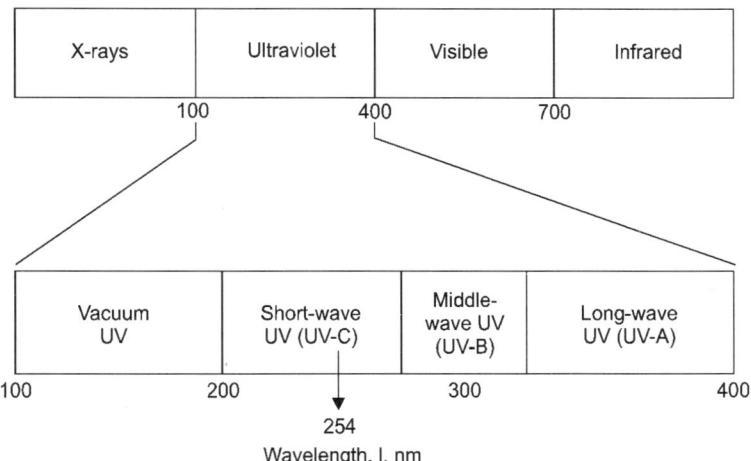

Fig. 10.4. Understanding UV light: (a) the electromagnetic spectrum; (b) output from low-pressure UV lamp; and (c) output from medium-pressure UV lamp.

Radiation just long enough to be outside the visible range is referred to as infrared radiation. Light in the UV spectrum is often further sub-divided into four segments, vacuum UV, short-wave UV (UV-C), middle-wave UV (UV-B) and long-wave UV (UV-A). In simple terms these classifications can be described as follows:

1. Both UV-A and UV-B activate the melanocytes in the skin to produce melanin ('a tan').
2. UV-B radiation also causes 'sunburn'.
3. UV-C radiation is absorbed by the DNA and is the most likely of the three to cause skin cancer.

If electromagnetic radiation is thought of as photons, then the energy associated with each photon is related to the wavelength of the radiation:

$$E = \frac{hc}{\lambda}$$

... (10.11)

where,

E = the energy in each photon, J.
h = Planck's constant (6.6×10^{-34} J·s).
c = speed of light, m/s.
λ = wavelength of radiation, m.

As a general rule, the more energy associated with each photon in electromagnetic radiation, the more dangerous it is for living organisms. Thus, visible and infrared light have little affect on organisms, whereas both X-rays and gamma rays can be quite dangerous. Beyond these broad considerations, there are other factors that determine the fraction of the UV spectrum that is effective in disinfection. The portion of the UV spectrum that is more effective in disinfection is called the 'germicidal range'. On the lower end, the germicidal range is limited by the absorption of UV radiation by water. As wavelengths decrease, water becomes an increasingly efficient barrier for UV. For practical purposes, vacuum UV, the fraction of UV with a wavelength below 200 nm, cannot penetrate water. So radiation having a wavelength of 200 nm or less is not considered germicidal. It is also well established that UV inactivates micro-organisms by transforming their DNA. This transformation cannot happen unless the UV is at a wavelength at which DNA will absorb it and this absorption does not occur above wavelengths of approximately 300 nm. Therefore the germicidal range for UV is between approximately 200 and 300 nm.

Sources of ultraviolet light

The UV disinfection units currently used in the water industry employ three different types of lamp technology: (i) low-pressure, low-intensity lamps; (ii) low-pressure, high-intensity lamps (also called low-pressure, high-output lamps); and (iii) medium-pressure, high-intensity lamps.

Low-pressure, low-intensity lamps are most common. The design of these lamps closely approximates that of the common fluorescent light bulb. Low and medium-pressure, high-intensity lamps are new technologies that are able to achieve a higher UV output in an equivalent space. Of the three technologies, medium-pressure UV has the greatest output. The spectrum of the UV light output by both types of low-pressure lamps is essentially the same, a very small amount of the light energy emanating at a wavelength of 188 nm and the vast majority of it emanating at a wavelength of 254 nm. The spectrum of the UV light output by medium-pressure lamps includes a number of wavelengths.

Reactivation

Reactivation is a more important consideration in UV disinfection than it is with disinfection by other methods. It is important to note that most forms of life evolved with some exposure to the sun and that sunlight includes significant amounts of UV irradiation. As a result, the process of evolution has addressed UV-induced damage by generating mechanisms for repairing the damage it causes. These mechanisms fall into two basic classes: (i) photo-reactivation; and (ii) dark repair. Photo-reactivation only takes place in the presence of light whereas dark repair has no such requirement. Organisms capable of dark repair generally show much greater UV resistance; however, understanding the importance of photo-reactivation requires that special tests be conducted, evaluating samples with and without light exposure to understand its effects. Certainly when water is being disinfected for discharge into the environment, only the net inactivation after photo-reactivation is important. Even in the case of drinking water systems, where light exposure is often more limited, the most conservative approach is to consider photo-reactivation as well.

Photo-reactivation was discovered by Kelner (*E. coli*) and Dulbecco (bacteriophage). It is now understood to involve the cleaving of the nucleic acid dimers with the enzyme DNA photolyase. The enzyme first adsorbs to the dimer and then cleaves it with the assistance of photons in the visible wavelength. Eventually, it will probably be possible to determine if an organism is capable of photo-repair by using its genetic fingerprint to map its position on the evolutionary tree. It is not safe to assume that any organism is incapable of photo-repair; unless through testing it has been demonstrated to be the case. Even some viruses have been shown to be capable of photo-repair, apparently taking advantage of enzymes in the host organism following infection. Fortunately, based on preliminary work done with *C. parvum*, it appears that this organism is not capable of photorepair.

Influence of water quality

The quality of the water being treated can have an important influence on the performance of UV disinfection systems. The two most important impacts stem from the action of dissolved and suspended substances.

Dissolved substances

Pure water absorbs light in the lower UV wavelengths. A number of dissolved substances also have important influence on the absorption of UV irradiation as it passes through the water on its way to the target organism. Among the more significant are iron, nitrate and natural organic matter. Chlorine, hydrogen peroxide and ozone can also have important effects.

Particulate matter

Particulate matter can also interfere with the transmission of UV light. Particulates are an aspect of water quality that can be particularly important where UV disinfection is concerned.

The impact of particulate matter is important with regard to the disinfection of secondary effluent from waste-water treatment facilities, especially when doses above 300 J/m^2 are employed. At these doses and higher, a tailing phenomenon is observed that can be directly related to the number of particles that have coliform organisms associated with them. Models have been developed that do a good job of characterising this phenomenon.

Influence of dispersion

Ultraviolet disinfection systems, particularly medium-pressure systems, are characterised by overall contact times that are much shorter than other kinds of disinfection systems. In these systems short circuiting and dispersion are difficult design issues. Designing these systems to achieve good performance requires a greater appreciation of the factors that influence dispersion and short circuiting than is required for the design of most other disinfection systems. The issues are the same as those discussed earlier with contactors for disinfection with chlorine, chloramines, chlorine dioxide and ozone; however, with UV disinfection contactors, the time spent in transition zones becomes much more important.

In chlorine contactors, for example, inlet conditions can have a big influence on performance. If the contactor is designed with a sufficiently long aspect ratio, good performance can be achieved in spite of non-ideal inlet conditions. In many UV reactors, the zones of flow transition can dominate most of the contact time. A further complication in UV reactors is that the fluence (light intensity) varies throughout the reactor. So the UV dose that an organism receives is not only a function of the length of time the organism spends in the reactor and the amount of light being emitted by the UV lamps but also of the specific path the organism takes as it makes its way through the reactor. Thus, the issue is not just the contact time the organism receives, but its cumulative exposure to UV.

Equipment configurations

A UV disinfection system consists of: (i) the UV lamps, (ii) transparent quartz sleeves that surround the UV lamps, protecting them from the water, (iii) the structure that supports the lamps and sleeves and holds them in place, (iv) the power supply for the system; and, in many cases, (v) the cleaning system to maintain transparency of the quartz sleeves. By themselves, UV lamps, which use an electrical arc, are not electrically stable because their electrical resistance decreases as their current increases. As a consequence, the electrical system must be ballasted to limit the current to the lamp. Disinfection systems are also classified by whether they operate as open-channel gravity flow systems or as closed-vessel pressurised systems. Open-channel systems are more common in waste-water disinfection, but closed-vessel systems are receiving a great deal of attention in the disinfection of drinking water.

Open-channel systems

Open-channel designs are available for low-pressure, low-intensity and low pressure, high-intensity UV systems. In low-pressure systems, the UV lamps are retained in modules or racks that are placed in the flow channel. Designs are available with lamps placed horizontally parallel to the flow and with lamps placed vertically perpendicular to the flow. Conventional low pressure systems are typically designed so that they can easily be removed and cleaned and most low-pressure, high-intensity and all medium-pressure systems are provided with mechanical or mechanical/chemical self-cleaning systems. These cleaning systems are necessary because the latter system operates at such high temperatures that salts with inverse solubility precipitate, fouling the outer surface of the quartz sleeve and reducing net UV output. These systems are always designed with parallel channels and the following are some important design tips:

1. Always provide stand-by banks and stand-by channels.
2. It is important to provide a reliable method for ensuring that the lamps will remain submerged. Weighted flap gates, sharp-crested weirs and automatic level controllers are often used.
3. Positive controls should be used to ensure flow is equally distributed between units (e.g. flow division via weirs).
4. Inlet conditions to these channels must be designed carefully to ensure that short circuiting does not occur.

Closed-vessel systems

Whereas most low-pressure systems are designed with open-channel flow, many low-pressure, high-intensity and medium-pressure systems are designed using closed vessels. These closed-vessel systems have the advantage that they can (and usually do) operate under pressure and this feature makes them particularly attractive in upgrades and retrofits because it is not necessary to 'break head' to use them. These systems, particularly the medium-pressure systems, operate with such a small number of lamps that a great deal more care is required to ensure that short circuiting does not occur. Biodosimetry standards are evolving that will ensure that the individual units perform as specified, but care must be taken to ensure that influent and effluent conditions do not contribute to short circuiting. Some important design tips for closed-vessel system are as follows:

1. All UV reactor designs should have their performance certified via a standard biodosimetry test conducted according to a widely accepted guidelines.
2. Always provide stand-by reactors.
3. Make sure that it is not possible for these units to fire up when dry or become dry while lit.
4. Positive controls should be used to ensure flow is equally distributed between units.
5. Inlet conditions to reactors must be carefully designed to ensure that short circuiting does not occur.

Chapter 11

Estimating the Risk of Infectious Disease Associated with Pathogens in Drinking Water

INTRODUCTION

Pathogenic micro-organisms frequently can be found in untreated water and, as a result, can cause infections which may lead to illness and even to death among those people who ingest that water. To reduce these risks, drinking water often is treated on a community basis. However, micro-organisms can escape even properly operated, community-based drinking water treatment processes, resulting in the presence in treated drinking water of pathogenic organisms belonging to three categories: bacteria, protozoa, and viruses. These residual microbial contaminants can be sufficient to cause appreciable levels of infectious disease among consumers of the water. The potential reality of these risks is borne out both by epidemiological studies and by published theoretical calculations.

The modelling equations presented in this chapter represent an example of how this infectious-disease risk can be estimated and the extent to which the risk can be lessened by reducing the level of micro-organisms in water prior to its ingestion. Two validations of these models were performed and presented in this chapter to serve as examples of how the risk estimation technique is employed and to show how closely it approximates the actual risk which those micro-organisms in drinking water impose upon the consumers of that water. These validation exercises were based on two published epidemiological studies by Smith and others in which these authors examined the relationship between gastrointestinal disease in humans and the community-supplied treated drinking water which those people ingested.

The first epidemiological study by Smith and others indicated that individuals ingesting conventionally treated, community-distributed tap water had an overall annual risk of 0.76 for gastrointestinal illness. This contrasted with an annual gastrointestinal illness risk of 0.50 among a control group of consumers who drank water that originated from the same community distribution system but had gone through an in-home filtration system prior to ingestion. The difference in annual risk thereby attributed to micro-organisms in the tap water was 0.26. Smith and others had not included data regarding the load of either pathogenic bacterial, protozoan, or viral contaminants in the community-supplied treated drinking water when they published that first epidemiological study. However, based on microbial occurrence data also published by Smith and others for a roughly similar time frame, estimated annual ingested dosages of pathogenic viruses and protozoa per individual consumer were determined. Those dosages then were multiplied by literature-derived estimates of the probability of infection per ingested organism and for the probability of illness per incidence of infection. These values for probability of infection and probability of illness came from studies in which human volunteers knowingly had been fed micro-organisms belonging to the three different categories. The validation exercises presented in this chapter did not include estimations for the risk of illness progressing to death, because mortality information was not included in the epidemiological study results that had been published by Smith and others. The validation exercise (validation 1) presented in this chapter for the first epidemiological study by Smith and others suggested that of the 0.26 annual risk of illness associated with infectious pathogenic organisms in the community-distributed tap water, viruses in the drinking water could have accounted for approximately 80.2 per cent of the observed cases of illness, while protozoa of the genera *Cryptosporidium* and *Giardia* could have accounted for approximately 1.0 and 2.6 per cent, respectively, of the observed annual risk of illness in that first epidemiological study by Smith and co-workers.

Several years later, after efforts to improve the efficiency of that community's drinking water treatment facilities, Smith and others performed a second epidemiological study, during which they found that individuals ingesting conventionally treated, community-distributed tap water had an overall 0.66 annual risk for incidence of gastrointestinal illness. This contrasted with an annual illness risk of 0.58 among a control group of consumers who drank either bottled water or tap water that originated from the same community distribution system but had gone through an in-home filtration system prior to ingestion. The difference in annual risk thereby attributed to micro-organisms in the tap water during the second epidemiological study was 0.08. The validation exercise presented in this chapter for the second epidemiological study of Smith and others used virus and protozoan data directly published as a part of that second study. This validation suggests that of the 0.08 annual risk, viruses in the drinking water could have accounted for approximately 85.9 per cent of the observed cases of illness, while protozoa of the genera *Cryptosporidium* and *Giardia* could have accounted for approximately 1.5 and 4.2 per cent, respectively, of the observed illnesses. The data which were published by Smith and others concerning bacterial concentrations in the tap water represent indicator bacterial groups rather than known waterborne bacterial pathogens. As such, the only way to estimate the proportion of observed infectious disease which would have been attributable to bacterial pathogens in the water during the time of the epidemiological studies is by subtraction. Knowing the total risk due to pathogens in the drinking water and then subtracting the estimates for risks attributable to viruses and protozoa results in the estimated proportion of illnesses due to bacterial pathogens in the tap water. This process reveals that for the two epidemiological studies published by Smith and others, bacterial pathogens could have accounted for approximately 16.1 per cent of the observed cases of illness attributable to micro-organisms in the community-distributed tap water for the first epidemiological study and correspondingly could have accounted for 8.4 per cent of the observed cases of illness for the second study.

The approach presented in this chapter for modelling the risks of acquiring infectious disease from contaminants in drinking water can be used to estimate the existing level of danger to a community's population. The accuracy of disease risk estimates relies on knowing the levels of pathogens in the water being ingested. Alternatively, by determining a target value for the maximum desired disease risk to a community and by knowing the levels of pathogenic micro-organisms in the source water, this modelling approach could be used to estimate the level of community-based water treatment which would be needed to decrease the disease risk to that target level.

NATURE OF THE HAZARDS ASSOCIATED WITH PATHOGENIC MICRO-ORGANISMS IN AQUATIC ENVIRONMENT

Most of the micro-organisms present in aquatic environments seem to have no effect on human health. However, some clearly do represent a public health risk and for this reason are considered to be disease causing and are referred to as pathogens. The true definition of the term 'pathogen' is 'something that causes pathological changes' in the appearance of cells and tissues. Often the 'pathogen' is the entire organism viewed as an infectious agent. In other cases, however, the pathological change is due to a toxic substance produced by a noninfectious organism, and some toxins, such as those produced by algae and cyanobacteria, can bioaccumulate within the food chain to extremely high concentrations. This chapter addresses organisms which are infectious for humans, and while some of these represent aquatic contaminants that come from either human or animal reservoirs, others represent organisms which are naturally present in the environment.

ROUTES OF PATHOGEN ACQUISITION

Humans can acquire infections from water by any of three major routes: ingestion, body surface contact, and inhalation. Table 11.1 lists examples of the types of infectious diseases which are associated with these major routes. The route addressed in this chapter is that of ingestion.

DEFINING THE MAIN VARIABLES INCORPORATED IN RISK ESTIMATES FOR THE LEVEL OF INFECTIOUS DISEASE ASSOCIATED WITH INGESTION OF WATER

Five main variables must be considered when estimating the risk of acquiring infectious disease by ingestion of water. These are the concentration of pathogenic organisms in water which are infectious by the route of ingestion, the amount of water ingested by an individual per a unit time, and the probabilities of infection, illness, and death associated with ingesting those pathogenic organisms.

Defining the Exposure Level

Two of the five main variables, the concentration of pathogenic organisms in the water which are infectious by the route of ingestion and the amount of water which an individual ingests per unit time, act together to represent the level of an individual's exposure to the hazard. The risk estimation technique presented in this chapter could be used for any period. However, the examples presented will employ a period of 1 year, calculated as 365.25 days. The value used as an estimate for the volume of water ingested on a daily basis per individual consumer is 2 litres per day. This volume is the estimate recommended by the US Environmental Protection Agency for use in calculating the risks from ingestion of water.

Defining the Probability of Infection

The values used for probability of infection are calculated as the inverse of the minimum or minimal infectious dose. The probability of infection thus represents the estimated likelihood that ingestion of a single measurable bacterium, protozoan, or virus will result in infection. The values for probability of infection are based on published results from studies during which human volunteers were fed different dosages of infectious micro-organisms. For the purpose of this chapter, the minimum infectious dose is defined as the smallest number of micro-organisms which could be found to produce infection when ingested by any of the volunteers, with smaller doses not found to produce infection. In most clinical studies, infection is defined serologically after the fact. The test for infection involves examining a pair of samples of serum collected from each individual volunteer. The first serum sample is collected at the time when that individual ingests the organism of interest, and the second serum sample is collected at some time between 4 and 6 weeks following that ingestion. If the titer (concentration) of the antibodies in the serum which were able to react with antigenic material from the strain of micro-organism fed to that individual increased at least twofold during the 4- to 6-week time interval, then that individual is considered to have been infected by the ingested organism.

Table 11.1. Examples of infectious diseases associated with waterborne pathogens.

Exposure route	Type of disease	Causative micro-organism(s)[a]
Ingestion (includes contaminated drinking water and other beverages plus ice- and water-associated contamination of foods)	Bacterial	
	Enteric fever	*Salmonella* (especially *S. enterica* serovar Typhi, which causes typhoid fever)
	Enteritis	*Campylobacter, Shigella* (causes bacterial dysentery), *Vibrio* (especially *V. cholerae*, which causes cholera)
	Febrile syndrome	*Francisella tularensis*
	Septicemia	*Vibrio vulnificus*
	Protozoan	
	Enteritis	*Cryptosporidium parvum, Entamoeba histolytica* (causes amebic dysentery), *Giardia lamblia*
	Viral	
	Encephalitis	*Enterovirus*
	Gastroenteritis	*Astrovirus, Calicivirus, Coronavirus, Rotavirus*
	Hepatitis	*Calicivirus, Hepatovirus*
	Meningitis	*Enterovirus*
Body surface contact (usually associated with either recreational or aquatic occupational activities)	Bacterial	
	Enteritis	*Vibrio cholerae*
	Nephritis	*Leptospira interrogans*
	Wound infections	*Vibrio parahaemolyticus, Vibrio vulnificus*
	Metazoan	
	Worm infestation	*Schistosoma*
	Protozoan	
	Encephalitis	*Naeglerio*
	Enteritis	*Entamoeba histolytica*
	Viral	
	Encephalitis	*Enterovirus*
	Gastroenteritis	*Astrovirus, Calicivirus, Coronavirus, Rotavirus*
	Meningitis	*Enterovirus*
	Pharyngocon junctival fever	*Mastadenovirus*
Inhalation	Bacterial	
	Pneumonic fever	*Legionella pneumophila*

[a] If an organism is indicated by both its genus and species names, the disease association is with that particular species. If only a genus name is given, then the disease association is with more than one species belonging to that genus.

Some researchers who perform risk estimations would prefer to use a mathematical approximation technique based on curve fitting to modify the findings from feeding studies. However, as explained in Table 11.2, this type of curve fitting approach seems to produce 'estimated' minimum infectious dose values which are higher than the actual minimum infectious dose values published by the authors of the feeding studies. The use of such curve fitting could then cause an underestimation of the true risk associated with ingesting infectious organisms. To avoid this problem of underestimating risk, the values for probability of infection used in this chapter will be those listed in Table 11.3, and they directly represent the actual minimum infectious doses as published by the authors of the respective feeding studies.

Table 11.2. How the use of curve fitting can alter value estimates for the minimum infectious dose of a micro-organism.

Micro-organism examined for infectivity	Experimentally determined minimum infectious dose[a]	Minimum infectious dose as estimated by curve fittings[c]	Underestimation of risk which would result from using curve-fitting values[d]
Echovirus 12	17	500.0	29.4
Poliovirus 1	20	110.2	5.5
Poliovirus 3	0.5[b]	1.93	3.9
Rotavirus	0.9	1.62	1.8
Giardia	10.0	49.27	4.9

[a] Lowest viral dose found capable of causing infection when ingested by human volunteers. The number of viruses was determined by infectivity in cultured cells. The number of *Giardia* cells was determined by direct microscopic enumeration.

[b] Actual dose was 1 $TCID_{50}$, divided by 2 to yield an estimated $TCID_{100}$ as reported in this table, using the technique described by Hurst.

[c] This column contains estimated minimum infectious dose values derived by performing curve-fitting operations on the actual experimental data. The curve-fitting values listed were presented as the estimated concentration of micro-organisms per litre of drinking water which would give an annual risk of one infection per 10,000 individuals and assumed that each individual ingested 2 litres of water per day. Therefore, backtracking to obtain the estimated minimum infectious dose values which had been derived by curve fitting was done by multiplying the published values by 7.3×10^6, which was calculated as (2 litres of water/day) × (365.25 days/year) × (10,000 individuals).

[d] If the calculation of risk were done using values derived by curve fitting in place of the actual experimentally determined values, an underestimation of risk would result. The underestimation of risk (as a factor of error) which would be associated with using those curve-fitting values was calculated as follows: (the estimated minimum infectious dose as determined by Regli through curve fitting) ÷ (the experimentally determined minimum infectious dose initially published for that organism).

Table 11.3. Probability that a human will become infected on ingesting an enteric micro-organism[a].

Type of micro-organism examined (category of estimate)	Probability of infection per ingested micro-organism[f]
Bacteria[b]	
Enteric pathogenic bacteria (overall estimate)	0.00001
Salmonella enterica serovar Typhi (this species in general)	0.00001
Vibrio cholerae (type O1 strains)	0.001
Vibrio cholerae (type non-O1 strains)	0.00001

(Contd ...)

Type of micro-organism examined (category of estimate)	Probability of infection per ingested micro-organism[e]
Protozoa[c]	
Enteric pathogenic protozoa (overall estimate)	0.067
Cryptosporidium parvum (this species in general)	0.033
Giardia lamblia (this species in general)	0.1
Viruses[d]	
Enteric pathogenic viruses (overall estimate)	0.5
Enterovirus (this genus in general)	0.5
Enterovirus human echovirus 12 (this species in general)	0.059[f]
Enterovirus human poliovirus 1 (this species in general)	0.5
Enterovirus human poliovirus 3 (this species in general)	2.0[f]
Rotavirus (this genus in general)	1.1[f]

[a] Probabilities were determined by volunteer feeding studies. The success of infection was determined by testing the sera of the volunteers before and after those individuals were dosed with micro-organisms. The values listed in this table are medians based on data published by Hurst. When values for the same genus or species of micro-organism were available from more than a single study, an overall estimate was derived to represent that genus or species by calculating the median of the pertinent values. Likewise, overall estimates for any particular group (e.g. enteric pathogenic bacteria) of micro-organisms were derived by calculating the median of the values available from studies in which members of that group had been examined.

[b] The number of bacterial organisms was determined by culture.

[c] The number of protozoa was determined as either cysts (for *Giardia*) or oocysts (for *Cryptosporidium*) by direct microscopic enumeration.

[d] The number of viruses was determined by infectivity assay in cultured cells.

[e] Probability of infection associated with each micro-organism ingested. This calculation is performed as $1 \div$ minimum infectious dose.

[f] For this virus type, the number of virus particles required to cause an infection of cultured cells is greater than the number of virus particles required to cause infection of a human. Thus, the value of the probability of a human developing an infection from this virus type is higher than the titer obtained by cell culture assay of the virus.

Defining the Probability of Illness

Illness is defined as the demonstration of symptoms characteristic of disease, and not all infected individuals show symptoms. The values used in this chapter for probability of illness are listed in Table 11.4 and were calculated as the number of individual volunteers who developed symptoms of illness following infection by ingesting a particular micro-organism divided by the total number of individual volunteers who were determined to have been infected when fed that micro-organism. Thus, the probability of illness is the estimated likelihood that an incidence of infection caused by a bacteria, protozoan or virus will progress to illness.

Defining the Probability of Death

The probability of death is the estimated likelihood that a case of illness induced by a bacterium, protozoan, or virus will progress to cause death. The values used in this chapter for probability of death are listed in Table 11.5 and were calculated from epidemiological data as the number of infected individuals who died from illness induced by a particular micro-organism divided by the total number of individuals who were made ill by that micro-organism.

Table 11.4. Probability that infection of a human by an enteric micro-organism will result in illness[a].

Type of micro-organism examined (category of estimate)	Probability of illness per incidence of infection
Bacteria	
Enteric pathogenic bacteria (overall estimate)	0.4928
Salmonella enterica serovar Typhi (this species in general)	0.4928
Vibrio cholerae (type O1 strains of this species)	0.2727
Vibrio cholerae (type non-O1 strains)	0.5333
Protozoa	
Enteric pathogenic protozoa (overall estimate)	0.4386
Cryptosporidium (this genus in general)	0.5538
Cyclospora (this genus in general)	0.2200
Giardia (this genus in general)	0.4617
Giardia (species of this genus other than *G. lamblia*)	0.5011
Giardia lamblia (this species in general)	0.4617
Viruses	
Enteric pathogenic viruses (overall estimate)	0.5769
Enterovirus (this genus in general)	0.486
Enterovirus human coxsackievirus A21 (this species in general)	0.9231
Enterovirus human echovirus 12 (this species in general)	0.0508
Rotavirus (this genus in general)	0.5769

[a] Probabilities were determined by volunteer feeding studies. The success of infection was determined by testing the sera of the volunteers before and after those individuals were dosed with micro-organisms. Illness was defined as the demonstration of any symptoms. When values for the same genus or species of micro-organism were available from more than a single study, an overall estimate was derived to represent that genus or species by calculating the median of the pertinent values. Likewise, overall estimates for any particular group (e.g. enteric pathogenic bacteria) of micro-organisms were derived by calculating the median of the values available from studies in which members of that group had been examined.

Table 11.5. Probability that a human made ill by an enteric micro-organism will die as a result of that illness.

Type of micro-organism examined (category of estimate)	Probability of death per incidence of illness
Bacteria	
Enteric pathogenic bacteria (overall estimate)	0.0180
Campylobacter (this genus in general)	0.0003
Escherichia coli (pathogenic strains, this species in general)	0.1000
Escherichia coli (type O111 strains)	0.1111
Escherichia coli (type O157:H7 strains)	0.0582
Salmonella enterica serovar Typhi (this species in general)	0.0368
Shigella flexneri (this species in general)	0.0016
Vibrio cholerae (this species in general)	0.0195
Vibrio cholerae (type O1 strains)	0.0764
Vibrio cholerae (type non-O1 strains)	0.0137

(Contd ...)

Type of micro-organism examined (category of estimate)	Probability of death per incidence of illness
Protozoa	
Cryptosporidium parvum (this species in general)	0.0001
Viruses	
Enteric pathogenic viruses (overall estimate)	0.0094
Calicivirus hepatitis E virus (this species in general)	0.0280[b]
Enterovirus (this genus in general)	0.0104
Enterovirus (subset 'coxsackieviruses' of this genus)	0.0075
Enterovirus (subset 'echoviruses' of this genus)	0.0028
Enterovirus (subset 'polioviruses' of this genus)	0.0183
Enterovirus human coxsackievirus A2 (this species in general)	0.0050
Enterovirus human coxsackievirus A3 (this species in general)	0.0741
Enterovirus human coxsackievirus A4 (this species in general)	0.0052
Enterovirus human coxsackievirus A6 (this species in general)	0.0098
Enterovirus human coxsackievirus A9 (this species in general)	0.0026
Enterovirus human coxsackievirus A10 (this species in general)	0.0171
Enterovirus human coxsackievirus A16 (this species in general)	0.0012
Enterovirus human coxsackievirus A19 (this species in general)	0.1667
Enterovirus human echovirus 6 (this species in general)	0.0029
Enterovirus human echovirus 9 (this species in general)	0.0027
Enterovirus human poliovirus 1 (this species in general)	0.0505
Enterovirus human poliovirus 2 (this species in general)	0.0182
Enterovirus human poliovirus 3 (this species in general)	0.0148
Hepatovirus hepatitis A virus (this species in general)	0.0060
Rotavirus (this genus in general)	0.0055

[a] Probabilities were determined by epidemiological data. Illness was defined as the demonstration of any symptoms, and death was defined as cessation of self-directed metabolic activity. When values for the same genus or species of micro-organism were available from more than a single study, an overall estimate value was derived to represent that genus or species by calculating the median of the pertinent values. Likewise, overall estimates for any particular group (e.g. enteric pathogenic bacteria) of micro-organisms were derived by calculating the median of the values available from studies in which members of that group had been examined.

[b] The value cited here for probability of death represents data reported for adult males. Values for females in their third trimester of pregnancy range from 0.17 to greater than 0.25.

CALCULATING THE RISK ESTIMATION

Format of the Equations

Equations 11.1 to 11.4 present the format used in this chapter for estimating the human health risk of infection, illness, and death on an annual basis per individual consumer.

$$
\begin{array}{l}
\text{Number of} \\
\text{organisms} \\
\text{ingested} \\
\text{annually}
\end{array}
=
\begin{array}{l}
\text{Number of} \\
\text{organisms} \\
\text{per litre} \\
\text{of water}
\end{array}
\times
\begin{array}{l}
\text{Number of} \\
\text{litres} \\
\text{ingested} \\
\text{per day}
\end{array}
\times
\begin{array}{l}
\text{365.25 days} \\
\text{of water} \\
\text{ingestion} \\
\text{per year}
\end{array}
\qquad \dots (11.1)
$$

$$
\begin{array}{l}
\text{Annual} \\
\text{risk of} \\
\text{infection}
\end{array}
=
\begin{array}{l}
\text{Number of} \\
\text{organisms} \\
\text{ingested} \\
\text{annually}
\end{array}
\times
\begin{array}{l}
\text{Probability} \\
\text{of infection} \\
\text{per ingested} \\
\text{organism}
\end{array}
\qquad \dots (11.2)
$$

$$
\begin{array}{l}
\text{Annual} \\
\text{risk of} \\
\text{illness}
\end{array}
=
\begin{array}{l}
\text{Number of} \\
\text{organisms} \\
\text{ingested} \\
\text{annually}
\end{array}
\times
\begin{array}{l}
\text{Probability} \\
\text{of infection} \\
\text{per ingested} \\
\text{organism}
\end{array}
\times
\begin{array}{l}
\text{Probability} \\
\text{of illness} \\
\text{per incidence} \\
\text{of infection}
\end{array}
\qquad \dots (11.3)
$$

$$
\begin{array}{l}
\text{Annual} \\
\text{risk of} \\
\text{death}
\end{array}
=
\begin{array}{l}
\text{Number of} \\
\text{organisms} \\
\text{ingested} \\
\text{annually}
\end{array}
\times
\begin{array}{l}
\text{Probability} \\
\text{of infection} \\
\text{per ingested} \\
\text{organism}
\end{array}
\times
\begin{array}{l}
\text{Probability} \\
\text{of illness} \\
\text{per incidence} \\
\text{of infection}
\end{array}
\times
\begin{array}{l}
\text{Probability} \\
\text{of death} \\
\text{per incidence} \\
\text{of illness}
\end{array}
\qquad \dots (11.4)
$$

Validation of the Risk Estimation Technique

For the purpose of this chapter, validation of Eqs 11.1 to 11.4 and of the values for probability of infection and illness, presented in Tables 11.3 and 11.4, respectively, will be done using the findings from studies performed by Smith and others. These efforts by Smith and others include two epidemiological studies in which the occurrence of gastrointestinal illness was correlated with whether individuals ingest within their homes, community-treated tap water as provided directly from the community distribution mains or the same water which had been treated within their homes with reverse-osmosis water purification units. Reverse-osmosis units, when operating correctly, remove from water all micro-organisms, including viruses. During their first epidemiological study, Smith and others did not include microbiological data for the water. It was possible, however to derive appropriate microbiological data from other published studies performed by the same researchers. The second epidemiological study did include microbiological data.

Equations 11.1 through 11.4 do not include the fact that people often ingest water which has been subjected to drinking water treatment processes that would reduce its microbial concentration. If the values available for the number of organisms per litre of water represent untreated water and if the consumers are drinking untreated water, then Eqs 11.1 through 11.4 can be used directly without need for correction. Likewise, these equations can be used directly without need for correction if the consumers are drinking treated water and the values available for the number of organisms per litre of water represent treated water. However, if the consumers are drinking treated water and the values available for the number of organisms per litre represent untreated water, a mathematical correction will have to be made to account for the level of treatment. That correction is to multiply the estimated level of risk by a factor of [1 – (percentage of microbial reduction achieved by the water treatment process)]. These corrections will be noted, where appropriate, in the validation exercises presented below.

For the purpose of these validation exercises, the risks of infection and illness will be calculated separately for viruses capable of infecting cultured mammalian cells (and thus presumably capable of infecting humans), oocysts of *Cryptosporidium*, and cysts of *Giardia*, because the information from

Smith and others was made available in that manner. Estimations of the risks for infection and illness due to bacterial pathogens will be done as back calculations and are necessarily based on an assumption that any risk not accounted for by the enumerated viruses or protozoa was due to bacteria. This assumption is required since Smith and others reported data for the levels of indicator bacteria, which generally are not pathogenic for humans, rather than reporting the levels of pathogenic bacteria. These validation exercises do not include estimations of the number of deaths due to gastrointestinal illness which may have occurred during the epidemiological studies, because those authors did not supply information on the causes of death of the individuals who made up the different study groups.

Caveat Emptor Principle

Several important issues must be recognised and their implications must be understood when risk assessment estimations for infectious diseases are made. Six of these are listed below. The first three of these issues specifically apply to estimations for drinking water. The last three apply not only to drinking water but also to any infectious disease risks regardless of the source of the pathogenic organisms.

1. Estimations of the levels of micro-organisms in samples of water are inherently underestimations. This is due to the deficiencies of the detection methodology. Among these deficiencies, it is important to note that techniques for concentrating micro-organisms from environmental samples are less than 100 per cent efficient; many of the types of micro-organisms present in water samples cannot easily be cultured by the methods available; and some micro-organisms which normally would be culturable, if their metabolic condition were healthy, will not readily be detectable if their environmental exposure has resulted in metabolic shutdown associated with starvation.

2. To some extent, estimations of disease risk based on the levels of microbial contaminants in the municipal water distribution mains will be overestimations. This is because immediately prior to consumption, some of the water which people ingest is treated by heating techniques that will reduce the level of viable microbial contaminants in the water. This includes the heating of water to produce food items that are consumed hot, such as broths or soups, as well as the heating used in preparation of beverages like tea and coffee. However, heating water and then maintaining it warm for a long time, as occurs with water-heating tanks in homes and commercial buildings including nursing facilities, can change the risk in that additional pathogenic organisms such as bacteria of the genus *Legionella* can grow in water that is kept warm.

3. To directly estimate the level of disease risk associated with pathogenic bacteria present in water, it is best to enumerate the bacteria which actually are pathogenic. Most 'indicator bacteria' are not considered pathogenic, so information regarding the level of their presence in water cannot directly be used for making risk assessment estimates.

4. If death does not occur as the outcome of an initial infection by a particular strain of a specific micro-organism, then each subsequent reinfection of the same person by the same strain may carry a decreased risk that the infection would progress to illness.

5. Immunosuppression can increase the probability that an infection will have a severe outcome. Some natural conditions, such as pregnancy, and abnormal conditions, such as malnutrition, can result in suppression of immune system function. The result of this immunosuppression is that when individuals become infected, they may have a greater probability of death as the result of that infection. A notable example of this effect is that human illness due to hepatitis E virus normally has a 0.028 probability of death (which also can be stated as 2.8 per cent or

1 death/36 individuals). However, the probability of death for women in their third trimester of pregnancy at the time when they become ill from this virus ranges from approximately 0.17 (17 per cent or 1/6) to greater than 0.25 (25 per cent or 1/4). As another example, Hoxie observed that the overall risk of death from waterborne *Cryptosporidium*, associated with a large general population of humans who experienced an outbreak of infectious illness caused by that genus in Milwaukee, Wis., was 0.0001 (0.01 per cent or 1/10,000). However, the risk of death from *Cryptosporidium* when underlying malnutrition is present has been observed to be 0.14 (14 per cent or 1/7).

6. Concurrent infections (incidences when a person is infected simultaneously with more than one pathogenic organism) can drastically increase the probability of death associated with any one of the involved pathogens. As an example, the risk of death from *Cryptosporidium* when a person concurrently has the viral infection measles has been observed as 0.20 (20 per cent or 1/5), representing a 2000-fold increase over the rate of death from *Cryptosporidium* infections in the general population (serial no. 5 above). By itself, measles virus is almost never fatal; however, it also seems to be the most powerfully immunosuppressive virus known to infect humans.

Chapter 12

Methods of Enumeration and Activities of Micro-organism

INTRODUCTION

Methods employed in environmental microbiology are as diverse as microbial world. Some methods can be employed for more than taxonomic groups of microbes as cultural methods for bacteria and fungi with necessary modifications. In this chapter, various cultural, physiological, immunological, and nucleic acid based methods are discussed.

CULTURAL METHODS

Cultural Methods for Enumeration of Bacteria

The microbial ecologist is often interested in isolating or enumerating viable micro-organisms present in an environmental sample. This information may help establish the level of soil fertility or degree of contamination of soil or water by metals, toxic organics, or pathogens. In addition, cultural methods can be used to evaluate the diversity of microbial communities or quantitate a specific organism of interest. However, unlike the assay of pure culture samples, culture techniques involving the diverse microbial communities found in the environment are complex, and the numbers obtained need to be qualified by the culture technique that is used. To determine the appropriate culture method, it is necessary to define which specific micro-organism or group of micro-organisms is to be enumerated. The methods for enumerating bacteria, fungi, algae, and viruses are very different. In addition, within each group special enumeration techniques may be needed, as in the case of anaerobic or autotrophic bacteria. Precise, standard methods are necessary for assaying indicator organisms such as fecal coliforms or pathogens such as *Salmonella* spp. The type of environmental sample under analysis must also be considered when enumerating micro-organisms. Different techniques may be necessary for extraction of the organisms from the sample and, depending on the number of organisms of interest present in the particular sample, dilution or concentration of the sample may be necessary. Selective or differential growth media may also be necessary to observe and small population of a specific organism when an abundance of other organisms are also present in the sample. For example, it is usually necessary to enumerate *Salmonella* in sewage samples in which a large number coliforms are present.

Enumeration and isolation techniques

Extraction of cells from soil

For accurate viable plate counts from soil it is necessary to have efficient extraction of the micro-organisms that are attached to soil particles or are present in the pores of soil aggregates. Methods often employed include hand or mechanical shaking with or without glass beads, mechanical blending, and sonication. Different extracting solutions may be used depending on the pH and texture of the soil. A surfactant such as Tween 80 (Difco) may be used, often with the use of a dispersing agent such as sodium pyrophosphate.

Dilution step

The extraction step is followed by serial dilution of the sample to separate the micro-organisms into individual reproductive units. Sterile water, physiological saline, and buffered peptone or phosphate solutions are a few of the solutions commonly used for this step. Although water is convenient and commonly used, it is not preferred because it does not prevent osmotic shock during the dilution process.

Concentration step for water samples

Some environmental samples, such as marine water or drinking water, contain low bacterial numbers and require concentration rather than dilution before enumeration. In this method, a specified volume of water is filtered through a membrane using a vacuum. The bacteria are trapped on the membrane, which is placed on the agar medium or a cellulose pad soaked in medium to allow growth of individual colonies. This is the basis of the membrane filtration technique. Different volumes of water may need to be filtered to obtain the correct concentration of bacteria on the membrane for isolation and counting purposes. In this situation, it is critical to select a type and size of membrane appropriate to the bacteria to be collected.

Plating methods

After dilution or concentration, the sample is added to petri dishes containing a growth medium consisting of agar mixed with selected nutrients. Two different methods are used for application of the diluted sample to the growth medium. In the spread plate method, a 0.1 ml aliquot of selected dilutions of the sample is uniformly spread on top of the solid agar with the aid of a sterile glass rod. Alternatively, in the pour plate method, 1 ml aliquots of appropriate sample dilutions are mixed with molten agar (45°C) in a petri dish and allowed to solidify. The spread plate technique is advantageous in that it allows colonies to develop on the surface of the agar, making it easier to distinguish different micro-organisms on the basis of morphology.

It also facilitates further isolation of the colonies Fig. 12.1. The spread plate method generally gives bacterial counts that are higher than the pour plate method (for the same size of inoculant), perhaps because of improved aeration and desegregation of clumps of bacteria with this method.

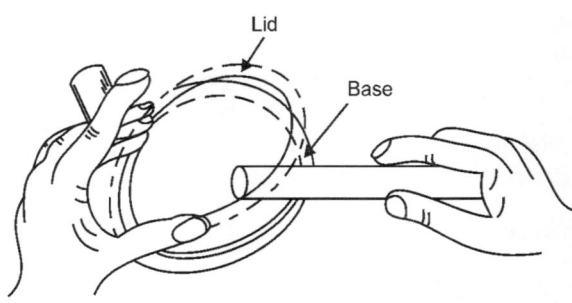

Fig. 12.1. Pour plate technique.

After plating, the samples are incubated under specified conditions, allowing the bacteria to multiply into macroscopic, isolated colonies known as colony-forming units (CFUs). Because it is assumed that each colony-forming unit originates from a single bacterial cell, it is critical for the organisms to be separated into discrete reproductive units in the dilution step prior to plating. In reality, however, colonies may arise from chains or clusters of bacteria, resulting in underestimation of the true bacterial number. The total number of bacteria of interest is calculated from the number of colonies found on a specific dilution. The range of colonies acceptable for counting is 30–300 on a standard 150 mm diameter agar plate. Below 30, accuracy is reduced; above 300, accuracy increases only slightly and, in fact, numbers may be reduced by overcrowding and competition between organisms growing on the plate.

The process of isolating micro-organisms from environmental samples often necessitates further isolation of individual colonies arising from the spread or pour plate method to allow further characterisation or confirmation. In this technique, a sterile inoculating loop is used to pick colonies from the original agar plate and the loop of bacteria is streaked to dilution on a new agar plate.

The standard plate count technique for the enumeration of micro-organisms is one of the oldest and most widely used techniques in microbiology. Despite its popularity, the dilution and plating technique has been subject to much scrutiny and criticism almost since its inception. One of the main criticisms is that only a small fraction of the total population, as observed microscopically, can be cultured on laboratory media. It is well documented that only 1–10 per cent of the number of cells observed with direct microscopic counts can be recovered as viable bacteria using plate count techniques. This large discrepancy between techniques may be due to several factors. First of all, environmental samples with low substrate levels may be populated with nutrient-starved or oligotrophic organisms, which are very slow growing or nonculturable and so do not appear as colonies in a reasonable time frame. Second, viable organisms may become nonculturable because of stress imposed on them during the extraction and dilution process. Another reason is that competition that occurs among bacteria to form colonies may suppress the growth of some microbes. In addition, all microbes have different optimal and minimal growth conditions with respect to nutrient and electron acceptor requirements. Thus, some microbes grow very slowly or not at all on the plating medium. Because most natural environments are oligotrophic, the nutrient status of the plating medium is an important point to consider when performing viable counts. Many soil isolates grow better on a nutrient-poor medium than on a rich medium.

Most probable number technique

The most probable number (MPN) technique is sometimes used in place of the standard plate count method to estimate microbial counts in the environment. In this method the sample to be assayed is dispersed in an extracting solution and successively diluted, as in the plate count. This method relies on the dilution of the population to extinction followed by inoculation of 5–10 replicate tubes containing a specific liquid medium with each dilution. After incubation, the tubes are scored as +/– for growth on the basis of such factors as turbidity, gas production, and appearance or disappearance of a substrate. Scoring a tube positive for growth means that at least one culturable organism was present in the dilution used for its inoculation. The number of positive and negative tubes at each dilution is used to calculate the number present in the original sample through the use of published statistical MPN tables or computer programmes designed to simplify the analysis. MPN tables can be found in American Public Health Association.

MPN methodology is useful because if allows estimation of a population of bacteria based on a process-related attribute such as nitrification by *Nitrosomonas* and *Nitrobacter* spp. or nitrogen fixation by *Rhizobium* spp. It is mandatory to use MPN analysis when enumerating a micro-organism requiring broth enrichment prior to culturing. It is also essential when assaying many foods, sediments, and soils. However, the MPN technique is very labour-intensive and results are usually less precise than those obtained with direct plating methods.

Cultural Media for Bacteria

General media used for culturing bacteria

Media for the cultivation of micro-organisms must contain substances that will support their growth, and the media available are as diverse as the micro-organisms themselves. The major components of microbiological media include: (i) a source of carbon for incorporation in biomass, such as glucose for heterotrophic bacteria or CO_2 for autotrophic bacteria, (ii) nitrogen, which is needed for growth and commonly supplied as ammonia, proteins, amino acids, peptones, or extracts from plants or meat, (iii) buffers to maintain a suitable pH, and (iv) growth factors such as defined trace minerals or metals, or undefined factors such as those found in extracts made from the environmental samples themselves. Agar is the most common solidifying agent used in media. It is a polysaccharide of an extract from marine algae. Agar exists as a liquid at high temperatures but solidifies on cooling to 38°C. Although agar ideally does not supply nutritional value to the medium, variations in growth may be observed with different types of agar. Bacteria may also be cultured in a broth medium lacking this solidifying agent. Many media also contain selective components that favour the growth of specific organism while inhibiting the growth of nontarget organisms.

While cultural media generally are used to enumerate or isolate specific micro-organisms, they also can be used in metabolic fingerprinting analysis to allow identification of microbial isolates. Such systems include API strips and Enterotube which are routinely used for clinical identification of gram-negative micro-organisms. For environmental microbiology, the most commonly use of the metabolic fingerprinting systems is Biolog. Biolog is used for identification of single isolates as well as for analysis of community composition. The basis for the Biolog system is a 96-well microtiter plate where 92 of the wells contain different carbon substrates. Each well is inoculated with the same isolate or community sample. If the substrate is utilised, the well turns purple. The plates can be read either manually or automatically using a plate reader. The results are then compared (in the case of isolates) with a database provided by Biolog. One of the drawbacks of this system for environmental isolates is that the database is not yet large enough to provide identification for most isolates. In the case of community samples, Biolog is used to provide a comparison of the community before and after a perturbation or simply to monitor the community for a period of time.

Heterotrophic plate counts

Heterotrophic bacteria obtain energy and carbon from organic substances, and it is often desirable to obtain a count of the total heterotrophic population from soil or water. Heterotrophic plate counts give an indication of the general 'health' of the soil as well as an indication of the availability of organic nutrients within the soil. Two basic types of media can be used for this analysis: nutrient-rich and nutrient-poor media. Examples of nutrient-rich media are nutrient agar, peptone-yeast agar, and soil extract agar amended with glucose and peptone. These media contain high concentrations of peptone, yeast, and/or extracts from beef or soil.

Culturing specific microbial populations

It is often necessary to detect or enumerate a specific population of micro-organisms, or even a very specific bacterial isolate, from the total population of bacteria found in an environmental sample. This may necessitate culturing with one or more specialised media and often requires that a specified sequence of steps be performed to ensure maximum culturability of the target environmental organism. We will now define the media involved in these steps and explain why they are important.

Pre-enrichment or resuscitation medium is a liquid medium that allows the micro-organisms in the environmental sample to begin to actively metabolise and increase in number. There are two purposes for a pre-enrichment medium. First, damaged cells are given time and the necessary nutrients to repair and grow. This step is important because direct inoculation of a sample into a selective enrichment medium (discussed next) may result in the death of some bacteria. This is especially critical for sublethally injured organisms, which may not be recoverable under selective conditions. The second purpose of pre-enrichment is to increase the number of cells. This step is crucial when trying to enumerate low numbers of a target organism in environmental samples.

Enrichment medium is a liquid medium that promotes the growth of a particular physiological type of micro-organism present in a mixture of organisms while suppressing the growth of competitive background flora. It aids in the detection of the desired environmental isolate when the specimen contains a high population of normal flora. Enrichment media may be elective or selective. Elective enrichment medium allows growth of a single or limited type of bacteria based on a unique combination of nutritional or physiological attributes. Selective enrichment medium involves the use of inhibitory substances or conditions to suppress or inhibit the growth of most organisms while allowing the growth of the desired organism. A detailed description of selective agents is given next.

A selective plating medium is a modification of an agar medium to suppress or prevent the growth of one group of organisms while fostering the growth of the desired organism. Antibiotics are among the most widely used and effective selection agents. A specialised isolation medium contains formulations that meet the nutritional needs of specific groups of organisms, such as *Staphylococcus* or *Corynebacterium*, thereby allowing differentiation and identification.

A differential medium contains ingredients to allow distinction of different microbes growing on the same medium. It includes an indicator, usually for pH, to distinguish certain groups of organisms on the basis of variations in nutritional requirements and the production of acid or alkali from various carbon sources.

Identification of Fecal Coliforms from water samples using the membrane filtration technique

Coliform bacteria are nonpathogenic bacteria that occur in the feces of warm-blooded animals. Their presence in a water sample indicates that harmful pathogenic bacteria may also be present. Coliforms are found in numbers corresponding to the degree of pollution, are relatively easy to detect, and are overall hardier than pathogenic bacteria. For these reasons, coliforms are important is that researchers may look for the presence of these 'indicator organisms' in an environmental sample to assess water quality prior to or in place of culturing other organisms.

Enumeration of Salmonella from sewage samples using the MPN technique

Salmonella is an enteric bacterium that is pathogenic to humans and causes a wide range of symptoms, primarily gastroenteritis. It can be transmitted by drinking improperly disinfected water or contaminated recreational water, but infection in the United States is primarily due to foodborne transmission because *Salmonella* infects both beef and poultry. Municipal sewage sludge contains many micro-organisms, including *Salmonella*. Some cities now apply treated sewage to agricultural soil to improve nutrient quality. Thus, it has become necessary to monitor the sewage for indicator organisms and pathogens such as *Salmonella* to assess health risks to the general population. As you will see from the following example, culture methods for *Salmonella* are varied, complex, and time consuming and require final confirmatory tests (Fig. 12.2).

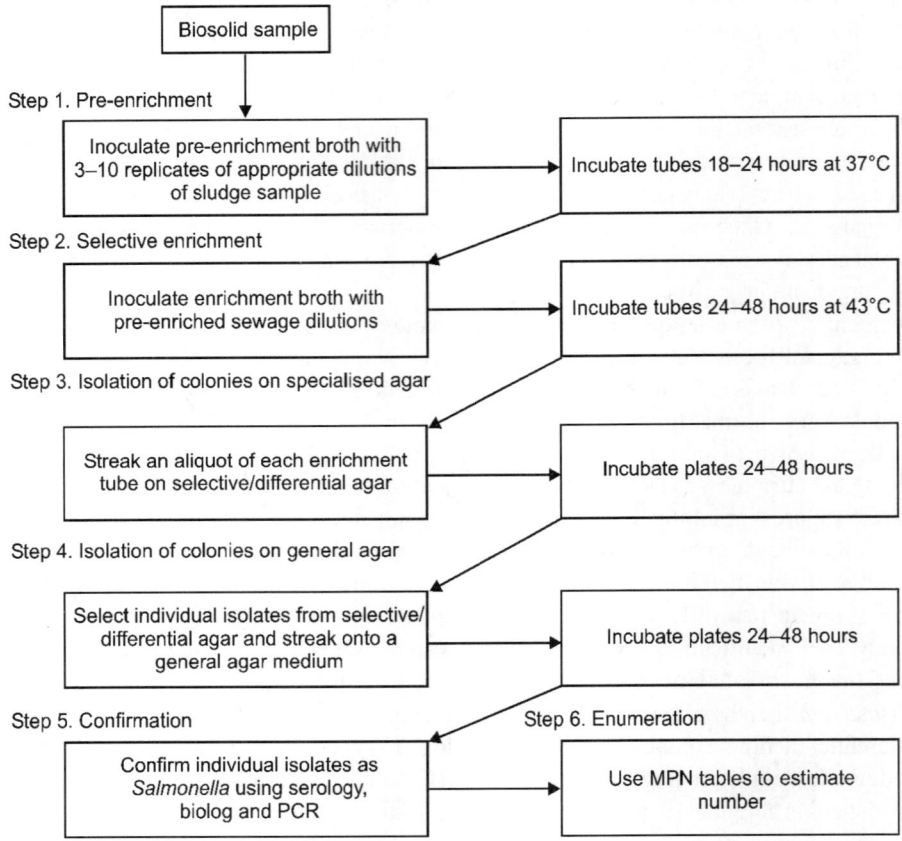

Fig. 12.2. Protocol for the detection of *Salmonella* from sewage sludge.

Isolation of Fluorescent Pseudomonads from soil

Pseudomonas are gram-negative aerobic chemoheterotrophs and are commonly found in the environment. They exhibit diverse enzymatic systems and are capable of conducting many biochemical transformations. Fluorescent pseudomonads are characterised by the production of siderophores. Siderophores are compounds that are produced under low levels of iron. They have the ability to chelate the iron and transport it into the microbial cell. These microbes also have a yellow-green pigment that diffuses through agar media during growth and fluoresces under ultraviolet light. Pigment production is enhanced by iron deprivation.

Isolation of nitrifying organisms from soil

Nitrosomonas and *Nitrobacter* are chemoautotrophic organisms found in soil and water and are responsible for the oxidation of ammonium to nitrite (*Nitrosomonas*) and nitrite to nitrate (*Nitrobacter*).

This process, known as nitrification, is important because it can affect plant growth beneficially, but nitrate also contributes to potable water contamination. Direct plating techniques are difficult, because even with use of strictly inorganic substrates in the medium, slight amounts of organic material introduced by inoculation with the environmental sample allow growth of faster growing heterotrophic organisms. One approach used to overcome this problem is a lengthy serial enrichment technique that includes soil enrichment, an initial culture enrichment step, a final enrichment step, isolation on agar, and rigorous purity check.

Isolation and enumeration of 2,4–D degrading bacteria in soil

Researchers are often interested in determining whether bacteria that perform a known metabolic function exist in an environmental sample. As an example, the research community is very interested in the fate of pesticides that are introduced into the environment. The herbicide 2,4-D (2,4-dichlorophenoxyacetic acid) is one such pesticide that has been commonly used in fate and ecology studies. A cultural medium used for enumeration and isolation of viable 2,4-D degraders is a selective, differential enrichment medium called eosin-methylene blue (EMB)-2,4-D agar. This agar contains minimal salts and 2,4-D as the carbon source.

Cultural Methods for Fungi

Fungi are ubiquitous in nature and can be found in samples taken from soil, sediments, and aquatic environments such as lakes, ponds, rivers, marine water, waste-water, and well water. They are heterotrophic organisms, mostly aerobic or microaerophilic in nature. Fungi exist in a variety of morphological and physical states, which makes them difficult to quantitate and identify by cultural techniques. Cultural methods for fungi are similar to those for bacteria but must be modified to restrict bacterial growth. This is normally done by the addition of antibiotics or dyes, or by lowering the pH of the medium. Cultural methods for fungi are normally used to obtain pure fungal isolates. Cultural methods for fungi are not, however, appropriate for quantitative dilution and plating analyses, because counts can be highly biased by the presence of spore-forming fungi. In this case, colonies can arise from spores, and the number of culturable colonies obtained will not be a true reflection of the number of colonies in the environmental sample.

Fungi are often isolated on a nonselective agar medium which allows the isolation of the maximum number of fungal taxa from the sample under study.

General cultural practices include the use of nitrate as a source of nitrogen and adjustment of the medium to an acidic pH. As already mentioned, it is often necessary to suppress the growth of bacteria to aid in fungal enumeration by the addition of growth inhibitors such as crystal violet or Rose Bengal or a broadspectrum antibiotic such as streptomycin. A lowered incubation temperature of $4°$–$10°C$ also helps fungi outcompete bacteria on the growth medium. The optimal number of colonies to count per plate is usually about 100.

Cultural Methods for Algae and Cyanobacteria

Algae are unicellular or multicellular phototrophic eukaryotic micro-organisms that occur in fresh and marine water and moist soil. In contrast, the so-called blue-green algae are not true algae—rather they belong to a group of bacteria known as cyanobacteria. The majority of soil algae are obligate photoautotrophs, using light to manufacture organic compounds from inorganic nutrients (phototrophic eukaryotes). Thus, their nutritional requirements include water, light, oxygen, carbon dioxide, and inorganic nutrients. A small population of algae are photoheterotrophic, requiring organic compounds for growth. Algae and cyanobacteria can be enumerated by dilution and plating techniques as well as MPN assay with some suggested modifications. Because of the filamentous nature of some algal species, it is sometimes necessary to use more forceful methods for extraction than with bacteria, prior to dilution and culturing techniques.

When isolating algae from environmental samples, the approach may be to select for a specific alga or to obtain a quantitative index of the entire community. In the latter case, an MPN technique using a growth medium consisting of a soil-water mix is often used. Many media are available for specific enrichment and plating, including Bold's Basa medium for green and yellow-green microalgae.

Cell Culture-Based Detection Methods for Viruses

Cell cultures are initiated by dissociating small pieces of tissue into single cells by treatment with a proteolytic enzyme and a chelation agent (generally trypsin and EDTA). The dispersed cells are then suspended in cell culture medium composed of a balanced salt solution containing glucose, vitamins, amino acids, a buffer system, a pH indicator, serum (usually fetal calf), and antibiotics. The suspended cells may be placed in flasks, tubes, or petri dishes, depending on the needs of the laboratory. Plastic flasks and multiwell plates are especially prepared for cell culture use and are currently the most common type of container used. The cells attach to the surface of the vessel and begin replicating. Replication ceases when a single layer or monolayer of cells occupies all the available surface of the vessel.

Unfortunately, not all animal viruses will grow in the same cell line, and others, including Norwalk virus, have never been grown in cell culture. Viruses require specific receptors for attachment and replication with the host cell, and if these are not present, no replication takes place. Cell lines commonly used to grow human viruses are shown in Table 12.1.

Table 12.1. Commonly used continuous cell cultures for isolation and detection of enteric viruses.

			Virus						
			Coxsackie	Coxsackie				Hepatitis	
Cell culture line	Adeno	Astro	A	B	Echo	Polio	Rota	A
BGM	$-^a$	−	−	+	+	+	−	−
BSC-1	−	*	−	*	*	+	*	*
CaCO$_2$	+	+	*	+	+	+	+	+
Hep-2 (HeLa)	+	*		+	*	*	*	*
RD	*	*	+	−	+	*	*	
RfhK	*	*	*	*	*	*	*	+

[a] +, growth and/or production of cytopathogenic effect (CPE); −, no growth and/or production of CPE; *, no data.

The two most common methods for detecting and quantifying viruses in cell culture are the cytopathogenic effect (CPE) method and the plaque-forming unit (PFU) method. Cytopathic effects are observable changes that take place in the host cells as a result of virus replication. Such changes may be observed as changes in morphology, including rounding or formation of giant cells, or formation of a hole in the monolayer due to localised lysis of virus-infected cells. Different viruses may produce very individual and distinctive CPE).

Generally, laboratory strains of viruses have been selected for their ability to grow in cell culture easily and rapidly with the production of CPE. In contrast, viruses isolated from environmental samples often come directly from infected animals or humans and do not grow as rapidly in cell culture. Often, 'blind passages' of cell culture showing no CPE are passed on to fresh monolayers of cells before a CPE is observed.

PHYSIOLOGICAL METHODS

In this section, various methods of mearurements of the activity of micro-organisms in pure and environmental sample have been discussed.

Device for the measurement of biological activities and/or physiological magnitudes, includes a measuring cell, equipped with a chamber suitable to receive micro-organisms to be analysed, and one or more probes opening into the chamber. The probes are connected to measuring resources and resources for processing the electrical signals emitted by the probes. The measuring cell includes a vertical well, equipped in its lateral wall with holes uniformly distributed around the well in order to allow the probes to open into the well. The probes are sealingly attached to the well, and rest in the support resources, and suitable to receive a cup whose cross section is homothetic to the section of the well. The cup is equipped on its lateral wall with holes which are located opposite to the holes of the well when the cup is positioned in the well.

Microbial Activity Measurements in Pure Culture

The activity of micro-organisims (in pure culture) can be measured in terms of appropriate parameters. This includes: (i) substrate disappearance, (ii) terminal electron acceptors, (iii) cell mass, and (iv) carbon dioxide evolution.

Substrate disappearance

For heterotrophic activity substrate is carbon-based, and such substrate can be measured in different ways depending on the chemistry of substrate molecule. Accordingly, these heterotrophic substrates are measured by using UV spectrophotometer, fluorimeter, high-performance liquid chromatograph (HPLC), gas chromatograph (GC), and mass spectrophotometer. For chemoautotrophic substrates, as oxidation of ammonia (nitrification), and sulphur (sulphur oxidation), different methodologies are employed. For nitrification ^{15}N-labelled products are measured by mass spectrophotometer, whereas sulphate formed during sulphur oxidation are measured by atomic absorption spectrophotometer.

Electron acceptor

An electron acceptor is a chemical entity that accepts electrons transferred to it from another compound. It is an oxidising agent that, by virtue of its accepting electrons, is itself reduced in the process.

Terminal electron acceptor

A terminal electron acceptor is a compound that receives or accepts an electron during cellular respiration or photosynthesis. All organisms obtain energy by transferring electrons from an electron donor to an electron acceptor. The process starts with the transfer of an electron from an electron donor. During this process (electron transport chain) the electron acceptor is reduced and the electron donor is oxidised.

Examples: Examples of acceptors include oxygen, nitrate, iron (III), manganese (IV), sulphate, carbon dioxide, or in some micro-organisms the chlorinated solvents such as tetrachloroethylene (PCE), trichloroethylene (TCE), dichloroethene (DCE), and vinyl chloride (VC). These reactions are of interest not only because they allow organisms to obtain energy, but also because they are involved in the natural biodegradation of organic contaminants. When clean-up professionals use monitored natural attenuation to clean up contaminated sites, biodegradation is one of the major contributing processes.

Inner cell mass

In early embryogenesis of most eutherian mammals, the inner cell mass [abbreviated ICM and also known as the embryoblast or pluriblast (the latter term being applicable to all mammals)] is the mass of cells inside the primordial embryo that will eventually give rise to the definitive structures of the fetus. This structure forms in the earliest steps of development, before implantation into the endometrium of the uterus has occurred. The ICM lies within the blastocoele (more correctly termed 'blastocyst cavity', as it is not strictly homologous to the blastocoele of anamniote vertebrates) and is entirely surrounded by the single layer of cells called trophoblast.

Carbon dioxide evolution

An alkaline trap, usually composed of NaOH or any other strong basic solution can be used to trap CO_2 produced during mineralisation. The trapped biocarbonate can be detected by using titrimetric, gravimetric, or conductimetric measurements. The most common method is to quantify trapped CO_2 by titration with standardised acid solution.

The use of ^{14}C-radiolabelled substrates allows very sensitive and specific measurements of $^{14}CO_2$ evolution. This technique is often used to evaluate the rates of biodegradation of organic contaminants. A biometer flask can be used, and the alkali within the sidearm trap is assayed for radioactivity using liquid scintillation counting.

Microbial Activity Measurements in Environmental Samples

Microbial activity in a pure culture, though can be easily determined by measurements of cell number (culturable plate or direct counts) or biomass (turbidity or protein content), these values are not realistic for environmental samples. In environmental samples, any such value is obtained for all forms of organisms present in the sample rather than an individual as in case of pure culture. Realistic information on microbial activity is in fact provided by environmental samples. In samples, activity is quantified in terms of respiration or synthesis of cellular macromolecules, which provide more direct reflection of levels of microbial activity within a sample.

Measurement of respiration gases (CO₂ and O₂)

As described earlier for pure cultures CO_2 and TEA (terminal electron acceptor) can be measured in environmental samples, both in laboratory as well as field conditions (i.e. *in situ*). In controlled conditions of a laboratory, a sample of porous medium is typically incubated in a sealed, airtight enclosure, usually referred to as a microcosm. Field studies can be performed by placing a field chamber over a plot of surface soil and using this setup to make *in situ* experiments. In microcosm, flux of CO_2 and/or O_2 within the headspace atmosphere is determined. Gas samples from headspace can be withdrawn using a gastight syringe and concentration of these gases measured by gas chromatography. Alternatively, CO_2 can be trapped in a basic solution using trap (biometer flask used for pure cultures). Gas chromatography is suitable method for field studies also.

Respiration measurements have several applications in environmental microbiology. These include the assessment of soil 'health' or condition as reflected by basal rate determinations of microbial activity in the samples; indictors of rates of biodegradation of pollutants, as petroleum hydrocarbons in contaminated sites; microbial biomass determinations; biological oxygen demand (BOD); and better understanding of the working of mixed microbial mats communities (biofilms).

Incorporation of radio-labelled tracers into cellular macromolecules

Protein or nucleic acid estimations can be used to monitor the increase in biomass of a bacterial population. In environmental samples biomass, specifically of bacteria is measured by incorporating radiolabelled tracer molecules into cellular macromolecules. Such tracers include nucleoside thymidine labelled with tritium (^{3}H) which is incorporated into DNA, and the amino acid leucine, labelled with either ^{3}H or carbon-14 (^{14}C) which is incorporated into protein.

Adenylate energy charge (AEC)

In terms of determining microbial activity, the relative abundance of ATP compared with its precursors, ADP and AMP indicates a rapid rate of ATP formation. The ATP/ADP/AMP ratio provides a biochemical basis for assessment of physiological and nutritional status of organisms. A measure of AEC ratio is the weighted ratio of cellular adenylates:

$$AEC = \frac{ATP + 1/2\ ADP}{ATP + ADP + AMP}$$

High AEC values (>8.0) reflect an active community, intermediate values (0.4 to 0.8) reflect cells in resting state, and low values (<0.4) reflect a high proportion of dead or moribund cells. Estimation of AEC values from environmental samples can be used as an indicator of bacterial biomass. Quantification of ATP from samples needs an extraction procedure, followed by concentration of cellular components into a buffer. The ATP is then determined directly using the luciferin-luciferase assay. ADP and AMP in the sample are then converted to ATP by enzymatic reactions and quantified.

Enzyme assays

All enzyme assays measure either the consumption of substrate or production of product over time. A large number of different methods of measuring the concentrations of substrates and products exist and many enzymes can be assayed in several different ways. Biochemists usually study enzyme-catalysed reactions using four types of experiments:

Initial rate experiments: When an enzyme is mixed with a large excess of the substrate, the enzyme-substrate intermediate builds up in a fast initial transient. Then the reaction achieves a steady-state kinetics in which enzyme substrate intermediates remains approximately constant over time and the reaction rate changes relatively slowly. Rates are measured for a short period after the attainment of the quasi-steady state, typically by monitoring the accumulation of product with time. Because the measurements are carried out for a very short period and because of the large excess of substrate, the approximation free substrate is approximately equal to the initial substrate can be made. The initial rate experiment is the simplest to perform and analyse, being relatively free from complications such as back-reaction and enzyme degradation. It is, therefore, by far the most commonly used type of experiment in enzyme kinetics.

Progress curve experiments: In these experiments, the kinetic parameters are determined from expressions for the species concentrations as a function of time. The concentration of the substrate or product is recorded in time after the initial fast transient and for a sufficiently long period to allow the reaction to approach equilibrium.

Transient kinetics experiments: In these experiments, reaction behaviour is tracked during the initial fast transient as the intermediate reaches the steady-state kinetics period. These experiments are more difficult to perform than either of the above two classes because they require rapid mixing and observation techniques.

Relaxation experiments: In these experiments, an equilibrium mixture of enzyme, substrate and product is perturbed, for instance by a temperature, pressure or pH jump, and the return to equilibrium is monitored. The analysis of these experiments requires consideration of the fully reversible reaction. Moreover, relaxation experiments are relatively insensitive to mechanistic details and are thus not typically used for mechanism identification, although they can be under appropriate conditions.

Enzyme assays can be split into two groups according to their sampling method i.e. continuous assays, where the assay gives a continuous reading of activity, and discontinuous assays, where samples are taken, the reaction stopped and then the concentration of substrates/products determined. Continuous assays are most convenient, with one assay giving the rate of reaction with no further work necessary. There are many different types of continuous assays.

Spectrophotometric assays

In spectrophotometric assays, you follow the course of the reaction by measuring a change in how much light the assay solution absorbs. If this light is in the visible region you can actually see a change in the colour of the assay, these are called colorimetric assays. The MTT assay, a redox assay using a tetrazolium dye as substrate is an example of a colorimetric assay.

Direct versus coupled assays

Coupled assay for hexokinase using glucose-6-phosphate dehydrogenase. Even when the enzyme reaction does not result in a change in the absorbance of light, it can still be possible to use a spectrophotometric assay for the enzyme by using a coupled assay. Here, the product of one reaction is used as the substrate of another, easily-detectable reaction.

Fluorimetric assays

Fluorescence is when a molecule emits light of one wavelength after absorbing light of a different wavelength. Fluorometric assays use a difference in the fluorescence of substrate from product to measure the enzyme reaction. These assays are in general much more sensitive than spectrophotometric assays, but can suffer from interference caused by impurities and the instability of many fluorescent compounds when exposed to light.

Calorimetric assays

Chemiluminescence of Luminol Calorimetry is the measurement of the heat released or absorbed by chemical reactions. These assays are very general, since many reactions involve some change in heat and with use of a microcalorimeter, not much enzyme or substrate is required. These assays can be used to measure reactions that are impossible to assay in any other way.

Chemiluminescent assays

Chemiluminescence is the emission of light by a chemical reaction. Some enzyme reactions produce light and this can be measured to detect product formation. These types of assay can be extremely sensitive, since the light produced can be captured by photographic film over days or weeks, but can be hard to quantify because not all the light released by a reaction will be detected.

The detection of horseradish peroxidase by enzymatic chemiluminescence (ECL) is a common method of detecting antibodies in western blotting. Another example is the enzyme luciferase, this is found in fireflies and naturally produces light from its substrate luciferin.

Light Scattering assays

Static light scattering measures the product of weight-averaged molar mass and concentration of macromolecules in solution. Given a fixed total concentration of one or more species over the measurement time, the scattering signal is a direct measure of the weight-averaged molar mass of the solution, which will vary as complexes form or dissociate. Hence the measurement quantifies the stoichiometry of the complexes as well as kinetics. Light scattering assays of protein kinetics is a very general technique that does not require an enzyme.

Discontinuous assays

Discontinuous assays are when samples are taken from an enzyme reaction at intervals and the amount of product production or substrate consumption is measured in these samples.

Radiometric assays

Radiometric assays measure the incorporation of radioactivity into substrates or its release from substrates. The radioactive isotopes most frequently used in these assays are ^{14}C, ^{32}P, ^{35}S and ^{125}I. Since radioactive isotopes can allow the specific labelling of a single atom of a substrate, these assays are both extremely sensitive and specific. They are frequently used in biochemistry and are often the only way of measuring a specific reaction in crude extracts (the complex mixtures of enzymes produced when you lyse cells). Radioactivity is usually measured in these procedures using a scintillation counter.

Chromatographic assays

Chromatographic assays measure product formation by separating the reaction mixture into its components by chromatography. This is usually done by high-performance liquid chromatography (HPLC), but can also use the simpler technique of thin layer chromatography. Although this approach can need a lot of material, its sensitivity can be increased by labelling the substrates/products with a radioactive or fluorescent tag. Assay sensitivity has also been increased by switching protocols to improved chromatographic instruments (e.g. ultra-high pressure liquid chromatography) that operate at pump pressure a few-fold higher than HPLC instruments.

Factors to control in assays

Some of the factors to control in assays are discussed below:

1. Salt concentration: Most enzymes cannot tolerate extremely high salt concentrations. The ions interfere with the weak ionic bonds of proteins. Typical enzymes are active in salt concentrations of 1–500 mM. As usual there are exceptions such as the halophilic (salt loving) algae and bacteria.

2. Effects of Temperature: All enzymes work within a range of temperature specific to the organism. Increases in temperature generally lead to increases in reaction rates. There is a limit to the increase because higher temperatures lead to a sharp decrease in reaction rates. This is due to the denaturating (alteration) of protein structure resulting from the breakdown of the weak ionic and hydrogen bonding that stabilise the three dimensional structure of the enzyme. The 'optimum' temperature for human enzymes is usually between 35 and 40°C. The average temperature for humans is 37°C. Human enzymes start to denature quickly at temperatures

above 40°C. Enzymes from thermophilic archaea found in the hot springs are stable up to 100°C. However, the idea of an 'optimum' rate of an enzyme reaction is misleading, as the rate observed at any temperature is the product of two rates, the reaction rate and the denaturation rate. If you were to use an assay measuring activity for one second, it would give high activity at high temperatures, however if you were to use an assay measuring product formation over an hour, it would give you low activity at these temperatures.

3. Effects of pH: Most enzymes are sensitive to pH and have specific ranges of activity. All have an optimum pH. The pH can stop enzyme activity by denaturating (altering) the three dimensional shape of the enzyme by breaking ionic, and hydrogen bonds. Most enzymes function between a pH of 6 and 8; however pepsin in the stomach works best at a pH of 2 and trypsin at a pH of 8.

4. Substrate saturation: Increasing the substrate concentration increases the rate of reaction (enzyme activity). However, enzyme saturation limits reaction rates. An enzyme is saturated when the active sites of all the molecules are occupied most of the time. At the saturation point, the reaction will not speed up, no matter how much additional substrate is added. The graph of the reaction rate will plateau.

5. Level of crowding: Large amounts of macromolecules in a solution will alter the rates and equilibrium constants of enzyme reactions, through an effect called macromolecular crowding.

IMMUNOLOGICAL METHODS

Immunology is the study of the immune system of higher organisms in relation to disease. Specifically, immunology can be defined as the branch of biology that is concerned with the structure and function of the immune system, the bodily distinction of self from nonself, and the use of antibody based laboratory techniques or immunoassays. In general, the immune system of higher organisms can be broken down into two primary response systems that work together to create immunity. The two primary response systems are the cell-mediated and the antibody-mediated responses. The cell-mediated response is produced when sensitised white blood cells or lymphocytes directly attack material, which has been determined to be foreign to the body. The antibody-mediated response involves the transformation of a subset of lymphocytes into cells that produce and secrete specific antibodies against these foreign objects. These two immune responses are triggered when foreign material is introduced into the host as depicted in Fig. 12.3.

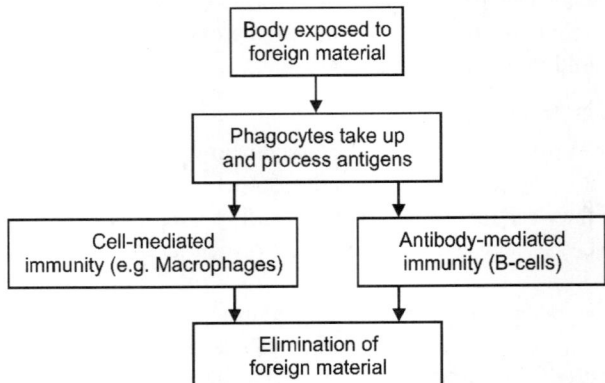

Fig. 12.3. Flow chart showing the two primary immune response systems that comprises the host response to foreign materials. Together these two branches of the immune system work together to create immunity. The cell-mediated response is designed to directly attack and destroy material determined by the body to be non-self. The antibody-mediated response is the branch of the immune system involved in the formation of antibodies.

Environmental Microbiology does not deal with all aspects of immunology or the immune responses *per se* but instead adapts immunology-based research technologies or immunoassays for the study of micro-organisms in association with the environment. The primary immunologic-based tool used in environmental microbiology is the antibody. In this section, introduction to antibodies is given, with respect to structure of antibodies, the various classes of antibodies [immunoglobulin G (IgG), IgA, etc.], and the interaction of antibodies with foreign objects (antigens). Following this introduction, several of the basic immunological methodologies (immunoassays), that are widely used in environmental microbiology, are discussed. These immunoassays include fluorescent immunolabelling, enzyme-linked immunosorbent assay (ELISA), magnetic bead antigen capture, Western immunoblotting, immunoaffinity chromatography, and immunoprecipitation. Finally, in order to provide perspective and illustrate how these immunoassays can be used in the field of environmental microbiology, an example of each immunoassay is provided in relation to current research topics such as bioremediation and pathogen detection.

Antibody

Antibodies are protein complexes produced by the immune system of higher life forms that help defend the host against foreign invasion. When a host is challenged by foreign material (bacteria, virus, toxins, etc.) the first response of certain host immune cells called macrophages is to engulf these invaders (antigens) and process them biochemically. This biochemical processing essentially creates a blueprint that is used for the development of an immune response that results in the production of antibodies. The unique feature of antibodies produced in response to an antigen is that they are synthesised in such a way that they are highly specific for that antigen. Thus, they can chemically react and bind only with that particular antigen, neutralise it, and aid in its destruction and removal from the body.

There are five different classes of antibodies or immunoglobulins (Igs): IgA, IgD, IgE, IgG and IgM. These immunoglobulins (antibodies) differ in many ways including their overall structures. The most common type of antibody used for immunoassays is the IgG class of immunoglobulins (Fig. 12.4). IgG antibodies are Y-shaped proteins composed of four protein chains that are joined together by disulphide linkages.

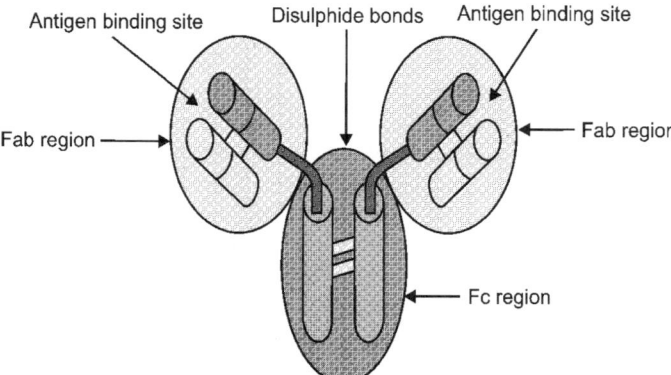

Fig. 12.4. Schematic representation of an IgG antibody, showing the various regions associated with the antigen/antibody interaction. There are two antigen-binding fragments (Fab), which interact with the antigen. There is also one crystal fragment (Fc), which is the part of the antibody recognised by the host immune system as self. There are two light chains joined to two heavy chains by disulphide bonds and the two heavy chains are in turn joined to each other in a similar fashion.

Antibody specificity

Specificity for a particular antigen is one characteristic that makes immunology-based methodologies such valuable tools. In essence, once they are produced, antibodies are very precise in recognition of the particular antigen. This discrimination is based on the molecular structure of the antigen binding sites located on the Fab portion of the antibody and on the epitopes or chemically reactive sites of the antigen. Antigen-antibody binding is the result of specific chemical interactions (i.e. charge-charge, dipole-dipole, hydrogen bonding, and van der Waals) that occur between the antigen and amino acid residues of the antibody that are located on the Fab region.

Antibody affinity

Affinity is defined as the attraction between an antibody and an antigen. More specifically, affinity is a measure of the strength of this interaction and is usually expressed as an interaction or association constant. Quantitatively, affinity is the sum of the chemical bonds that form between the antigen and the antibody. These are usually relatively weak interactions such as hydrophobic interactions and hydrogen bonds. Even though individually such chemical bonds are relatively weak, collectively they form very strong and tight interaction.

Polyclonal and monoclonal antibodies

In the past, injecting a laboratory animal with the antigen of interest produced particular antibodies needed for environmental and other research purposes. In response to immunisation, the animal produces antibodies that can be collected (in serum) directly from the blood of the animal. These blood-derived antibodies are termed polyclonal antibodies because they are not derived from a single B lymphocyte but are instead a product of many different B cells reacting in slightly different ways to the same antigen. As a result, this method yields a very impure product that contains a wide variety of antibodies, proteins, and blood factors.

To avoid problems associated with use of such antibody mixtures, scientists developed the technology to produce monoclonal antibodies. A monoclonal antibody is an antibody that is the product of a single B cell. Production of monoclonal antibodies involves the *in vitro* combination of two types of cells. The first type of cell is a B cell that produces a single, unique antibody. The second type of cell is an immortalised myeloma cell: a cancer cell that is able to thrive and multiply *in vitro*. The specific antibody-producing B cell is fused with the myeloma cell to form a hybrid cell called a hybridoma. This hybridoma combines the characteristic of 'immortality' with the ability to produce the desired specific antibody in high concentrations and in pure form. As a result of the development of monoclonal antibody technology, scientists are able to produce large amounts of pure and highly specific antibodies.

Antiglobulins

Antiglobulins, as indicated earlier, are antibodies that are specific (usually targeting the Fc portion) for another individual antibody. Usually, antiglobulins are developed to recognise a whole antibody class for a specific organism, e.g. a mouse. Because antibodies are large proteins with complex structures, they have the potential to be seen as antigens if they do not have the 'self' recognition sites common to the host that produced the antibody.

Immunoassays

Immunoassays are analytical methods used for the detection and/or quantitation of the antigen-antibody interaction. For the most part, the types of immunoassays used in environmental microbiology are based on quantitation or detection of antigens as opposed to characterisation of the antigens. That is, we are usually interested in using immunoassays to determine how much antigen is in an environmental sample and not in characterising an antibody/antigen interaction or the role of the antigen in disease or in the immune response. However, in order to quantitate or detect the antigen there must be a way to visualise the antigen-antibody interaction. This visual signal is produced by the attachment of specific signal molecules to the antibodies or antiglobulins used to detect the antigen within an environmental sample.

For almost all types of immunoassays, attachment of a signal molecule to the antibody and/or antigen is very important. Many types of signal molecules are used in immunoassays, including iodine, enzymes, fluorochromes, and radioisotopes. These signal molecules produce a visual signal that allows quantitation of the specific antibody-antigen interaction being investigated.

As just mentioned, attachment of an antibody or antigen to a signal molecule is an almost universal way to allow for visual detection via immunoassays. There are also two universal formats for immunoassays. These are termed direct labelling and sandwich (often termed indirect) labelling. Essentially with direct labelling, the primary antibody (antibody specific for the target) has the signal molecule attached to it allowing for one-step detection. Sandwich or indirect labelling involves two steps. The first is the attachment of a primary antibody to the target, and the second is the attachment of a secondary (antiglobulin) antibody to the primary antibody. In indirect labelling, the secondary antibody has the signal molecule attached. Both methods work well though both have advantages and disadvantages. With direct labelling, the binding and signal are usually more specific because there is a smaller signal to background noise ratio. However, the use of sandwich labelling allows for one labelled antibody to be used with many different primary antibodies provided they are all of the same type, meaning that each primary antibody does not have to be labelled separately. For instance, if you use a mouse to produce monoclonal IgG primary antibodies against 4 different protozoa (*Microsporidia*, *Giardia*, *Cryptosporidium*, and *Entamoeba*), then you can use the same labelled antiglobulin to bind to each of these. Because conjugating signal molecules to antibodies is tedious and often difficult, and because a wide range of antiglobulins conjugated to various signal molecules are available commercially, this may be the format of choice. However, if you desire a one-step assay that is slightly more specific, then the use of direct labelling is often preferred.

Fluorescent immunolabelling

Technique

Fluorescent immunolabelling (immunofluorescence) is the use of fluorescent signal molecules conjugated to antibodies to interact with and subsequently indicate the presence of a particular antigen by the production of fluorescent light. The basic procedure for immunofluorescence microscopy, for example, is to attach the sample antigen to a microscope slide, add a fluorescent chemical/antibody conjugate specific to the antigen, and view the sample under a microscope equipped with a fluorescent light source. When viewing fluorescence-labelled samples under the microscope, the labelled antigen appears bright green against a dark background. Immunofluorescence microscopy is one of the easiest and most widely used immunoassays. Immunofluorescence microscopy was first used in environmental microbiology for the study of *Rhizobia* in soil.

Application

One of the main uses for immunofluorescence is in the detection of protozoan parasites in water. The protozoan parasites *Giardia* sp. and *Cryptosporidium* sp. are two of the main causes of diarrhea in humans and many animals and are transmitted by the fecal-oral route, often through the contamination of surface waters.

Enzyme-linked immunosorbent assays

The enzyme-linked immunosorbent assay (ELISA) is a very sensitive laboratory method used to detect the presence of antigens. There are many different approaches to ELISAs. They are typically performed as a direct or an indirect sandwich method but call also be performed as a competitive assay. For any ELISA procedure the antigens of interest are concentrated (if necessary) and solubilised in an appropriate buffer. For a direct sandwich ELISA, a primary antibody is attached to a microtiter plate, microcentrifuge tube, or other solid support. The antigen is then added and allowed to incubate in order to bind with the antibody (antigen capture). After the antigen is bound, a second antibody called the signal antibody is added. These secondary antibodies are conjugated to a signal molecule, for example, the enzyme alkaline phosphatase. This secondary antibody then binds to the antigen (antibody capture), which is already bound to the primary antibody. Substrate is then added that causes a colour change in response to the presence of the signal molecule. This colour change is usually in proportion to the amount of antigen present; thus, the assay becomes quantitative. This makes it possible to quantify the amount of antigen present in a given sample. Once a signal is produced it can be used to visually score the results based on the colour change or an automated plate reader can be used. Plate readers provide highly sensitive detection of low-level signals and can determine accurately the strength of a given signal in comparison with a standard curve.

Application

In many environmentally relevant processes such as the degradation of pollutants in water treatment plants and bioreactors, the concentration of certain bacterial species within biofilms is very important. Biofilms are specialised environments where micro-organisms are firmly attached to surfaces and to one another by exopolymeric substances. Previously, the enumeration of specific organisms within biofilms was done using cultural counts, most probable number, or immunofluorescence microscopy.

Advantages and disadvantages

As mentioned, there are many advantages of using ELISA over other detection or quantification methods. ELISA is sensitive and can be quantitative when used in conjunction with standard curves. Disadvantages are similar to all antibody-based methods and are related to cross-reactivity and non-specific signal production. ELISAs must also be optimised to provide consistent results especially when using environmental samples.

Competitive ELISA

In competitive ELISA, both a labelled control antigen and a sample, that contains an unknown quantity of unlabelled antigen are added to a sample well coated with antibody. In this assay, the sample is added first and the bound antibody captures the unlabelled antigen. The labelled antigen is then added and is captured by any remaining antibody binding sites. If all sites are taken, no labelled antigen will bind and no signal will be emitted.

Advantages and disadvantages

The advantage of competitive ELISA lies in its ability to detect extremely low antigen concentrations. This is related to the inverse relationship between target concentration and signal strength described previously. The disadvantage is that these assays require a great deal of optimisation and must be used in conjunction with a plate reader.

Immunomagnetic separation assays

Magnetic immunoseparation is an antigen capture methodology that uses antibodies conjugated to paramagnetic beads to attach to, concentrate, and purify antigens. Immunomagnetic separation is rapidly becoming one of the more popular approaches to the specific manipulation of micro-organisms, proteins, and nucleic acids. The increase in popularity of this approach is due to its ease, mild conditions, low cost, and the ability to automate the process. In its most simplistic form, an immunomagnetic separation is accomplished with antibody-coated magnetite beads and a magnet. Essentially, the antibodies coated on the magnetic beads bind with antigens in solution and are then separated from the solution using a magnet (Fig. 12.5).

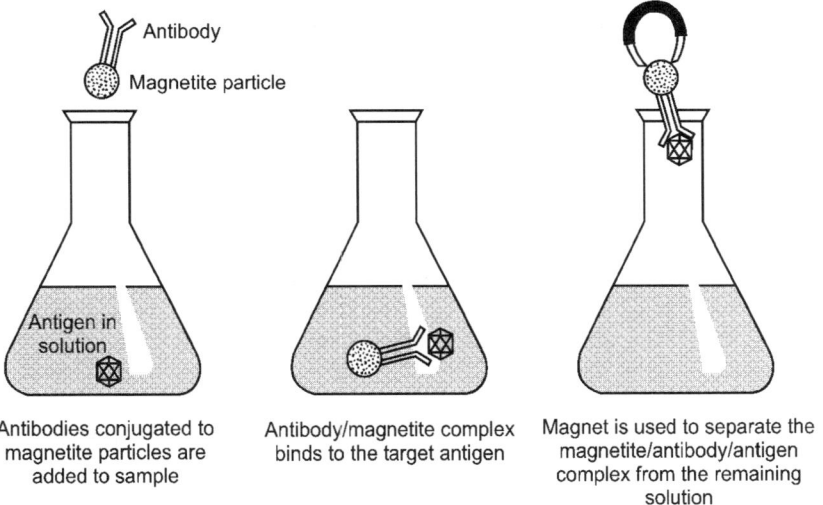

Antibody

Magnetite particle

Antigen in solution

Antibodies conjugated to magnetite particles are added to sample

Antibody/magnetite complex binds to the target antigen

Magnet is used to separate the magnetite/antibody/antigen complex from the remaining solution

Fig. 12.5. This is a schematic representation of the principle of immunomagnetic separation. In the first flask is a solution containing the antigen (virus). Antibodies specific to the virus are added to this solution. These antibodies, which have been conjugated (attached chemically) to a paramagnetic particle such as magnetite, then bind the virus. When a magnet is applied to the solution the magnetic particle is attracted to the magnet along with the attached antigen (virus). Immunomagnetic separation is a valuable tool for concentrating and purifying antigens from complex solutions.

Application

Immunomagnetic separation was used to recover thermophilic sulphate-reducing bacteria from oil field waters below oil production platforms in the North Sea. These bacteria can proliferate in oil field waters and cause considerable problems for oil companies during oil recovery, so the ability to detect their presence is of great value to the petroleum industry.

Western immunoblotting assays

Western immunoblotting is a three step, binding assay used to identify the presence of target antigens in a complex mixture of many other nontarget antigens such as might be found in environmental samples. This assay can be done with simple dot blot hybridisation with a labelled antibody or an electrophoretic separation followed by hybridisation with the labelled antibody. In dot blot hybridisation, environmental samples are added directly to an immobilising nitrocellulose membrane, followed by immunolabelling and signal detection. In the second technique a sample of antigen is added to a gel and separated by size using electrophoresis. After electrophoretic separation, the sample is transferred to an immobilising nitrocellulose membrane. This membrane is then incubated with enzyme-labelled or radiolabelled antibodies, that specifically bind to the antigen. After incubation, a substrate for enzyme-labelled or photographic film for radiolabelled substrate is used to detect the presence of the target antigen. Either method (dot blot or electrophoretic separation) indicates the presence and relative quantity of an antigen. If a separation step is used, this also allows molecular size determination of the antigen, which aids in confirming its identity.

Advantages and disadvantages

Obvious advantages of immunoblotting lie in its ability to specifically detect and quantify a particular antigen (target) within heterogeneous matrices. Further, immunoblotting can detect extremely low levels of target antigen. However, this procedure can be very time consuming and is subject to the problems inherent to all antibody-based methods.

Immunoaffinity chromatography assays

Affinity chromatography is a very powerful method used in purification and concentration of antigens. In affinity chromatography, the antibody is chemically bound to an inert support matrix (usually a glass, latex, or plastic bead) in a chromatography column. The sample containing the antigen is eluted through the column, and the antigen is selectively retained within the column while the sample passes through. After the sample is run through the column, the purified antigen is eluted, usually by changing the pH of the column, which causes the antigen to detach from the antibody. The antigen can then pass out of the column and be collected in highly purified form. This process provides a very efficient means of both concentration and purification. Immunoaffinity chromatography offers several advantages compared with conventional purification techniques. Not only is the process selective and efficient, it also enables the processing of large-volume samples with relatively few steps.

Immunocytochemical assays

An immunocytochemical assay is used for the detection and determination of the cellular localisation of target antigens. The purpose of an immunocytochemical assay is to determine where target antigens are localised within a particular cell. For instance, you can determine whether the target antigen (a particular protein for example) is localised in the cytoplasm or on the cell surface. In many cases, light microscopes are used to determine the location of antigens within a eukaryotic cell. However, immunocytochemical assays more often involve the use of electron microscopes to increase resolution and magnification of the area being studied.

Immunoprecipitation assays

Immunoprecipitation is a methodology that uses the antigen-antibody reaction in solution to semiquantitatively determine the amount of antigen or antibody in a sample by determining the amount of precipitation or clumping of the antigen-antibody complex. Immunoprecipitation can be used to determine the concentration of low levels of antigen or can be used to quantify or titer antibodies or antigens. Immunoprecipitation can also be used to determine the optimal concentration ratio for an antibody and antigen. Most commonly, a series of reaction tubes are set up, each of which contains a constant titer of antibodies. Antigen is then added in increasing concentration to consecutive tubes. In the initial tubes, where the lowest concentration of antigen has been added, there is no obvious precipitation. As the antigen concentration is increased, the formation of antigen-antibody complexes increases until a visible precipitate is formed. As the antigen concentration is further increased it will eventually exceed the concentration of antibody present and the amount of precipitate will decrease again (Fig. 12.6).

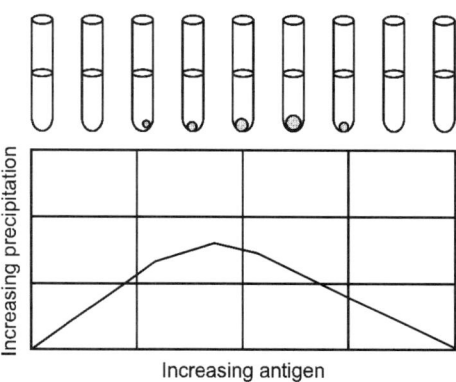

Fig. 12.6. This is a schematic representation of an immunoprecipitation assay. The test tubes at the top all have constant concentrations of antibody and increasing concentrations of antigen starting from the left. In the middle tubes there is optimum antigen-antibody interaction and precipitation of the antibody and antigen occur. As indicated by the graph on the bottom there is a point where the highest amount of precipitation forms. This type of assay is used to determine optimum antibody to antigen ratios for immunoassays. It is also useful for quantitating, either antibody or antigen concentrations in solutions.

Application

This technology has been utilised in environmental microbiology to determine the mechanism for the inhibition of certain economically important fungal plant pathogens by another nonpathogenic fungus. It was hypothesised that the fungus *Talaromyces flavus* can control the proliferation of several fungal plant pathogens including *Sclerotinia sclerotionum*, *Rhizoctonia solani*, and *Verticillium dahliae* by the production of hydrogen peroxide. *T. flavus* produces hydrogen peroxide as a product of glucose metabolism in the presence of the enzyme glucose oxidase. To determine whether hydrogen peroxide was responsible for the inhibition, cultures containing the pathogens were incubated with and without glucose oxidase. The cultures without glucose oxidase showed a high percentage of germination, whereas cultures containing glucose oxidase exhibited very low germination (inhibition). To confirm these results, the cultures with glucose oxidase were subjected to immunoprecipitation. Antibody to the glucose oxidase was added to the cultures, effectively removing the glucose oxidase from the culture. After immunoprecipitation of the glucose oxidase, the fungal pathogens recovered and showed high levels of germination. Germination was subsequently halted again when more glucose oxidase was added. This research effectively showed that glucose oxidase is the enzyme produced by *T. flavus* that controls the plant pathogens.

NUCLEIC ACID-BASED METHODS OF ANALYSIS

The advent of molecular biology has created a new array of methodologies for examining micro-organisms in the environment. The premise of these methodologies relies on basic concepts that have created some very powerful tools. Techniques such as gene probing and polymerase chain reaction (PCR) have made possible a very specific and sensitive evaluation of the microbial world. Microbiologists are now able to use a small sample of microbial nucleic acids to identify unculturable bacteria, track genes, and evaluate genetic activity in the environment.

Nucleic Acid-Based Methods

Gene probes and probing

Gene probes are an application of nucleic acid hybridisation. Typically, probes are small pieces of DNA known as oligonucleotides, which are complementary to the target sequence of interest, that are marked or labelled in some way in order to make them detectable. Probes have been used in environmental microbiology to examine soil microbial diversity, to identify a particular genotype, and to test for virulence genes of suspected pathogens that have been isolated from water.

Gene probe methodology takes advantage of the fact that DNA can be denatured and reannealed. To make a gene probe, the DNA sequence of the gene of interest must be known. This gene may be unique to a particular microbial species, in which case the sequence may be useful for the specific detection of that organism. Alternatively, the gene may code for the production of an enzyme unique to some metabolic pathway, and a gene probe constructed from such a sequence may indicate the potential activity of a group of bacteria in a soil or water sample. This kind of probe can be defined as a functional gene probe. For example, gene probes can be made from sequences that code for enzymes involved in nitrogen fixation, and such a probe could then be used to estimate whether a particular soil contained any bacteria with nitrogen-fixing genes. Additional probes could later be constructed to determine whether such organisms were in fact nitrogen-fixing *Rhizobium* or *Azospirillum* spp. or even cyanobacteria. The gene could even be universal to all bacteria, thus allowing detection of all known bacteria.

The basic strategy in the construction of a gene probe is to obtain the sequence of the target gene and then to select a portion of this sequence for use as a probe. The size of the probe can range from 18 base pairs to as many as several hundred base pairs. The probe is then synthesised and labelled in such a way that it can be detected after it hybridises to the target sequence.

Colony lifts or hybridisation

Gene probes can also be used to detect a specific gene sequence within bacterial colonies on a petri plate containing a mixed population of bacteria through use of a process termed colony hybridisation or lifts. To perform a colony hybridisation, a piece of filter paper is lightly pressed onto the petri plate so that some bacterial cells from each colony adhere to the paper. The cells are lysed directly on the filter paper, and the DNA is fixed to the filter. The filter is then probed and detected as described above. After this procedure, only the colonies that contain the specific DNA sequence give a radioactive signal. Because the original petri plate contains all the intact colonies, the viable colony of interest can now be identified and retained for further study (Fig. 12.7).

Southern and Northern hybridisations

Another application of gene probe technology is to identify a target sequence of interest via Southern or Northern hybridisations, which are also known as blots. Southern blotting is a technique used to identify a DNA sequence. For example, it may be important to know whether a gene is plasmid or chromosomally borne. To determine whether the sequence is plasmid borne, all the plasmids within the strain can be extracted and separated by gel electrophoresis. The plasmid DNA is then transferred onto a special membrane by blotting, and the membrane is subsequently probed. Once again, only the DNA molecules that contain the target sequence hybridise with the probe, thus allowing detection of those plasmids containing the target sequence.

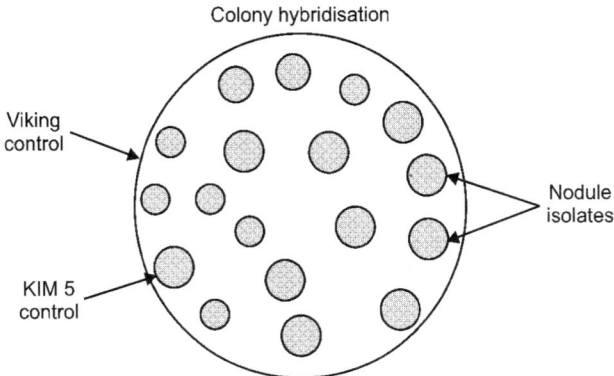

Fig. 12.7. A colony lift from a petri plate containing a mixed population of the nitrogen-fixing bacteria rhizobia isolated from root nodules. The gene probe was constructed from a unique plasmid associated with the isolate known as KIM 5.

Microarrays

A microarray refers to a set of DNA probes, usually purified PCR products from DNA or genomic clones, deposited on a solid support (generally a glass microscopic slide) with a spot density of several hundred individual spots per cm². Arrays have been used primarily for gene expression profiling. For example, RNA is extracted from cells which have been exposed to stress and cells that have not been exposed to stress. A cDNA copy of the mRNA is labelled enzymatically by the inclusion of a fluorochrome labelled nucleotide during an amplification step. The array is hybridised with the target, the labelled cDNA copy of the mRNA, and the hybridisation intensity is acquired by laser scanning and CCD camera detection. Since it is known where on the array that the individual DNA probes are located, the result is a global fingerprint or signature of the genes that are expressed as a result of a specific stress. This technology has been used in the pharmaceutical industry to look at drug responses. However, the applications of this technology seems limitless. In environmental microbiology microarrays have the potential to allow screening of a drinking water supply for several hundred pathogens with one assay.

Polymerase chain reaction

The PCR, a technique used to amplify the amount of target DNA up to 10^6-fold or more, has revolutionised molecular biology methodologies. First discovered in 1985, PCR has become a key protocol in many biological laboratories, including those concerned with environmental microbiology. The ability to amplify small amounts of DNA that may be present in the environment makes detection very sensitive. For example, the detection of 100 *Salmonella* cells in 1 gram of soil using traditional cultural techniques would be very difficult. However, amplifying the 100 cells to the equivalent of 10^6 cells makes detection much easier. PCR is a relatively simple enzymatic reaction that uses a DNA polymerase enzyme to copy a target DNA sequence repeatedly during a series of 25–30 cycles. During each cycle of PCR, the amount of target DNA is doubled, resulting in an exponential increase in the amount of DNA.

Recombinant DNA techniques

Recombinant DNA technology or DNA cloning has been widely used in environmental microbiology to examine the genetic structure of bacterial populations. Cloning has enabled scientists to find new or closely related genes, as well as characterise and identify unculturable or unknown isolates. Cloning may also be used to examine the activity of specific genes.

A clone is a foreign DNA fragment that is replicated in a host organism after being shuttled in by a cloning, vector such as a plasmid. Cloning basically involves three steps: (i) choosing the source of DNA for cloning, (ii) producing a collection of DNA fragments that can be inserted into a vector and in many cases ultimately creating a cDNA library, and (iii) screening for the desired sequence of interest. Thus, cloning results in a population of organisms that contain recombinant DNA molecules. Following screening for the target sequence of interest, a particular clone can be propagated to amplify the recombinant molecules, resulting in a large mass of the DNA sequence of interest.

SECTION IV

Waste-water Treatment

13. **Detection of Viruses in Environmental Water,
    Sewage and Sewage Sludges**   205

14. **Detection of Bacterial Pathogens in Waste-water and Sludge**   220

Chapter 13

Detection of Viruses in Environmental Water, Sewage and Sewage Sludges

INTRODUCTION

A great many different groups of viruses are found in environmental waters. These include the many types of viruses whose hosts are natural aquatic organisms. There also are groups of viruses present in environmental waters which represent exogenous contaminants, whose hosts are nonaquatic plants and animals. Those water-associated viruses which are of greatest concern from the viewpoint of human public health are viruses which replicate in cells of the human gastrointestinal tract. These are referred to as human enteric viruses, and they are shed in fecal material. The principal transmission route for enteric viruses is fecal-oral, meaning that they can cause illness when a susceptible host ingests fecally contaminated water or food. Septic tank effluents and waste-water from human populations doubtlessly contribute many of the human enteric viruses that are found in surface waters and groundwaters. The presence of human enteric viruses in surface water can also result from human recreational activities performed either in or around those bodies of water.

At least some of the human enteric viruses can cross infect, meaning that they are capable of also replicating within animals other than, humans. As an example, acquisition of the disease poliomyelitis, caused by a member of the genus *Enterovirus*, has been documented in a wild population of chimpanzees that were living in relative proximity to a human community. Other genera of enteric viruses which should be considered to have the potential for cross-infection of related mammalian host species are *Orthoreovirus*, *Rotavirus*, and *Mastadenovirus*. Likewise, some enteric viruses of animals likely can cross-infect humans. Sources of such viruses are fecal wastes from animals, including farm waste-water which can drain into surface waters, overland runoff containing animal fecal wastes, and fecal wastes directly deposited in the water by animals. It is difficult, and in some instances it may be impossible, to determine the host species from which many of these potentially cross-infective enteric viruses originate. Thus, for the purpose of protecting public health, the presence in water of any virus capable of replicating in mammalian host animals or cultured mammalian cells is presumed to represent a pathogenic hazard to humans.

Enteric viruses are capable of surviving improperly operated conventional drinking-water treatment processes and may, therefore, be found in potable water. This knowledge has resulted in the development of methods for detecting both human enteric viruses and also fecally associated bacteriophages in water. These bacteriophages are not considered to be enteric in nature, because they infect the intestinal bacteria rather than the animal host, and correspondingly they themselves do not pose a public health threat. Bacteriophages are considered unsuitable in terms of meeting regulatory requirements for assessing enteric virus pollution due to problems with lack of specificity and sensitivity for many of the assay techniques and the lack of an association between detection of bacteriophage and disease occurrence in human populations. Instead, these bacteriophages have been studied as potential indicators of the presence of fecal material and on the premise that they may prove useful as indicators by which to gauge the fate of human enteric viruses within the natural environment. They also may prove useful as indicators of the removal or destruction of human enteric viruses during the course of water treatment processes. Selectively choosing the bacterial hosts employed in the assay process, including the decision as to whether the bacterial host strains used would be considered indigenous or non-indigenous to the environment being sampled, is an approach which may be used to detect specific bacteriophage populations. Conversely, microscopic methods and even automated techniques such as flow cytometry have been used to detect total bacteriophage populations in samples of water.

The levels of bacteriophages in environmental waters, particularly those that receive sewage effluent, are often sufficiently high that the water can be directly assayed for the presence of these bacteriophages without need for resorting to concentration techniques. This represents a desirable advantage, since viral concentration techniques have a measure of inherent inefficiency, which results in some loss of virus. Also, some virus inactivation may occur during concentration processes as a result of such factors as changes in pH and the addition of chelating chemicals like EDTA. The levels of human enteric viruses present in environmental waters often are so low as to require that they be concentrated from the water in order to perform a successful, cost-effective assay. This chapter describes techniques that can be used for concentrating human enteric viruses from environmental water, drinking water, raw waste-water, waste-water sludges, and waste-water effluents.

MECHANISMS INVOLVED IN VIRAL CONCENTRATION METHODS

A large variety of approaches have been used for concentrating viruses contained in water samples. These can be separated into five major categories based upon the mechanisms involved.

Passive Adsorption

Passive adsorption relies upon unaided entrapment of viruses via adsorption or absorption into pads of gauze and cotton, used in stationary or flowthrough configurations. This is followed by expression of the entrapped fluids, either with or without the supplemental use of an eluant solution, to recover the viruses.

Directed Adsorption

Directed adsorption entails adsorption of viruses to the surface of filter materials or granular solids, facilitated in many instances by pretreatment of the water. Processes which have been evaluated as pretreatments include using charged resins to remove dissolved organics from the water, adjustment of the water sample pH, and addition of salts to the water. Efforts have also been made to pretreat the adsorbent material itself in order to enhance the efficiency of viral adsorption. Examples of the latter approach include the binding of metal precipitates or charged polymers to the matrix of microporous filters. Subsequent recovery of the bound viruses from the filter materials or granular solids used as adsorbents is achieved by exposing the adsorbents to an eluant solution.

Ultrafiltration

Ultrafiltration involves retention of viruses in the original water sample during a reduction of its volume achieved by pore size exclusion. This generally is accomplished by either recirculating the water, under pressure, through hollow-fibre filters, or over the surface of flat-sheet filters in either recirculating or nonrecirculating systems. In some cases an eluant is subsequently passed through the same concentration unit to facilitate virus recovery.

Direct Physico-chemical Flocculation and Phase Separation

Direct physico-chemical flocculation and phase separation are representative of processes by which chemicals are used that cause a water sample to partition into two parts. One of those parts will contain the viruses, which in effect results in concentration of the viruses. This category of viral concentration methodology relies upon (i) physico-chemical flocculation (the viruses' associating with a precipitate which forms when chemicals are added to the water); (ii) polymeric phase separation (the viruses' selectively partitioning with one of two aqueous phases which are generated and gravimetrically separate when soluble polymeric substances are added to the water sample); and (iii) hydro-extraction (placing the water sample into a tubing made of water-permeable membrane, with that tubing in turn packed in a hydroscopic polymeric substance which draws water out through the pores of the membrane, to result in the viruses and only a small amount of residual water being retained within the tubing).

Affinity Chromatography

Affinity chromatography describes the process whereby viruses are retained during passage of the water sample through a column of polysaccharide gel particles which bear covalently linked antibodies that specifically recognise and bind a particular virus type. Viruses are then released from the column by alterations in the composition or ionic strength of the buffer fluid used for maintaining hydration of the polysaccharide gel.

DIRECTED ADSORPTION, ELUTION AND RECONCENTRATION OF VIRUSES

Directed adsorption has generally replaced the other types of concentration techniques for use in recovering viruses from large sample volumes of water. The different types of solid adsorbents that have successfully been utilised for this can be divided into two groups: filters and granular solids. The types of filters that have been tested are composed either of barn fibre wound around a hollow core to form a depth filtration cartridge, sheet filter materials configured either as flat layers or in the form of cartridges composed of pleated sheets of filter material, and cartridges prepared as hollow tubes of filter material. Viruses adsorb onto the filter matrix during the passage of virus-containing water samples through these filters. Recovery of the adsorbed viruses is normally achieved by subsequently passing an eluant through the filters or by dissolution of the filter if it is made of alginate. Granular solids generally are used as viral adsorbents in one of three modes: batch utilisation by which the granules are mixed into the water sample and then recovered by using either filtration or magnetic attraction, the latter requiring that the granules be magnetic; as layers of granules either supported on or sandwiched between sheets of non-virus-adsorbing flat filter material through which the water sample is passed; or as columns or fluidised beds of granules exposed to the water sample using a flowthrough configuration. Subsequent desorption of viruses from the granular solids occurs during exposure of the granules to an eluant using one of these same three modes.

Of the many different types and configurations of virus adsorbents, the ones that currently seem to be preferred for use in recovering viruses from large volumes of water are wound-yarn cartridge filters; pleated-sheet cartridge filters based upon either glass microfibre or nylon, including the use of sheet filter material types that are modified to have a more electropositive charge; and columns of glass powder. Filters seem to offer some advantage over the use of glass powder columns in that the filters are easily transported to the field (even by postal shipment) and readily used in the field, after which, if packed properly, the filters may be shipped back to the processing laboratory with very little concern about detrimentally affecting either the filter or the adsorbed viruses. This chapter presents the use of cartridge filter-based methods for concentrating viruses from environmental water, drinking water, and sewage effluent. The methods presented in this chapter for isolating viruses from raw sewage (raw waste-water) and waste-water sludges effectively utilise a process of directed adsorption and elution but differ in that they rely upon the waste-water solids to serve as an *in situ* adsorbent.

The adsorption of viruses onto filters, and presumably other solid adsorbents as well, is governed by both electrostatic interactions, which predominate at lower pH levels, and hydrophobic interactions, which predominate at higher pH levels. The chemical composition of the solid adsorbent and the fluid in which the viruses are suspended will influence the extent to which either attraction or repulsion occurs between the viruses and the adsorbent. Two other important factors that relate to virus adsorption from water include the relative proximity and amount of contact time allowed between the adsorbent and those viruses contained in the water sample. The latter factors are influenced by the rate at which the water being sampled flows either past or through the solid adsorbent and the ratio of the diameter of the virus particle to the diameter of any pores in the adsorbent material through which the water may pass. Virus adsorption can be inhibited by the presence of added proteins in the input virus suspension, possibly as a result of competition between the proteins and viruses for those adsorption sites available on the solid matrix. Other types of soluble organics such as humic and fulvic acids may also interfere with virus adsorption. Means of facilitating the adsorption of viruses onto solid matrices include first removing dissolved organic materials from the water samples by passing the water through a resin column and then adding salts, including chlorides of either sodium, magnesium, or, more effectively, aluminum. Virus adsorption to filters can also be facilitated by adjusting (usually lowering) the pH of a water sample, a process that can be performed by adding HCl either through a batch operation or using an in-line injector configured ahead of the filter. A pH of 3.5 to 4.0 seems preferred for use with negatively charged filters, whereas near neutrality is preferred for the more positively charged filters. There may also be optimal pH levels for adsorption of viruses to the different types of granular solids.

The adsorption process is reversible, and the reverse process is termed elution. Elution is accomplished by exposing the adsorbent to an aqueous solution termed an eluant, following which the eluant is termed an eluate and contains at least some of the viruses which had been adsorbed to the solid matrix. Eluants can be divided into at least two categories based upon their mode of action. The first category includes proteinaceous materials which simply compete with the proteins of the virus particles for binding sites on the adsorbent. The second category consists of compounds that alter the favourability of adsorption. Beef extract is now the most predominantly used eluant material of the first category and is recommended in this chapter, although other protein products may prove to be suitable substitutes. The second category of eluants includes solutions that contain various active substances, among which are chaotropic agents like glycine or trichloroacetic acid, detergents like Tween 80, and EDTA, which serves as a chelating agent for metal cations. Both categories of eluant may utilise elevated pH, which would help to decrease the electrostatic attraction between the viruses and adsorbents. Eluants containing viruses which have been desorbed from filters generally are reduced in volume by a secondary, or second-step, concentration technique prior to their being assayed for the presence of viruses. The secondary concentration method customarily used for beef extract-based eluants is organic flocculation as introduced by Katzenelson and co-workers, which is performed by lowering the pH of the eluate, during which the proteinaceous material supplied by the beef extract spontaneously precipitates. Viruses contained in the beef extract become associated with this precipitate, which can subsequently be collected by centrifugation and dissolved in a small amount of higher-pH salt-based buffer, such as sodium phosphate or glycine. A pH of approximately 3.5 appears optimal for recovering viruses from beef extract eluates via organic flocculation.

Sampling Apparatuses and Processing Equipment

A variety of sampling apparatuses may be needed when detecting viruses in environmental waters. These include not only the usual sampling jars or bottles but often also buckets; carbuoys, pumps, cartridge filters and their holders, and hoses. If it is necessary to store a volume of sample water which is too large to fit in a standard jug or carbuoy, then even a new sterilised plastic garbage can with a tightly fitting cover can be used as a water storage container. Whenever possible, it is preferable that containers used for water samples be made of polypropylene, which can reduce unintentional adsorption of viruses to container walls. Alternatively, it is suggested that the apparatus be made of either stainless steel, brass, or a chemically resistant polymer. This will aid in terms of ruggedness for use in field operations, cleanability, and resistance to the chemical corrosion that can be caused by salt or extremes of pH.

It is necessary to neutralise any chemical disinfectant present in the water being sampled to prevent disinfectant related die-off of the viruses. Neutralisation must be done before the water sample is processed on-site or transported for off-site (i.e. laboratory) processing. This necessity often occurs with drinking water and waste-water. The chemical disinfectants most commonly used by water treatment facilities are chlorine, dioxide, and monochloramine, any of which can be neutralised by the addition of sodium thiosulphate. If concentration of viruses from such water samples is to be performed in the field, then a compact portable chlorine test kit can be used to confirm that, by adding the prescribed quantity of sodium thiosulphate solution, you have successfully neutralised all free chlorine or other chlorine compounds present in the water sample. If concentration of viruses from water samples in the field is to be done using cartridge filters, then a portable pH meter may be needed, since it might be necessary to adjust the pH of the water samples prior to filtration. When using virus-adsorbing filters that are considered electropositive in nature, you generally will only need to adjust the acidity of water whose pH is above approximately 7.5. If negatively charged filters are used, then the sample water, must be adjusted to a pH of approximately 3.5 and supplemented with aluminum chloride prior to filtration. Positively charged filters thus offer a tremendous advantage in terms of their ease of use. However, the positively charged, filters are more expensive and do not work well for concentrating viruses from saline waters. You should use only negatively charged filters for concentrating viruses from saline waters.

Cartridge Filtration Apparatus

The types of apparatus used for concentrating viruses from large volumes of environmental waters, drinking water, or sewage effluent by means of cartridge filtration are not standard equipment for most environmental microbiology oratories. Thus, the equipment used with the cartridge filters will be described in some detail. Figure 13.1 presents the type of equipment configuration used for concentrating viruses by means of cartridge filters. A lightweight portable electric or gasoline-powered water pump fitted for use with either brass or stainless steel quick-disconnect plumbing adapters or garden hose couplings can be used to supply pressure for passing water or sewage effluent through cartridge filters. Such a pump will not be needed if the water or waste-water is being supplied directly via a tap from a pressurised source. Figure 13.1 also shows two filter holders designed for use with standard 10-inch-long cartridge water filters, fitted for use with either quick-disconnect plumbing adapters or garden hose couplings. The first holder contains a non-virus-adsorbing prefilter that may be needed to prevent clogging of the virus-adsorbing filter, which is contained in the second filter holder. If it is necessary to modify the water pH or to add aluminum chloride to aid viral adsorption, then an in-line injector can be configured between the pump (or pressurised tap) and the first filter holder. An alternative to using an injector is to first pump the water sample into a large temporary storage tank, make any necessary chemical adjustments (supplementations) to the water while it is in that tank, and then pump the adjusted water from the tank through the filters. The filter holders should have clear base sections so that the filtration process can be visually monitored, particularly for air pockets which occasionally develop within the filter holders and tend to impede the water flow. Suitable holders should be available from the manufacturers and distributors from which the cartridge filters are obtained. Water being processed through these filters flows from the outside of the filter, through the filter material, into the hollow core. It is helpful if the top of each filter holder is fitted with a stainless steel finger-operated air pressure release valve connecting to the inflow side of the filter holder. The installation of air pressure release valves in this manner allows easy elimination of air pockets from the filter holders. Figure 13.1 also shows a portable water meter fitted for use with either quick-disconnect plumbing adapters or garden hose couplings attached to the outlet end of the second filter holder. Suitable meters are available from plumbing suppliers and used for metering the amount of water which is passed through the filters. Four lengths of fibre-reinforced garden hose, likewise fitted for use with either quick disconnect plumbing adapters or garden hose couplings, are used for connecting the pump, filter holders containing their cartridge filters, and water meter to form the virus concentration apparatus shown in Fig. 13.1.

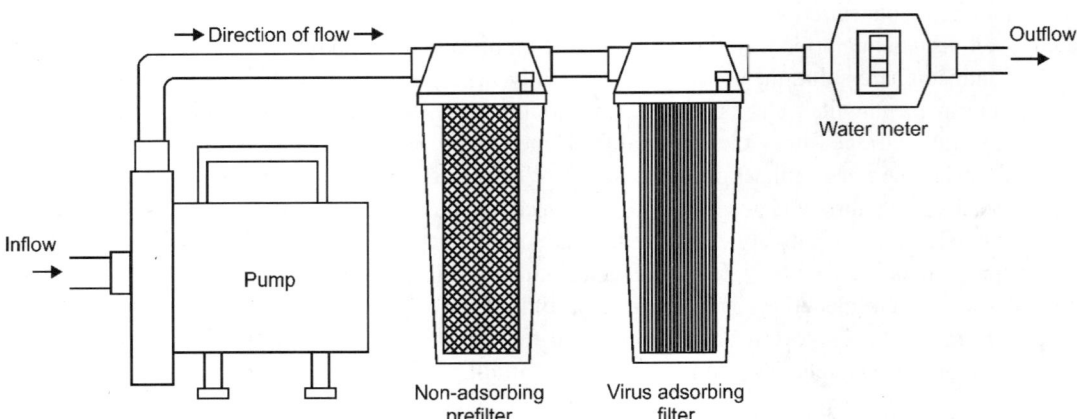

Fig. 13.1. Diagram of apparatus used for concentrating viruses using cartridge filters.

The fourth piece of hose is used to direct outflow from the water meter. Also, one length of strong-walled water supply hose, fitted for use with either quick-disconnect plumbing adapters or garden hose couplings, will be required for leading water into the pump. Standard garden hose may not be suitable for this usage since many types of garden hose would tend to collapse inward during operation of the pump. It may be helpful to have a stainless steel or brass wire strainer fitted to the intake end of the water supply hose when concentrating environmental waters by pumping directly from their source through the filters. Use of a strainer in this instance will help to prevent the water supply hose from clogging. The suggested maximum force for driving a water sample through these filters is 30 lb/in^2. If the pump being used is particularly powerful, it may be necessary to install a flow-restricting device ahead of the first filter holder. An ice chest containing either wet ice or commercial ice packs will be needed for transport (or shipment via overnight delivery service) of the used virus-adsorbing filter, inside a polypropylene container of eluant, to the laboratory where the elution and secondary concentration steps are to be performed.

Sterilisation Requirements

All reagents that will be used during any of these methods must be sterilised prior to their use. Solutions of HCl or NaOH which are at least 1 M in strength are considered to be self-sterilising, as are solutions of at least 0.5 per cent (wt/vol) sodium hypochlorite or calcium hypochlorite. Containers used to store water or waste-water samples should be presterilised. All parts of equipment that will come into contact with the samples should likewise be visibly clean and sterile. The only parts of a pump or fluid injector which need to be sterile are those which come into contact with the water sample. These parts can be treated using a hypochlorite solution sterilisation technique. The choice of methods to be used for sterilising different items of equipment will depend upon the size of the equipment and the item's resistance to heat. Many types of apparatus can be sterilised in the laboratory before their use in the field, either by autoclaving, ethylene oxide gas, or, if accessible, gamma irradiation. All of the cartridge filters described in this study can be presterilised within their holders using ethylene oxide gas treatment before they are transported to the field. Alternatively, some of the filters have sufficient resistance to heat and steam that they can be sterilised by autoclaving.

Techniques for Field Sterilisation of Sampling Equipment

Field sterilisation of metal or borosilicate glass objects can be performed using flame from a portable gas torch (such as propane or butane) to heat the sample collection objects to red incandescence. Objects less resistant to high heat, such as non-borosilicate glass and some polymer materials, can be surface sterilised by dousing or immersing them in commercial 95 per cent ethanol and then igniting the alcohol with a flame. Nearly all objects described in this chapter, with the exception of filters and liquids, can be subjected to field sterilisation techniques which consist of chlorine-based chemical treatment using solutions such as calcium hypochlorite or sodium hypochlorite followed by the use of sodium thiosulphate to neutralise the residual disinfectant activity.

If the water sample is to be placed into a storage container for any reason, then that container should be visibly clean, must be sterile, and preferably should not previously have been used for any purpose other than the storage of water samples. Large (up to 50 gallon and some even larger) polyethylene containers with lids are available from scientific supply companies and can be used for storage of water samples. In an emergency situation, as mentioned earlier in this chapter, a new plastic garbage container with a lid can be used for this purpose. Field sterilisation of containers can be done by first completely filling them with clear water and then adding calcium hypochlorite solution (0.5 per cent, wt/vol) to the water at a rate of 3.8 ml per gallon. A 10 per cent solution of standard household liquid bleach (sold as 5.25 per cent sodium hypochloride by weight) can be used as a substitute for the calcium hypochlorite solution. The hypochlorite solution should be throughly mixed into the water, and this water should then be allowed to remain in the storage container for at least 15 minutes. The chlorinated water should then be emptied from the storage container, and the container should be rinsed very thoroughly with clear (lacking obvious turbidity) water and again completely filled with clear water. Sterile sodium thiosulphate solution (50 per cent, wt/vol) should be added to this second filling of water at a rate of

10 ml per gallon. The sodium thiosulphate solution should thoroughly be mixed into the water, and the water should then be allowed to remain in the storage container for at least 5 minutes. The container should then be emptied and filled with the intended water sample.

All of the equipment used in the virus concentration apparatus (Fig. 13.1), except for the filters, can be field sterilised using this type of chlorine-based technique. The apparatus (shown in Fig. 13.1) should be disassembled prior to treatment by the field sterilisation technique in order to make certain that all surfaces that may later come into contact with the sample water will have received thorough exposure to the chlorine disinfectant and that all residual hypochlorite is then neutralised before the water sample is processed through the system. Hoses and assembled filter holders (without filters), with their air pressure release valves open, should be carefully submerged in a solution of clear water containing hypochlorite prepared as described above for treatment of storage containers. Be certain that there are no pockets of trapped air in either the hoses or filter holders, as this may preclude thorough treatment. Keep the hoses and filter holders completely submerged in this chlorinated water for a minimum of 15 minutes, and then remove and drain the hoses and filter holders. Next, similarly treat the hoses and filter holders by submersion for a minimum of 5 minutes in clear water containing sodium thiosulphate solution prepared as described above for treatment of sample storage containers. The hoses and filter holders should be drained and can then be used immediately, or, if they are to be used at some later time, their openings should be wrapped with sterile aluminum foil or some other type of sterilised covering to prevent contamination. Field sterilisation of pumps and injectors can be done by first continuously recirculating clear water containing hypochlorite solution through them for at least 15 minutes. Residual chlorinated water should then be drained from the pump or injector and its necessarily attached hoses. Clear water containing sodium thiosulphate solution, likewise prepared as described above, should then be recirculated through the pump or injector and attached hoses for at least 5 minutes. Residual water containing thiosulphate solution should then be drained from the pump or injector and attached hoses. The pump, injector, and hoses can then immediately be used to process a water sample, or, if not used immediately, their openings should receive a sterile covering as described above to prevent contamination. When properly performed, this chlorine-based field sterilisation technique should be sufficient to destroy the infectiousness of viral contaminants on the treated surfaces of the equipment. This technique may not be sufficient to destroy some encysted protozoans which are resistant to chlorine.

DETECTION OF VIRUSES

Detecting viruses in water involves a three-stage process. The first stage consists of either collecting a grab sample which will be processed in the laboratory or performing a field operation in which the water being sampled is passed through cartridge filters, with those filters usually then returned to the laboratory for processing. The grab sampling approach is likely to be utilised when either the level of viruses or the content of solids is relatively high, as is the case for raw waste-water or waste-water sludges. If the level of viruses and the concentration of solids are expected to be relatively low — as is the case with most environmental surface water, 'groundwater', treated drinking water, and waste-water effluents—then the cartridge filtration approach is likely to be used. The second stage consists of laboratory elution processing, performed by either of two methods. The first method is used for unfiltered grab samples and consists of concentrating viruses onto the solids naturally present in a sample and then eluting viruses from those solids. The second method, used for cartridge filters, consists of eluting adsorbed viruses from those filters. Usually either type of eluate will be further reduced in volume through a subsequent concentration process. The third stage consists of assaying viruses contained in the eluates.

Viral Concentration Techniques

The viral concentration techniques presented in this chapter are categorised according to the type of environmental sample being examined. These are environmental waters (fresh or saline), treated drinking water, waste-water effluent, raw waste-water, and waste-water sludges.

Environmental water, drinking water and sewage effluent

The procedure described below can be used to concentrate suspected human enteric viruses from environmental surface waters, including brackish and marine waters; groundwater; tap water; and sewage effluents. The produced concentrated sample materials can then be examined for the presence of viruses. The viruses are first adsorbed onto cartridge membrane filters. A beef extract-based eluant fluid is then used to desorb viruses from the filters. Viruses contained in the eluate are then secondarily concentrated by low-pH organic flocculation. The volume of water sample that is processed when looking for the presence of human enteric viruses depends upon the level of viruses likely to be present in the water. For sampling sewage effluents, the minimum useful sample volume may range from 15 to 30 litres. For general surface waters, a minimum useful volume would probably be 100 litres, with 150 to 200 litres preferred. For water that has very low turbidity, the apparatus described here could be used for processing volumes as great as 1000 litres, as may be necessary when examining either groundwater or tap water that has been processed by sand filtration.

If the intended water sample consists of waste-water effluent, tap water, or some other water that may have received a chemical disinfectant, then the collected water sample should be dosed with sodium, thiosulphate solution (50 per cent, wt/vol) at a rate of 10 ml per gallon, and a sample of the dosed water should be checked to be certain that all residual disinfectant activity has been neutralised. This neutralisation of residual disinfectant activity must be done before the water is either stored as a bulk sample for processing off-site or processed on-site for concentrating any viruses which the water may contain. If possible, the water sample should be processed by the virus concentration technique immediately upon its collection. If it is necessary to store a water sample before it is processed, then the storage container must first be adequately treated by a sterilisation technique, and after the sample water is added, the container should be kept covered to prevent contamination. The sample should also be kept in a cool location to reduce viral inactivation until such time as the water sample can be processed. Prolonged storage of water samples should be done in a refrigerated room (no more than approximately 4°C) and preferably for no longer than two days.

For each water sample that is to be processed using cartridge filtration you will need a virus-adsorbing filter. You also may need a prefilter (if the water has such a high turbidity that it cannot easily be passed directly through the virus-adsorbing filter). Neither of these filter types should be reused. The prefilter should be a 3 μm-porosity (nominal porosity) wound polypropylene yarn filter of the type available as stock number M27R10S from Process Filtration Division, Parker Hannifin Company (Lebanon, Ind.). When using cartridge filters that are considered to be positively charged for the purpose of concentrating viruses from nonsaline environmental waters, it should not be necessary to chemically modify the waters pH prior to passing them through the filters except for lowering the pH of naturally alkaline waters to below the range of 7.5 to 8.5. This is the maximum pH range which should be used when adsorbing viruses onto the positively charged filters. The recommended positively charged virus-adsorbing filters are the pleated 0.45-μm-porosity glass and cellulose fibre membrane type, presently available as the Zeta Plus Virosorb 1MDS filter cartridge from Cuno Inc. (Meriden, Conn.), and the pleated 0.20-μm-porosity charged nylon membrane type, presently available as a Posidyne N 66 filter cartridge from Pall Trinity Micro Corporation. Both of these virus-adsorbing filters are positively charged relative to untreated cellulose ester or glass microfibre filters and are, therefore, suitable for concentrating viruses from water at most ambient pH levels (approximately 5 to 7.5). If the 1MDS type of charged fibre filter is used, then it is preferable to use dilute hydrochloric acid to adjust the pH of any alkaline water samples to less than 7.5 before filtration. The charged nylon filter type may be preferable for use with water samples of moderately alkaline pH level, as these filters can be used effectively for concentrating viruses from water at pH levels up to approximately 8.5. A possible alternative choice for the use of a positively charged filter would be the cellulose, glass, and melamine matrix filter presently available as the Zeta Plus MK (Micro Klean) filter cartridge from Cuno Inc. You can also use a more electronegatively charged glass microfibre filter, of the type available as a 0.25-μm-porosity Duo-Fine series filter cartridge from Vivendi Water Company, Filterite Division. In fact, the use of this type of filter is preferred when concentrating viruses from either brackish or marine water, as the more positively charged filters seem to be inefficient when used on saline waters. When you use the more negatively charged filters, such as the Duo-Fine series filter cartridge, as your virus-adsorbing filters, it will be necessary to both adjust the pH of the sampled water to 3.5 by addition of hydrochloric acid and add aluminum chloride before passing the sample water through the filter. The recommended final concentration of aluminum chloride varies depending upon the nature of the water sample, from 0.0005 M (freshwater) to 0.0015 M (full seawater).

The various reagents, namely, aluminum chloride, hydrochloric acid, and sodium thiosulfate, which may need to be added to the water sample before the sample passes through the filters, may be prepared as stock reagent solutions and injected into the flowing sample in between the pump (or pressurised source) and the prefilter. Alternatively, a batch adjustment technique can be used. For this technique, the water sample is collected into a sterilised temporary storage container, chemical supplementation of the water is performed while the sample is in that storage container, and the water is then pumped through the filters.

Assemble the cartridge filter virus concentration apparatus as shown in Fig. 13.1, making certain to properly align the pump, filter holders containing their respective filters, and water meter with respect to their indicated directions of water flow. If injectors will be used for chemical supplementation, they should be installed ahead of the prefilter. Be certain that the tops of the filter holders are fully tightened to their bottom sections, that the filter holders contain the correct filters, and that the air pressure release valves on the filter holders are closed. Connect one end of the water supply hose to the pump inlet, and place the other end of this hose into the water being tested, whether it is being taken directly from an environmental water source, a tap water source, or from a storage container prepared as described above. Note the reading on the water meter (flow meter), as these meters usually cannot be reset; in order to know how much water sample you have filtered it will be necessary to read the meter both before and after the filtration. Turn on the pump and let the sample water pass through the filters and meter. It may be necessary to occasionally release trapped air pockets from the filter holders. The presence of air pockets that form within the filter holders can easily be seen by looking at the clear bottom parts of the holders. Such air pockets should be released, as they otherwise may reduce the rate at which water is able to flow through the filters. To release air, first make sure that the holder in question is being held upright so that the trapped air can pass through the release valve. While keeping the holder upright, slightly open the valve until the air has passed through, and then close the valve to keep the water sample from leaking out. If the holder does not contain a release valve, it is sometimes possible to eliminate the trapped air by rapidly inverting the filter holder. This manoeure is done by first placing the filter holder into an upright vertical position and then quickly inverting the filter holder, followed by rapidly restoring the holder to its normal upright position. It may be necessary to repeat this 'flipping' manoeuvre. Alternatively, while making certain to hold the filter holder upright, the top of the filter holder may be loosened slightly where the top connects with the bottom by means of screw threads, to release trapped air. The top and bottom sections should then be tightened back together when water rather than air begins to seep out between the screw threads. After the water sample has been filtered, disassemble the apparatus and allow any remaining water to drain from the filter holders.

There are several procedures available for eluting viruses from the virus-adsorbing cartridge filters. This chapter describes two which allow elution to begin in the field, with eluant presterilised ahead of time in the laboratory. For each water sample to be processed, you will need 1600 ml of filter eluant solution consisting of sterile, pH 7.0, 3 per cent (wt/vol) beef extract solution contained in a separate watertight, 1-gallon-capacity, wide-mouth, screw-cap autoclavable container. A polypropylene container is preferred for this purpose because viruses do not readily adsorb to polypropylene. The eluant solution can be sterilised by autoclaving it inside these polypropylene containers. Containers of this type are available as stock number 2121–0010 from Nalge Nunc International. The eluant (beef extract) solution should consist of 48 grams of microbiological grade beef extract powder. The type of beef extract powder chosen should be one which produces an easily visible amount of precipitate when, dissolved in distilled water, its pH is adjusted to 3.5. It may be helpful if at least one-half of this powdered beef extract is of the type sold as Bacto Beef Extract, Desiccated, by BD Biosciences.

This will help to assure that later on an easily visible precipitate will be produced during subsequent secondary concentration of the filter eluant. Performing either of these two elution procedures can be done more easily if a supporting device is made to maintain in an upright position the filter holders containing the filters. We have seen these types of supporting device made in many, and sometimes imaginative, ways. A simple suggestion is to take a short three-legged wooden stool and cut two round holes in the seat, with the diameter of those holes being slightly greater than the diameter of the filter holders. Placing the filter holders into these holes will then maintain the filter holders upright. Because the three legs will serve as a tripod, this type of support should be able to sit in a steady manner on nearly any horizontal surface.

The first field elution procedure requires that the filter holders then be opened carefully. The virus-adsorbing membrane filter should aseptically be placed into the container of beef extract solution (eluant). If a prefilter was used, both it and one-half of the eluant should be placed into an additional sterile 1-gallon-capacity polypropylene container. These filters should be kept in their containers of eluant and, along with wet ice or ice packs, immediately placed into the insulated container. The insulated container of samples should then be sealed and immediately be transported to the laboratory or else shipped there, preferably by an overnight delivery service. These filters must be kept in their containers of eluant until such time as the filters are processed by the virus elution technique. The virus-adsorbing filter (and prefilter, if one was used) and filter eluant should be processed as soon as possible (preferably immediately) upon their receipt by the laboratory, using the elution and secondary concentration processes described below. If immediate elution is not possible, then the filters (and also any prefilters that have been used) should be stored under refrigeration at a temperature not greater than about 4°C until such time as the elution and secondary concentration processes can be performed. Storage of the filters inside their respective containers of eluant is done to prevent drying of the filters, since drying causes most waterborne viruses to die off very quickly. Please note that excessive bacterial growth in the filters and containers of eluant may result if the total period of transport and storage is longer than two days.

Therefore, if lengthy storage must be done, then storage should be performed at $-70°C$, although with the understanding that some viruses will lose viability when frozen. Frozen storage at a relatively warmer temperature, such as $-20°C$, would be less preferred because viruses tend to die off more rapidly in parallel with the increased sublimation which occurs at this relatively warmer temperature. In particular, members of the family *Inoviridae* seem extremely susceptible to loss of infectivity when frozen, even at temperatures of about $-70°C$.

To continue this first elution procedure, open the container of eluant and aseptically remove the virus-adsorbing filter from its container of eluant and place the filter into a filter holder that has been presterilised. The top of the filter holder should be fully tightened to its bottom section, and the air pressure release valve should be closed. The outlet side of the filter holder should then be connected by a hose to the outlet of a stainless steel pressure can (1-gallon-capacity, of the type sold by Millipore Corporation, Bedford, Mass.) which will hold the eluant. If a prefilter was used, then the prefilter likewise should be placed into a presterilised holder, and both holders (respectively containing the virus-adsorbing filter and prefilter) should be connected together in series so that the filters which they contain will be eluted simultaneously. Next, while continuously stirring the full 1600 ml volume of filter eluant, adjust it to a pH of 9.5 by dropwise addition of 1 M sodium hydroxide and hydrochloric acid as necessary. The eluant then should be placed into the pressure can and from there be passed through the filter (and prefilter, if one was used); three times sequentially. The flow direction of the eluant; during this elution process should be opposite to the flow direction used for the water sample during the virus adsorption. Air pressure supplied to the inlet side of the pressure can at approximately 15 lb/in^2 will provide the necessary force for driving the eluant through the filters. After passing through the filter (and prefilter, if used) the eluate then should be collected into a sterile polypropylene container. This elution procedure represents a reversal of the virus adsorption technique described above. The eluted filters should then be discarded. Any eluant remaining in the filter holder (or both holders, if a prefilter was used), pressure can, or hoses should then be added to the eluate that has passed through the filters. At this point, the pH of the eluate should immediately by neutralised by the dropwise addition of 1 M hydrochloric acid.

Alternatively, the second elution procedure begins in the field by reassembling the virus concentration apparatus as shown in Fig. 13.1 but without either the pump or the water meter. A hand-pumped, positive-pressure spray vessel fitted with quick-release connectors is connected in place of the water meter in such a way as to provide air pressure into the outflow side of the holder which contains the virus-adsorbing filter. In this manner, the direction of flow of the eluant will be the reverse of the flow direction used during the virus adsorption process. This type of spray vessel is available at many hardware stores that sell products for use in homes and gardens. It is important to note here that the eluant solution is not to be placed into the spray vessel, as the spray vessel will be used only to provide a gentle source of air pressure to facilitate the elution process. The 1600 ml volume of 3 per cent beef extract eluant contained in the polypropylene container is then adjusted to pH 9.5. As much of this beef extract eluant solution as can fit (approximately 900 ml) then is poured into the filter housing which already contains the virus-adsorbing filter. Using air pressure provided by the spray vessel, this eluant is then forced from the filter holder, through the prefilter in its housing if a prefilter was used, and collected in the polypropylene container. Correct operation of this elution process is performed by gently pumping the spray vessel until the pressure increases and the beef extract solution flows evenly through the filter holder. This process should then be repeated five more times, each time pouring as much of the eluant solution as will fit into the holder which contains the virus-adsorbing filter and using air pressure provided by the spray vessel to gently force that eluant from the filter holder with the eluant collected in the polypropylene container. Then, adjust the beef extract eluant back to pH 7 and return it to the laboratory in the polypropylene container. This will provide an elution which numerically is roughly equivalent to passing all 1600 ml of eluant three times through the virus-adsorbing filter as was described above for the first elution procedure.

Back in the laboratory, a portion of the pH neutralised filter eluate can be removed to a separate container, and this portion of the eluate can then be saved and this subsequently assayed for the presence of bacteriophage. The enteric viruses contained in the filter eluate tend to have a greater resistance to brief low pH exposure than do many type of bacteriophage. Enteric viruses contained in the filter eluate (produced by following either of the two elution procedures described herein) then can be subjected to a secondary, or second-step, concentration procedure. This second concentration step, as described by Katzenelson, represents a 20-fold concentration of the filter eluate. To begin the Katzenelson procedure, lower the pH of the filter eluate to 3.5 by dropwise addition of 1 M hydrochloric acid with continuous stirring. During this pH adjustment the eluate should become cloudy as an organic precipitate forms. The pH 3.5 eluant should be stirred for 30 minutes and then centrifuged for 10 minutes at $3000 \times$ gram to collect the precipitate. The supernatant resulting from this centrifugation should be discarded, and the precipitate from all 1600 ml of eluant should be dissolved in place (within the centrifuge bottles) using a total volume of 80 ml of sterile, 0.15 M Na_2HPO_4 prepared in distilled water. The pH of the dissolved precipitate, now referred to as final virus concentrate, should then be checked and if not already between pH 7.0 and 7.2 adjusted to within that range by adding 1 M sodium hydroxide or 1 M hydrochloric acid. Antibiotics can be added to the final virus concentrate if desired. Suggested final antibiotic concentrations would be 200 U/ml for penicillin G, 200 µg/ml for streptomycin sulphate, and 2.5 µg/ml for amphotericin B. These concentrates can then be stored at $-70°C$ prior to assay if the viruses of interest are stable when frozen in this solution. Otherwise, the concentrates should be stored under refrigeration (no greater than about 4°C). These same storage conditions should be used for any pH-neutralised filter eluate that was kept for bacteriophage analysis.

The same types of filter material used in fabricating the cartridge filters are usually available commercially as circular flat-sheet filters which can be used with other types of commercial filter holders. These flat-sheet filters and filter holders are often suitable for processing smaller volumes of sample and can be eluted by passing a correspondingly smaller volume of the same type beef extract-based eluant through the filters.

Raw waste-water

The following procedure was developed for the detection of viruses in 8-litre-volume samples of raw sewage. Following collection of the sample, adjust it to a final concentration of 0.05 M $MgCl_2$ by addition of a 4 M $MgCl_2$ stock solution. The sample should then be stored overnight under refrigeration (preferably at 1°C) to allow the solids to settle, following which approximately the top one-half of the supernatant volume is decanted and discarded. The remainder of the sample is then centrifuged in aliquots for 20 minutes at $9500 \times$ gram in order to pellet the solids. All of the pelleted solids from the sample are then combined and resuspended to approximately 120 ml with phosphate-buffered saline (pH 7.2, 0.01 M phosphate, prepared using sodium phosphates) containing 2 per cent (vol/vol) fetal bovine serum. The resuspended solids are then extracted with an equal volume of chloroform (trichloromethane) and centrifuged at $1200 \times$ gram for 30 minutes to separate the two phases. The upper (aqueous) phase from this extraction is collected and centrifuged for 20 minutes at $9500 \times$ gram to remove residual suspended solids. Then, the supernatant from this last 9500 × gram centrifugation step is adjusted to a final concentration of 3.0 per cent (wt/vol) beef extract by addition of a 30 per cent (wt/vol) beef extract solution and is subjected to concentration by organic flocculation at pH 3.5 using the Katzenelson procedure, as defined earlier in this chapter, and the low-pH precipitate is dissolved in 10 ml (final volume) of 0.15 M Na_2HPO_4. The dissolved precipitate should be adjusted to pH 7.0 to 7.2 if it is not already within that range. Also, the small amount of residual suspended solids which had been present in the upper phase following chloroform extraction should be eluted by mixing them in a 5 ml volume of buffered 10 per cent beef extract elution solution (containing, per litre, 100 grams of commercial powdered beef extract, 13.4 grams of $Na_2HPO_4 \cdot 7H_2O$, and 1.2 grams of citric acid), followed by centrifugation for 10 minute at 2800 × gram to repellet the solids and filtration of the resulting supernatant through a sterile 0.25 µm-pore-size (rated by absolute filtration) microbiological filter. In order to yield a final concentrated sample, the dissolved precipitate resulting from organic flocculation of the chloroform-extracted supernatant should be combined with the filtered eluate from the residual solids. The volume of the final concentrated sample ranges from approximately 11 to 14 ml (because there is some volume loss associated with processing the eluant of the residual suspended solids), representing an overall sample concentration factor of approximately 6000-fold. These concentrated samples can then be stored at −70°C prior to assay if the viruses of interest are stable when frozen in this solution. Otherwise, the concentrated samples should be stored under refrigeration (preferably at not more than approximately 4°C). The beef extract used should be of a type that produces a visible amount of precipitate when adjusted to pH 3.5.

Waste-water sludge

Viruses can be isolated from waste-water sludges using the following procedure. An appropriate sample size would be approximately 500 ml of primary waste-water sludge, or 2,000 ml of secondary (mixed-liquor, activated) waste-water sludge. Each sludge sample should be adjusted to pH 3.5 with 1 M HCl and then 0.05 M $AlCl_3$ added to a final concentration of 0.0005 M. This can be done while stirring the sludge in a large beaker. The adjusted sample should then be centrifuged at 1400 × gram for 15 minutes, and the supernatant should be discarded. Resuspend the solids pellet in an amount of buffered 10 per cent beef extract elution solution equal to five times the pellet volume. The beef extract used should be of a type that produces a visible amount of precipitate when adjusted to pH 3.5. The solids pellet should be stirred in the eluant solution for 30 minutes. The resuspension should then be centrifuged at 7000 × gram for 30 minutes, and the supernatant yielded from this centrifugation should then be passed in series through sterilised filters capable of effectively removing particles of diameters to less than 0.25 µm based upon nominal porosity (a suggestion being AP40, followed by either AP25 or AP20 and then AP15 series glass fibre filters from Millipore Corporation, obtainable as 142 mm-diameter disks) to remove remaining sludge solids and contaminating bacteria. The resulting filtrate can be assayed for bacteriophages. If the goal is to detect enteric viruses, then the filtrate should be diluted with a volume of distilled water equal to 2.3 times the filtrate volume (to achieve a final effective beef extract concentration of 3 per cent by weight) and concentrated using the Katzenelson technique as defined above. The supernatant resulting from the Katzenelson procedure is then discarded and the precipitate is resuspended using a volume of 0.15 M Na_2HPO_4, equal to 1/20 the volume of the diluted filtrate. These concentrates can be stored at −70°C prior to assay if the viruses of interest are stable when frozen in this solution. It has been noted that at least some types of viruses will die off even when stored at −70°C in these processed eluates. If there is suspicion regarding the stability of viruses when they are frozen in any type of sample, then storage should be done under refrigeration (recommended to be approximately 4°C or less) and for the least possible amount of time. In general, the colder the storage temperature is, the greater the stability of the viruses will be (except for those few virus groups whose members are extremely susceptible to the effects of freezing).

VIRAL ASSAY TECHNIQUES

The infectivity of enteric viruses contained in environmental samples can be examined by inoculating the sample into cultures of either human or animal cells that are prepared in the laboratory as opposed to inoculating them into live animals. The simplest way to score the results from this type of test relies upon looking for virally induced changes in the inoculated cells, either in the form of a cytopathogenicity assay technique (looking for virusinduced cytopathogenic effects) or a plaque formation assay technique (looking for the development of focal areas of cell death). Methods for performing these two types of assays and also molecular approaches for detecting enteric viruses contained in environmental samples. It may prove necessary to use toxicity reduction methods when assaying concentrated environmental samples in cultured human or animal cells.

Chapter 14

Detection of Bacterial Pathogens in Waste-water and Sludge

INTRODUCTION

The number and variety of bacteria present in waste-water and associated solids are legion. Their sources are the excrement of humans and animals, other waste materials that find their way into domestic sewage, and the microbial flora in the source water. This great diversity and the associated variety of required growth conditions hamper attempts to isolate, identify, and enumerate most bacterial members of this microcosm. The presence and number of pathogenic bacteria that might be present in waste-water and biosolids are a function of the disease morbidity in the community from which the waste materials are derived and the degree of sewage treatment received. Relative to the total number of bacteria present the pathogens will normally represent but a minor part. In most instances these pathogens will play a passive role in the dynamics of the microbial ecosystem; the waste environment is hostile, and as a result, the number of pathogens present tends to decrease over time. The isolation, identification, and enumeration of pathogens from this milieu are replete with all the difficulties listed above.

Representative genera of bacterial pathogens that might be found in domestic waste-water and sludge include *Salmonella, Shigella, Vibrio, Escherichia, Campylobacter*, and *Yersinia*. The genus *Salmonella* includes two species and more than 1,800 serovars; the genus *Shigella* is represented by 4 species encompassing 34 serovars; *Vibrio cholerae* O1 and non-O1 can be present; the pathogenic varieties of *Escherichia coli* include at least 5 serovars; *Campylobacter jejuni* is the most common *Campylobacter* found in waste-water, followed by *Campylobacter coli*; and there are reported to be more than 50 serovars of *Yersinia enterocolitica*.

Because of the difficulties in the isolation and detection of bacterial pathogens in waste-water and sludges the use of surrogate (indicator) bacteria has been standard practice in water quality monitoring. Historically, the coliform group, the enterococci, and *Clostridium perfringens* have, in descending priority, been the bacterial indicators of choice. In all these cases the indicator bacteria are assumed to be indigenous to feces, and thus their presence in environmental samples is indicative of fecal contamination. The presence, and in some instances the absence, of these indicators is not an absolute indication of the presence of bacterial pathogens. Rather, they indicate the potential for the presence of pathogens because of the likelihood that infectious feces are present in waste-water or sludge. The bacterial indicators' greatest weakness as a public health monitoring tool for water and waste-water is their greater sensitivity to disinfection relative to viruses and, in particular, the cysts of protozoan parasites. In these and other instances, the absence of indicator bacteria is not a guarantee that other more resistant microbial forms are not present. Because of these problems there have been ongoing efforts to find better indicators for the presence of microbial pathogens in environmental samples. The ideal would be to monitor for the presence of all microbial pathogens that might be present in a liquid or solid sample, an ideal not likely to be realised anytime in the foreseeable future.

There are a number of instances in which there is a need to make direct measurements for the presence and number of bacterial pathogens in water and solids. This approach can be most useful in epidemiologic studies of waterborne disease; in the development of the relationship between indicator numbers and specific pathogen concentrations, which could be of aid in standard setting; in determining the efficacy of water and solid treatment processes in the reduction of pathogens; and in situations in which the sanitary significance of high indicator numbers is in question. This latter situation has been observed in composting sewage sludge in which coliform after growth can occur, the significance of which is determined by the direct measurement of *Salmonella* species. If a specific pathogen type were to be used as an indicator of pollution, one would still have the problem of selecting what would be the ideal surrogate. At the present time the majority of work done thus far on measuring bacterial pathogens in water and solids has been directed towards salmonellae and they have become a 'standard pathogen' by consensus.

TRADITIONAL METHODS FOR THE DETECTION AND ENUMERATION OF BACTERIAL PATHOGENS

To date, methods for the detection and enumeration of bacterial pathogens from waste-water and sludge have used a cumbersome approach which includes enrichment, isolation, and identification. The number of bacterial pathogens found in waste-water and sludges is usually lower than the number of nonpathogens present. For example, the number of salmonellae detected in activated sludge and digested sludge ranges from none detected to a most probable number (MPN) of 400/g (dry weight) of solids. The numbers reported in raw sewage range from 7 to 8000 per 100 ml. Because of these small numbers some form of enrichment, which may be proceeded by a sample concentration step, is required.

Determining the concentration of bacterial pathogens in waste-water and biosolids is a cumbrous task. The traditional method of choice is to estimate the numbers of bacteria by the MPN method. The MPN is determined by placing a series of dilutions of a waste-water sample, sample concentrate, or sludge solids extract into tubes of enrichment broth and determining if the target bacterium has grown in any of the inoculated tubes. This requires that each tube (dilution) be screened for the presence of the pathogen by completing the isolation and identification steps.

Sampling Requirements, Transportation, and Preservation

Raw (untreated) sewage does not require the concentration of large-volume samples (typically 250 ml to 1 litre) because of its high levels of bacteria. Sterile glass or polypropylene bottles can be used to collect samples. All samples should be refrigerated or stored on ice and processed in the laboratory as soon as possible, i.e. within 24 hours of collection.

The detection of bacterial pathogens in treated sewage effluent (primary, secondary, or better process) may require the collection of larger sample sizes (1 to 10 litres), depending upon the final quality of the waste-water effluent. If the collected waste-water represents an effluent which has been treated by a tertiary or better process, large samples (10 to 20 litres or more) must be collected. For molecular investigations, depending upon the final water quality, bacteria may be concentrated directly onto filters in the field for recovery of their nucleic acid. Furthermore, if a chlorinated effluent is discharged, the sample must be dechlorinated by adding to the sample bottle an appropriate concentration of sodium thiosulphate usually enough to give a final concentration of 100 mg/litre prior to sterilisation. Sewage sludge samples should be collected. The tested sample is composed of subsamples taken at various locations throughout the same pile or source of sludge. All samples should be refrigerated or held on ice and processed within 24 hours of collection. Results are reponed as MPN per gram (dry weight) of sample.

Waste-water samples generally are concentrated by filtration or centrifugation of known volumes of sample, whereby the bacteria present are captured on a filter or in a centrifuged pellet. Filtration using 0.45-μm-pore-size membranes is most applicable to clean (filterable) water. In the case of sewage, the utility of filtration is restricted because clogging limits the sample volume. If necessary, coarser filters such as spun glass can be employed although they will be less efficient as bacterium collectors. In the case of waste-water and biosolids extracts centrifugation can he employed, but there will be some obvious limitations in the size of the sample that can be examined. A nonquantitative pseudofiltration method has been used in which gauze pads (sanitary napkins work well) are suspended in sewage flow for a period of time, after which the entire pad is placed in a suitable enrichment broth. Bacteria can be isolated from these pads up to four times more frequently than from grab samples. This method can also be used to detect enteric viruses in waste-water. Other concentration methods using diatomaceous earth, fibreglass, and membrane filtration are described separately. Methods that use concentration prior to enrichment are problematic in that these methods concentrate all manner of materials and microbes as well as the target pathogen; thus, while concentrating the sample one also significantly increases the amount of interference that will be encountered in subsequent steps of the isolation and identification process.

Enrichment and Isolation

The enrichment process employs a broth medium that allows the target bacteria to multiply to numbers large enough to facilitate their isolation and identification. In many instances the medium to be used has been chosen directly from those used in clinical laboratories where isolation would be from the stools of infected individuals. In these samples one would expect large numbers of pathogens. The enrichments are devised to limit the growth of interfering bacteria, such as coliforms and *Proteus* species, and encourage the growth of the pathogen. Frequently, the formulations used seem less inhibitory to the target than to the unwanted bacteria, although, at the same time, they are often not optimum for the growth of the pathogen. In the case of stools from infected individuals, and in many cases infected food, the target bacteria are robust enough and present in sufficient numbers that they multiply despite the suboptimal growth conditions. In waste-water, sewage solids, and treated effluents the pathogenic bacteria are in a hostile environment, are not robust, and are present in small numbers. In these instances, the types of selective enrichment media and techniques used in the clinical setting may be too restrictive for successful enhancement. Less restrictive or nonrestrictive media can be used to allow the growth of the target pathogen, and other bacteria, followed by secondary enrichment in a restrictive medium. After enrichment for the appropriate time and temperature, material from the broth is streaked or plated onto a solid medium for bacterial isolation. There are a large variety of such media which are selective, differential, or both. Typical colonies are selected and further identified using standard biochemical (triple sugar iron agar, lysine iron agar, urea agar, etc.) and immunological methods.

Salmonella Detection

Because of the great diversity of pathogens that might be found in waste-water and solids, the diversity of media and incubation regimes required, and, in most instances, the paucity of information available, the following discussion is limited to the salmonellae. Most of the available literature, and reported experience in the sphere of waste-water and sludge, have been directed towards the detection and enumeration of salmonellae. Much of the methodology used has been adapted from food microbiology, an area of intense activity in the detection of salmonellae.

A representative list of selective enrichment media is presented in Table 14.1. These media fall into three ingredient categories: (i) selenium based, (ii) tetrathionate based, and (iii) malachite green-magnesium salt based (Rappaport-Vassiliadis broth [RVB]). The conditions prevailing in these enrichment media do not present optimum growth conditions for salmonellae and are selective in that they inhibit the growth of interfering bacteria, such as *Proteus* species and coliforms, but the media provide opportunity for the ascendancy of any salmonellae that might be present. A number of additional ingredients have been incorporated into *Salmonella* cultivation media in an attempt to enhance selectivity. These include brilliant green dye, cystine, sulphapyridine, and novobiocin. This latter antibiotic, in concentrations up to 80 µg/ml, has been found to be very useful in the suppression of interfering bacteria and is incorporated in the RVB. Because these enrichment media can exert a significant inhibitory effect upon stressed salmonellae, some form of pre-enrichment, such as buffered peptone broth, can be used prior to inoculation into a selective medium. This is the procedure followed when the tetrathionate brilliant green broth of Hussong is used.

Table 14.1. List of selective enrichment media used for the detection of *Salmonella* in waste-water and sludge.

Enrichment medium
Selenite-F broth
Selenite brilliant green
Selenite brilliant green sulpha broth
Selenite cysteine broth
Selenite dulcitol broth
Tetrathionate broth
Tetrathionate brilliant green broth
RVB

The incubation temperature can have a significant impact on the recovery efficiency of a selective medium. Harvey and Price in their review of salmonellae isolation methods, indicate that the enhancement of *Salmonella* recovery by using incubation temperatures between 40° and 43°C has been recognised since the turn of the century. The use of elevated temperature must be matched with the enrichment system used. These authors point out that the results obtained using elevated temperatures can vary among laboratories, probably because of a lack of standardisation of methods. In general, elevated temperature does not enhance the usefulness of selenite brilliant green broth or tetrathionate-based medium samples. These authors also point out that elevated temperature methods may not detect *Salmonella enterica* serovar Typhi. The elevated temperature regime works well with RVB and is used in the secondary enrichment scheme of Hussong and with the use of selenite dulcitol broth.

Salmonella Isolation Media

There are a number of formulations of isolation agar for the detection of *Salmonella*. The major components for six of the most common agars are shown in Table 14.2. We have included a seventh, less common medium, modified semi-soft Rappaport-Vassiliadis (MSRV) medium, because of our success in isolating *Salmonella* from waste-water and biosolids using this formulation. These agars are selective and differential. Selectivity is brought about through the use, singly or in combination, of brilliant green dye, bile salts, or desoxycholate. Variations on this theme include the use of bismuth sulphite along with brilliant green in bismuth sulphide agar and the use of malachite green (a homolog of brilliant green), a relatively high concentration of magnesium chloride, and novobiocin in MSRV medium. Differentiation among colony types is, for the most part, based on carbohydrate fermentation and associated pH change, as shown by indicator dyes. For example, the fermentation of lactose, present in five of the seven agars, is an attribute of coliforms but not *Salmonella*; colonies of the former will assume the characteristic colour of the pH indicator used. A good illustration of the use of these activities to differentiate pathogens from nonpathogens is shown in the make-up of xylose lysine desoxycholate (XLD) agar. As listed in Table 14.2, the medium contains three sugars: lactose, saccharose, and xylose, with the last in a smaller concentration than those of the first two. The pH indicator used is phenol red, which is yellow in acid. Coliform colonies will be yellow because of fermentation of all the carbohydrates present; nonpathogen, nonlactose fermenters, such as *Proteus*, will be detected by the fermentation of the other carbohydrates present; while salmonellae, which can ferment xylose, will neutralise the acid produced through the decarboxylation of lysine, resulting in a colony of red colour. The presence of excess lactose and saccharose prevents lysine-decarboxylase-active coliforms from neutralising the acid produced. The production of hydrogen sulphide is indicated by reaction with ferric salts in the medium to produce red-black or yellow-black colonies, the former being common to *Salmonella* species. In contrast, bismuth sulphite agar, which contains only dextrose, relies on the production of hydrogen sulphide with the concomitant precipitation of iron sulphide, producing a black metallic sheen on suspect *Salmonella* colonies. The medium relies on the inhibitory effect of brilliant green to control the growth of interfering coliforms.

Table 14.2. Major components of selected *Salmonella* isolation agars.

Class	Ingredient	Agar[a]						
		SS	*XLD*	*XLBG*	*BGA*	*HEA*	*BSA*	*MSRV*
Carbohydrate	Dextrose	−	−	−	−	−	+	−
	Lactose	+	+	+	+	+	−	−
	Saccharose	−	+	+	+	+	−	−
	Xylose	−	+	+	−	−	−	−
Amino acid	Lysine	−	+	+	−	−	−	−
Inhibitors	Bile salts	+	−	−	−	+	−	−
	Brilliant green	+	−	+	+	−	+	−
	Desoxycholate	−	+	−	−	−	−	−
	Bismuth sulphate	−	−	−	−	−	+	−
	Malachite green	−	−	−	−	−	−	+
	MgCl$_2$	−	−	−	−	−	−	+
	Novobiocin	−	−	−	−	−	−	+

[a] SS, Salmonella shigella agar; XLD, xylose lysine desoxycholate agar; XLBG, xylene lysine brilliant green agar; BGA, brilliant green agar; HEA, Hektoen enteric agar; BSA, bismuth sulphite agar; MSRV, modified semi-soft Rappaport-Vassiliadis agar. A plus or minus sign indicates the presence or absence, respectively, of ingredient in agar.

MSRV agar does not contain any carbohydrate but relies on the motility of salmonellae as a differential characteristic. The use of motility as a means of separating salmonellae from background bacteria has been reported by numerous authors throughout the years. More recently, DeSmedt and DeSmedt and Bolderdijk have developed a modification of the Rappaport-Vassiliadis enrichment medium in semi-soft (0.8 per cent) agar. Drops of enrichment culture are placed onto the agar, the sample is incubated at 42°C, and motility (cloudy growth) away from the spot is considered to be an indication of the presence of salmonellae. This limits the test to those *Salmonella* that are motile; however, there are very few nonmotile *Salmonella*, notably *S. enterica* serovar Pullorum.

The considerable variety of enrichment and isolation media available pose a problem for the investigator who wishes to select the most efficacious method for the isolation of salmonellae from waste-water and biosolids. Part of this decision will be made based upon the investigator's past experience and sample matrix. Morinigo evaluated the use of various enrichment media and isolation agars for their effectiveness in the isolation of salmonellae from environmental samples. In one study of river water, they examined 10 different enrichment media, which included RVB with four concentrations of novobiocin, selenite cysteine at two different incubation temperatures, and selenite-F with and without novobiocin at two incubation temperatures. All of the samples were pre-enriched in buffered peptone broth and isolated onto XLD and brilliant green agars. They concluded that the best recoveries of *Salmonella* were made using RVB containing novobiocin at a concentration of 10 µg/ml. In another study, these authors again evaluated this enrichment media with the inclusion of tetrathionate broth, using laboratory cultures of bacteria including *Salmonella* and species known to interfere with the isolation of salmonellae. They found that selenite-containing media were less effective at restraining the growth of gram-positive, interfering bacteria. Salmonellae grew poorly in tetrathionate broth and media containing brilliant green.

Stressed (stressed by exposure to seawater) salmonellae did well in selenite-F and selenite cysteine media and in RVB. When natural water samples were used, RVB was deemed the best for enrichment. These authors also evaluated a great variety of isolation agars, excluding MSRV, for their ability to grow *Salmonella* as well as to inhibit interfering bacteria. Laboratory cultures of bacteria, both stressed and unstressed, were used. The results indicated that XLD agar was of equal selectivity to the best of the others and was best for the recovery of stressed *Salmonella*. From their work it appears that RVB with novobiocin is the enrichment medium of choice, and XLD agar is an acceptable isolation agar for use with polluted water samples.

There has been an increased interest in the ability to isolate *Salmonella* from sewage sludge, particularly with the adoption of a *Salmonella* standard for treated sludge that can be applied, unrestrictedly, to land. Yanko evaluated five different procedures for the isolation and enumeration of salmonellae in biosolids: two standard methods which use dulcitol selenite and tetrathionate enrichments, respectively, followed by the use of XLD and brilliant green isolation agars; Kenner and Clark's method using dulcitol selenite enrichment incubated at 40°C followed by subsequent isolation on XLD agar; the method of Hussong which uses a pre-enrichment in buffered peptone at 36°C followed by a secondary enrichment in tetrathionate brilliant green broth incubated at 43°C and isolation on xylose lysine brilliant green agar; and a method using selenite brilliant green enrichment and incubation at 37°C with confirmation on XLD agar. This evaluation was conducted on digested sludge, activated sludge, and composted sludge. The Hussong and selenite brilliant green methods recovered more salmonellae than did the other three, and the first two of the five methods were determined to be inadequate for use in compliance testing.

In none of these evaluations was MSRV agar used. In one laboratory three methods have been compared for the enumeration (MPN) of salmonellae in biosolids and raw sewage. These methods include selenite brilliant green enrichment and selenite dulcitol enrichment, both followed by isolation on XLD agar, and a nonselective enrichment in tryptic soy broth, followed by isolation on MSRV agar. All the enrichments were incubated at 35°C for 24 to 48 hours. The MSRV agar was then spotted with up to 0.2 ml (1 drop per spot) of enrichment and incubated for 24 hours at 42°C. At the end of this period, the MSRV agar plates were examined, and motility away from the spots was recorded. Material taken from the edge of a motile spot was either restreaked onto XLD agar or used directly for biochemical screening. Experience has shown that this material contains, with few exceptions, a pure culture of *Salmonella*. This procedure is two to three times more effective than the other two methods tested. In no instance have we recovered *Salmonella* by either of the first two methods and not made a recovery in the MSRV agar. The US environmental protection agency has recently completed an evaluation of this method for monitoring biosolids for salmonellae and found the use of MSRV to be superior to the methods now being used. Round-robin testing is to be conducted in the near future. One reason for the success of this method may be the use of a large inoculum (0.2 ml) of enrichment culture relative to that used on a streak plate. Harvey and Price point out that *Salmonella* growing in an enrichment culture tend not to be uniformly distributed, and a streak plate may well come up negative even when *Salmonella* are present in the enrichment. Our observations seem to confirm this phenomenon, since frequently not all of the six spots on a positive MSRV agar plate are motile. The use of a nonselective enrichment medium may also be of value in resuscitating stressed *Salmonella*.

Other Bacterial Pathogens

The recently published literature on the detection of bacterial pathogens other than *Salmonella* species in waste-water and sludge is relatively sparse. The cultural method for the detection of *Vibrio cholerae* is well established. The use of alkaline peptone water as an enrichment followed by isolation on thiosulphate citrate bile salts sucrose agar and confirmation of typical colonies by biochemical and serological tests is an effective method. In one laboratory there has been good success in isolating non-O1 *V. cholerae* from estuarine sediments using this procedure.

Reports on the detection of *Shigella* have not been found. Research information compiled suggests that methods similar to those for the detection of *Salmonella* should be employed for the detection of *Shigella*, but it should be noted that the methodology is qualitative and the sensitivity is low. Reports on the isolation and enumeration of *Yersinia* in waste-water and sludge were not found.

Methods for the isolation and enumeration of thermophilic *Campylobacter* species from the water environment are reported by Arimi and Jones. These authors report successful isolation and MPN enumeration using Preston Campylobacter selective enrichment broth incubated at 43°C and isolation on Preston Campylobacter selective agar incubated under microaerophilic conditions.

It is known that *Aeromonas* spp. are associated with human enteric disease. Because of this understanding, the distribution of these autochthonous aquatic bacteria in surface water, waste-water, and sludge has been enumerated and reported. The most recent literature contains descriptions of methods that may or may not use an enrichment step. Nishikawa and Kishi describe a qualitative method using an alkaline peptone broth enrichment followed by isolation on bile salts brilliant green starch agar. Amylase-positive colonies on the latter medium were considered as positive presumptive *Aeromonas* isolates. Further identification was based upon biochemical differentiation. Poffe and Op de Beek reported good results in the isolation and enumeration of *Aeromonas hydrophila* from sewage and sludge using direct detection methods. In the case of waste-water, they employed the membrane filter method using mA agar, which contains ampicillin, trehalose, and desoxycholate. In the case of sludge samples suspensions of various dilutions were spread on the surface of mA agar. In both instances the suspect colonies are yellow, a result of trehalose fermentation. These species are further identified using biochemical differentiation. Ribas compared four similar media for their usefulness in the detection of *Aeromonas* spp. in polluted water. The media were of various compositions containing ampicillin, complex carbohydrate, and, in one case, glutamate; mA agar was included.

The authors concluded that all of the media gave similar results but that the starch glutamate ampicillin was the simplest to use. The US environmental protection agency is evaluating a method for the enumeration of *Aeromonas* species in finished water. This is a membrane filter method using ampicillin-dextrin agar containing bromthymol blue. *Aeromonas* species produce yellow colonies on this agar. The applicability of this method to waste-water is unknown.

In one laboratory there have been success with a presence/absence method in which samples are filtered through a 0.45-μm-pore-size cellulose acetate filter which is subsequently immersed in alkaline peptone water at 35°C for 24 hours. This enriched material is then spread onto Rimmler-Shotts agar. *Aeromonas* isolates produce yellow colonies.

There has been an increasing interest in the occurrence of *Helicobacter pylori* in the aquatic environment. The role of drinking water in the transmission of this bacterium to humans remains problematic. At the present time, there are no culture methods for the isolation or enumeration of these micro-organisms from water and waste-water. Their presence in these media has been confirmed by molecular biological (PCR) methods.

MOLECULAR BIOLOGICAL METHODS

The use of molecular techniques is growing rapidly in the environmental microbiology field. In the last 10 years molecular techniques have transitioned from the research laboratory to waste-water agencies. There can be unsurmountable difficulties concerning the use of culture methods for the isolation of certain pathogenic bacteria from waste-water and sludges.

The primary advantage of the application of molecular methods to the detection of pathogenic micro-organisms from environmental samples is the ability to specifically and rapidly detect the organism of interest without having to actually isolate it on growth media. It may even be possible to apply molecular methods to detect target organisms that exhibit characteristics commonly associated with viability (the presence of RNA species). Given the large background of competing bacteria, these methods hold promise for the future of monitoring for either specific or selected groups of pathogenic bacteria in waste-water and sludges.

Polymerase Chain Reaction (PCR)

In the section that follows, it is assumed that the reader is familiar with the basic theory behind PCR. The PCR involves a few critical steps; the following are highlights of those procedures.

Primer selection

PCR primers can be chosen as specific sequences adjacent to predetermined target sequences, or a group of random hexamers can be used as primers to create PCR-amplified molecules that contain the target sequence. If specific primer sequences are to be chosen, one can determine the preferred primer sequences either by searching the literature or nucleic acid libraries or by sequencing areas of DNA or

RNA of the bacteria of interest. Primer sets can be highly specific for only one particular gene that, one hopes, is unique to a single organism, or they can be universal, amplifying related sequences that cross taxonomic family or kingdom boundaries. In addition, depending upon the amplification conditions, multiple primer sets may be simultaneously run in a nested form or as a suit of primers. Target sequences of high tertiary complexity (e.g. superhelices) may be more difficult to amplify due to the structural geometry of the finished sequence. It is important to limit the size of the final target (<3 kbp) because, as the PCR is used to replicate longer and longer pieces of DNA, there is greater chance of error (e.g. in base pair matching, the formation of 'primer dimers', etc.).

PCR conditions

PCR is performed using a thermal cycler, an instrument that is capable of rapidly cycling temperature changes over a wide range, generally from 4 to 100°C. Thermal cycler settings must be optimised for each primer set and desired product and often are determined empirically through trials. In addition, the various commercial thermal cyclers may perform slightly differently. Therefore, in order to keep thorough quality control, it is imperative that PCR conditions be optimised on one instrument and subsequently reproven if the investigator switches to another brand or type of cycler. The optimal annealing temperature is critical and can be determined either empirically through a series of experiments or by consulting computer-based software which can predict melting and annealing temperatures based upon the base pair sequences.

Molecular probes and detection techniques

There are many methods available for the detection of the PCR product. One of the easiest methods is by gel electrophoresis, whereby the products of the PCR amplification are separated into bands based upon their size. Some drawbacks to this method include: (i) if the target was present in an extremely low concentration even an amplified product may be difficult to visualise, (ii) other amplification products may interfere with the resolution of the electrophoretic bands, and (iii) if only a single band is generated for a target organism, then a single electrophoretic image may not be definitive proof of a successful amplification of specific target. The first of these two problems can be revolved by using specific complementary probes, which can be labelled either isotopically or nonisotopically. One may then visualise extremely low concentrations of product and avoid interfering bands which do not bind to the probe.

As an aid in solving the third problem, it is advisable to amplify more than a single target site for any organism of interest, since the presence of multiple products from the same organism provides a more specific test for that organism.

Isotopic versus nonisotopic techniques

For many years isotopically labelled (e.g. ^{32}P-labelled) probes were the method of choice given their sensitivity for detecting nucleic acid sequences. Methods using isotopically labelled probes can measure as little as femtogram (10^{-15} gram) quantities of product (4–6). However, because of the necessary precautions associated with the handling and cleanup of radioactive material, many investigators now prefer to use nonisotopic detection procedures whenever possible. Biotin-strepavidin, digoxigenin, or fluorescent labels can be used for obtaining colorimetric results.

Methods for nonisotopically labelled probes may not be as sensitive as those which employ isotopically labelled probes, but they are improving. The final product from the reaction can be detected by immobilising the target sequences onto nylon filters and hybridising them in place with a known, labelled probe, or the known probe can be immobilised and used to capture the labelled target material through hybridisation as the sample is placed into contact with the filter.

The specificity of binding between the nucleic acid probe and the complementary target nucleic acid sequences versus other sequences which are similar, but not identical, is known as stringency and is dictated by the chemical conditions and temperature under which the hybridisation is performed. The selection of hybridisation conditions and the format of test presentation (Southern hybridisation, dot blots, etc.) will vary depending upon the make-up of the probe and the typical yield of the PCR. Consequently, the best hybridisation conditions must be determined empirically for each complementary set of nucleic acid primer sequences.

Direct probing

If the micro-organisms of interest are in sufficient quantity, PCR amplification of the target nucleic acid sequences may not be necessary. Recovery of total nucleic acid from an environmental source such as sewage, sludge, or filter concentrates may yield sufficient material for direct probing. Sommerville and Knight have demonstrated that some pathogens may be present in sufficient quantity but may not be culturable. In these instances, total nucleic acids were harvested, and known nucleic acid probes specific for the bacterial pathogens were used as a detection system.

Quantification

It is possible to use PCR to quantify the number of bacteria present in the sample. Internal nucleic acid amplification controls can be run simultaneously, and the intensity of the hybridisation product between the probe and the PCR amplification can be compared with that of the product of hybridisation with known quantities of target nucleic acid. Picard prepared multiple PCR tubes containing undiluted and diluted DNA prepared from soil extracts and used the resulting test reactions as a basis for constructing an MPN result. Yamashiro used a technique based on the method of Holland, whereby the PCR-amplified product is fluorescently labelled as the PCR proceeds. To arrive at a quantitative result, the increasing concentration of fluorescently labelled target DNA is compared to the increasing concentration of fluorescently labelled control DNA within the same reaction tube. This procedure has become commercialised as 'TaqMan'.

Quality control

There are several quality control measures that can be employed to test for a successful PCR and hybridisation. If the same source of waste-water or sludge is to be repeatedly tested, spiking samples with either the organism in question or target DNA will provide information on the efficiency of sampling techniques and the effects of inhibitory compounds.

In addition to a spiked study, every reaction tube must contain a known piece and quantity of control DNA that will indicate whether PCR had been inhibited and to what extent. Specificity of the hybridisation can be measured by using known labelled DNA that differs from the target by one or two base pairs. Therefore, the level of stringency first must be empirically determined for each primer set, and then the established amplification regimen can be rigorously followed. Sensitivity of the test results can be determined by adding either target DNA or the test organism in the desired waste-water or sludge matrix in such a manner as to conduct dilution to extinction experiments.

Sample preparation

Currently, the primary obstacle to the application of molecular methods to environmental microbiology is the presence of naturally occurring compounds, such as humic acids and minerals, which may be inhibitory to the molecular procedures. Therefore, removing these inhibitory agents is of primary importance when designing sample preparation and concentration techniques. The following discussion will highlight a variety of methods that have been employed to isolate the DNA or RNA of the target bacteria in question.

DNA is a double-stranded nucleic acid which can form very complex superhelical structures. These structures can be quite recalcitrant to environmental effects. Treated waste-water may contain disinfectants which are indiscriminate oxidising compounds and can damage DNA molecules. Some treated sludges can have extremely high pHs from liming processes, which can denature DNA. Therefore, if molecular techniques are to be applied to lime-added sludges, it may be necessary to adjust (lower) the pH of suspensions before testing.

Use of RNA as a molecular target is important because it may be helpful as a qualitative measure of viability (i.e. by measuring the presence of mRNA or rRNA). In addition, RNA is far more abundant within bacterial cells and will be easier to detect with fewer cycles of amplification in a PCR. Reverse transcriptase PCR has been used for the detection of enteric viruses in environmental waters and waste-waters, and techniques used to preserve viral RNA should be applicable to bacterial RNA. Although RNA is more plentiful than DNA, RNA is a single-stranded nucleic acid which has been shown to be quite labile in environmental matrices. Therefore, it is important that steps be taken to improve the survival of RNA molecules during sample manipulation.

As stated above, one of the major obstacles in amplifying bacterial nucleic acids from environmental sources is the presence of inhibitory agents such as humic acid. Since environmental matrices can be complex and undefined, it is difficult to apply a 'one cleanup step, fits all' procedure. For example, approaches used for the recovery of bacterial nucleic acids from waste-water differ from those used for sludge. Palmer used the following techniques to recover bacterial (*Legionella* spp.) nucleic acids from waste-water and reclaimed waste-water. First, the sample was concentrated onto Teflon filters. The filter was then vortexed in a lysis buffer and heated to release the bacterial DNA. The nucleic acids were further purified by centrifuging through a Spin-X filter. A carrier reagent (homopolymer A RNA) was added to aid subsequent precipitation of the DNA in isopropanol. Following centrifugation, the DNA pellet was washed twice in isopropanol. Finally, the pellet was resuspended in PCR water, and an aliquot was submitted to amplification by PCR. Although one manufacturer (PE Applied Biosystems) of an environmental PCR kit uses Durapore filters, other investigators have reported that greater yields were achieved using Fluoropore filters.

Environmental water samples may also he centrifuged to concentrate bacteria into a pellet, and the pelleted cells are then lysed in a silica-guanidinium thiocyanate lysis buffer followed by several washing steps. These crude extracts are sufficiently concentrated to be added directly to a PCR amplification.

For sludges, Tsai reported that rapid freeze-thawing of samples followed by a phenol-chloroform extraction and cleanup through a Sephadex-200 spin column provided adequate recovery of target bacterial DNA. Typically, small volumes (about 25 to 30 μl) of the extracted nucleic acids are used in 100-μl reaction mixes; therefore, rare targets may be missed. Analysis of larger-volume samples by PCR has been described by Stewart and Abbaszadegan, who applied it to the detection of viruses in the environment. The basic volumes were adjusted to accommodate 100 μl of extract for a total volume of 300 μl. Because the larger volumes may introduce more inhibitory compounds, Stewart and Abbaszadegan employed a cleanup procedure through a Sephadex G-100 column (for humic acids) and through a column of Chelex-100 (for ionic inhibitors).

Recently, magnetic capture beads have been employed in a wide variety of clinical, food, and environmental applications for the recovery of target nucleic acids, which were subsequently amplified by PCR. For example, Hultén used paramagnetic beads in an attempt to isolate *H. pylori* from environmental samples to reduce the effects of background chemicals and micro-organisms on the PCR. This technique holds promise as a mechanism for capturing the target organisms and their associated nucleic acid, while incorporating wash steps which help to remove potential inhibiting agents.

Fluorescent Antibodies (FA) Techniques

Fluorescently labelled antibodies specific for a target organism have been used to screen waste-water and reclaimed waste-water for target micro-organisms. Since there are many micro-organisms in waste-water and sludge, there may be some that share similar antigenic reaction sites (epitopes) with which the antibody stain could cross-react. Therefore, when searching through a diverse number of bacteria, it is best to limit the occurrence of cross-reactivity by using monoclonal fluorescent antibodies (FA). The biggest advantage to using monoclonal FA stains is that they can be used to produce a rapid and relatively inexpensive test (once the monoclonal antibody has been made). In the case of organisms such as *H. pylori*, which have yet to be successfully cultured from the environment, FA techniques offer a means by which to rapidly screen samples for the presence of this organism. The disadvantages of the use of monoclonal FA stains for environmental samples are that: (i) there remains a probability that cross-reactivity may occur, (ii) the FA stain can be used to detect the presence of the target organism but does not measure viability, (iii) the array of antigens exhibited by a bacterium may differ depending on the conditions in its environment, and for this reason, depending upon the physiological state of the bacteria used to prepare the monoclonal cell line (which, in turn, produces the antibodies used in the detection reaction) the chosen monoclonal antibody or antibody mixture may not recognise the key target bacterial epitopes that are expressed on bacteria in environmental waters and (iv) the number of bacteria required for microscopic detection must be substantial, and especially in environmental samples, the level of test organisms may be below this minimum number. A negative test result, therefore, should be reported as 'none detected' rather than as 'none present'.

Enzyme-Linked Immunosorbent Assays (ELISA)

Enzyme-linked immunosorbent assays (ELISA) are also antibody based. Brigmon applied an ELISA to detect the presence of *S. enterica* serovar Enteritidis in raw sewage, sludges, and waste-water. The advantage associated with applying the ELISA technique directly to environmental samples is that it can provide a mechanistic measure (via a commercial device termed a microtiter dish reader) with only minor manipulation of the sample. However, one of the major disadvantages to this technique is the lack of sensitivity. Brigmon reported that a minimum of 10^5 *S. enterica* serovar Enteritidis cells/ml were required to generate a clear response above background. Tamanai-Shacoori demonstrated that PCR was superior to ELISA for screening environmental samples for enterotoxigenic *E. coli*. Given the current standard of <3 salmonellae per 4 grams of sludge, this technology has not yet reached the sensitivity required for performing the testing for regulatory compliance. Other disadvantages of this technique include those outlined for FA stains.

Thus, there is a growing interest in the direct measurement of specific bacterial pathogens in environmental samples. This interest is stimulated by concern for the reliability of the use of standard indicator bacteria in setting public health standards, the need for better epidemiologic information relative to the use of reclaimed waste-water and the disposal of biosolids, and the evaluation of pathogen reduction in waste treatment processes. Methods for the isolation and enumeration of pathogenic bacteria in waste-water and sludge are not well developed. There are a myriad of difficulties ranging from the huge diversity of micro-organisms involved to the considerable variety of methodological approaches one can pursue. At this time, development of methods for the detection and enumeration of salmonellae is the most advanced, followed by an increasing development of cultural methods for *Campylobacter* and *Aeromonas* species among others. The more recent developments in molecular biological methods offer great possibilities for the future.

SECTION V

Aquatic and Extreme Environment

15. **Basic Concepts of Aquatic and Extreme Environment**   235

16. **Primary Productivity and Producers**   261

17. **Bacterial Secondary Productivity**   279

18. **Community Structure: Bacteria and Archaea**   289

19. **Protistan Community Structure**   298

20. **Phagotrophy in Aquatic Microbial Food Webs**   312

21. **Biogeochemical Cycling**   321

Basic Concepts of Aquatic and Extreme Environment

INTRODUCTION

Aquatic microbiology, in broad terms, is the study of micro-organisms and microbial communities in water environments. Aquatic environments occupy more than 70 per cent of the earth's surface. Although most of this area is occupied by oceans, there is also a broad spectrum of other aquatic environments including estuaries, harbours, major river systems, lakes, wetlands, streams, springs, and aquifers. Water is essential for life and may arguably be our most important natural resource. Aquatic environments, in addition to providing water for drinking, provide necessary resources for agriculture, mining, power generation, semiconductor manufacturing, and virtually every other industry. Thus, protection and preservation of these environments are vital for the continuation of life as we know it. Microbiota are the primary producers in the aquatic environment (and are responsible for approximately one half of all primary production on the earth). They are also primary consumers. The microbiota that inhabit aquatic environments include bacteria, viruses, fungi, algae, and other microfauna. Identifying the microbiological composition of the aquatic environment and determining the physiological activity of each component are the first steps in understanding the ecosystem as a whole. The approach taken in the first part of this chapter is to define the organisation, composition, and functioning of the community structural habitats that are characteristic of aquatic environments. These habitats are the planktonic, sediment, microbial mat, and biofilm communities. The second part of this chapter contains descriptions and general microbial characteristics of the three main aquatic environments: (i) inland surface waters (lakes, revers, streams), (ii) seas (oceans, harbours etc.), and (iii) groundwater. Finally, examples, of extreme enable environments are given along with features that enable micro-organisms to survive in these locations.

MICROBIAL HABITATS IN THE AQUATIC ENVIRONMENT

Planktonic Environment

Plankton refers to the microbial communities suspended in the water column. Photoautotrophic organisms within this community including both eukaryotes (algae) and prokaryotes (cyanobacteria) are collectively referred to as phytoplankton. Suspended heterotrophic bacterial populations are referred to as bacterioplankton, and protozoan populations make up the zooplankton. Together these three group, of organisms make up the microbial planktonic community. Fig 15.1 shows the relationship and inter-dependence of the various microbial components within a general planktonic food web. Phytoplankton are the primary producers in the food web, using photosynthesis to fix CO_2 into organic matter. In the planktonic microbial community, this primary production is the major source of organic carbon and energy, which is transferred to other trophic levels within the web. The organic compounds produced by phytoplankton can be divided into two classes, particulate or dissolved, depending on their size. Particulate organic matter (POM) compounds are large macromolecnles such as polymers, which make up the structural components of the cells, including cell walls and membranes. Dissolved organic matter (DOM) is composed of smaller compounds, such as amino acids, carbohydrates, organic acids, and nucleic acids, which are rapidly taken up by microbes and recycled.

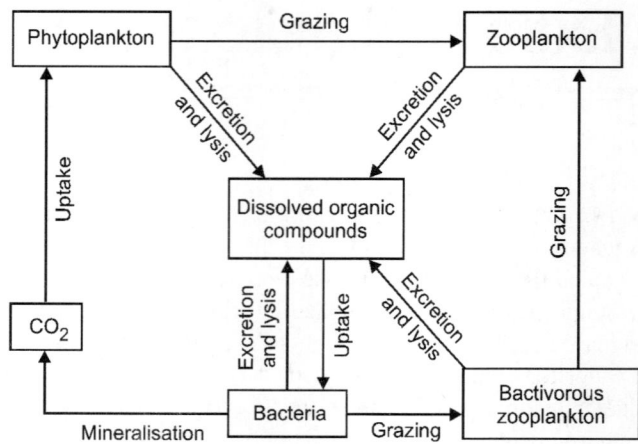

Fig. 15.1. The microbial loop in the planktonic food web. The microbial loop represents a pathway in which the dissolved organic products are efficiently utilised. The role of bacterioplankton is to mineralise important nutrients contained within organic compounds and to convert a portion of the dissolved carbon into biomass. Grazing by bactivorous protozoans provides a link to higher trophic levels.

Primary production

The amount of primary production within a given water column depends on a variety of environmental factors. These factors include the availability of essential inorganic nutrients, particularly nitrogen and phosphorus; water temperature; and the turbidity of the water, which affects the amount of light transmitted through the water column. Open oceans have relatively low primary productivity because of low levels of the essential nutrients nitrogen and phosphorus. The exceptions are areas where currents cause up-welling of water from the bottom of the ocean, bringing with it nutrients from the deep sea. Coastal areas are productive because of the introduction of dissolved and particulate organic material from river outflows and surface runoff from the terrestrial environment. Higher nutrient loading also results from the decomposition of aquatic plants rooted in shallow waters. Freshwater lakes, like the open seas, are often low-productivity environments, particularly those with large, deep, nutrient-poor (oligotrophic) bodies of water. In contrast, smaller and shallower freshwater bodies tend to be nutrient rich (eutrophic). Nutrient loading, which causes eutrophication of lakes, can be a natural evolutionary process or the result of human activity. Sources of natural nutrient loading include terrestrial runoff, rivers that feed into the lake, and plant debris (such as leaves). Nutrient loading resulting from human activities includes the disposal of municipal waste-water and runoff of fertilisers from agricultural fields. Both of these nutrient sources contain high levels of nitrogen and phosphorus, the nutrients that are most often limiting in the aquatic environment.

Secondary production

In a typical food web, phytoplankton (primary producers) are consumed by microfauna (zooplankton), which in turn are consumed by progressively larger organisms, such as fish or other filter feeders. This is called the grazing food chain. However, the actual transfer of carbon and energy between trophic levels is much more complex than what is implied by the grazing food chain. A substantial portion of the carbon fixed by photosynthesis (>50 per cent) is released into the water column in the form of dissolved organic matter (Fig. 15.1). This DOM is rapidly utilised by heterotrophic bacteria (bacterioplankton) present within the planktonic community. The utilisation of dissolved organic compounds released from phytoplankton or from other sources (see later) is a pathway in the aquatic food web referred to as the microbial loop. In this loop, bacterioplankton mineralise a portion of the organic carbon into CO_2 and assimilate the remainder to produce new biomass. This production of bacterial biomass is referred to as secondary production. Secondary production is a major pathway for the utilisation of photosynthates, as well as a pathway for the transfer of carbon and energy to higher trophic levels in the aquatic environment. Thus, the microbial loop serves in the efficient utilisation of DOM released into the water column by various mechanisms.

The source of the DOM pool in the planktonic environment is primarily phytoplankton, but zooplankton and bacterioplankton also contribute. Aquatic fauna contribute to the dissolved organic carbon pool through excretion and the lysis of dead cells. Among the phytoplankton, it is known that both 'healthy' cells and 'stressed' cells (those under some form of environmental stress) can release DOM into the water column. Another suggested mechanism is that 'sloppy' feeding habits of aquatic animals that prey on phytoplankton may allow a portion of the DOM to be released into the water column. Finally, dissolved organic compounds are also released during the lysis of phytoplankton and bacterioplankton by viruses.

Benthic habitat

The benthos is a transition zone between the water column and the mineral subsurface. This interface collects the organic material that settles from the water column or that is deposited from the terrestrial environment. The interface is a diffuse and non-compacted mixture of organic matter, mineral particulate material, and water. This zone is characterised by a dramatic increase in the concentration of micro-organisms (as much as five orders of magnitude) compared with the planktonic environment. The actual microbial concentration, however, depends on the availability of organic material and the availability of oxygen. Below the interface zone, microbial numbers may decline because of oxygen depletion. Anaerobic microbial processes, that utilise nitrate, sulphate, or iron as terminal acceptors characterise deeper sediment layers, and in even deeper layers there is a methanogenic consortium.

The benthic habitat is an important feature of the aquatic environment. The cycling of essential nutrients, such as N, and S, in this area is dependent on a combination of aerobic and anaerobic microbial transformations (Fig. 15.2). The benthic environment can support and often favours the formation of conjoined aerobic and anaerobic microenvironments. Oxygen depletion by microbial activity at the organic rich interface creates anaerobic microenvironments that support the activity of facultative and strictly anaerobic microbes. Specific physiological groups of micro-organisms are strategically positioned relative to the oxic-anoxic interface to take advantage of such microenvironmental niches. In this way, the sediment zone acts to support a physiologically diverse aquatic microbial community. For example, fermentative bacteria metabolise DOM into organic acids, such as acetic acid, and CO_2, Organic acids can act as electron donors for a group of strictly anaerobic bacteria, which utilise CO_2 as the final electron acceptor in anaerobic respiration, thus generating methane (CH_4). The methanogenic activity in turn supports the activity of the methane-oxidising bacteria, which, under aerobic conditions, can utilise methane and other one-carbon compounds as an energy source, regenerating CO_2. Methanotrophic activity is localised at the sediment water interface zone in order to use CH_4 released from the anaerobic zone and the oxygen available in the water column. This is an example of biogeochemical cycling on the scale of a small habitat.

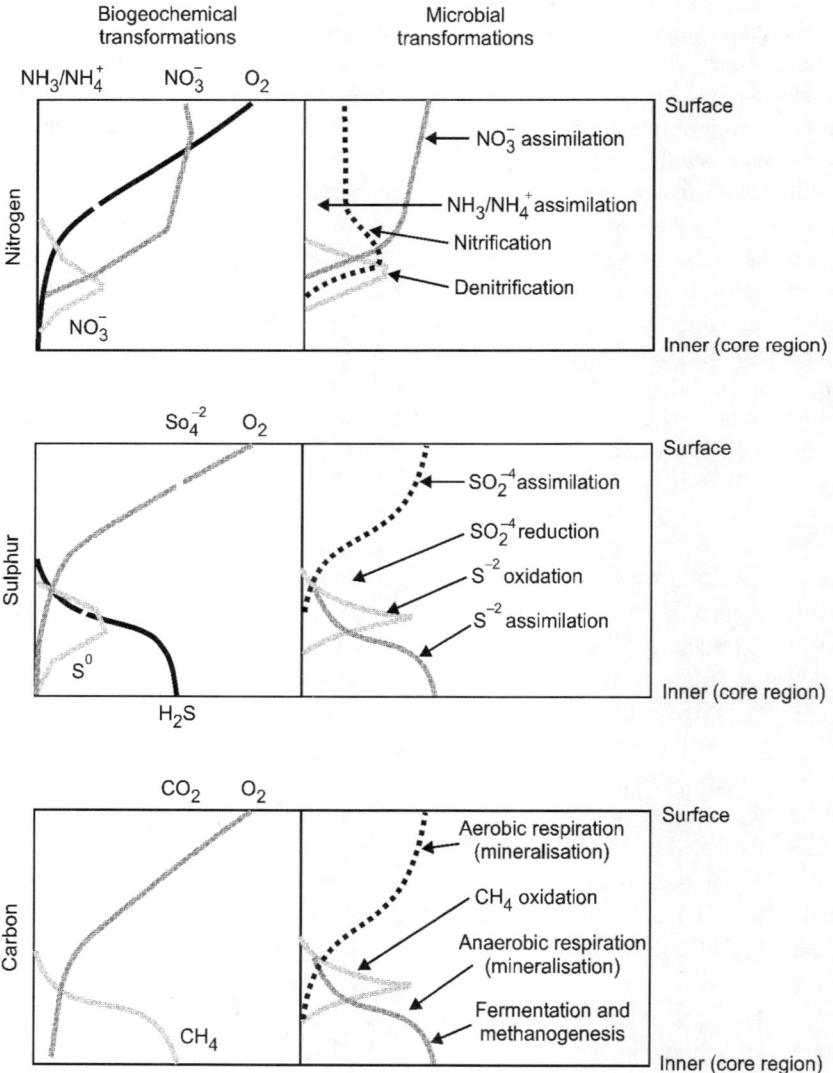

Fig. 15.2. Biogeochemical profiles and major carbon, nitrogen, and sulphur transformations that can be predicted for environments in which oxygen levels are highest at the 'surface' layer and are depleted by microbial activity to create anoxic conditions in the 'inner' region.

The decomposition of organic material in the sediment layer also generates ammonia from organic debris. The fate of ammonia is controlled by: (i) the assimilation of ammonia as a source of essential nitrogen by planktonic and sediment micro-organisms, and (ii) nitrification of ammonia as an energy source by chemoautrophic micro-organisms (*Nitrosomonas* and *Nitrobacter* species). These organisms sequentially oxidise ammonia to nitrate (NO_3^-), a process known as nitrification. Nitrification is often localised at the sediment-water interface because it relies on the presence of oxygen and the release of ammonia by the decomposers. The control of ammonia compounds can be important, especially in alkaline environments, where the undissociated NH_4OH form can be toxic to aquatic animals. The activity of the ammonia-oxidising or nitrifying bacteria, *Nitrosomonas* and *Nitrobacter*, is highly sensitive to the presence of certain DOM, including naturally occurring and industrial chemicals. Therefore, the inhibition of nitrification (ammonia oxidation), which can be detected by an accumulation of ammonia or NO_2, provides a sensitive indicator of the environmental impact of certain toxic pollutants.

Microbial Mats

The previous section described the organisation of the sediment-planktonic interface into microenvironments in which the combined aerobic and anaerobic activity is able to support a diversity of microbial populations. Microbial mats are an extreme example of an interfacial aquatic habitat in which many microbial groups are laterally compressed into a thin mat of biological activity. These groups interact with each other in close spatial and temporal physiological couplings. The thickness of mats ranges from several millimeters to a centimeter in depth. They are also vertically stratified into distinct layers. Microbial mats have been found associated with environments such as the surface-planktonic interface of hot springs, deep-sea vents, hypersaline lakes, and marine estuaries. By supporting most of the major biogeochemical cycles, the mats are considered largely self-sufficient.

Stal describes an example of a laminated form of a microbial mat. Cyanobacteria (photosynthetic prokaryotes) occupy the upper zone of the mat, where they have access to sunlight. This primary production by cyanobacteria is the most important source of organic compounds within the microbial mat community. The photosynthetic activity of the cyanobacteria creates an oxygenic environment in the upper layer of the mat. Photosynthetic activity creates oxygen-supersaturated conditions during the day, but at night, in the absence of sunlight, microbial respiration rapidly depletes all the available oxygen. An important process in the decomposition of DOM released by the cyanobacteria involves the activity of sulphate-reducing bacteria. Anaerobic respiration by sulphate-reducing bacteria has been considered a strictly anaerobic process; however, in microbial mats, sulphate-reducing bacteria are active in the oxygenic upper layer. Considering that anoxic conditions develop rapidly at night, there may be a temporal separation between oxygenic photosynthesis and anaerobic activity. It is also possible that a high demand for oxygen by heterotrophic activity can create microenvironments in which anaerobic activity is supported. This type of temporal and spatial separation of aerobic and anaerobic activity is also important in nitrogen cycling. A major source of nitrogen for the mat community is the fixation of atmosphere nitrogen by bacteria that function only under strictly anaerobic conditions. The reduced nitrogen and sulphide products of anaerobic activity provide energy sources for the nitrifying and sulphur-oxidising bacteria, which are present in close association with the cyanobacteria and are dependent on oxygen to complete the sulphur and nitrogen cycles.

A layer of sediment rich in oxidised iron may form directly underneath the cyanobacterial layer. The origin of the oxidised iron layer is not known, but it appears to form a barrier between aerobic and anaerobic phototrophic activities. Purple sulphur bacteria can form a distinctive layer just below the aerobic-anaerobic interface. The purple sulphur bacteria are also photosynthetic, but compared with the cyanobacteria, these bacteria have a less developed photosynthetic process and they are not capable of the photolysis of water. The purple sulphur bacteria utilise a reduced sulphur compound (sulphide) as an electron donor. The sulphide is provided by the activity of the sulphate-reducing bacteria. Purple sulphur bacteria are readily visible in microbial mats, or other anaerobic environments, because of their conspicuous purple pigmentation. Below the purple sulphur bacteria is an extensive black layer enriched by the precipitation of iron sulphide (FeS).

The microbial mats are unique communities because the interdependent microbial components form clearly stratified and distinctively coloured zones. Mats are often found in extreme environments or in environments where conditions fluctuate rapidly. The cyanobacteria are known to be tolerant of extreme conditions, such as high temperatures or highly saline waters, and thrive in locations where competition from other microbial groups and predation by grazing organisms are limited by the inhospitable environment. The microbial mats are also of evolutionary significance. Fossilised microbial mats, known as stromatolites, dating back 3.5 billion years were among the first indications of life on earth. At that time, the earth's atmosphere lacked oxygen, and the stromatolites from that era were probably formed with anoxygenic phototrophic bacteria (purple and green sulphur bacteria).

Biofilms

A biofilm is a layer of organic matter and micro-organisms formed by the attachment and proliferation of bacteria on the surface of an object. In most cases this surface is submerged in nonsterile water or surrounded by a moist environment. Solid surfaces suitable for bacterial colonisation in the natural environment include inert surfaces, such as rocks and the hulls of ships and even living surfaces, such as the submerged portions of aquatic plants. Biofilms are characterised by the presence of bacterial extracellular polymers, which can create a visible 'slimy' layer on a solid surface. The secretion of extracellular polysaccharides (glycocalyx) provides a matrix for the attachment of bacterial cells and forms the internal architecture of the biofilm community. The exopolymer matrix is also an integral component influencing the functioning and survival of biofilms in hostile environments. Biofilms have been extensively studied for their role in nutrient cycling and pollution control within the aquatic environment, as well as for their beneficial or detrimental effects on human health.

Biofilm development is a complicated issue and is usually initiated by the attachment of bacteria to a solid surface (Fig. 15.3). Permanent attachment of bacteria requires two stages: (i) reversible attachment, which is a transitory physico-chemical attraction, and (ii) irreversible attachment, which is a biologically mediated stabilisation reaction. Reversible attachment is a function of the initial attraction of a bacterium to a surface and is controlled by several competing forces, including hydrophobic, electrostatic, and van der Waals forces. The van der Waals forces attract an object to a solid surface; while hydrophobic and electrostatic forces can be attractive or repulsive, depending on the surface properties of the bacterium and the solid surface. The properties of a solid surface may be modified by the presence of an adherent conditioning film of organic compounds. Bacterial attachment is promoted when DOM of limited water solubility coats the surface of solid substrates. The source of these hydrophobic organic compounds can be the decomposition of organic material, excretion by living organisms, or lytic products from dead organisms. The organic substrates attached to solid surfaces are a source of nutrients for bacteria, especially in oligotrophic environments where nutrients are limited. In specific cases, it is believed that the initial contact between the cell and the organic conditioning film is made using microbial appendages, such as pili or flagella.

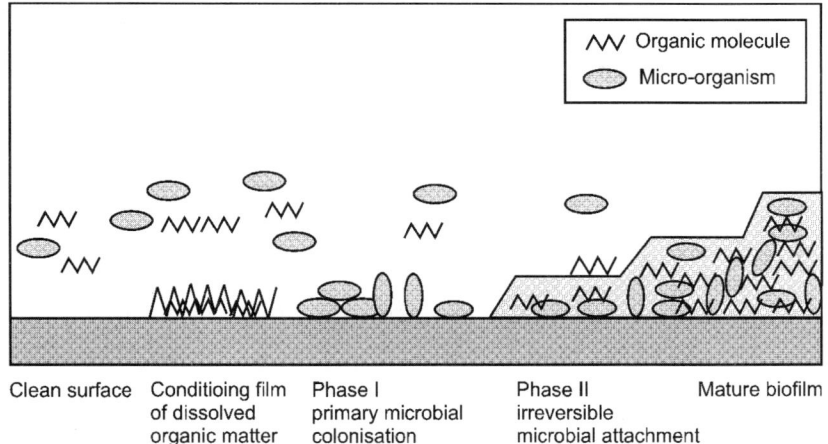

∧∧ Organic molecule
Micro-organism

| Clean surface | Conditioing film of dissolved organic matter | Phase I primary microbial colonisation | Phase II irreversible microbial attachment | Mature biofilm |

Fig. 15.3. Representation of biofilm formation. Dissolved organic molecules of a hydrophobic nature accumulate at the solid surface-water interface and form a conditioning film. Bacteria approach the solid surface because of water flow and/or active motility. The initial adhesion (phase I) is controlled by various attractive or repulsive physico-chemical forces leading to passive, reversible attachment to the surface. An irreversible attachment is a biological, time-dependent process related to the proliferation of bacterial exopolymers forming a chemical bridge to the solid surface (phase II). By a combination of colonisation and bacterial growth, the mature biofilm is formed. It is characterised by cell clusters surrounded by waterfilled voids.

The initial reversible attachment of organisms to surfaces can, with time, become permanent. Irreversible attachment is initiated by the excretion of extracellular polymers by the reversibly attached bacteria. The extracellular polymers create a matrix that surrounds the cell and forms a strong chemical bridge to the solid surface. The fibrous, anionic extracellular matrix can have several functions for the adherent cells. The matrix can act as an ion-exchange resin to filter and collect essential nutrients. It can protect the attached community from environmentally stressful conditions such as desiccation or changes in pH and temperature. It can also provide some protection of the cells from predation by protozoans. Attachment has been reported to increase, inhibit, or have no effect on the growth rate of attached cells when compared with planktonic cells. The positive effect of attachment reported for some bacteria may be the result of indirect factors such as an increase in the local concentration of nutrients or the protective effects of the biofilm exopolymeric matrix. It has also been suggested that attachment may activate the expression of specific genes that are beneficial to the micro-organism in the altered physic-ochemical conditions of the biofilm environment.

The transition from initial microbial attachment to the development of a mature biofilm is due not only to the proliferation of the attached cells but also to the continued deposition of cells from the bulk fluid. Examination of mature biofilms using microscopic techniques has revealed a complex organisation. The biofilm cells have been shown to be organised into column-shaped clusters embedded within the extracellular polymer matrix and surrounded by large void spaces. The void spaces form channels, which can carry limiting nutrients, such as oxygen, into the exopolymeric matrix. The presence of void spaces increases the biofilm surface area and the efficiency with which nutrients and gases are transferred between the biofilm and the surrounding water. The exact nature of the biofilm architecture depends on numerous factors, including the type of solid surface, the microbial composition of the biofilm, and environmental conditions.

Attachment and biofilm formation as a survival strategy are well illustrated by the distribution of bacteria in the nutrient-poor environment of high alpine streams. Bacteria are predominantly found attached to rock surfaces, where they can take full advantage of the continuous renewal of nutrients provided by water flow, and thus act as a biological filter to remove DOM from flowing waters. The filtration of DOM from the water by these attached communities represents a water purification system in natural environments. This system has long been exploited for use in purifying water from municipal (sewage) or industrial sources. In systems known as trickling filters and fluidised bed reactors, the biomass of the biofilm is maximised by providing a porous network of solid supports for bacterial attachment. The porous network is meant to provide an expansive area for biofilm development and increase the area of contact between the biofilm and DOM in the waste-water stream. Biofilter systems have also been used for the degradation of a wide range of pollutants, including hydrocarbons, pesticides, and industrial solvents, that may be present in waste-water streams.

Microbial colonisation at a solid-liquid interface can be beneficial, as in nutrient cycling and water purification processes. However, in some instances, biofilm development can be detrimental, and there is a need to control the growth of biofilm cells using antibacterial substances. For example, biofilm control is required in industries that make extensive use of pipelines. Many industries use water pipelines in cooling towers or heat exchange structures, and the presence of a thick biofilm inside these pipes can lower the flow capacity of the pipes and decrease heat-exchange efficiency. Compared with planktonic cells, biofilm formations are more resistant to antibacterial substances, such as antibiotics and disinfectants. The reduced efficacy of these agents may be due to their inability to penetrate the extracellular matrix material or to an altered physiological state of the attached bacterial cells. The resistance of bacteria in biofilms can present problems in the use of antibiotics to treat diseases and in the use of chemical disinfectants to sterilise medical devices, such as surgical tools or medical implants. Control of biofilm development in drinking water distribution systems is a well-documented problem. Attached bacteria grow by utilising low concentrations of DOM present in water distribution systems. These biofilms can harbour opportunistic pathogens and require high doses of disinfectant for their control. The high concentration of disinfectant, in turn, can cause public and environmental health problems. The development of innovative technologies has focused on ways to enhance the effectiveness of disinfectants so that lower doses can be used or to remove the disinfectant from the water supply after the biofilm has been destroyed. Alternative approaches include the impregnation of materials, such as plastics, with biocides or antibiotics so that colonisation of the surface is inhibited.

The support of mixed microbial groups, which interact as a consortium within the architecture of the biofilm, provides a distinct advantage for the survival of diverse physiological types within the biofilm environment. The advantage of adherence of microbial consortia is evident with the biodegradation of cellulose fibres. Bacteria that possess cellulolytic capabilities are able to attach directly to the cellulose fibres. These cells are then surrounded by other cells capable of utilising the glucose monomers released by the depolymerisation of cellulose. Another good example is the positioning of different physiological types in separate aerobic or anaerobic microenvironments that are engaged in complementary metabolic activities.

The products of one form of metabolism are exchanged across the aerobic-anaerobic interface, providing the essential requirements for the growth of a different physiological type. The contribution of biofilm structure to the corrosion of metal surfaces is an example of this level of organisation.

Aquatic Environments

Freshwater environments

Freshwater environments are inland bodies of water (such as springs, rivers and streams, and lakes) that are not directly influenced by marine waters. The science that focuses on the study of freshwater habitats is called limnology and the study of freshwater micro-organisms is microlimnology. There are two types of freshwater environments. The first is standing water, or lentic habitats, such as lakes, ponds, and bogs. The second is running water, or lotic habitats, including springs, streams, and rivers. These freshwater environments have very different physical and chemical make-ups and correspondingly characteristic populations and communities of micro-organisms. In addition, no single description of these microbial communities can accurately describe each and every case. For instance, the microbial community in a lake in Egypt is not the same as the microbial community in one of the Great lakes in the northeastern United States. In this section we define various freshwater environments and provide ranges of numbers and the types of populations of the microbes that inhabit them.

Springs

Springs form wherever subterranean water reaches the earth's surface. There are many types of springs. For instance, melted snow and ice in mountainous regions feed cold springs such as scree springs. In contrast, groundwater springs are fed by surface-linked aquifers. Warm or hot springs originating from volcanic areas or great depths are known as thermal springs. A good example are the geysers found in Yellowstone National Park. Springs are also characterised with respect to their chemical properties, especially if they have a distinctive mineral or chemical composition. Examples include sulphur and magnesium springs, acid springs, and radioactive springs. Micro-organisms, especially bacteria and algae, are often the only inhabitants of springs.

In general, photosynthetic populations dominate spring environments with photosynthetic bacteria and algal communities ranging from 10^2 to 10^8 organisms/ml. These primary producers are present in highest concentrations (10^6 to 10^9 organisms/ml) along the shallower edges of the spring and in association with rock surfaces, where light is available and inorganic nutrients are in highest concentrations. Although heterotrophs are also present, because the nutrient level, especially DOM, is low, numbers are usually quite low (10^1 to 10^6 organisms/ml) compared with other surface waters. As they mature and die, photosynthetic populations provide the initial source of organic matter for downstream heterotrophic populations. However, the largest portion of DOM found in surface freshwater originates from surrounding terrestrial sources. This organic input, which originates from sources such as plant exudates, dead plants, animals, and microbial biomass, is transported into lotic habitats by mechanisms such as terrestrial runoff, seepage, and wind deposition. Thus, we have the image of spring water starting with very low concentrations of DOM and heterotrophs. The DOM and the heterotrophic populations steadily increase as the spring moves away from the source as inputs of terrestrial organic matter and microbial biomass continues to accumulate.

Rivers and streams

Springs, as they flow away from their subsurface source, merge with other water sources to form streams and rivers that eventually flow into other bodies of water such as lakes or seas. These water sources can be other springs, terrestrial runoff, and even hyporheos. A hyporheos is defined as an interface where the subsurface water table meets and interacts with the streambed. It can provide an active exchange interface between the surface water and subsurface aquatic environments.

As a stream progresses and becomes larger, it tends to accumulate organic matter and heterotrophic populations. These heterotrophic populations are inoculated from the surrounding terrestrial environment, and the profile of micro-organisms in river water often resembles associated terrestrial microbial communities.

Most physical characteristics of rivers and streams, such as temperature, volume, velocity, and chemical make-up, are determined by the geographic and climatic conditions of the area through which they flow. For instance, mountain streams, because of the steep terrain through which they flow, have fast currents and low temperature, whereas rivers that flow through plains may flow very slowly and have higher temperatures. Rivers are usually not very deep (only a few metres), but larger rivers can have pools over 50 m deep. The volume of rivers and streams is highly dependent on seasonal fluctuations. All of these variables influence the microbial populations found in rivers and streams. For instance, current affects the development of planktonic communities. In lotic habitats these communities develop only if the local current is extremely slow or nonexistent. In most cases, stable and localised planktonic communities in rivers are found only in pools.

Streams, like most surface water environments, contain primary producer communities, especially when light can penetrate to the bottom of the stream. Photosynthetic populations range from 10^0 to 10^8 organisms/ml and tend to be present as attached communities associated with biofilms because of the flowing nature of the water column. Phytoplankton (free-living) communities also exist in streams, but because of the constant water movement, they are not spatially stable populations.

As streams evolve into rivers, they tend to acquire more DOM. The increase in DOM limits the penetration of light and consequently begins to limit photoautotrophic populations. However, heterotrophic populations begin to increase in response to increased DOM. Because of their flow patterns, stream and river waters are for the most part well aerated. Therefore, heterotrophic populations are predominantly aerobic or facultatively anaerobic. In general, the concentration of heterotrophs in streams and rivers ranges from 10^4 to 10^9 organisms/ml, with microbial numbers increasing as DOM increases. Sewage outflows are areas where this is especially evident. Downstream of these outflows, heterotrophic populations often increase two to three orders of magnitude. Although isolated pools that form in rivers act as DOM and POM sinks and support fairly stable heterotrophic planktonic communities, the only truly stable populations in the lotic habitats of streams and rivers are the biofilm and sediment (benthic) communities.

Lakes

Physical and chemical characteristics

Lakes vary in depth from a few metres to more than 1000 metre. Lakes can also vary considerably in surface area, from small ponds of only a few square metres to huge lakes that cover areas of up to 1,00,000 km². Although often regarded as lentic or nonflowing environments, lakes have inflows and outflows, wind-generated turbulence, and temperature-generated mixing, all of which create a dynamic environment. Salt lakes, such as the Great Salt Lake in Utah, are distinguished by a high salt content and are examples of extreme environments. Other lakes are also characterised by chemical composition, such as bitter lakes that are rich in $MgSO_4$, borax lakes that are high in $Na_2B_4O_7$, and soda lakes that are high in $NaHCO_3$. Lakes are by far the most complex of the freshwater environments. As a result, the microbial communities and their interactions are equally complex and diverse.

Lakes are divided into subsections based upon morphometric (depth, dimension, geology of shores, currents, etc.) and physico-chemical (temperature, pH, oxygen content) parameters (Fig. 15.4). The edge of the lake, where sunlight can penetrate to the bottom, is known as the littoral zone. The air-water interface including the upper few millimetres of the water column is known as the neuston layer.

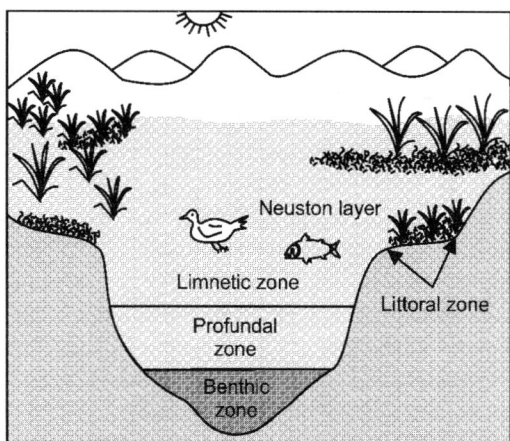

Fig. 15.4. Schematic representation of a typical lake showing common designations based on sunlight. Other designations for zones are based on features such as temperature, oxygen concentration, and pH. However, the most common are those shown here primarily because of the controlling influence sunlight has on these environments.

The general structure of the neuston is shown in Fig. 15.5. The neuston is known to accumulate nutrients. The very top layer of the neuston is a thin lipid layer (10 nm deep) that is created because it is energetically favourable for nonpolar organic molecules to align at the air-water interface. Adjacent to this is a slightly thicker layer (100 nm) containing proteins and polysaccharides. Together, these layers form a thin gel-like matrix at the air-water interface. Bacteria attach to this organic layer in a firm but reversible manner.

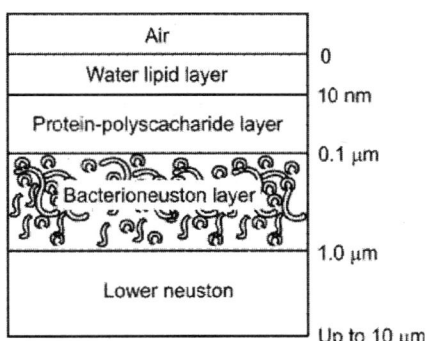

Fig. 15.5. Schematic representation of the neuston. This is the upper layer of aquatic environments and can range from 1 to 10 μm in depth. Most scientists consider the neuston an extreme environment because of many factors, including intense solar radiation, large temperature fluctuations, and the natural accumulation of toxic substances including chemicals, organic matter, and heavy metals. The upper layer that interacts with the atmosphere consists of a water-lipid mixture that has increased surface tension. Below this is a layer of organic matter that accumulates from organic matter rising up the water column.

Thus, we also have the image of the neuston layer as a biofilm, where organic molecules 'condition' the air-water interface, allowing bacteria to attach. The limnetic zone refers to the surface layer of open water away from the littoral zone where light readily penetrates. The area below the limnetic zone, where light intensity is less than 1 per cent of sunlight (the light compensation point), is known as the profundal zone. Finally, the benthic zone consists of the lake bottom and the associated sediments.

Temperature is very important, especially in lakes, giving rise to another classification scheme. The three regions in this scheme are the upper zone or epilimnion, the lower zone or hypolimnion, and the thermocline, which is a middle zone characterised by a rapid change in temperature (Fig. 15.6). Because water is most dense at 4°C, temperature-induced density stratification occurs at the thermocline in the summer and the winter. In the summer, the epilimnion, which is heated by sunlight, is typically warm and oxygen rich. This zone is usually characterised by intensive primary productivity that can deplete the epilimnion of mineral nutrients, resulting in nutrient-limiting conditions. The characteristics of the epilimnion are reversed in the hypolimnion, which has low temperature and oxygen levels, lack of light penetration, and a high mineral nutrient content. This stratification would tend to make lakes very static, but as fall and winter approach, the warm waters of the epilimnion cool until they reach the temperature, and consequently the density of the hypolimnion. When this happens the thermocline breaks down and allows mixing of the epilimnion and the hypolimnion. In the winter a layer of ice forms at the top of the lake and the epilimnion is formed in the region of 0°C (ice layer) to 4°C. The hypolimnion remains at 4°C or warmer, and again a thermocline is formed and no mixing occurs. In the spring, as the lake thaws and the two zones reach a similar temperature, mixing occurs once again. In essence, the turnover and mixing of these two layers allow reoxygenation of the hypolimnion and replenishment of mineral nutrients in the epilimnion.

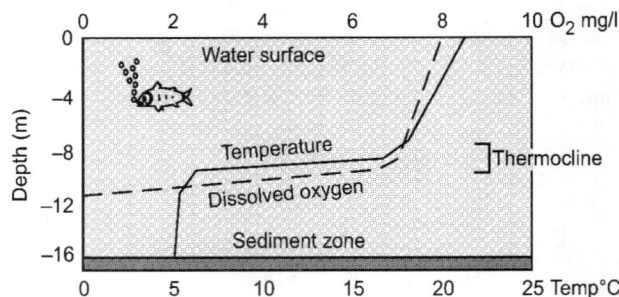

Fig. 15.6. Idealised profiles of temperature and oxygen in a temperate region, eutrophic lake. Stratification is due to thermal warming of the upper layers in the summer months. Cooling of the upper layer in the fall and early winter breaks the mixing barrier and allows the sediment zone to be reoxygenated.

Microbial characteristics

Lakes are the most extensively studied aquatic environments and in many cases are the most complex microbially. Lakes contain extensive primary and secondary productive populations that interact dynamically. In the littoral zone the planktonic community is composed predominantly of algae and secondarily of cyanobacteria. Even the attached communities are dominated by the presence of filamentous and epiphytic algae. Thus, the primary productivity in the littoral zone is high. The limnetic habitat is also dominated by phytoplankton, which form distinct community gradients based upon the wavelength and the amount of light that penetrates to a given depth. Figure 15.7 shows characteristic photosynthetic organisms (phytoplankton) and their light absorption spectra. As an example of how these microbial community gradients become established, consider *Chlorobium*, a green sulphur bacterium. *Chlorobium* can utilise longer wavelengths of light than many other phototrophs. They are also anaerobic organisms, requiring H_2S rather than H_2O for photosynthesis. Thus, they have a competitive advantage in establishing a niche at depths lower in the water column or even in the surfaces of sediments, where only small amounts of light penetrate, little or no oxygen is present, but hydrogen sulphide is available.

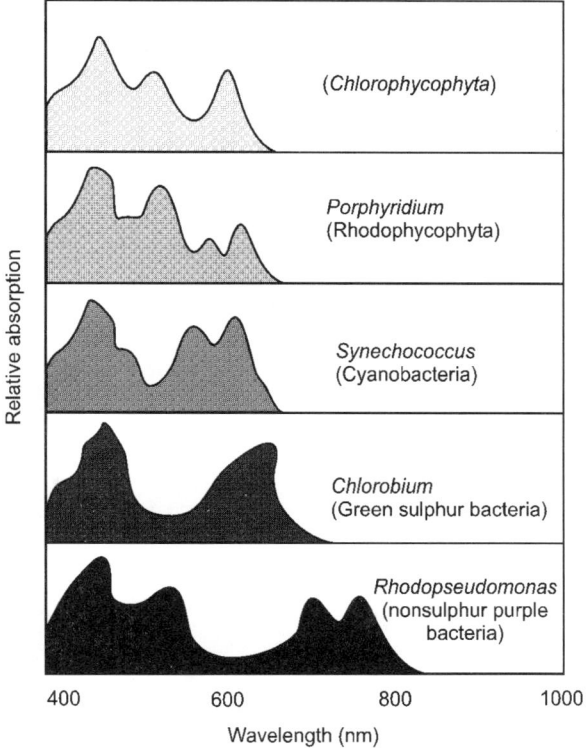

Fig. 15.7. Graph showing the light absorbance spectrum of common phytoplanktonic algae and photosynthetic bacteria. It can be seen that each of these groups has a different profile. This enables groups to take advantage of their niche. In general organisms that are capable of utilising longer wavelengths are found deeper in the water column. Thus, they do not have to compete with organisms higher in the water column that absorb the shorter wavelengths.

In addition to their phototrophic populations, lakes have extensive heterotrophic communities. The amount of secondary production is directly related to primary production. In general, secondary production in the photic zone is 20 to 30 per cent of primary production. Heterotrophic concentrations vary with depth, but there are three areas that generally have elevated numbers of heterotrophs. The first is the neuston layer, where accumulations of proteins and fatty acids create localised eutrophic conditions of which heterotrophic organisms take advantage. The second area where heterotrophic populations are markedly higher is the thermocline, which is just below the zone of highest primary productivity. This is because organic debris tends to settle and accumulate at the thermocline as it does in the neuston layer. The third area of a typical lake that is characterised by higher concentrations of heterotrophs is the upper layer of the benthos. The populations here are primarily anaerobic. In comparing oligotrophic and eutrophic lakes, there are some striking differences. Figures 15.8(a) and (b) compare the major bacterial populations found in typical oligotrophic and eutrophic lakes along with their relative concentrations. In terms of primary productivity, oligotrophic lakes have higher rates (20 to 120 mg carbon/m^3/day) than eutrophic lakes (1 to 30 mg carbon/m^3/day).

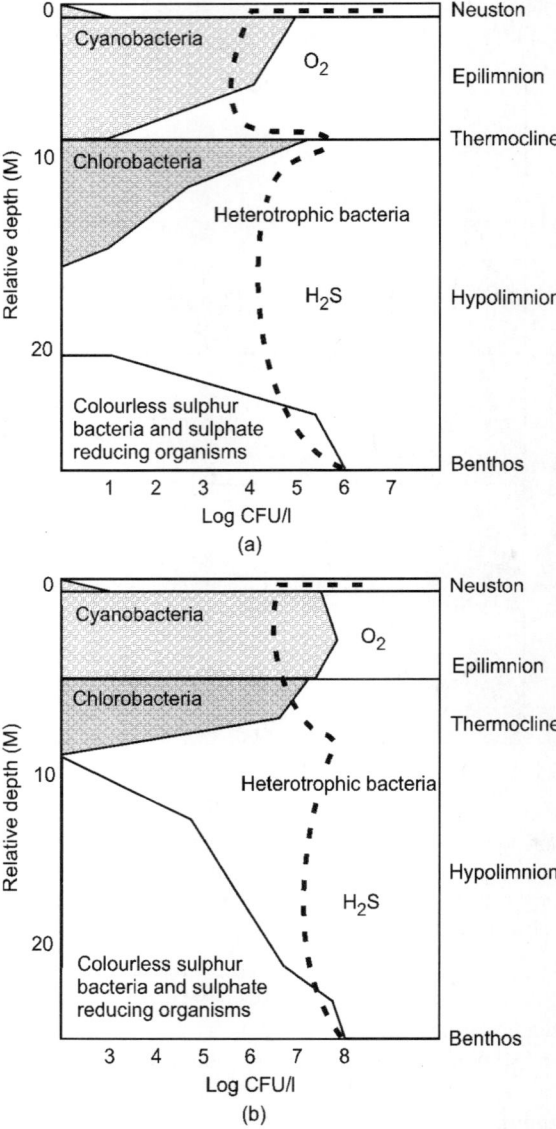

Fig. 15.8. (a) Schematic representation of bacterial distribution in a typical oligotrophic lake. Notice especially the distribution and concentrations of the photosynthetic populations. Also note the lower concentration of heterotrophs in the upper zone, where cyanobacteria predominate. The large increase in the heterotrophic population between the epilimnion and the hypolimnion is related to the presence of a zone where organic matter accumulates. This area is known as a thermocline and is a zone where the sunlight-warmed surface water (less dense) and the deeper colder water (more dense) meet, forming a density gradient where organic matter accumulates. (b) Schematic representation of a typical eutrophic lake. The figure shows the same groups of organisms as in (a) indicating the localisation and relative concentrations throughout the water column. Notice that both the photosynthetic and the heterotrophic populations are considerably higher in a eutrophic lake.

This is because eutrophic lakes have much higher levels of organic matter, which causes turbidity and interferes with light penetration. As might be expected, in terms of secondary productivity, eutrophic lakes have much higher rates (190 to 220 mg carbon/m^3/day) than oligotrophic lakes (1 to 80 mg carbon/m^3/day).

Apart from their bacterial and algal populations, streams, rivers, and lakes also contain fungal, protozoan, and viral populations, which interact and contribute to the functioning of the food web. Fungi are rarely planktonic but they serve as parasites of planktonic algae, preventing overpopulation and allowing light to penetrate farther into the water column. Some fungi have simple lifestyles, colonising surfaces and often forming fungal lawns, whereas other fungi such as *Zoophagus insidians* have a more complex lifestyle. *Z. insidians* live on filamentous green algae in rivers and lakes. The fungi have long hyphae, which branch off and extend into the water column. The hyphae actually act like fishing lines because when they are touched by the cilia of rotifers they quickly secrete a sticky substance, essentially hooking or attaching to the animal. At this point, and with great rapidity, the hyphae grow into the mouth of the rotifer, where a fungal mycelium forms, absorbing the contents of the animal's body.

Viruses in freshwater environments can be very abundant and can utilise bacteria, cyanobacteria, and microalgae as their hosts. Viruses affect both population dynamics and community composition in the planktonic environment. The impact of viral infection is a complex issue and an area of current study. Much attention has been focused on the interaction between bacterioplankton and viruses. The total viral population can exceed that of the bacterioplankton by two orders of magnitude, and viral population densities tend to fluctuate with the population densities of their hosts. Viruses that infect bacterioplankton (bacteriophage) also vary in host specificity. In other words, some bacteriophage can be infective to a variety of bacterial species and genera, whereas others may be infective to only one particular species. Virus-induced lysis of bacterioplankton can account for approximately 20 to 50 per cent of bacterial mortality. The DOM released by the lysis of microbial cells is an important nutrient source for noninfected bacteria, as discussed.

Protozoa are also important predators of aquatic micro-organisms, bacteria and algae in particular. As bacterial and algal concentrations increase, they provide an abundance of food for protozoa, whose populations begin to increase. As the food source is used up, the numbers of protozoa are no longer supported by bacterial and algal biomass and so begin to decline. As protozoal numbers fade, this allows a renewed increase in bacterial and algae concentrations. It should be noted that protozoan populations are several logs lower than bacterial numbers. They are able to affect the numbers of bacteria and algae because each protozoan is able to consume hundreds of bacteria and algae per day. It should also be noted that viral concentrations tend to mimic protozoan ones. Thus, it can be assumed that together the protozoan and viral populations help to control the concentrations and biomass of the bacterial and algal communities, providing population balance in freshwater ecosystems.

Brackish Water

Brackish water is a broad term used to describe water that is more saline than freshwater but less saline than true marine environments. Often these are transitional areas between fresh and marine waters. An estuary, which is the part of a river that meets the sea, is the best known example of brackish water. Estuaries are highly variable environments because the salinity can change drastically over a relatively short distance. Dramatic change can also occur at a given point in the estuary as a function of the time of day or season of the year. For example, at high tide, the salt content at a given spot in the estuary will increase as ocean water moves into the area. In contrast, seasonal increases in freshwater due to rainfall or snowmelt will decrease the salinity at a given point in the estuary. The variation in salinity can range from 10 to 32 per cent with the average salinity of freshwater being 0.5 per cent. In order to survive in these environments, microbes and plants in an estuary must be adapted to the fluctuations in salinity. Despite this, estuaries are very productive environments. Specific examples of highly productive brackish water environments are mangrove swamps such as those found in the everglades of Florida. The salinity in these swamps is usually very close to that of seawater. Mangrove swamps are named for the characteristic trees that grow in the saturated soils of the swamp. These trees have specially adapted roots that grow up from the water to allow gas exchange above the water so that the trees can continue to obtain oxygen and 'breathe'. A second important adaptation is that mangrove trees inhibit salt transport into the roots to avoid salt stress. Mangrove swamps are an important transition community because they help filter contaminants and nutrients from the water, they stabilise sediments, and protect the shoreline from erosion. They also provide an active habitat for more than 250 animal and 180 bird species.

In general, estuarine primary production (10 to 45 mg carbon/m^3/day) is not always enough to support the secondary populations. Estuaries tend to be turbid because of the large amount of organic matter brought in by rivers and the mixing action of tides. As a result, light penetration is poor. Numbers of primary producers are variable, ranging from 10^0 to 10^7 organisms/ml, and these populations also vary considerably in relation to depth and proximity to existing littoral zones. Despite low primary productivity, because availability of substrate is not limited, heterotrophic activity is high, ranging from 150 to 230 mg carbon/m^3/day. Local runoff and organic carbon are brought in abundantly by the rivers that flow into the estuaries. In fact, the supply of nutrients can be so great that in many cases estuaries can actually become anoxic for whole seasons during the year. As a result of the steady and abundant carbon supply, numbers of secondary producers fall into a much narrower range from 10^6 to 10^8 organisms/ml.

Marine Water

Physical and chemical characteristics

Marine water environments, like those of lakes, are highly diverse. Marine water is characterised by salinity between 33 and 37 per cent and can range in depth up to 11,000 metres in the deepest of ocean trenches. Oceans are not static in size or shape; considerable mixing (especially of the surface layers) and movement are caused by the action of tides, currents, temperature upwelling, and winds. Because the oceans are so expansive and their surface area is so great, the effect of sunlight is important. The ocean is divided into two zones, the photic zone, through which light can penetrate, and the lower aphotic zone. Light is able to penetrate to a depth of 200 metres, depending on the turbidity of the water. In coastal areas, where the amount of suspended particulate matter in the water is high, light may penetrate less than 1 metre.

Other designations for classifying zones in marine environments are based on habitats. Four major habitats are important from a microbiological standpoint. At the surface of the sea (air-water interface) is the habitat referred to as the neuston (Fig. 15.5). The pelagic zone is a broad term used to describe the water column or planktonic habitat. The pelagic habitat is subdivided on the basis of the precise depth in the water column. The habitat in the upper 100 metres of the water column is known as the epipelagic zone (i.e. the photic zone). A large proportion of the organisms in the epipelagic zone are photosynthetic. Further depths are designated as mesopelagic, bathypelagic and abyssopelagic habitats. Finally, the benthopelagic zone (benthos) is the sea-sediment interface. Apart from the pelagic zone and the neuston layer, the third major habitat is the epibiotic habitat, which refers to surfaces on which attached communities occur. The fourth is the endobiotic habitat, which pertains to organisms found within the tissues of other larger organisms such as fish. One interesting endobiotic bacterium, *Epulopiscium fishelsoni*, is discussed later in this chapter.

Microbial characteristics

The ocean contains diverse microbial habitats. Also, depending on location (compare the Mediterranean with the Antarctic ocean), markedly different microbial populations predominate. As a rule, especially in deep waters, microbial concentrations are highest within the neuston and drop markedly below this region. Immediately below the neuston, the numbers are on average close to 10^7 organisms/ml and decrease by more than a log at a depth of 100 metres. As a result of organic input from the terrestrial environment, total bacterial numbers are on average one order of magnitude higher in coastal water than in the open ocean. This is especially true for harbours and near populated areas. If we consider the vertical distribution of the heterotrophic populations we find that, as in lake environments, numbers increase at the thermocline that exists below the zone of primary production (50 metres). As mentioned lake thermoclines are also regions of high heterotrophic activity because of the accumulation of organic matter. At greater depths, the numbers of heterotrophs quickly diminish until, at a depth of 200 metres, concentrations are very low. Heterotroph numbers increase again immediately above the ocean floor.

Thermocline-induced stratification is not as dramatic in coastal waters because of the mixing of water by winds, currents, and temperature. For this reason, bacterial numbers are uniform at all depths except when the weather is very calm for long periods of time. In addition, seasonal fluctuations in bacterial numbers, which are not observed in the open ocean, are common near the coast. In general, there are two times of the year when there is an increase in bacterial populations in coastal waters, late spring–early summer and late summer–early fall. These are also times when the phytoplankton are most active. Thus, primary production is intimately tied to the overall productivity of the ocean. Figure 15.9 shows the relationship between chlorophyll *a*, a photosynthetic pigment, and heterotrophic numbers. It can be seen that an increase in phototrophs is followed by an increase in heterotrophs. This tends to be true in many aquatic environments where the concentration of primary producers ranges from 10^0 organisms/ml in some benthic environments to 10^8 organisms/ml in surface zones. Heterotrophic populations generally vary considerably with depth as well, although not as much as the photosynthetic populations. The general range is from 10^1 organisms/ml in zones characterised as oligotrophic to 10^8 organisms/ml in zones where organic matter is present. Oceans have profiles similar to those presented in Figs 15.8(a) and (b), depending on whether the marine environment is oligotrophic like the open ocean or eutrophic like coastal waters, especially coastal waters where sewage outpours may be present.

Phytoplankton abundance and community compositions, as mentioned above, vary depending on the season. Interesting phenomena related to phytoplankton in the ocean are algal blooms. These occur when a eutrophic water body and the appropriate environmental conditions (usually warm, sunny, and calm) occur. In these nutrient-rich and agreeable environmental conditions, it is common for certain algae or cyanobacteria to proliferate rapidly, resulting in blooms. Such blooms are often a natural part of the yearly cycling of lake and ocean ecosystems. On the other hand, when uncharacteristically high eutrophication events occur, algal blooms can adversely affect the water quality. Such water tends to be scummy and smelly, making it unpleasant for recreation and even dangerous for fishing, boating, or swimming. In the worst cases, algal blooms may be composed of algae that produce potent toxins. The 'red tide' is an example of such an event. This is an immense algal bloom that is composed of red-pigmented dinoflagellates. These algae produce powerful toxins that are taken up by fish, causing extensive fish kills. These toxins can also affect marine birds, marine mammals, and even humans who consume fish caught in these waters. Types of human illness caused in humans by algal toxins include skin rashes, eye irritation, vomiting, diarrhea, fever, and joint pain.

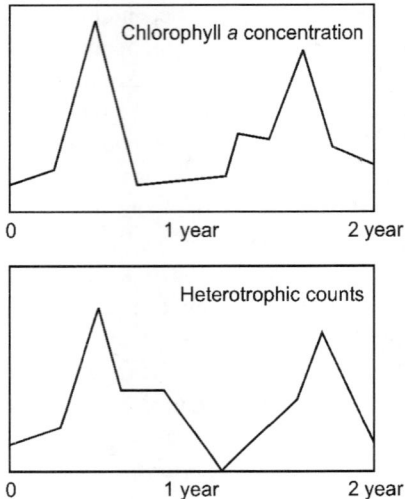

Fig. 15.9. Diagram of the interrelationship between the concentration of chlorophyll *a*, a photosynthetic pigment, and heterotroph concentration. The concentration of chlorophyll in water is related to the amount of primary production. This in turn influences the amount of secondary production by heterotrophic populations. In this figure, it can be seen that as the chlorophyll *a* concentration increases, it is closely followed by an increase in heterotrophic populations. Thus, secondary production is intimately tied to primary production.

Fungi, protozoa, and viruses in their roles as predators and parasites are also important components of marine waters. Bacteriophage in particular are highly prevalent, often occurring in concentrations one to two logs higher than the bacterial populations at any given location in the water column. Viruses that infect phytoplankton are very important from an organic cycling standpoint because the organic matter released by phytoplankton after they are infected and lysed promotes secondary productivity. Secondary producers and even algae and larger marine animals such as fish and crabs are also targets of marine viruses. Such viruses can be responsible for considerable economic losses for the fishing industry.

Protozoa also act as important bacterial predators in the marine environment. In addition, fungi have recently been found to be widely distributed in the marine habitat. Marine fungi feed on both plants and animals in addition to bacteria and algae. A relatively small number of genera are found in the marine environments, but these can often be present at surprisingly high concentrations. Certain species of fungi can be isolated with concentrations up to 10^3 to 10^4 organisms/ml, especially in carbon-rich areas of the water column and in organic-rich benthic habitats. Another surprising fact is that fungi have been found at depths well below 1000 metres.

Deep-sea hydrothermal vents

In 1977 geologists first described deep-sea hydrothermal vents (Fig. 15.10). These are areas on the ocean floor where, driven by magma-derived hydrothermal convection, hot water laced with minerals flows up through cracks and fissures. The cracks, which are known as hydrothermal vents, often have a buildup of chemicals resembling chimneys surrounding them. Water, reaching temperatures up to 400°C, is emitted from these vents at rates of 1 to 2 m/sec. This mineral-rich, hydrothermal water forms a dark cloud of mineral precipitates as it mixes with seawater. Thus, these hydrothermal vent chimneys are known as 'black smokers'. It was surprising to find that in this environment, which has no light and extremely high temperature and pressure, whole self-contained ecosystems consisting of microscopic and macroscopic life have developed.

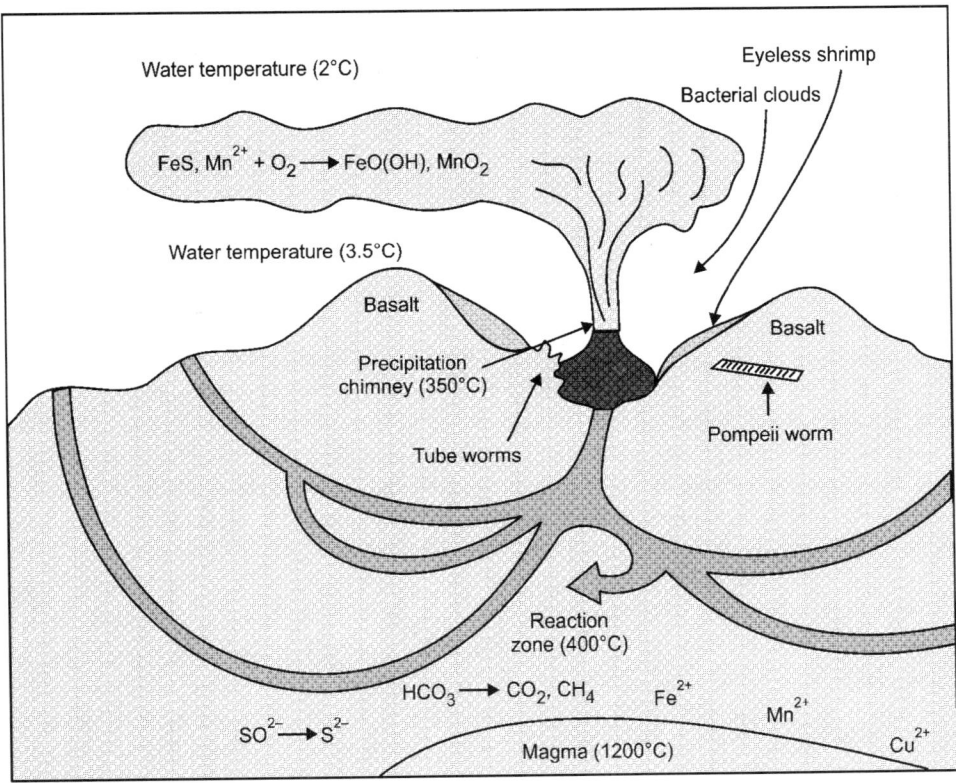

Fig. 15.10. Schematic representation of a hydrothermal vent community. This figure shows a black smoker rising from the ocean floor and a plume of chemical-rich superheated water rising from it. Also depicted are characteristic organisms, including bacterial clouds that primarily consist of heterotrophic bacteria. Other creatures that are often found in hydrothermal vent communities are shown, including tube worms, eyeless shrimp, and Pompeii worms. Each of these is reliant on microbial primary or secondary productivity. The tube worms, as described in the text, rely on sulphur bacteria for primary production of carbon which is shared symbiotically with the worm. The eyeless shrimp consume sulphur bacteria from microbial mats around vents, and Pompeii worms rely on enzymes produced by thermophilic bacteria for their own thermal protection.

There is currently vigorous debate about the actual source of primary organic carbon and the process for its introduction into these ecosystems. There are three main schools of thought, each of which is based as much on theory as on actual scientific data. The hypothesis that currently has the most credibility is the chemoautotrophic theory. Chemoautotrophs utilise CO_2 as a major and usually exclusive source of carbon. In this hypothesis, primary organic carbon production in hydrothermal vent communities is based on bacterial use of H_2S, H_2 and CH_4 as electron donors, and CO_2 as a carbon source. Thus, according to this hypothesis, the entire food web in a hydrothermal vent community is based on chemoautotrophy, not photoautotrophy as in surface environments.

Another hypothesis for the formation of primary organic carbon is based on an abiotic process called organic thermogenesis. This hypothesis followed from the observation that organic-free rocks initially collected near hydrothermal vents, were able to synthesise sugars and amino acids from para-formaldehyde and urea at high temperatures in the presence of carbonates. This observation led Ingmanson and Dowler to think that the production of carbon in hydrothermal vent communities is the result of the combination of physico-chemical factors found in the vent environment. Specifically, these factors include: a concentrated source of energy in the form of heat; a considerable amount of ionising radiation is present; there is a ready supply of all the required precursors (e.g. CO_2, N_2, NH_3, H_2, CH_4); and the advective flow around vents provides rapid cooling and the subsequent quenching of thermogenic reactions necessary for organic synthesis reactions. Once CO_2 is converted to simple sugars in this thermogenic process, chemoheterotrophs can utilise the carbon to build biomass. Thus, thermogenesis results in primary production by heterotrophic bacteria.

The third major theory, called the advective plume hypothesis, is that primary production in hydrothermal vents is based on photoautotrophy. This theory is strange on the surface because such deep-sea vents are essentially isolated from solar radiation. The theory, however, is based on the assumption that settling of organic carbon from near the ocean surface occurs. This carbon, which settles from near-surface phototrophic layers, is concentrated around hydrothermal vents by advection (uprising in the water column due to heat), which essentially draws water and DOM in from relatively great distances. In this case, primary production would also result from heterotrophic activity. In summary, it may very well be that all three of these hypotheses are valid and each mechanism contributes to some extent to primary production in the hydrothermal vent community.

The vent communities support macrofauna that rely on the bacterial populations as a source of organic carbon. There are at least three major mechanisms for transfer of this bacterial carbon and energy to the next trophic level. The first is an endosymbiotic relationship between vent bacteria and an invertebrate that has been dubbed 'tube worm'. Tube worms are large tube-shaped creatures that grow from the seafloor. These worms have no mouth, gut, or any other digestive system and depend completely on bacteria for their nutrition. Instead of consuming the bacteria, the worms have interior surfaces that are colonised by massive quantities (3×10^{11} bacteria per ounce of tissue) of sulphur-oxidising chemoautotrophs. The worm's body is filled with blood containing large amounts of haemoglobin that binds H_2S. The blood transports the H_2S to the bacteria, which oxidise it and fix CO_2 into organic compounds that nourish the worm. The second method is termed microbial gardening. In this case, bacterial cultures are maintained by mussels and other invertebrates on specialised appendages such as tentacles and gills. These invertebrates periodically harvest and consume the bacteria, retaining small inocula to initiate the next crop. The third mechanism for carbon transfer to higher trophic levels is direct consumption of free-living bacterial cells, filaments, or mats. Crabs, amphipods, predatory fish, and even other micro-organisms, including bacteria, have been observed to feed directly on the chemoautotrophic or chemoheterotrophic primary producers.

Free-living bacteria around hydrothermal vents have been seen in such great numbers that the area was described as a bacterial snowstorm. These concentrations range from 10^4 to 10^8 bacteria/ml, depending on the proximity to the mouth of the vent, with the highest concentrations occurring within 1 metre of the vent mouth. In addition to the clouds of bacteria, bacterial mats several inches thick are found surrounding the vents. Thus, extensive bacterial communities are found in this extreme environment. The cellular mechanisms that help these micro-organisms survive in this high-pressure, high-temperature environment are described further. Primary production around hydrothermal vents when all forms of chemoautotrophy are considered is estimated to be on the order of 2 mg carbon per litre, although this may be somewhat conservative.

Subterranean Water

The groundwater environment is found inland in the subsurface zone and includes shallow and deep aquifers. The characteristics and microbial communities of the groundwater environment have been discussed. Briefly, micro-organisms are the sole inhabitants of these environments and bacteria are the dominant type of microbe present. In groundwater, unlike other aquatic environments that have substantial planktonic populations, most of the bacterial populations are attached or only transiently suspended. In general, levels of microbial activity are low, especially in intermediate and deep aquifers. This is due to low nutrient levels. Many subsurface environments may even be considered extreme from a nutrient perspective.

ENVIRONMENTAL DETERMINANTS THAT GOVERN EXTREME ENVIRONMENTS

There are two definitions of extreme environments. The first characterises an environment as extreme if the environmental conditions are at one of two extremes (high or low). These environmental conditions can include pH, temperature, salinity, pressure, and nutrients. The second definition refers to environments in which conditions select for extremely low microbial diversity. In general, growth and multiplication are the primary factors that influence the survival of any organism and its progeny. Extremophile is a term applied to organisms that have successfully adapted to environments where it is difficult or impossible for other organisms to survive. Thus, extremophiles have been selected over time for characteristics that allow them to grow and multiply in a variety of extreme environments. This section will briefly describe some of these environments and the physiological adaptations used by extremophiles to compete or survive in their particular niche.

Air-Water Interface

The air-water interface is a unique habitat that is often considered an extreme environment for many reasons, including high levels of solar radiation; accumulation of toxic substances (e.g. heavy metals, pesticides); large temperature, pH, and salinity fluctuations; and competition. As a rule, the air-water interface, also referred to as the neuston, actually contains higher concentrations of organisms than other layers of the water column. There is some debate, however, about the percentage of viable bacteria in the neuston. Indeed, most reports suggest that the numbers may be higher but the ratio of metabolic activity to total counts is lower in the neuston than in the planktonic habitat. The neuston accumulates nutrients and especially attracts nonpolar organic and inorganic molecules, which form a film at the air-water interface. Bacteria attach to this layer because it concentrates nutrients. However, in addition to the nutrient accumulation, the neuston tends to accumulate toxins. Among these toxins are nonpolar organic molecules, including pesticides such as DDT and petroleum hydrocarbons, as well as metals such as Cd, Cu, Mn, Hg, Pb, Se and Cr. Thus, the micro-organisms that inhabit the neuston have developed unique metabolic, genetic, and functional strategies that allow them to survive the extremes found in this environment. These strategies include the use of pathways that catabolise toxic compounds and provide resistance to metals that accumulate at the interface. Some microbes have developed efficient DNA repair mechanisms to combat DNA damage caused by exposure to ultraviolet radiation. Finally, neuston inhabitants also need to be able to respond and quickly adapt to variations in environmental conditions.

High Temperature

There are many examples of environments with extreme temperatures. Environments with high temperatures (>70°C) include terrestrial and submarine hot springs, some of which can reach temperatures of 100°C, and hydrothermal vents, which can reach temperatures in excess of 300°C. Such high temperatures are inhospitable for most forms of life except for certain bacteria and archaebacteria. Most notable of these organisms are the genera *Thermus, Methanobacterium, Sulfolobus, Pyrodictium,* and *Pyrococcus.* The last two of these genera are especially adapted to high temperatures and can live at temperatures in excess of 100°C. Another species of thermotolerant bacteria, namely *Thermus aquaticus,* is especially renowned because of its thermotolerant DNA polymerase. This enzyme has been patented and is used around the world in the polymerase chain reaction (PCR).

Many mechanisms allow micro-organisms to survive at temperatures that would normally denature proteins, cell membranes, and even genetic material. In terms of proteins, thermophiles have an increased number of salt bridges (cations that bridge charges between amino acid residues). These bridges help the protein to remain folded even at high temperatures. In terms of cell membranes, thermophilic eubacteria have increased amounts of saturated fatty acids in their membranes that allow the membranes to remain stable at high temperatures. Extreme thermophiles, almost all of which are archaebacteria, have an entirely different cell membrane composed of repeating units of the five-carbon compound isoprene (Fig. 15.11) that are ether linked to glycerol phosphate. Finally, in terms of DNA, thermophiles contain special DNA binding proteins that arrange the DNA into globular particles that are more resistant to melting. Another factor that is common to all hyperthermophiles tested so far is a unique DNA gyrase. This gyrase acts to induce positive supercoils in DNA, theoretically providing considerable heat stability.

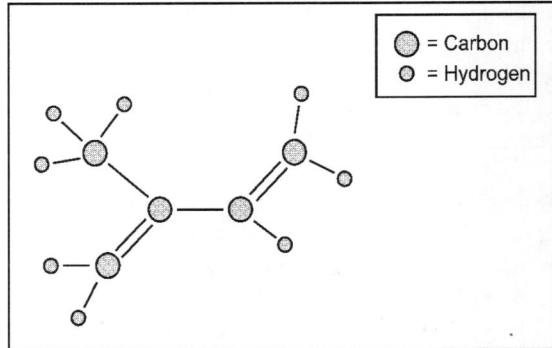

Fig. 15.11. Molecular structure of isoprene, monomer subunit from which archaebacterial cell membrane components are assembled.

There are numerous biotechnological applications for enzymes isolated from thermotolerant micro-organisms and the number of applications is growing rapidly, especially in commercial industry. One example already mentioned is the thermostable DNA polymerase used in the PCR. Other examples include proteases, lipases, amylases, and xylanases that are used in the agricultural, paper, pharmaceutical, water purification, bioremediation, mining, and petroleum recovery industries.

High Solute

Halotolerant, or high-salt-tolerant, organisms require salt concentrations for growth that are substantially higher than that found in seawater (Table 15.1). One of the best known examples of this type of environment is Utah's Great Salt Lake; another is the Dead Sea, which lies between Israel and Jordan. *Halobacterium* and *Haloanaerobium* are two examples of halotolerant bacteria. In addition to bacteria, some algae and fungi are known to be halotolerant. In general, halotolerance is not a requirement for high solute concentration but is specific for Na^+. The main mechanism of salt tolerance displayed by bacteria is internal sequestration of high concentrations of a balancing solute to equal the salt concentration found external to the cell. Examples of these balancing solutes include K^+, which is important in halotolerant bacteria, and glycerol, which is important in halotolerant eukaryotes. A second mechanism of salt tolerance of halophiles involves proteins that are acidic and typically have low proportions of nonpolar amino acids. Thus, for these proteins to be active, high salt concentrations are needed to balance their charge and acidity. Because of these macromolecular modifications, halotolerant bacteria are usually unable to survive in environments lacking high salt concentrations. Thus, many are considered obligate halophiles.

Table 15.1. Ranges of microbial salt tolerance.

Category	Salt range (metre)	Optimum salt concentration (metre)
Nonhalophile	0–1.0	<0.2
Slight halophile	0.2–2.0	0.35
Halophile	1.4–4.0	2.5
Extreme halophile	2.0–5.2	>3.0

Low pH

Acidic environments, such as acid hot springs, the gastrointestinal tract, mining waste streams, acid mine waste-water, and various mineral oxidising environments, are populated by bacteria such as *Thiobacillus*, which are discussed extensively in relation to acid mine drainage. Other examples of acidophiles are *Clostridium acetobutylicum* and *Sarcina ventriculi*, which are obligate anaerobes that ferment sugars. In addition to bacteria, some fungi, algae, and protozoa are also known to be acid tolerant. Strategies used by micro-organisms to deal with high or low pH values usually involve modifications of the cell membrane. The first of these modifications is to the structure of membrane components to allow them to be acid tolerant. This includes the incorporation of very long chain dicarboxylic fatty acids, which make up more than 50 per cent of the membrane fatty acids. These specialised fatty acids help inhibit acid hydrolysis of the membrane. The second modification involves control of ion transport across the membrane. By controlling ion transport, these organisms can maintain an internal pH in the range between 5 and 7, even though the external environment can have a pH less than 2.

High Pressure

Deep-sea environments are characterised by high pressure and cold temperatures. Microbes that live in this environment are called barophiles. Barophiles have developed unique mechanisms that allow them to tolerate the high pressures of more than 1000 bars found in deep-sea trenches (normal atmospheric pressure is 1 bar). Bacteria retrieved from depths greater than 2000 m actually grow better under high pressure than at normal atmospheric pressure. In addition to being pressure tolerant, barophiles are for the most part also psychrophilic, meaning that they grow better at low temperature. Finally, these microbes are adapted to darkness, as they live at depths that are essentially isolated from solar radiation.

Organisms that live under these deep ocean conditions, like other extreme microbes, have developed unique mechanisms for survival. Many of these mechanisms involve changes in macromolecular structure and function. For example, long-chain polyunsaturated fatty acids are found in high concentration in the membranes of barophiles. These modified fatty acids maintain the membrane in a fluid state under a pressure and temperature that would otherwise tend to gel or crystallise them. Interestingly, many of the mechanisms used by these organisms are the opposite of the mechanisms utilised by thermotolerant organisms. Although little is actually known experimentally about the mechanisms involved in replication, transcription, and translation in barophiles, it appears that they involve control of salt concentrations.

No Nutrient

Another extreme environment, which is of great importance in today's world, is ultrapure or nutrient-free water. Ultrapure water is used in the semiconducter, medical, and many other industries. Contamination of ultrapure water by micro-organisms can be devastating to these industries. For instance, microbial contamination of ultrapure water can cause flaws in the crystal design of computer chips, lowering the efficiency of the chips. Because there are virtually no nutrients in ultrapure water, it is considered an extreme environment. Indeed, very few organisms are able to survive, let alone proliferate, in ultrapure water. One of these is the bacterium *Caulobacter* and another is *Pseudomonas fluorescens*. In distilled water, for instance, even the limited exchange of nutrients (CO_2) from the atmosphere provides enough nutrients to allow limited growth. Springwater bottling plants are also of concern because the low-nutrient spring water is able to provide an even more favourable environment for proliferation and colonisation of many of these organisms, even opportunistic pathogens such as *Pseudomonas aeruginosa*.

AQUATIC MICROBES IN THE NEWS

Giant Marine Bacteria

A number of aquatic microbes are unique, or interesting for one reason or another. One of these is a bacterium known as *Epulopiscium fishelsoni*. *E. fishelsoni* is a symbiotic bacterium isolated from the gut of a Red Sea surgeonfish. One of the seminal laws in microbiology, and life in general, is that for every rule there is an exception. *E. fishelsoni* provides exceptions for two of the main rules governing bacteria. *E. fishelsoni* is 0.6 mm long and 0.08 mm in diameter. So, unlike every other known bacterium, this organism does not require a microscope to be visualised; it can be seen with the naked eye. In addition to their large size, these bacteria appear to have viviparous progeny, as opposed to undergoing typical bacterial division. Two small 'baby bacteria' develop inside the cytoplasm of the mother cell and are released through a slit in one end of the cell. This phenomenon is thought to have evolved from *Epulopiscium* ancestors as a modification of sporulation. Thus, instead of forming inert endospores, as their relatives do, these prokaryotes form internal vegetative progeny, which are then released from the cell. Another organism, *Metabacterium polyspora*, which has been isolated from the gastrointestinal tract of guinea pigs, has an intermediate stage of propagation between that of endospore-forming *Clostridium* sp., and that of the internal viviparous reproduction of *E. fishelsoni*. This organism is able to produce multiple endospores by asymmetric division at both poles of the cell and by symmetric division of the endospores at an early stage of their development. *Epulopiscium* is still unique, however, because of its ability to give birth to live offspring in a process that mimics to that in higher life forms.

The initial work on *E. fishelsoni* was done by Fishelson, who initially thought this organism was a protozoan. Later, using molecular methodologies, such as PCR and *in situ* hybridisation, the genes encoding the ribosomes of this organism were characterised and it was found that the genetic sequence was that of a prokaryote. Phylogenetic analysis has further classified this organism as a relative of the clostridia. *E. fishelsoni* offers unique opportunities for molecular and cellular microbiology as well as for environmental microbiology. Because of its large size, it is feasible that intracellular probes can be inserted within the bacterial cell without disrupting its cellular machinery to allow the study of cellular processes such as membrane transport, cellular responses to environmental stimuli, and even genetic regulation processes. Such experiments are not as easy to perform with typical bacteria, which have only one millionth the cellular volume of *E. fishelsoni*.

Aquatic Microbes: Food for the Future

Spirulina is a cyanobacterium (phytoplankton) that is rapidly becoming a popular dietary product. *Spirulina* is a traditional food for several cultures and can be found growing naturally in warm-water alkaline volcanic lakes. Scientific studies suggest that *Spirulina* protein may be one of the highest quality types of protein available, because these cyanobacteria are up to 70 per cent protein by weight and their protein contains all of the essential amino acids. In addition to being a high quality protein source, *Spirulina* is considered to be a vitamin and mineral 'gold mine'. *Spirulina* has more beta-carotene than carrots, 28 times as much iron as beef liver, the highest whole-food concentration of vitamin B_{12} and even high levels of glycogen (a natural source of energy).

NASA has considered use of *Spirulina* as a food of choice for space flight. The United Nations (UN) considers *Spirulina* as a possible solution to global protein shortages. To this end, *Spirulina* farms are being constructed in underdeveloped countries. The per-acre protein yield of *Spirulina* is 10 times that of soyabeans and 200 times that of beef (Table 15.2). The human body is also able to assimilate over 95 per cent of the protein derived from *Spirulina*, whereas we can assimilate only about 20 per cent of the available protein found in beef. Studies also suggest the possibility that consumption of these cyanobacteria can strengthen the immune system, enhancing the ability to fight infection. These other health benefits may be related to the presence of phytonutrients. Phytonutrients are special compounds produced naturally by plants and other organisms that are able to enhance and strengthen the body's natural resistance to disease, support cardiovascular health, lower cholesterol, aid in digestion, and even act as natural antioxidants. As a result, *Spirulina* is used as a supplement in many dietary supplements, including protein drinks.

Table 15.2. Comparison of *Spirulina* aquatic farming with other important agricultural products.

Protein source[a]	Land area needed (m^2)	Water needed (I)	Total energy input required (10^7 kilojoules)	Total energy output generated (10^7 kilojoules)
Spirulina	0.6	2100	3.5	23
Soyabeans	16	9000	11.7	13.8
Corn	22	12,500	55	16.5
Grain-fed (feedlot) beef	190	10,5000	456	16

[a] Values are related to the production of 1 kg of protein.

To meet the increasing demand for *Spirulina*, the number of *Spirulina* farms are proliferating. One example is California's Earthrise Farms. At Earthrise Farms, *Spirulina* is grown in numerous specially designed pond systems, each of which is larger than a football field. These growing ponds function like immense chemostats, maintaining a constantly growing pure culture of the cyanobacterium. One of the primary concerns in growing cyanobacteria commercially is the need to maintain a pure culture. This is done using specially designed pond systems and carefully balancing the pond's environmental conditions to favour the growth of *Spirulina* over other, potentially toxic organisms. Thus, these ponds are supplied with purified mineral- and nutrient-rich water and are kept aerated by mechanical paddles. To enhance growth rates further, carbonated water is bubbled through the ponds to increase the amount of carbon dioxide available for photosynthesis. Unlike other crops that are harvested once or twice a year, *Spirulina* can be harvested 24 hours a day during peak growing seasons. After harvesting, *Spirulina* cells are dried by a flash evaporation system and packaged for sale. If processed and packaged appropriately, *Spirulina* can be stored at room temperature for 5 years or more without degradation of the nutrients in the product.

Finally, *Spirulina* production has environmental advantages over normal crop production. Because *Spirulina* is an aquatic micro-organism, it does not require fertile land and hence does not cause soil erosion or groundwater contamination. Table 15.2 shows the amount of land required by other major food sources to produce 1 kg of protein compared with that required to produce 1 kg of protein from *Spirulina*. As a further example of the efficiency of farming this cyanobacterium, consider that production of 1 kg of corn results in loss of 22 kg of fertile topsoil. In addition, *Spirulina*, in spite of being an aquatic micro-organism, uses less water per kilogram of protein produced than other crops (Table 15.2). Thus, *Spirulina* requires less room, soil, and water than traditional agricultural production of protein.

Chapter 16

Primary Productivity and Producers

INTRODUCTION

Primary production is the production of organic compounds from atmospheric or aquatic carbon dioxide, principally through the process of photosynthesis, with chemosynthesis being much less important. All life on earth is directly or indirectly reliant on primary production. The organisms responsible for primary production are known as primary producers or autotrophs, and form the base of the food chain. In terrestrial ecoregions, these are mainly plants, while in aquatic ecoregions algae are primarily responsible. Primary production is distinguished as either net or gross, the former accounting for losses to processes such as cellular respiration, the latter not.

Primary productivity is the rate at which energy is converted by photosynthetic and chemosynthetic autotrophs to organic substances. The total amount of productivity in a region or system is gross primary productivity. A certain amount of organic material is used to sustain the life of producers; what remains is net productivity. Net marine primary productivity is the amount of organic material.

PROCESS OF PRIMARY PRODUCTION AND RELEVANT MICRO-ORGANISMS

Primary production is the biochemical conversion of inorganic into organic matter in aquatic ecosystems. The rate of primary production (pirmary productivity) and the subsequent accumulation of organic matter are key indicators of trophic state and water quality. Except for systems dominated by macroalgae and macrophytes, the bulk of production of organic matter at the base of planktonic and benthic food webs is mediated by microbial photoautotrophy or phototrophy (photosynthetically mediated CO_2 fixation) and chemolithotrophy or chemoautotrophy (chemically mediated CO_2 fixation). In these processes, light- or chemical-derived energy is used to reduce CO_2 to organic carbon. The basic stoichiometry of CO_2 fixation is:

$$n\,CO_2 + 2n\,H_2O \longrightarrow n\,(CH_2O) + O_2 + n\,H_2O$$

The manner in which autotrophs derive and donate reducing power to drive this process differs between physiological groups. Phototrophs utilise light-driven photolysis of simple, readily oxidizable compounds as a source of electrons and protons to generate reducing power (NADPH) and energy (ATP) for CO_2 fixation. The two modes of phototrophy are oxygenic and anoxygenic photosynthesis. In oxygenic photosynthesis, the dominant form of primary production in oxygenated surface waters, water (H_2O) fills this role. The photolysis of H_2O results in the formation of protons, electrons, and molecular oxygen (O_2), as outlined below:

$$2H_2O \xrightarrow{\;h\nu\;} 4H^+ + 4e^- + O_2$$

This is the light reaction of photosynthesis. Combining the reducing power of the light reaction with the CO_2 fixation, or the dark reactions, of photosynthesis yields the following overall biochemical reaction:

$$6CO_2 + 12H_2O \xrightarrow{hv} C_6H_{12}O_6 + 6O_2 + 6H_2O$$

Oxygenic photosynthesis is conducted by all chlorophyll a-containing micro-organisms. These include the prokaryotic blue-green algae (cyanobacteria) and eukaryotic microalgal groups shown in Table 16.1. Anoxygenic photosynthesis is confined to the photosynthetic bacterial groups outlined in Table 16.1. H_2S is the source of protons and electrons, with S being the product:

$$2H_2O \xrightarrow{hv} 4H^+ + 4e^- + 2S$$

Anoxygenic photosynthesis is restricted to O_2 devoid water columns or surface sediments. Anoxygenic photosynthesis was, in all likelihood, the prevalent form of aquatic primary production during the O_2- free Precambrian period (~2.5 billion years before present). Oxygenic photosynthesis corresponded to the appearance of cyanobacteria during the late Precambrian. Present-day dominance by oxygenic phototrophs has led to the confinement of anoxygenic phototrophs to stratified, illuminated bottom waters in lakes, near-surface anoxic sediments, nutrient-enriched waste-water systems (e.g. sewage treatment ponds and animal waste lagoons), laminated microbial mats and biofilms containing O_2-free microenvironments, and endosymbiotic (O_2-free) habitats.

Chemolithotrophy takes place in oxic and anoxic waters. Here, the oxidation of a range of inorganic and organic compounds provides reductant and energy for CO_2 fixation. Examples of aerobic chemolithotrophs commonly encountered in the water column and oxic surface sediments include nitrifying bacteria (*Nitrosomonas* and *Nitrobacter*), which oxidise ammonium (NH_4^+) or nitrite (NO_2^-) and reduce CO_2 (Table 16.1). Representative facultative anaerobic and microaerophilic chemolithotrophs include the nonpigmented sulphide (S^{2-}) oxidising bacteria (*Beggiatoa*) and the sulphur (S^0) oxidisers (*Thiobacillus*) (Table 16.1).

ECOLOGICAL AND BIOGEOCHEMICAL SIGNIFICANCE

Primary productivity is a measure of ecosystem fertility and trophic state and as such serves as a prime indicator of either natural or anthropogenic alteration in the productive capacity and associated water quality of impacted waters. The rates of CO_2 fixation and O_2 production and consumption are critical facets of ecosystem metabolism. The acceleration of primary production, or eutrophication, is generally accompanied by increases in phototrophic and heterotrophic biomass. Excessive production may manifest itself as unsightly, odouriferous, and toxic microalgal blooms, where production exceeds the ability of the food web to consume the vast amounts of producer biomass. The accumulation of ungrazed microalgal biomass often enhances microbial decomposition (O_2 consumption), which may have undesirable consequences, including hypoxia (stressfully low dissolved-O_2 concentrations) or anoxia (no detectable O_2) which is fatal to a broad spectrum of fauna and flora.

Short-term (hourly to weekly) changes in primary productivity reflect diel and seasonal changes in ecosystem nutrient inputs, physical alterations (e.g. changes in optical properties, vertical mixing, and water residence time), and shifts in plant community composition and activity. Long-term (interannual, decadal, millennial, and geological) changes may be indicative of ecological, geochemical, and climatic changes. Analysis of long-term data sets has demonstrated the utility of primary-productivity measurements as indicators of either natural or anthropogenic ecosystem alterations.

Table 16.1. Major functional groups of aquatic microbial autotrophs, categorised according to environmental requirements and oxic or anoxic characteristics[a].

Oxygenic phototrophs

    Prokaryotic

        Cyanobacteria (P, B)

        Photosynthetic bacteria (P, B)

        Erythrobacter (O_2 tolerant, contains bacteriochlorophyll *a*)

    Eukaryotic

        Chlorophytes (green algae; P, B), chrysophytes (including diatoms; P, B), cryptophytes (P), dinoflagellates (colourless and pigmented; P), euglenophytes (colourless and pigmented; P, B), prasinophytes (mainly P)

Anoxygenic phototrophs (prokaryotic: also grow photoheterotrophically)

    Sulphide and sulphur as sole electron donor for photosynthesis

        Chromatium, Chlorobium, Thiocapsa, Thiopedia, Thiospirillum (P only in meromictic waters, B) *Ectothiorhodospira* (P, B, high salinity and alkaline requirements)

        Prosthecochloris (B, strictly marine)

        Rhodobacter (P, B)

        Rhodopseudomonas, Heliobacter (mainly B)

        Chloroflexis (B, thermophilic)

Aerobic chemolithotrophs (prokaryotic)

        Colourless sulphur-oxidising bacteria

        Thiobacillus (mainly B)

        Thiovulum (B, microaerophilic)

        Beggiatoa (B, microaerophilic)

        Thiosphaera (mainly B, microaerophilic)

    Iron- and manganese-oxidising bacteria

        Gallionella (B, acidophilic)

        Planctomyces (P, B)

        Leptothrix (mainly B)

    Nitrifying bacteria

        Nitrosomonas, Nitrosococcus (P, B, ammonia oxidisers)

        Nitrobacter, Nitrococcus (P, B, nitrite oxidisers)

---

[a] P, planktonic; B, benthic.

Productivity measurements have been coupled to other indicators of ecological and biogeochemical change. These include (i) stable C, N, S, and O isotope analyses of sources, fluxes, and fates of nutrients, (ii) biochemical indicators and markers of plant and microbe community structure and function (e.g. diagnostic pigment analyses and cell wall carbohydrate, lipid, and protein structural markers), (iii) molecular phylogenetic indicators (16S rRNA and other diagnostic genes such as *nifH*, *nitr*, and *RbcI*), and (iv) fluorescence identification and enumeration techniques (e.g. flow cytometry). In addition, there are a suite of microscopic identification and enumeration techniques available, including various cytological stains, nucleic acid probes, immuno- and autofluorescence, and microautoradiography. These techniques add to the sensitivity, specificity, relevance, overall utility of productivity measurements as indicators of ecosystem structure, function, and change. The use and application of these ancillary techniques is discussed in this chapter.

METHODS FOR MEASURING PRIMARY PRODUCTION IN AQUATIC HABITATS

Assessments of aquatic primary production are based on measuring the rates of consumption of reactants or formation of products of autotrophy. The most commonly examined reactants are CO_2 (ΣCO_2) and H^+ (i.e. pH), while the products include O_2 and changes in pH. Primary-production measurements are frequently supplemented with measurements of nutrient (e.g. N and P) uptake to obtain parallel estimates of biomass production and nutrient cycling.

Changes in biomass over time have also been used to approximate primary production. This is accomplished either by microscopic determination of cell numbers or by spectrophotometric or fluorometric determination of chlorophyll *a* content. Cell counts or chlorophyll *a* levels are converted to cellular carbon (C) by multiplying these parameters by cellular conversion factors (e.g. number of cells or amount of chlorophyll *a* multiplied by the amount of carbon per cell per unit of chlorophyll *a* or biovolume). Biomass has also been estimated dierectly as C or N, by combusting microbial biomass and measuring C and N particulate content with a C, H, N analyser.

Other techniques that have been used to assess primary production include monitoring changes in the stable-isotope composition (under natural-abundance or enrichment conditions) of C, N, S, and O contained in microbial biomass. One example is the use of ^{13}C for measuring CO_2 fixation and characterising specific CO_2 fixation pathways (i.e. Calvin cycle versus C-4 and other pathways). This approach is based on the knowledge that micro-organisms discriminate against heavy-isotope forms of each element during uptake, assimilatory, and growth reactions. Details of these techniques are provided in Table 16.2 CO_2 fixation can be routinely assessed by: (i) measurements of CO_2 uptake by gas analysis, or (ii) use of isotopic tracer techniques, specifically measuring $^{14}CO_2$ uptake.

Table 16.2. Commonly used methods for determining primary production in planktonic and benthic aquatic environments.

Method
^{14}C technique; planktonic systems
^{14}C technique; benthic systems
O_2 Winkler method; planktonic systems
O_2 Winkler method; benthic systems
O_2 microelectrode method; benthic and planktonic systems

CO$_2$ Uptake by Gas Analysis

CO_2 uptake can be assessed directly by gas chromatographic or infrared absorption analyses of CO_2 consumption. either in the headspace or in the soluble phase of gas-tight vessels. This approach is most often used in a laboratory or in manipulated field experiments (e.g. sealed chambers or mesocosms), where access to analytical instrumentation is readily available.

Procedure

Populations and communities of interest are dispensed into a gas-tight, optically transparent vessel (normally borosilicate or Pyrex glass, polycarbonate). The vessel aqueous and gas phases may be altered as desired. For the gas phase, the vessel may be flushed with gas mixtures, either initially or continuously.

Vessels are incubated under different illumination regimes (including a dark treatment). At prescribed intervals, gas and/or aqueous sub-samples (0.1 ml to several millilitres) are withdrawn through a serum stopper placed in the vessel. Vessels can also be sampled on a continuous gas flow basis.

Analysis of $[CO_2]$ or $[\Sigma CO_2]$ is done either by infrared [IR] absorption (nondispersive IR detector) or by gas chromatography (thermal conductivity detector). The sensitivities of these instruments are comparable. If CO_2 or ΣCO_2 are the only species to be analysed, IR analysis is the most practical (especially if flowthrough procedures are used). If additional gases (e.g. O_2 and N_2) are to be analysed, gas chromatography is preferred. CO_2 is analysed by sampling the gas phase directly or by sparging the aqueous phase with an inert gas (He, Ar or N) and passing the evolved CO_2 through appropriate instruments. To determine ΣCO_2, a 0.2 ml sample of the aqueous phase is acidified in 5 to 10 ml of 30 per cent H_3PO_4 to volatilise all inorganic C species (CO_2, HCO_3^-, and CO_3^{2-}), sparged with inert gas, and subjected to IR analysis. Instrument sensitivity and linearity are determined by making up a set of CO_2 or dissolved inorganic carbon (using $NaHCO_3$ or Na_2CO_3) standards covering the range of concentrations encountered in samples.

Discussion of technique

The gas analysis technique is most applicable to laboratory studies, where constant access to instrumentation can be ensured. If dense and active cultures (i.e. single populations) or microcosms (i.e. mixed communities) are available, this is a very effective, easily executed, and unequivocal techique. However, the ^{14}C technique is often preferred in ecosystem-level (i.e. water column or whole lake/ marine environments) studies.

While both techniques measure CO_2 uptake as an estimate of primary production, they do not necessarily account for identical processes and as a result may yield contrasting results. This is because the measure of CO_2 'uptake' by gas analysis includes simultaneous production (e.g. respiration and decomposition) and autotrophic consumption of CO_2. Thus, like the oxygen method of assessing primary production, this is a net measure of CO_2 flux. The ^{14}C method, on the other hand, measures only the consumption of CO_2 (as $^{14}CO_2$). Small amounts of refixation of respired or excreted ^{14}C may be possible, especially during lengthy (>3 hours) incubations.

The CO_2 uptake and production technique is suitable for both planktonic and benthic studies. Limitations are largely due to physical (structure and size) constraints. In particular, analyses of benthic samples can be problematic. Maintaining the structural integrity of sediments, microbial biofilms, and mats is critical. Disruption of these samples can lead to altered diffusivity and structural changes along vertical biogeochemical gradients. Such alterations can significantly impact measurements of primary production and respiration, which in turn affect CO_2 flux. In addition, laminar flow and other forms of turbulence can profoundly affect CO_2 flux at the sediment-water interface. Therefore, samples must be stirred during measurements.

Uptake of ^{14}C-Labelled CO_2

The radioactive (β-emitting) form of carbon, ^{14}C, is a long-half-life (5760-year) isotope, commonly used to quantify CO_2 uptake and its incorporation into organic matter. The ^{14}C method is based on several assumptions: (i) biochemically, $^{14}CO_2$ behaves similarly to the stable, dominant form of carbon, ^{12}C, but there is a small (6 per cent) isotopic discrimination against the heavier ^{14}C, and (ii) the chemical form of ^{14}C administered (usually as $NaH^{14}CO_3$) rapidly (within a few seconds after mixing) equilibrates with other forms of inorganic C (CO_2, CO_3^{2-}) in accordance with pH and $[\Sigma CO_2]$. ^{14}C should be administered in trace quantities ($[\Sigma CO_2]$ remains unaltered by ^{14}C additions). Because of its high sensitivity, ease of use and deployment in the field, and relatively low cost, the HC method is widely used for measuring primary production in natural waters.

Procedure

Planktonic samples

Water samples are collected (preferably with a nonmetallic sampler, such as a Van Dorn or Niskin bottle), and dispensed in 100 to 250 ml incubation bottles. Transparent Pyrex, borosilicate, or polycarbonate bottles with gas-tight seals are preferred. If the bottles are incubated outdoors, they should be kept shaded during filling and processing, to prevent photo inhibition. For field assessments of primary productivity, at least triplicate light and single dark (in opaque bottles) samples should be incubated at each sampling location. Bottles should be stored in a light-tight box until collection is complete. Parallel water samples for ΣCO_2 ($CO_2 + HCO_3^- + CO_3^{2-}$) by alkalinity and pH measurements or by direct gas analysis are collected in well-sealed glass bottles or vials (completely filled). All samples should be stored in a cool, dark place until analysis.

After sample collection, $NaH^{14}CO_3$ is added to each bottle, using a preset repeating pipetter or syringe equipped with disposable tips or needles. From 0.5 to several millilitres of stock ^{14}C solution (3 to 10 $\mu Ci\ ml^{-1}$) is added per bottle. Commercial sources of $NaH^{14}CO_3$ which specific activities of 20 to 100 $\mu Ci\ \mu mol$ are available. At the above-mentioned dilutions, trace quantities of ΣCO_2 (i.e. $<0.1\ \mu M$) are added. Stock solutions can be made up and stored in several ways: (i) commercially available $NaH^{14}CO_3$ solutions (usually shipped in sterile deionised water) can be diluted with sterile water (at pH 7.5 to 8.5) and stored refrigerated in a well-sealed bottle; (ii) alternatively, the diluted stock can be dispensed in glass, break-neck ampules (i.e. 'gold seal' or 'blue seal' types), which are then sealed and autoclaved for long-term storage at room temperature.

Standard radioactivity protection measures, including wearing disposable gloves and laboratory coats and protecting the work area with disposable protective paper or plastic covers ('Benchkote' or equivalent material), should be used. A well-labelled set of glassware, pipettes, filtration funnels, forceps, etc. should be dedicated for repeated use with ^{14}C. Appropriate disposal and handling of solid and liquid waste is required.

After ^{14}C additions, bottles are sealed tightly and agitated by repeated inversion (5 to 10 times) to ensure thorough mixing of the isotope. Dark bottles must be truly dark (they should be covered with layers of foil if necessary), and (for *in situ* studies) the bottles are placed at the locations and depths from which they were collected. They should be incubated for 2 to 4 hours.

To terminate the incubation, the bottles are collected and placed in a well-insulated, light-tight box for rapid, transportation to the laboratory. If the time between sample collection and processing is longer than 1 hour the samples should be 'iced down' to restrict biological activity, (i.e. respiration and decomposition of ^{14}C-labelled cell constituents). Bottle contents (either partial or entire volume) are filtered, under gentle vacuum (2 to 5 lb in^{-2} or 200 mm Hg), on 0.45 μm-pore-size nitrocellulose (HA Millipore or equivalent) or fibreglass (Whatman GFF or equivalent) filters. Filter funnels are rinsed with small quantities of prefiltered (0.45 μm-pore-size filters) water from which samples were obtained. The filters are then removed and placed face up in a well-ventilated area.

Following drying, filters are placed in a well-sealed plastic container with a removable lid (20 by 30 by 10 cm polyethylene storage containers are suitable). A small (~50 ml) wide-mouth beaker half-filled with concentrated (fuming) HCl is placed in the box, and the filters are exposed to HCl fumes for at least 1 hour. This treatment eliminates adsorbed and abiotically precipitated inorganic ^{14}C from filtered material. The filters are then allowed to thoroughly vent and dry (usually overnight). They are placed in liquid scintillation (LS) vials to which a chemically compatible, high-efficiency LS cocktail is added. A variety of suitable cocktails are available, the most popular of which are those composed of biodegradable, nontoxic solvents. Radioactivity is determined by liquid scintillation counting (LSC). Typically, ^{14}C counting efficiencies, determined from a set of quenched standards, range from 90 to 95 per cent for ^{14}C in filtered materials.

Water samples stored for ΣCO_2 measurements are analysed (preferably on the same day) either by IR absorbance gas chromatography, or titration accompanied by pH determination.

The primary productivity (CO_2 fixed per unit time) is calculated as follows:

$$\text{Primary productivity (mg of C m}^{-3}\text{ h}^{-1}) = \frac{(A_l - A_d) \times {}^{12}C \times 1.06 \times 1000}{{}^{14}C \text{ added} \times T}$$

where A_l is the mean counts (dpm) in light bottles, corrected for quenching and instrument background; A_d is the quench and background corrected counts (dpm) in dark bottles; ^{12}C is the total inorganic carbon (ΣCO_2) available, in milligrams of carbon per litre; 1.06 is a correction for the 'isotope effect'; 1000 is the conversion factor from litres to cubic metres; ^{14}C added is the total activity of ^{14}C added, in dpm; and T is the incubation time, in hours.

Primary productivity can also be calculated per square metre, utilising integration techniques. It can additionally be 'normalised' per milligram of chlorophyll *a* and per microeinstein of photosynthetically active radiation (PAR; 400 to 700 nm for oxygenic phototrophs; 400 to 850 nm for anoxygenic phototrophs) per square metre (or micromoles of photons per square metre per second).

Benthic samples

Primary productivity of benthic microalgal mats and submersed biofilms can also estimated using the ^{14}C technique. A modification of the methods described for planktonic samples is used. Mat or biofilm sub-samples are first obtained by cutting or coring small (~1 to 2 cm^2) pieces from natural assemblages. Whole cores, incubated intact, can also be used. Because of a high degree of heterogeneity in mats and biofilms, a large number of replicate (minimally triplicates) light samples is recommended. Replicated light and single dark subsamples are placed in cleaned 22 ml borosilicate LS vials having polypropylene-lined caps. Dark vials can be wrapped with foil. Then 20 ml of sample water and 0.2 ml of NaH^{14}CO$_3$ (total activity, 2 to 4 μCi; specific activity, 40 to 100 μCi μmol^{-1}) are added. The vials are sealed, gently mixed, and placed on their sides, with the mat surface facing up, under natural illumination and temperatures. The vials are agitated to ensure a uniform distribution of isotope, nutrients, and gases. Dissolved inorganic carbon (ΣCO_2) is determined by IR or gas chromatographic analyses of acidified, sparged ambient water.

For whole cores, NaH^{14}CO$_3$ can be directly injected into cores, which are then left to incubate under the conditions described above. Cores are normally incubated under natural illumination and temperature.

Incubations are terminated by removing mat and core samples from vials and placing them face up in a fuming HCl atmosphere for at least 30 minutes. The samples are then air dried and processed for LSC, using a tissue-solubilising, biodegradable cocktail (e.g. Cytoscint cocktail). Quenching (due to pigments and humic and other coloured substances in sediments) may reduce LSC efficiencies. This is corrected for by generating a quench curve where various amounts of unlabelled sediments are amended with known quantities of calibrated [^{14}C] hexadecane or toluene.

The formula described for planktonic samples can be applied to benthic microalgal communities. Results are usually expressed as milligrams of C per unit surface area (square centimeter or square metre), by correcting for the surface area of mat or biofilm samples used, per unit time. Data can be normalised per amount of chlorophyll *a* and/or PAR flux.

Discussion of technique

The ^{14}C technique is the technique of choice when large numbers of samples must be rapidly analysed and when extensive spatial-temporal (synoptic) sampling of lakes, rivers, estuaries, or oceanic regions is required. Since its introduction to marine and freshwater productivity studies in the 1950s, this technique has been the dominant method for assessing primary production in natural waters. Despite its popularity, the ^{14}C method is not without technical and interpretational problems, which are summarised below.

Sample collection and preparation

Water samples should preferably be collected with nonmetallic samplers (e.g. Van Dorn or Niskin bottles, plastic hose, and buckets) using 'clean techniques', as far away as possible from sampling platforms or vessels. Heavy metals, hydrocarbons (fuel and oil) and other anthropogenic contaminants (lubricants, paints, solvents, etc.) associated with sampling can alter the photosynthetic rates of the resident microflora, especially in oligotrophic waters.

The $NaH^{14}CO_3$ stock solution should be made up with either distilled deionised water or seawater free of above-mentioned contaminants. When trace metal contamination is of concern, stock solution water is initially 'cleaned up' by passing it through a Chelex column. The stock pH must be sufficiently alkaline ($>pH\ 8.0$) to guard against ^{14}C losses to the atmosphere (as $^{14}CO_2$) during preparation and dispensing.

Once samples are collected and transferred to incubation vessels, they are placed in a light-shielded environment as soon as possible. Photoinhibition, photorespiration, and (under extremely high light conditions) photooxidation of microflora may otherwise result. One should be particularly concerned about this problem in high-irradiance environments, including alpine, low-latitude tropical and subtropical regions, and intertidal and shallow-water benthic regions, and when high reflectivity is encountered (on ice, snow, sand, etc.). Photoinhibition can lead to underestimates (10 to 40 per cent) of primary productivity.

If *in situ* incubations are undertaken, the investigator is faced with several choices. If samples are incubated at the depth from which they were collected, they are essentially held under constant light and turbulence conditions during the incubation period. This is the simplest and most commonly used means of incubating ^{14}C samples. In nature, phytoplankton seldom resides under static conditions. For example, in well-mixed, illuminated near-surface waters (i.e. epilimnia in freshwater and mixed layers in estuarine and marine waters), phototrophs experience a highly dynamic, transient light regime during daytime.

Incubations conducted under varied light regimes yield primary-productivity estimates differing from those derived under static conditions. This disparity holds true even when multiple-depth static incubations are integrated to yield an areal estimate (per square metre) of primary productivity. The differences between static and dynamic incubations are particularly large in highly turbid waters where vertical light extinction gradients are very steep in the mixed layer. To compensate for this, various transient light incubation schemes have been devised. These include on-deck incubators, which rotate samples through variable light regimes, and *in situ* variable light incubators, such as rotating wheels containing a range of light transmittance screens. Sedimentation effects in the incubation vessels may also affect productivity measurements. To compensate for this, bottles should be agitated to most closely mimic natural turbulence.

Applying the ^{14}C method is particularly challenging in benthic systems. Because $^{14}CO_2$ needs to diffuse and thoroughly disperse in benthic core and mat samples, adequate time must be allotted for them to do so. One approach is to preincubate both light and dark samples with $^{14}CO_2$ under complete darkness prior to initiating measurements of photosynthesis. This will enable ^{14}C to diffuse into the sediment matrix. Stirring and agitation of samples during incubation also help minimise this problem. If benthic samples are highly productive, dissolved inorganic carbon limitation may occur. This creates the possibility of establishing a diffusional barrier to inward movement of dissolved inorganic carbon from the overlying water column, which will almost certainly lead to underestimates of productivity based on the ^{14}C method. The potential problem this imposes on the ^{14}C method should be evaluated before this technique is used on uncharacterised sediments.

Choosing appropriate incubation periods is also important. No single incubation time can be prescribed for all aquatic ecosystems. As a rule, 2 to 4 hours is recommended, because this time frame should be long enough to allow for sufficient isotopic labelling of cell pools and constituents critical to the primary production process while being short enough to avoid significant ^{14}C losses due to excretion, cell death and lysis, grazing, etc. all of which ultimately affect quantification of primary productivity. There may also be significant seasonal differences in the above-mentioned sources of error. It is suggested that the investigator start with this time frame and adjust it according to sensitivity needs.

The time of day when productivity measurements are made can be critical to the determination of daily and longer-term production. It has been shown that primary productivity 'tracks' irradiance. Productivity and PAR curves are not necessarily superimposable, however. Studies on a variety of marine and freshwater ecosystems have shown that the highest rates of primary productivity occur during mid to late morning, prior to the period of maximum irradiance. Fee and others have modelled primary productivity, based on knowledge of such deviations from the diel PAR flux curve. In addition to diel variability in the relationship between PAR flux and productivity, there are day-to-day, seasonal, and spatial differences in environmental variables regulating primary productivity.

Following incubation, samples should be processed expeditiously. Unless it is necessary for specific applications (e.g. autoradiography, sample vouchers, and radioimmunological and molecular samples), one should avoid the use of preservatives, such as formaldehyde, glutaraldehyde, and Lugol's iodide. Preservation steps can lead to loss of radioactivity from samples; this is particularly true if the samples are washed or treated with other solutions prior to filtration, drying, and further handling. If required, samples can be quick-frozen and freeze-dried, but only if this is the last step prior to determining radioactivity.

Planktonic samples are usually concentrated by filtration prior to ^{14}C determination. One should determine appropriate filtration vacuum conditions under which successful concentrations can be achieved without loss of cellular labelled constituents. Appropriate vacuum settings depend on phytoplankton community composition and physiological state. A gentle vacuum, on the order of ~200 mm Hg, is recommended if the above-mentioned variables are unknown. The choice of filter is also critical. One should choose a filter that quantitatively retains all relevant phototrophs at .any time of the year and at any location. From a practical standpoint, it is probably best to choose the smallest-pore-size filter capable of passing the volume of water incubated.

Filtration prevents ^{14}C-labelled dissolved organic carbon, which may have been produced, during the incubation, from being measured. As an alternative to filtration that would include nonvolatile ^{14}C-labelled dissolved organic carbon production, one can collect 5 to 20 ml of incubation water and perform direct ^{14}C counting by LSC. Prior to counting, samples are acidified (0.1 N HCl), sparged with air or N_2 to expel non-assimilated $^{14}CO_2$ and neutralised (with dilute NaOH) prior to counting. This approach is most appropriate for productive waters, where large incubation volumes are not required. Relatively low LSC efficiencies (<85 per cent) may result, because large amounts of water (which is a strong quenching agent) must be mixed with the LS cocktail. An LS cocktail chemically compatible with the sample and yielding maximum HC counting efficiency is preferred.

Data interpretation

Interpretation of data is one of the more critical aspects of the ^{14}C technique. Compared to other methods of assessing productivity, specifically the O_2 method, the ^{14}C technique yields results that range from general agreement to serious underestimate. Some of the disagreement can be attributed to various methodological problems, including those for determining cellular ^{14}C losses. Other discrepancies may be due to the biochemistry of respective measurements. Because it accounts for the balance between CO_2 uptake, assimilation, and losses via respiration, excretion, and lysis, the ^{14}C method with dark bottle correction represents net primary production i.e. primary production that is retained in algal cells and not released as dissolved organic matter. In contrast, the O_2 method accounts for total photosynthetic production of O_2 including or minus O_2 consumption (depending on the calculation used) via dark-mediated respiration. It does not account for other biomass losses incurred during excretion, lysis, death, grazing, and additional light-mediated respiratory processes such as photorespiration. As such, the O_2 method estimates either gross primary production or net community production.

In practice, when these methods are compared, the results are variable. The greatest disparity between these methods occurs when primary productivity is closely linked, in time and space, to higher trophic levels. At the level of microbial interactions, autotrophy is closely linked to heterotrophy. Carbon and other nutrient (N and P) transfers can occur within minutes between host autotrophs to epiphytic heterotrophs. Such close metabolic coupling of primary and secondary producer components of the microbial food web is referred to as the microbial loop if significant C transfer occurs within the ^{14}C incubation period, it could lead to overestimates of 'net' primary production, since both primary producers and secondary producers (microheterotrophs) are filtered for analysis. Under these closely coupled conditions, the ^{14}C method may therefore more reflect total microbial community production. Similarly, if micro- or mesozooplankton grazing of phytoplankton takes place during the incubation period, transfer of fixed ^{14}C can occur, leading to overestimates of net primary production. Conversely, respiration by primary producers and their grazers can lead to the loss of fixed C. If recently synthesised ^{14}C-labelled organic compounds are respired, it is possible that a portion of ^{14}C fixed can be lost and/or recycled during incubation. This may lead to underestimates of primary production. Grazers can be eliminated prior to incubation to assess the quantitative significance of trophic transfer and respiratory losses during incubatons.

Oxygen-based methods

Oxygen evolution has long been used as a benchmark measurement of photosynthesis and, for natural microbial assemblages, primary productivity. The O_2 method is often preferred over other, less direct, primary-production measurements. One of its main advantages is that it measures photosynthetic performance based on a key product of the light reaction of photosynthesis. Hence, measurements can be directly related to the capture of radiant energy, electron flow, and the generation of reductant. The O_2 technique is suitable for determining the productivity of complex microbial assemblages, where parallel measurements of O_2 evolution and its utilisation serve as useful indicators of community metabolism. Oxygen can be easily and accurately measured without a need for radioactive materials and sophisticated detection equipment. As discussed above, diffusion limitations with the ^{14}C technique can seriously compromise the use of this method in laminated benthic systems. Such limitations are minimised with O_2 measurements (especially using microelectrodes), as outlined below. In moderate to high productivity systems, where sensitivity is not limiting, the O_2 method is often preferred.

Oxygen evolution is routinely measured in two ways: (i) determining changes in dissolved oxygen in bottles or other vessels by wet chemistry with titration (i.e. the Winkler method) or by spectrophotometry and (ii) electrochemical, macro- or microelectrode determinations. The Winkler, spectrophotometric and electrochemical techniques should yield identical results, although in the electrochemical technique, small amounts of O_2 are consumed during measurement. Supply needs and measuring equipment are relatively simple and portable, making this an excellent shipboard or field technique. In addition, there are novel, experimental (i.e. nonroutine) approaches, such as the use of stable oxygen isotope uptake measurements.

Winkler method

A divalent manganese solution is added to water samples, and then alkaline iodide is added. Under these conditions, dissolved O_2 is rapidly reduced and precipitated as manganese oxide. This step 'fixes' O_2 prior to analysis. The sample is then acidified in the presence of iodide, and the oxidised manganese is converted back to its divalent state, while iodine (equivalent to the original amount of dissolved O_2 present) is liberated. The iodine is quantified either titrimetrically or spectrophotometrically.

Procedure

Duplicate 60 to 300 ml Pyrex, ground-glass-stoppered light and dark biological oxygen demand (BOD) bottles are normally used to hold samples. Quartz bottles may be substituted if the effects of UV radiation are assessed. The bottles are flushed several times with sample water and filled, making sure not to generate bubbles during filling (bubbling will affect O_2 solubility and hence concentration). This is accomplished by inserting a filling tube (Tygon or equivalent material) to the bottom of the bottle and rapidly filling the bottle, allowing it to overflow. The ground-glass stopper is then quickly inserted. The sample is stored in a cool, light-tight box until all samples are collected.

After incubation, reagents are added through a pipette in the following order: 3 M manganese chloride (600 grams of $MnCl_2 \cdot H_2O$ is dissolved in distilled water to make 1 litre) (solution 1); alkaline iodide solution (8 N NaOH plus 4 N NaI), which is made by dissolving 320 grams of NaOH and then 600 grams of NaI in distilled water to make a volume of 1 litre (solution 2); and 10 N sulphuric acid (280 ml of concentrated H_2SO_4 is added to 500 ml of distilled water) (solution 3). Each solution should be allowed to cool to room temperature before use.

The following procedure is for 125 ml bottles. A 1 ml volume of solution 1 and a 1 ml volume of solution 2 are added to the bottles. The bottles are stoppered and the contents are mixed well (by inverting the bottles at least 10 times). Following this fixation step, the bottles can be kept cool and dark for at least 1 to 2 hours prior to analysis. When analysing, the precipitate should be allowed to settle at least halfway down the bottle.

For titrimetric determination of $[O_2]$, the iodine liberated is titrated with a 0.01 N standardised thiosulphate solution, yielding $[O_2]$ in milligrams per litre. The solutions are as follows. For solution A, 2.9 grams, of analytical grade $Na_2S_2O_3 \cdot 5H_2O$ and 1 gram of Na_2CO_3 are dissolved in 1 litre of distilled water. A 0.2 ml volume of CS_2 is added as preservative. For solution B, a 0.01 N potassium iodide standard is made up by adding 0.3567 gram of air-dried (105°C) analytical grade KIO_3 to 250 ml of distilled deionised water (the solution is warmed if necessary to dissolve KIO_3). For solution C, a 1 per cent starch solution is made up (it is solubilised in dilute NaOH if necessary and neutralised with dilute HCl). A 1 ml volume of solution 3 (10 N sulphuric acid) is added. Within 1 hour of sample acidification, 50 ml is transferred (using a pipette) into a 125 ml Erlenmeyer flask containing a magnetic stir bar. The solution is titrated immediately with 0.01 N thiosulphate which has previously been dispensed in a burette, while stirring, until the solution is slightly yellow. Then 0.5 ml of starch solution is added, which should turn the sample blue. The titration is continued slowly and carefully, until the blue colour disappears. N_2-flushed blanks are titrated, and the blank value is subtracted to obtain a corrected titration volume (V) in millilitres. $[O_2]$ is calculated as follows:

$$[O_2] \text{ (mg/litre)} = 0.1016 \times f \times V \times 16$$

To determine the calibration factor, f, 1 ml of concentrated sulphuric acid (solution 3) is added to a 125 ml BOD bottle filled with distilled water, and the solution is mixed thoroughly. A 1 ml volume of the $MnCl_2$ solution (solution 1) is added, and the solution is mixed again. A 50 ml volume is transferred into the titration flask, and 5 ml of the 0.01 N KIO_3 standard (solution B) is added. The solution is mixed gently for 2 minutes and titrated with thiosulphate. Using V as the titration volume, $f = 5.00/V$. The calibration factor should be determined in triplicate.

For spectrophotometric determination of $[O_2]$, the precipitate formed (after adding solutions 1 and 2 to sample) should be allowed to settle halfway down the sample bottle. A 1 ml volume of solution 3 is added, the bottle is stoppered, and the solution is mixed again. At least 5 minutes should be allowed for the precipitate to thoroughly dissolve. A small aliquot (5 to 10 ml) is withdrawn from the bottle and diluted to 100 ml with distilled water. It is important to avoid violent mixing, to minimise oxidation of iodine by exposure to air. Simply inverting the mixing flask will provide adequate mixing. The extinction of iodine (E_s for samples and E_b for blank) is measured at 287.5 nm in a 1 cm quartz cuvette, using a UV spectrophotometer. An N_2 flushed (to remove O_2) water sample is run in parallel as a blank. To quantify $[O_2]$ using the spectrophotometric technique, a calibration factor must first be determined by the standard titration technique discussed above. One can also refer to a table to determine standard O_2 saturation (percentage, or milligrams of O_2 per litre) at specific temperatures (at sea level) and salinity conditions. This yields an O_2 solubility factor (F_s) based on the following formula:

$$F_s = [O_2]/(E_s - E_b)$$

Discussion of technique

The Winkler technique is best suited for moderately to highly productive waters, characterised by relatively high O_2 production and consumption rates. Measurements are most easily conducted on planktonic samples, where the sampling volume is not restricted. The technique can also be applied to sediments, biofilms, mats, and epibiotic communities. In these systems, the method is most effectively deployed by placing domes and other enclosures directly over the surface to be examined. Specially designed domes, equipped with electric stirrers and ports for withdrawing sub-samples, are routinely deployed for assays of benthic primary production, respiration, and (by subtracting respiration from production) net community metabolism. During measurements of benthic metabolism, corrections must be made for metabolism of the overlying water column.

Surficial O_2 production and consumption measurements suffer from methodological shortcomings, some of them similar to those incurred with ^{14}C-based methods. Perhaps most serious are diffusional problems. Oxygen, like CO_2, is a gas and is transported along diffusional gradients. In the case of very high rates of O_2 production or consumption, the rate of diffusional transport can control the rate at which O_2 is emitted from or absorbed by surficial and subsurface boundary layers. If the surface is submersed, diffusional barriers are most pronounced. Unless turbulence (as stirring or shaking) can overcome this limitation, under estimates of O_2 production and/or consumption may result. Potential underestimation due to transport of photosynthetically produced oxygen to lower depths in sediments may constitute an additional problem. Diffusional problems can be minimised by stirring or agitating samples. Care must be taken, however, to avoid disturbance and re-suspension of sediments.

One way to minimise diffusional problems is to employ microelectrode measurements of $[O_2]$. Microelectrodes are small enough to minimise the impact of diffusional gradients.

An interpretational limitation of the O_2 method is that, unlike the ^{14}C method, it does not account for primary productivity of anoxygenic phototrophs (e.g. photosynthetic bacteria) or chemolithotrophs (e.g. nitrifying bacteria and sulphur oxidising bacteria). It does account for respiration by these and all other microorganisms (which the ^{14}C method is incapable of doing).

Electrochemical techniques

Oxygen can be measured electrochemically, using several types of electrodes. The most widely used are cathode and Clark-style combination electrodes. In the cathode-style electrode, a membrane-coated platinum cathode coated with a gold surface (to sense O_2) is embedded in a thin glass or steel cannula housing. A voltage of -0.75 V (relative to a calomel reference electrode) is applied to the cathode. The current resulting from the reduction of O_2 at the gold surface is directly proportional to $[O_2]$ in the sample. Cathode-style electrodes can be constructed with very small ($\sim$5 μm) sensing tips. Microelectrodes with small sensing tips have a high degree of spatial resolution. Furthermore, the O_2 diffusion path to the tip is fast, leading to short response times. The 90 per cent response time (i.e. the time required to approach equilibration of external and internal $[O_2]$ conditions) is typically <0.5 second.

In Clark-style or combination O_2 electrodes, the cathode and reference electrodes are combined, but physically separated, in the same electrode. The cathode and electrode are both immersed in an electrolyte (1 M KCl), which additionally serves to shield the cathode. The tip of the electrode is sealed with an O_2-permeable membrane. Clark-style electrodes vary in size and shape, depending on the scale, sensitivity, stability, response time, and durability, desired. The larger BOD electrodes are typically used for environmental measurements (e.g. on a sampling line) or measurements to be made on large containerised water samples (BOD bottles, domes, and other enclosures). The response time is on the order of a few seconds to 30 seconds. If high resolution, high sensitivity, and short response time are desired, smaller-diameter electrodes are called for. Minielectrodes have sensing tips of 0.5 mm to several millimeters in diameter, while microelectrodes have tips as small as 5 μm. The response time of Clark-style microelectrodes is similar to that of cathode microelectrodes.

$[O_2]$ is determined by measuring changes in the current flowing through the electrode. In large Clark-style electrodes, a digital or analogue ammeter serves as the detector. With all microelectrodes, the current changes are very small, and hence a very sensitive ammeter, capable of measuring down to 1 pA (10^{-12} A), is used. Portable picoammeters are commercially available for field studies.

Mini- and microelectrodes are typically utilised for small-scale pO_2 measurements in biofilms, benthic boundary layers, sediments, and microbial mats, where strong, microscale (micrometer) O_2 gradients are an integral component of primary production and nutrient cycling dynamics. Details of the construction and applications of mini- and micro-Clark-style electrodes are provided by Revsbech and Jorgensen.

The procedure briefly outlined below is applicable to the use of fabricated mini- and microelectrodes for measuring pO_2 in surficial environments.

Oxygen regimes and changes therein are analysed by mounting the micro- or minielectrode on a micromanipulator. The micromanipulator is positioned above the sample in such a way that the electrode can be rapidly inserted into the sample and $[O_2]$ measurements can be made at discrete intervals. O_2 profiles are obtained by advancing the electrode downward into the sample at 0.25 to 0.5 mm intervals. Measurements are usually made under both illuminated and dark conditions. Once stable O_2 conditions occur, the next interval is measured. The most common way to determine photosynthetic O_2 production is to use the light-dark shift method. In this method, samples are exposed to the irradiance of interest. The microelectrode is inserted to the sampling location. $[O_2]$ is determined. Photosynthesis is then stopped by darkening the sample. This allows O_2 consumption and diffusional exchange to continue at the presumed original rate (prior to darkening). Since there is no photosynthtic O_2 production to balance consumption, $[O_2]$ decreases in proportion to the photosynthetic rate just prior to darkening. The initial slope (measured either on a recorder or on a data logger) of the decrease in $[O_2]$ after darkening is equal to the $[O_2]$ produced by photosynthesis.

Electrodes are standardised and calibrated by measuring current changes in O_2-depleted (N_2- or Ar-flushed) and saturated (by bubbling with air) solutions of ambient water.

Discussion of techniques

Electrochemical measurements of $[O_2]$ have several advantages over 'wet-chemistry' Winkler techniques. They are rapid and easily executed and, in the case of mini- and microelectrodes, circumvent diffusional artifacts and limitations. All electrodes require calibration against the Winkler technique. Some microelectrodes are sensitive to illumination, and this must be compensated for by measuring current changes in response to light and dark conditions in media free of micro-organisms (e.g. distilled water). Oxygen microelectrodes are also sensitive to temperature changes, and this must be assessed using sterile or micro-organism-free solutions.

Because they are very thin and are constructed in large part of glass, mini- and microelectrodes are prone to breakage. Breakage is a particularly serious problem in sediments and microbial mats, where impenetrable heterogeneous substrates (sand, gravel, mucilage, slimes, etc.) can hinder their use. In addition, microelectrodes are sensitive to fouling, chemical poisoning (e.g. by sulphide), and changes in sensitivity and response time, all of which necessitate extensive and repeated standardisation. Unless investigators have ready access to the technology for constructing microelectrodes, replacement of broken electrodes, calibration, and standardisation are expensive, time-consuming aspects of these techniques.

Possible interpretive problems include extrapolating or integrating from the short timescales required to make measurements (minutes) to the longer timescales required to assess productivity on a community and ecosystem level (hours to days). In addition, microelectrodes can potentially disrupt biofilms and mats during measurements, leading to invasion of O_2 into previously hypoxic and anoxic microzones.

Electrochemical techniques require electronic amplification (of current output) and detection and measurement accessories; the, degrees of sophistication and costs of these vary considerably.

Complementary techniques and their applications

Standard measurements of primary productivity can be complemented by modifications, ancillary techniques, and procedures which enhance physiological, ecological, and taxonomic interpretations of productivity measurements. Measurements of photopigments are highly informative in that they can serve as surrogates for rate measurements of productivity and as complementary biomass measurements. In addition, the ^{14}C method lends itself to specific techniques capable of assessing species-, group-, and community-specific rates of primary production. Two such techniques, microautoradiography and radiolabelling of diagnostic microalgal pigments, can enhance the specificity and dimensionality of primary-productivity measurements.

Pigment-based estimates of primary productivity

Changes in net phototrophic biomass over time can be used as an indirect estimate of primary productivity. This approach has been used when direct physiological-rate measurements are not available. The simplest and often most effective estimate of phototrophic biomass is one based on determining the cellular photopigment content. The most common pigment assessed is chlorophyll *a*, which is present in all oxygenic phototrophs. Other pigments, specific for taxonomic groups (carotenoids, phycobilins) (Table 16. 3), can be used to partition production among the groups. All pigment-based approaches rely on the assumption that the cellular ratio of photopigments to dry weight or carbon content is reasonably constant This assumption may not be valid in communities experiencing environmental change and physiological stress. Pigment-based measurements are net estimates, since they simultaneously account for synthesis and loss of photopigments. Synthesis is largely due to growth and to production of new biomass. Losses and changes in cell pigment synthesis may be attributable to death, consumption, advective processes (sinking, sedimentation, and horizontal transport due to flushing), and changes in physiological state (i.e. photoadaptation, photobleaching, and photooxidation).

Table 16.3. Diagnostic photopigments used as indicators of major phytoplankton functional groups.

Pigment	Phytoplankton functional group
Chlorophylls	
a	All groups
b	Chlorophytes, euglenophytes, prasinophytes, prochlorophytes
c_1, c_2, c_3	Bacillariophytes, chrysophytes, dinophytes, prymnesiophytes, raphidiophytes, xanthophytes
Carotenoids	
Alloxanthin	Cryptophytes
19′-Butanoyloxyfucoxanthin	Dinophytes, prymnesiophytes, raphidiophytes
Crocoxanthin	Cryptophytes
Dinoxanthin	Dinophytes, prymnesiophytes
Echinenone	Cyanobacteria, prochlorophytes
Fucoxanthin	Bacillariophytes, chrysophytes, dinophytes, prymnesiophytes
19′-Hexanolyloxyfucoxanthin	Dinophytes, prymnesiophytes
Lutein	Chlorophytes, prasinophytes
Myxoxanthophyll	Cyanobacteria
Peridinin	Dinophytes
Zeaxanthin	Cryptophytes, cyanobacteria, prasinophytes, prochlorophytes
Phycobilins	
Allophycocyanin	Cyanobacteria
Phycocyanin	Cyanobacteria, cryptophytes
Phycoerythrin	Cyanobacteria, cryptophytes, rhodophytes

The most common measures of chlorophyll *a* are by fluorometric emission and spectrophotometric absorbance of intact cells or extracted pigment (usually in 90 per cent acetone or 90 per cent ethanol or methanol) at either 440 or 663 nm and 750 nm (turbidity blank). Detailed methods for microalgal pigment extraction and spectrophotometric analyses are given by Strickland and Parsons, and applications to aquatic production studies are addressed by Millie Jeffrey, and Wright. Fluorometric analyses of chlorophyll a (excitation maximum ~400 nm; emission maximum ~660 nm) are also commonly used in limnological or oceanographic studies. Fluorometry is more sensitive than and not as prone to optical interference as spectrophotometry. Fluorometry can additionally be applied *in situ*, by placing fluorometers and data-logging and telemetry devices in the water column.

The above-mentioned methods are applicable to both planktonic and benthic systems. For benthic measurements, accumulations of chlorophyll degradation products (e.g. pheophytin) can be significant. This can lead to interpretational problems, even if acidification corrections for pheophytin are made. High-performance liquid chromatography (HPLC) has, however, proven effective in rapidly separating photosynthetic pigments, prior to spectrophotometric and fluorometric detection. When coupled to real-time multiwavelength rapidscanning capabilities, such as those provided by a photodiode array spectrophotometer (PDAS), HPLC offers the ability to rapidly and accurately distinguish and quantify pigments and a number of their degradation products within monotypic and mixed algal assemblages. In this manner, certain taxa known to contain unique pigments and combinations of pigments (such as carotenoids, phycobilins, specific combinations of chlorophylls a, b, and c, and chlorophyll derivatives [Table 16.3]) may be identified and potentially quantified. When coupled with autoinjectors and driven by state-of-the-art computer software, HPLC-PDAS is very amenable to large-scale production characterisation and monitoring programmes (the National Science Foundation's Global Ocean Flux Studies and long-term Ecological Research Programmes, the EPA's Toxics and Lakes Survey and Coastal Indicator Programmes, and the National Oceanic and Atmospheric Administration's Coastal Ocean Programme).

Photopigment radiolabelling

Photopigment radiolabelling is a relatively new, potentially useful technique for measuring C-specific growth rates among diverse phototrophic taxa. This method, while neither a direct nor a surrogate measure of primary productivity, provides additional dimensionality to primary-productivity measurements. The conventional ^{14}C method gives a single production value for the entire community. In contrast, the photopigment-radiolabelling technique allows simultaneous measurement of specific growth rates among individual taxa. The method is a useful extension of the ^{14}C technique. When algae are exposed to ^{14}C, the labelled C is incorporated into 3-phosphoglyceric acid, passes through the Calvin-Benson cycle, and eventually passes into a pool of lower-molecular-weight intermediates. These labelled compounds enter the various photopigment biosynthetic pathways. After a sufficiently long incubation (2 to 24 hours), the specific activity of photopigment C will equal the total phytoplankton carbon pool. The C-specific growth rate (μ, where $C_p = C_{p,0} e^{\mu t}$ [Cp is C-specific biomass at time t, $C_{p,0}$ is C-specific biomass at time zero, and t is time]) is a more useful parameter for describing the physiological responses of functional groups of phototrophs, in terms of C assimilation and allocation into biomass. A comparison of μ values obtained under different experimental conditions allows for predictions of the ultimate fate of community components over time.

Using HPLC coupled to PDAS detection of pigments, photopigments are separated on a sequential series of C_{18} HPLC columns, to optimise the separation of diagnostic chlorophyll and carotenoid photopigments. The sequence is as follows: a monomeric Rainin Microsorb-MV and two polymeric

Vydac 201TP reverse-phase C_{18} columns using a binary gradient elution system. Monomeric columns provide strong retention and high efficiency, while polymeric columns select for similar compounds with minor differences in molecular structure and shape. On-line PDAS produces real-time three-dimensional chromatograms of absorbance spectra for each photopigment. Computer software allows data from individual chromatograms to be stored and processed for qualitative analyses and comparisons with samples of known absorption spectra and peak retention times. C-specific growth rates are determined using published methods. The ^{14}C specific activity (dpm) of individual pigments is quantitatively determined using an in-line radioactivity (β emission) detector. Radioactivity (dpm) peaks are overlaid onto pigment chromatograms to identify labelled pigments. Values of community-specific growth rate (μ) and C biomass are determined using equations given by Goericke and Welschmeyer and Redalje. The growth rates and C biomass of phylogenetic groups (cyanobacteria, diatoms, dinoflagellates, chlorophytes, etc.) are estimated from the activities of labelled diagnostic pigments.

Microautoradiography

Samples that have been incubated with $^{14}CO_2$ can be analysed for single-cell, taxon, and community-specific photosynthetic potential and rates using microautoradiography. This method is based on the fact that ^{14}C assimilated by micro-organisms can be detected by radiationsensitive, silver halide emulsions which are placed over radioactive organisms. Following exposure (in complete darkness) and processing using standard photographic techniques, microautoradiographs can be viewed microscopically. Exposed silver halide crystals appear as optically dense silver grains, which can be photographed and quantified. Image analysis has proven useful for quantification.

^{14}C-labelled micro-organisms are either filtered onto membrane filters or settled onto glass slides after which radiosensitive liquid or film emulsions are applied. Following exposure and processing, both the micro-organisms and their radioactivity (which appears as exposed silver grains in the developed emulsion) can be examined microscopically and photographed. Radioactivity can be quantified (either as silver grains or as tracks of grains) if a uniformly thick film or liquid emulsion is applied. Below is a short description of the technique as applied to planktonic samples.

Samples that have been incubated with ^{14}C are either fixed in 2 per cent borate-buffered formaldehyde (pH 7 to 8) (for delicate cells) or left unfixed and gently filtered (~200 mm Hg) through 0.45 μm-pore size Millipore HA filters. The filters are then rinsed with prefiltered, unlabelled sample water to remove excess radiolabel. For marine samples, salts must be removed from filters, or they will crystalise on drying and interfere with microscopic viewing. This is accomplished by rinsing the filters twice with small quantities (10 to 30 ml) of 0.01 N phosphate-buffered saline, which removes salts while avoiding lysis of fragile cells. The filters are then either quick-frozen and/or air dried (for at least 4 hours) and optically cleared by placing them, face up, on clean microscope slides and carefully passing them over the mouth of a 250 ml beaker containing boiling (fuming) acetone. This step will simultaneously clear the filters and attach them to the slides.

The following steps are conducted in the darkroom, either under a dark-red 20 W or less safelight (at least 1 m away from the samples) or in complete darkness, which ensures the lowest background. The slides are dipped in a melted nuclear track emulsion. Kodak NTB-2, a moderately sensitive, low-background, reusable emulsion, is widely recommended. The stock emulsion is kept in a light-tight container in a refrigerator at ~4°C. In darkness, it is melted at 40°C and diluted 1:1 with water for use with grain density microautoradiography. If quantitative auto radiography is desired, the application of thick emulsions combined with track autoradiography, or stripping film emulsions (Kodak AR-10), is recommended. All emulsions should be thoroughly melted at 40°C (a water bath is used in the darkroom and light-tight containers containing emulsion are heated in an oven) before use. In both thick-layer (undiluted emulsion) track and diluted thin-layer grain density techniques, dipped and jelled slides are air dried for 30 minutes and exposed in light-tight slide boxes containing packets of silica gel desiccant. Exposure times typically vary from 2 to 10 days.

Exposed track and thin-layer microautoradiographs are developed in Kodak D-19 for 5 and 2 minutes, respectively, transferred to a stop bath for 2 minutes. They are then fixed for 5 minutes in Kodak rapid fixer or 15 per cent NaS_2O_3 and rinsed for 15 minutes in a water bath with gentle flowing water. All slides are then air dried overnight. Microautoradiographs are viewed microscopically, usually with phase-contrast oil immersion optics. For thin-layer autoradiographs, immersion oil can be directly applied to the dried emulsion. Track autoradiograph emulsions are allowed to swell by the addition of 30 per cent glycerine followed by the application of a coverslip. Immersion oil can then be applied to the coverslip.

While microautoradiography can overcome and circumvent the technical limitations of other techniques, it is time-consuming, requires experience and patience, and is limited by microscopic resolution. It can also be prone to interpretational differences among investigators.

Regional and global studies of primary production: the utility of remote sensing

Increasingly, humans are affecting aquatic primary production. The most perceptible and problematic symptoms of this trend are anthropogenic nutrient-enhanced eutrophication and associated increases in harmful algal blooms in lakes, estuaries, and coastal waters. Such ecosystem-level and larger-scale (regional and global) events range from days to several weeks, depending on the timing of and relative enrichments from direct and indirect (run-off) nutrient inputs. In addition, expanding industrial, agricultural, and domestic discharges of toxic xenobiotic substances are suspected of altering the rates of primary production and community composition. The functional overlay of growth and community-altering anthropogenic inputs in the world's aquatic ecosystems over a wide range of spatiotemporal scales necessitates short as well as long-term detection and characterisation of productivity responses, often integrated over large areas.

These informational needs are clearly beyond the scope of limited field surveys using the techniques discussed above. However, they constitute tasks ideally suited for multispectral (to detect diagnostic phytoplankton pigments such as chlorophylls, carotenoids, and phycobilins) aircraft and satellite-based remote sensing. The National Aeronautics and Space Administrations's SeaWiFS satellite-based remote-sensing platform is capable of discrete spectral analyses with sufficient sensitivity and resolution. SeaWiFS includes six narrow visible bands (412, 443, 490, 510, 555, and 670 nm) and two broader near-IR bands (765 and 865 nm). All eight bands are recorded in 10 bit resolution to take advantage of their high signal-to-noise ratios (up to 1000). This makes for a very high spectral resolution remote-sensing platform ideal for detecting and discriminating primary-production contributions among phylogenetically diverse algal groups whose diagnostic pigments can be individually characterised.

The spatial and temporal coupling of nutrient and toxic inputs with patches of pigmented phototrophs is critical in determining primary production, changes in algal biomass, bloom dynamics, and CO_2 flux in these waters. In addition, the contemporaneous impacts of autochthonous (internal; upwelling, regenerated N supplied to surface waters via mixing) and allochthonous (external; run-off, riverine and atmospheric) nutrient- and toxic-mediated primary production should be assessed. Evolving remote-sensing techniques will help clarify and evaluate these nutrient production interactions on scales appropriate for each system in question.

Chapter 17

Bacterial Secondary Productivity

INTRODUCTION

Planktonic heterotrophic bacteria (bacterioplankton) are now recognised to be a large and metabolically active group that contribute significantly to the biomass and to the flow of carbon in aquatic systems. Bacterial cell densities often exceed 10^9 cells litre^{-1}. Bacterial biomass in marine systems is generally greater than that of zooplankton and can exceed phytoplankton biomass in the open sea.

Estimates of bacterial biomass and growth rate show that bacterial biomass turns over rapidly (in a matter of hours). Comparison of bacterial biomass and production rate with corresponding phytoplankton biomass and production rate indicates that bacteria consume a substantial fraction (20 to 40 per cent) of the carbon fixed by phytoplankton. In unproductive aquatic systems, bacterial respiration can exceed phytoplankton net production. Bacteria are now considered major secondary producers, since they convert dissolved organic matter (DOM) derived from primary producers into an abundant biomass. The consumption of bacterial biomass by specialised predators (ciliates and flagellates) may be an important pathway for the transfer of DOM to metazoan food webs.

This view of bacteria has emerged as a result of the use of modern methods for measuring the abundance and growth rates of bacteria in environmental samples. Bacterial abundance is currently determined from the direct enumeration of bacterial cells by using nucleic acid-specific fluorochromes and epifluorescence microscopy. The contributions of bacteria to overall metabolism in ecosystems have been determined primarily from estimates of bacterial biomass production (BBP). If the assimilation efficiency of the bacterial population is known, BBP rates can be used to calculate the total utilisation of organic matter by bacteria and consequently to determine the geochemical significance of bacterial metabolism. The amount of bacterial biomass potentially available to grazers and thus to higher trophic levels can also be determined from rates of BBP. In addition, changes in the rates of BBP can be used as an indicator of the response of bacteria to spatial and temporal fluctuations in environmental conditions. Currently, the most widely used methods of estimating BBP involve the measurement of various aspects of population growth such as the increase in cell numbers when predators are removed, changes in the frequency of dividing cells (FDC), and increases in the rate of synthesis of cellular constituents such as nucleic acids and proteins. System-wide estimates of BBP determined using these independent methods all show that bacterial production is high and consumes a substantial fraction of the algal primary production.

Even though there is general agreement regarding the importance of bacteria in aquatic ecosystems, the methods of assessing BBP have remained controversial and there is little agreement about the best method. This disagreement is based largely on uncertainties regarding the aspects of the taxonomy and metabolism of aquatic bacteria necessary to calibrate the methods and thus to determine the aspect of growth which is the best indicator of BBP. We know very little about the species composition and metabolic diversity of natural assemblages of bacteria. Traditional methods of laboratory microbiology, which depend on the isolation of single bacterial strains, are inadequate bacteria in aquatic systems. Only a small fraction of the bacteria in natural aquatic systems can be cultured. Therefore, current approaches to estimation of BBP have focused on determining the growth of all heterotrophic bacteria rather than on attempting to assess the growth rates of individual species. To use this synecological approach, assumptions must be made regarding the overall requirements and response of the entire bacterial assemblage. Furthermore, little is understood about the growth state and the growth environment of aquatic bacteria. For example, methods which measure the rate of synthesis of cellular constituents assume that bacteria undergo balanced, steady-state growth. Under these conditions, the rates of synthesis of all cellular constituents are equal. However, there is evidence that unbalanced growth is common in natural populations of bacteria. During unbalanced growth, the rates of synthesis of different macromolecules will yield different estimates of growth. Theoretically, determinations of BBP can also be based on the utilisation of DOM. However, in practice this approach can be inaccurate because bacteria in aquatic systems consume DOM which occurs in very low concentrations and the chemical composition of this DOM is largely unknown.

Additional challenges to developing accurate and precise methods of measuring BBP involve the possible artifacts caused by the manipulation of natural samples. Bacteria occur in association with other organisms and non-living particles of similar size, making it difficult to isolate them by size fractionation. Bacteria can also respond rapidly to changes in their environment; thus, containment (bottle effects) and long incubation periods may significantly modify species composition and growth.

To properly assess the contribution of bacterial metabolism to the flow of organic matter in ecosystems, BBP needs to be converted into units of carbon or nitrogen production. Because of the difficulties in isolating bacteria from surrounding debris, these values can seldom be measured directly. Carbon production is generally obtained by applying a conversion factor which translates cell production into carbon production. These conversion factors are typically derived by comparing the volume of cells to carbon content in mixed cultures of bacteria under laboratory conditions. Published values of the volume-to-biomass ratios vary by a factor of 14 and thus conclusions derived from BBP are very sensitive to the choice of conversion factor used. This problem is compounded by the need for other conversion factors which translate a given measure of growth (e.g. DNA or protein synthesis) into cells produced.

Given these uncertainties and methodological limitations, it is remarkable that current BBP estimates based on a variety of independent methods show strong correlations to algal production and biomass. This suggests that when compared over large time and space scales, these methods reflect actual rates of BBP and thus are adequate to study system-wide contributions by bacteria. However, as our knowledge of the ecology of aquatic bacteria increases, questions concerning the dynamics and controls of BBP have become more detailed and demand greater accuracy and precision of the methods. Intercalibration studies have shown that differences between methods can be significant, making the study of the dynamics and controls of bacteria over small time and space scales problematic. Because of the various uncertainties of the current methods of measuring BBP, their adequacy in attempts to accurately estimate bacterial processes and rates has been questioned. However, for many ecological questions, the current methods are still a valuable tool. For example, understanding the response of bacteria to changing conditions requires high precision but knowledge of the absolute value of BBP is not needed. Additionally, there has been substantial research on the causes of differences among methods, and uncertainties can be reduced by experimental validation of the assumptions of the chosen method in the system being studied.

Because current methods of BBP suffer from many uncertainties and limitations, the choice of method must follow careful consideration of the question being addressed and the degree of accuracy that is satisfactory. For example, estimates of protein synthesis may be a better indicator of the amount of biomass available for consumption by grazers than the rate of cellular division because cells may increase in mass but not divide. The investigator must also be prepared to invest time and effort to test, in a given system, the validity of the assumptions of the chosen method.

A common approach, given the difficulties in determining the accuracy of each method, has been to simultaneously measure BBP by using independent methods. Agreement between these methods provides confidence that observed variations reflect real changes in rates of bacterial production. Because each method has its own assumptions and targets a different aspect of bacterial growth, differences in BBP estimates may reveal variations in environmental factors or changes in the growth state of bacteria. In this context, current methods of measuring BBP can complement each other and provide useful information regarding the growth state and the environment of bacteria.

Of the various methods available to determine BBP, those that employ radiolabelled precursors to estimate the rate of synthesis of nucleic acids and proteins have become the most widely used and will be the focus of this chapter. The rate of synthesis of these macromolecules closely reflects cell division and cellular growth. Advantages of these procedures include high specificity for bacteria, high sensitivity, small sample volumes (<100 ml), and short incubation periods (<1 hour). Disadvantages include the need to test for the specificity of the label for the chosen macromolecule, the need to determine possible sources of external and internal isotope dilution, and the need for conversion factors to obtain cell or carbon production estimates from incorporation rates. Currently, the three methods most commonly used are thymidine (TdR) incorporation into DNA, leucine (Leu) incorporation into protein, and adenine incorporation into RNA. When bacterial growth is balanced, rates of incorporation into RNA, DNA, and protein will give equal estimates of growth. However, uncoupling of macromolecular synthesis (unbalanced growth) occurs when bacteria shift growth rates. Periods of unbalanced growth may consequently reflect the response of bacteria to changes in environmental conditions. Simultaneous measurement of the rates of synthesis of DNA, RNA, and protein are therefore not redundant and can provide information regarding the growth state of bacteria in nature.

Methods to determine BBP have been reviewed extensively, and a manual giving step-by-step instructions for several methods is available. Because adenine is also incorporated by algae and size fractionation is required to obtain estimates of BBP, only the procedures for TdR and Leu are described in detail. In addition, methods for the empirical determination of a conversion factor from TdR or Leu incorporation to cells produced is presented, as well as several procedures designed to test various assumptions of these methods. Alternative methods to determine BBP which do not rely on the uptake of radiolabelled compounds are also briefly discussed.

THYMIDINE (TdR) INCORPORATION INTO DNA

Rationale, Advantages and Disadvantages

Of the various methods available to measure bacterial production, TdR incorporation into DNA has become the most widely used in both freshwater and marine systems. This method estimates the rate of bacterial DNA synthesis and consequently the rate of cell division. Nucleotides are normally synthesised *de novo*, but in bacteria they can also be produced via a salvage pathway. dTMP is synthesised *de novo* from dUMP by the enzyme thymidylate synthetase. In the salvage pathway, dTMP is formed through the phosphorylation of exogeneous TdR by the enzyme thymidine kinase. The TdR method assumes that TdR is added in sufficient concentration to inhibit *de novo* biosynthesis of nucleotides and promote the salvage mechanism. The added TdR is also assumed to be taken up rapidly by bacteria, to remain stable during uptake, and to be rapidly converted into nucleotides.

The advantages of the TdR method include: (i) high specificity for heterotrophic bacteria, (ii) high sensitivity and precision; and (iii) ease of use in the field. Only bacteria have the transport mechanisms for assimilating TdR and the enzyme thymidine kinase needed for nucleotide synthesis via the salvage pathway. Other organisms such as eukaryotes or cyanobacteria do not appear to incorporate TdR into DNA. The TdR method requires minimal manipulation of water samples. Short incubation periods (typically <1 hour) are sufficient, and only nanomolar concentrations of TdR are needed. The procedures are relatively simple and do not require complex equipment, and the method can be completed quickly and is easily adapted to field use. Furthermore, the filters containing the incorporated TdR can be stored for subsequent extraction and determination of radioactivity. The basic TdR procedure has been modified for determining BBP in sediments, and marine snow.

Extensive use and testing in many aquatic environments has raised concerns related to the accuracy of the method. The major concerns are as follows: (i) TdR is not always incorporated into DNA, and other macromolecules are also labelled; (ii) the conversion factors required to obtain cell or carbon production from TdR incorporation vary widely and are difficult to derive empirically; (iii) *de novo* synthesis may not be entirely suppressed by added TdR, leading to underestimates of DNA synthesis; (iv) microautoradiographic studies show that not all the bacteria present incorporate TdR; it is not clear if these unlabelled bacteria are dormant or moribund cells, actively growing bacteria with unique metabolisms, or bacteria that lack TdR transport systems; and (v) bacteria in anaerobic environments such as chemolitotrophic and sulphate-reducing bacteria do not appear to incorporate TdR, and thus the TdR method will underestimate BBP in anaerobic waters. The problems associated with the molecular specificity of TdR and with the conversion factor have been extensively studied. Isotope dilution caused by *de novo* synthesis and the fraction of the bacterial population incorporating TdR in various environments has not been investigated in as much detail.

The ^{3}H radioactive label of TdR often appears in macromolecules other than DNA, such as RNA and proteins, indicating catabolism of the added TdR. The fraction of the total incorporated radioactivity which appears in non-DNA macromolecules can be very variable and may lead to serious overestimates of DNA synthesis. To circumvent this problem, several purification procedures have been developed to isolate the DNA fraction. Although differences in the degree of TdR catabolism reported can be traced to differences in the extraction method used, examination of the degree of TdR catabolism obtained using individual methods suggests that the extent of this metabolism varies predictably with experimental and environmental conditions. For example, nitrogen starvation in seawater cultures of bacteria causes the utilisation of TdR for protein synthesis. Differences in TdR catabolism have been observed over a salinity gradient and with depth. The patterns of TdR metabolism in San Francisco Bay could be related to differences in the source of carbon for the system and to differences in bacterial species composition. Consistent patterns of TdR metabolism within a system may reflect an environmental or physiological basis. Thus, measurements of TdR catabolism can give insights into environmental conditions. Determination of the extent of TdR catabolism has become an important component of the TdR method. The procedures used to isolate DNA are relatively simple and should be performed routinely. However, if the percentage of the total TdR that is incorporated into DNA is constant in a given system, it may not be necessary to perform the DNA isolation for every sample.

Another major concern regarding the TdR method is the need for conversion factors to translate TdR incorporation rates (picomoles per litre per hour) to cells or carbon produced (cells per litre per hour or micrograms of carbon per litre per hour). Rates of TdR incorporation are first converted into cells produced by using a theoretically or empirically derived factor. The theoretical factor is calculated from: (i) the percentage of thymidine in bacterial, DNA, (ii) the DNA content per cell, (iii) assumptions regarding the extent of extracellular isotope dilution; and (iv) the effective inhibition of *de novo* synthesis (intracellular isotope dilution). The value of these factors is generally obtained from the results of laboratory studies and varies from 0.4×10^{18} to 0.5×10^{18} cells per mol of TdR incorporated. A major drawback of the theoretical conversion factor is that the information used in its computation is derived from studies using laboratory monocultures of bacteria.

These values may not be representative of natural bacteria or of natural assemblages of bacteria. In studies using natural assemblages of bacteria, the amount of DNA per cell was quite variable, with smaller cells generally containing more DNA per unit volume. Furthermore, other assumptions related to the theoretical factor, such as the extent of isotope dilution by *de novo* synthesis, are very difficult to determine directly.

The difficulties of the theoretical conversion factor can be circumvented by deriving an empirical conversion factor. The empirical conversion factor uses natural assemblages of bacteria from the study area to calibrate the TdR method and is derived by comparing, under controlled conditions, TdR incorporation with increases in bacterial numbers. An added advantage of these empirical conversion factors is that all possible relationships between TdR incorporation and cell production (such as TdR catabolism) are, in theory, included in the conversion factor. Published empirical conversion factors vary widely and are generally higher than the theoretical factors. The most commonly used empirical factor is 2×10^{18} cells per mol of TdR, which is the median of 97 studies. Variations in the empirical conversion factor may be caused by differences in the method used to analyse the data or by differences in the experimental procedures.

For example, high conversion factors relative to the commonly used factor of 2×10^{18} cells per mol of TdR may be due to underestimates of TdR incorporation caused by the lack of uptake saturation. If uptake is not saturated, *de novo* synthesis leads to isotope dilution. Bell reported that variations in the conversion factors could be reduced by using higher concentrations of added TdR. Thus, when the TdR method is used, preliminary experiments should be performed to determine the TdR concentration at which maximal uptake is achieved (saturation concentration). Conversion factors determined using the same method can vary seasonally and spatially. These variations may reflect changes in environmental parameters such as nutrient availability.

For many questions regarding the ecological role of bacteria, BBP in units of carbon is required. Once TdR incorporation values are converted to cell production, another factor is applied to translate this information into units of carbon produced. This factor is determined by comparing cell volumes to carbon content in cultures of bacteria.

A wide range of biomass conversion factors are available. The choice of conversion factors and their combination is likely to significantly affect the calculated magnitude of bacterial carbon production based on any method using incorporation of radiolabelled compounds. Because of the various factors available to convert TdR incorporation rates into cell or carbon production, data are generally reported in units of TdR incorporation (typically picomoles per litre per hour).

Procedures for Estimating BBP from Incorporation Rates of TdR

The basic TdR procedure examines [methyl-^{3}H]TdR incorporation into a macromolecular fraction containing DNA, RNA, and protein, which is isolated using a cold trichloroacetic acid (TCA) extraction. For each water sample, replicate samples (usually 10 ml) are incubated with [methyl-^{3}H]TdR (specific activity, 40 to 80 Ci mmol^{-1}; usually a final concentration of 20 nM). Abiotic absorption of radioactivity is measured in replicate samples in which bacteria have been killed with TCA or formalin. After an incubation period (30 to 60 minutes), the samples are placed in an ice-cold water bath for 1 minute, and then 1 ml of ice-cold 50 per cent TCA is added to reach a final concentration of 5 per cent. The samples are cooled for an additional 15 minutes. This step kills the bacteria and lyses cell membranes, allowing the removal of the TdR taken up but not incorporated into macromolecules. Following this extraction, samples are filtered through 0.45 µm-pore-size cellulose acetate membrane filters. Incubation vials are rinsed with 2 ml of ice-cold 5 per cent TCA. Filters with the bacteria are then rinsed with the TCA used to rinse the vials and with three additional 1 ml aliquots off ice-cold 5 per cent TCA. To remove TdR incorporated into cells but not into DNA (e.g. into lipids), the filter is rinsed further with five 1 ml aliquots of 80 per cent ethanol. Radioactivity in the filter is measured using a scintillation counter. Quenching during the radioassay, caused by the filter, is reduced by dissolving the filter with 1 ml of ethyl acetate for 30 minutes prior to the addition of scintillation solution.

Several procedures are available for the isolation of DNA if incorporation into other macromolecules is an issue. These include an acid-base hydrolysis and a phenol-chloroform extraction. To purify DNA using the phenol-chloroform procedure, lysed samples are filtered through cellulose nitrate membrane filters instead of cellulose acetate filters, since the latter are dissolved by phenol-chloroform. Following the incubation period, incorporation is stopped by addition of formaldehyde to a final concentration of 2 per cent. RNA is extracted first by adding 5 N NaOH to reach a final concentration of 0.25 N and incubating the sample at 20 to 25°C for 60 minutes. The sample is then acidified with 100 per cent TCA (1 to 2 ml/10 ml sample) to a pH of approximately 1 and stored on ice for 15 minutes. The sample is filtered, and any protein remaining on the filter is extracted by rinsing with 5 ml of a 50 per cent (wt/vol) phenol-chloroform solution (50 grams of phenol in 100 ml of chloroform), leaving radiolabelled DNA on the filter. A final rinse is done with five 1-ml portions of ice-cold 80 per cent ethanol, and the filter is prepared for radioassay as described above.

Exact sample volumes, incubation times, and final TdR concentration may vary depending on the environment (freshwater, saltwater, oligotrophic, eutrophic, etc.) and are derived from preliminary experiments. The appropriate concentration of TdR is determined from saturation experiments where replicate water samples are incubated with various concentrations of TdR (e.g. 5, 10, 15, 20, and 25 nm). The rate of TdR incorporation is then plotted against the amount of added TdR, and the lowest concentration at which incorporation is saturated (i.e. no increase in incorporation observed) is chosen. The length of incubation is determined by incubating replicate samples over various time intervals (e.g. 10, 20, and 30 minutes) and determining if the relationship between the rate of incorporation and incubation time is linear. To avoid bottle effects, the shortest incubation time which yields a linear, measurable rate is chosen.

The extent of extracellular isotope dilution is determined indirectly using a standard addition method. Replicate samples are incubated with a given amount of radioactive TdR (e.g. 5 nm), but with different concentrations of unlabelled TdR (e.g. 0, 5, 15, 25, and 35 nm). Increasing amounts of added unlabelled TdR dilute the radioactive TdR and yield decreasing incorporation rates. In the absence of external concentrations of the substrate, the plot of reciprocal incorporation rate against total added substrate (labelled plus unlabelled) should pass through 0. However, if additional TdR is present in excess of what is added experimentally, the plot will not pass through 0 and the ambient concentration of TdR is determined by extrapolating the linear plot to reciprocal incorporation rate equals 0 ($y = 0$).

The proportion of bacterial cells incorporating TdR can be determined using microautoradiography. In this technique, filters containing bacterial cells previously incubated with TdR are placed faced down on a radioactivity-sensitive emulsion coating one surface of a microscope slide. The filter is then removed, and the bacteria are embedded in the emulsion. Cells which incorporated TdR 'expose' the underlying emulsion, producing silver grains. The entire slide is developed by normal photographic means and additionally stained with DNA-specific fluorochromes so that silver grains and bacteria can be identified simultaneously. The preparation is then observed under bright-field illumination to count the cells with associated silver grains. By using epifluorescence, the total bacterial population is determined by counting the fluorescent cells in the same sample.

LEUCINE INCORPORATION INTO PROTEIN

Rationale, Advantages and Disadvantages

The Leu method estimates the rate of bacterial protein synthesis by measuring the incorporation rate of radioactive Leu into proteins which are isolated using a hot-TCA extraction. Protein synthesis is a good indicator of BBP because proteins account for a large percentage (~60 per cent) of bacterial biomass in natural assemblages and their synthesis consumes a large percentage of cellular energy. Protein synthesis can also be estimated by using $^{35}SO_4^{2-}$ as a radiolabelled precursor, but [^{3}H]Leu is the preferred substrate because algae also utilise $^{35}SO_4^{2-}$, and the high concentration of sulphate in marine environments can result in significant isotope dilution. The rate of Leu incorporation is a good indicator of the rate of protein synthesis because Leu constitutes a fairly constant fraction of bacterial protein and is not transformed into other amino acids. Kirchman demonstrated that the same as those described above for the TdR procedure. Leu specificity for proteins can be tested by comparing the radioactivity on the filter with that in the hot-TCA filtrate, which contains nucleic acids. The Leu method is very specific for proteins, and nearly all the Leu taken up is assimilated into the protein fraction. If experimental evidence shows that this is the case in the study area, this procedure can be simplified by substituting a cold-TCA extraction for the hot-TCA extraction. In this case, the procedures are identical to those of the TdR method. Leu incorporation can be translated directly to carbon production by using a theoretical conversion factor of 3.1 kg of C mol^{-1}. Alternatively, an empirical conversion factor can be determined using mixed cultures of bacteria from the study area.

The basic Leu procedure has been modified by using centrifugation instead of filtration to separate the bacterial proteins. In this modified method, the use of smaller sample volumes (1 to 2 ml) and the lack of a filtration step result in a faster, cheaper procedure and reduced radioactive waste. Briefly, samples are incubated with radiolabelled Leu in microcentrifuge tubes. Following incubation of the samples and removal of unincorporated Leu, the samples are centrifuged and the resulting pellets are treated with hot TCA to isolate the proteins. Scintillation fluid is then added directly to the microcentrifuge tube prior to radioassay. The Leu method involving centrifugation has been further modified for use in freshwater and sediment. The efficiency of protein precipitation in freshwater samples is increased by including a cold-TCA extraction step and by adding humic extracts or NaCl that act as co-precipitants for protein. In sediment samples, the pellets formed by centrifugation are treated to form a gel matrix, which prevents the sediment from settling during the radioassay.

Because simultaneous measurements of protein and DNA synthesis can provide complementary information regarding the growth of bacteria, a method has been developed to concurrently measure Leu and TdR incorporation in the same sample. To achieve this, [^{3}H]TdR and [^{14}C]Leu are added to a single sample and the cold-TCA-insoluble material is collected. [^{14}C]Leu is used because [2-^{14}C]TdR can undergo catabolism and still label DNA owing to the retention of the label in uracil. A final concentration of 20 nM Leu (10 nM [^{14}C]Leu [specific activity, about 300 mCi mmol^{-1}] and 10 nM nonradioactive Leu) is used. Termination of incubation, cold-TCA extraction of macromolecules, subsequent filtration, and preparation of filters for radioassay are the same as described above for the basic Leu procedure. This dual-label method has been modified and used successfully with sediments.

EXPERIMENTS TO DETERMINE EMPIRICAL CONVERSION FACTORS FOR THE TdR AND Leu METHODS

Cultures of mixed assemblages of bacteria are prepared by inoculating filter-sterilised sample water with small volumes of prefiltered sample water (1 part prefiltered to 9 parts sterilised). Prefiltering is accomplished using a 0.6 to 2.1 µm-pore-size track-etch polycarbonate filter and reduces the proportion of grazers, phytoplankton, and detritus in relation to bacteria. The exact pore-size used will vary depending on the environment. For eutrophic systems, using a very small-pore filter may exclude a significant portion of the bacteria, whereas for oligotrophic systems, a large-pore filter may be ineffective in removing bacterivores. Sterile filtration is accomplished using 0.22 µm-pore-size Nuclepore filters. Adding the prefiltered sample to sterilised water from the same site promotes bacterial growth, reduces grazing by bacterivores escaping the prefiltration step, and reduces viral encounters. Gentle vacuum (<100 mmHg) is used throughout to avoid the release of DOM due to cell lysis. Inoculated samples (1 to 4 litres) are kept in the dark and incubated at the *in situ* water temperature. Subsamples (10 to 50 ml) are taken every 2 to 4 hours for a total of 24 hours and assayed for both bacterial abundance and TdR (and/or Leu) uptake by the methods described above.

There are several methods to compute a conversion factor from these experiments. The simplest and most commonly used method (integrative method) involves dividing the increase in cell numbers by the substrate incorporation integrated over time. The change in cellular abundance is simply the final abundance minus the initial abundance.

The total substrate incorporated is computed by integrating the area under the curve described by the plot of substrate incorporation against time. A drawback of this approach is that only two values are used to determine the increase in cell numbers. Alternatively, the data can be analysed using an exponential-growth model which considers all points. The conversion factor is computed by the equation $CF = \mu e^B / e^b$, where μ is the growth rate determined from changes in cell abundance over time, B is the y intercept of ln (cells) plotted against time, and b is the y intercept of ln (substrate incorporation) plotted against time. The conversion factor yields values in units of cells. To obtain carbon production, an additional factor must be applied.

ADENINE INCORPORATION INTO DNA AND RNA

Rationale, Advantages and Disadvantages

The adenine method has been applied to many water column environments and sediments but has not been used as extensively as the TdR method. The adenine method estimates the rates of both DNA and RNA syntheses. Adenine is incorporated into bacteria and into eukaryotic algae and is designed to measure the growth of the entire microbial community. Estimates of BBP using this method, therefore, require a size fractionation step to isolate the bacteria.

An advantage of this method is the ability to simultaneously measure RNA and DNA synthesis in the same incubation. Relative changes in these rates can yield information regarding the extent of unbalanced growth in mixed assemblages of bacteria. In this method, it is possible to correct for isotope dilution by measuring the specific activity of the nucleic acid precursors. Adenine is incorporated into nucleic acids through various precursor molecules (ATP, dATP, ADP, and AMP). The existing pools of these compounds, in addition to inputs via *de novo* synthesis, dilute the specific activity of the added radioactive adenine ([2-^{3}H]adenine). In the adenine method, the immediate nucleoside triphosphate precursors are isolated and assayed for their specific activity.

Briefly, the procedure for the adenine method involves incubating replicate water samples with [^{3}H]adenine. Following incubation, subsamples are extracted for isolation of nucleic acids. RNA, DNA, and the precursor ATP are separated, purified, and radioassayed. From this information, rates of RNA and DNA synthesis corrected for isotope dilution are computed. If a time course experiment is performed where replicate samples are incubated over different times, the individual rates of DNA and RNA synthesis are determined. DNA synthesis can be converted to BBP if the C/DNA ratio is known.

OTHER METHODS

Several other methods for the estimation of BBP which do not require radiolabelled compounds are available. The FDC method, for example, estimates the growth rate of bacteria from an empirically determined relationship between growth rate, temperature, and FDC. By using epifluorescence microscopy, dividing cells are identified as cells visibly showing invagination. Advantages of this method are the lack of incubation and the ability to determine the bacterial biomass from the same sample. Disadvantages are the need to determine the relationship between temperature, growth rate, and FDC for each environment investigated, the time and tedium involved in microscopic examination of all samples, and the subjectivity involved in the identification of dividing cells.

Another method that does not rely on radiolabelled substrates uses the TdR analogue 5-bromo-2'-deoxyuridine (BrdU) to estimate DNA synthesis. In this method, BrdU incorporation into bacteria is detected and measured by a chemiluminescence immunoassay. Throughout the development and testing of this novel procedure, BrdU incorporation rates agreed well with TdR incorporation rates. The BrdU method may be a good alternative method to estimate DNA synthesis when the use of radioactive TdR is impractical.

Other methods focus on determining increases in bacterial numbers after grazers have been eliminated by size fractionation or by use of eukaryotic inhibitors. The overlap of the sizes of bacteria and their predators and the possible control of bacteria by virus action complicates this method. Furthermore, metabolic inhibitors and antibiotics which are used to selectively inhibit groups of organisms are not always effective. Finally, the long-term incubations that are required to observe changes in bacterial numbers can introduce significant artifacts such as changes in DOM. Some of these bottle effects can be minimised by using dialysis bags, which allow the exchange of dissolved materials but maintain the bacterial population isolated from predators.

To sum up the importance of heterotrophic bacteria as decomposers of organic matter and as secondary producers in aquatic systems is widely recognised. While many methods have been developed to successfully study the contributions of bacteria at the ecosystem level, these methods have been difficult to apply to the understanding of the dynamics and controls of BBP over short timescales. A major difficulty in refining the current methods and developing new ones has been the dearth of knowledge regarding the species composition of natural assemblages of bacteria. Current methods of BBP determination must make assumptions regarding the overall characteristics of the community under consideration. Consequently, important information is lost and it is difficult to accurately calibrate the methods. Knowledge of the species composition and metabolic diversity of aquatic bacteria is sorely needed.

The most commonly used methods for the estimation of BBP are TdR incorporation into DNA, Leu incorporation into protein, and adenine incorporation into DNA and RNA. These methods can be easily used to determine BBP *in situ*, and information derived from their use has contributed significantly to our current view of bacteria in aquatic systems. Nevertheless, there are several concerns associated with these methods, such as the specificity of the label for bacteria and for the target molecule, dilution of the specific activity of the added substrate by extracellular and intracellular pools, and the need for conversion factors to determine carbon production. These uncertainties introduce ambiguities which make the study of issues related to the control and fate of bacterial production problematic. For example, the various conversion factors needed to obtain carbon production are not easily derived, and the ranges associated with the choice of factors and their combination may exceed the experimental errors of the methods. It is, therefore, difficult to compare BBP over small time and space scales and to accurately assess the role of bacteria in carbon flow.

Given the current state of knowledge, the simultaneous measurement of several independent indices of BBP is recommended. Simultaneous measurements help reduce the methodological limitations of individual methods. For example, because the Leu method provides carbon production estimates without the need for a carbon-per-cell conversion factor, simultaneous measurements of TdR and Leu provide estimates of both carbon and DNA synthesis. Furthermore, because balanced growth is probably not common in nature, simultaneous measurements of DNA, RNA, and protein can provide important complementary information regarding the growth state and growth environment of bacteria. Evaluation of the most appropriate combination of methods must follow careful consideration of the question being addressed and thorough testing of methodological assumptions in the study area.

Chapter 18

Community Structure: Bacteria and Archaea

INTRODUCTION

Community structure is generally considered to be information related to the types of organisms present in an environment and the relative proportions of those types. The various ways to determine such structure could easily fill a whole book; therefore, this brief chapter will basically provide an outline of the various approaches that are used to determine community structure and some guidelines regarding which methods may be most appropriate for specific aquatic environments and specific scientific questions.

Community structure analysis might be performed for a number of purposes, and the most appropriate techniques to use will vary with the goals of the study. Certain techniques may allow study of subsets of the microbial community in great detail but at the same time may entirely miss other community members. For example, the portion of the community that is performing a certain function, such as photosynthesis, might be of interest, and this might be studied by analysing extracted photosynthetic pigments or possibly by flow cytometric analysis of pigment contents of individual cells. Several of the techniques described below rely on the ability to identify particular preselected components of the microbial community. However, in addition to learning about known types of micro-organisms, a study of the general community structure can also potentially find previously unknown microbial types, and some of these may be the most interesting. As might be expected, when novel organisms are found, it takes considerable time to characterise such organisms.

MICROSCOPY

The oldest method for obtaining information on microbial community structure is to examine a sample with a microscope and characterise the microbes by their morphology. With some aquatic environments, many types of organisms can in fact be identified this way. An example is microbial mats: numerous types have morphologies which are readily recognisable, particularly by electron microscopy. Many colonial micro-organisms have distinctive colony morphologies that can be recognised by standard light microscopy or even with the naked eye. However, this approach is risky in that some types may appear similar by convergence and may thus be misidentified. More to the point, in most aquatic environments, the vast majority of bacteria have nondistinct and/or variable morphologies, so this method is inappropriate for them. Related to identification by morphology is the identification of certain kinds of organisms by their natural fluorescence properties as examined by epifluorescence microscopy. For example, methanogens contain the F420 cofactor, which fluoresces green when excited with light at

420 nm wavelength. More commonly, such identification may be based on photosynthetic pigments (often in conjunction with general size and morphology information) that can yield distinctive colours of fluorescence. For example, marine *Synechococcus* cells containing phycoerythrin have a distinctive golden fluorescence when excited with blue light and a bright orange-red fluorescence when excited by green light. These pigment 'signatures' can be more precisely determined by flow cytometry, which provides quantitative fluorescence (and other) information on many thousands of individual cells in a short time. In fact, with this approach, the presence in the ocean of large quantities of tiny marine prochlorophytes that had evaded detection by standard visual epifluorescence microscopy techniques was discovered.

CULTURING

The most common traditional method of determining community structure involves culturing the organisms from the habitat in question and identifying the cultures by standard techniques. If the organisms of interest are culturable, then this approach may be suitable. It goes without saying that the culturable organisms must be viable in order to be detected, and such viable counts are often thought to avoid the problem of counting inactive organisms that may be of less interest. However, there are several important caveats. First, even for culturable organisms, it has been suggested that many individuals may be viable but still not culturable. A second and bigger concern is that culture conditions that have typically been used to perform such counts usually recover on the order of 1 per cent or less of the total number of organisms in aquatic habitats, even though the majority of organisms in such habitats can be shown to be metabolically and synthetically active. The reasons for this inability to culture the organisms are somewhat speculative and are generally thought to be the result of culture conditions that do not adequately mimic the natural growth conditions. Possibilities include wrong substrates, trace metal sensitivity, excessive storage product formation (leading to cell damage), and viral infection, among others. Recent attempts to use dilution cultures (i.e. dilute a sample with filtered seawater so that only a few bacteria are in each tube) have greatly increased the percentage of bacteria that can be cultured, but for unknown reasons many such cultures stop growth when abundances reach about 10^5 cells per ml. At this concentration, the classical identification tests cannot be readily performed, but it may be possible to coax such organisms into growth under richer conditions. While this may assist identification, the organisms may have substantially different physiological properties after such a process.

In any case, many aquatic microbiologists are concerned about accepting the results of culture-based approaches for determining community structure in a comprehensive manner, and such studies are rarely seen in the literature.

IMMUNOLOGICAL APPROACHES

Immunological approaches have been used primarily to characterise and count nitrifying bacteria and cyanobacteria. To prepare the antibodies for such a study, first the organisms must be cultured and then the culture is used to vaccinate an animal. The antibodies are purified from serum and then labelled with a fluorescent tag, such as fluorescein. The antibodies are mixed with a sample, unattached antibodies are rinsed away, and the cells are then observed by epifluorescence microscopy. Of course, this method presupposes that one must have cultures of the organisms in question, so this approach would miss organisms that are resistant to cultivation. Also, some knowledge of the cross-reactivity of the antibodies to other organisms is required. Often, antibodies can be specific for a particular species or strain. This approach has the potential to identify nonculturable members (viable or not) of a particular serotype, which is functionally defined as the type(s) to which the antibody binds. It has also been found to be useful for organisms that might be cultured but grow slowly such as nitrifiers.

LIPID ANALYSIS

Different groups of micro-organisms have different types of lipids, and this fact has been used extensively for microbial identification. For environmental work, lipid analysis has been used as an indicator of community structure, with certain classes of lipids (or particular lipids) being used as markers for certain groups. In general, the lipids are extracted in organic solvents and analysed by gas chromatography.

The approach provides information not only about cell type but also about nutritional status or stress. Because most such analytical methods require a substantial number of microbes of each type in each sample, this approach has been used primarily for sediments. However, with the appropriate instrumentation, it is now possible to extend such work to planktonic environments.

LOW-MOLECULAR-WEIGHT RNA PROFILES

The molecular size distribution pattern of low-molecular-weight RNA (including tRNAs and 5S rRNA) is thought to be unique within narrow phylogenetic groups of micro-organisms. This concept has been applied to natural aquatic planktonic communities by Hofle and Brettar; the whole community RNA profile is examined on an electrophoretic gel and compared with profiles of other communities and of standard cultures.

As long as the bands on the gel are well resolved, patterns of similarity and dissimilarity between natural communities can be seen. It is also possible to detect patterns that suggest the presence of specific organisms or groups, and bands may be excised and sequenced for partial identification.

DNA-DNA HYBRIDISATION

There are situations in which it is desirable to know the community structure to learn if two microbial communities contain the same or different organisms, yet the quantitative information on species compositions is not of particular interest.

That is, the question is simply whether two communities are the same or different. In such situations, it is possible to perform a DNA-DNA hybridisation assay with total DNA extracted from the two communities. In this approach, extracted prokaryotic DNA is labelled by nick translation, and samples are compared two at a time (reciprocal hybridisations). The results are scored by expressing the cross-hybrids as a percentage of the self-hybrids on the same filter.

This percentage is expected to be the sum total of the shared common fractions between the two filters. For example, if sample 1 has 10 per cent species A, 40 per cent species B, and 50 per cent species C, and sample 2 has 25 per cent species A, 30 per cent species C, and 45 per cent species D, the shared common fraction is $10 + 30 = 40$ per cent.

Tests with mixtures of pure-culture DNAs have shown that the results are usually as expected, with 100 per cent hybridisation of identical or nearly identical samples, ranging down to about 5 to 10 per cent hybridisation between samples sharing few or no species (the 5 to 10 per cent represents well-conserved DNA sequences that cross-hybridise between distantly related organisms).

This approach has been used to see broad differences between ocean basins, between depths at stratified locations, and over seasonal scales at the same location, and it has also shown similarities between some communities over time and space.

OTHER DNA-BASED APPROACHES

Another type of DNA-DNA hybridisation is analysis of reassociation rates of single-sample DNA that is melted (strands separated). The rate is related to the complexity (or one may say diversity) of the DNA and thus can be an index of the diversity of the sample. In other words, it indicates the number of species present but not what types they are. Torsvik examined DNA extracted from soil micro-organisms in this way, and the very slow reassociation of most of the DNA was interpreted as indicating the presence of about 4000 completely different genomes in a 30-gram sample from a deciduous forest. This was 200 times higher than the diversity determined from standard plate counts and graphically demonstrates the remarkably high potential overall diversity of microbial communities and its undersampling by conventional means. However, it should be pointed out that other fingerprinting methods have indicated that the large majority of some soil communities is made up of only perhaps a few dozen distinguishable taxa (see the terminal restriction fragment length polymorphism [T-RFLP] method below).

A method to 'fingerprint' a microbial community based upon the quantitative distribution of genomes with different per cent G + C contents was recently described. In this approach, community DNA is centrifuged in a density gradient that separates the DNA on the basis of per cent G + C content, yielding a profile. Although the per cent G + C is not an unambiguous identifier for particular groups, the profile of DNA along the per cent G + C gradient can strongly indicate the presence of certain groups (prompting probe analysis for verification; see below). Furthermore, profiles from different samples can indicate differences between the microbial communities over time and space.

16S rRNA-BASED APPROACHES

The difficulty of culturing bacteria and the ordinary requirement that a culture be available to identify a species have presented a major dilemma to microbial ecologists. About 25 years ago, Pace and Olsen presented the elegant idea that cultures are not necessary to identify the organisms present in a natural habitat. The idea came against the backdrop of the increasing use of molecular phylogeny to help define microbial systematics. It was becoming clear that the nucleotide sequences in molecules like 16S rRNA are very powerful tools in determining phylogenies and consequently in microbial systematics. Large databases of such sequences have been made available. The new idea was to use molecular biological techniques to obtain 16S rRNA sequences directly from the organisms freshly collected from the natural habitat without culturing them. These sequences could then be compared with those in the databases and with each other to learn how they fit into the microbial phylogenetic framework. Even if the sequence is unknown from previous work, it can be placed in relation to those of known organisms and other sequences from nature. A further major benefit is that the sequences can be used to make probes for quantitative composition analyses of microbial communities. This is proving to be a very powerful approach and is being augmented by inclusion of 23S rRNA data as well. A major aspect of the power of this approach is that the data are in the form of sequences that are universally understood and readily analysed. The 16S rRNA sequence analyses can be done in a few ways. If what types of organisms are present in a sample is of interest, it is best to use the approaches that do not restrict the results to certain groups. The initial method proposed by Pace called for extraction of DNA from all the micro-organisms, fragmentation of that DNA with a restriction enzyme, and ligation to DNA from a bacteriophage (e.g. lambda) to make a library containing bits of DNA from all of the original organisms in the sample. That library is then screened for fragments coding for 16S rRNA by low-stringency hybridisation to rRNA from a culture (or from cultures representing very broad groups, such as the domains bacteria, archaea, and eucarya). The low stringency, combined with the moderately high level of conservation of this molecule, is expected to allow virtually any 16S rRNA to be detected. Screening with another rRNA, rather than with universally conserved oligonucleotide probes, is preferable because of the large number of false positives expected as a result of random matches between the oligonucleotide and the myriad genes from the natural community (in our laboratory, we found such false positives were very common). The positive clones are then sequenced, and the sequences are aligned to sequences from a database and analysed for phylogenetic relationships by a computer. This library approach was used successfully by Schmidt with marine plankton.

Ward took a different approach to cloning 16S rRNA sequences of micro-organisms from a well-studied hot spring at yellowstone national park. They extracted RNA from the natural sample and performed reverse transcription with universal primers to make cDNA that was then cloned and sequenced. The length of the cloned products was usually a few hundred bases, which is suitable for identification and some general phylogenetic analyses. The results yielded numerous clone types that were not the same as the organisms that had previously been cultured from that hot spring and had been thought to be the dominant types (by microscopical observation as well as culture work). This has been seen as strong evidence that the culture-based approach finds only a subset of the natural diversity and that morphological identification of even distinctive organisms can be deceptive.

One major alternative related approach to cloning 16S rRNA genes has also been used with great success: the PCR. With PCR methods, DNA is extracted from freshly collected organisms, PCR is performed, and with the primers of choice (see below), the PCR products are ligated into a phage or plasmid vector. These products are cloned by standard techniques, and the clones are sequenced and analysed phylogenetically. This approach is particularly suited to planktonic communities because there is usually very little DNA to work with. For example, with a typical bacterial abundance of 10^6 bacteria per ml and a typical bacterial DNA content of a few femtograms per cell, there is on the order of a few micrograms of bacterial DNA per litre. The amplification inherent in PCR means that one can begin with 1 ng of total genomic DNA (representing roughly 10^6 bacteria) and end up with micrograms of amplified 16S rRNA genes. This is more than enough to work with.

The existence of regions of the 165 rRNA molecule that are essentially invariant among all known organisms means that universal primers can be used in PCR; the longest distance between such primers is about 860 bases, which is adequate (but not ideal) for most phylogenetic analyses. The universal nature of the primers allows the broadest coverage, but it can also be a problem. For example, the high copy number of nucleus-encoded 16S rRNA genes in eukaryotes may cause eukaryotic PCR products to swamp those of prokaryotes if they are together in the sample. One solution is to try to remove eukaryotes by filtration, as can be done in marine plankton; Fuhrman used glass fibre filtration that removed essentially all of the eukaryotes but only about 10 per cent of the prokaryotes. Alternatively, one can choose more-specific primers to target specific groups. An example of this might be to use bacteria-specific primers to avoid amplifying eukaryotic genes when eukaryotes and prokaryotes may not be easily separated; however, it must be remembered that chloroplasts and mitochondria contain 'bacterial' 16S rRNA sequences, so these primers do not completely avoid interference from eukaryotes. On the positive side, there are nearly universal bacterial primers that allow cloning of almost the whole 16S rRNA gene (nearly 1500 bases), maximising the available phylogenetic information. However, the so-called universal primers should be used with care, in part because they are not exactly universal. Even with some ambiguous bases, such primers have mismatches with certain known groups, and updated databases should be checked before this work is embarked on. For example, the nearly universal bacterial primer set does not match the *Planctomyces* sequence well, and therefore it may be suspected that other unknown groups may be missed. The commonly used primer at *Escherichia coli* position 1492, often treated as if it is universal, is not really so. While the PCR annealing conditions can be adjusted to allow priming in the presence of some mismatches, there still may be some rRNA genes that do not amplify well, and this possibility must be considered in interpreting the results.

Among results from PCR-based cloning and sequencing of 16S rRNA genes from marine plankton, Giovannoni used moderately specific PCR primers designed primarily for cyanobacteria, yet still found a novel proteobacterial group (SAR11) in addition to a cyanobacterial group. Fuhrman used universal primers with microbial DNA collected from the deep-sea and the euphotic zone and found novel groups of both archaea and bacteria. DeLong examined 'marine snow' with bacterial primers and found that the clones were distinctly different from those collected from surrounding free-living bacteria. As was found by Schmidt, who used a phage library, very few of these marine clones (with the exception of cyanobacteria) were closely related to previously known cultures. The SAR11 group was found at several marine locations and depths. Some of the clones were so distant as likely to be considered separate phyla from those previously studied. DeLong used archaeal primers and found two distinct archaeal groups in coastal temperate and polar waters; these were in the same groups as found by Fuhrman at different locations and with universal primers. When specific primers are used, even a relatively minor component can be amplified, detected, and studied. As an example, ammonium-oxidising bacterial 16S rRNA genes were specifically amplified from plankton by Voytek and Ward.

The PCR methods have been used with samples from sediments and microbial mats, as well as deep-sea holothurian guts, often showing unexpected bacterial and archaeal diversity. Although such material is far more concentrated than plankton (more organisms per unit volume), making it easier to obtain enough material for analysis, there are many substances that can interfere with molecular analyses, so the DNA may need extensive purification. Also, such organisms may be difficult to extract. A further consideration with at least some such samples is that they should be frozen or extracted immediately upon sampling; this is because it has been found that storage of sediments (especially if the sediments were initially anaerobic but are stored under aerobic conditions) can lead to significant and rapid shifts in species compositions. Such rapid potential changes seem most likely in rich material with rapid potential growth.

Some studies have used information about the 16S rRNA clones short of partial or full sequences. These include restriction fragment length polymorphism analysis, which can be useful in grouping clones together. However, such analysis is based on only a few base positions in the sequence, so it can lack resolution between closely related groups, particularly when universal primers are used for the initial amplification. On the other hand, it is possible to do the PCR with more-specific primers, followed by restriction analysis, to indicate rather rapidly the presence of particular groups or types of interest and to compare different samples with respect to these groups. Some fairly specific analyses might be possible by this approach with judicious selection of primers and restriction enzymes. Even with universal primers, the results can have resolution adequate for many types of studies.

A related and powerful application has been the use of fingerprinting methods that give a snapshot of the entire microbial community at once, with the ability to tentatively identify different components. These methods include T-RFLP and length heterogeneity PCR (LHPCR). In T-RFLP, PCR is performed with one fluorescent primer and one nonfluorescent primer, and the products are cut with a restriction enzyme and analysed for the size of the fluorescent end products. The presence or absence of restriction sites, as well as positions and lengths of insertions and deletions, leads to different-length products. The result looks like a chromatogram, with peaks representing different taxa. In LHPCR, the length variations of the entire PCR product (due to insertions and deletions) permit separations. In both these methods, it is possible for multiple taxa to have the same length of detected product, so 'identification' of a particular peak on the basis of database information is not definitive. In general, the T-RFLP approach has the potential for greater phylogenetic resolution than LHPCR, as the presence or absence of restriction sites, plus the ability to choose different enzymes, adds variety.

Different types of fingerprint analyses are denaturing gradient gel electrophoresis (DGGE) and thermal gradient gel electrophoresis (TGGE), which are ways to separate similar-length nucleic acid molecules on the basis of small differences in the sequences. Muyzer and colleagues, have described the use of gradient gel electrophoresis analysis of PCR products to separate different components of the PCR mixtures from natural samples. Typically, extracted DNA from an environmental sample is amplified with bacterial primers, and the products are separated by DGGE. One can get an idea of the broad diversity (number of different bacterial types) of a sample by examining the number of different bands in such an analysis. It is also possible to use probes to characterise individual bands, or the bands may be excised, cloned, and sequenced (or sometimes directly sequenced) for a detailed phylogenetic analysis. This latter ability to cut out the bands for characterisation is a particular advantage of DGGE over T-RFLP or LHPCR, in that the identity of individual components can be verified. However, the T-RFLP and LHPCR methods yield specific size information on the products, suitable for database analysis and comparison to known sequences (for presumptive identification), whereas DGGE banding patterns are much harder to standardise. Also, in a recent comparison with marine communities, T-RFLP was found to detect more taxa than DGGE, although this may be specific to the particular conditions of the study.

A non-16S rRNA method that deserves mention here is PCR from repetitive sequences in DNA to yield electrophoretic banding patterns (genomic fingerprints) that can be distinct for particular groups or strains. Such methods may be applicable in the future for characterising community structure, particularly for simple communities with little complexity.

There is still a question about possible biases in the molecular methods: have we replaced culturing biases with unknown biases? Potential biases could arise at a few stages. In the extraction stage, one can check microscopically to see that substantially all the cells have lysed, and this has been done with some of the extraction techniques with aquatic samples. The PCR may introduce biases due to variations in primer binding or extension efficiency. There are also possible biases in the cloning step, as it is known that some sequences clone more readily than others. These possible problems indicate that caution is still in order when data on the relative amounts of different clones in libraries are being interpreted, and it cannot be assumed that clones are found in proportion to their natural abundance. There is also some concern about chimeras possibly being formed during PCR amplification, and clone libraries must be checked for these. However, even if there are biases, they are probably quite unrelated to culturing biases, so these approaches are yielding much new information on what organisms are present. Can we yet say if cultures are representative of 'typical' aquatic prokaryotes? Comparisons of cultivation and cloning results have been inconclusive. Suzuki compared 127 cultivated organisms and 58 16S rRNA clones from the same Oregon coastal sample and reported little overlap, also noting that even the culturable marine organisms are poorly represented in sequence databases. Pinhassi took a different approach, using whole-genome hybridisation (from 48 culturable organisms) to examine community DNA from the northern Baltic Sea. He reported significant genomic overlap, suggesting that the cultures are representative of a large part of the native community. One group readily found in 16S rRNA clones and also cultures is the so-called marine alpha proteobacteria. On the other hand, some readily cloned sequences, such as those of SAR11 and the marine archaea, have not yet been cultured. Therefore, it appears that cultures may represent important components of native communities but may also miss important ones as well.

PROBES

In the quantitative analysis of community structure, oligonucleotide probes are powerful tools that avoid possible biases in cloning and yield a more direct measure of the target groups of interest. One way this can be done is with oligonucleotide probes hybridised to bulk nucleic acid extracted from the aquatic habitat in question. Probe sequences are determined from sequence databases and can be universal or specific to certain domains, groups, or even some species. RNA is generally preferred as the target instead of DNA because DNA is likely to yield far more false positives (unintentional hits) since mixed genomic DNA from innumerable species will have an immense variety of genes. RNA also has the benefit of being present in ribosomes and is thus much more abundant as a target and is also related to the cellular growth rate (although the exact relationship in natural communities is still uncertain). Quantitative bulk hybridisations to extracted RNA from aquatic habitats have been reported by Giovannoni and DeLong, showing the relative abundances of the SAR11 cluster and archaea, respectively. It should he noted that in order to best standardise such probe binding, it is ideal to have a culture of the organism or group in question to determine the relative binding of different probes empirically.

Although quantitative hybridisation can be readily used with natural samples, it is not the fraction of RNA coming from particular groups but instead the proportions of individuals in those groups that are usually of interest. For such work, it is ideal to tag each cell type with specific probes that allow visual identification. The preferred mode of observation has been epifluorescence microscopy with fluorescent oligonucleotide probes for rRNA. Flow cytometry has also been used to automate the analysis with success. This area has blossomed very recently. However, it has been found to work best with relatively rich environments, probably because slowly growing cells in relatively oligotrophic environments have few ribosomes and thus fluoresce dimly. Nevertheless, some approaches have yielded useful data with difficult samples like marine plankton. Lee found that multiple probes yield enough fluorescence for standard visual observation of about 75 per cent of the bacteria in marine plankton. Fuhrman found that about 75 per cent of the DAPI (4′,6′-diamidino-2-phenylindole)-countable cells from coastal marine plankton may be seen even with single fluorescent probes when video image intensification is used to boost the brightness of the images, and about 8 per cent fluoresced with an archaea-specific probe. Ouverney and Fuhrman found that typically 90 to 95 per cent of the total DAPI-countable cells can be tagged with a universal 16S rRNA probe after a marine plankton sample is treated for an hour with the antibiotic chloramphenicol. This ostensibly stops protein synthesis (and prevents changes in community composition) while allowing rRNA synthesis to continue, apparently permitting particularly strong binding of probes to the RNA. Bright fluorescence can also be obtained with long probes containing multiple fluorochromes, such as polyribonucleotides; however, with relatively conserved targets such as 16S rRNA, such long probes are best for tagging broad rather than narrow phylogenetic groups. Enzymatic amplification is another way to produce a bright signal with relatively few probe molecules, by using an enzyme-linked probe that reacts with a substrate to deposit brightly fluorescent product near the site of the probe.

There is also the possibility of using *in situ* PCR to produce fluorescent products within particular target cells, although the ability to use this as a general method in natural communities is still under development. Laser confocal microscopy is another promising approach with excellent sensitivity that allows for visualisation of dim fluorescence and also permits examination of microbes in complex matrices. This is clearly an area that will progress rapidly.

PROBE-MICROAUTORADIOGRAPHY COMBINATION

A very recent development is the ability to combine microautoradiographic characterisation of the activity of single cells with 16S rRNA probes to identify those cells. With such methods, the activities of individual cells within natural mixed communities can begin to be measured, while at the same time it can be learned which taxa are responsible for each measured property. This is a natural extension of community structure analysis. This approach thus greatly reduces the need to separate and/or cultivate an organism to learn what it is doing in its natural habitat. Such an approach is a particularly powerful means to dissect the 'black box' of natural microbial communities, going far beyond simply determining the taxonomic breakdown of community composition. There is little doubt that in the future, such single-cell probe approaches will be common tools in studies of microbial community structure.

Chapter 19

Protistan Community Structure

INTRODUCTION

Protistan assemblages of aquatic ecosystems have become the focus of a concerted research effort in aquatic ecology. One stimulus for this work has been the recognition that phototrophic protists (the microalgae) constitute a major fraction of the primary productivity within aquatic ecosystems. Another incentive has been the realisation that protozoa (heterotrophic protists) play a pivotal role in the flow of energy and elements in these communities. Studies of the abundance, biomass, and trophic activities of protists have now been conducted in a wide range of aquatic ecosystems. In addition, many laboratory studies have examined the general biology and physiology of various protistan species grown under carefully controlled conditions. The synthesis of the information into useful models of how protistan assemblages are structured and how they function in nature has advanced considerably during the last decade.

Protozoa typically are defined as heterotrophic, eukaryotic, single-celled organisms that exist by absorbing dissolved organic substances or (more commonly) by engulfing prey (phagocytosis). It is important to recognise that the term protozoa now has more historical significance than phylogenetic or, ecological meaning. For example, many 'protozoan' taxa occur among the 'algal' taxa. Numerous species of heterotrophic chrysomonads and dinoflagellates exist. These species are closely related to chloroplast-bearing species based on ultrastructural features and DNA sequence data but obviously have quite different modes of nutrition than their photosynthetic counterparts. In addition to the existence of apochlorotic 'algal' species, chloroplast-bearing genera exist within the chrysophyte, dinoflagellate, prymnesiophyte, cryptophyte, and euglenophyte algae that are capable of phagotrophy in addition to photosynthesis. This 'mixotrophic' behaviour obscures the distinction between traditional definitions of algae and protozoa.

There are also difficulties in the classification of some ciliated protozoa as phototrophs or heterotrophs. Some ciliate species ingest and digest algal prey but are able to retain the chloroplasts of these prey in a functional state, thereby providing those ciliates with limited photosynthetic ability. Photosynthesis in these 'green ciliates' contributes significantly to the overall nutrition of the protozoan and also forms a notable fraction of the primary productivity of some planktonic communities.

The close phylogenetic affinities of some flagellated protozoa with algal taxa, as well as the mixed nutrition of many protists, indicate the artificiality of the historical distinction between the algal and protozoan taxa. Therefore, use of the term 'protist' in reference to both chloroplast-bearing forms (i.e. algae) and heterotrophic forms (protozoa) has gained popularity in recent years.

ASSESSING PROTISTAN COMMUNITY STRUCTURE

Protistan Species Diversity

Complete taxonomic characterisation of the protistan assemblage has been achieved for few, if any, aquatic communities. There are at least three major, interrelated reasons for this situation: (i) the large number of protistan species present in most aquatic environments, (ii) the tremendous mendous range in size and abundance of these species, (iii) the disparate methodologies that are required for sampling and identifying them. Surmounting the difficulties posed by these issues is central to improving the present state of our knowledge concerning the structure of natural assemblages of aquatic protists.

Literally thousands of species of protists have been described, and many of these species have sympatric distributions. In addition, free-living protists range in size from approximately 2 μm flagellates up to some species of radiolaria that can form cylindrical gelatinous colonies ~1 cm in diameter and >1 m in length. Similarly, relative abundances of protistan species within a habitat may vary by several orders of magnitude. These immense ranges of size and abundance make it necessary to apply a number of sampling techniques in order to adequately sample all of the protistan species in an environment. The unique physical and chemical characteristics of different aquatic environments (e.g. planktonic versus benthic environments) also contribute to the varied protocols that are necessary to sample protistan assemblages. Even within an environment, sampling protocols must be adapted for particular groups of protists. For example, enumerating species of benthic protozoa among the sedimentary particles in which they exist has been a long-standing problem in assessing protistan species diverisity in those environments. Various methods for extracting and concentrating protozoa from sediments have relied on the mobility of the community in response to changing salinity, extraction by centrifugation, or enrichment culture. Such approaches have resulted in reasonable estimations of the protozoan diversity of some sediment environments, but the success of these methods is usually group-specific. The extraction of protozoa by the sea-ice method may work well for highly mobile species such as benthic ciliates but this method may be less useful for slower-moving forms such as small amoebae. For the latter forms, enrichment cultivation appears to be the most appropriate method.

Adjustments to sampling protocols are also necessary for sampling different protistan groups within plankton communities. Sample volumes of 200 to 500 ml are usually sufficient in oceanic waters for flagellated protists (typical abundances are hundreds to thousands per millilitre), and volumes of 0.5 to 2 litres are usually sufficient for ciliated protists (typical abundances are tens to thousands per litre), but actinopods and foraminifera usually must be concentrated using plankton nets or filters. These latter techniques, however, are damaging to delicate species of planktonic ciliates. Common methods for sampling planktonic protists have been recently reviewed.

Preservation, fixation, and other manipulations are necessary prerequisites for the identification of most protistan species once appropriate samples have been collected. Notable exceptions to this generalisation are the 'naked' amoebae (primarily the *Gymnamoebae*) in which some of the characteristics that are essential for proper identification are present only in living specimens). For the remaining protistan groups, correct preservation is dependent on the protistan group under consideration. For flagellated (and often ciliated) protists, aldehyde fixatives (formaldehyde and glutaraldehyde) are commonly employed, and osmium tetroxide is often added when electron microscopy is planned. A variety of fixatives have been developed for ciliated protozoa, most of which are usually employed in combination with post-fixation staining methods that are used to visualise cytological features of the cells.

The preservation of some protistan taxa requires special consideration. A preservative that does not promote the dissolution of skeletal structures must be employed for species which possess such structures because the latter are important diagnostic criteria (e.g. for the identification of some actinopods and foraminifera). Careful adjustment of the pH of the preservative is necessary to prevent dissolution of foraminiferan tests, while addition of strontium is necessary to prevent dissolution of acantharian skeletons. When these requirements conflict, subsamples must be preserved separately for the different groups. For example, planktonic foraminifera (which require alkaline pH) typically would be preserved differently than planktonic ciliates which are often preserved in acid Lugol's solution.

The identification of protistan species in mixed natural assemblages depends on criteria that are often as different as the methodologies used to sample and preserve these assemblages. Therefore, taxonomic expertise among protistologists is often limited to one of the major groups (e.g. diatoms, dinoflagellates, amoebae, ciliates, or foraminifera) or some portion of one of these major categories.

Ciliates typically possess morphological features that provide sufficient taxonomic criteria for identifying species by light microscopy. Cell size, shape, location and characteristics of the oral area, presence of a lorica, and particularly the arrangement of the somatic ciliature are useful features for species identification. Ciliates are often easier to identify than many of the flagellated and amoeboid protists because of the presence of these features, and extensive species lists exist for various environment.

Flagellated protists typically possess fewer morphological features that can serve as useful taxonomic criteria when they are observed by light microscopy. Cell size and shape, chloroplast arrangement, and flagellation are important criteria for identification by light microscopy. Some diagnostic features, however, are only visible using electron microscopy (e.g. flagellar mastigonemes and body scales). Electron microscopy is often necessary for distinguishing the numerous genera and species of small heterotrophic flagellates (<10 μm). The need to establish these features using electron microscopy makes it difficult to process large numbers of samples. Moreover, many of the latter taxa have not been adequately described. There is considerable uncertainty about the validity of numerous genera and thus the true species diversity of small heterotrophic flagellates in many environments.

The amoeboid protists are a polyphyletic collection of species, and the methods of identification applied to these species are heterogeneous. The naked amoebae are identified based on features of the living organisms, i.e. cell size and shape during locomotion, arrangement and type of pseudopodia, and morphology of the floating form, etc. The requirement for live material for species identification has made the determination of species diversity of natural assemblages of amoebae a difficult topic, but the taxonomy, distribution, and general ecology of these species is slowly emerging. Identification of the many types of testate amoeboid protists (testacea, foraminifera, radiolaria, and others) is based on the skeletal structures that are present in many of these species and on features of their cellular organisation. The presence of a rigid skeletal structure in many of these species makes it possible to use plankton nets or screens for collecting and concentrating these specimens from the plankton or sediment.

Difficulties associated with sampling and identifying the entire spectrum of protists (as described above) in natural communities hampers documentation of the true protistan species diversity of any natural ecosystem. Exceptions to this generality might be found in environments where protozoan diversity is greatly reduced due to severe environmental factors such as anaerobic conditions, but it is safe to generalise that the vast majority of studies of natural communities have underestimated total protistan species diversity.

However, the species diversity of particular taxa of protists (e.g. the ciliates, flagellates, and amoebae) has been more accurately determined. The most complete information exists for plankton communities where extensive lists of ciliated protozoa, chloroplast-bearing flagellates, and skeleton-bearing sarcodines (foraminifera and actinopods) have been obtained.

Protistan Abundance and Biomass

Identification of the protistan species present in an aquatic environment provides useful but limited information on their potential contribution to the structure and function of the total biological community because of the tremendous size range and varied trophic activities of protistan species. A much broader understanding of their importance can be obtained by combining species lists with estimates of abundance and biomass. Most modern methods for collection and identification of protists have been designed with this goal in mind. Generally accepted methods are now emerging which minimise problems associated with loss of cells during collection, enrichment, preservation, and sample processing for specific groups of protists, thus allowing accurate estimates of protistan abundance.

As with species identification, estimating population abundances of amoebae is a difficult problem. The amorphous shapes of these micro-organisms in preserved samples make them difficult to recognise and enumerate in natural samples, and their association with detritus and other particles also can obscure them from view. The few abundance estimates that are available for these species have been obtained using a most-probable-number culture technique that relies on the growth of the amoebae in serial dilutions of the water or sediment samples.

Protistan abundance measurements can be used to calculate total protistan biomass (typically expressed in units of carbon) using measurements of abundance, cell volume, and empirically derived carbon:volume conversion factors. Cell volume measurements obtained from microscopical studies are combined with abundance estimates to calculate the volume of particular protistan taxa, and carbon:volume conversion factors are then applied to calculate the carbon content. Carbon:volume conversion factors must take shrinkage due to fixation and the variable vacuolar space of protists into account. Shrinkage due to fixation can be both taxon and size specific. Typical values for converting carbon to volume range between 160 and 360 fg of C μm^{-3} for flagellated protists, whereas a value of 190 fg of C μm^{-3} is typical for ciliated protists. Carbon:volume conversion factors for larger sarcodines (acantharia, radiolaria, and foraminifera) are based on aspects of the cells that are resistant to net collection. A method has been proposed for estimating the cell volume of naked amoebae that directly relates the diameter of the nucleus to total cell volume. Commonly used conversion factors have been summarised and reviewed recently.

Describing Protistan Community Structure

The term community structure implies that organised relationships exist between protists and other micro-organisms within natural ecosystems. Indeed, the niche concept has been applied to protistan assemblages with the implication that the number of protistan species in an environment is indicative of the number of unique ecological roles for protists in these assemblages. Unfortunately, it is unrealistic to consider all protistan species in a community as separate entities at this time because of the great species diversity of these assemblages, the limited ecological information available on the realised niches of many protistan species, and the extreme difficulty in obtaining species identifications and abundance or biomass information for all protistan species in an assemblage. For these reasons, various simplifying groupings of protists have been employed as a way of reducing the complexity of protistan assemblages into manageable (and measurable) quantities.

Various means of simplifying protistan community structure have been used. The most popular approaches have grouped protists by trophic mode (phototrophic versus heterotrophic), size, and prey type (for heterotrophs) in keeping with the trophic-level concept of Lindeman. Trophic categories must be somewhat more flexible than simply phototrophy versus heterotrophy because of the common behaviour of mixed nutrition among protists. Nevertheless, aggregation of species into 'trophospecies' is still a useful and necessary procedure for partitioning the assemblage in order to allow investigations of energy and elemental flow through aquatic communities in models of manageable size.

For heterotrophic protists, it is a common procedure to group species according to the type of prey that they consume. Bacterivorous flagellates and ciliates in plankton communities or bacterivorous flagellates and amoebae on suspended particles may be grouped together to represent a major sink for bacterial biomass in the plankton. Similarly, ciliate species may be grouped into bacterivorous, herbivorous, or predacious specie. Such 'feeding guilds' ignore some of the details of protistan feeding behaviour (such as omnivory), but they are useful for reducing the complexity of the assemblage. Feeding guilds are often treated as single species in biological or biogeochemical models of ecosystem function.

The organisation of protists by size is logical for two reasons. Allometric dependence of growth and metabolism can be used to constrain the potential contribution of a particular size range of protists to biogeochemical cycles. In addition, predator-prey relationships are typically size dependent, with larger predators consuming smaller prey. For example, many heterotrophic flagellates that are 2 to 20 μm in size consume bacteria and cyanobacteria that are <2 μm in size, and many ciliate species that are 20 to 100 μm in size consume algae and protozoa that are <20 μm in size.

While this generalisation is realistic, it is not absolute. Many species of heterotrophic dinoflagellates consume diatom prey that are considerably larger then themselves by employing a pseudopodial 'feeding veil'. Similarly, some planktonic sarcodines (acantharia, radiolaria, and foraminifera) are capable of consuming metazoan prey considerably larger than themselves due to the production of a sticky pseudopodial network that entangles and immobilises prey items.

Notwithstanding these exceptions, size-dependent grazing models are the most common manner of organising protistan populations into manageable units for inclusion into models of elemental flow in aquatic ecosystems. The aggregation of species into groups within models probably reduces the predictive capabilities of these models, but their outcomes thus far appear to be in reason agreement with field data. It remains to be seen how the reduction of species diversity in these models will affect predictions of the response of the community to internal and external perturbations, but the gradual disaggregation of these models into more ecologically relevant compartments should provide insight into this issue.

An example from the plankton

A hypothetical example indicates the analytical approaches for examining protistan community structure and the limitations of these approaches. A species list of protists that is representative of an oceanic plankton community is given in Table 19.1. This assemblage is not meant to be complete but rather is indicative of the breadth of protistan sizes and nutritional modes in this type of ecosystem. Pertinent information on cell size, photosynthetic and phagotrophic ability, prey type(s), and typical abundances are also provided. The species in this assemblage have been arranged according to major taxonomic (i.e. phylogenetic) affinities.

As shown in Table 19.1 taxonomic groupings of the protists correspond poorly to the nutritional modes and ecological roles of the species. Reorganisation of the same species into groups based on the nutritional modes of the species provides a very different classification of this assemblage Table 19.2. This reorganisation indicates the classical dichotomy between phototrophs and heterotrophs, but it also indicates the more recent realisation that many protistan species have mixed modes of nutrition. Note that the latter characteristic results in some of the species occurring in more than one category.

Table 19.1. Species list showing the range of protistan diversity of an oligotrophic oceanic environment[a].

Protistan taxon	Avg. size (μm)	Chloroplasts present	Phagotrophy	Probable prey[b]	Representative abundances (litre⁻¹)
Flagellated or nonmotile protists					
Dinoflagellates					
Protoperidinium sp.	55	No	No	Dia	10^2
Gymnodinium sp.	50	Yes	Yes	C, Din, Dia	2×10^2
Prorocentrum micans	25 × 40	Yes	No		10^3
Ornithocercus magnificus	40	No[c]	Yes	Dia, Sf	10^2
Chrysophytes or chrysomonads					
Paraphysomonas imperforata	7.0	No	No	B, Cc, Sf	10^5
Ochromonas sp.	6.0	Yes	Yes	B, Cc	10^5
Prymnesiophytes					
Chrysochromulina ericina	6.0	Yes	Yes	B, Cc	10^4
Chlorophytes					
Nanochloris atomus	4.0	Yes	No		10^5
Bacillariophytes (diatoms)					
Minutocellus polymorphus	3.0	Yes	No		10^5
Coscinodiscus concinnus	75 × 200	Yes	No		10^{-2}
Ditylum brightwellii	20 × 100	Yes	No		10^2
Rhizosolenia clevei	200 × 500	Yes	No		10^{-3}
Ethmodiscus rex	1000	Yes	No		10^{-4}
Choanoflagellates					
Diaphanoeca grandis	2.5	No	Yes	B	10^5
Amoeboid protists					
Gymnamoebae (naked amoebae)					
Platyamoeba weinsteini	3 × 12	No	Yes	B	10^0
Flabellula citata	4 × 30	No	Yes	B	5×10^{-1}
Foraminifera					
Globigerina bulloides[h]	700 [f]	No	Yes	Omi, Mz	10^{-3}
Globigerinoides sacculifer[h]	700 [f]	No[d]	Yes	Omi, Mz	10^{-3}
Acantharea					
Amphilonche elongata[h]	50 × 400	No[d]	Yes	Omi	10^{-1}
Spumellarian radiolaria					
Thalassicolla nucleata[h]	1000[f]	No[d]	Yes	Omi, Mz	10^{-4}
Collozoum caudatum[h]	200[f,g]	No[d]	Yes	Omi, Mz	10^{-5}
Ciliated protists					
Tintinnids					
Tintinnopsis parva	20 × 40	No	Yes	B, Ce, Sf	10^2

(Contd ...)

Protistan taxon	Avg. size (μm)	Chloroplasts present	Phagotrophy	Probable prey[b]	Representative abundances (litre[-1])
Oligotrichs					
Strombidium sulcatum	25 × 50	Yes[e]	Yes	B, Cc, Sf	10^2
Loboea strobila	50 × 150	Yes[e]	Yes	B, Cc, Sf	10^2
Hypotrichs					
Euplotes woodruffi	65 × 120	No	Yes	C	5×10^{-1}
Hymenostomatids					
Uronema marinum	10 × 20	No	Yes	B, Cc	10^3

[a] This list is only a partial list of representative species to exemplify the breadth of trophic modes in a real assemblage. Species are organised according to major taxa. Pertinent ecological information and realistic abundances are provided based on literature values.
[b] Dia, diatoms; C, ciliater; Din, dinoflagellates; Sf, small flagellated protists; B, bacteria; Cc, chroococcoid cyanobacteria; Omi, omnivorous on prokaryotic and eukaryotic unicells; Mz, metazoan zooplankton.
[c] Species harbouring extracellular, symbiotic cyanobacteria that contribute to the photosynthetic nutrition of the host.
[d] Species harbouring intracellular, symbiotic dinoflagellates that contribute to the photosyntheitic nutrition of the host.
[e] Species that retain functional chloroplasts from ingested prey.
[f] Size does not take into account extensive pseudopodial network.
[g] Colonial species with colonies up to 1 cm in width and 1 m in length.
[h] Considers adult specimens only.

Table 19.2. Planktonic protistan species list descriped in Table 19.1 arranged according to trophic category[a].

Phototrophs

    Gymnodinium sp.

    Prorocentrum micans

    Ochromonas sp.

    Chrysochromulina ericina

    Nannochloris atomus

    Minutocellus polymorphus

    Coscinodiscus concinnus

    Ditylum brightwellii

    Rhizosolenia clevei

    Ethmodiscus rex

Mixotrophs

    Phagotrophic algal species

        Gymnodinium sp.

        Ochromonas sp.

        Chrysochromulina ericina

    Chloroplast-retaining species

        Strombidium sulcatum

        Loboea strobila

(Contd ...)

Symbiont-bearing species
    Ornithocercus magnificus
    Amphilonche elongata
    Thalassicolla nucleata
    Collozoum caudatum
Heterotrophs
  Bacterivores
    Paraphysomonas imperforata
    Ochromonas sp.
    Chrysochromulina ericina
    Diaphanoeca grandis
    Platyamoeba weinsteini
    Flabellula citata
    Amphilonche elongata
    Thalassicolla nucleata
    Collozoum caudatum
    Tintinnopsis parva
    Strombidium sulcatum
    Loboea strobila
    Uronema marinum
  Herbivores
    Protoperidinium sp.
    Gymnodinium sp.
    Ornithocercus magnificus
    Paraphysomonas imperforata
    Amphilonche elongata
    Thalassicolla nucleata
    Collozoum caudatum
    Tintinnopsis parva
    Strombidium sulcatum
    Loboea strobila
    Globigerina bulloides
  Carnivores
    Paraphysomonas imperforata
    Globigerina bulloides
    Amphilonche elongata
    Thalassicolla nucleata
    Collozoum caudatum
    Euplotes woodruffi

[a] Note that some species occur in more than one category.

The collection of species in this assemblage also demonstrates the enormous breadth of cell sizes that can be displayed by protistan assemblages (Fig. 19.1). The size range is not necessarily restricted for any particular trophic mode. In this assemblage heterotrophs range from 2.5 μm to >1 mm in size, phototrophs range from 3.0 μm to >1 mm, and mixotrophs range from 6.0 μm to >1 mm if one, includes symbiont-bearing sarcodines in this last category. Commonly employed plankton size class designations, are also shown in this figure. These designations correspond to organisms that are 2 to 20 μm (nanoplankton), 20 to 200 μm (microplankton), 0.2 to 20 mm (mesoplankton), and 20 to 200 mm (macroplankton) in their longest dimension. Protists occur in all of these size classes, as indicated in Fig. 19.1, although the majority of these species are in the nano- and microplankton size classes.

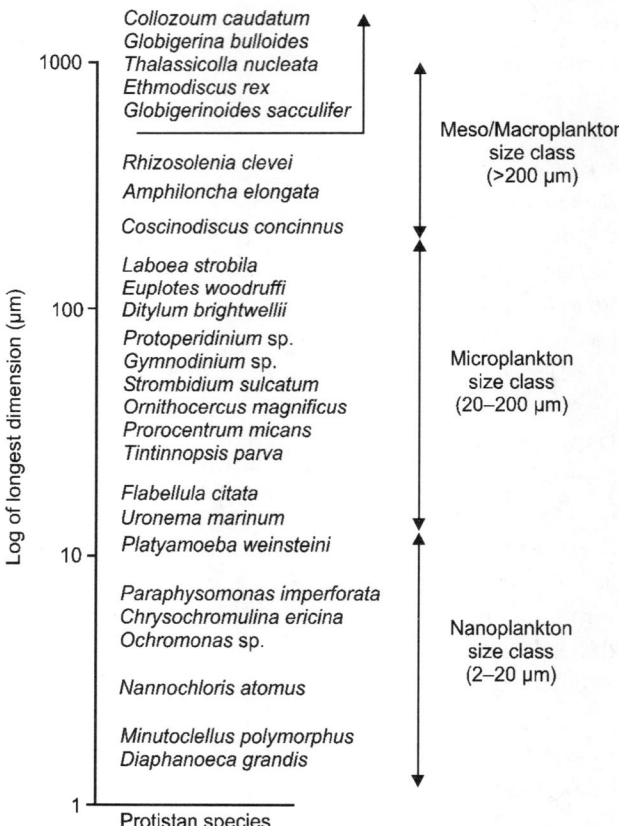

Fig. 19.1. Approximate sizes (longest dimension) of the planktonic protistan species listed in Table 19.1. Commonly employed size class designations are shown on the right. Note that the sizes of these protists span more than 3 orders of magnitude. Upward-pointing arrow under *Globigerinoides sacculifer* indicates that the group of five species enclosed by the arrow can be larger than 1000 μm.

One generality that is clear from Table 19.1 is that small planktonic protists are typically more abundant than large species are. This relationship is shown clearly by 1 to 2 orders of magnitude differences in abundances when the individuals in a community are tallied according to plankton size class Fig. 19.2(a). Phototrophic, heterotrophic, and mixotrophic protists all show large differences in abundance among size classes. However, the abundances of phototrophic, mixotrophic, and heterotrophic protists within a size class are often similar. The large disparity that is apparent when comparing the abundances of protists in different size classes Fig. 19.2(a) is greatly reduced when the total volume of living protists is analysed Fig. 19.2(b). The small cell size of the nanoplanktonic protists is generally balanced by the high abundances of these species, while the low abundances of the larger protists are balanced by their large cellular volumes. These general relationships of protistan abundance and biovolume are consistent with data from natural assemblages of nanoplanktonic and microplanktonic protists.

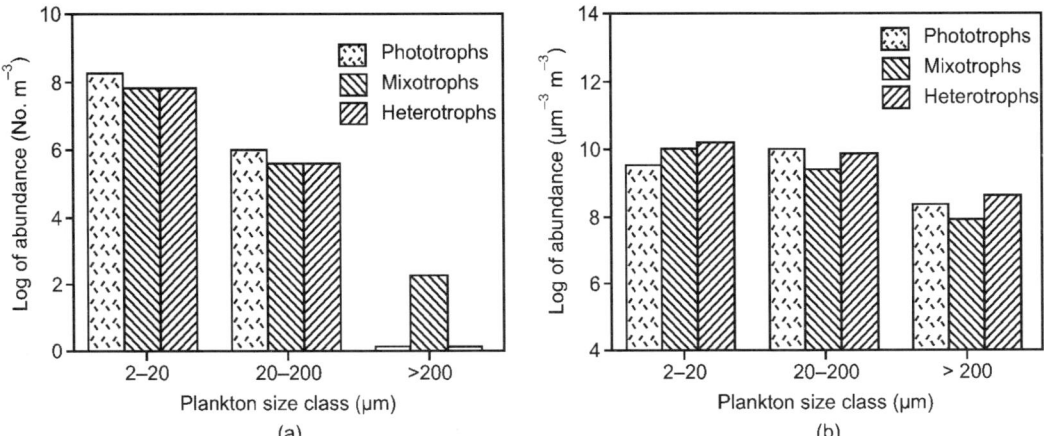

Fig. 19.2. Abundance and biovolume relationships of the protistan assemblage listed in Table 19.1. The species have been grouped according to size class and according to trophic mode (phototrophic, mixotrophic, heterotrophic).

The information summarised in Tables 19.1 and 19.2 and Figs. 19.1 and 19.2 can be used to construct a typical box model depicting the flow of materials from producers through consumers for this hypothetical protistan community (Fig. 19.3). The species have been grouped according to their nutritional modes, sizes, and approximate predator-prey relationships. The presumed directions of energy and material flow (i.e. from producers to consumers and from small organisms to large consumers) in the hypothetical community are indicated. The development of working models that accurately describe energy and elemental flow within these communities and the incorporation of microbial processes into classical models of aquatic community structure and function have been among the major goals for work on modelling microbial loop processes. Models such as the one shown in Fig. 19.3 are appropriate for these purposes be cause they attempt to reduce a complex assemblage of micro-organisms to a manageable number of trophic 'compartments' and trophic interactions. These models, therefore, are strongly influenced by methodologies available for identifying protistan species (or trophospecies) and for investigating their trophic interactions.

The model in Fig. 19.3 might adequately describe energy or elemental flow in this hypothetical protistan assemblage if the biomass and flow parameters of the model could be determined. However, this type of depiction of community structure still has some inherent flaws. As referred to earlier, predator-prey relationships that are not size dependent are difficult to represent and measure. Energy is depicted as moving from smaller to larger size classes in this model, but this representation is incorrect for species such as *Protoperidinium* sp., which can graze on diatoms larger than itself, and for the sarcodines

Globigerinoides sacculifer, Thalassicolla nucleata, Collozoum caudatum, and *Globigerina bulloides,* which can consume metazoan prey. The doubleheaded arrows connecting these latter compartments indicate the potential for the flow of energy in either direction. In practice these measurements are difficult to make.

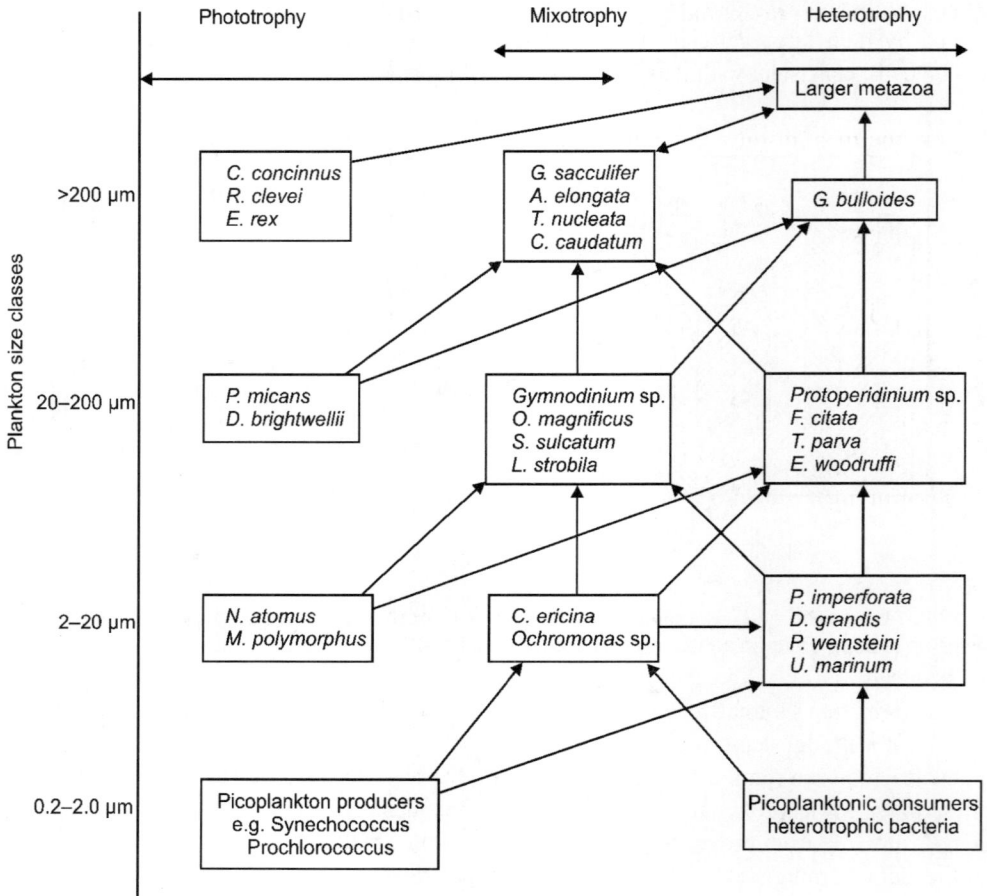

Fig. 19.3. Box model approximating the major trophic interactions within the protistan assemblage listed in Table 19.1. The species have been grouped according to known or presumed size-dependent and trophy-dependent relationships. Arrows indicate the directions of energy flow in predator-prey interactions.

Similarly, selective grazing and omnivory are difficult to incorporate into this type of model. For example, *Euplotes woodruffi* is a predacious ciliate that feeds primarily on other ciliates (in this assemblage it might feed on *Uronema marinum*). On the other hand, *Tintinnopsis parva* may accept a variety of small protists and other micro-organisms as prey. The distinction between these two rather different nutritional modes has been forfeited by placing them into the same trophic compartment. Clearly, if the goal of this modelling exercise were to understand the factors affecting the success or failure of either of these two species in plankton communities, then this model would be unsatisfactory. It is for reasons such as this last example that the appropriate conceptualisation and representation of protistan community structure must take into account the goal of the investigator.

TEMPORAL AND SPATIAL CHANGES IN COMMUNITY STRUCTURE

The most significant differences in the species composition and trophic relationships of protistan communities exist between different aquatic environments. However, there is also a rapidly increasing data base on changes in community structure over seasonal and shorter time scales. These latter changes appear to be most significant in temperate and polar climates.

Freshwater versus Marine Ecosystems

Probably the most distinct difference between freshwater and marine protistan communities is the restriction of the larger sarcodines (acantharia, radiolaria, and foraminifera) to brackish and marine ecosystems. In tropical and subtropical oceanic plankton communities, adult sarcodines are often the most conspicuous macroscopic organisms in surface waters, while swarmer cells and juvenile specimens of these species contribute to the entire size spectrum of protozooplankton. In marine sediments, benthic foraminifera can constitute an important component of faunal assemblages.

In contrast to large differences in the assemblages of sarcodine species in freshwater and salt water, there appears to be a fair degree of similarity with respect to types of ciliates and flagellates in these environments. The ecological roles of small protozoa in freshwater and marine plankton communities appear to be similar and related to bacterial production over very broad scales of examination. Most ciliates in the freshwater and marine plankton also appear to play grossly analogous ecological roles as consumers of small prokaryotes and eukaryotes. Mixotrophic (phagotrophic) algae exist in both freshwater and marine water, as do species of chloroplast-retaining ciliates. These generalisations do not necessarily mean that the same species of flagellates or ciliates occur in both ecosystems, but rather that similar ecological niches have been filled by protistan species in these different environments.

Benthic versus Pelagic Ecosystems

Although there are species of protists that are commonly found in both benthic and pelagic environments, there are clearly numerous species within these assemblages that are uniquely suited for one environment or the other. Morphological adaptations of ciliates to life between sediment particles in the benthos has resulted in the evolution of cell forms that allow movement through this medium. Common adapotations include cylindrical or flattened shapes, flexible cell walls, and patterns of ciliature that allow 'crawling' along surfaces and grazing on prey loosely associated with particulate material. Some species permanently attach to surfaces. In contrast, ciliates in pelagic environments (e.g. choreotrichs) tend to have more rounded shapes and patterns of ciliature that afford rapid swimming behaviour and feeding on suspended particles.

There is great chemical and physical diversity among and within benthic environments as a consequence of sediment grain size, organic loading, oxygen gradients, etc. The number of micro-environments at one locale may be considerable. Spatial heterogeneity, the remoteness of many benthic environments, and difficulties associated with sampling, concentrating, and observing benthic protists continues to limit our knowledge of the protistan fauna of many benthic ecosystems. There are extremely few observations of the protistan fauna of the deep ocean benthos.

The amoebae are particularly well suited for existence in benthic environments. Locomotion and feeding of these species take place on particles. Therefore, benthic environments tend to support significant assemblages of amoebae. Amoebae occurring in the plankton are generally assumed to be associated with suspended particulate material or with the air-water interface.

Among the larger sarcodines, there are clear differences between pelagic and benthic assemblages. Foraminifera occur in both environments, but the species occurring in these two environments are different. The planktonic species are restricted to pelagic, oceanic ecosystems while benthic species are common in ecosystems ranging from salt marshes to abyssal depths. Most radiolaria (polycystines and phaeodaria) and acantharia are restricted to pelagic, oceanic ecosystems. There are relatively few exceptions to these generalities, making the larger sarcodine fauna of benthic and pelagic ecosystems quite distinct.

The contribution of phototrophic protists to flagellate assemblages in surface waters of pelagic ecosystems clearly differentiates them from flagellate assemblages of benthic environments in which phototrophs may be reduced or absent. However, the heterotrophic flagellate assemblages of these two habitats also differ in composition. Many flagellated protozoa occur in both environments, but species that are capable of particle attachment or movement along surfaces (e.g. bodonid flagellates) tend to predominate in benthic environments. Forms that feed on suspended bacteria (e.g. chrysomonad flagellates and choanoflagellates) tend to predominate in pelagic ecosystems.

Pelagic environments generally are considered more homogeneous ecosystems than benthic ecosystems, but there are clearly sources of heterogeneity in the plankton. Epibiotic (and possibly enteric) protistan assemblages have not been adequately studied, but they contribute to protistan species diversity in the water column. Suspended particles also create unique microhabitats in pelagic ecosystems for some protozoan species that are more characteristic of the benthos. Macroscopic detrital aggregates in marine planktonic ecosystems (so-called 'marine snow') may create a 'false benthos' for benthic species by creating microenvironments with elevated abundances of bacteria and other prey. Similar 'oases' for unique protozoan assemblages in the plankton may be established using artificial foam substrates. It has been demonstrated that colonisation and species succession of protozoa on these natural and artificial substrates may follow a pattern similar to that of the colonisation of oceanic islands by higher organisms.

Depth and Seasonal Distributions

The seasonality of algal species composition and abundance in pelagic environments is well-known. Distributions of protistan algae associated with depth have also received considerable attention. Although changes in total protozoan abundance or biomass associated with changes in season or depth have been documented, there is relatively little information on changes in species composition or community structure or function associated with changes in depth. This paucity of information is not surprising given the logistical problems associated with the collection of long-term data sets or multiple samples from considerable depth.

Most studies to date have been restricted to a particular group of heterotrophic protists because of either methodological approach or taxonomic expertise. Often these investigations have reported only changes in broad taxonomic or ecologically relevant categories (heterotrophic flagellates, mixotrophic flagellates, and ciliates, etc.). For example, depth and seasonal changes in abundance have been reported in a variety of marine and freshwater environments for flagellated and/or ciliated protozoa. More detailed data on the spatiotemporal distributions are available only for specific taxa for which identification is more straightforward.

It is difficult to generalise concerning changes in the community structure of protozoan assemblages as a function of season from these scattered reports. For temperate communities, seasonal changes in species composition and winter reductions in the intensity of grazing activity are likely, but the extent of these changes remains largely undetermined for most environments. Temperature is a strong controlling influence on processes within the microbial loop of temperate ecosystems, but diverse heterotrophic and phototrophic protistan assemblages abound even in extremely low-temperature environments such as marine habitats of Antarctica. In the latter environment, rather unique assemblages of phototrophic and heterotrophic protists have adapted to existence within and on the bottom of the sea ice and in the water column.

The vertical distributions of protozoa typically demonstrate greater overall abundances in surface waters relative to abundances at depth. These distributions of abundance are clearly related to the production of organic material in surface waters. Fine-scale vertical distributions, however, can be complex. Elevated abundances of protozoa have been observed at the air-water interface, at oxic-anoxic boundaries within water columns and at subsurface biological features such as deep chlorophyll maxima. Vertical distributions of protists in the sediments typically are related to physical and chemical gradients within the benthos. The exploitation of these chemicals and/or physical features within the benthos and water column can increase the diversity of protistan assemblages of an environment by providing unique microhabitats for the growth of species able to exist there.

MOLECULAR APPROACHES TO STUDYING PROTISTAN COMMUNITY STRUCTURE

Problems associated with determining species identity, abundance or biomass, and trophic activity of protistan assemblages in aquatic ecosystems continue to hamper in-depth analyses of the structure and function of these communities. As described above, classical methods of identification are time-consuming and often do not provide quantitative information on the occurrence of species. New conceptual and methodological approaches will be necessary to deal with these recalcitrant problems.

Conceptual approaches and methodologies from molecular biology and immunology offer hope for addressing some of these problems. For example, rRNA-targeted oligonucleotide probes and PCR-based methods for determining the presence of specific micro-organisms are becoming commonplace in medical and environmental bacteriology. Until recently, the application of these approaches to protistan ecology had been confined largely to investigations designed to determine the presence or absence of species of interest to human health or species that might have adverse environmental impact. Applications to species of purely ecological significance, however, are beginning. In some cases these approaches are indicating previously undocumented diversity among small protistan assemblages or new insights into the biogeography of these species. Immunological methods for determining the presence and abundance of specific microbiological taxa also have been developed. This approach has been successfully applied to phototrophic protists.

The application of molecular and immunological approaches to protistan groups with difficult taxonomic features may be particularly helpful in the future. For example, species of heterotrophic flagellates or amoebae which presently involve nonquantitative or impractical approaches such as electron microscopy or the observation of living specimens might be particularly suited for this approach.

Biochemical markers to indicate the presence, abundance, and activity of heterotrophic protists also may provide new methods for examining natural assemblages of protists. Detailed pigment analyses have provided useful insights into the contribution of microalgal taxa to total algal biomass. Lipid biomarkers have been applied to obtain information on the biomass and nutritional status of bacterial assemblages and to study eukaryotic decomposers. Analogous methodologies for assessing 'community-level' features of heterotrophic protistan assemblages would be useful for investigating natural, mixed assemblages of protists.

Chapter 20

Phagotrophy in Aquatic Microbial Food Webs

INTRODUCTION

Determination of predator-prey interactions, and of pathways of energy and elemental flows within microbial food webs, is vital to understanding the structure and functioning of ecosystems. Methods to assess specific trophic pathways, e.g. bacterivory and herbivory, have been applied to *in situ* microbial communities in both marine and freshwater systems. Most approaches used so far entail assumptions that have been demonstrated to be invalid in certain cases. In order for grazing assays to yield interpretable results, it is important for alternate mortality processes to be assessed via judicious control treatments. For example, spontaneous lysis or viral lysis of bacterial cells may confound results in long-term prey disappearance methods used to estimate protistan bacterivory; thus a nongrazer control treatment should be included in the experimental design. There are few methods for the evaluation of food resources for phagotrophic protists other than bacteria or phytoplankton, e.g. other heterotrophic protists or nonliving organic matter. Application of molecular techniques to identify specific phylogenetic groups of prey, to identify protistan grazers, or to develop new approaches for analysis of predator-prey interactions has only just begun. This field is a rapidly expanding area within aquatic microbial ecology; additional methodological approaches, and further application of existing approaches, should yield major new insights into microbial trophodynamics.

TROPHIC INTERACTIONS AMONG AQUATIC MICROBES

The focus on grazing interactions—who eats whom—in aquatic ecosystems has expanded during the past two decades to include microbial predators, phagotrophic protists, in addition to the usual macroscopic consumers such as copepods and fish. Textbook notions about linear food chains and clearly distinct trophic levels do not easily apply to microbial predators. For instance, mixotrophy, a combination of trophic modes, is common among protists. Examples of mixotrophy include photosynthetic nanoflagellates and dinoflagellates that ingest bacteria, algae, and heterotrophic protists and heterotrophic ciliates and dinoflagellates that temporarily 'enslave' chloroplasts from algal prey. Realisation of the great diversity among aquatic microbes, in terms of size, trophic modes, and phylogenetics, has also resulted in changes in our conceptualisation of microbial trophic interactions. For instance, in open ocean gyres, prochlorophytes, 1-µm-sized phototrophic bacteria containing divinyl chlorophyll *a*, are an abundant and often significant component of both the phytoplankton and the bacterioplankton. In these marine systems, bacterivorous flagellates play a dual role as herbivores and bacterivores. Another case in which ideas have changed has to do with dinoflagellates. Long assigned to the phytoplankton, dinoflagellates are now recognised as having an equally important role as consumers of a wide range of microbial prey. Not only are half of all dinoflagellates species nonpigmented, and thus heterotrophic, but many species of autotrophic dinoflagellates are also phagotrophic.

In the 'microbial loop' concept of Azam, as formalised by Ducklow bacteria, bacterivorous flagellates, and ciliate consumers of the flagellates function as regenerative adjuncts to the classic metazoan food web. This simplistic model has by now evolved to the more expansive concept of a microbial food web. The microbial food web encompasses all microbial cells: phytoplankton, bacteria, and phagotrophic protists and carries out the functional roles of primary production, microbial heterotrophy, and nutrient regeneration. The overall microbial food web, and not just phytoplankton, should be considered as the resource base for metazoans in aquatic ecosystems. It is also characterised by complex interactions between microbes, and between microbes and their physical and chemical environment, that are not yet well understood.

Little is yet known about the ecology of individual species of free-living phagotrophic protists, for at least two reasons: (i) taxonomic identification of protists in natural aquatic systems is difficult. Work on the phylogenetic affinities of free-living protists has lagged both taxonomic studies of multicellular zooplankton and molecular genetic studies of phylogenetic groups of bacterioplankton; and (ii) the focus in most studies of trophic roles of protists has been in the area of systems ecology rather than of population and community ecology. Heterotrophic microbes are responsible for a large share of overall carbon utilisation, respiration, and regeneration of mineral nutrients in ecosystems. Thus, most research in this field has considered functional groups of microbes rather than individual populations. However, Verity and Smetacek argued that understanding how aquatic ecosystems function will remain incomplete so long as we do not understand the life histories of their major biotic components. Clearly, there is much work still to be done with respect to the ecology of phagotrophic protists.

Developing methods to elucidate specific trophic links within microbial food webs continues to be a challenge. We will outline the techniques that have been used to investigate the processes of bacterivory and of herbivory in microbial food webs and also briefly discuss approaches used to study other trophic pathways. Current methods are still not entirely satisfactory. This area of research remains methods limited.

BACTERIVORY

The question of the fate of bacterial production within aquatic food webs has been of long-standing interest. Heterotrophic bacteria process a large fraction (30 to 60 per cent) of primary production in both freshwater and marine systems. Phagotrophic protists, predominantly <10-μm-sized heterotrophic flagellates but also phototrophic flagellates and ciliates, have been identified as significant consumers of bacterioplankton in aquatic systems. Although most work on bacterivory has been done in pelagic systems, there are a few studies of bacterivory in aquatic sediments. Loss rates due to viral lysis can rival protistan bacterivory, but more work is needed to determine the extent to which lysis is important as a source of bacterial mortality across a range of aquatic systems.

The rationale and results of methods used to quantify rates of protistan bacterivory (Table 20.1) depend on the idea that all of the cells in a bacterial assemblage are equally liable to be grazed. This is certainly not so. It has been shown that bacterivorous flagellates preferentially clear larger-sized and dividing bacteria, bacteria with highly active electron transport systems, and motile bacteria. Flagellate grazing rates can be two- to three-fold higher on these classes of bacteria compared to rates of grazing on smaller, less active, or nonmotile cells. Very large and filamentous bacteria can also show grazing resistance; this is particularly characteristic of freshwater systems. Experimental studies have demonstrated that bacterivorous flagellates can discriminate among different bacterial strains or selectively graze bacteria based on phylogenetic affiliation. New approaches additionally suggest that there is a large spectrum of physiological states of bacterial cells in natural waters from highly metabolically active, to alive but inactive, to dead. Except for a study showing selective grazing on bacteria with highly active electron transport systems, the extent to which protistan grazers discriminate among physiological categories of bacteria is unknown.

Table 20.1. Summary of methods used to study protistan grazing on heterotrophic and autotrophic bacteria in natural microbial communities in marine and freshwater systems.

Method[a]	Advantages	Disadvantages
Change in prey abundance with and without predation Selective filtration Selective inhibition Progressive dilution	Not technically difficult, can estimate both growth and grazing rates	Manipulation artifacts, disruption of predator-prey feedbacks, time-consuming microscopy unless flow cytometry is used
Ingestion of labelled prey Fluorescent beads FLB or heat-inactivated, or live-stained bacteria Radiolabelled bacteria	Short-term incubations, can identify which protists are bacterivores	Preparation of labelled prey, manipulation artifacts, labelled prey not 'natural', incomplete separation of radiolabelled bacteria from protist grazers
Disappearance of labelled prey FLB Radiolabelled bacteria or *E. coli* minicells	Simplified microscopy for FLB, no microscopy for radiolabelled prey	Preparation of labelled prey, manipulation artifacts, labelled prey not 'natural'
Acid lysozyme activity Cleavage of peptidoglycan analog, MUF-chitotriose, by enzymes at pH 4.5	*In vitro* assay, no incubation of live organism, high-resolution sampling	Uncertainty about specificity of enzyme activity, must be calibrated with other method

[a] Papers with method protocols or applications of methods in laboratory or field studies are cited in parentheses.

In light of these considerations, methods used to quantify protistan bacterivory (Table 20.1) should be applied with caution. Measurement of this parameter is, however, useful. Estimates of bacterivory provide an independent lower estimate of bacterial cell production and an upper bound on transfer of bacterial carbon to larger organisms in the food web via bacterivorous protists. Bacterivory techniques segregate into two general approaches: (i) manipulation to reduce grazing and then monitoring change in bacterial abundance during long-term (> 1 day) incubations; and (ii) addition of surrogate prey particles, followed by determining uptake of the particles by protists during short-term (<1 hour) or long-term (>1 day) incubations. Determining the *in situ* activity of protistan digestive enzymes that specifically degrade bacterial cell walls has also been proposed.

Manipulation Methods

The basic rationale for these methods is that the grazing impact of bacterivores is eliminated (or substantially minimised) and then changes in the abundance of bacterial prey are compared in treatments with and without grazing. The earliest of such methods was physical separation of bacterial prey from their predators via size-selective screening. Bacterial abundance was monitored over 24 hours in unfiltered samples (with grazers) and in samples screened through either 1 µm- or 3 µm-pore-size filters. Wright and Coffin reported that the concentration of bacteria increased linearly in the 1 µm-pore-size-filtered sample and showed a slower increase, or a decline, in abundance in 3 µm-pore-size-filtered water and in unfiltered samples. Large numbers of heterotrophic flagellates passed through the 3 µm-pore-size filter, and some flagellates were also able to pass through the 1 µm-pore-size filter.

Concerns about physical manipulation of samples via screening and incomplete separation of predator and prey led to attempts to quantify *in situ* bacterivory via selective metabolic inhibition of either the prokaryotic prey or the eukaryotic predator. As in the size fractionation method, changes in bacterial cell abundance in treatments with and without addition of inhibitory chemicals are monitored over time periods of 12 to >24 hours. The main assumption inherent in this method is that the chemicals added specifically inhibit only the group of interest and do not affect the growth rate or feeding rate of the components of the planktonic assemblage. Fuhrman and McManus used ampicillin at 5 mg litre^{-1} to inhibit bacterial growth, assuming that eukaryotic predators were not affected. Newell used a combination of cycloheximide and thiram to inhibit bacterivorous protists and assumed that bacteria were unaffected. However, such assumptions were not entirely justified. Working with microbial communities in freshwater sediments, Tremaine and Mills reported apparent incomplete inhibition of protists by cycloheximide as well as decreased bacterial growth rate in the presence of the antibiotic. Taylor and Pace examined effects of several eukaryotic inhibitors on both phototrophic and heterotrophic protists and concluded that use of metabolic inhibitors to evaluate bacterivory did not produce reliable results. Sherr evaluated a number of prokaryotic and eukaryotic chemical inhibitors and found that most were not sufficiently selective. Only combinations of cycloheximide (200 mg litre^{-1}) plus colchicine (100 mg litre^{-1}) for protists and of vancomycin (200 mg litre^{-1}) plus penicillin (1 mg litre^{-1}) for bacteria resulted in acceptable selective inhibition of the two target assemblages. Liu reported that kanamycin at 1 mg litre^{-1} specifically inhibited the growth of the marine prokaryotic phototrophs *Prochlorococcus* and *Synecococcus* spp. but not growth of heterotrophic bacteria or of phagotrophic protists.

A third manipulation approach to bacterivory was adapted from the Landry and Hassett dilution method, which was initially developed to quantify microzooplankton grazing on phytoplankton. The dilution method involves establishing a series of mixtures of whole water and of 0.2 µm-pore-size-filtered water in order to progressively decrease grazing impact on prey populations. In theory, prey cells in diluted samples will show decreased grazing mortality in proportion to the dilution but will have the same intrinsic growth rate. Monitoring growth of prey over 24 hours in the dilution series allows calculation of both grazing rate and growth rate. The dilution method has been used to estimate bacterivory in Hawaiian coastal waters and protistan grazing mortality on coccoid cyanobacteria at several sites in the northwest Atlantic Ocean. Tremaine and Mills did a limited test of the assumptions of the dilution method applied to analysis of bacterivory and concluded that the approach was valid for the eutrophic freshwater habitat examined.

Several problems are associated with the manipulation-incubation methods: (i) separation of predator and prey varies in degree depending on the microbial community involved but is rarely complete; (ii) manipulation can cause experimental artifacts in the behaviour of the predator and prey assemblages, e.g. screening can cause mechanical damage to protists; (iii) long-term incubations of small water samples may cause shifts in the original species composition of predator and prey assemblages, i.e. growth of some species may be inhibited, and growth of others favoured, by the experimental conditions; and (iv) separation of predator and prey or artificial decrease of grazing rates can disrupt positive feedback interactions between microbial assemblages. For example, excretions of bacterivorous protists provide their bacterial prey with dissolved organic substrates and inorganic nitrogen and phosphorus nutrients. Sherr found lower bacterial growth rates in experimental manipulations in which protist grazing was reduced via addition of eukaryotic cell inhibitors. Addition of ammonium to such treatments resulted in enhanced bacterial growth, suggesting that bacteria were limited by the availability of inorganic nutrients excreted by protists. There is a substantial literature on requirements of heterotrophic bacteria in both marine water and freshwater systems for inorganic nitrogen and phosphorus.

Prey Addition Methods

Methods based on quantifying protistan ingestion of added, surrogate prey particles have advantages compared to the approaches described above: (i) they avoid the experimental artifacts inherent in manipulation methods; (ii) they can yield results with much shorter incubation times (<1 hour) compared to manipulation experiments (12 to >24 hours); (iii) they allow detection of the particular protists that are ingesting the prey of interest and can be used to quantify protist-specific feeding rates; and (iv) they can be successfully used in sediments. Two approaches to prey addition have been employed: short-term incubations in which appearance of labelled prey in protists is monitored and long-term incubations in which the rate of disappearance of labelled prey is monitored. To minimise alteration of protist clearance rates, labelled prey should be added at concentrations of <50 per cent of natural bacterial abundance. McManus and Okubo suggested that tracer concentrations of surrogate particles should not exceed 10 per cent of prey abundance. However, if surrogate particles are added at low concentrations, estimation of protistan ingestion by direct observation of prey in food vacuoles becomes increasingly difficult and the disappearance of added particles must be monitored over longer time periods to obtain interpretable results.

Surrogate prey particles used to determine bacterivory have included: (i) fluorescently labelled prey, either bacterium-size fluorescent plastic beads or fluorescently labelled bacteria (FLB); and (ii) radioisotopically labelled prey, either laboratory cultured bacteria, *Escherichia coli* minicells, or natural bacterioplankton assemblages. In the case of radioisotopically labelled prey, extensive microscopy is not required as it is in the other approaches. Bacterivory assays using fluorescently labelled tracer particles have revealed that ciliates as well as flagellates can be significant bacterivores in some aquatic systems and that many phototrophic flagellates are also bacterivorous. Specific protocols for estimating bacterivory via addition of fluorescently labelled particles or of radiolabelled minicells or bacteria are available.

A major assumption inherent in grazing methods based on addition of labelled prey is that predators do not show significant selection for or against the added prey compared with their rates of feeding on natural, unlabelled prey. This assumption has been tested and found to be invalid in certain respects. Some species of bacterivorous protists have lower feeding rates on plastic microspheres than on FLB. A criticism of the FLB grazing method of Sherr is that the labelled bacteria are heat inactivated and thus may not be an adequate surrogate for live bacteria. Gonzalez concluded that if more than a few per cent of natural bacterioplankton were motile, use of heat-inactivated FLB would result in underestimation of rates of bacterivory.

Use of live-stained FLB has been proposed to achieve more accurate estimates of bacterial mortality due to protist grazing. Landry developed a method to live-stain bacteria with fluorescein isothiocyanate by treatment with dithioerythritol, a compound that breaks sulphur bonds in the bacterial cell wall, enhancing binding of the fluorochrome. Epstein and Rossel suggested cyanoditolyl tetrazolium chloride for preparation of live FLB. CTC is a redox compound that can be used in place of oxygen in electron transport the systems. In its oxidised state, CTC is nonfluorescent and easily diffuses into bacterial cells; after reduction in the electron transport system, CTC precipitates in the bacterial cytoplasm as a red-fluorescent fluorochrome. Epstein successfully used CTC-labelled cultured bacteria to estimate rates of bacterivory by protists in marine sediment communities.

Notwithstanding the problems inherent in using surrogate prey, the FLB uptake-disappearance approach has been widely used to evaluate grazing within natural microbial assemblages on both heterotrophic bacteria and cyanobacteria. FLB have also been used to evaluate rates of digestion of bacterial prey by protists. Use of flow cytometry in place of microscopy to follow decrease in FLB concentration due to grazing increases the precision of the method. Gonzalez has suggested a method for estimating *in situ* bacterivory based on determination of the fraction of active bacterivorous protists (fraction of protists with ingested FLB).

Radiolabelled bacteria can be readily prepared from living cultured bacteria or *in situ* bacterioplankton. Hollibaugh pioneered the use of bacteria prelabelled with tritiated thymidine in grazing experiments. Lessard and Swift modified this technique to determine relative grazing by specific components of the >20 μm-sized protistan assemblage on [³H]thymidine-labelled heterotrophic bacteria and on ¹⁴C-labelled phototrophic cells. Servais combined size-selective filtration with long-term disappearance of [³H]thymidine-labelled bacteria in order to discriminate between bacterial mortality due to protist grazing and that due to viral lysis or other factors. The potential problem of growth of labelled natural bacteria in disappearance experiments was addressed by the protocol of Wilkner, in which nongrowing *E. coli* minicells the size of natural bacteria were labelled with [³⁵S]methionine for use in grazing assays. Nygaard and Hessen subsequently developed a method for ¹⁴C-labelling of bacterial cells via growth on ¹⁴C-labelled protein hydrolysate and reported higher ingestion rates of radiolabelled cells compared with the ingestion rates of heat-inactivated FLB.

A problem with the use of [³H]thymidine-labelled bacteria as prey is that bacterivorous protists do not efficiently assimilate the labelled DNA, so that most of the radiolabel ingested prey is excreted into the medium. Another concern with respect to use of radiolabelled bacterial cells in grazing experiments is effective separation of label ingested by bacterivores from label present in unconsumed bacterial cells. Zubkov and Sleigh proposed a method that overcomes these problems by using dual labelled ([³H]thymidine and [¹⁴C]leucine) cultured bacteria. Since virtually all of the ³H in the DNA of grazed bacteria is released to the medium, the rate of decrease of ³H in the particulate fraction is a measure of the rate of disappearance of labelled bacteria due to grazing. However, a fraction of the ¹⁴C-labelled protein in ingested bacteria is retained during biosynthesis of macromolecules by protists, based on their carbon growth efficiencies. Thus, comparing the proportional losses of ³H and of ¹⁴C from the particulate fraction over time can be used to estimate the growth efficiency of the bacterivores.

Digestive Enzyme Activity Approach

Gonzalez suggested that *in situ* bacterivory might be assessed by determining the activity of hydrolytic enzymes present in the food vacuoles of phagotrophic protists that would be specific for bacterial prey. The basis of the method is that protistan digestive enzymes can be distinguished from extracellular hydrolytic enzymes of bacteria or from extracellular or intracellular hydrolytic enzymes of other eukaryotic organisms by pH optima of enzymatic activity. In theory, the digestive enzymes of protists should have peak activity at a pH of 4 to 5, the pH of their early-stage food vacuoles, while other types of hydrolytic enzymes would have optimal activity at higher pH. The assay of Gonzalez was based on measurement of activity of lysozyme, an enzyme that hydrolyses the cell walls of eubacteria, in sonicated seawater samples at acid pH. Lysozyme activity was assessed by rate of enzymatic cleavage of the fluorochrome methylumbelliferone (MUF) from the substrate MUF-triacetylchitotriose, which serves as an analog of peptidoglycan, a major structural component of bacterial cell walls. MUF exhibits low fluorescence when bound to substrates and high fluorescence after cleavage.

In cultures of bacterivorous flagellates, the rate of production of MUF fluorescence showed two peaks, one at pH 4.5, presumably due to flagellate digestive enzymes, and one at pH 7 to 8 presumably due to bacterial ectoenzymes, with little overlap. Cultured phytoplankton exhibited only minor lysozyme activity at pH 4.5. Acid lysozyme activity was calibrated against rates of bacterial ingestion via the FLB uptake method. Results of the two methods were significantly correlated ($r^2 = 0.98$) over a wide range of rates of bacterivory (10^3 to 10^8 bacteria ingested ml^{-1} hr^{-1}). Positive correlations between the two indices of bacterivory have been found *in situ*: both for Oregon shelf waters and offshore gyre waters as well as for surface waters in the Arctic Ocean. An advantage of the digestive enzyme activity approach is that it allows higher-resolution sampling in time and, space compared with incubation approaches. The enzyme assay must be calibrated with an other method of estimating bacterivory, and, in our experience, there does not appear to be a universal calibration factor.

Vrba proposed an alternative substrate, MUF-β-N-acetylglucosaminide, as an indicator of digestive enzyme activity associated with *in situ* protistan bacterivory. However, a number of species of phytoplankton show high levels of glucosaminidase activity at both pH 4.5 and pH 6 to 8. Thus, outside of controlled laboratory experiments, enzyme activity based on cleavage of MUF-β-N-acetyl-glucosaminide could be attributable to multiple sources and not strictly to bacterivorous protists.

PROTISTAN HERBIVORY

Phagotrophic protists are increasingly recognised as significant grazers of phytoplankton in aquatic ecosystems. In marine systems, the microzooplankton, mainly <200-μm-sized protists, often grazes 50 to 100 per cent of daily phytoplankton production. Protistan grazers expel both particulate and dissolved digestion products into the environment. Development of methods to assess protistan herbivory is critical in determining the structure and functioning of pelagic food webs. However, less work has been done in this regard (Table 20.2) compared to the number of approaches developed to investigate protistan bacterivory.

Table 20.2. Summary of methods used to study protistan herbivory in natural microbial communities in marine systems.

Method[a]	Advantages	Disadvantages
Progressive dilution Change in pigments or in cell abundance measured over 24 hours in a series of dilutions to decrease grazing impact	Commonly used, can obtain both prey growth and grazing rates, technically simple	Doesn't always work, labour intensive, manipulation artifacts, disruption of predator-prey feedbacks and nutrient regeneration
Selective filtration Size fractionation separates phytoplankton from protist grazers	Technically simple, can obtain growth and grazing rates	Manipulation artifacts, incomplete separation of prey and predator, disrupts feedbacks
Ingestion of labelled prey Fluorescently labelled, or radiolabelled cells	Can get cell-specific uptake rates for particular types of predator and prey	Preparation of labelled prey, labelled prey may not he 'natural', indirect estimates of community grazing rates

[a] Papers with method protocols or applications of methods in laboratory or field studies are cited in parentheses.

The most widely used method for quantification of protist grazing on phytoplankton is the dilution method of Landry and Hassett discussed above. The advantages of this approach are that the method is straightforward and that both grazing mortality (g) and intrinsic phytoplankton growth rate (μ) are obtained. Change in abundance of phytoplankton can be determined either by chlorophyll-*a* concentration, by cell counts via microscopy, or by flow cytometry. In addition, by analysing photopigments via high-performance liquid chromatography at the beginning and end of the incubations, grazing on specific taxonomic components of the phytoplankton assemblage can be assessed. Disadvantages are that the dilution method requires manipulation which can disrupt predator-prey feedback (e.g. regenerative) processes and long-term incubation (24 hours) which may create experimental artifacts. When the method is used in low-nutrient environments, a separate dilution series with addition of macronutrients is recommended as part of the experimental protocol. Carrying out a dilution assay is sufficiently time consuming that at most only one experiment can be performed in a day.

The assumption that the cell-specific clearance rates of herbivorous protists are not affected by dilution can be violated in eutrophic systems where protists experience saturated feeding at *in situ* prey abundance. To directly address the problem of variation in cell-specific grazing rate with dilution, Landry modified the dilution assay protocol by adding FLB or fluorescently labelled algae (FLA) to the dilution series. Proportional decrease of added labelled prey cells over the duration of the dilution experiment is then used to determine relative grazer activity, and the net phytoplankton growth is regressed against relative grazing rather than against the dilution factor as in the original protocol. This modification of the method allows direct measurement of effect of dilution on grazing, thus making application of the assay less ambiguous.

Verity used a size fractionation-incubation method to quantify grazing rates of microzooplankton 10 to 202 μm on <10 μm-sized phytoplankton in coastal marine waters. This approach is limited to situations in which there is a clear size difference between phytoplankton and the dominant protistan herbivores. Subsequently, in the open North Atlantic, Verity compared estimates of microzooplankton herbivory made via the dilution method and via the size fractionation approach. In this study, the dilution method yielded consistently higher estimates of grazing mortality, suggesting an incomplete separation of predator and prey in the size fractionation experiments.

Other approaches to determination of protist grazing within natural planktonic communities appear to be useful only for specific predator-prey interactions. Campbell and Carpenter showed that use of prokaryotic inhibitors could be used to measure protist grazing on coccoid cyanobacteria. The dual-label radiotracer method of Lessard and Swift can be used to evaluate cell-specific grazing rates of ciliates and heterotrophic dinoflagellates on phytoplankton. Rublee and Gallegos adapted the heat-inactivated 5-([4,6-dichlorotriazin-2-yl]amino)fluorescein (DTAF) staining method of Sherr to prepare FLA for *in situ* grazing experiments in coastal marine waters. Sherr further tested this labelled prey addition method and concluded that for short-term uptake assays it was necessary to add fluorescently labelled phytoplankton at abundances higher than generally present in mesotrophic systems. Labelled prey could be added at tracer concentrations only in applications of the method in which long-term (>12 to 24 hours) disappearance of the prey is analysed.

The method of Rublee and Gallegos produces fluorescently labelled algal cells which have been shrunk and hardened by heating, and which are no longer motile in the case of phytoflagellates. One solution has been to use live algal cells which have distinctive fluorescence signatures, for example cryptophytes. Putt developed a live-staining method for algal cells using the fluorochrome hydroethydine. She reported higher ingestion rates of live-stained phytoplankton compared to heat inactivated and stained phytoplankton. However, in practice hydroethydine-stained cells lost fluorescence upon fixation and could only be visualised by the pink to brown colour of stained algae in transmitted light microscopy. Hydroethydine staining also decreased the motility of phytoflagellates. A more promising compound for producing live-stained fluorescently labelled eukaryotic cells, including both phototrophic and heterotrophic protists, is the dye 5-chloromethylfluoroscein diacetate. CMFDA is membrane permeable. Once inside the cell, the compound is cleaved by esterases and fluoresces bright green when stimulated with blue light and is precipitated by thiol groups so that it is membrane impermeable. Fluorescence is not diminished by preservation. Li demonstrated ingestion of CMFDA-stained ciliates and phyto-flagellates by autotrophic dinoflagellates in the Chesapeake Bay. Quantification of protistan grazing on autotrophic picoplankton or on eukaryotic cells via evaluation of digestion rate of ingested prey cells appears to be useful in laboratory experiments of phagotrophic protists feeding on selected prey, but it has not been applied in field studies.

OTHER TROPHIC PATHWAYS IN MICROBIAL FOOD WEBS

Phagotrophic protists ingest other types of food in addition to bacteria and phytoplankton. Alternate food resources include copepod eggs and nauplii, other heterotrophic protists, sperm cells, viral particles, and high-molecular-weight compounds. Fluorescently labelled viral particles and colloidal poly-saccharides and proteins have been used to demonstrate ingestion of submicrometer nonliving material. However, there is still virtually no information on the quantitative importance of alternate foods for phagotrophic protists in marine and freshwater environments.

Of particular interest is assessing the significance of the protist-protist trophic link in marine and freshwater food webs. From the early conceptualisations of the role of heterotrophic microbes in marine ecosystems to more recent ideas about the structure of pelagic microbial food webs, a major component of the microbial loop concept has been that bacterivorous flagellates are consumed in turn by larger protists. There is supporting evidence for this trophic link. Both Ohman and Snyder and Verity reported that cultured ciliates were able to grow on a diet of heterotrophic flagellates. Working in a shallow bay in the Adriatic Sea, Solic and Krstulovic monitored the growth of heterotrophic flagellates during 24 hour incubations in seawater samples prefiltered through either 8 µm-pore-size polycarbonate filters or a 100 µm mesh screen. These authors reported that on an annual basis, phagotrophic protists in the 8 to 100 µm-size fraction consumed nearly 100 per cent of the biomass production of <8 µm-sized flagellates in the bay. We have prepared CMFDA-stained bacterivorous flagellates by incubating cells for 2 to 3 hours at 0.5 mM final concentration of the reagent and have observed ingestion of the labelled flagellates by a heterotrophic dinoflagellate, *Gymnodinium* sp., by a tintinnid, *Tintinnopsis* sp., as well as by a variety of protists in Oregon coastal waters. Feeding by predatory protists on phagotrophic ciliates has been studied by prelabelling prey ciliates with fluorescent microspheres that accumulated in food vacuoles. This area of research remains ripe for further investigation.

Chapter 21

Biogeochemical Cycling

INTRODUCTION

In ecology and earth science, a biogeochemical cycle is a circuit or pathway by which a chemical element or molecule moves through both biotic (bio-) and abiotic (geo-) compartments of an ecosystem. In effect, the element is recycled, although in some such cycles there may be places (called reservoirs) where the element is accumulated or held for a long period of time. The elements, chemical compounds, and other forms of matter are passed from one organism to another and from one part of the biosphere to another through the biogeochemical cycles.

All chemical elements occurring in organisms are part of biogeochemical cycles. In addition to being a part of living organisms, these chemical elements also cycle through abiotic factors of ecosystems such as water (hydrosphere), land (lithosphere), and the air (atmosphere); the living factors of the planet can be referred to collectively as the biosphere. All the chemicals, nutrients, or elements—such as carbon, nitrogen, oxygen, phosphorus—used in ecosystems by living organisms operate on a closed system, which refers to the fact that these chemicals are recycled instead of being lost and replenished constantly such as in an open system.

BIOGEOCHEMICAL CYCLES

What happens to the vast array of organic matter that is produced on the earth during photosynthetic processes? This material does not keep accumulating rather, it is consumed and degraded, and a delicate global balance of carbon is maintained; carbon dioxide is removed from the atmosphere during photosynthesis and released during respiration. This balance is a result of the biologically driven, characteristic cycling of carbon between biotic forms such as sugar or other cellular building blocks and abiotic forms such as carbon dioxide. Cycling between biotic and abiotic forms is not limited to carbon. All of the major elements found in biological organisms (Table 21.1), as well as some of the minor and trace elements, are similarly cycled in predictable and definable ways.

Taken together, the various element cycles are called the biogeochemical cycles. Understanding these cycles allows scientists to understand and predict the development of microbial communities and activities in the environment. There are many activities that can be harnessed in a beneficial way, such as for remediation of organic and metal pollutants or for recovery of precious metals such as copper or uranium from low-grade ores. There are also detrimental aspects of the cycles that can cause global environmental problems such as the formation of acid rain and acid mine drainage, metal corrosion processes, and formation of nitrous oxide, which can deplete the earth's ozone layer. As these examples illustrate, the microbial activities that drive biogeochemical cycles are highly relevant to the field of environmental microbiology. Thus, the knowledge of these cycles is increasingly critical as the human population continues to grow and the impact of human activity on the earth's environment becomes more significant. The biogeochemical cycles pertaining to carbon, nitrogen, sulphur and phosphorus are discussed here under.

Table 21.1. Chemical composition of an *E. coli* cell.

Elemental breakdown	% Dry mass of an E. coli cell
Major elements	
Carbon	50
Oxygen	20
Hydrogen	8
Nitrogen	14
Sulphur	1
Phosphorus	3
Minor elements	
Potassium	2
Calcium	0.05
Magnesium	0.05
Chlorine	0.05
Iron	0.2
Trace elements	
Manganese	All trace elements combined
Molybdenum	comprise 0.3 per cent of
Cobalt	dry weight of cell
Copper	
Zinc	

Gaia Hypothesis

In the early 1970s, James Lovelock theorised that earth behaves like a superorganism, and this concept developed into what is now known as the Gaia hypothesis. To quote Lovelock, 'living organisms and their material environment are tightly coupled. The coupled system is a superorganism, and as it evolves there emerges a new property, the ability to self-regulate climate and chemistry'. The basic tenet of this hypothesis is that the earth's physico-chemical properties are self-regulated so that they are maintained in a favourable range for life. As evidence for this, consider that the sun has heated up by 30 per cent during the past 4–5 billion years. Given the earth's original carbon dioxide-rich atmosphere, the average surface temperature of a lifeless earth today would be approximately 290°C (Table 21.2). In fact, when one compares earth's present-day atmosphere with the atmospheres found on our nearest neighbours Venus and Mars, one can see that something has drastically affected the development of earth's atmosphere. According to the Gaia hypothesis, this is the development and continued presence of life. Microbial activity, and later the appearance of plants, have changed the original heat-trapping carbon dioxide-rich atmosphere to the present oxidising, carbon dioxide-poor atmosphere. This has allowed earth to maintain an average surface temperature of 13°C, which is favourable to the life that exists on earth.

Table 21.2. Atmosphere and temperatures found on Venus, Mars and Earth.

Gas	Planet			
	Venus	*Mars*	*Earth without life*	*Earth with life*
Carbon dioxide	96.5%	95%	98%	0.03%
Nitrogen	3.5%	2.7%	1.9%	9%
Oxygen	Trace	0.13%	0.0	21%
Argon	70 ppm	1.6%	0.1%	1%
Methane	0.0	0.0	0.0	1.7 ppm
Surface temperature (°C)	459	−53	290 ± 50	13

How do biogeochemical activities relate to the Gaia hypothesis? These biological activities have driven the response to the slow warming of the sun resulting in the major atmospheric changes that have occurred over the last 4–5 billion years. When Earth was formed 4–5 billion years ago, a reducing (anaerobic) atmosphere existed. The initial reactions that mediated the formation of organic carbon were abiotic, driven by large influxes of ultraviolet (UV) light. The resulting reservoir of organic matter was utilised by early anaerobic heterotrophic organisms. This was followed by the development of the ability of microbes to fix carbon dioxide photosynthetically. Evidence from stromatolites suggests that the ability to photosynthesise was developed at least 3.5 billion years ago. Stromatolites are fossilised laminated structures that have been found in Africa and Australia. These structures were formed primarily by cyanobacteria that grew in mats and entrapped or precipitated inorganic material as they grew. The evolution of photosynthetic organisms tapped into an unlimited source of energy, the sun, and provided a mechanism for carbon recycling, i.e. the first carbon cycle (Fig. 21.1). This first carbon cycle was maintained for approximately 1.5 billion years. Geologic evidence then suggests that approximately 2 billion years ago, photosynthetic micro-organisms developed the ability to produce oxygen. This allowed oxygen to accumulate in the atmosphere, resulting, in time, in a change from reducing to oxidising conditions. Further, oxygen accumulation in the atmosphere created an ozone layer, which reduced the influx of harmful UV radiation, allowing the development of higher forms of life to begin.

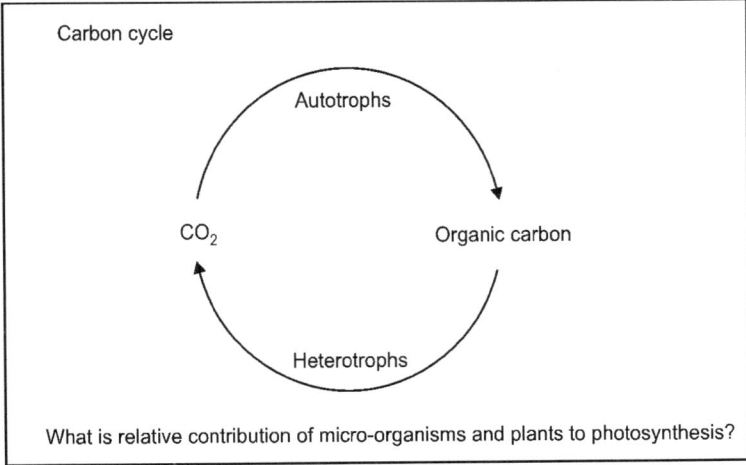

Carbon cycle

Autotrophs

CO_2 ⟶ Organic carbon

Heterotrophs

What is relative contribution of micro-organisms and plants to photosynthesis?

Fig. 21.1. The carbon cycle is dependent on autotrophic organisms that fix carbon dioxide into organic carbon and heterotrophic organisms that respire organic carbon to carbon dioxide.

At the same time that the carbon cycle evolved, the nitrogen cycle emerged because nitrogen was a limiting element for microbial growth. Although molecular nitrogen was abundant in the atmosphere, microbial cells could not directly utilise nitrogen as N_2 gas. Cells required organic nitrogen compounds or reduced inorganic forms of nitrogen for growth. Therefore, under the reducing conditions found on early earth, some organisms developed a mechanism for fixing nitrogen using the enzyme nitrogenase. Nitrogen fixation remains an important microbiological process, and to this day, the nitrogenase enzyme is totally inhibited in the presence of oxygen.

When considered over this geologic time scale of several billion years, it is apparent that biogeochemical activities have been unidirectional. This means that the predominant microbial activities on earth have evolved over this long period of time to produce changes and to respond to changes that have occurred in the atmosphere, i.e. the appearance of oxygen and the decrease in carbon dioxide content. Presumably these changes will continue to occur, but they occur so slowly that we do not have the capacity to observe them. However, one can also consider biogeochemical activities on a more contemporary time scale, that of tens to hundreds of years. On this much shorter time scale, biogeochemical activities are regular and cyclic in nature, and it is these activities that are addressed in this chapter. On the one hand, the presumption that earth is a superorganism and can respond to drastic environmental changes is heartening when one considers that human activity is effecting unexpected changes in the atmosphere, such as ozone depletion and build-up of carbon dioxide. However, it is important to point out that the response of a superorganism is necessarily slow (thousands to millions of years), and as residents of earth we must be sure not to overtax earth's ability to respond to change by artificially changing the environment in a much shorter time frame.

CARBON CYCLE

Carbon Reservoirs

A reservoir is a sink or source of an element such as carbon. There are various global reservoirs of carbon, some of which are immense in size and some of which are relatively small (Table 21.3).

Table 21.3. Global carbon reservoirs.

Carbon reservoir	Metric tonnes carbon	Actively cycled
Atmosphere		
CO$_2$	6.7×10^{11}	Yes
Ocean		
Biomass	4.0×10^{9}	No
Carbonates	3.8×10^{13}	No
Dissolved and particulate organics	2.1×10^{12}	Yes
Land		
Biota	5.0×10^{11}	Yes
Humus	1.2×10^{12}	Yes
Fossil fuel	1.0×10^{13}	Yes
Earth's crust[a]	1.2×10^{17}	No

[a] This reservoir includes the entire lithosphere found in either terrestrial or ocean environments.

The largest carbon reservoir is carbonate rock found in the earth's sediments. This reservoir is four orders of magnitude larger than the carbonate reservoir found in the ocean and six orders of magnitude larger than the carbon reservoir found as carbon dioxide in the atmosphere. If one considers these three reservoirs, it is obvious that the carbon most available for photosynthesis is in the smallest of the reservoirs, the atmosphere. Therefore, it is the smallest reservoir that is most actively cycled. It is small, actively cycled reservoirs such as atmospheric carbon dioxide that are subject to perturbation from human activity. In fact, since global industrialisation began in the late 1800s, humans have affected several of the smaller carbon reservoirs. Utilisation of fossil fuels (an example of a small, inactive carbon reservoir) and deforestation (an example of a small, active carbon reservoir) are two activities that have reduced the amount of fixed organic carbon in these reservoirs and added to the atmospheric carbon dioxide reservoir (Table 21.4).

Table 21.4. Net carbon flux between selected carbon reservoirs.

Carbon source	Flux (metric tonnes carbon/year)
Release by fossil fuel combustion	7×10^9
Land clearing	3×10^9
Forest harvest and decay	6×10^9
Forest regrowth	-4×10^9
Net uptake by oceans (diffusion)	-3×10^9
Annual flux	9×10^9

The increase in atmospheric carbon dioxide has not been as great as expected. This is because the reservoir of carbonate found in the ocean acts as a buffer between the atmospheric and sediment carbon reservoirs through the equilibrium equation shown below.

$$H_2CO_3 \rightleftarrows HCO_3^- \rightleftarrows CO_2$$

Thus, some of the excess carbon dioxide that has been released has been absorbed by the oceans. However, there has still been a net efflux of carbon dioxide into the atmosphere of approximately 7×10^9 metric tonnes/year. The problem with this imbalance is that because atmospheric carbon dioxide is a small carbon reservoir, the result of a continued net efflux over the past 100 years or so has been a 28 per cent increase in atmospheric carbon dioxide from 0.026 per cent to 0.033 per cent. A consequence of the increase in atmospheric carbon dioxide is that it may contribute to global warming through the greenhouse effect. The greenhouse effect is caused by gases in the atmosphere that trap heat from the sun and cause the earth to warm up. This effect is not solely due to carbon dioxide; other gases such as methane, chlorofluorocarbons (CFCs), and nitrous oxide add to the problem.

Carbon Fixation and Energy Flow

The ability to photosynthesise allows sunlight energy to be trapped and stored. In this process carbon dioxide is fixed into organic matter (Fig. 21.1). Photosynthetic organisms, also called primary producers, include plants and micro-organisms such as algae, cyanobacteria, some bacteria, and some protozoa. As shown in Fig. 21.2, the efficiency of sunlight trapping is very low; less than 0.1 per cent of the sunlight energy that hits the earth is actually utilised. As the fixed sunlight energy moves up each level of the food chain, up to 90 per cent or more of the trapped energy is lost through respiration. Despite this seemingly inefficient trapping, photoautotrophic primary producers support most of the considerable ecosystems found on the earth.

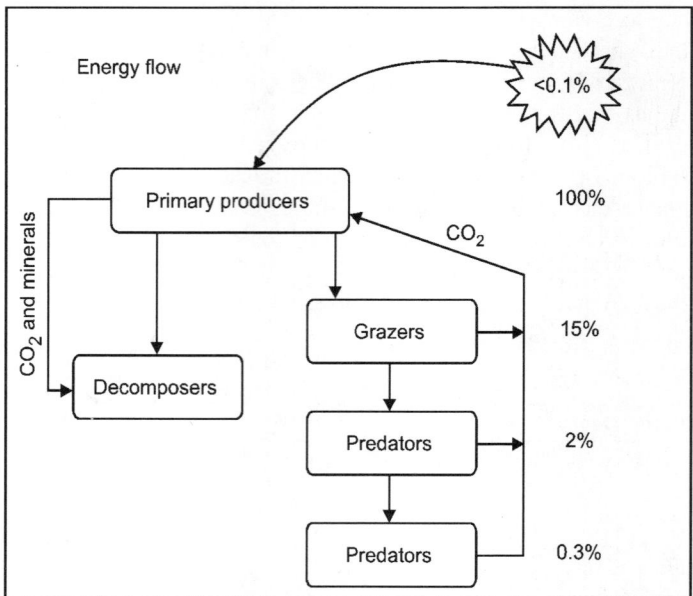

Fig. 21.2. Diagram of the efficiency of sunlight energy flow from primary producers to consumers.

Productivity varies widely among different ecosystems depending on the climate, the type of primary producer, and whether the system is a managed one (Table 21.5). For example, one of the most productive natural areas is the coral reefs.

Table 21.5. Net primary productivity of some natural and managed ecosystems.

Description of ecosystem	Net primary productivity (g dry organic matter/m^2/year)
Tundra	400
Desert	200
Temperate grassland	Up to 1500
Temperate or deciduous forest	1200–1600
Tropical rain forest	Up to 2800
Cattail swamp	2500
Freshwater pond	950–1500
Open ocean	100
Coastal seawater	200
Upwelling area	600
Coral reef	4900
Corn field	1000–6000
Rice paddy	340–1200
Sugarcane field	Up to 9400

Managed agricultural systems such as corn and sugarcane systems are also very productive, but it should be remembered that a significant amount of energy is put into these systems in terms of fertiliser addition and care. The open ocean has much lower productivity, but covers a majority of the earth's surface and so is a major contributor to primary production. In fact, aquatic and terrestrial environments contribute almost equally to global primary production. Plants predominate in terrestrial environments, but with the exception of immediate coastal zones, micro-organisms are responsible for most primary production in aquatic environments. It follows that micro-organisms are responsible for approximately one half of all primary production on the earth.

Carbon Respiration

Carbon dioxide that is fixed into organic compounds as a result of photoautotrophic activity is available for consumption or respiration by animals and heterotrophic micro-organisms. This is the second half of the carbon cycle shown in Fig. 21.1. The end-products of respiration are carbon dioxide and new cell mass. An interesting question to consider is the following: if respiration were to stop, how long would it take for photosynthesis to use up all of the carbon dioxide reservoir in the atmosphere? Based on estimates of global photosynthesis, it has been estimated that it would take 30 to 300 years. This illustrates the importance of both legs of the carbon cycle in maintaining a carbon balance.

The following sections discuss the most common organic compounds found in the environment and the microbial catabolic activities that have evolved in response. These include organic polymers, humus and C1 compounds such as methane (CH_4). It is important to understand the fate of these naturally occurring organic compounds. This is because degradative activities that evolved for these compounds form the basis for degradation pathways that may be applicable to organic contaminants that are spilled in the environment. But before looking more closely at the individual carbon compounds, it should be pointed out that the carbon cycle is actually not quite as simple as shown in Fig. 21.1. This simplified figure does not include anaerobic processes, which were predominant on the early earth and remain important in carbon cycling even today. A more complex carbon cycle containing anaerobic activity is shown in Fig. 21.3. Under anaerobic conditions, which predominated for the first several billion years on earth, some cellular components were less degradable than others.

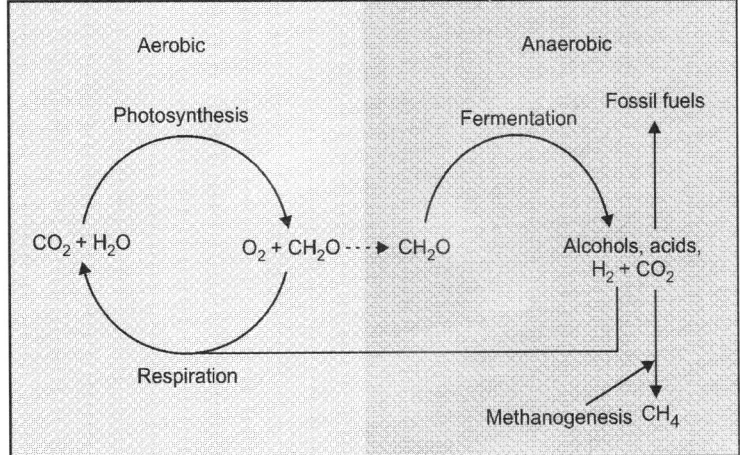

Fig. 21.3. The carbon cycle, showing both aerobic and anaerobic contributions.

This is especially true for highly reduced molecules such as cellular lipids. These components were, therefore, left over and buried with sediments over time and became the present-day fossil fuel reserves. Another carbon compound produced under anaerobic conditions is methane. Methane is produced in soils as an end product of anaerobic respiration. Methane is also produced under the anaerobic conditions found in ruminants such as cows.

MICROBIAL NITROGEN CYCLING

Nitrogen availability is a key factor regulating the biological productivity of many aquatic ecosystems, and bacteria have long been recognised as important agents affecting N pools through various transformations. The assimilation into organic form and subsequent release of inorganic N, as performed by a broad array of prokaryotic and eukaryotic organisms, comprise the inner core of the N cycle in nature. However, it is the uniquely bacterial processes of N_2 fixation, nitrification, and denitrification that define the broader cycle and can affect directly the availability and form of N within particular ecosystems. The relative importance of each of these processes can vary greatly among systems.

Hence, there is sustained interest in determining the importance of these processes in the environment, and a variety of procedures are currently in use. This section provides an update and synopsis of current procedures for determining N_2 fixation, nitrification, and denitrification in aquatic environments.

N_2 Fixation

Biological N_2 fixation is the reduction of N_2 gas to ammonium as catalysed by nitrogenase, an enzyme system found in a physiologically diverse array of eubacteria and *Archaea*. As such, N_2 fixation provides a primary input of usable N to the biosphere and may be a key process in N-limited environments. Nitrogen fixation occurs throughout a variety of freshwater and marine habitats. Several recent reviews on aquatic N_2 fixation are available.

Assessing populations

Growth on N-free media has long been used as *prima facie* evidence of N_2 fixation, and early researchers in the field relied on such procedures in attempting to enrich and enumerate N_2-fixing populations in nature. Because of the physiological diversity of N_2 fixers, a variety of media and conditions appropriate for aerobic and anaerobic photoautotrophs and chemoautotrophs, as well as aerobic and anaerobic heterotrophs, must be employed, depending on the circumstances. However, because of pervading contamination of many ingredients into bacteriological media with low levels of combined N and the possibility of uptake of atmospheric NH_3, growth on N-free media is generally not considered definitive evidence of N_2 fixation.

The nitrogenase enzyme has been highly conserved. Antibodies to components of the protein complex have been developed which show broad reactivity with nitrogenases from diverse sources. Such antibodies, linked to fluorescent reporters, have been used to detect and enumerate diazotrophs in marine seagrass and mat systems.

Nucleotide probes for genes encoding the structural proteins of nitrogenase (*nif*H and *nif*K) have been developed and used to identify diazotrophs in a variety of marine substrates. More recently, the application of denaturing-gel gradient electrophoresis has allowed for sensitive detection of sub-components of the diazotrophic populations as determined by subtle differences in *nif*H sequences. These approaches hold great promise for developing routine procedures for detecting and enumerating diazotrophs in aquatic environments.

Assessing activity

Assessing the rate of N_2 fixation in the environment has been the subject of intensive study and method development over the past several decades. As for the determination of any activity or process in the environment, there are a variety of generic issues and caveats which need to be considered.

The earliest method for detecting active N_2 fixation in natural samples was to monitor increases in organic N, typically using the Kjeldahl method (Table 21.6). However, this is a very insensitive approach to observe small increases above a large background of organic N pre-existing in the sample. Furthermore, it cannot clearly discriminate increases in N due strictly to N_2 fixation from those which may result from assimilation of pre-existing inorganic forms of N.

Table 21.6. Summary of common assays for microbial N cycle reactions[a].

Method	Substrate(s)	Analyte(s)	Detection	Comments
N_2 fixation				
Growth without combined N	N_2	Biomass, PON, protein	–	Insensitive, often false positives
Isotopic				
Direct tracer	$^{13}N_2$	^{13}N-PON, NH_4	Gamma counter	Requires cyclotron; isotope has 10-min half-life
	$^{15}N_2$	^{15}N-PON, NH_4	IRMS or ES	
Isotope dilution	$^{15}NH_4$	$^{15}NH_4$	IRMS	Follow decrease in $^{15}N/^{14}N$ in NH_4 pool
Natural abundance	NA	$\delta^{15}N$-PON	IRMS	
Enzymatic				
C_2H_2 reduction	C_2H_2	C_2H_4	FID-GC	Sensitive, field capable
			Photoacoustic detector	Requires conversion factor
Nitrification				
Isotopic				
Direct tracer	$^{13}NH_4$, NO_2	$^{13}NO_3$	Gamma counter	Requires cyclotron; isotope has 10 minutes half-life
	$^{15}NH_4$, NO_2	$^{15}NO_3$	IRMS or ES	
Isotope dilution	$^{15}NO_3$	$^{15}NO_3$	IRMS	Follow decrease in $^{15}N/^{14}N$ in NO_3 pool
Natural abundance	NA	$\delta^{15}NH_4$, $\delta^{15}NO_3$ $\delta^{15}N_2O$	IRMS	
Dark $^{14}CO_2$ uptake	$^{14}CO_2$	^{14}C-POC	Scintillation counter	$\pm$ Nitrification inhibitors and requires conversion factor
Enzymatic				
Inhibitor	N-serve ATU, CH_3F	NH_4, NO_2, NO_3	Spectrophotometry Ion chromatography	$\pm$ Nitrification inhibitors

(Contd ...)

Method	Substrate(s)	Analyte(s)	Detection	Comments
Dentrification				
Direct				
N_2 flux	–	N_2	TCD-GC	Long assay, stringent isolation of sample from air, potential for leakage artifacts
	–	N_2/Ar ratio	QMS, IRMS	Improves precision
Isotopic				
Direct tracer	$^{13}NO_3$	$^{13}N_2$	Gamma counter	Requires cyclotron; isotope has 10 minutes half-life
	$^{15}NO_3$	$^{15}N_2$	IRMS or ES	
Natural abundance	NA	$\delta^{15}NH_4$, $\delta^{15}NO_3$, $\delta^{15}N_2O$	IRMS	
Enzymatic				
C_2H_2 blockage	C_2H_2	N_2O	ECD-GC	Highly sensitive

[a] Abbreviations: NA, not applicable; IRMS, isotope ratio mass spectrometry; N-serve, nitrapyrin; FID-GC, flame ionisation gas chromatography; ATU, allylthiourea; ECD-GC, electron capture gas chromatography; TCD-GC, thermal conductivity gas chromatograph; QMS, quadrupole mass spectrometer, ES, emission spectrometer.

$^{15}N_2$ Direct tracer: When first introduced, the $^{15}N_2$-isotope tracer procedure provided the first sensitive, direct method for quantitatively assessing rates of N_2 fixation in natural samples and remains the procedure of choice for definitively demonstrating and quantitating N_2 fixation.

The assay in outline is very straightforward and follows standard stable-isotope-tracer protocols. Briefly, a sample is amended with enriched $^{15}N_2$ (available commercially in up to 99 atom% $^{15}N_2$ or generated in the laboratory or field from ^{15}N-enriched ammonium salts) and incubated for a period after which the biological material is collected and analysed for increases over natural abundance due to the uptake of the enriched $^{15}N_2$. For aquatic samples, assay protocols often call for removal of atmospheric N_2 by sparging or vacuum before introduction of the $^{15}N_2$. This is done in order to maximise the enrichment of the actual substrate pool and thereby provide a discernible enrichment in product over a relatively brief period of sample exposure and containment.

Enrichment in the sample is determined either by isotope ratio mass spectrometry (IRMS) or emission spectrometry. Biological samples are typically converted to N_2 gas by Dumas combustion for either instrumental method, either off-line or on-line in carbon-nitrogen analyser/flow-through mass spectrometers. Dedicated ^{15}N emission spectrometers are considerably less expensive than IRMS spectrometers and have the advantage of requiring a substantially smaller sample mass (<1 to 10 µg of N) than single-collector mass spectrometers (a mass spectrometer with a single Faraday cup collector, which analyses samples by scanning several masses, usually requires >500 µg of sample). However, emission spectrometers are typically far less precise than those for IRMS and therefore require higher enrichment in product pools to discern activity. Instruments for multiple-collector IRMS are now available at reasonable cost and can obtain accurate isotope ratios on much smaller samples (<10 µg) than are required for single-collector instruments. Furthermore, the high precision of the triple-collector systems has allowed researchers to reproducibly detect small increases in ^{15}N in natural plankton assemblages

(1 to 10 $\delta^{15}N$ units, where $\delta^{15}N = [(^{15}N:^{14}N_{sample}/^{15}N:^{14}N_{atmosphere} - 1) \times 1000])$ after 4 hours of exposure to 10 atom% $^{15}N_2$, without the need to remove ambient N_2.

Acetylene reduction method

The introduction of the C_2H_2 reduction method in 1967 prompted a revolution in studies of N_2 fixation by providing a highly sensitive, convenient, and readily available procedure for determining nitrogenase activity. The procedure was designed for field studies, and the original description included assays of lake cyanobacteria.

Briefly, a sample of interest is incubated in the presence of C_2H_2, typically at a level of about 0.1 to 0.2 atm pC_2H_2 (1 atm = 101 kPa). Nitrogenase preferentially reduces C_2H_2 at these levels to C_2H_4. C_2H_2 and C_2H_4 can be easily separated and quantified by flame ionisation gas chromatography. In active samples, C_2H_4 production can be detected within minutes of C_2H_2 addition. Experimental set-up and sample preparation (injection of C_2H_2 into the gas phase of a sealed sample) are trivial and simple. Sample throughput is very rapid (about 1 minute per analysis). Portable gas chromatographs suitable for this analysis are readily available and may be operated in remote locations as long as power and the requisite compressed gases (typically N_2, air, and H_2) are available. Acetylene may be obtained in small, portable cylinders or generated in the field from CaC_2.

Where broad-based field studies are required or in experimental studies requiring multiple treatment levels, the C_2H_2 reduction procedure is the method of choice. Moreover, in certain systems, it may be run concurrently with C_2H_2 block assays of denitrification. However, the C_2H_2 reduction procedure is indirect in that it measures the reduction of a substrate analog of N_2 and ideally assumes that the activity observed would have been directed to N_2 reduction in the absence of C_2H_2. While good correspondence has been found between rates of C_2H_2 reduction and direct determination of $^{15}N_2$ fixation for a broad variety of systems, large divergences have been noted with other systems. Departures from the theoretical value of 3:1 have often been related to the relative extent of nitrogenase-catalysed H_2 production, a natural function of nitrogenase. Hence, when quantitative estimates are required, it is strongly advised to perform direct inter-comparisons with $^{15}N_2$ reduction.

Natural-abundance determinations

The natural-isotope-abundance approach is rapidly gaining broad use for assessing aquatic N_2 fixation. The nitrogenase reaction, unlike many of the other N transformation, results in very little isotope fractionation and thereby leaves a very distinct signature in some ecosystems. Thus, biological systems (e.g. N_2-fixing cyanobacteria) that are directly dependent upon N_2 gas for a substantial portion of their N nutrition should have N isotopic signatures very close to that in N_2 (0.3662 atm% ^{15}N or a $\delta^{15}N$ close to 0 [see above for definition of $\delta^{15}N$]), compared to systems dependent on N arising from deep oceanic pools or terrestrial run-off. Such anomalies have been observed in studies of marine ecosystems but have not generally been interpreted with respect to a possible diazotrophic source.

The method depends on the collection of a sufficient mass of material from a pool of interest (e.g. surface particulate matter, isolated organisms, or soluble pools), off-line (Dumas) or on-line (continuous-flow IRMS) combustion of the nitrogen in that material to N_2, and high-precision determination of the relative masses of the $^{28}N_2$, $^{29}N_2$ and $^{30}N_2$ peaks. Usually, the highest precision can be achieved by off-line combustion and dual-inlet mass spectrometry. Studies of the natural abundance of N implicating diazotrophy in marine systems have focused on particulate matter suspended near the surface in the oligotrophic ocean, planktonic cyanobacteria and zooplankton (J. Montoya, E. Carpenter, and D. Capone, unpublished data) in these systems, and components of coral reef ecosystems. Through large regions of the temperate and boreal ocean, NO_3 pools are generally enriched in ^{15}N near the surface as phytoplankton takes up NO_3, discriminating against the heavier isotope. In the oligotrophic oceans with chronically low NO_3 concentrations (typically at the limit of detection), it has been difficult to collect sufficient dissolved NO_3 without an imposed artifact to obtain sufficient material for mass spectrometry. However, recent advances are overcoming this limitation, and several studies have noted the opposite trend for NO_3 in these systems; i.e. NO_3 pools near the surface are relatively depleted of the heavier isotope, indicating the input of lighter N presumably through dinitrogen fixation.

Emergent and innovative procedures

With the development of molecular tools to detect nitrogenase (*nif*) genes in the environment, there arises the possibility of detecting mRNA from *nif* genes as well. This capability could provide direct evidence of active synthesis of the genes, rather than of the potential for N_2 fixation. Furthermore, using *in situ* PCR or fluorescent *in situ* hybridisation, one may determine the location and distribution of cells induced for N_2 fixation in natural samples. Quantitative PCR methods are developing rapidly, and it may soon be feasible to quantitate gene copies and, ideally, the relative proportions of different components of the diazotrophic population in different marine systems in particular environments. *nif*H gene detection may also prove useful in paleoecological reconstructions.

Geochemical approaches have long been used to infer broad-scale spatial or temporal trends in biogeochemical processes. A recent analysis of the relative regeneration rates of nitrogen and phosphorus in discrete layers of the oceans, using a derived parameter termed N-star (N^*), has revealed large areas in the oligotrophic oceans where there appears to be an excess of nitrogen regeneration over phosphorus. This excess has been related to the presence of significant populations of N_2 fixers in surface waters that eventually sediment through the water column and are degraded.

Nitrification

Nitrification is the oxidation of more reduced species of inorganic N, namely NH_4^+ and NO_2^-, to more oxidised forms, and thus is a crucial component of the N cycle. While nitrification has been found in some heterotrophs, in the environment it is generally attributed to two specialised types of aerobic autotrophic eubacteria, the ammonium oxidisers and the nitrite oxidisers. Nitrifying bacteria appear to be ubiquitous in soils, as well as in marine and freshwaters and sediments, although their activity is not always simple to demonstrate directly.

Assessing populations

While nitrifiers have been grown in culture for over a century, they are generally regarded as slow growing and difficult to culture. Nitrifier populations in aquatic environments may be enumerated by most-probable-number-type procedures, using media designed to support dark autotrophic growth in air with NH_4^+ or NO_2^- as the electron donor. However, growth yields are poor, and this approach is generally not held to be a quantitative means of accurately assessing nitrifier populations in nature. Ward pioneered the application of immunofluorescent detection in aquatic systems (earlier developed for soil systems by Schmidt), having developed antisera to several strains of marine nitrifiers. Ward's studies detected far more numerous populations of nitrifiers than previously noted by conventional procedures. However, the polyclonal antibodies used are generated against cell surface antigens, often have a high degree of strain specificity, and may still underestimate *in situ* populations.

Nucleotide probes developed to the genes of functional enzymes or to 16S ribosomal DNA sequences unique to nitrifiers are being widely and effectively used to identify and localise nitrifiers in environmental matrices without the bias inherent in traditional procedures. As for diazotrophs (see above), use of denaturing gradient gel electrophoresis is facilitating detailed studies of the *in situ* diversity of these organisms.

Assessing activity

At the simplest level, nitrification may be detected by the production of NO_3^- and NO_2^- from N_4^+, and simple spectrophotometric procedures may suffice (Table 21.6). In highly oligotrophic environments such as open ocean waters, high-sensitivity methods (1 to 100 nM range) for NO_3^- and NH_4^+ are also available. In freshwater environments, ion chromatographic methods with UV detection are a good alternative for NO_3^- and NO_2^- detection. However, the success of this approach depends on a relatively high rate of nitrification and detects only a net rate over any concurrent consumption of the N oxides produced (either by dissimilatory or assimilatory NO_3^- reduction). In a system with balanced nitrification and NO_3^- uptake, changes in the NO_3^- pool may be undetectable, and other procedures are required. Henriksen has described an assay of nitrification potential for sediments in which a high (1 mM) amendment of NH_4^+ is added, and NO_3^- production is monitored. N_2O, which may be sensitively detected by electron capture gas chromatography, is also formed during nitrification and has been used as a diagnostic for nitrification in certain environments.

[15]N Direct-tracer methods

Both direct ^{15}N-tracer and ^{15}N isotope dilution methods have been used to obtain quantitative estimates of nitrification in various aquatic ecosystems. As mentioned above, while access to appropriate instrumental facilities is often a constraining factor, highly sensitive IRMS has become more affordable and available.

Studies of open-ocean and estuarine waters have used the direct tracer approach. Briefly, a small enrichment of the ambient NH_4^+ pool (typically a 10 per cent increase) is made with highly enriched [15]NH_4^+. After a period of incubation, samples are taken for analysis of the ^{15}N enrichment in the NO_2^- and NO_3^- pool, along with changes in concentration by appropriate methodology. The [15]NO_2^- pool may also be enriched and flux into the NO_3^- traced.

For isotopic analysis, NO_2^- is typically trapped as a diazo complex and extracted. NO_3^- can be reduced to NO_2^- and then trapped as above, and each can be dried and combusted on- or off-line to N_2 before introduction into the mass or emission spectrometer. Alternatively, NO_3^- and NO_2^- may be reduced to NH_4^+ by Devarda's alloy or $TiCl_3$ with the NH_4^+ captured by microdiffusion or other methods). Methods have recently been developed for direct microbiological conversion of very low levels of NO_3^- or NO_2^- to N_2O in advance of mass spectrometry.

The isotope dilution approach, which provides simultaneous estimates of both nitrification and nitrate consumption (combined assimilatory and dissimilatory), has also been used in water column and sediment systems. For this method, one labels the product pool, NO_3^-, and monitors the decrease in isotope ratio of that pool relative to the initial enrichment.

This provides a measure of the rate of NO_3^- production, on the assumption that new product will initially be unlabelled. A distinct advantage of isotope dilution approaches is that, apart from a small addition (typically 10 per cent) to the product rather than the substrate pool, rates of increase in the product pool may be estimated even in systems near steady state, where consumption is in close balance with production. Furthermore, when combined with measures over time of the NO_3^- pool, one can accurately estimate both its production and consumption.

Inhibitor-based nitrification assays

Nitrification has been particularly amenable to the 'inhibitor approach' because of the sensitivity of the steps of nitrification to a variety of metabolic inhibitors. Inhibitors, if appropriately applied, can disrupt steady-state conditions and reveal production or consumption of nitrification substrates, intermediates, or end products. Besides relevant N species, these may also include functions specific to autotrophic metabolism (e.g. dark $^{14}CO_2$ uptake). In contrast to ^{15}N methods, which require careful pool separation and sample preparation, inhibitor-based assays are often more rapid and sensitive.

In general, the ammonium oxidation step has received the most attention in the application of inhibitors. One of the earliest compounds to be used as a nitrification inhibitor in an ecological context was N-serve (nitrapyrin or 2-chloro-6-trichloromethyl pyridine), a compound developed to minimise ammonium fertiliser loss from agricultural systems. Inhibition of ammonium oxidation should result in a cessation of NO_2^- and NO_3^- production, with a concomitant decrease in NH_4^+ consumption, as was reported in several early studies.

Typical concentrations used in sediments have been about 10 to 20 µg/ml. N-serve requires an organic solvent, typically acetone, and therefore solvent controls should be employed in addition to inhibitor-free, solvent-free controls. Furthermore, N-serve may sorb to particles and is susceptible to degradation, which may result in decreasing effectiveness with time. Hall proposed use of allyl-thiourea (ATU) as a water-soluble alternative to N-serve. He found somewhat higher rates of nitrification in a direct comparison, with some evidence of N-serve and solvent inhibition of nontarget processes. Hall recognised the problems of slow diffusion of these inhibitors to sites of nitrification.

Most recently, gaseous inhibitors of ammonium oxidation have been promoted. Gaseous inhibitors have the advantage for sediment systems of diffusing more rapidly through assay systems than organic compounds such as N-serve or ATU. Sloth found 1 kPa (0.01 atm) C_2H_2 to be an effective inhibitor of nitrification in sediments. Greater ammonium efflux from sediments occurring in the presence of C_2H_2 is taken as the measure of nitrification. C_2H_2 is of course also an inhibitor of N_2 fixation and of the final step in denitrification, and this may need to be considered. Similarly, Miller has proposed CH_3F (at $pCH_3F = 0.1$ atm) as an effective and specific inhibitor of nitrification, which apparently does not interfere with dissimilatory NO_3^- reduction and denitrification. Moreover, CH_3F has the added benefit of inhibiting nitrifier-specific N_2O production. A comparison of both methods found insignificant differences with respect to stimulating ammonium flux from sediments. It should be noted that, whereas C_2H_2 is readily available or easily generated from CaC_2, CH_3F is relatively difficult to obtain and can be quite costly.

Coupling inhibition of ammonium oxidation with determination of dark $^{14}CO_2$ uptake has been used to detect nitrifier activity. This approach provides a very sensitive radioisotopic method, but requires knowledge of a conversion factor if N transformation estimates are the objective (reported values of N oxidised to C assimilated range from 4 to over 100). Jones described an alternative radioisotope approach based on the ammonium oxidiser-specific (via ammonium monooxygenase) conversion of ^{14}CO to $^{14}CO_2$ in the presence and absence of N-serve.

Nitrite oxidation can also be specifically inhibited, and this offers the advantages of monitoring the NO_2^- pool, for which the spectrophotometric procedure is quite sensitive and for which there is generally no appreciable background.

Belser and Mays found chlorate at about 10 mM to be effective and were able to determine NO_2^- oxidation rates in soils and sediments. However, a subsequent report by Hynes and Knowles pointed out several potential problems with this approach.

Emergent and innovative approaches

As molecular approaches continue to improve, their application in nitrification studies will promote novel insights into the structure and functioning of nitrifier communities. Routine application of quantitative PCR is not far off and will permit sensitive detection of dilute populations. Improvement in RT-PCR will foster our understanding of physiological and environmental controls on nitrification, particularly with the advent of gene chip technologies.

Microprobes have been developed which allow very fine (less than millimeter)-scale resolution of NO_3^- profiles in sediments. The dynamic aspects of these profiles in sediments were experimentally demonstrated by the introduction of an inhibitor (C_2H_2) into the system under study. The coupling of microprobe technology with molecular tools such as fluorescent *in situ* hybridisation gene probes and *in situ* PCR will provide a completely novel level of resolution and understanding of the structure of and controls on these communities.

While becoming more routine, the full promise of the ^{15}N natural abundance approach in nitrification studies remains to be realised. Under NH_4^+-replete conditions, nitrification can impose a very large (up to 20 per 25.4 μm) fractionation between the NH_4^+ and NO_3^- pools. For instance, clear evidence for nitrification has been obtained in Chesapeake Bay waters where increases in sub-pycnocline $\delta^{15}N$ values of NH_4^+ have been directly correlated with increases in NO_2^- concentrations. Natural abundance data have also been used to infer a nitrification source for N_2O in open ocean waters.

Dissimilatory Nitrate Reduction and Denitrification

Dissimilatory nitrate reduction, the reduction of NO_3^- to NO_2^- in anaerobic respiration, is performed by a wide variety of facultative anaerobes when O_2 is unavailable; biological denitrification is the reduction of nitrogen oxides, such as NO_3^- and NO_2^-, as respiratory electron acceptors to gaseous end products, either N_2O or N_2. In general, denitrification is performed by a subset of the facultative anaerobes that reduce nitrate. Denitrification represents a sink for combined nitrogen within ecosystems. In aquatic ecosystems, denitrification is generally associated with anaerobic zones or with chemical interfaces where NO_3^- may be supplied from adjacent NO_3^--rich zones or areas sustaining aerobic nitrification. In some aquatic ecosystems, denitrification has been implicated in promoting N limitation. Dissimilatory NO_3^- reduction to NH_4^+ (NO_3^- fermentation), which conserves combined N within an ecosystem, can at times equal or exceed denitrification. Recently, a biological pathway of anaerobic nitrification ('anammox') has been identified.

Assessing populations

As for N_2 fixation and nitrification, enrichment and enumeration of denitrifiers has long been used to evaluate the presence of such populations *in situ*. Media appropriate for organotrophs, NO_3^- as an electron acceptor, and anaerobic conditions are generally employed. Production of gas in anaerobic enrichments with NO_3^- is often taken as positive for denitrifiers, and such a test may be coupled with most-probable-number analysis. However, as for a broad range of bacteria, heterotrophic denitrifiers have diverse organic substrate requirements, and it is unlikely that any particular media would satisfy the specific requirements of all such that a quantitative procedure based on enrichment could be developed.

Immunofluorescence methods based on cell surface antigens of cultured denitrifiers have been developed and applied. However, while the antibodies provide information on the cultured strain and closely related sero-types *in situ*, they appear to be highly strain specific and unsuitable as a means of assessing the broader denitrifying population. Similarly, antibodies generated to nitrite reductase (NiR) are similarly strain specific, despite the apparent similarity among denitrifiers. Probes to functional genes appear to hold greater promise for *in situ* evaluation of denitrifier populations.

Assessing activity

Because of the perceived importance of denitrification in many aquatic ecosystems, considerable effort has been placed on developing and improving procedures to provide quantitative estimates of this process. As for nitrification, many of our initial insights into denitrification came from observations of spatial and temporal patterns of NO_3^- or NO_2^- using standard colourimetric procedures or more recently by ion chromatographic approaches (Table 21.6). Observations of NO_3^- anomalies in waters and sediments, i.e. deficits in NO_3^- concentrations over that predicted by simple diffusion, mixing and nutrient regeneration, or excesses of N_2 relative to Ar, provide geochemical evidence for denitrification in various water bodies and sediment systems. Occasionally, transients of NO_2^- or excesses (or deficits) of N_2O have also been taken as evidence of denitrification.

Direct N_2 flux

Field efforts have sought to demonstrate short-term rates of denitrification in many aquatic systems, typically using an incubation format. As for nitrification and with similar provisos, one may monitor decreases in NO_3^- pools as a measure of nitrate reduction. However, this will provide an estimate only of net NO_3^- flux and will not discriminate between assimilatory and dissimilatory consumption.

Determination of N_2 flux has long been employed as a quantitative measure of denitrification in respirometric studies of bacterial physiology. Aquatic studies have also used N_2 flux from natural samples as an estimate of *in situ* denitrification rates. N_2 is generally determined on subsamples of headspace gas after equilibration or sparging by thermal conductivity detection following gas chromatographic separation. However, the large natural background of N_2 provides a formidable barrier to detecting small increases. Furthermore, leakage of ambient N_2 into assay vessels or sampling devices (e.g. syringes) may be erroneously construed as denitrification. Seitzinger developed procedures for stringent containment of sediment samples, removal of ambient N_2, and incubation of samples in N_2-sparged seawater. Incubation periods of over a week are required for establishing rates of denitrification, and the procedures are not amenable for broad-scale sampling, replication, or experimentation. Nowicki has provided some improvements in the method which provide means of obtaining estimates over shorter (<1 week) time periods. She also introduced the use of a control using deoxygenated water over the sediments, which eliminates NO_3^- arising from *in situ* nitrification. Devol has developed automated procedures for obtaining samples for N_2 analysis at depth and running those samples by gas chromatography without contamination artifacts.

The analysis of N_2 relative to the conservative tracer Ar has been used to detect excesses of N_2 resulting from denitrification. However, the availability of affordable, bench top quadrapole mass spectrometers suitable for gas analysis through membrane inlet systems is providing a new stimulus for direct gas flux studies. Kana and others have reported on an improved N_2/Ar method applicable for real-time analysis of denitrifying sediments. More recently Kana and others applied this method to estuarine sediments with promising results.

$^{15}N_2$ Tracer methods

Direct tracer procedures have been widely applied in denitrification studies. The bulk of these studies have used ^{15}N isotopes, although some research on denitrification has been accomplished with the short-lived radioisotope ^{13}N. Tracer studies offer the advantage of unambiguous evidence of the presence of a pathway, although quantitation may be confounded by a number of factors, such as artificially elevating substrate pools or disturbing natural gradients.

Koike and Hattori introduced procedures to determine NO_3^- reduction and nitrification rates in sediments by isotope dilution of the NO_3^- pool and denitrification and NO_3^- reduction to NH_4^+ by direct tracer studies of $^{15}NO_3^-$ conversion to $^{15}N_2$ or $^{15}NH_4^+$, respectively. The latter direct tracer procedure was estimated to be about 100-fold more sensitive than the isotope dilution approach. NH_4^+ may be captured for subsequent conversion to N_2 and mass spectrometry by steam distillation or direct microdiffusion. A variety of alternative procedures for collection of NH_4^+ is available including precipitation, organic complexation and extraction or trapping on C-18, or trapping on zeolite resin.

The problem of elevated NO_3^- concentrations was addressed by Oren and Blackburn, who applied a kinetic approach to the problem. Nishio and others subsequently described a flowthrough method for determining denitrification near the sediment-water interface. $^{15}NO_3^-$ is fed into the waters overlying sediment contained in a core, and the enrichment in the $^{15}N_2$ pool in the effluent is monitored. However, this can provide only an estimate of denitrification, depending on an overlying water source, which may be only a fraction of total denitrification. Parallel assays with $^{15}NH_4^+$ can provide an estimate of nitrifier-dependent denitrification. A recent innovation is the use of isotope-pairing analysis to estimate from tracer $^{15}NO_3^-$ assays alone the extent of denitrification that is dependent on external NO_3^- and that derived from *in situ* NO_3^- production via nitrification.

Acetylene blockage method

The observation that C_2H_2 blocked the final step of the denitrification pathway led rapidly to development of a convenient and sensitive procedure for aquatic denitrification. C_2H_2, in blocking N_2O reductase, focuses the products of denitrification into N_2O, easily detected by electron capture gas chromatography and for which there is no appreciable background. C_2H_2 is typically added at about 10 to 20 per cent of the gas phase of the assay vessel or injected into cores in C_2H_2-saturated waters. Over time, samples are taken and analysed for the appearance of N_2O. Both N_2 fixation and denitrification may be assayed simultaneously.

This C_2H_2 blockage procedure has found wide use in diverse habitats. However, under conditions of low ambient NO_3^- or where NO_3^- derives primarily from nitrification (which is inhibited by C_2H_2) or where S^{2-} is present, the procedure likely underestimates *in situ* denitrification. In conjunction with C_2H_2 blockage, Joye and Paerl have used additions of millimolar NO_3^- and glucose along with chloramphenicol to prevent *de novo* synthesis of denitrifying enzymes and to provide an estimate of potential denitrification. However, chloramphenicol may interfere with the activity of some of the denitrifying enzymes. In systems in which NO_3^- levels are relatively high (>10 μM), denitrification is uncoupled from nitrification, and denitrification over the length of the incubation is unlikely to deplete the NO_3^- pool, the C_2H_2 blockage procedure remains an appropriate and useful approach.

An alternative to the C_2H_2 blockage assay is the N_2O reductase assay first proposed by Sherr and Payne and refined by Miller. In this procedure, solutions containing known quantities of N_2O (5 to 20 kPa) are injected into samples, and the consumption of N_2O is determined over time (relative to that of killed or C_2H_2-inhibited controls).

Emergent and innovative approaches

As mentioned above, current efforts are being directed towards development of molecular methods to detect, enumerate, and establish the diversity of denitrifying populations in the environment. In contrast to nitrifiers, the phylogenetic diversity of denitrifiers probably precludes use of rRNA-based probes for assessing denitrifying populations at large. However, functional genes are providing a useful target for such studies. There are two major known forms of nitrite reductase (NiR), those containing *cd1* heme type cytochrome (NiRS) and non-heme Cu-containing (NiRK) forms. An oligonucleotide probe developed to a portion of the *cd1*-type NiRS of *Pseudomonas stutzeri* was more broadly reactive with other cytochrome *cd1*-containing strains and may prove more useful in evaluating denitrifier population sizes in environmental samples. Other reports of probe development for NiRK and NiRS have been forthcoming. Braker and others have recently examined the relative dominance of NiRK and NiRS clones from marine sediments. Scala and Kerkoff have identified diverse nitrous oxide reductase genes in continental shelf sediments. The prospect for discriminating active from inactive denitrifiers using mRNA also exists.

Microprobe technology continues to advance, and as mentioned above, microprobes for N_2O and NO_3^- have been described which allow fine-scale determination of profiles and temporal dynamics of each pool and thereby provide an ability to observe the results of nitrification and denitrification with minimal sample perturbation.

Other Aspects

Bacteria are involved in a variety of other relevant N transformations, including the key processes of uptake of inorganic N and its incorporation into organic form and the degradation of organic forms, returning N to inorganic pools. Considerable research has proceeded on both fronts. Research using tracer methodology, with size fractionation and specific inhibitors, has demonstrated important roles of heterotrophic bacteria in the uptake of limiting nutrients in lacustrine and marine environments.

With respect to the regeneration of nitrogenous organics, the $^{15}N_4^+$ isotope dilution procedure has been used widely and indicates that the regeneration of inorganic N is often dominated by heterotrophic bacteria in many systems. Shallow sediments are particularly important sites of N regeneration, and procedures for sediment studies are well developed. Degradation of proteins and peptides has been a particular recent focus, and recent observations of novel cell surface hydrolytic enzymes in phytoplankton have promoted renewed investigations of the sources of regenerated ammonium. Various fluorescent analogs of model peptides have been used to estimate potential rates of protein hydrolysis.

With the concern about radiatively important trace gases ('greenhouse gases'), there is renewed interest in sources and sinks of N_2O in aquatic ecosystems. N_2O can arise from both nitrification and denitrification, and it may be consumed by denitrification (including via nitrifier-catalysed denitrification). There remains considerable controversy over the relative roles of denitrifiers and nitrifiers in N_2O cycling. Efforts to resolve the source of N_2O in short-term assays with inhibitors have often been confounded by the nonspecificity of commonly used inhibitors (e.g. C_2H_2). In this regard, the CH_3F nitrification inhibitor described by Miller and others reportedly is specific in inhibiting nitrifier but not denitrifier N_2O production.

New pathways of N transformations continue to be found, and novel observations are often abetted by innovations in methods or instrumentation. In the late 1970s, with the advent of the $^{15}NO_3^-$ isotope dilution procedure, the existence and importance of the dissimilatory pathway to NH_4^+ was first recognised. While N_2O production from nitrification has been known since the 1960s, its importance in particular ecosystems was not appreciated until a decade later with the widespread use of electron capture gas chromatography. Most recently, direct ^{15}N tracer studies have provided evidence for a novel pathway of anaerobic nitrification. Aerobic denitrifiers have been isolated, and their importance in the environment remains to be resolved. Similarly, heterotrophic nitrifiers are widely distributed in soils, but little is known of their contribution to nitrification in aquatic environments. Much remains to be discovered of the role of micro-organisms in the marine N cycle.

SULPHUR CYCLING

Sulphur (S) is an important element biochemically and geochemically. It constitutes ~1 per cent of the dry mass of organisms where it serves many structural and enzymatic functions. S also acts as a significant electron donor and acceptor during many bacterial metabolisms. S can be found in a range of valence states from the highly reduced sulphide (–2) to the most oxidised form in sulphate (SO_4^{2-}) (+6). There are several intermediate valence forms of S that can act as both electron donors and acceptors depending on environmental conditions, the most notable being elemental sulphur (S^0) and thiosulphate ($S_2O_3^{2-}$). Microbial S transformations are closely linked with the carbon cycle in which S reduction coupled with organic matter utilisation is a major mineralisation pathway in anoxic habitats, while S oxidations can occur aerobically and anaerobically, some of which are autotrophic and/or phototrophic. Many S compounds are highly reactive, and micro-organisms often must compete with abiotic reactions including the oxidation of reduced S by O_2. This competition makes the study of certain physiological types of S bacteria difficult. In addition, the S cycle is complicated further by the reactivity of sulphide with metals and the oxidation of metal sulphides by bacteria. In general, the high chemical and biological reactivity of S compounds results in a tight coupling of the oxidative and reductive portions of the S cycle in aquatic habitats, particularly at the redoxcline where S cycling can be extremely rapid.

Micro-organisms of the S cycle are extremely diverse. The anaerobic sulphate-reducing bacteria (SRB), which are unique physiologically and genetically, are represented by several genera, most of which were discovered in the last 20 years. Once thought to be restricted primarily to sulphate respiration or fermentation, it has been shown that SRB are capable of utilising iron and manganese as electron acceptors. O_2 reduction has been demonstrated, but O_2-dependent growth has not been confirmed. SRB are rather nutritionally diverse with the known suite of electron donors increasing each year. Chemolithotrophic S oxidation is mediated aerobically by colourless S bacteria, some purple S bacteria, and SRB.

Anaerobically, nitrate-respiring chemolithotrophs oxidise sulphide, and both oxygenic and anoxygenic phototrophic bacteria use sulphide as an electron donor for photosynthesis. Intermediates produced during sulphide oxidation can be oxidised, reduced, or fermented (disproportionated). Although S reduction and oxidation are often studied separately for convenience, it is becoming clear that in many instances both processes must be considered simultaneously.

Sample Collection and Handling

For enumeration, isolation, and rate measurements, it is important to process samples quickly after they are obtained and to maintain samples near ambient temperatures. For anaerobes, collect and handle samples anoxically. Typical water samplers include Niskin bottles and bags. To prevent contamination by O_2 that diffuses through or from plastic, water samples should be transferred to glass bottles by allowing them to overflow before sealing. If rubber stoppers are used, they should be the type that retard O_2 diffusion. Butyl rubber is best, and thick butyl stoppers (e.g. Bellco 2048–11800) can be penetrated many times with needles without risk of O_2 contamination. Since many S compounds are reactive with rubber stoppers, and most stoppers release some S compounds, Teflon-lined butyl septa (e.g. Wheaton 224168) are a useful alternative. However, they usually cannot withstand repeated needle penetration.

Sediments are obtained using grabs or corers. Ideally, for rate measurements, use small diameter (3 cm) whole core (~15 cm long) incubations in which amendments of dissolved constituents are introduced axially via microlitre syringes through predrilled holes filled with a sealant. Whole cores should have the overlying water removed prior to incubation to prevent depletion of O_2 in surficial layers due to stagnant water. In aquatic sediments where O_2 penetration is less than 10 to 15 mm, most of the sulphide-oxidising activity takes place in the upper 20 to 30 mm so cores for S oxidation measurements should sample the upper 3 to 5 cm. Sediments removed from cores should be handled in N_2-filled glove bags or boxes. However, most relatively active sediments can withstand short exposures to laboratory air, i.e. the few minutes required to transfer samples from cores to incubation vessels. Cores should be used within 24 hours after collection and should not be exposed to temperatures above ambient since even short exposures to elevated temperatures will increase rates.

Sedimentary pore water samples for chemical measurements can be obtained in several ways: (i) centrifugation of sediments under N_2 or by centrifugation (2 minutes at 8000 × gram) of small volumes in a microcentrifuge filter unit (pore size, 0.45 µm; Gelman Sciences, Ann Arbor, Mich.); (ii) squeezing of samples under inert gas flow (N_2) using pre-fabricated squeezers; (iii) *in situ* dialysis samplers referred to as 'peepers', which must equilibrate with pore water for ~2 weeks (these cannot be used when sediments are subjected to tides; peepers should also be preincubated in an O_2-free atmosphere for 2 weeks prior to deployment to remove O_2 that dissolves in plastic and will diffuse into anoxic sediment; and (iv) *in situ* 'sippers', which are lysimeters with porous Teflon collars which allow for the removal of pore water in a few minutes under a slight vacuum. In all cases, samples must not come in contact with O_2, and except for peeper samples which do not have to be filtered, pore waters must be filtered anoxically. When sampling sites such as marshes or seagrass beds that contain macrophytes, avoid sampling techniques that disturb the sediments since damaged roots leak dissolved material. Sippers are well suited for vegetated sediments.

Sulphate Reduction

SRB are a specialised group of anaerobic bacteria that are responsible for the dissimilatory reduction of SO_4^{2-} to sulphide that is linked to organic matter oxidation. They are important in the anaerobic degradation of organic matter in most aquatic habitats, where they are situated at the bottom of the anaerobic food chain. SRB are also a major source of sulphide, which is highly reactive, geochemically important, and used by S-oxidising bacteria as described below. Because of the abundance of SO_4^{2-} in seawater, dissimilatory SO_4^{2-} reduction is considered a major process in marine sediments, yet it is also important in the decomposition of organic material in anoxic freshwater habitats. Although the reduction of SO_4^{2-} is considered to be the classic role of SRB in the environment, these bacteria are capable of diverse metabolisms including metal and O_2 reduction, metal methylation and demethylation, organic fermentations, use of xenobiotics and petroleum by-products, S disproportionation, and the utilisation of various intermediate redox states of S, including sulphonates. It is generally accepted that SRB oxidise products of fermentative bacteria such as fatty acids, alcohols, some aromatic acids, a few amino acids, and hydrogen. The suite of substrates used varies among genera. For detailed information on SRB, various reviews are available.

Rate measurements

Rates of SO_4^{2-} reduction have become routine measurements in studies of the biogeochemistry of anoxic aquatic environments. Rates can be determined by monitoring the loss of SO_4^{2-} in samples incubated anaerobically over time (jar experiments) or by determining the rate of production of reduced ^{35}S end products generated during the reduction of [35]SO_4^{2-} in incubated samples. The ^{35}S technique is preferable since it uses a short incubation period (1 day or less) and undisturbed sediment. However, if one is unable to use the radiotracer technique, a jar approach provides an estimate of SO_4^{2-} reduction. Minimise manipulation in all cases, especially the mixing of oxic and anoxic layers in sediments.

Jar technique

Homogenise a sediment sample and either distribute sub-samples among several vessels which are sacrificed over time or maintain one vessel in which sub-samples are removed over time. For the former, use centrifuge tubes (15 to 50 ml) so pore water can be extracted easily. When using plastic, preincubate tubes in an inert atmosphere for ~2 weeks, and incubate sample-filled tubes similarly since many plastics allow O_2 to pass and they bleed O_2. Diffusion-retardant plastic bags can be used without additional precaution. At selected time intervals, extract and filter pore waters and measure SO_4^{2-} content (analytical techniques are described below). Determine the wet and dry weight of known volumes of sediment so rates can be expressed on a sediment volume basis. Incubation times should be kept to a minimum and a linear decrease in SO_4^{2-} content is desirable since nonlinear losses are difficult to quantify. For nearshore marine sediments, incubation times range from a few days to weeks. Since freshwaters contain much less SO_4^{2-}, incubation times are short (a few days to <2.0 hours).

Consumption of SO_4^{2-} is calculated from the slope of the concentration over time ($\delta[SO_4^{2-}]/\delta t$). The consumption rate is either expressed in amount (micromoles or millimoles) per volume (e.g. cubic centimetres of sediment or millilitres of pore water), or per mass (grams of sediment) per unit of time (minutes, hours, or days). Rates are usually expressed as nanomoles per cubic centimetre of sediment per day or integrated over depth as micromoles per square metre.

[35]S Technique

The radiotracer technique is the most widely used method for measuring rates of sulphate reduction. [35]SO_4^{2-} is introduced in tracer amounts in small volumes, and ^{35}S end products are isolated. Unlike the jar technique, which measures SO_4^{2-} loss, the accumulation of labelled, end products is very sensitive, so incubation times are short and undisturbed samples can be used. Early studies measured the accumulation of acid-volatile sulphides only. However, since the discovery that significant quantities of reduced ^{35}S are recovered as nonacid soluble phases (i.e. S^0 and, FeS_2), these are included by using a chromium reduction procedure. The rapid isotopic exchange of ^{35}S among reduced pools precludes the use of this technique as a measure of the rate of production of specific reduced S species. Therefore, rates are determined from the sum of all the inorganic reduced S compounds. The following is a one-step distillation procedure for measuring rates of SO_4^{2-} reduction.

Radiotracer (2.0 µl of [35]SO_4^{2-} in weak acid or sterile water; ~1.0 µCi) is introduced, via microlitre syringe through longitudinal holes filled with a sealant, into small diameter (~3 cm) short (15 cm) cores while withdrawing the syringe. It is also possible to take horizontal subcores in syringes which are sealed and injected. For sample that are difficult to subcore, like salt marsh peats that contain thick rhizomes, large diameter cores are sliced into small sections that are introduced into 5 ml syringes without homogenising. To determine the SO_4^{2-} content of the sediments, additional cores are sectioned horizontally and pore water is extracted by centrifugation. After incubation in the dark from 2 to 24 hours at *in situ* temperature, cores are cut into segments that are mixed with 20 per cent zinc acetate and then frozen ($-20°C$) to fix sulphides and terminate activity. Incubation time depends on the SO_4^{2-} content of the sediment with shorter incubation periods used when SO_4^{2-} levels are low. Try to avoid consuming more than 10 per cent of the ambient SO_4^{2-} during incubation. Thawed samples are centrifuged, and the activity of the remaining $^{35}SO_4^{2-}$ is determined by scintillation counting. When using individual subcores, the ^{35}S activity of this supernatant is not required as long as the activity of the injected ^{35}S is known. The pellet is washed twice with seawater to remove excess $^{35}SO_4^{2-}$, transferred to a reaction vessel, and mixed with 10 ml of ethanol. The vessel is connected to a condenser and contains ports for introducing reagents and a constant stream of N_2 gas. The top of the condenser is connected to two traps in series each containing 10 ml of a solution of 6 per cent zinc acetate and 10 per cent ammonium hydroxide. Two drops of Antifoam B (Baker Chemicals) are added to the traps to prevent foaming, which can be excessive for samples with relatively high reduced S levels. The reaction vessel is flushed with N_2 for 20 minutes, and then 40 ml of $CrCl_2$ solution and 20 ml of concentrated HCl are added. The contents are gently boiled for 45 to 60 minutes with a constant stream of N_2. The total reduced S is converted to H_2S that is swept into the traps and collected as ZnS. Over 98 per cent of the ^{35}S is collected in the first trap. Half of the trap material (~5.0 ml) is mixed with a gel-forming scintillation cocktail and counted. If desired, the sulphide content of the remaining ZnS can be determined colourimetrically for a measure of the pool size of total reduced inorganic S.

The reduced chromium solution is prepared by percolating 1.0 M $CrCl_3 \cdot 6H_2O$ (pH 1.0) through a column filled with amalgamated zinc granules (grain size, 0.5 to 1.5 mm). The zinc is amalgamated by soaking briefly in a solution of $HgCl_2$ (~0.25 M), transferred to a glass column (inside diameter, ~2 cm), and washed with 3 volumes of 0.1 M HCl. The $CrCl_3$ solution, which is dark green, changes to a bright blue colour when reduced ($CrCl_2$) and is collected within syringes or in serum bottles previously flushed with N_2. It is important that only bright blue reduced chromium be used in reaction vessels. If the solution elutes from the column as even slightly greenish blue, it should be discarded. Only freshly prepared $CrCl_3 \cdot 6H_2O$ solutions should be run through zinc columns since solution a few days old will not reduce adequately. Also use only $CrCl_3 \cdot 6H_2O$ which is a fine green powder. $CrCl_3 \cdot 6H_2O$ in large chunks that exhibit a slight bluish hue will never completely reduce, even if recently manufactured.

An alternative to the active distillation procedure was developed by Ulrich, in which the chromium reduction reaction is employed in a passive mode. Briefly, samples (<2 ml) incubated with ^{35}S and fixed as described above are introduced anoxically into 120 ml serum vials and sealed with thick butyl rubber stoppers. Each serum vial contains a test tube with 2.5 ml of anoxic 10 per cent zinc acetate as a sulphide trap. Reduced chromium solution (8 ml) followed by anoxic concentrated HCl (4 ml) are injected into each vial, which are then placed on a rotary shaker. After ~30 hours, the zinc acetate-filled test tubes are removed and homogenised by both vortexing and sonicating, and the radioactivity of the ZnS precipitate is then determined by scintillation counting. All solutions must be prepared and stored anoxically. This passive technique does not quantitatively recover S in S^0, so care should be taken when analysing samples in which S^0 is a significant portion of the S generated during the incubation. In addition, calcium carbonate-containing sediments rapidly neutralise the acidic chromium solution, preventing the reduction of S species. This neutralisation may be overcome by using small sediment samples. However, sediments rich in calcium carbonate cannot be analysed with this technique unless the quantities of reagents are increased and excess pressures in vials are released.

Microscale distribution of sulphate reduction activity can be assessed using Ag foil coated with $^{35}SO_4^{2-}$ is washed with acetone followed by application of ^{35}S (use carrier-free $^{35}SO_4^{2-}$) and air-drying overnight. The coated foil is then placed on a fresh sediment sample and incubated. After rinsing off residual sulphate, activity is determined by measuring $Ag^{35}S$ formed during SO_4^{2-} reduction using autoradiography or gel scanners (e.g. bio-rad molecular imager system GS-525) that are able to detect ^{35}S on the foil surface.

The rate of sulphate reduction is calculated as:

$$\text{Rate} = \frac{[SO_4^{2-}] \cdot (^{35}S\text{-TRIS}) \cdot 1.06}{(^{35}SO_4^{2-} + ^{35}S\text{-TRIS}) \cdot t} \, \text{nmol} \, SO_4^{2-} \, \text{cm}^{-3}$$

(whole sediment) day^{-1}

where, $[SO_4^{2-}]$ is the SO_4^{2-} concentration in nanomoles per cubic centimetre (calculated from pore water values and water content of sediment determined from wet and dry weights), ^{35}S-TRIS is the total radioactivity of ZnS, $^{35}SO_4^{2-}$ is the total radioactivity of the supernatant (the $^{35}SO_4^{2-}$ left after incubation), t is the incubation time in days, and 1.06 corrects for isotope fractionation.

Cultivation, enrichment, enumeration and isolation

SRB are readily isolated from many environments including freshwater and salt water, soils, oil-bearing shales and strata, intestinal contents, sewage, and paper mill effluents. In addition to the common mesophilic forms, thermophilic and hyperthermophilic forms have also been isolated. Although sulphate reduction has been shown to occur in the oxic layer of microbial mats, and SRB can survive brief exposure to air, they have so far been cultivated only under anoxic, reducing conditions. The physical removal of O_2 from a medium, by boiling or sparging with N_2, is often sufficient to permit growth of robust strains since sulphide may be carried over in the inoculum. However, the use of a prereduced, 'poised' medium is usually necessary, particularly for fastidious strains or when SRB are present in low numbers as would be obtained from a dilution series. The greatest variety of SRB has been isolated using a bicarbonate-buffered medium that contains sulphate, an electron donor, and sulphide as the reductant. Most probable number (MPN) enumeration of SRB has been reported to have been improved through the use of natural sediment or sludge media combined with the inclusion of $^{35}SO_4^{2-}$ as a tracer.

Composition and preparation of media

The following general-purpose media can be used to cultivate most types of mesophilic or moderately thermophilic SRB, although no one particular formulation is suited to all types. As with all anaerobes, cultures are incubated in the absence of air by filling the containers as completely as possible or using a headspace filled with anoxic gas. Use septa or stoppers made from butyl rubber. Incubation temperatures are typically 30°C for mesophilic strains and 55°C for thermophilic strains.

Postgate's media

Postgate's media are useful for routine cultivation of *Desulphovibrio* and *Desulphotomoculum* spp. and are relatively simple to prepare. They may be solidified with 10 grams of agar per litre. N_2 may be used as the headspace gas.

Postgate's medium B

KH_2PO_4	0.5 g
NH_4Cl	1.0 g
$CaSO_4$	1.0 g
$MgSO_4 \cdot 7H_2O$	2.0 g
Sodium lactate	3.5 g
Yeast extract	1.0 g
Ascorbic acid	0.1 g
Thioglycolic acid	0.1 g

FeSO$_4$	0.5 g
Tap water	1.0 litre

Adjust pH between 7.0 and 7.5 and autoclave under N$_2$. Add NaCl to 2.5 per cent for marine strains or use seawater. This medium forms a precipitate that is useful for long-term maintenance of cultures, since it is thought to provide a microhabitat. The ingredients may be prepared as stock solutions. However, the reductants thioglycolate and ascorbate must be freshly added, as they readily deteriorate in air. The completed medium should be used as soon as possible; deterioration of the reductants is indicated by the transient appearance of a purple colour. The medium diagnostically blackens upon growth of SRB.

Postgate's medium C

KH$_2$PO$_4$	0.5 g
NH$_4$Cl	1.0 g
NaSO$_4$	4.5 g
CaCl$_2$·6H$_2$O	0.06 g
MgSO$_4$·7H$_2$O	0.06 g
Sodium lactate	6.0 g
Yeast extract	1.0 g
FeSO$_4$	0.004 g
Sodium citrate·2H$_2$O	0.3 g
Distilled water	1.0 litre

Adjust pH to ~7.2, autoclave under N$_2$. For marine strains prepare with NaCl at 2.5 per cent. The medium may be cloudy after autoclaving. However, this is generally a clear medium, the iron being chelated by the citrate, and is useful for large-scale cultures of *Desulphovibrio* or *Desulphotomaculum*.

Widdel and Pfennig medium

Widdel and Pfennig medium, a defined medium described in detail by Widdel and Hansen, was used to isolate most of the newly identified genera of SRB. All SRB cultivated to date can be grown on a variation of this medium. The basal salts solution is adjusted with respect to NaCl, MgCl$_2$·6H$_2$O, or CaCl$_2$ for preparation of a freshwater, brackish, or marine medium, respectively. Electron donors are selected according to the physiology of the strain to be grown or the type to be enumerated. The medium is assembled under 80 per cent N$_2$–20 per cent CO$_2$ from separately prepared solutions added to the basal salts solution in the order given below. To make a solid medium it is recommended that the agar be washed three times in distilled water to remove inhibitors and growth substrates, added to the basal salts (10 grams litre^{-1} final concentration), and then autoclaved. The sterile solution must be kept at ~50°C to keep the agar molten while the other additions are made.

Solution A, basal salts	*Freshwater*	*Brackish*	*Marine*
Na$_2$SO$_4$	4.0 g	4.0 g	4.0 g
H$_2$PO$_4$	0.2 g	0.2 g	0.2 g
NH$_4$Cl	0.25 g	0.25 g	0.25 g
NaCl	1.0 g	7.0 g	20.0 g
MgCl$_2$·6H$_2$O	0.4 g	1.2 g	3.0 g
KCl	0.5 g	0.5 g	0.5 g
CaCl$_2$·2H$_2$O	0.1 g	0.1 g	0.15 g

| Resazurin (0.1%) | 1.0 ml | 1.0 ml | 1.0 ml |
| Distilled water | 950 ml | 950 ml | 950 ml |

The volume of water used is adjusted for subsequent additions; the amounts of salts given are for 1 litre of the completed medium. The salts are added to stirred water to prevent the formation of precipitates. The solution is prepared in a stoppered vessel for making the subsequent additions and dispensing the completed medium. After autoclaving, the solution is cooled under N_2-CO_2 and the additions made under the anaerobic gas with mixing on a magnetic stir plate. The final pH of the medium should be between 7.0 and 73.

Solution B, trace elements

HCl (66%, 8 M)	12.5 ml
$FeSO_4 \cdot 7H_2O$	2100 mg
H_3BO_3	30 mg
$MnCl_2 \cdot 4H_2O$	100 mg
$CoCl_2 \cdot 6H_2O$	190 mg
$NiCl_2 \cdot 6H_2O$	24 mg
$CuCl_2 \cdot 2H_2O$	2 mg
$ZnO_4 \cdot 7H_2O$	144 mg
$Na_2MoO_4 \cdot 2H_2O$	36 mg
Distilled water	987 ml

The solution is autoclaved under N_2. Add 1.0 ml litre of medium[-1].

In addition, some strains require additional trace elements (prepared in 100 ml of distilled water): NaOH, 40 mg; $Na_2SeO_3 \cdot 5H_2O$, 0.6 mg; $Na_2WO_4 \cdot 2H_2O$, 0.8 mg. The solution is autoclaved under nitrogen. Add 1.0 ml litre of medium[-1].

Solution C, bicarbonate (1.0 M)

$NaHCO_3$	8.4 g
Distilled water	100 ml

The solution is saturated with CO_2 by sparging or by shaking in a tightly closed bottle with one-third of the capacity left as headspace that is replenished several times with CO_2. The solution may be filter sterilised or autoclaved in bottles with secured seals under CO_2. Add 30 ml litre of medium[-1].

Solution D, organic substrates

These are prepared separately as neutralised solutions (usually 1.0 to 2.0 M) and autoclaved under N_2. One is selected for use depending upon the physiology of the strain to be grown or the type to be enriched.

The range of substrate concentrations used is given below.

Acetate	10–20 mM
Benzoate	2–5 mM
Butyrate	5–12 mM
Caproate	2–5 mM
Ethanol	10–20 mM
Formate	10–20 mM

Lactate	10–40 mM
Propionate	10–20 mM
Succinate	10–20 mM
Pyruvate (filter sterilised)	10–40 mM

Add 5 to 15 ml of the concentrated stock solution litre of medium^{-1}.

Solution E, vitamins

These are prepared separately for stability, filter sterilised, and kept in the dark.

Mixed vitamins

4-Aminobenzoic acid	4 mg
D(+)-biotin	1 mg
Nicotinic acid	10 mg
Calcium D(+)-pantothenate	5 mg
Pyridoxine dihydrochloride	15 mg
Sodium phosphate buffer (10 mM, pH 7.1)	100 ml

Add 1.0 ml litre of medium^{-1}.

Thiamine

Thiamine chloride	10 mg
Sodium phosphate buffer (25 mM, pH 3.4)	100 ml

Add 1.0 ml litre of medium^{-1}.

Vitamin B$_{12}$

Cyanocobalamin	5 mg
Distilled H$_2$O	100 ml

Add 1.0 ml litre of medium^{-1}.

Solution F, sulphide

Na$_2$S·9H$_2$O	4.8 g
Distilled water	100 ml

Sodium sulphide crystals deteriorate in air. Clean crystals are selected or rinsed with distilled water if needed. The water is sparged with and kept under N$_2$ while the crystals are being dissolved. The solution is autoclaved under N$_2$ and added at 7.5 ml litre of medium^{-1}.

The more fastidious species of SRB are stimulated by the addition of sodium dithionite, which further reduces the medium. This reductant is added after the medium has already been reduced with sulphide. Aqueous sodium dithionite rapidly breaks down with exposure to O$_2$ and must be prepared with O$_2$-free water under N$_2$.

A small tube with a side arm to introduce a sterile stream of N$_2$ is recommended for preparation and storage of dithionite, which will remain stable for about 5 days if not exposed to air. Since dithionite crystals are usually sterile due to their toxicity, an estimated solid amount may be added directly from the reagent bottle to the medium using a sterile spatula fashioned from a platinum wire.

Dithionite solution

O$_2$-free distilled water (autoclaved under nitrogen)	10 ml
Na$_2$S$_2$O$_4$ (weighed aseptically)	0.2 g

Add 1.0 ml litre of medium^{-1}.

Inoculation, transfer and storage

Use a 1 to 10 per cent (vol/vol) inoculum (the higher concentration is used for slower-growing strains). Inoculations can be made with a syringe needle through septum seals or with a pipette for stoppered cultures. When using a pipette, the tip is placed beneath the surface of the medium to limit exposure of the inoculum to air and the headspace is flushed with the anaerobic gas as the stopper is replaced. Transfers are performed on a 6- to 12-week basis; a stringiness becomes apparent due to lysis of cells in aged cultures. For storage, cultures are removed from the incubator prior to attaining stationary phase and stored at a 4° to 6°C in the dark. Care should be taken as some strains may expire if refrigerated. Postgate's medium B is suited to the storage of *Desulphovibrio* and *Desulphotomaculum* species. For long-term maintenance, cultures containing 5 to 10 per cent dimethyl sulphoxide may be kept under liquid nitrogen.

Enumeration and isolation, MPN and solid media

Dilution of samples into liquid media can be used to enumerate SRB by the MPN technique. Containers should be refrigerated until the samples are processed, and samples are diluted in poised media and shaken to dislodge and separate cells, since many SRB grow in clumps. Growth of SRB is confirmed by the blackening of media that contain an iron indicator, such as Postgate's medium B, or by growth in the media of Widdel and Pfennig medium. With the latter, growth is indicated by the appearance of turbidity, which can be measured by increase in absorbance at 600 nm against an uninoculated blank and should be confirmed by chemically testing for SO$_4^{2-}$ reduction or sulphide production by one of the assays given below. Transfer of an inoculum from a culture obtained with the highest dilution of the series to a medium solidified with agar can be used to obtain an isolate in pure culture.

Enumerations can also be performed by counting the formation of colonies in solid media. The sample is serially diluted into tubes held at 40°C that contain the liquefied medium. A pipette is used to make a transfer and briefly distribute the inoculum, the tip being kept beneath the surface of the medium. The agar can be solidified as a layer over the inside of the tube (roll tubes) or as a plug.

Tubes are incubated in an inverted position so that condensation collects away from the forming colonies and can easily be removed. As described above, colonies of SRB blacken in the presence of the iron indicator. In media without an indicator the colonies may contain a slight yellowish, reddish, or grayish pigment.

Pure cultures can be obtained from well-separated colonies picked from the agar with a drawn pasteur pipette. The colony mass is suspended in a few millilitres of anoxic saline or medium and used to inoculate a small volume (~10 ml) of medium, some being reserved for microscopic inspection.

Tests for purity

Microscopically observe the cell suspension for uniform morphology. It may be necessary to repeat selection of a colony. Postgate offers the following additional tests. Contamination of a culture by an aerobe can readily be determined by plating onto nutrient agar containing glucose and peptone; no colonies should appear after the plate is incubated in air at 30°C. To test for contaminant anaerobes, prepare 25 ml of molten peptone-glucose agar (pH 7.0 to 7.6) and add sterile $Fe(NH_4)_2(SO_4)_2$ to 0.05 per cent. Cool the medium to 40°C, dispense into tubes, and make a dilution series from the test culture. Allow the agar to solidify as a plug and incubate under N_2. Black colonies should appear; colonies that are not black are contaminants.

Molecular techniques for density and diversity measurement

It is now well understood that only a very small percentage of the bacteria observed in an environmental sample can be cultivated. In recent years, molecular methods based on PCR, gene cloning, and hybridisation probes have been developed to examine natural populations of bacteria directly without cultivation. SRB have historically been difficult to isolate and identify with culture techniques. Most genera have been described only fairly recently. Yet, it has been determined that culturable numbers of SRB, based on MPN methods using the defined medium of Widdel and Pfennig, could account for measured rates of sulphate reduction. Nonetheless, strains continue to be isolated that are morphologically, physiologically, and phylogenetically unique. Similarly, 16S rRNA genes are frequently cloned from environmental samples, indicating that many lineages of SRB have not yet been cultured. In addition to rRNA sequences, use of the bisulphite reductase gene and hydrogenase gene sequences have provided phylogenetic information on natural communities of SRB.

rRNA-based methods

rRNA-based methods are based largely on oligonulceotide probes designed to hybridise to 16S rRNA sequences and take advantage of the agreement between the phylogeny and physiology of the gram-negative mesophilic SRB. Oligonucleotide probes may be used to measure the rRNA abundance of a targeted group by hybridisation to nucleic acids extracted from a natural microbial community and immobilised on a membrane support. The probes may also be used to directly observe and enumerate targeted cells in a sample by whole-cell hybridisation and serve as PCR primers for the amplification of 16S rDNA of a targeted group from an environmental sample.

rRNA-targeted oligonucleotide hybridisation probes have been developed and described for many of the gram-negative SRB. These probes have been used to examine communities and *in situ* distribution of SRB in various habitats including sediments, activated sludge, biofilms, microbial mats, the water column, and the rhizosphere of aquatic plants. Understanding the distribution of SRB in microbial mat and biofilm studies in particular has been greatly aided through the use of micro-electrodes to measure gradients of O_2 and H_2S.

DNA-DNA hybridisations

A technique termed reverse sample genome probing (RSGP) has been developed and used to study SRB in oil field production waters, soil, and mining wastes. In RSGP, genomic DNAs from pure cultures are applied to membranes and serve as standards against which radio-labelled DNA obtained from a microbial community, either by direct extraction or selective enrichment, is hybridised. The pure cultures may be obtained from collections or isolated from the environment under investigation. The genomes used as standards are chosen on the basis of having no or limited cross hybridisation. RSGP was shown to generate reproducible fingerprints of microbial communities of SRB. Quantitation of hybridisation signals indicated enrichment of SRB in biofilms on metal surfaces that had contact with production water. A DNA-DNA hybridisation method using rDNA-targeted oligonucleotide probes to estimate SRB cell numbers has also been described.

Sulphur Oxidation

Sulphide is produced from degradation of sulphur-containing organic matter and by dissimilatory sulphate reduction. Microbial metabolism of sulphide competes with chemical oxidation, either by O_2 or Fe^{2+}. During biological and chemical conversion of sulphide (oxidation state, −2) to SO_4^{2-} (+6), a variety of intermediates (−1 to +5) are formed. Some of these intermediates, such as polysulphides, react chemically with organic matter, during which organic S compounds can be formed. Metabolic pathways discussed here include those that are of particular environmental importance. Under oxic conditions, chemolithotrophic sulphide oxidation takes place by colourless S bacteria (thiobacilli and filamentous sulphide oxidisers) and some purple S bacteria and SRB. In the absence of O_2, nitrate-respiring chemo-lithotrophs engage in sulphide oxidation, and both oxygenic and anoxygenic phototrophic bacteria use sulphide as an electron donor for photosynthesis. Intermediates that are produced during sulphide oxidation can either be oxidised or sometimes reduced by SRB or, alternatively, fermented (disproportionated). Various reviews provide detailed information on oxidative processes in the S cycle.

During chemolithotrophic metabolism, sulphide is respired with O_2 or nitrate. Since chemical oxidation competes with microbial oxidation, thiosulphate ($S_2O_3^{2-}$) is often used as a sulphide analog since the former does not chemically oxidise spontaneously. Chemolithotrophic sulphide oxidation plays an important role in hydrothermal vent environments, in certain symbiotic relationships, and, most importantly, in (marine) sediments and water columns of stratified lakes.

Rate measurements

Rate measurements can be determined by O_2, nitrate, and/or sulphide (or $S_2O_3^{2-}$) consumption. Problems introduced by chemical reactions of sulphide and O_2 or metals such as iron require one to exercise caution. Killed controls of the use of inhibitors are necessary to account for abiotic loss factors. When adding sulphide or $S_2O_3^{2-}$ it is important to employ concentrations as close to those *in situ* as possible. Sulphide can be especially toxic to bacteria at elevated concentrations, so its use in excess may cause consumption rates that deviate from those *in situ*.

For water samples, 5 to 50 ml is placed in a chamber or reaction vessel. The endogenous O_2 consumption is determined while the sample is stirred. A small volume (10 to 100 µl) of a concentrated sulphide or $S_2O_3^{2-}$ stock solution is added to the reaction vessel, and the disappearance of O_2 is monitored over time. Alternatively, the disappearance of sulphide or $S_2O_3^{2-}$ is monitored by measuring the concentration in subsamples.

Autoclaved or filter-sterilised controls are treated identically to determine abiotic loss factors. Preparation of stock solutions of sulphide and $S_2O_3^{2-}$ (50 to 500 mM) is described below. The analytical techniques for measuring these compounds and nitrate and O_2 are described below.

Intact core measurements are the preferred method for measuring S transformations. A major problem with sediments is abiotic reactions, such as iron and manganese sulphide mineral formation and oxidation of reduced S by O_2. The substrate (preferably $S_2O_3^{2-}$ to minimise abiotic consumption) is introduced with a glass syringe by axial micro-injection. Concentrated stock solutions are added in 5 to 25 µl amounts to avoid changing pore water content. Multiple cores are incubated simultaneously, and individual ones are sacrificed at each time point. Upon sampling, the cores are sectioned and the pore water concentration of the substrate is measured as described below. To prevent use of $S_2O_3^{2-}$ by SRB, molybdate (5 to 10 mM, final concentration Na_2MO_4) can be injected simultaneously as a 'specific' inhibitor. Since this method is elaborate and makes the use of replicates impractical, it may be necessary to repeat the experiment twice or three times.

Measurements can also be made using sediments that are slurried (e.g. 1 part sediment plus 1 to 3 parts sterilised water or artificial medium of similar ionic strength) and placed in a vessel (100 to 500 ml) or serum bottle (20 to 100 ml), where the consumption of sulphide, $S_2O_3^{2-}$, O_2, or a nitrate is monitored in aliquots collected over time. The slurry should contain a gas phase (N_2 for anoxic conditions) and be stirred or shaken, and temperature should be controlled. In slurry experiments, sulphide and O_2 can be measured directly with electrodes placed in the reaction vessel.

Consumption of sulphide or $S_2O_3^{2-}$ can be calculated from the initial slope of the concentration over time ($\delta[\text{sulphide}]/\delta t$ or $\delta[S_2O_3^{2-}]/\delta t$). For water samples this can be normalised easily to volume; for sediments the consumption rate is either expressed in amount (micromole or millimole) per volume (e.g. cubic centimetre of sediment or millilitre of pore water) or per mass (gram of sediment) per unit of time (minute, hour, or day). Another approach is to calculate rates of sulphide or $S_2O_3^{2-}$ consumption by assuming steady state in which the SO_4^{2-} reduction (i.e. sulphide production) rate equals the sulphide consumption rate, or the $S_2O_3^{2-}$ oxidation rate equals $1/2 \times SO_4^{2-}$ reduction rate $\times 0.6$ (which assumes that 60 per cent of the sulphide proceeds through $S_2O_3^{2-}$).

Radiotracer studies and isotope exchange

The use of [^{35}S]sulphide for measuring S oxidation creates difficulties because, in addition to chemical reaction with O_2 and Fe^{2+}, rapid isotope exchange occurs among reduced S species (i.e. $H^{35}S^-$, $^{35}S^0$, and $Fe^{35}S$), presumably via polysulphide ($^{35}S_n^{2-}$). However, [^{35}S]sulphide synthesised from $^{35}S^0$ has been applied successfully for short-term sediment incubations. In general, the use of $^{35}S_2O_3^{2-}$ is preferable. Incubation times must be short (<1 hour) to prevent metabolism of the S compounds formed. Radionuclides are applied as described above for whole-core incubations, and reactions are terminated by slicing and then homogenising sections of cores in ice-cold 1 per cent zinc acetate. Pore water is retrieved by micro-centrifugation (2 minutes at 8000 × g at 4°C) and filtration through a 0.2 μm-pore-size nylon filter (Alltech, Deerfield, Ill.), and after dilution in a $S_2O_3^{2-}$ carrier solution (0.1 mM), $S_2O_3^{2-}$, polythionates, and SO_4^{2-} are separated by ion chromatography. The eluted fractions containing these S compounds are collected in scintillation vials, and activity is determined using scintillation cocktails designed for aqueous samples (e.g. Ecolume or Ecolite; ICN Biomedicals, Inc.).

Cultivation, enrichment and isolation

Cultivation and enumeration of sulphur oxidisers are most successful in CO_3^{2-}-buffered media. *Thiomicrospira* spp. from marine environments are especially sensitive to elevated phosphate concentrations. Success of isolation procedures can be enhanced by mixing mineral media with filtered (pore size, 0.45 μm) and autoclaved water from which the inoculum is taken (1:1, vol/vol). Gradient cultures (two-layered agar systems, in which the bottom compartment contains sulphide and the upper compartment consists of mineral agar in which O_2 can freely diffuse) are useful for the isolation and cultivation of filamentous S bacteria (*Beggiatoa*). Cultivation in chemostats is also a powerful tool to determine sulphide oxidation rates.

Medium for S oxidisers

Mineral medium base

NH$_4$Cl	0.2 g
CaCl$_2$ · 2H$_2$O	0.225 g
KCl	0.2 g

| MgCl$_2$·6H$_2$O | 0.2 g |
| KH$_2$PO$_4$ | 0.02 g |

Add 1.0 ml of Widdel and Pfennig's trace element solution (listed above for SRB). For *Thiobacillus* and *Thiomicrospira* spp. add FeSO$_4$·7H$_2$O (0.1 to 1 mg), and for marine isolates add NaCl (25 grams). Dissolve salts for 1.0 litre in 800 ml and autoclave. Prepare and autoclave each of the following separately (g litre^{-1}): Na$_2$CO$_3$, 20; Na$_2$S$_2$O$_3$, 248; Na$_2$S·7-9H$_2$O, dissolve 24 grams litre^{-1} in boiling 2 per cent Na$_2$CO$_3$ solution, close cap, and autoclave; CH$_3$COONa, 8.1; 1 M HCl; 1 M NaOH; if desired, add yeast extract to final medium. Add Na$_2$CO$_3$ stock to medium base (1:9, vol/vol) and either Na$_2$S$_2$O$_3$ (1:9, vol/vol) or Na$_2$S (1:99, vol/vol); if required, add CH$_3$COONa (1:9, vol/vol) and yeast extract (1:99, vol/vol). Adjust final volume to 1.0 litre.

Adjust pH to 7.5 to 8.0 with sterile HCl or NaOH. For aerobes, transfer complete medium into an Erlenmeyer flask with a cotton plug (do not shake because this will cause the pH to change since the medium is CO$_3^{2-}$ buffered). For anaerobes, use bottles or tubes sealed with butyl rubber stoppers or septa. Culturing of phototrophs requires incandescent light (a 15 or 40 watt light bulb placed 10 to 20 cm from the culture).

Inocula and maintenance

Inocula that have a high success rate can be obtained from a variety of environments including marine sediments, estuaries, stratified lakes, and sulphur-containing (thermal) springs. Storage time of the inoculum should be kept to a minimum; when sediments are used, allow air exchanges for aerobes, restrict exposure to O$_2$ for microaerophiles (denitrifying *Thiomicrospira* spp.), and keep O$_2$-free for anaerobes (purple and green sulphur bacteria). Isolation can be direct (sample is transferred to a bottle or Erlenmeyer flask) or after serial 10-fold dilutions. Alternatively, a reaction vessel through which a continuous low flow (dilution rate of 0.005 to 0.02 hr^{-1}) of medium is fed (simple chemostat) yields good results for slow-growing species such as certain (denitrifying) thiobacilli and *Thiomicrospira* spp. Phototrophs are isolated after dilution in liquefied agar, and single colonies are picked and transferred to fresh (liquid) medium.

Enumeration

S oxidisers are enumerated using MPN techniques. A water sample of known size is diluted by 10 × several times to obtain a series ranging from 10^{-3} to 10^{-9} that of the original concentration. Sediment samples are diluted 100 times and sonicated for approximately 30 to 60 seconds (40 W) prior to the same treatment. Each dilution step is used as an inoculum (1:9, vol/vol) in fresh medium to which a pH indicator is added (e.g. bromocresol purple; 0.4 gram litre^{-1}). Acetate, or any desired organic substrate, can be added in addition to CO$_3^{2-}$ to include heterotrophic S$_2$O$_3^{2-}$ utilisers. Since a wide variety of heterotrophs use S$_2$O$_3^{2-}$ but produce polythionates, it is important to include a pH indicator. Three or five tube series can be used.

Positive tubes are scored and transferred to fresh medium to confirm growth. Positives from the greatest dilutions are used to isolate organisms that are likely to be important in the environment. Epifluorescence microscopy can be used to estimate cell counts in the inoculum. Molecular approaches described above for SRB can similarly be used to study S oxidisers.

For example, oligonucleotides specific for *Thiomicrospira* spp. have been used as PCR primers and probes for investigating the distribution and diversity of this group in coastal sediments and at hydrothermal vents.

Phototrophic processes

Sulphide and $S_2O_3^{2-}$ are electron donors for purple S and green S bacteria. Certain cyanobacteria can also use these electron donors, but this is only environmentally significant in a few lakes. In addition to sulphide oxidation experiments discussed above, chlorophyll (Chl), bacteriochlorophylls (BChl), and other pigments can be used as biomass indicators, and light of specific wavelengths can be used to selectively stimulate or inhibit certain species. Cyanobacteria have additional pigments, phycobiliproteins, which have peak absorbance at 550 to 650 nm. Accessory pigments (carotenoids) of anoxygenic phototrophs typically absorb at wavelengths between 400 and 550 nm. Purple non-S bacteria contain spirilloxanthin, okenone, or lycopene, while green non-S bacteria contain carotene, chlorobactene, or isonieratene. By using certain light sources, either group of bacteria (oxygenic versus anoxygenic) can be selected in enrichment cultures. Photopigments can be measured spectrophotometrically after extraction in methanol with or without hexane. In methanol, absorption maxima (nm) and coefficients $(g^{-1} \cdot litre \cdot cm)$ of Chl$a$, BChl$a$, BChl$b$, BChl$c$, BChl$d$, BChl$e$, and BChl$g$ are 665 and 745; 771 and 84.1; 794 (unstable in air); 668 and 86.0; 654 and 82.3; 646 and 82.3; and 765 (coefficient in methanol unknown), respectively. Different BChls can be further differentiated upon acidification, yielding bacteriopheophytin with different absorption maxima. High performance chromatography with fluorescence detection enables photopigment quantification as well. Photopigment concentrations can be used to calculate protein, which may be useful for specific rate measurements *in situ*. Although the specific photopigment concentration typically increases with decreasing light intensity, a value of 22 to 30 mg of protein µg of BChl^{-1} is typically found.

Organic sulphur oxidation

Dimethyl sulphide (DMS), dimethylsulphoxide (DMSO), and methane thiol (CH_3SH) are consumed by a variety of thiobacilli, anoxygenic phototrophs, saltmarsh ascomycetes, and SRB. Metabolism can be assessed as described above for other rate measurements. Stock solutions of oxidisable substrates should be prepared fresh. DMS is prepared gravimetrically. CH_3SH is prepared from dimethyl disulphide by cleaving with tributylphosphine. The concentration of these organosulphur compounds can be determined by gas chromatography (GC) with flame ionisation detection (FID) or flame photometric detection (FPD). FPD is S specific and therefore more sensitive, than FID, and measurements of volatile S compounds using GC-FPD are facilitated by doping the FPD fuel line with low levels of an S gas (i.e. carbonyl sulphide, sulphur dioxide, or carbon disulphide). Purge and trap systems can further lower detection limits. Care should be taken with CH_3SH, which rapidly binds to sediments, glass, and rubber, and appropriate controls are necessary to account for abiotic losses. DMS can be oxidised photochemically and biologically to DMSO, which can be measured as DMS after chemical reduction with a stabilised titanium solution. When performing experiments with organosulphur compounds, use Teflon-lined stoppers and heat glassware overnight (300° to 450°C) prior to use.

Inhibitors

Molybdate (MoO_4^{2-}) has been used as a 'specific' competitive inhibitor of SO_4^{2-} reduction, which is useful for determining the direct role of SRB in biogeochemical transformations. In marine samples, MoO_4^{2-} concentrations of 20 to 28 mM are applied, although a much lower (5 mM) concentration has almost the same effect. For freshwater samples, less MoO_4^{2-} is required (<2 mM). In addition to SRB, MoO_4^{2-} inhibits S metabolism by thiobacilli and phototrophs by 15 and 30 per cent, respectively. All of these organisms contain ADP and/or ATP sulphurylases, which may be affected by MoO_4^{2-}. Other group

VI oxyanions, such as WO_4^{2-} and SeO_4^{2-}, have also been used to inhibit SO_4^{2-} reduction but are not quite as thorough. However, these latter compounds may prove useful for some applications since they interact less with free thiols than MoO_4^{2-}. Anthraquinone derivatives like 1,8-dihydroxy-anthraquinone (DHAQ) can also be used to selectively inhibit SO_4^{2-} reduction.

Analyses

Dissolved sulphide is measured colourimetrically or with ion-specific electrodes. Both techniques have a detection limit of 1 to 10 μM. Alternatively, samples can be acidified in closed serum bottles and the headspace can be assayed for H_2S by GC with FID or FPD. GC techniques have very low detection limits (nM) and are not necessary for most applications. Microelectrodes can be used for measuring sulphide microprofiles. $S_2O_3^{2-}$ is determined colourimetrically after cyanolysis, which can also be used to measure polysulphides. The detection limit with this technique for both compounds is 10 μM. It is important to use pore water as a reagent blank to account for possible background colour. Alternatively, anion-exchange chromatography and UV detection can be used to measure $S_2O_3^{2-}$ using samples that are filtered through 0.2-μm-pore-size nylon filters (Alltech) (limit of detection, 1 to 5 μM).

High-performance liquid chromatography with fluorometric detection can also be used to measure oxidised and reduced S species. Nitrate is measured colorimetrically after reaction with sulphanilamide (1 to 5 μM detection limit) or in freshwater samples with an ion-specific electrode (30 μM, limit of detection). Nitrate can also be measured by ion chromatography. O_2 is measured with an O_2 monitor consisting of a polarographic electrode (Clark-type; Yellow Springs International, Ohio) attached to a strip chart recorder.

PHOSPHORUS CYCLING

Phosphorus is considered to be a major growth-limiting nutrient in aquatic systems, and, unlike the case for nitrogen, there is no large atmospheric source that can be made biologically available. All living organisms require phosphorus for growth and metabolism. It is essential in both cellular energetics (ATP) and cellular structure (DNA, RNA, and phospholipids). As such, the phosphorus cycle has been extensively studied and is of great interest in the aquatic environment.

In contrast to the microbial cycles of nitrogen and sulphur, phosphorus, in general, does not commonly undergo any oxidations or reductions and remains combined with oxygen in its pentavalent state as phosphate (PO_4^{3-}). It is thought that some sediment micro-organisms may have the ability to utilise phosphate as a terminal electron acceptor in the absence of oxygen, nitrate, or sulphate and produce phosphine (PH_3). This process is, however, of little ecological significance, and we will limit, our discussion to the cycling of phosphate.

The aquatic cycling of phosphorus is essentially the conversion of phosphorus from the organic to the inorganic state and *vice versa*, with micro-organisms playing a key role in both transformations. There are two pools of phosphorus that must be considered: the large, slowly cycled pool of geologic phosphate contained in the earth's crust and sediments and the much smaller but rapidly cycled pool of biologically active phosphate. In the water column, this smaller pool exists predominantly as dissolved organic and inorganic phosphates, living and dead biological materials, and some suspended particulate and sediment inorganic phosphates. Although micro-organisms play a key role in the release of phosphorus from the large geologic pool by mineralisation and dissolution through the production of organic acids, the primary role of micro-organisms is in mediating the cycling of the small biologically active pool. It is this pool of phosphate and the methods used to examine it that will be discussed here. In addition, this section is limited to a discussion of phosphorus cycling in the water column.

Since phosphorus is a limiting nutrient in most of the systems studied, its examination is often difficult because of the small quantities actively being cycled. Many of the traditional analytical methods are not sensitive enough to be used or require such large sample volumes that they are inappropriate. For this reason and because there are readily available radioactive phosphorus isotopes, ^{32}P and ^{33}P, many researchers have relied on the use of radioisotopes for the examination of phosphorus cycling. In addition, there are several nonradioactive methods for the examination of phosphate availability, based on the production of the enzyme alkaline phosphatase or 5′-nucleotidase. The magnesium-induced co-precipitation (MAGIC) technique for the measurement of nanomolar quantities of phosphate, developed by Karl and Tien, may help to solve some of the analytical problems, but it is unlikely to replace the use of radioisotopes since the pool of available P is cycled so rapidly in oligotrophic environments.

Measurement of Phosphorus Uptake and Incorporation by Using ^{32}P and Differential Filtration

The method described below relies on the use of radioactive $^{32}PO_4^{3-}$ and requires that certain precautions be taken. Briefly, the procedure consists of the addition of trace amounts of $^{32}PO_4^{3-}$ to a water sample; after various periods of incubation, the sample is filtered through a series of filters with different pore sizes to obtain both the turnover times and the incorporation rates of various size fractions. The use of radioactive phosphate allows the determination of phosphate uptake at concentrations at or near ambient concentrations. In general, phosphate concentrations below 0.03 μM are not easily quantifiable by colourimetric chemical analysis; concentrations of phosphate in aquatic systems are often below this concentration. If this is the case or if one wishes to determine which size fraction is incorporating phosphorus from the water column, the use of ^{32}P is appropriate.

Due to the speed at which trace quantities of phosphate are taken up by microbial communities, this procedure is generally divided into two phases. In the first phase, the objective is to collect as many samples as possible within the first 10 minutes for the calculation of phosphate turnover time (as defined by Lean and others). After this intensive sampling, a second series of samples is differentially filtered for size fractionation, if these results are desired. Since turnover times of less than 10 minutes are common, it is important to have all the necessary materials set up before the procedure is begun. Initially the assay is conducted using a 0.2 μm-pore-size filter only, and the results are calculated on the basis of the percentage of the ^{32}P remaining in the filtrate. Since it is easier to count the ^{32}P retained on the filters and hence calculate the activity remaining in the filtrate, the filtrate is discarded and the filters are retained for counting.

For this procedure, duplicate 250 ml water samples are placed in 500 ml polyethylene, glass, or polycarbonate bottles. Plastic bottles are recommended because of the fragility of glass, even though they can represent a radioactivity decontamination problem. Samples are placed in a water bath at field temperature. The initial uptake of phosphate is usually restricted to bacterial activity; however, if it is suspected that photosynthetic organisms play a significant role, incubations should be done under both light and dark conditions.

After a sufficient interval has been allowed for the samples to reach equilibrium with the incubation conditions, generally in the range of 30 to 60 minutes, a calculated amount of carrier-free $^{32}PO_4^{3-}$ (Amersham or DuPont, NEN Research Products) is added to give between 20,000 and 50,000 cpm/ml. This amount should be calculated before the experiment is begun and is based on the decay rate and isotope preparation data provided by the manufacturer. Since ^{32}P decays to form ^{32}S, the volume added is unimportant, because it does not affect the final P concentration, but should generally be less than 0.1 ml because of the acidity of the phosphate solution (generally as orthophosphoric acid in dilute HCl). After the isotope is added, the contents of the bottle are mixed vigorously without shaking, so that changes in dissolved-oxygen concentration are minimised and to avoid the formation of radioactive aerosols, and the bottle is placed in the water bath.

Sub samples of 1 to 10 ml are then removed at intervals of approximately 0.5, 1, 2, 4, 8, 10, 20, 40, 60, and 120 minutes and filtered through a 25 mm-diameter, 0.2 µm-pore-size filter. Before the isotope is added, the water to be examined should be checked for its filtration time. As a general rule, the filtration process for a 10 ml sample should take less than 30 seconds. If a longer time is required, smaller subsamples should be used.

The use of subsamples smaller than 1.0 ml is not recommended, since reproducibility seems to suffer. In general, nuclear track emission (Nuclepore or equivalent) filters work better than depth retention filters. It is critical that accurate times be recorded and that the precise time of the end of filtration be noted. For this purpose, each bottle should have a separate stopwatch assigned to it and the filtration process should be observed so that the time can be recorded as soon as the surface of the filter is dry. The accuracy of the results depends highly on the precision of the time determination and reproducibility of subsample collection.

After 1 or 2 hours, additional 10 ml subsamples are filtered onto 0.2, 1.0, 5.0, and 12 µm-pore-size (Nuclepore or equivalent) filters to determine the size dependence of phosphate uptake. The use of Nuclepore or equivalent filters ensures that little water will remain in the filter matrix. This is important since any remaining water retained by the filter will give inaccurately high values. As the filters are collected, they are placed into 20-ml scintillation vials and covered with a sufficient amount of scintillation cocktail or water.

If water is used (recommended, since problems associated with the disposal of waste are minimised), it is important that the filters be placed flat on the bottom of the vial, filtration surface up, and covered with 2 ml of water. The use of water makes the counting efficiency dependent on the geometry of the sample, since water has no scintillation properties. This reproduction of the geometry is not necessary if a scintillation cocktail is used.

For the determination of total activity, samples along with 2.0 ml of the unfiltered water are placed in a liquid scintillation counter and assayed for radioactivity. For calculation of the turnover time, the counts retained on the filter are subtracted from the total counts, and the natural logarithm of the percentage of the ^{32}P in the filtrate is plotted against time (Table 21.7 and Fig. 21.4). The initial linear portion of the slope is used to approximate the uptake rate constant, and the turnover time of phosphate is the reciprocal of the rate constant (Fig. 21.4).

Usually, short turnover times indicate phosphorus-limiting conditions and long turnover times indicate that some other factor limits microbial activity or system productivity.

Just as the turnover time gives information about the rate of phosphate uptake, the result of the size fractionation gives us information about the type of micro-organism responsible for the phosphate uptake. To calculate these values, the counts from the sequential filtration steps are subtracted from each other to give the percentage of phosphate incorporated by the 0.2 to 1.0, 1.0 to 5.0, 5.0 to 12.0, and >12.0 µm fractions.

The value for the material collected on the 1.0 µm-pore-size filter is subtracted from the value for the material collected on the 0.2 µm-pore-size filter to give the 0.2 to 1.0 µm fraction, the value for the material collected on the 5.0 µm-pore-size filter is subtracted from the value for the material collected on the 1.0 µm-pore-size filter to give the 1.0 to 5.0 µm fraction, etc.

Table 21.7. Rate of $^{32}PO_4^{3-}$ incorporation by an Everglades National Park, Fla., water sample.

Sample collection time (minutes)	cpm[a] on filter	% of total cpm on filter[b]	% of total cpm in filtrate[c]	Ln % of total cpm in filtrate
0.80	14,736	25.23	74.77	4.31
4.20	35,369	60.56	39.44	3.67
7.80	43,992	75.33	24.67	3.21
11.20	47.464	81.27	18.73	2.93
17.80	49,511	84.78	15.22	2.72
30.90	52,675	90.20	9.80	2.28
40.60	54,462	93.26	6.74	1.91
49.30	53,888	92.27	7.73	2.04
68.50	54,369	93.10	6.90	1.93
77.50	55,218	94.55	5.45	1.70
109.00	55.426	94.91	5.09	1.63

[a] Total cpm added was 58,400.

[b] This value is given by (filter cpm/total cpm) × 100.

[c] This value is given by 100 − % of total cpm on filter.

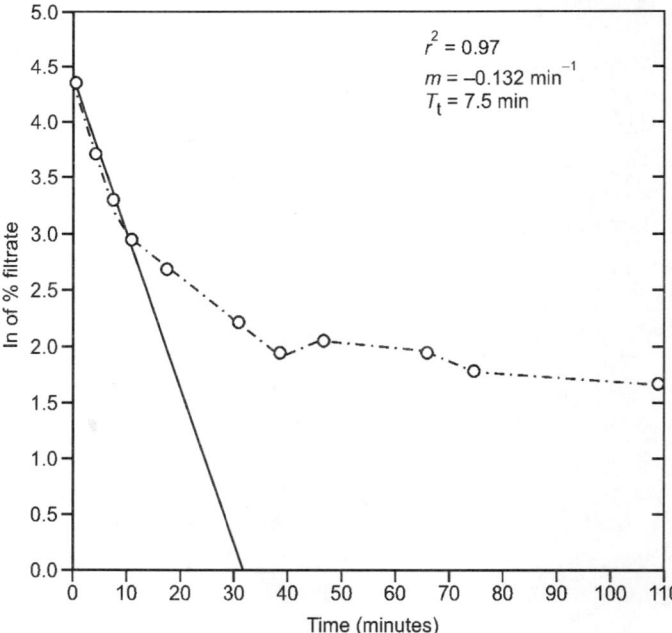

Fig. 21.4. Incorporation of $^{32}PO_4^{3-}$ into particulate material. The solid line shows the initial slope for turnover time calculation; m, slope; T, turnover time.

The counts retained on the 12.0 μm-pore-size filter represent the fraction incorporated into particles of >12 μm. Results are generally presented as the percentage of the total amount of phosphate incorporated (counts on the 0.2 μm-pore-size filter). However, if the initial concentration of phosphorus is known, the result can be expressed in absolute terms. An example of the type of distribution expected is shown in Fig. 21.5. For interpretation, it is generally assumed that the 0.2 to 1.0 μm fraction represents bacteria, that the 1.0 to 5.0 μm fraction represents a mixture of bacteria and phytoplankton, and that both the 5.0 to 12.0 μm and >12.0 μm fractions represent phytoplankton. These size classifications are highly dependent on the system being examined, and microscopic examination of the retained material should be performed for a more precise determination. In addition, the determination of chlorophyll or other specific pigments can help interpret size classes.

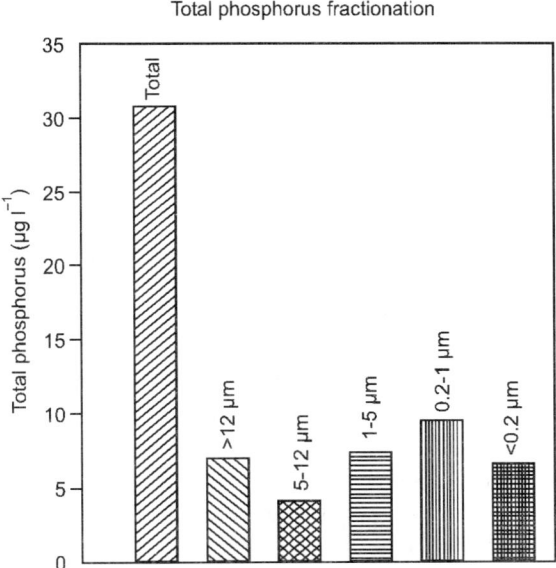

Fig. 21.5. Total phosphorus size fractionation in a water sample.

The use of $^{32}PO_4^{3-}$ has many other applications, several of which require only minor modification of the foregoing procedures. Rate constants such as K_m (the Michaelis-Menten constant) and V_{max} (the maximum velocity of the reaction at substrate saturation) can be calculated by using several different concentrations of cold (nonradioactive) phosphate to dilute the activity of the $^{32}PO_4^{3-}$. Generally, total inorganic phosphate concentrations must be kept below 2.0 μM for this procedure. ^{32}P can also be used as a tracer of phosphate movement through the food web. This can be accomplished by conducting the differential filtration described above at different time intervals and observing the change in the percentage of P contained in each fraction. This approach could also be used to examine phosphorus incorporation by larger organisms and macrophytes.

The procedure described above is generally limited in use only by the precautions necessary for the safe handling of ^{32}P. Care must be taken that the samples collected for ^{32}P uptake and incorporation are homogeneous and do not contain large particles or rapidly settling sediments. If the samples used contain these materials, the subsamples will not be uniform and the results could be affected. The samples should not be refrigerated and should be assayed as soon as possible after collection. In general, samples more than 6 to 12 hours old should not be used. The effects of storage are, however, easily determined by experimentation.

Interpretation of the data can be more difficult if there are high concentrations of particulate inorganic phosphates. Without chemical analysis of the incorporated phosphate, it is not possible to be certain that all incorporation was a consequence of the conversion of inorganic phosphates into organic phosphates. Results from many systems have indicated that much of the material is incorporated into microbial cells, but caution is advised when making this assumption.

Alkaline Phosphate as an Indicator of Phosphorus Limitation

The use of $^{32}PO_4^{3-}$ as described above yields potential rates for the microbial conversion of inorganic phosphorus to organic phosphorus. It cannot establish if phosphorus is limiting. Alkaline phosphatase, on the other hand, can serve both as an indicator for the potential for conversion of organic phosphorus into inorganic phosphorus and as a test for phosphate limitation. The importance of alkaline phosphatase in the cycling of phosphorus in natural waters is paramount. Phosphorus occurs in nature in the form of organic and inorganic phosphates, but almost all osmotrophic organisms utilise only inorganic phosphate for growth. All organisms possess alkaline phosphatase, but only bacteria, fungi, and perhaps some algae excrete the enzyme outside their cells, thus participating in the remineralisation and dissolution of organic phosphates. Since inorganic phosphate is almost always found in growth-limiting concentrations, one of the most important microbial activities is the conversion of organic phosphates to inorganic phosphates through the action of phosphatases.

There are two types of phosphatases, acid and alkaline, identified by their pH optima. These enzymes function by hydrolysing organic phosphate esters as well as inorganic pyrophosphate and certain other inorganic phosphates. Phosphatases are present in all organisms, but only organisms that possess an extracellular (usually periplasmic in gram-negative bacteria) phosphatase are able to mineralise external sources of organic phosphate. Without alkaline phosphatase, organic phosphorus would act as a terminal sink for inorganic phosphorus and productivity in most systems would be limited by external phosphorus inputs from continental weathering. Phosphatase production and the various genetic and kinetic control mechanisms are in themselves an area of active study; indeed, some of the mechanisms for cell regulation make phosphatases excellent bioindicators. However, for purposes of this section, the determination of extracellular phosphatases will be discussed as a method.

What makes alkaline phosphatase so appropriate is that micro-organisms do not produce extracellular alkaline phosphatase in the presence of excess dissolved phosphate. This allows the use of alkaline phosphatase as a sensitive indicator for phosphate limitation. Alkaline phosphatase has been used successfully as a bioindicator in several systems, but because of the nature of the regulation of alkaline phosphatase, the environments for which it is best suited are oligotrophic. Extracellular production of alkaline phosphatase responds to increased levels of dissolved inorganic phosphate on the order of minutes to hours rather than days, making it suitable for the temporal variations of phosphorus concentration often found in natural systems.

In environments in which the concentration of dissolved inorganic phosphate is high and phosphate is not growth limiting, the assay of extracellular alkaline phosphatase may not be appropriate. If, however, there is an active cycling between the organic and inorganic forms of phosphate, alkaline phosphatase activity assays may prove useful.

The most commonly used procedures for assaying alkaline phosphatase activity are based on the use of an artificial substrate such as *p*-nitrophenyl phosphate (PNPP) or 3-*o*-methylfluorescein phosphate (MFP). PNPP is a colourless compound that is cleaved by alkaline phosphatase to yield PO_4^{3-} and *p*-nitrophenol, which is yellow and can be detected spectrophotometrically. MFP is hydrolysed by alkaline phosphatase to yield PO_4^{3-} and 3-o-methylfluorescein (MF), which is highly fluorescent and is detected fluorometrically. The use of MFP gives sensitivity nearly three orders of magnitude greater than that obtained with PNPP, and detection can be in the nanograms-per-litre range. Because of this and several other difficulties pertaining to the use of PNPP, the procedure for the use of MFP will be described here. However, in addition to MFP, another fluorogenic substrate, 4-methylumbelliferyl phosphate, can be used, requiring only a modification of the excitation and emission wavelengths. For the MFP procedure, it is very important that caution be used to ensure that the conditions of the assay are standardised. Temperature and pH both affect the fluorescence of MF as well as the activity of alkaline phosphatase. A saturating concentration of MFP dissolved in a buffer solution is added to a water sample. The sample is incubated, and alkaline phosphatase present in the samples cleaves the phosphate from the MFP, leaving MF. The concentration of MF, assessed fluorometrically, is proportional to the alkaline phosphatase activity of the sample. In this assay, duplicate 3 ml subsamples are pipetted into disposable cuvettes (four clear sides, polystyrene) and 30 µl of MFP reagent is added to each and mixed by using disposable plastic transfer pipettes. The MFP reagent is prepared by dissolving 53 mg of anhydrous MF (Sigma Chemical Co.) in 100 mM Tris buffer (pH 8.7) (Sigma). The final concentrations of MFP are 10 mM in the stock reagent and 100 µM in the assay mixture. The final concentration of Tris buffer in the assay mixture is 1 mM. This stock MFP reagent can be stored in the freezer at –20°C in small plastic vials (2 to 3 ml) and thawed prior to each use. The reagent stored in this manner has a useful life of more than 1 year.

The pH of the MFP reagent generally depends on the pH of the sample and the type of phosphatase, acid or alkaline, being assayed. Once again, it is important to remember that the objective of this procedure is to assay enzyme activity under standard conditions and that the greatest sensitivity will be obtained under conditions favouring the activity of alkaline phosphatase, not microbial growth. It is also important to note that the concentration of the Tris buffer may have to be increased or the pH of the samples may have to be adjusted if samples with high acidity or alkalinity are used. For most natural waters, the MFP reagent described here will be adequate.

After the sample is mixed, the fluorescence is immediately measured by using a fluorometer with an excitation wavelength of 430 nm and an emission wavelength of 507 nm and recorded. The subsamples are then incubated for 2 hours at 25°C in the dark. The temperature of incubation is unimportant as long as the conditions are standardised. However, again it is important to remember the objective of optimum conditions for enzyme activity. After this period, the cuvettes are again assayed fluorometrically at the same wavelengths, and the results are recorded. The amount of MF produced in 2 hours is quantified by subtracting the initial value from the final value, and the result is compared against a standard curve. Different incubation periods can be used, but the amount of MFP converted to MF should never exceed 10 per cent (i.e. 10 µM) of the initial concentration of MFP.

If it does, the assayed value for alkaline phosphatase activity will be lower than the actual value since the MFP may no longer be at a saturating concentration. As an alternative to using a standard fluorometer, sample volumes can easily be scaled to use 96 well microtiter plates and a fluorescence plate reader with the appropriate excitation and emission filters. We use a CytoFluor series 4000 multiwell plate reader (PerSeptive Biosystems, Framingham, Mass.). The use of microtiter plates and the associated small sample size (200 µl) results in higher variability in the fluorescence reading, and thus the mean of eight subsamples is necessary to achieve the precision of the results with duplicate 3 ml subsamples using a conventional fluorometer.

To prepare the standard MF curve, it is important to ensure that the standards are made up in the same buffer at the same pH, concentration, and temperature as the assay mixture. Working standards are prepared by using a 1 mM solution of MF in methanol (stock stored in freezer at –20°C) diluted into 1 mM Tris buffer (pH 8.7). Concentrations of 0, 1, 2.5, 5, and 10 μM are used to generate the final curve. If care is not taken to ensure that the pH and temperature are the same as in the assay mixture, the results will be incorrect, since MF fluorescence is highly dependent on these factors.

When samples for alkaline phosphatase activity are collected, it is important that sediments and large particles be excluded, since they will cause a positive interference and lead to difficulty in data interpretation. In general, samples should not be refrigerated but, rather, stored in the dark at ambient or assay temperature. It is important that the samples be assayed as soon as possible, but if necessary, they can be stored for up to 24 hours with little change in activity. An example of the results obtained by this procedure is shown in Fig. 21.6. Although the results of alkaline phosphatase assays are often expressed in absolute units (micromoles per litre per hour), these results are generally not strictly comparable from one environment to another. Results are best viewed in the context of relative values when one is working with a series of similar matrix samples.

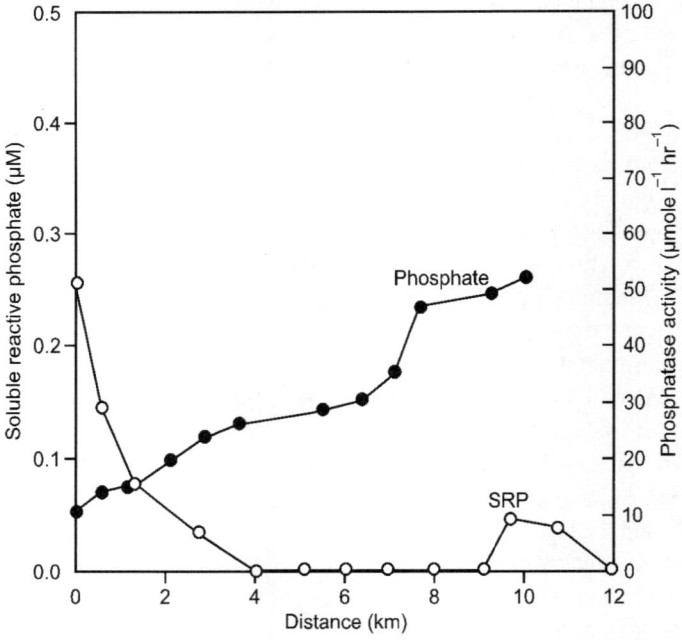

Fig. 21.6. Alkaline phosphatase activity and soluble reactive phosphate (SRP) concentrations in water samples from a transect.

In the example in Fig. 21.6, the low alkaline phosphatase activity at 0 km indicates an excess of dissolved inorganic phosphate and nonlimiting conditions while the value of 53 at 11 km indicates a situation in which phosphate is in limited supply. Thus, high alkaline phosphatase activity indicates phosphate limitation while low activity represents the opposite. In some situations, alkaline phosphatase activity can be affected by the number of bacterial cells present in the sample. This can be of concern when low alkaline phosphatase activities are associated with small cell numbers rather than low concentrations of alkaline phosphatase per cell. When examining waters with small bacterial numbers one should exercise caution when interpreting results. The opposite result can occur with large bacterial numbers, but this is much less common.

In addition to the procedure described above, MFP can be used to estimate the actual rate of organic phosphate mineralisation. This is done by assaying the rate of MF formation at several different concentrations of MFP below saturation, applying first-order rate kinetics. The most appropriate method for interpreting these results when the concentration of organic phosphate cannot be measured accurately is that proposed by Wright and Hobbie and discussed by Button. MFP can also be applied to sediments and soils with only minor modifications, usually involving centrifugation or filtration to remove interference with the fluorescence measurements. Other enzymes, such as 5'-nucleotidase, are also important in the regeneration of phosphate. Intracellular production of alkaline phosphatase is also related to nutrient limitation and can be used to examine phosphate cycling by larger organisms, particularly the algae associated with surfaces. In summary, the use of MFP is a simple, sensitive, and inexpensive method to examine phosphate mineralisation and limitation.

SECTION VI

Soil, Subsurface and Landfill

22. **Groundwater Chemistry and Subsurface Microbiology**   365

23. **Microbial Ecology of Groundwater Systems**   371

24. **Abundance and Distribution of Bacteria in the Subsurface**   393

25. **Microbiological Sampling of Subsurface Environments**   410

26. **Microbial Acclimation to Groundwater Contamination**   422

Groundwater Chemistry and Subsurface Microbiology

INTRODUCTION

This chapter shows that the history of subsurface microbiology has been one of convergence between the different disciplines of geology and microbiology. Observations of groundwater chemistry had suggested, from at least the 1920s, the possible importance of microbial processes. However, an observational approach was not sufficient to demonstrate that this was, in fact, the case. Rather, rigorous sampling methods and experimental techniques were required to demonstrate adequately the role of microbial processes.

On the other hand, experimental methods themselves had significant disadvantages, such as extreme cost, disturbance of sediments, and an inherent small scale that was inappropriate for large hydrologic systems. In the end, an integration of microbiology and the geosciences has proved to be the most effective means of studying subsurface microbiology.

This integrated approach, which requires that microbiologists and geologists understand each other's unique and important perspectives, holds much promise for unravelling important environmental problems in terrestrial subsurface environments.

Given the very different histories and different traditions of geology and microbiology, it is not particularly surprising that they have stayed largely separate during much of the twentieth century. As the scopes of each discipline enlarged over time, however, it was inevitable that they would eventually reach a point of intersection. The issue that provided this point of intersection was the chemical quality of groundwater.

Characteristically, however, each discipline approached the groundwater quality issue from different viewpoints. Geoscientists were initially concerned with understanding the natural water chemistry of pristine aquifer systems. It was observations associated with oil field brine geochemistry that first suggested that microbial processes in deep subsurface environments could be of major importance. This, of course, is perfectly in keeping with the observational tradition of the geosciences.

Microbiologists were also drawn into the study of subsurface environments by water-quality issues. In this case, however, concern about chemical contamination due to waste-disposal practices was the primary motivating factor.

Furthermore, development of asceptic drilling techniques and experimental evidence for the presence and activity of subsurface micro-organisms was required before the presence of subsurface micro-organisms were widely accepted in the microbiological community. This, in turn is in keeping with the experimental nature of the microbiological sciences.

SUBSURFACE MICROBIOLOGY AND THE GEOSCIENCES

In 1919, a geologist named Sherburne Rogers, working for the US geological survey, published a comprehensive study of the Sunset-Midway oil field in California. As part of this study, analyses of the brines associated with the hydrocarbons were made. These analyses pointed out that the oil field brines had a significantly different composition than shallower groundwater in the area. Specifically, shallow groundwaters were characterised by relatively high sulphate and relatively low bicarbonate concentrations. The brines associated with oil, on the other hand, showed high concentrations of bicarbonate but an almost total absence of sulphate.

Rogers interpreted this as evidence that, where groundwater and hydrocarbons coexisted, the oxidation of hydrocarbons resulted in the reduction of sulphate with the production of bicarbonate. Furthermore, Rogers suggested that sulphate-reducing bacteria were involved in these reduction processes. At the time, however, there was no direct evidence that such micro-organisms were actually present in hydrocarbon reservoirs.

Over the next 50 years or so in the United States, the idea that bacteria are present in deep subsurface environments and that they have important effects on groundwater chemistry was gradually broadened.

Zobell suggested that sedimentary organic material subjected to 'dynamochemical processes' (i.e. increased temperature and pressure) would progressively eliminate carboxyl groups and produce carbon dioxide abiotically. This carbon dioxide could then react with carbonate minerals to form the observed high-bicarbonate groundwater. Foster's paper was highly cited in the geochemical literature and was enormously influential. Consequently, the previous tradition that microbial processes were important in groundwater chemistry began to be replaced by a conviction that inorganic processes were predominant.

A study of the microflora of the groundwaters of the artesian basins in western Siberia lowland has shown that sulphur oxidising and denitrifying bacteria that oxidise sulphur are widespread in these deep artesian aquifers, even at depths exceeding 1.8 km.

One reason that consideration of microbial processes fell out of favour in the United States during the 1960s and 1970s was that groundwater chemists focused on other equally important lines of inquiry. The concept of hydrogeochemical facies showed that water chemistry evolution followed regular, predictable patterns that depended on groundwater flow paths. Back's observations were crucial because, for the first time, it could be shown that the chemical composition of groundwater was not random (as it had often seemed) but that it followed regular, predictable patterns. Subsequent studies by Garrels and Mackenzie showed that basic concepts of inorganic chemistry and phase equilibria could explain many of the patterns that were observed.

SUBSURFACE MICROBIOLOGY AND MICROBIAL ECOLOGY

Soil microbiology has been an important subdiscipline of microbiology in the early twentieth century. However, soil microbiology traditionally has been preoccupied with soil fertility as it relates to agricultural practice. Consequently, prior to 1970, there was virtually no microbiologic investigations that extended below the root zone of plants.

In the 1970s, however, it became increasingly evident that certain waste-disposal practices were contaminating subsurface environments below the root zone. There was considerable concern about the effects of these waste-disposal practices on groundwater quality and this concern began to draw microbiologists toward studying subsurface environments. In 1973 two environmental protection agency microbiologists, William Dunlap and James McNabb, driven by concern about contaminated aquifers, prepared a literature search that turned up many of the early references from the geologic and petroleum literature.

At this point, the historical differences between microbiology and geology began to make an impact. The problem was that the early studies by Rogers and Cedarstrom were entirely observational and thus, by microbiologic standards, unreliable.

Even early studies that included normal microbiological methods seemed to be of dubious reliability. Normal microbiologic standards of aseptic technique did not appear to have been followed in most of these studies.

Furthermore, the possibility that micro-organisms could have been introduced to the deep formations through the drilling process or through the finished wells made the microbiologists deeply suspicious of the early work from the petroleum industry.

Microbiologists, who were trained to rely on rigorous experimental procedures, began working on ways to evaluate the possibility that drilling procedures or wells introduced micro-organisms into previously sterile subsurface environments. One particularly clever study was described by Olson. The investigators were studying groundwater from the deeply buried Madison aquifer of Montana, which produced fairly hot (47–54°C) groundwater. It was found that sulphate reducing bacteria were present that would grow at 50°C but not at 22°C. Accordingly, this suggested that the micro-organisms were adapted to the temperature of the aquifer and therefore were certainly indigenous to it. Water from wells was used by a number of microbiologists to isolate micro-organisms from pristine and contaminated aquifers.

The development of aseptic procedures for sampling deeply buried sediments was of particular importance in studying subsurface microbiology. Dunlap described a method by which a bore-hole was drilled with augers, sediment samples were collected with a split-spoon sampler, and the outer (i.e. contaminated) portion of the core removed with a paring device. With some modifications, these sampling procedures were used to collect aseptic samples for a variety of experiments and procedures by Wilson. These aseptic coring procedures have been widely accepted and are standard techniques for sampling shallow water table aquifers.

Auger drilling is not suitable for coring sediments more than about 30 metres deep. For such deeper sediments, mud-rotary, or air-drilling methods are necessary. This greatly complicates the process of obtaining uncompromised sediment samples.

Chapelle used a combination of mud-rotary drilling and split-spoon coring to obtain sediments as deep as 200 metres. Phelps described sampling procedures using mud-rotary coring. Development of these sampling procedures made it possible to investigate microbial processes in deeply buried sediments without being restricted to water samples from wells.

Many of the microbiologists who were involved in early investigations of subsurface environments were microbial ecologists by training. It is not surprising, therefore, that many of these studies focused on ecological issues, such as microbial diversity and nutritional status. The most important focus, however, was on the ability of subsurface micro-organisms to degrade potential pollutants.

Consideration of microbial ecology issues, however, pointed out some limitations of relying solely on asceptically cored sediments for investigating subsurface microbiology. Specifically, coring deep holes using mud-rotary methods was extremely expensive. Obtaining enough core holes to cover adequately the large areas of most deep aquifer systems was an economic and practical impossibility. This led microbiologists to consider the use of groundwater chemistry in ecological studies of the deep subsurface.

Again, however, microbiologists brought a different perspective to the problem. It was well-known from studies of surface aquatic sediments that certain unstable intermediate products of microbial metabolism, such as hydrogen gas and organic acids, provided information as to the distribution of terminal electron-accepting processes. This raised the possibility that measuring certain metabolites in groundwater could provide information as to the distribution of microbial processes in the subsurface. The use of hydrogen concentrations as a tool for delineating iron-reducing, sulphate-reducing, and methanogenic zones of aquifers was first proposed by Lovley and Goodwin. This technique, which has been used in subsequent hydrologic studies, is an effective combination of both microbiologic and hydrologic methods.

In addition to helping delineate the distribution of metabolic processes in subsurface environments, groundwater chemistry has helped in studying the microbial ecology of subsurface environments. Because of the nutrient-limited nature of most aquifer systems, it was expected that rates of microbial metabolism would be very low in the deep subsurface.

This was confirmed by experiments with cores of deep sediments that showed the turnover of radiolabelled carbon compounds was relatively low. However, because the sediments had been disturbed by coring and because artificial nutrients had been used in the assays, it was difficult to extrapolate these experimental results to *in situ* rates of metabolism. Chapelle and Lovley used a combination of radiotracer experiments and geochemical modelling to delineate actual *in situ* rates of microbial metabolism in a deep aquifer system.

This approach again showed that a combination of microbiologic and geochemical techniques could be effective in addressing issues of microbial ecology.

SUBSURFACE MICROBIOLOGY AND CONTAMINANT BIODEGRADATION

Throughout most of the nineteenth and twentieth centuries, the principal method of disposing of municipal, industrial, and military wastes was to simply dump them into holes in the ground and forget about them. However, by the 1960s, it was becoming increasingly apparent that these waste-disposal strategies were causing serious environmental damage. Concern over environmental pollution was also raised by incidents like the 1967 sinking of the supertanker *Torrey Canyon*, which spilled millions of barrels of crude oil into the English Channel. At that time, it was widely feared that such spills could result in permanent and irreversible damage to the environment. Public awareness of industrial pollution due to sloppy waste-disposal practices crystallised in the early 1970s with the discovery that an elementary school in Love Canal, New York, had been built on top of an industrial dump. Furthermore, toxic chemicals had seeped through the foundations of the school and had exposed hundreds of children and teachers to potentially poisonous vapours.

The public outcry over the Love Canal incident led to congressional action in the form of the Resource Conservation and Recovery Act (RCRA) of 1976. This act mandated the end to unregulated disposal of hazardous wastes. Later, Congress enacted the Comprehensive Environmental Response Compensation and Liability Act (CERCLA) in 1980 to deal with the problem of industrial wastes that had accumulated over the last hundred years. This act established a billion dollar fund to help clean up heavily polluted sites. Because of the huge amounts of money involved, CERCLA quickly acquired the nickname 'Superfund', which it has been called ever since.

The establishment of Superfund created a new industry for cleaning up environmental pollution. At first, it was widely assumed that physical methods, such as simply digging up old waste drums and transporting them to more secure repositories, would be sufficient to clean up most sites. Soon, however, it became apparent that much contamination had accumulated in soils or in underlying groundwater systems. Cleaning up this kind of contamination proved to be much more difficult then had originally been anticipated.

Underground storage tanks, which were commonly used to store and distribute petroleum products and which often leaked, were found to be one of the most common sources of groundwater pollution (Council on Environmental Quality, 1981). However, methods for remediating such contamination were problematic. By far the most common remediation strategy was to pump contaminated groundwater out of the ground and treat it to remove contaminants. Many of these pump-and-treat systems, as they came to be called, did not consider the possibility that the contaminants were being degraded by microbial processes. The reason for this was that, in the early 1980s, little was known about the microbiology of shallow aquifers. Early studies of soil microbiology had indicated that numbers of bacteria dropped off sharply with depth, and it was widely assumed that shallow aquifers were sterile. With the increasing concern about groundwater contamination, however, the microbiology of shallow aquifers became an active topic of investigation. Several of these studies focused on the ability of subsurface micro-organisms to degrade petroleum hydrocarbons. In particular, Wilson showed that toluene, a common component of gasoline, was easily degraded by micro-organisms in shallow aquifers.

Throughout the 1980s, laboratory studies of microbial activity in shallow aquifer systems, and the ability of those micro-organisms to degrade petroleum hydrocarbons, was a topic of intense inquiry. In addition, numerous field studies showed that microbial oxidation of petroleum hydrocarbons was clearly reflected in groundwater chemistry. These field studies were important because they suggested that petroleum hydrocarbons were being oxidised under anaerobic conditions. Throughout most of the 1970s, it was widely believed that molecular oxygen was necessary to biodegrade petroleum hydrocarbons. In fact, the first micro-organism capable of completely mineralising an alkyl benzene under strictly anaerobic conditions was only first isolated in 1988. Laboratory studies have now shown that petroleum hydrocarbons are degraded under nitrate-reducing, Fe(III)-reducing, sulphate-reducing, and methanogenic conditions. For many years, the anaerobic oxidation of petroleum hydrocarbons was controversial, largely because aerobic microbial processes had been the focus of the early studies of the marine environment. Nevertheless, the importance of anaerobic oxidation processes in the biodegradation of petroleum hydrocarbons became firmly established.

But petroleum hydrocarbons were only part of the contamination problem. Industrial solvents like trichloroethene (TCE) and perchloroethene (PCE) had been used since the early 1900s and huge amounts of these chemicals had been simply dumped into seepage basins and unlined landfills. Because solvents like TCE and PCE are entirely synthetic compounds, with no natural analogues, it was widely believed in the 1970s that micro-organisms could not degrade them. But in 1985, Wilson showed that aerobic methane-oxidising bacteria were capable of co-oxidising TCE to carbon dioxide. Later, it became evident that chlorinated ethenes were also subject to reductive biodegradation, with chlorine atoms being sequentially stripped off of the ethene molecule to form dichloroethene (DCE) and chloroethene (which is usually called vinyl chloride, or VC). Later still, it was discovered that VC and DCE were also subject to oxidation by micro-organisms.

Starting from the initial assumption that chlorinated ethenes were not biodegradable at all, it is now evident that there are a number of microbial processes that biodegrade these compounds in groundwater systems, and that they are very important in restricting contaminant transport. This experience has been repeated with a number of so-called 'recalcitrant' contaminants. Polychlorinated biphenols, explosives such as TNT, and gasoline additives such as methyl-*tert* butyl ether (MTBE) are other examples of compounds that were initially thought to be completely inert, but which now are known to undergo microbial degradation in groundwater systems. This, of course, is the normal progression of scientific inquiry. Understanding the behaviour of human-produced contaminants in groundwater systems requires both an observational approach (looking for evidence that degradation is occurring *in situ*) and an experimental approach (showing that a particular microbial process occurs under controlled laboratory conditions). This takes time. From the time that TCE was first identified as an important contaminant of groundwater systems (~1970) it took 15 years to identify a cometabolic microbial degradation process, and another 10 to fully appreciate the importance of reductive and oxidative degradation processes. As new human-produced contaminants are introduced in the future, it is safe to assume that an understanding of their microbial degradation processes will lag in a similar fashion.

Chapter 23

Microbial Ecology of Groundwater Systems

INTRODUCTION

Ecology is that branch of biology dealing with relationships between organisms, and relationships between organisms and their physical surroundings. An ecological community consists of all the organisms—microbes, plants, and animals—that live in an area. A community, together with the physical environment to which it is tied, is called an ecosystem. The major focus of ecological research is to determine how communities are structured, how species in the community interact with each other, and how communities interact with their physical environment.

In order to simplify the investigation of communities, ecologists usually study some subset of the organisms present in a given environment. For example, an ecologist might be mainly concerned with a particular taxonomic group such as birds. Thus, one might read about 'bird communities'. Similarly, microbial ecologists are mainly concerned with relationships between micro-organisms and focus on 'microbial communities'. Given this considerable simplification, determining how organisms in particular communities interact with each other and their environment might appear to be perfectly straightforward. In practice, however, there are often considerable technical and interpretive problems to be overcome.

Consider, for example, a particularly simple ecological question—one concerning blackbirds in marshy environments of the northwestern United States. It is observed that during the mating season, part of a particular marsh is populated exclusively by yellow-headed blackbirds and another part of the marsh populated only by red-winged blackbirds. The ecological question is, what factor or combination of factors lead to the observed segregation of the two populations?

In order to answer this question, the ecologist would begin by making observations on how the two species interact. In this case, the ecologist would observe that in the spring, male redwinged blackbirds arrive in the marsh several weeks before its cousin, the yellow-headed blackbird. The redwing males promptly stake out all of the best nesting sites, which consist of cattail clumps surrounded by deep water. This ensures a constant supply of insects to feed upon while the nestlings are growing. However, when the larger male yellowheads arrive in a couple of weeks, they forcibly evict the redwings from the best sites. This forces the redwings to retreat to less desirable sites. By the time the females arrive and mating starts, the yellowheads inhabit one part of the marsh—the part with the most food—totally excluding redwings. For their part, the redheads inhabit an entirely different, less productive, part of the marsh. The two populations do not overlap.

Given these observations, it is perfectly clear how the yellowheads exclude redwings—they kick them out. Less clear is why yellowheads do not also inhabit the poorer nesting sites as well and thus totally exclude the redwings. To answer this question, the ecologist would have to observe the rate at which yellowheads successfully raise their chicks as a function of the productivity of the nesting site. It turns out that the physical size of the yellowheads, in addition to providing the advantage in one-on-one encounters with redwings, also carries a price. This price is that they are not able to successfully raise their young in nesting sites that do not produce sufficient food. The young yellowheads simply require too much food. The redhead, however, with their smaller size can successfully support offspring in the poorer nesting sites. The redwings exploit an environment closed to the yellowheads, proving that competition can cut two ways. The yellowheads have a competitive advantage in direct confrontation between the two species. The redheads, on the other hand, have a competitive advantage in exploiting available habitat.

Having made these observations, the next challenge to the ecologist is to determine the evolutionary factors that led to the observed relationship between the two species. The yellowheads are bigger, perhaps a result of natural selection favouring the ability to oust redwings from high-quality areas. On the other hand, the size of the yellowheads might reflect a completely different factor, such as a relatively low temperature in its winter range, or perhaps their size makes them less vulnerable to predators. In this case the yellowhead's ability to outcompete redwings for productive nesting sites is unrelated to the actual competition between the species. For their part, are the redwings smaller due to natural selection for inhabiting poorer nesting sites? Alternatively, does the smaller size of the redwings reflect something like migration habits or food availability in their winter range? These kinds of questions, central to the study of ecology, are very difficult to answer with confidence and require careful study. The relationship between the two blackbird species used here as an example, for instance, is still a matter for investigation and debate.

Given the inherent difficulties associated with unravelling relationships between species such as birds — species that can be directly observed — the formidable difficulties faced by the microbial ecologist studying subsurface environments can be more readily appreciated. The microbial ecologist faces the same difficult questions concerning relationships between micro-organisms and their environments, but in this case it is very difficult to directly observe either the micro-organisms or the environments that they inhabit. Furthermore, the sheer number of microbial species inhabiting many subsurface environments can be daunting. It has been only recently that microbial ecologists have begun to examine relationships between micro-organisms in groundwater systems. Nevertheless, the social problems that arise in studying these unique ecosystems place subsurface microbial ecology as an important subdiscipline of both microbiology and ecology. This chapter surveys our present understanding of microbial ecology of groundwater systems.

SCOPE OF SUBSURFACE MICROBIAL ECOLOGY

The microbial ecology of many natural environments has been carefully investigated over the years. Examples of environments that have received particular attention include soils and aquatic sediments. It is no accident that these environments have been selected for intensive study. The microbial ecology of soils is directly related to the agricultural productivity of farmland. Understanding the microbial ecology of soils, and understanding how microbial populations adjust to different agricultural practices has helped boost agricultural production in the twentieth century. Thus, the microbial ecology of soils has been closely studied because it is a problem with major economic applications.

Similarly, the microbial ecology of aquatic sediments has received considerable attention because of practical reasons. In the nineteenth and early twentieth centuries, rivers and streams in the United States and Europe were routinely used for direct waste disposal. It was observed, however, that if raw sewage, to use one example, was placed in a river, the resulting pollution decreased rapidly downstream. Furthermore, it was observed that microbial processes in the fluvial sediments were largely responsible for consuming the dissolved organic carbon and nutrients in the sewage, thus renovating the river. Understanding the nature of these pollution-attenuating processes as well as understanding their rates and capacities was therefore an important practical problem. Also, much of the worldwide cycling of carbon and nutrients occurs in aquatic sediments. In order to understand global carbon balances, for example, it is necessary to understand carbon cycling in aquatic sediments.

The microbial ecology of subsurface environments has emerged as an important topic because of similar practical considerations. In the 1970s it became apparent that there was widespread contamination of shallow water-table aquifers due to a variety of human activities, including waste disposal, agricultural practices, and fuel storage. Gradually, it became clear that subsurface micro-organisms played an important role in the mobility and fate of many contaminants. As such, understanding the ecological interactions between micro-organisms in the subsurface assumed much more importance than had been placed on it earlier in the twentieth century.

The scope of microbial ecology as applied to the terrestrial subsurface is closely tied to the occurrence of groundwater. While groundwater by definition can be produced only from aquifers, confining beds (also called aquicludes) play an important role in shaping groundwater flow patterns and have important affects on groundwater chemistry. Thus, the microbial ecology of subsurface environments has come to include both water-bearing aquifer, sediments and water-retarding aquiclude sediments. In addition, unsaturated sediments overlying water-table aquifers are generally included in the scope of subsurface microbial ecology. To some degree, this introduces some overlap with traditional soil microbial ecology. However, subsurface microbial ecologists tend to focus on unsaturated sediments below the root zone, whereas soil microbial ecologists focus on the root zone. The questions considered in the microbial ecology of the terrestrial subsurface are similar to those considered by aquatic sediment microbial ecologists. Energy, most of which comes originally from the sun, is trapped by photosynthetic plants in organic carbon compounds and flows through an ecosystem. The subsurface microbial ecologist seeks to elucidate the pathways through which the flow of carbon and energy passes, and the processes that control the flow. Closely linked to the flow of carbon are the flow of nutrients, such as phosphorus and nitrogen, and the flow of electron acceptors that are coupled to carbon oxidation.

One important regulating mechanism of carbon flow in subsurface environments are microbial food chains. It is impossible for any single microbe to have the enzymatic capability for degrading all of the organic carbon compounds present in sediments. Thus, it is common for particular microbes to extract energy from organic carbon by performing specialised metabolic processes and transforming the carbon. The transformed carbon compounds may then be used by other microbes as primary substrates.

A simple example of how microbial food chains may regulate carbon and energy flow in a water-table aquifer is shown in Fig. 23.1. In this example, the chain begins with the fermentation of lignin under anaerobic conditions, perhaps by *Clostridia*, to produce molecular hydrogen and carbon dioxide. Some of the hydrogen and carbon dioxide produced may diffuse upward to an oxygenated zone near the water table, and be utilised by hydrogen-oxidising bacteria. Under anaerobic conditions in which mineral electron acceptors such as nitrate, Fe(III), or sulphate are absent, methanogenic bacteria take up a portion of the hydrogen and carbon dioxide, and produce methane. Some of the methane produced in this manner may then diffuse upward to oxygenated water near the water table and is oxidised to carbon dioxide.

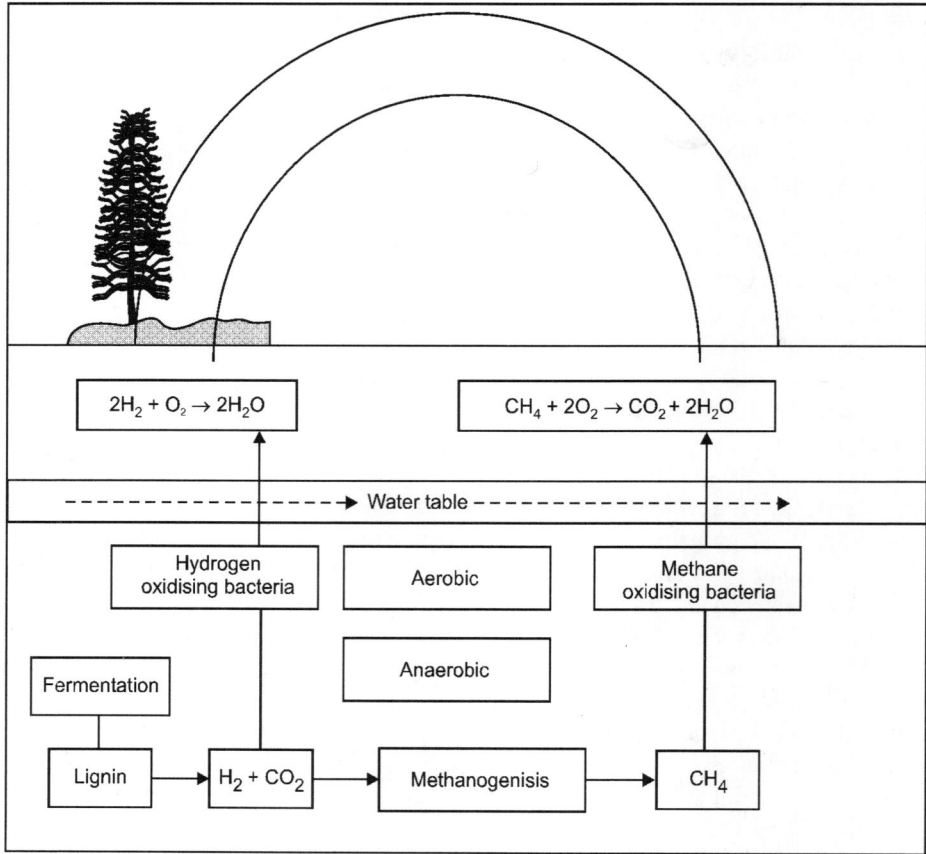

Fig. 23.1. Schematic diagram showing how carbon and energy flow in a shallow water table aquifer proceeds via food chains.

Because sediments typically contain many different kinds of organic compounds that are potential substrates for microbes, it is commonly observed that numerous food chains operate parallel to each other. For example, the lignin present in sediments typically consists of many classes of complex organic compounds, including aliphatic side chains and carbohydrates linked to aromatic molecules. Metabolising these different compounds requires specific enzymatic capabilities, not all of which are present in particular fermentative micro-organisms. Thus, a number of different species may be present, which not only attack certain classes of compounds but also produce different fermentation products. These different fermentation products in turn, form different food chains.

The summation of these ongoing, parallel food chains give rise to the community structure of a subsurface environment. This community structure may be fairly simple, with only a few species actively metabolising in a given system. Alternatively, the community structure may be very complex, with hundreds of strains of microbes involved in organic matter degradation. Determining the community structure of aquifer systems is a major goal of the subsurface microbial ecologist.

Also of primary importance to the microbial ecologist is determining the trophic levels associated with subsurface environments. In classical ecology, the concept of a trophic pyramid is widely used

(Fig. 23.2a). In this model, the base of the pyramid is made up of photosynthetic primary producers such as plankton or plants. The next level of the pyramid contains the primary consumers, principally plant-eating herbivores. At the same trophic level are the decomposers, micro-organisms (fungi, bacteria) that oxidise the organic matter. The next trophic level is filled by secondary consumers (carnivores) that feed on the primary consumers and decomposers. There may be additional trophic levels filled by progressively higher animals. Subsurface environments are fundamentally different from many ecosystems in that the number of trophic levels present is greatly restricted. For example, there can be no photosynthetic primary production in subsurface environments isolated from sunlight. It is possible for primary production to be based on chemolithotrophy, and this occurs in some subsurface environments. However, that is the exception rather than the rule. In most subsurface systems, there is a greatly attenuated trophic structure (Fig. 23.2b). Because there is no primary production and no primary consumers, decomposers, largely bacteria but also including fungi, form the base of the trophic pyramid. Superimposed on the decomposer level is a secondary consumer level. This trophic level is largely filled by protozoa that may actively graze on the decomposers, although it is possible that some predatory bacteria such as *Bdellovibrio* are present in some systems.

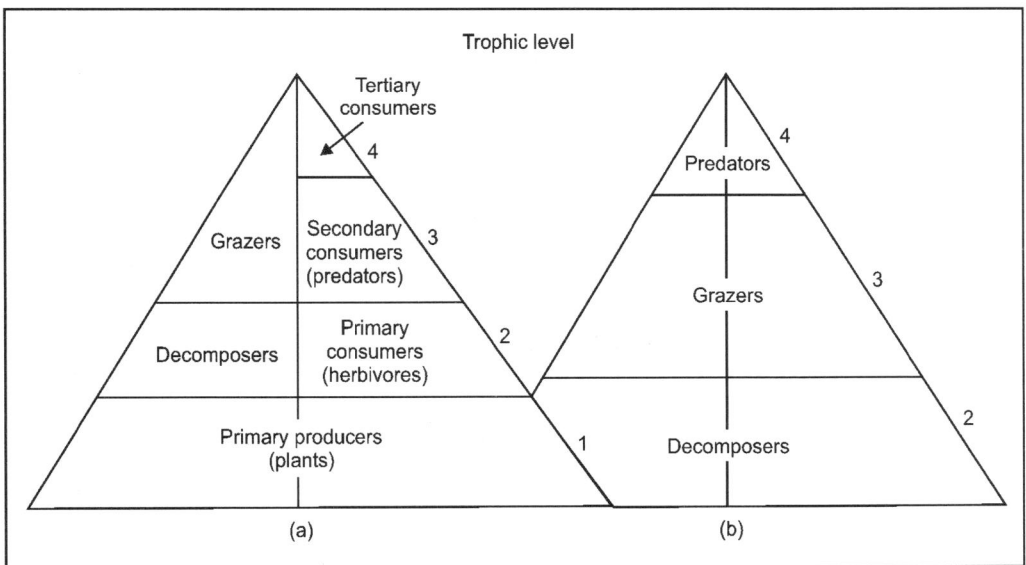

Fig. 23.2. The trophic pyramid model (a) as typically used in classical ecology, and (b) those trophic levels present in groundwater systems.

There may be, in some systems, higher trophic levels present. An example of these higher trophic levels in groundwater systems would be the blind fish that inhabit some carbonate aquifers. These fish feed on protozoa that in turn feed upon the decomposers. Understanding the energy flow, food chains, community structure, and trophic levels of different subsurface environments are important goals in subsurface microbial ecology. The motivation for achieving these goals is varied and depends upon the systems under consideration. For example, in a contaminated shallow water table aquifer, the effects of protozoan predation on decomposer populations — and therefore rates of contaminant decomposition — might be an important consideration. For a clastic aquifer system consisting of alternating beds of sand

and clay, the important ecological question might be how community structure differs between the two sediment types and how these differences impact carbon and energy flow. The focus of an ecological study for any subsurface environment therefore depends largely on the questions being asked. The nature of the relevant questions is then a major factor in selecting appropriate methods of study.

METHODS IN SUBSURFACE MICROBIAL ECOLOGY

Micro-organisms can only be directly observed using a microscope. In order to achieve the resolution required to observe bacteria using either light or electron microscopy, only a very small portion of the microbial habitat can be observed at the same time. Aquifer systems that compose the habitat of subsurface micro-organisms are often hundreds of square miles in area and may be hundreds of feet thick. Clearly, if the decision is made to directly observe bacteria in an aquifer, it will not be possible to observe the aquifer itself in a meaningful way. This difference in scale between the micro-organisms and their habitat is a basic conflict in studying the microbial ecology of subsurface environments. In practice, this conflict can be partially resolved by adjusting the scale of the observation technique used to the scale of the feature of interest. To use an example familiar to geoscientists, one cannot observe cross-bedding by looking at thin sections of sandstone with a petrographic microscope. Such features can be observed only at a hand-specimen or outcrop scale. Conversely, one cannot observe individual sand grains in hand specimen; one must use a petrographic microscope. In general, therefore, the scale of observation must match the scale of the feature that is of primary interest.

This scale conflict is especially important in subsurface microbial ecology. In selecting a scale suitable for observing micro-organisms, little information concerning the microbial environment (i.e. the aquifer system) can be gleaned. Alternatively, if a scale suitable for observing a large aquifer system is selected, little direct information concerning individual micro-organisms can be gleaned. This scale conflict is a fundamental consideration in choosing the most appropriate method for studying a given environment. The most commonly used methods to study the microbial ecology of subsurface environments include:

1. Culture techniques (small-scale).
2. Direct observation (micro-scale).
3. Biochemical marker techniques (small-scale).
4. Activity measurements in microcosms (small-scale).
5. Groundwater geochemistry (large-scale).
6. Molecular techniques (small-scale).

Most of these techniques make observations at a small scale (i.e. samples from 1 to 1000 grams of aquifer material) or micro scale (i.e. samples less than 1 gram). The only large-scale observations (volumes of aquifer material greater than 10 square metres) that can be made are those using groundwater geochemistry.

As we shall see, each of these approaches and their corresponding scales have particular advantages and disadvantages that must be carefully evaluated by the investigator. As a practical matter, the selection of a method or combination of methods should reflect the questions being asked. If the presence of micro-organisms capable of particular metabolic functions (such as degradation of organic chemicals, for example) are of primary interest, culture techniques or activity measurements (small scale) are appropriate. On the other hand, some questions, such as determining overall rates of microbial metabolism in regional aquifer systems require the larger scale of geochemical techniques.

Culture Methods

Procedures for obtaining micro-organisms in pure culture and then analysing pure cultures for their biochemical properties are deeply embedded in the practice of microbiology. Many of the most important advances in medical and agricultural microbiology were made by pure culture methods. These methods can, in some instances, be usefully applied to the microbial ecology of subsurface environments. The basic idea is to isolate as many strains of bacteria as possible from sediments recovered aseptically from the subsurface, and characterise them on the basis of colony morphology, cell morphology, cell arrangement, gram reaction, and physiological/biochemical characteristics. This data can then be used qualitatively and quantitatively to characterise populations of micro-organisms and to compare populations from different environments.

Culture methods for investigating the microbial ecology of subsurface sediments have both advantages and disadvantages. The chief advantage is that culture methods can demonstrate the diversity of microbial communities. The principal conclusion was that the communities, both water-borne and sediment-bound, were fairly diverse. This result could not have been predicted *a priori*. This method also has the advantage of being able to compare microbial diversity between different horizons. For example, the community of sediment probe 2–04 was characterised by a very high percentage of gram-negative bacteria, which set it apart from other sediment communities. Thus, the culture method is also effective in documenting heterogeneity within aquifers. Culture methods have been extensively applied in other groundwater systems to document microbial diversity and heterogeneity.

Culture methods also have significant disadvantages, the most serious of which is the selectivity introduced by using culture media. Many of the bacteria present may not grow on the media and culture conditions offered, or grow so slowly as to be underrepresented.

In addition to the selectivity problem, there is the problem of strain duplication. How does one determine whether two strains of isolates are closely related or whether they simply came from two cells of the same strain? If all isolates are treated as different strains, then isolating a large number of cells of the same strain would create the appearance that many strains have similar properties. On the other hand, if one were to take the position that isolates with similar properties belong to the same strain, one takes the risk of obscuring the presence of closely related but different organisms.

Another problem with culture methods has to do with the 'activity' of bacteria in culture versus under *in situ* conditions. Just because a particular strain exhibits the ability to reduce nitrate in culture, for example, does not mean that particular strain is involved in nitrate reduction *in situ*. In general, it is not feasible to evaluate which processes are important *in situ* from culture data alone. For these important ecological questions, other techniques must be utilised.

Direct Observation

An important limitation of culture methods in subsurface microbial ecology is that they give little information as to processes that are active *in situ*. Other limitations are that culture methods give no information as to the niches inhabited by microbes or to their nutritional status. In order to deal with these problems, microbial ecologists have employed a number of direct observational techniques to address questions of community biomass, diversity, and nutritional status.

An example of how microbes present in aquifer sediments can be directly observed was given by Ghiorse and Balkwill. These investigators employed a two-step approach. First, they prepared smears of aquifer sediment on microscope slides and stained them with acridine orange (AO). Acridine orange binds with nucleic acids present in cells forming a fluorescent dye. When excited by ultraviolet light, the nucleic acid-acridine orange complex emits visible light that can be observed microscopically. In theory, the only light visible should be stained cells. In practice, however, acridine orange binds with mineral and organic particles present in the sediment, and these AO-mineral complexes may also fluoresce. Distinguishing between bacterial cells and sediment particles is often straightforward due to size and shape considerations. However, a measure of experience and skill is involved in such determinations.

The second step used by Ghiorse and Balkwill for directly observing aquifer micro-organisms was to separate the microbes from sediment particles, embed the microbes into epoxy, and observe them with transmission electron microscopy (TEM). Micro-organisms were released by placing sediment in a Waring blender with a solution of 0.1 per cent pyrophosphate and 3.0 per cent glutaraldehyde. The purpose of the pyrophosphate was to help dislodge the micro-organisms. The glutaraldehyde was added to fix the micro-organisms and prevent growth during the mounting procedures.

Using the sediment smear-acridine orange method, it was found that micro-organisms were not uniformly distributed in the sediment. Rather, bacteria tended to occur in microcolonies consisting of anywhere from three or four to hundreds of cells. Several other features of interest, such as bacteria present in filaments, were also observed. Both rod-shaped and coccoidal cells were observed as well as cells that exhibited features of eucaryotes. Additionally, cells that appeared to be in the process of dividing were observed. This indicated that the cells were actively growing *in situ* immediately before or during the sampling procedures.

Phospholipid Fatty Acid Analysis

Direct observational techniques for observing micro-organisms in deep subsurface sediments suffer primarily from two limitations. First, only a very small volume of aquifer material can be accurately screened due to the high magnification required to resolve microbes. Second, direct observations of micro-organism give only limited information as to the phylogenetic types present. It is feasible, as we have seen, to differentiate rod versus coccoid forms and to discriminate between gram-positive and gram-negative micro-organisms. However, in order to accurately evaluate the community structure present, it is desirable to have better resolution of different micro-organisms present. The basis of this technique is that different micro-organisms exhibit significant biochemical differences in the phospholipids and fatty acids (PLFAs) that comprise cell walls and internal storage granules. By directly extracting and characterising PLFAs, information concerning total biomass, cell wall type, presence or absence of eucaryotic organisms, and nutritional status of the micro-organism present may be derived.

Phospholipids are part of every bacterial membrane and maintain a relatively constant proportion of the cell mass. Furthermore, phospholipids are turned over fairly rapidly after cellular death. Because of these two features, the amount of phospholipid present in sediments is a potential measure of the living biomass present. Moreover, the fatty acid composition of the phospholipids can give informations as to the kinds of micro-organisms present. Eukaryotic micro-organisms, for example, are characterised by long-chain polyenoic fatty acids. The presence of such fatty acids in sediment is therefore a direct indicator of the presence of eukaryotic micro-organisms. Similarly, anaerobic micro-organisms form *cis*-vaccenic acids and therefore indicate the relative importance of anaerobic and aerobic metabolism in a given sediment.

Activity Measurements in Microcosms

Culture methods, direct observation, and biochemical marker techniques focus on the nature of organisms present and their physical and biochemical characteristics. Because the kinds of micro-organisms present in anyone environment are often so large, and the relationships between microbial species so complex, these methods are often quite cumbersome. An alternative to these methods, developed initially by soil microbiologists, is to focus on the net effects of microbial processes rather than on the kinds of organisms present. The guiding philosophy of this approach is that what the micro-organisms do is important ecological information and can supplement the findings of culture studies.

An example of how activity measurements can supplement the use of culture methods in studying the microbial ecology of aquifer systems was given by Chapelle. The system being studied was the carbonate Floridan aquifer and its overlying confining bed, the Hawthorn formation. Groundwater produced from the carbonate aquifer at this site contained up to 12 mmole/l of dissolved inorganic carbon—far more than could be accounted for by dissolution of carbonate minerals in a system closed to CO_2. The obvious hypothesis, therefore, was that CO_2 was being actively produced in the sediments of this system. The production of this CO_2 resulted in an open system with respect to CO_2 and vastly more carbonate material dissolution. In order to test this hypothesis, the first and most obvious question was ecological: are there micro-organisms present capable of producing CO_2?

To answer this initial question, culture methods similar to those employed by Kolbel-Boelke were used to show the presence of a viable bacterial community. Not surprisingly, the results of these culture studies documented the presence of a wide variety of aerobic and facultatively anaerobic bacteria, including strains of *Bacillus*, *Pseudomonas*, and *Micrococcus*. However, the fact that those micro-organisms could be coaxed into active growth on culture media in no way indicated that they actively produced CO_2 *in situ*, and that was the important question.

To address this question, samples of aquifer material were placed in sterile vials under aerobic and anaerobic conditions, and the production of CO_2 monitored with time (Fig. 23.3a). CO_2 was generated aerobically and somewhat slower anaerobically. However, sediment samples that had been sterilised by gamma radiation showed no CO_2 production.

While these data were consistent with the hypothesis that bacteria produced CO_2 in the sediments, the possibility that the sterilisation technique had in some way interfered with an abiotic CO_2-producing process could not be ruled out. For this reason, a second experiment was performed. In this experiment (Fig. 23.3b), sediment samples were initially incubated anaerobically for seven days. CO_2 production was very low, ruling out the possibility that inorganic degassing of the sediments was producing significant amounts of CO_2 in the experiments. After seven days, oxygen was introduced into the vials and CO_2 production rapidly increased. However, that did not rule out the possibility that inorganic oxidation of organic material was producing CO_2. In order to test this possibility, dicyclohexylcarbodiimide (DCCD), an inhibitor that uncouples chemiosmosis from ATP synthesis in bacteria, was added to the sediment. This immediately shut off CO_2 production and showed unequivocally that bacteria in the sediments were capable of active CO_2 production. Thus, a combination of culture data and measurements of potential activity (Fig. 23.3) neatly answered an ecological question posed by observing the groundwater chemistry of this system.

Aquifer materials placed into laboratory vessels such as test tubes or septated serum vials for measurement of microbial activity are referred to as microcosms. Microcosm is defined in the dictionary as 'a community or other unity that is representative of a larger unity'. The reasoning for microcosms is that by understanding the activity of a small portion of an aquifer, much can be learned about the aquifer as a whole. Microcosms have been used extensively in studies of subsurface microbial ecology and are especially useful in assaying indigenous microbial communities for specific activities. In the example given above, the activity of interest was potential bacterial CO_2 production in a pristine aquifer. Often, however, the activity of interest deals with the potential of microbial communities to metabolise particular xenobiotic organic compounds.

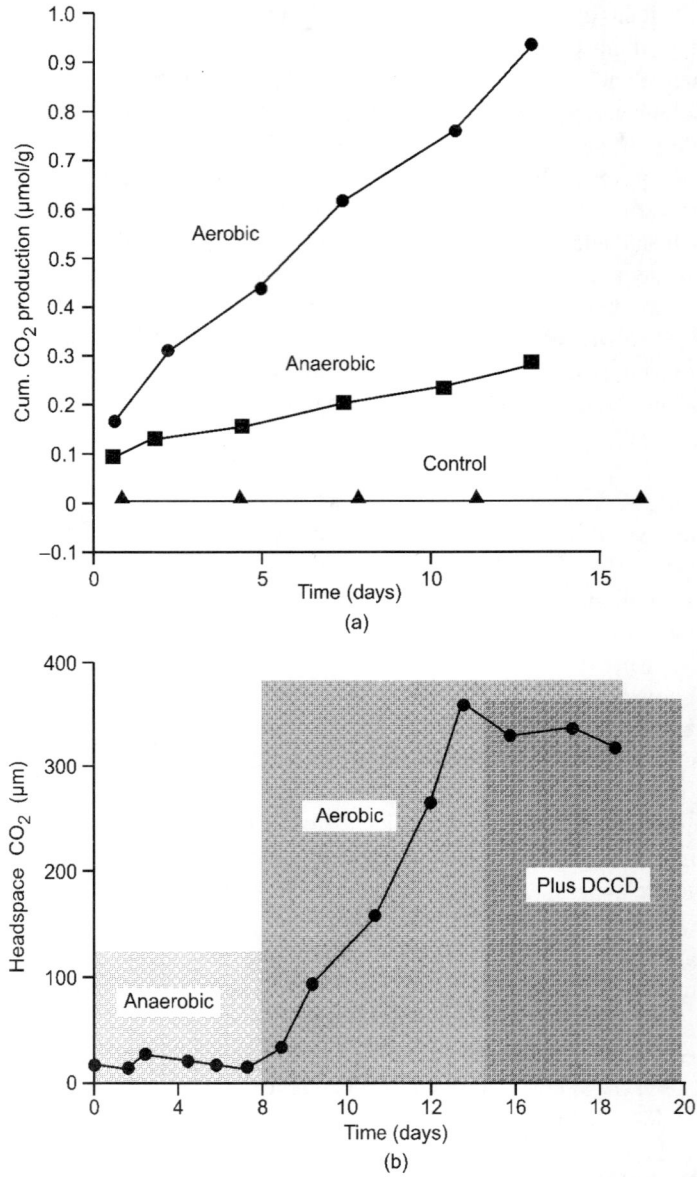

Fig. 23.3. Production of CO_2 from: (a) aerobic and anaerobic incubations of aquifer sediments; and (b) inhibition of CO_2 production by DCCD.

Measuring CO_2 production or organic compound degradation activity in microcosms is suitable for answering specific questions concerning metabolic potential of microbial populations. They are not suitable for determining what part of a microbial community is involved in the degradation processes. It is common practice, therefore, to establish the presence of particular metabolic activity using microcosms and, if necessary, follow up with culture methods for isolating particular micro-organisms capable of the activity.

Geochemical Methods

A feature shared by culture methods, direct observational methods, biochemical tracer methods, and microcosm methods is that the scale of observation is relatively small. Each of these methods is based on recovery of anywhere from one to a thousand grams of aquifer material. This scale is fixed largely by practical considerations of core recovery and sample handling in the laboratory. It is very difficult, however, to extrapolate meaningfully from such small-scale samples to the scale of regional aquifer systems, which may be hundreds of square kilometres in area. One way of enlarging the scale of subsurface microbiological observations is to use the chemistry of groundwater as a tracer for particular microbial processes.

The basic premise of using groundwater geochemistry in this fashion is that particular microbial processes impact the chemistry of groundwater in certain ways. For example, oxygen respiration in an aquifer closed to the atmosphere will result in the consumption of oxygen as groundwater flows downgradient. Similarly, sulphate reduction will consume sulphate along aquifer flowpaths. In addition to consuming oxidants or substrates, some microbial processes will mobilise chemical species. Ferric iron reduction will result in the production of dissolved iron and methanogenesis will produce methane. Thus, the accumulation or depletion of particular chemical species in groundwater can be used to indicate the presence or absence of certain microbial processes in aquifer systems.

As with other methods for studying the microbial ecology of subsurface aquifer systems, the use of groundwater geochemistry has advantages and disadvantages. The chief advantage is that groundwater chemistry provides a large-scale view of aquifers that cannot be obtained with other methods. Also, it is relatively inexpensive to sample water chemistry, whereas obtaining aseptic samples of deep sediments can be prohibitively expensive. However, groundwater chemistry can give no direct information as to the biomass, diversity, or biochemical potential of the organisms responsible for the processes under consideration. This can be a considerable disadvantage if, for example, the potential degradation of chemical contaminants is of primary interest. Furthermore, because groundwater chemistry integrates the effects of microbial processes, it is difficult to deduce the heterogeneity of microbial processes.

Molecular Microbial Ecology

One of the most frequent observations made in the microbial ecology of subsurface environments is that direct microscopic counts of bacteria commonly exceed plate counts by a factor of a hundred or more. Because plate counts are dependent on culturing and bacterial growth, this has widely been attributed to the presence of large numbers of bacteria that cannot be cultured. Studies from other natural microbial habitats (lakes, oceans, aquatic sediments) show a similar pattern, and this has been termed 'the great plate count anomaly' by Staley and Konopka. The problems this presents to the microbial ecologist have long been recognised. In particular, it implies that there are many more bacteria in the subsurface then are evident by culture methods. This, in turn, suggests that micro-organisms that perform important processes in the subsurface may not be recovered by culture methods.

The advent of molecular methods for studying and classifying bacteria has provided new methods for studying microbial ecology that are not dependent upon culture methods and, at least in principle, avoid culture bias.

A schematic diagram showing the general strategy for using molecular methods to study microbial ecology is shown in Fig. 23.4. First, samples of sediment or water are obtained, the bacteria present are lysed, and the nucleic acids from the total microbial community are extracted. Second, the DNA coding for ribosomal RNA (rDNA), usually about 1500 base pairs, is amplified using PCR with primers designed to isolate the 16S RNA genes. Alternatively, the rRNA present in the sample can be converted to DNA by means of reverse transcriptase, and the copied genes c rDNA amplified by PCR. Next, the amplified rDNA is either cloned using a plasmid cloning vector inserted into *E. coli* or specific rDNA sequences obtained suing denaturing gradient gel electrophoresis and reamplified with PCR. After sufficient DNA has been obtained, either by cloning or by DCCE-PC, the rDNA is sequence and the results compared to sequences in an rDNA database. From this it is possible to determine genera, and sometimes species, of micro-organisms present in an environment without having to use culture methods. Because of the lack of culture bias, it is possible to identify micro-organisms that would otherwise be missed.

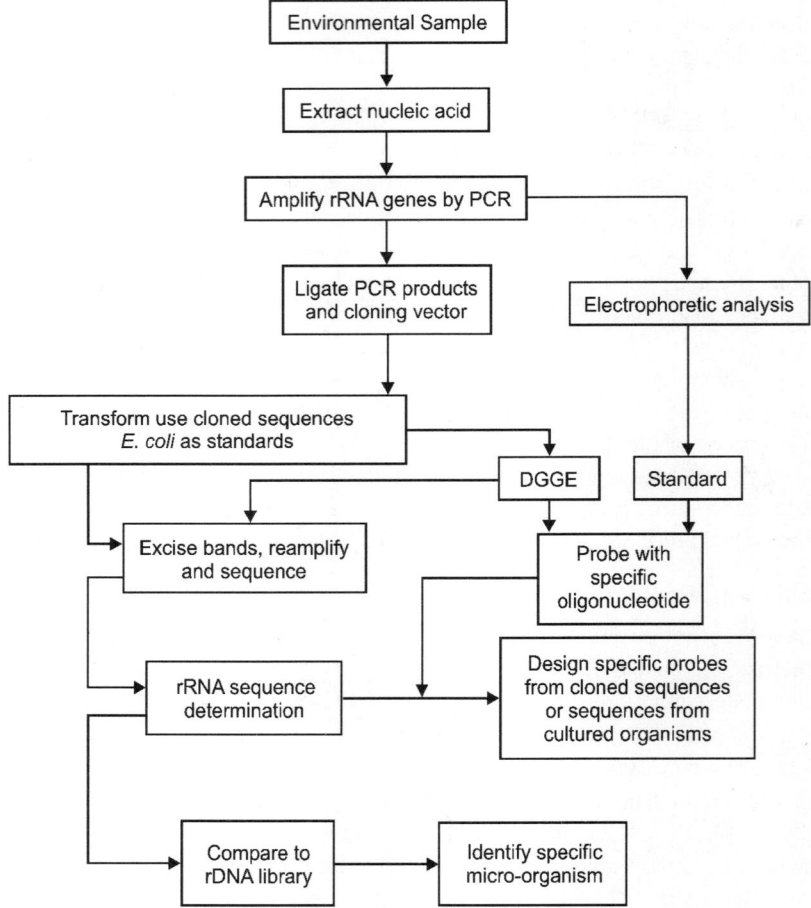

Fig. 23.4. Schematic diagram showing the general strategy for using various molecular methods to study the microbial ecology of environmental samples.

An example of how this technology has been applied to the microbial ecology of an aquifer system was given by Rooney-Varga. The aquifer being studied was a shallow glacial outwash aquifer located near Bemidji, Minnesota, that had been contaminated by a crude oil spill. Much of the risk associated with petroleum hydrocarbon-contaminated groundwater comes from benzene, a known carcinogen.

However, benzene is relatively recalcitrant under anaerobic conditions, and is often observed to migrate farther than other soluble contaminants. In accordance with this common observation, most sediments cored from Bemidji site do not contain micro-organisms capable of oxidising benzene. However, in some locations characterised by Fe(III)-reducing conditions, rapid oxidation of benzene was observed. The ecological question was, what micro-organisms are present in the benzene-degrading zone that are not present in the zones where benzene was not degrading?

To address this question, Rooney-Varga used polymerase chain reaction-denaturing gradient gel electrophoresis (PCR-DGGG), 16S rRNA sequence analysis, phospholipid fatty acid (PLFA), and most probable number polymerase chain reaction (MPN-PCR) methods to assess the ecological differences between benzene-degrading and benzene-nondegrading zones without the bias of culture methods. In MPN-PCR, the rDNA extracted from sediments is serially diluted and amplified by PCR. Those fragments that are present in greatest abundance appear in successively higher dilutions, and this can give a quantitative estimate of relative rDNA abundance.

PCR-DGGE patterns of DNA extracted from sediments differed markedly between the benzene-degrading and benzene-nondegrading zones. The diversity of bacteria appeared to be higher in the benzene-degrading zone, as indicated by the greater number of bands found in DGGE profiles. In addition, one distinct band present only in the benzene-degrading zone exhibited rDNA sequences characteristic of the Geobacteraceae family. This was consistent with the MPN-PCR analysis, which exhibited 10^7 copies of Geobacteraceae 16S rDNA in the benzene-degrading sediments as opposed to 10^4 copies in the benzene-nondegrading zone. Finally, PLFA analysis indicated greater numbers of bacteria in the benzene-degrading sediments, a higher degree of diversity, and higher concentrations of PLFA characteristic of known *Geobacter* species.

The obvious conclusion is that the benzene-degrading sediments were characterised by an abundance of bacteria associated with the family Geobacteraceae. Furthermore, because members of the Geobacteraceae are the only known micro-organisms that have been shown to degrade aromatic hydrocarbons under anoxic conditions, it suggested that Geobacteraceae were responsible for the observed benzene biodegradation. Because attempts to isolate this micro-organism were unsuccessful, however, this conclusion could not have been reached using traditional culture methods. While studies such as this using molecular techniques to answer ecological questions are in an early stage, it is likely that this approach will be increasingly used in future years.

MICROBIAL DIVERSITY AND NICHES IN AQUIFER SYSTEMS

A fundamental characteristic of aerobic aquifer systems is that they exhibit a high degree of morphologic and physiologic diversity in their bacterial flora. This was observed as early as the 1950s by German investigators using culture techniques. Subsequent studies, also using primarily culture techniques, have shown that diversity among aerobic and facultatively anaerobic heterotrophic bacteria appears to be the rule rather than the exception. Similar culture studies of anaerobic aquifer systems have not been performed as extensively. However, the results of White using biochemical marker techniques suggest that microbial diversity is also characteristic of anaerobic groundwater systems. In this section, the ecological implications of microbial diversity in groundwater systems is examined.

Measurement of Diversity

The term diversity describes heterogeneity within a system. As applied to ecology it refers to: (i) the total number of species present in an ecosystem, and (ii) the number of individuals of each species present in that ecosystem. It is possible to define a 'diversity index' based on these two quantities. The Shannon index is the most widely used measure of diversity in ecology and consists of one term for describing total number of species (S) and a second term to describe the distribution of that species within the community (Dists). Total diversity (Divtot) may thus be defined as:

$$Divtot = S + Dists$$

when dealing with macroscopic organisms, such as our earlier example of blackbirds, the information needed to quantify diversity may be obtained by direct observation. However, when dealing with microbes individuals can only be identified using culture techniques. The selectivity of culturing micro-organisms in the laboratory automatically introduces bias into measuring diversity. Even with the limitations imposed by the selectivity bias, it is evident that determining the number of individuals of each species present is not a practical undertaking.

Niches and Sources of Microbial Diversity

In natural systems, habitats are seldom uniform. Rather, there is considerable variation in the availability of nutrients and the suitability of living space. A niche is defined as the physical space and immediate environmental conditions within which an organism survives and reproduces. A fundamental niche is determined by the organism's genotype and is all of the conditions under which it may survive. The actual niche occupied by an organism, however, is generally limited by competition with other organisms and is referred to as the realised niche. There are a wide variety of niches available in saturated subsurface environments.

In general, one might expect a functional relationship between the number of exploitable niches in an ecosystem and the diversity of organisms. This depends, however, on how much each of the niches overlap and on the interaction between organisms. If there is significant overlap between niches and the organisms compete strongly, than a single organism may become dominant and the ecosystem exhibit low diversity. Alternatively, if competition is weak, there may be considerable diversity even if there is overlap of niches.

In groundwater systems, it would seem that some niches are largely isolated from others (i.e. a pore within a feldspar grain) and others with considerable overlap (a grain boundary between a quartz and feldspar grain). The observed high degree of diversity thus may represent the high number of available niches, the lack of competition, or both. It is likely that both of these factors enter in to the development of diversity in groundwater systems.

One possible explanation for the observed low microbial diversity found in groundwater pumped from aquifers as opposed to the high diversity found in sediments is that groundwater samples represent a single niche. This niche would be the pore water present in the relatively large, open pores between sand-sized mineral grains. Water in these large pores is subject to less surface tension forces than water in small pores and is preferentially drawn into pumping wells. The large pore niche, which would tend to be populated by free-living rather than attached micro-organisms, would be expected to exhibit much less diversity than the sediment as a whole. This, in turn, could explain the lower diversity in groundwater relative to sediments.

Stress and Microbial Diversity

In addition to responding to the availability of nutrients and the abundance of niches, microbial diversity also responds to environmental stresses. The impact of environmental stress on microbial diversity has been extensively studied in surface water systems. These studies show a measurable decrease in microbial diversity as the load of pollutant increased. For example, Larrick examined the diversity of heterotrophic bacteria as influenced by effluents from fossil fuel power plants. Total bacterial counts, as measured by number of colony-forming units (cfu), at his stations remained fairly constant with increasing stress from the effluents. However, the proportion of pigmented bacteria within the community correlated with physio-chemical stresses. At an unpolluted reference site, the proportion of pigmented bacteria was about 13 per cent of total cfu. At a heavily polluted ash basin, in contrast, pigmented bacteria made up almost 60 per cent of the cfu.

Decrease in diversity due to pollution stress is often, but not always, observed. For example, in a study of sewage effluent discharge to Lake Huron, Stevenson and Stoermer found high diversity of diatoms in areas receiving effluent. They hypothesised that species diversity is low in algal populations that either grow very rapidly (low stress) or populations that grow very slowly (high stress). Intermediate growth rates, however, seem to encourage greater species diversity. Thus, species diversity could be greatest at a moderate level of pollution stress. This example points out that there is not always a simple relationship between species diversity and environmental stress.

Systematic studies of how diversity in subsurface microbial communities vary with stress are largely lacking. Based on analogy with surface-water systems, however, several effects would be anticipated. In pristine systems, highly oligotrophic aquifers (high stress) might be expected to have lower diversity than moderately oligotrophic aquifers.

Similarly, the onset of pollution due to chemical spills or migration of landfill leachates (high stress) might decrease species diversity in a moderately oligotrophic (low stress) aquifer. This species-diversity decrease may coincide with an actual increase in cell numbers as a few well-adapted strains become dominant. Because changes in microbial diversity may either decrease or enhance the bioremediation potential of a system, this is a topic that deserves detailed study.

POPULATION INTERACTIONS

The ecology of microbial communities is strongly affected by interactions between populations that make up the community. The term 'population' in this context refers to organisms that are phylogenically related to each other. In subsurface environments that lack a community of higher plants and animal, the most important interactions are those between microbial populations. Interactions between microbial populations are extremely varied. Positive interactions are those that enhance the abilities of populations to survive within a community. In some cases, development of positive interactions allows the exploitation of resources that would otherwise not be available. This, in turn, allows some microbes to survive in environments that would otherwise be closed to them. Negative interactions are those that are in some way detrimental to certain populations. Negative interactions act as negative feedback mechanisms that limit population densities. These feedback mechanisms, in turn, adjust populations to the resources of the habitat. Negative interactions also tend to preclude the invasion of an established community by allochthonous populations and tend to maintain community stability.

Neutralism

The first kind of possible interaction is actually a lack of interaction. This is called neutralism. Neutralism occurs if, for some reason, populations are physically separated from each other. Furthermore, neutralism is more likely to occur under conditions of low population density that increase the chances for lack of contact. The most common example of neutralism would be the coexistence of a living bacterial population with endospores.

Because the bacteria cannot use resources needed by endospores (which need none) and do not prey on or otherwise degrade endospores, bacteria have no effect on the endospore population. Similarly, because the endospores do not use resources or attack bacteria, the endospore population does not affect the bacterial population.

In subsurface environments characterised by large numbers of potential niches and low population densities, neutralism is probably very common. For example, populations growing exclusively on dispersed microcline grains would have negligible interaction with populations growing exclusively on equally dispersed biotite grains. Similarly, populations inhabiting closed pore spaces would have virtually no interaction with those in interconnected pore spaces. There is no direct evidence for the presence (or absence) of neutralism in subsurface environments. This almost certainly reflects the lack of techniques for documenting such interactions.

Commensalism

In a commensal relationship between populations, one population benefits from the interaction and the other population is unaffected. This type of interaction between bacterial populations is also probably very common in subsurface environments. The best, and probably most widespread, example of a commensal interaction in subsurface environments is consumption of oxygen. In and near outcrop areas of aquifers, oxygen is supplied to the system by vertically percolating recharge. As groundwater flows downgradient, oxygen is systematically removed by aerobic and facultatively anaerobic bacteria. The removal of this oxygen by the aerobically respiring population creates the conditions required by obligately anaerobic bacteria.

Thus, the anaerobic bacteria benefit by the presence of aerobes in aquifer systems. Since aerobes and anaerobes inhabit different niches, however, the aerobes are neither positively nor negatively affected by the anaerobic population.

In some commensal interactions, one microbial population physically alters the habitat in such a way as to allow a second population to exist. Another possible way this occurs in subsurface environments is the presence of boring bacteria. These bacteria bore through solid grains, presumably to obtain substrates for growth. It is possible, however, for these borings to be opportunistically colonised by other bacteria as well. In this case, the opportunistic bacteria benefit with no corresponding benefit for the boring bacteria.

Another common commensal interaction is based on the production of specific growth factors. Some organisms produce specific compounds—vitamins or proteins, for example—and excrete them to the surrounding environment. These compound can then be taken up and used by other organisms to support growth. As long as these compounds are produced in excess and the producing organism derives no particular benefit in their removal, this interaction is commensal. It is common for this relationship to develop in communities of free-living bacteria and algae in aquatic environments. While there have been no reports of such relationships between subsurface micro-organisms, this probably occurs quite commonly.

Synergism and Symbiosis

Synergism and symbiosis refer to relationships that benefit both populations. The difference is that two synergistic populations, while benefiting from interaction, are capable of living without each other. Symbiotic populations, on the other hand, require the interaction to sustain life. The distinction between synergism and symbiosis has not always been made. Symbiosis was originally used to describe any close relationship between populations. In present usage, however, the term symbiosis is restricted to interactions that are obligatory.

In microbial ecology, a classic example of synergism is the relationship between *Streptococcus faecalis* and *Lactobacillus arabionsus*. *L. arabinosus* requires phenylalanine, which is produced by *S. faecalis*, for growth. Similarly, *S. faecalis* requires folic acid for growth, and *L. arabinosus* happens to produce folic acid. These two micro-organisms are quite capable of growing independently of each other, as long as these specific nutritional requirements are present in the media. However, if media are prepared that lack both folic acid and phenylalanine, *L. arabinosus* and *S. faecalis* can grow together in culture but they cannot grow separately.

Synergistic relationships are very important in the ecology of subsurface environments. The best example is the relationship between fermentative and respirative micro-organisms in anaerobic groundwater systems. Fermentative bacteria are able to directly utilise complex organic matter, carbohydrates and lignins, present in subsurface environments to supply energy for growth. However, because fermentative metabolism cannot completely oxidise organic compounds, fermentation products such as hydrogen and acetate still contain significant amounts of energy. Anaerobic respirative bacteria [Fe(III) reducers, sulphate reducers, methanogens] are incapable of directly metabolising carbohydrates or lignins, but they are capable of metabolising fermentation products. Thus, the presence of fermenting bacteria encourages the growth of respirative bacteria. Furthermore, since acetate and hydrogen are waste products, at least as far as the fermenters are concerned, their consumption by respirative bacteria encourages fermentation. Thus, both populations benefit.

The classic example of a symbiotic relationship between micro-organisms is the growth of lichens. Lichens consist of photosynthetic algae or cyanobacteria and heterotrophic fungi. The algae fix atmospheric carbon dioxide and in some cases atmospheric nitrogen that provides the substrate for growth of the fungi. In return, the fungi produce organic acids that help dissolve rock material allowing for further growth of the algae. These symbiotes inhabit extremely cold and dry environments of the Arctic desert. Individually, these species are not capable of surviving this extreme environment, whereas together they can.

Symbiosis has been closely studies by microbial ecologists for a number of reasons. One is that many ecologists feel that the formation of symbiotic relationships are a mechanism by which two populations unite to form new species. Many current theories of .microbial evolution point to the structural similarities between organelles of eukaryotic cells (mitochondria, chloroplasts, for example) and prokaryotic microbes. The consensus is that eukaryotic cells developed through the formation of symbiotic relationships between prokaryotic micro-organisms.

Competition

Different microbial populations in particular environments often strive to utilise the same resource. In this case, the relationship between the populations becomes competitive. The resource that is the object of competition can include nutrients, moisture, space, or any other commodity that is necessary for life functions and is in scarce supply.

Competitive interactions tend to separate populations in ecosystems. In microbial ecology there are numerous well-documented examples of population separation by means of competitive interaction. Perhaps the classic example is the separation of methanogenic and sulphate-reducing activity in aquatic sediments. It is ubiquitously observed in anaerobic marine sediments that a sulphate-reducing zone develops near the sediment-water interface and is underlain by a methanogenic zone.

In anaerobic sediments, both methanogenic and sulphate-reducing populations are supported by fermentation products such as acetate and hydrogen. These micro-organisms thus compete directly for the same substrates. Because sulphate reducers have a higher affinity for acetate and hydrogen than methanogenic bacteria, sulphate reducers outcompete methanogens. Furthermore, because the sulphate reducers lower concentrations of these substrates below the threshold required by methanogens, methanogens are effectively excluded from the environment. This is termed competitive exclusion. Sulphate reducers continue to dominate the ecosystem until concentrations of dissolved sulphate are exhausted. Once sulphate reducers can no longer respire, concentrations of hydrogen and acetate increase to levels required by methanogens, and methanogenic bacteria become dominant.

It is interesting to note that methanogens have both a commensal and competitive relationship with sulphate reducers. While methanogens cannot successfully compete with sulphate reducers for fermentation products, methanogens depend on the sulphate reducers to deplete pore water with respect to sulphate, thus providing the sulphate free environment they require.

Competitive exclusion is very important in the ecology of subsurface environments, and this is reflected in many commonly observed groundwater chemistry patterns. For example, one of the most frequently occurring groundwater problems is the presence of high concentrations of dissolved iron. Furthermore, it is often observed that these high concentrations of dissolved iron occur in discrete zones.

Similar patterns of competitive exclusion have been reported from contaminated aquifers as well. Most of the contaminated zone was characterised by high organic carbon concentrations and low sulphate concentrations. In these areas, methanogenesis was the predominant terminal electron-accepting process. In one localised area, however, the groundwater contained significant concentrations on sulphate and sulphate reduction was the predominant process. Significantly, while methanogenic bacteria were more numerous in the methanogenic zone than sulphate reducers, sulphate reducers were present. Similarly, while sulphate reducers were more numerous than methanogens in the sulphate-reducing zone, methanogens were also present.

Antagonism, Parasitism and Predation

When one population produces a substance that inhibits another population, the interaction is termed antagonistic. The population that produces the inhibitory substance thus gains a competitive advantage over the other. The production of antibiotics by some strains of bacteria and fungi are thought by some microbial ecologists to be an example of antagonism in nature. Curiously enough, however, it has not been unequivocally shown through ecological studies that antibiotic production actually confers a competitive advantage in natural systems. One reason for this may simply be that virtually all studies of antibiotic production are performed by medical microbiologists who are not specifically concerned with ecological interactions in nature.

Studies showing the presence or absence of antagonistic interactions between micro-organisms in groundwater systems have not been performed. It is notable, however, that numerous strains of bacteria characterised by antibiotic resistance have been isolated from subsurface environments. It is possible that this resistance reflects antibiotic production by some species in the subsurface, and thus may reflect antagonistic interactions. On the other hand, given the low population densities characteristic of subsurface environments, it is not clear that antibiotic resistance would confer much of a competitive advantage. As antibiotic production in different environments has been and will remain a topic of practical interest, studies on antagonistic relationships in subsurface environments are certainly warranted.

A parasitic relationship occurs when one population derives its nutritional requirements from hosts of another population. The parasitic population is clearly benefited by this relationship, generally to the detriment of the host population. Host-parasite interactions require long-term contact between the populations, and members of the parasitic population are generally much smaller than members of the host population.

The most common host-parasite interaction in microbial communities is between viruses and bacteria. Viruses are obligate parasites and are able to reproduce only inside other cells. Recent investigations have shown that a rather high percentage of planktonic marine bacteria harbour viruses. This shows, in turn, that parasitic interactions are rather common in that environment. Studies of viruses in aquifer systems have focused on the transport of human viruses rather than on viruses that parasitise indigenous bacteria. Thus, the extent that virus parasitism in growndwater systems is largely unknown.

When one population actively consumes another population, the interaction is termed predation. The predatory population derives nutrients from this interaction and is clearly beneficial. Conversely, such interaction is clearly detrimental to the prey population.

The most common predator-prey relationship in microbial ecology is the interaction between protozoa and bacteria. Many protozoa actively feed, or graze as it is often termed, on bacterial populations. Early studies of the microbial ecology of subsurface environments noted a lack of protozoa in groundwater systems, but subsequent studies have documented that protozoa are fairly common in aerobic aquifer systems. Furthermore, there is evidence that protozoa actively graze on bacterial populations in some systems. The observed higher numbers of protozoa in contaminated water table aquifers suggest that predation may act to limit bacterial degradation of contaminants in some systems. This important possibility needs to be explored in a variety of systems.

r AND K STRATEGIES IN MICROBIAL ECOSYSTEMS

Different populations in a microbial community employ different metabolic and reproductive strategies in their competition with each other. These different strategies allow different niches within a habitat to be successfully colonised. One scheme for classifying such strategies is to view them as falling along a gradient between pure r strategists and pure K strategists (Fig. 23.5). In this scheme, r strategists rely on high rates of reproduction to maintain themselves in the community. Similarly, r strategists tend to be characterised by catastrophic mortality events. The r strategists tend to have short life spans, high productivity-low efficiency utilisation of resources, high migratory tendency, and variable population size. Conversely, K strategists are characterised by lower reproductive rates, longer life spans, low productivity-high efficiency utilisation of resources, low migratory tendency, and stable population size. Clearly, most micro-organisms are neither pure r nor K strategists but fall somewhere between the two end members (Fig. 23.5).

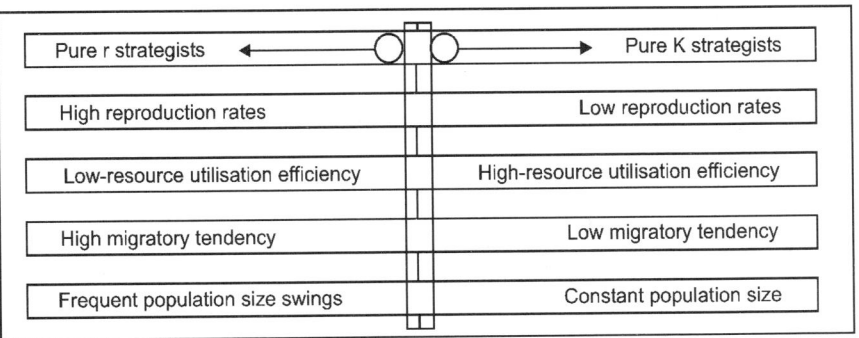

Fig. 23.5. Some of the characteristics of pure r and K strategists.

The possible effects of r or K strategies on species survival can be illustrated by considering the emergence of a new volcanic island in the Pacific ocean. The creation of land obviously creates new habitat that can potentially be exploited by plants and animals. In colonising this new habitat, r strategists have an immediate advantage. Plants that produce highly mobile seeds (palm trees, for example) reach the new habitat first. Once on the island, the plants that can reproduce the fastest are the ones that can claim the most space. Since resources (space) is not limiting, there is little initial advantage to use available resources efficiently. Clearly, r strategists have the advantage over K strategists when colonising new habitat. As time passes on the island, however, and more and more plants come and colonise, all of the available niches become crowded. Shortages of resources begin to appear and there might be mass mortality events among the r strategists. The ability to reproduce quickly becomes less and less important and the ability to utilise resources efficiently becomes more and more important. Plants with K strategy physiologic adaptations, adaptations that allow successful reproduction in space-limited conditions, have the advantage. Clearly, as time goes on, K strategists will gradually occupy most of the available niches.

The distinction between these two end-member strategies has been around a long time. Winogradsky, the great soil microbiologist of the early twentieth century, referred to some microbial species as being 'zymogenous', or opportunistic. There are obvious similarities between the concept of zymogenous and r-selected micro-organisms. Similarly, Winogradsky's concept of 'autochthonous', or indigenous, micro-organisms is analogous to that of K strategists.

r and K Strategies in the Aquifer Environment

At first glance, pristine subsurface environments would appear to be ideal for K-selected micro-organisms, since there has been long term stability in ambient conditions and nutrients are in limited supply. Indeed, available evidence suggests that K-selected micro-organisms dominate in subsurface environments. For example, population densities appear to be fairly constant spatially, reproduction rates are very low, and separation of populations by competitive exclusion mechanisms evident. Furthermore, the presence of fairly diverse microbial communities is consistent with the predominance of K strategies.

Interestingly, however, there is evidence that some micro-organisms in pristine subsurface environments retain some r-strategy characteristics. For example, in an experimental investigation Hirsch and Rades-Rohlkohl found that sterilised sediments exposed to groundwater were rapidly (within 12 weeks) colonised by bacteria. Furthermore, the diversity of the colonising micro-organisms was considerably less than the microbial diversity in the surrounding aquifer sediments. This evidence suggests that r strategists do exist in subsurface environments that are able to rapidly take advantage of new habitat.

Ghiorse and Wilson suggest an intriguing hypothesis that may explain the long-term presence of r strategists in subsurface environments. Ghiorse and Wilson were concerned with explaining why subsurface bacteria appear to divert considerable carbon resources to synthesise exopolysaccharides, a diversion that appears wasteful of scarce resources. They suggest that organic carbon particles in aquifers serve to absorb and concentrate dissolved organic carbon transported in from the overlying soil zone. As organic carbon absorption continues, the carbon particle becomes capable of supporting microbial growth. If a micro-organism comes in contact with this enriched particle, it is crucial to bind strongly to it in order to fully exploit the resource. This neatly explains the presence of exopolysaccharides, which enable the micro-organism to bind firmly to surfaces. After the organic resource is exhausted, the cells die or disperse, and the organic particle again begins to concentrate dissolved organic carbon. Ghiorse and Wilson thus envision a cycle of nutrient concentrations, colonisation by micro-organisms, exploitation, and finally dispersal (Fig. 23.6). Implicit in this cyclic model is the steady-state presence of r strategists capable of moving to and rapidly exploiting resources. Thus, the continued presence of *r*-selected micro-organisms in subsurface environments may not be as surprising as it might appear at first glance.

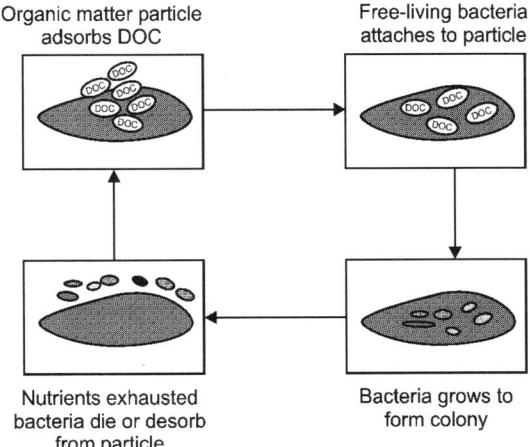

Organic matter particle
adsorbs DOC

Free-living bacteria
attaches to particle

Nutrients exhausted
bacteria die or desorb
from particle

Bacteria grows to
form colony

Fig. 23.6. Cycle of microbial colonisation and dispersal for r strategists in nutrient-poor groundwater systems.

Thus, the ecology of micro-organisms in subsurface environments may be studied using a number of different techniques, including culture methods, direct observations, biochemical marker methods, activity measurements, geochemical methods, and ecological modelling. Each of these methods has advantages and disadvantages. Ultimately, the methods used for a particular problem must reflect the nature of the questions being asked.

Subsurface environments present a number of different niches that may be exploited by subsurface micro-organisms. Groundwater in relatively large pores, for example, may support a free-living population of bacteria. Alternatively, bacteria may attach themselves to a variety of aquifer materials, including mineral grains and organic matter. The large number of potentially exploitable niches in aquifers apparently leads to highly diverse microbial populations in many groundwater systems.

Populations interact in a number of different ways in groundwater systems. Neutralistic interactions are not either positive or negative for different populations. This is the case for bacteria that depend on different carbon sources. In a commensal relationship, one population benefits from interaction, whereas the other is unaffected. In synergistic relationships, such as between fermenting and respiring bacteria, both populations benefit. Symbiosis refers to a relationship in which interaction is a prerequisite for life. Relationships between populations can also be competitive, which is the case when each population exploits the same resource.

Competitive relationships between different terminal electron-accepting micro-organisms, each of which attempts to use the same carbon sources, are very common in subsurface environments. In particular, the competitive exclusion of sulphate reduction by Fe(III)-reducing bacteria leads to very distinctive zones of high-iron groundwater in some systems.

Different microbial populations employ different metabolic and reproductive strategies in competing with each other. Pure *r* strategists rely on high rates of reproduction, with consequent less efficient utilisation of resources, to maintain themselves. Pure K strategists, on the other hand, rely on the efficient utilisation of resources while sacrificing a high reproduction rate to maintain themselves. The low-nutrient conditions characteristic of subsurface environments appears to favour efficient K strategists in many systems. However, there is evidence that fast-colonising r strategists are also present in many systems. This may reflect the way carbon is cycled in subsurface environments and allows subsurface micro-organisms to rapidly exploit niches as they become available. The term 'population' in this context refers to organisms that are phylogenically related to each.

Chapter 24

Abundance and Distribution of Bacteria in the Subsurface

INTRODUCTION

The study of microbial processes and how they affect groundwater geochemistry begins with consideration of the abundance and distribution of bacteria. That bacteria inhabit virtually all subsurface environments is not currently a matter of dispute. However, the types of bacteria present, their abundance, and the kinds of physiologic processes that they carry out differ widely from environment to environment.

In order to deal with bacterial abundance and distribution in subsurface environments, it is first necessary to draw some fundamental distinctions between the types of environments most commonly encountered. This requires some basis on which to classify subsurface environments.

CLASSIFICATION OF SUBSURFACE ENVIRONMENTS

One possible basis for classifying subsurface environments, and one that is often used informally, is simply one of depth. It seems reasonable that deep subsurface environments are fundamentally different than shallow environments. Closer inspection of this as a classification strategy, however, immediately points out several difficulties. First, deep and shallow are relative terms that have different meanings in different geologic settings. Deep to a soil scientist would mean depths on the order of metres, whereas to a petroleum geologist it would conjure thoughts on the order of 10,000 metres. Aside from the semantic problems, the terms 'deep' and 'shallow' give no information concerning the hydrologic setting of the environment.

In describing microbial processes in subsurface environments, the hydrologic setting is the most important frame of reference. The hydrologic setting is what determines such important factors as the amount of communication with the surface, rates of groundwater movement, and transport of dissolved substrates. Thus, classifying subsurface environments based on their hydrologic attributes is logical when considering microbial processes. Also, because the physics of groundwater flow does not vary between diverse hydrologic systems, such classification can be applied universally.

The most fundamental distinction between subsurface environments that may be drawn on the basis of hydrology is the difference between the saturated and unsaturated zones. The unsaturated zone is characterised by pore spaces that are incompletely filled with water. Pore space not filled with water is filled with gas. The significance of the unsaturated zone to hydrologists is that capillary forces between water and sediment particles prevent water from flowing to wells. Thus, wells completed in the unsaturated zone do not yield appreciable quantities of water. The amount of water present in the unsaturated zone varies widely and is highly sensitive to climatic factors.

Once the distinction between saturated and unsaturated hydrologic systems is made, the next step is to classify different types of saturated subsurface environments. Hydrologists have long recognised, on a qualitative basis, three categories of groundwater flow systems. These include an 'upper zone' of active flow strongly influenced by local precipitation events, a 'medium zone' of deeper flow only moderately affected by local precipitation events, and a 'lower zone' of relatively stagnant water unaffected by local precipitation. This 'zonation' was recognised entirely on an empirical basis and suffers from a vague notion of the importance of depth. As such, it offers little in terms of a classification scheme over just considering depth.

Three distinctly different types of flow systems develop for a wide variety of topographic configurations (Fig. 24.1). These different types of flow systems have the general characteristics:

1. A local system has its recharge area at a topographic high and its discharge area at a topographic low that are located adjacent to each other.
2. An intermediate system where recharge and discharge areas are separated by one or more topographic highs.
3. A regional system where the recharge area occupies the water divide and the discharge area occurs at the bottom of the basin.

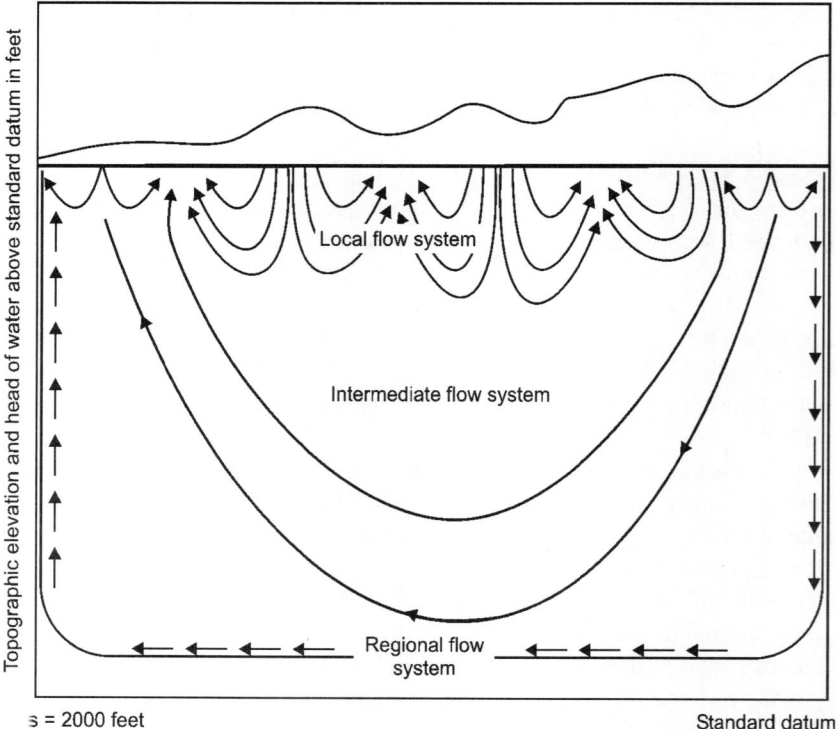

Fig. 24.1. Delineation of local, intermediate, and regional groundwater flow systems.

The classification of subsurface environments shown in Table 24.1 emphasises the different degrees of connection with the surface. Obviously, the unsaturated zone will be most closely connected with the surface, with water entering the system with every major precipitation event. Local flow systems also may be closely connected with the surface and may be strongly influenced by precipitation. Intermediate and regional systems, on the other hand, are much more isolated from events on the surface. The varying degrees of connection with the surface will obviously influence the kinds of microbial processes that may occur in different environments. For example, the unsaturated zone and local flow systems are often free to exchange gases with the atmosphere. Thus, a continuous source of oxygen is potentially available and aerobic bacteria may be able to continuously respire. In intermediate systems, however, the transport of dissolved oxygen is much more limited and anaerobic processes usually predominate. Regional systems are entirely isolated from atmospheric oxygen and anaerobic bacteria predominate.

Table 24.1. Classifications and hydrologic attributes of subsurface environments.

Connection with surface recharge	Flow rates	Oxygen status	Potential for microbial transport from surface	Example
		Unsaturated		
Extensive	Fast 0.5–5.0	Usually aerobic	High	Soil zones
		Saturated (Local)		
Extensive	Fast 0.5-5.0 ft./d	Often aerobic	High	Water table aquifer
		Saturated (intermediate)		
Small	Slow 0.5–5.0 ft./year	Generally ·anaerobic	Moderate	Confined aquifer
		Saturated (regional)		
Virtually nonexistent	Almost stagnant	Anaerobic	Virtually nonexistent	Deep basin/ petroleum reservoirs

UNSATURATED ZONE

An important hydrologic feature of the unsaturated zone is that it contains both water and gas in void spaces between sediment particles. More practically, water in the unsaturated zone is under a negative hydraulic pressure—that is, less than atmospheric pressure—and therefore water cannot flow into wellbores that penetrate it. The unsaturated zone is important from a hydrologic viewpoint because recharge to aquifers must first move through it. The rate that water moves through the unsaturated zone therefore directly controls the rate that an aquifer may be recharged. From a geochemical perspective, the unsaturated zone is important because gases are freely exchanged with water that ultimately recharges aquifers. Recharge water infiltrating through the unsaturated zone is readily charged with highly reactive gases such as oxygen and carbon dioxide due to the coexistence of the gas and aqueous phases. The carbon dioxide input is particularly important in studies of carbon isotopes in groundwater.

Microbial processes in the unsaturated zone are important for several reasons. First, bacterial processes such as oxygen consumption, carbon dioxide production, nitrification, and denitrification will directly impact the chemistry of infiltrating water reaching the water table. Secondly, anthropogenic compounds such as petroleum hydrocarbons, herbicides, pesticides, and a host of others must pass through the unsaturated zone in order to impact groundwater quality. Because many of these compounds are biologically active, bacterial processes in the unsaturated zone will affect their fate and transport.

The unsaturated zone is frequently divided into three components (Fig. 24.2). The first of these is the soil zone, generally a metre or two thick, which contains living roots and which supports plant growth. The porosity and permeability of the soil subzone is generally higher than that of underlying material. This underlying material, which varies in thickness from place to place, is often referred to as the intermediate zone (not to be confused with the 'intermediate' zones of saturated flow systems) and consists of sediments or rocks that have not been exposed to extensive pedogenic (soil-forming) processes. The boundary between the unsaturated zone and the saturated zone is termed the capillary fringe.

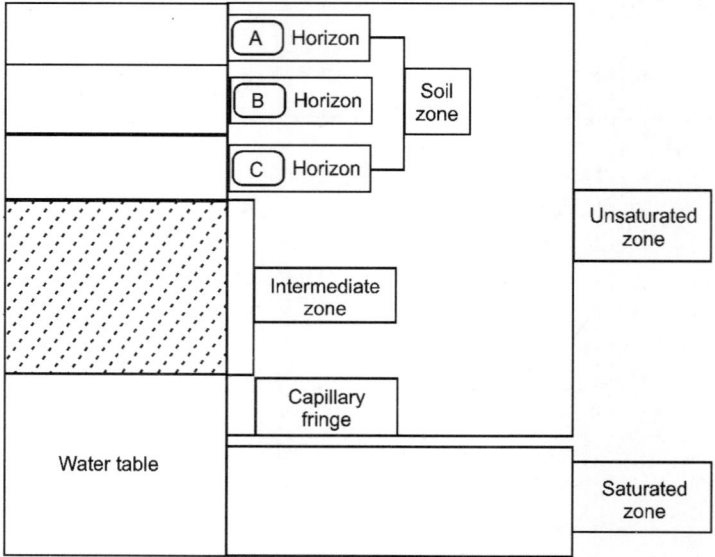

Fig. 24.2. Three components of the unsaturated zone.

Capillarity, the process that forms the capillary fringe, results from two forces: the mutual attraction (cohesion) between water molecules and the molecular attraction between water and solid mineral surfaces. Because of the forces, water will rise into small tubes to varying levels above water in a container. Pores in sediments are often of capillary size and therefore water is pulled from the saturated water table into the unsaturated zone. As a general rule, finer-grained sediments have thicker capillary fringes (up to 1 metre) than sandy sediments (generally 1 to 10 centimetres).

The abundance and distribution of bacteria in the unsaturated zone, particularly in soils, has been intensely studied over the last 150 years. Much of the impetus for this research has come from the recognition that microbial processes are directly involved in soil fertility and therefore agricultural productivity. For example, it had been known for hundreds of years that composting manure under aerobic conditions leads to the accumulation of nitrate, an important component of plant fertilisers. It was the famous soil microbiologist. Winogradsky who showed that this process is brought about by bacteria oxidising the ammonia present in manure to nitrates. Winogradsky isolated in pure culture a bacterium that carried out this nitrification process in 1891. At about the same time, Winogradsky isolated an anaerobic bacterium of the genus *Clostridium* capable of nitrogen fixation: the process of converting atmospheric nitrogen into organic nitrogen. In 1888, the Dutch bacteriologist Beijerinck isolated a nitrogen-fixing bacteria from the roots of legumes, a micro-organism later assigned to the genus *Rhizobium*. It is hard to overestimate the importance of these discoveries to modern agriculture, as they form the core of crop fertilisation and crop rotation strategies.

The traditional association of soil microbiology with soil fertility and crop production has led to a curious lack of attention to microbial processes in sediments just below the root zone. In fact, the unsaturated zone between the root zone and the water table is probably the least studied of any subsurface environment. In many areas, there are only a few metres between the root zone and the water table, and this lack of attention is understandable. However, in some parts of the world, and particularly in arid climates, there may be hundreds of metres of unsaturated material below the root zone. Just from the standpoint of sheer volume, the deep unsaturated zone in these areas bears consideration.

Unsaturated Zone as a Microbial Habitat

The unsaturated zone, by virtue of its proximity to photosynthetic organic carbon production at land surface, is the most biologically active and varied subsurface environment. By far the most biologically active part of the unsaturated zone is the soil.

There are five interactive factors that lead to the development of soil: climate, topography, parent material, time, and biologic processes. Soil microbiologists have long recognised that soil organisms play a significant role in the development of their own habitat. This is certainly true, to varying degrees, in other subsurface environments but is most easily observed in soils. Soil formation is initiated by chemical and physical weathering of rocks or sediments. The release of nutrients by these weathering processes allows colonisation by algae and lichens.

This initiation of primary production leads to the establishment of a heterotrophic bacterial population. In turn, these combined microbial processes speed the weathering processes by increasing the carbon dioxide partial pressure and by releasing organic acids. Once a suitable combination of weathered rock debris and organic matter is achieved, plants are able to initiate growth. This, in turn, speeds organic carbon production, increases the activity of heterotrophic bacteria, and results in more efficient weathering processes. The development of soil, therefore, may be viewed as a combination of processes that are linked in positive feedback loops. That is, as each process becomes more efficient, the efficiency of the other processes is enhanced.

The net result of soil formation processes is the establishment of a stable, well-defined soil profile at the top of the unsaturated zone (Fig. 24.2). The formation of clay-organic complexes results in the stabilisation of clay, sand, and silt particles into aggregates. Aggregate formation is initiated when polysaccharides from roots and micro-organisms combine with clays to form organic matter and mineral complexes. Micro-organisms live in the pores of the soil aggregates, generally attached to particle surfaces. The porosity and size of pore throats in the soil is an important structural feature because it forms living space and access to nutrients.

Moisture and gas content

The coexistence of aqueous and gas phases in the unsaturated zone has important consequences to microbial processes. All micro-organisms require water for cell maintenance and growth. The lack of water is, therefore, extremely detrimental to microbial activity. On the other hand, if pore space is completely filled with water, the transport of metabolically important gases such as oxygen is greatly decreased. Thus, too much water or too little water will both depress microbial processes.

Biomass Measurements in Soil Microbiology

Many of the methods commonly used in subsurface microbiology were developed by soil microbiologists. For this reason, a brief survey of some techniques commonly used in soil microbiology is of practical as well as historical interest.

Biomass is defined as the living part of soil organic matter exclusive of roots and soil animals larger than 5000 cubic micrometres in size. In soils, biomass consists of both bacteria and protozoa. Below the soil zone, however, biomass consists largely of bacterial cells. While protozoa are commonly found deeper in the unsaturated zone, their total biomass is small compared to that of the bacteria.

Measurements of biomass in soils has long been a preoccupation of soil microbiologists. The most important reason for this is that there is a direct relationship between soil biomass and soil fertility. There are many techniques for measuring biomass, each of which has certain advantages and disadvantages.

Direct microscopy

Winogradsky, the founder of soil microbiology, was the first investigator to successfully attempt direct microscopic examination of soils. The basic method developed by Winogradsky is simplicity itself: a microscope slide is placed into the ground, soil is packed carefully around it, and it is left for several weeks. After the incubation time has passed, the slide is removed and stained, and the micro-organisms that have adhered to the slide observed directly. This technique provides a qualitative view of microbial interaction *in situ*.

More recently, the scanning electron microscope (SEM) has been used to observe micro-organisms in soils. This technique works well for observing fungi and actinomycete structures, but is less useful for observing bacteria. This is because bacteria are difficult to unequivocally distinguish from small soil particles.

A common technique designed specifically for counting micro-organisms is epifluorescence microscopy. This technique is notable because it has been widely adapted to enumerating bacteria in samples from the deep subsurface. A sediment sample is dispersed in a blender or shaker, sometimes with the help of a detergent (tween 80), and a small volume of the suspension is filtered onto a membrane. The membrane is then stained with a fluorescent dye, and bacteria enumerated with an epifluorescent microscope. This type of microscope illuminates the stained sample with ultraviolet light, causing the fluorescent dye to emit visible light that can be observed by the microscopist. Acridine orange is the most widely used fluorescent stain.

Acridine orange binds to nucleic acids, RNA and DNA and the resulting complex fluoresces at a characteristic wavelength of UV light. Fluorescein isothiocyanate (ATC) binds to sulphydryl groups of proteins in cell walls and is sometimes used. After enumeration of the stained sample, the number of bacteria in the sample is calculated from the volume of sample filtered, the area of the filter, and the area of filter actually enumerated.

There are significant practical problems with obtaining reproducible counts of bacteria using the epifluorescence method. It is impossible, for example, to be sure that all bacteria have been desorbed from sediment grains by the shaking process. Also, acridine orange is a cationic dye and binds to negatively charged clay mineral surfaces. Thus, there is generally a large amount of interference in counting samples containing large amounts of clay. Despite these problems, direct counting procedures have been widely used.

Chemical techniques

The most widely used chemical technique for determining soil biomass is the soil fumigation method. Chloroform ($CHCl_3$) under pressure is forced through the soil in order to completely saturate it. The chloroform destroys the cell membrane of living cells allowing the cytoplasm to leak into the soil. After the fumigation has lysed the bacteria, the soil is inoculated with a small amount of soil containing living micro-organisms, and the mixture incubated for 10 days along with untreated controls. Biomass is then calculated as:

$$B_c = (CO_{2tr} - CO_{2un})/K_c$$

where, CO_{2tr} is the amount of carbon dioxide evolved from the chloroform-treated sediment, CO_{2un} is the carbon dioxide evolved from the untreated soil, and K_c is the percentage of biomass carbon mineralised to carbon dioxide (often assumed to be 0.41).

A chemical technique that has been used to measure biomass in estuarine and marine environments is to quantify the abundance of phospholipids. This method has been applied to biomass measurements in soils and in other subsurface environments. The phospholipid technique takes advantage of the fact that bacterial membranes (as well as the cell membranes of some eukaryotes) contain phospholipids and that these compounds are unstable once the cell dies. Thus, the phospholipid content of sediments is directly proportional to biomass. Phospholipids can be extracted from sediments using chloroform, hydrolysed into their various fatty acid components, and quantified using gas chromatography. Given the average amount of phospholipid present in bacterial cells (generally on the order of 50 micromoles/g bacteria), phospholipid content can be converted to bacterial numbers.

A significant advantage of the phospholipid method over other measures of biomass is that the kinds of fatty acids liberated from sediment samples gives some information about microbial diversity. This reflects the unique phospholipid content of each bacterial strain. Also, because some phospholipids are found only in prokaryotic organisms and some found only in eukaryotic organisms, the method can help distinguish the biomass component attributable to each class. More important, in deep subsurface sediments where eucaryotes are rare, the method gives information on bacterial diversity that cannot be gathered using culture techniques.

Another chemical technique for estimating biomass in soils is by ATP assay. In this method, the cells in a soil sample are lysed, often by chloroform fumigations as discussed earlier, and the ATP content determined. This determination is based on the reaction:

$$\text{D-Luciferin} + ATP + \text{Luciferase} + O_2 \longrightarrow AMP + CO_2 + \text{Luciferase} + \text{light}$$

The light released by this process can be measured by a photometer and related to ATP content by comparison to standard curves. This technique is widely known as the 'firefly' assay because firefly tails are a ready source of D-Luciferin and Luciferase.

Activity measurements

The essence of heterotrophic microbial metabolism is the oxidation of organic carbon compounds to carbon dioxide, with the micro-organisms using the energy obtained for cell maintenance and growth. Thus, measurement of carbon dioxide production from soils has been widely used as an indicator of microbial activity. As with methods for estimating biomass, this method has been widely adapted for activity measurements of subsurface environments other than the soil horizon. The most straightforward method of estimating carbon dioxide production from soils is to place a sample in a closed vial, and monitor CO_2 production with time. Carbon dioxide is most conveniently quantified by gas chromatography equipped with a thermal conductivity detector. However, CO_2 can also be stripped from incubation vials by a stream of nitrogen gas and trapped in a NaOH solution. The carbonate ions generated by the trapping procedure are then quantified by titration with $BaCl_2$ and precipitation of $BaCO_3$.

The accuracy and sensitivity of quantifying respiratory activity by measuring CO_2 production can be significantly enhanced by use of radiotracers. For example, amending a soil sample with ^{14}C-labelled glucose and quantifying the rate of $^{14}CO_2$ evolved is a powerful tool for measuring metabolic rates in soils. This radiotracer technique is particularly useful for determining relative decomposition rates of different organic compounds by indigenous micro-organisms. For example, numerous radiotracer studies have shown that aliphatic compounds degrade much faster than phenolic compounds. This has been interpreted as reflecting the relative difficulty soil micro-organisms encounter in degrading compounds containing relatively stable benzene rings. A significant problem with both the CO_2 and $^{14}CO_2$ methods is that they tend to overestimate respiration rates.

This may reflect soil disturbance due to sampling procedures, enhancement of activity by addition of artificial nutrients, or a host of other possibilities. For this reason, these procedures are most effective when used to determine the activities of different soils relative to each other. In this way, sampling and experimental protocols can be standardised, and the relative activities of soil samples determined fairly accurately.

Other measures of biomass and activity

A large amount of work has been invested into alternative methods of measuring biomass and microbial activity. One of these methods, that of measuring the activities of specific enzymes in soils, has been widely applied. Because micro-organisms catalyse a wide variety of metabolic functions enzymatically, the activity of a particular enzyme should reflect the rate of metabolism. Some of the enzyme activities used for this purpose include oxidoreductases such as catalase, dehydrogenase, glucose oxidase, and oxidase as well as hydrolases such as amylase and cellulase.

Dehydrogenases, enzymes used by microbes to transfer hydrogen ions, have been widely used as a direct measure of metabolic activity. Dehydrogenase activity is measured by determining rates of 2,3, 5-triphenyl tetrazolium chloride reduction to triphenyl formazan in soil under anaerobic conditions. The triphenyl formazan can then be extracted and measured spectrophotometrically. Significantly, a positive correlation between dehydrogenase activity and respiration rates determined by CO_2 or $^{14}CO_2$ production is not commonly observed. This has been interpreted as reflecting the nonactive state of much of the ambient soil microflora.

One variation of the enzyme activity method is to measure the products of an enzymatic process, such as hydrolosis, that may be carried out by a wide variety of enzymes. For example, fluorescein diacetate (FDA) is hydrolysed by many different enzymes, such as lipases, proteases, and esterases. Thus, the rate of FDA hydrolosis in soil slurries is a qualitative measure of a broad spectrum of enzymatic activity and thus bacterial activity.

The FDA hydrolosis technique has been applied to a number of soils. A notable example was given by Federle who showed that microbial activity was present well below the soil zone and pointed out that this activity could well impact the transport of contaminants. These authors also showed that FDA activity correlated well with the abundance of phospholipids, a measure of biomass. Another method long used in soil microbiology as a measure of microbial activity is heat output. This was first proposed by Hesselink van Suchtelen, and significant advances in the technique, termed microcalorimetry, have been made recently in applying this concept to soils.

Given the wide variety of methods available for measuring microbial activity in soils, it is reasonable to wonder how well these techniques agree with each other. In fact, the methods do not always agree as well as one might expect.

One such comparison was given by Sparling, who showed that heat output consistently correlated with respiration, ATP content, amylase activity, and biomass. Curiously, however, heat output did not correlate well with dehydrogenase activity. In addition, there was little correlation between the various activity measurements and soil properties such as pH, carbon content, or nitrogen content.

The lack of consistently strong correlations of the various measures of soil biomass and microbial activity troubled microbiologists for many years. Recently, however, a consensus has emerged that no one technique is capable of giving the answer as to soil biomass or microbial activity. Each of these techniques, however, is suited for answering particular questions. Thus, the technique selected generally depends more on the question being asked than on the 'accuracy' of the technique. If, for example, the question deals with relative respiration rates between the soil zone and sediments deeper in the unsaturated zone, CO_2 or $^{14}CO_2$ production rates are suitable techniques. If, on the other hand, the question is how much of the soil organic matter is tied up in microbial biomass, the chloroform fumigation or ATP techniques may be more appropriate. Finally, if the question deals with the diversity of the microbial community, the phospholipid technique is most appropriate. This generality applies to biomass and microbial activity measurements in all subsurface environments. The measurement technique largely depends on the question being asked.

Distribution of Bacteria in the Unsaturated Zone

The unsaturated zone is inhabited by bacteria, viruses, actinomycetes, cyanobacteria (formerly called blue-green algae), fungi, and algae as well as by microscopic arthropods. Bacteria, however, are the most numerous of the micro-organisms in this environment and their metabolism is the most broadly based. For these reasons, bacterial processes probably have the largest impact on the chemistry of infiltrating groundwater. No attempt will be made here to discuss the distribution of micro-organisms other than the bacteria.

Soil zone

A great deal is known about the types and distribution of bacteria in soils. Much less is known about bacterial abundance and distribution in the unsaturated zone below the soil. Very early on, it was observed that total numbers of bacteria decreased sharply with depth in the soil. Because of this, it was easy for soil microbiologists to conclude that bacteria ceased to exist in appreciable numbers below the soil zone. This reasoning, while inaccurate, was perfectly logical based solely on the distribution of bacteria in the soil zone.

For example, if one fits an exponentially decreasing function to a bacterial profile like that of Waksman, it is found that the numbers of bacteria approach zero at a depth of only a few metres. Based on this type of reasoning, it is easy to see why some soil microbiologists earlier in this century dismissed the possibility that bacteria could live deep in the unsaturated and saturated zones.

Intermediate unsaturated zone

While studies of bacteria in the saturated portion of the subsurface have become fairly common in the last twenty years, relatively little attention has been paid to the unsaturated zone. A notable exception to this was a study at a site in Oklahoma. These investigators carefully noted the saturation status of samples taken from the subsurface and reported them as being from the unsaturated zone, from the interface of the saturated and unsaturated zones (i.e. the capillary fringe), or from the saturated zone. Total counts of bacterial cells (determined by acridine orange direct count) were in the range of 10^6 to 10^7 cells per gram of dry sediment. The highest counts were found just below the soil zone and were somewhat lower in the capillary fringe. Interestingly, there was no significant change of bacterial numbers from the capillary fringe into the saturated zone.

Balkwill and Ghiorse also reported the total numbers of distinct colony types observed on several types of growth media. The number of distinct colony types observed to grow on dilute PTYG (peptone, tripticase, yeast extract, glucose) agar declined noticeably with depth. In contrast, samples plated onto SSA (surface soil extract agar) media showed no noticeable decrease in numbers of colony types with depth. While it is tempting (and probably inappropriate) to overinterpret these data, it does suggest that bacteria from deeper in the unsaturated and saturated zones grew most efficiently on carbon sources readily available from indigenous sources.

The types of bacteria identified by Balkwill and Ghiorse from the unsaturated and saturated zones were dominated by isolates of the genera *Pseudomonas* and *Arthrobacter*. Significantly, members of these genera are typically observed to be the dominant types of bacteria present in soils. It is entirely possible that the types of bacteria present in the unsaturated zone and shallow saturated zone largely reflects bacterial populations present in the overlying soil zones. However, data are presently insufficient to firmly decide this interesting question.

Deep unsaturated zone

Significant areas of the earth are characterised by arid or semiarid climates. These areas often exhibit unsaturated zones that are tens or even hundreds of metres thick. The microbiology of these deep unsaturated zones has not been extensively characterised. However, one such study has been reported by Colwell that focused on unsaturated sediments from the high desert of Idaho.

One unusual finding reported by Colwell is that of 32 strains of aerobic, heterotrophic bacteria recovered from these deep unsaturated sediments, 84 per cent were gram-positive. This again contrasts with studies of saturated subsurface sediments where gram-negative strains typically predominate. It is possible that a major component of the stress in this environment is the lack of moisture. If this is the case, the relatively thick peptidoglycan layer in the cell wall of gram-positive bacteria may confer upon them a competitive advantage over gram-negative bacteria. In any case, the observed presence of nearly atmospheric levels of gaseous oxygen in these deep unsaturated sediments certainly reflects the extremely low indigenous activity of micro-organisms in this environment.

LOCAL FLOW SYSTEMS

A local system of groundwater flow is defined as having its recharge area at a topographic high and its discharge areas at an adjacent topographic low (Fig. 24.1). From a practical perspective, local flow systems often correspond to shallow water table aquifers that are recharged directly from precipitation and that provide base flow to adjacent streams or rivers. Local flow systems are hydrologically very active.

Local Flow Systems as a Microbial Habitat

Local flow systems are very active hydrologically. This is to say that water from precipitation events quickly passes through the unsaturated zone, recharges local flow systems, and is readily discharged to surface streams or transpired by phreatophytic plants that tap the saturated zone. It is not uncommon for the water table of local flow systems to rise or fall several metres over time periods of only a week or two. It is generally considered that residence times of water in local flow systems are on the order of days to years.

The dynamic nature of local flow systems has important consequences as to the microbial habitat that is generated. Fast rates of recharge and discharge mean that local flow systems are relatively open to sources of nutrients from the surface or from the unsaturated zone. For example, dissolved inorganic carbon generated in the unsaturated zone may be quickly transported into local flow systems. Figure 24.3 shows a conceptual model of DOC production, transport through the unsaturated zone and delivery to the water table. DOC that is transported into local flow systems, while often consisting of high-molecular-weight compounds that are not easily assimilated by micro-organisms, may provide an important source of nutrients to bacterial populations.

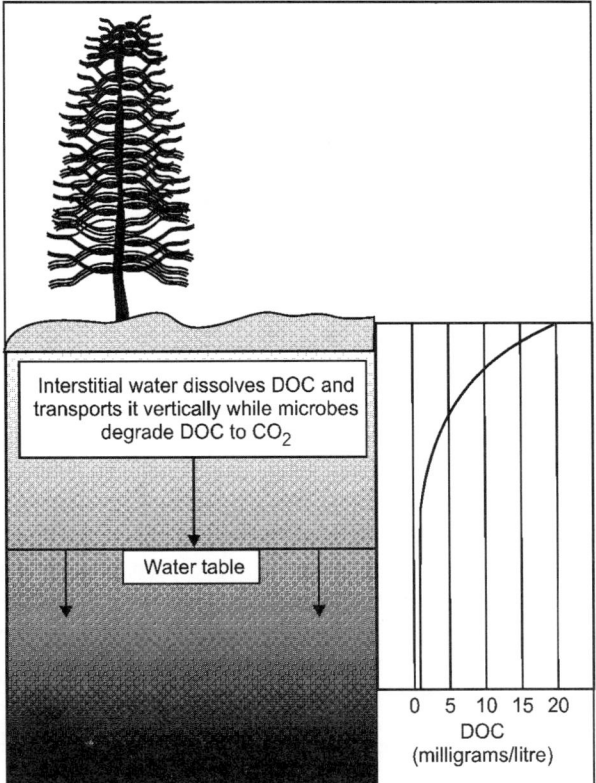

Fig. 24.3. The production of DOC in the soil zone and it is transport to shallow groundwater systems.

Fast rates of recharge and the relatively free exchange of gases within the unsaturated zone encourages the transport of dissolved oxygen to local flow systems. Because aerobic bacteria obtain more energy per mole of organic material oxidised than bacteria using other electron acceptors (such as ferric iron or sulphate), local flow systems are often dominated by aerobic bacteria. Such aerobic activity can be sustained only if there is a steady flux of oxygen into the system, and this requires the close connection with the unsaturated zone characteristic of local flow systems.

Recharge to local flow systems that passes through the unsaturated zone may also facilitate the transport of dissolved nitrogen species in the form of nitrate or ammonia. Because nitrogen fixation, ammonification, and nitrification are processes characteristic of the soil zone, recharge moving through soils to local flow systems is a potential source of organic nitrogen.

Because nitrogen is required for protein synthesis, it is an important nutrient requirement for bacterial growth. In many subsurface environments, nitrogen is the principal limiting factor for bacterial growth. In this respect, however, local flow systems have less limitation than intermediate or regional flow systems.

The proximity to land surface and the dynamic nature of groundwater movement in local flow systems makes them relatively accessible to chemical pollutants from human activities. These pollutants come from a variety of sources and include a dizzying variety of compounds. From the standpoint of a microbial habitat, such pollution is often a boon for bacterial growth, as it may provide carbon (gasoline, sewage wastes) or nitrogen (fertilisers, septic effluents) sources, each of which is often limiting in subsurface environments. Although precise figures are difficult to compile, it is probable that as much as 80 per cent or 90 per cent of all documented cases of groundwater pollution occur in local flow systems.

Distribution of Bacteria in Local Flow Systems

Systematic studies on the distribution of bacteria in shallow water table aquifers, herein termed local flow systems, are relative latecomers to subsurface microbiology. Whereas the distribution of bacteria in unsaturated sediments (the soil zone) and in regional flow systems (petroleum reservoirs) have been studied since the early twentieth century, studies on local flow systems began only in about 1980. To a large degree, this neglect simply reflects lack of pertinent questions concerning microbial processes in local flow systems.

In the case of the unsaturated zone (soils), the obvious connection between soil fertility and the abundance of soil microbes presented investigators with a wide range of important questions. Addressing these questions has spurred research in soil microbiology for 100 years. In the case of regional flow systems (largely petroleum reservoirs), it was recognised in the 1920s that bacterial processes affected petroleum production. This led to many studies on bacterial corrosion of well casings, biofilm plugging of reservoir rocks, and metabolism of hydrocarbons.

However, there was no similar impetus for studying microbial distributions in shallow water table aquifers. The increasing awareness that some human activities contaminated water table aquifers to the point of their having to be abandoned as sources of potable water changed this in the early 1970s. It was widely thought, for example, that simple dilution by uncontaminated water was the most practical method of renovating contaminated water supplies. Since many surface water bodies are completely turned over in a matter of days or weeks, the 'dilution solution' was deemed a workable remediation strategy. By this criterion, however, groundwater reservoirs could not be salvaged once contamination took place. Even local flow systems are recharged very slowly relative to surface water bodies, and the conventional wisdom was to simply write off contaminated groundwater as a loss.

INTERMEDIATE FLOW SYSTEMS

The defining characteristic of an intermediate flow system is that one or more topographic low areas separate recharge areas from discharge areas (Fig. 24.1). In practice, intermediate flow systems generally correspond to confined aquifer systems of moderate (<300 metres) depth. Because intermediate flow systems produce a high percentage of the groundwater used for human consumption, animal consumption, and irrigation in the United States, and other developing countries they are economically very important.

Intermediate flow systems differ from local flow systems primarily in the extent of connection with the surface and with the unsaturated zone. Because of this relative lack of connection, recharge rates are much lower than in most local flow systems.

Intermediate Flow Systems as a Microbial Habitat

The low rates of recharge and groundwater flow characteristic of intermediate flow systems creates a microbial habitat very different from that of local flow systems. Available evidence suggests that the delivery of DOC from the soil zone to local flow systems is fairly low. This being the case, delivery of DOC from surface soils to intermediate flow systems is often negligible. Because of this, microbial populations in intermediate flow systems are largely dependent upon organic material present in the sediments as a primary carbon source. These carbon materials are often refractory in nature, and this severely limits the metabolic and growth potential of microbes. Due to oligotrophic conditions, rates of bacterial metabolism in intermediate flow systems are slower than those in local flow systems.

In addition to being strongly carbon-limited, microbial processes in intermediate flow systems are limited by the availability of organic nitrogen. Delivery of dissolved nitrogen compounds to intermediate flow systems from soil water is limited for the same reasons that delivery of DOC is limited. Not surprisingly, therefore, concentrations of nitrogen species in groundwater of intermediate flow systems are generally less than 0.1 mg/l.

Distribution of Bacteria in Intermediate Flow Systems

The presence of bacteria in intermediate flow systems has been recognised for much of the twentieth century.

Studies of bacterial distributions in local flow systems had clearly shown that a large percentage of indigenous bacteria were attached to sediment particles rather than living free in interstitial water. However, by the late 1980s there were no systematic studies of bacteria in sediments of intermediate flow systems. The reason for this lack of attention, as was the case with local flow systems prior to 1970, was that there were no pressing scientific questions that required such investigations. In the case of local flow systems, this impetus was supplied by concerns about groundwater pollution. However, because of the low recharge rates, chemical contamination of intermediate flow systems was a rare occurrence. Thus, microbial studies of intermediate flow systems were not mandated by concern about groundwater pollution.

Not surprisingly, the aerobic activity of bacterial populations behaved in a similar manner. Hicks and Fredrickson investigated the metabolism of ^{14}C-labelled acetate, 4-methoxybenzoic acid, and phenol in cored sediments over a 21-day incubation period. Acetate and phenol were metabolised relatively easily in the sandy sediments, whereas 4-methoxybenzoate was metabolised less readily. In general, the highest rates of mineralisation were observed in the sandy sediments and the lowest in the clays for each compound.

Microbial Processes in Confining Beds

While it is generally observed that microbial activity in clayey confining beds is much less than that in more permeable aquifers, activity is often not completely absent. Furthermore, this activity can have important impacts on the chemistry of water in adjacent aquifer sediments in intermediate flow systems. The first systematic evaluation of microbial processes in confining beds, and how they affect groundwater chemistry, was given by McMahon and Chapelle. This study pointed out some unusual patterns in the concentrations of organic acids in aquifer and confining-bed pore water. Specifically, confining-bed pore water contained fairly high concentrations of acetate and formate (50–70 μM) compared to the fairly low concentrations found in the pore water of sandy aquifer sediments (1–10 μM). Furthermore, the clayey confining bed sediments showed much greater hydrogen (H_2) production than did sandy aquifer sediments. Since hydrogen, acetate, and formate are central to anaerobic food chains, these data were exactly opposite of what one might have expected given reported low rates of microbial activity in clayey sediments.

The reasons for the observed accumulation of organic acids in confining beds relative to aquifer pore water were clarified by means of radiotracer experiments. When confining-bed sediments were amended with ^{14}C-labelled glucose, a fermentable substrate, it was observed that most of the ^{14}C accumulated in formate and acetate rather than being oxidised to CO_2. In sandy aquifer sediments, however, all of the ^{14}C was oxidised rapidly to CO_2. In anaerobic sediments, complete oxidation of fermentable substrates such as glucose requires the symbiotic action of fermentative and respirative bacteria. Fermentative bacteria partially oxidise the substrate with the production of organic acids and hydrogen. These simpler compounds are then utilised by respirative bacteria. The observed accumulation of organic acids, therefore, indicated that fermentation outpaced respiration in clayey confining-bed sediments.

This finding has important implications for the microbial cycling of carbon in groundwater systems. The concentration differences of organic acids between confining beds and aquifers appears sufficient to drive a net diffusive flux of organic carbon to aquifer. This, in turn, provides a constant supply of simple organic substrates for respirative bacteria in the aquifers.

Furthermore, it was shown that a particular fermentation known as acetogenesis was an important microbial process in these sediments. Acetogenic bacteria are widely dispersed in the environment and have the ability to derive energy by converting molecular hydrogen (H_2) and carbon dioxide (CO_2) into acetate according to the stoichiometry:

$$4H_2 + 2CO_2 \longrightarrow CH_3COOH + 2H_2O$$

Like methanogenic bacteria, acetogens can use H_2 as an electron donor and CO_2 as an electron acceptor. However, with acetogens, acetate is the end waste product. This process was demonstrated by showing that radiolabelled $^{14}CO_2$ was converted to $^{14}CH_3COOH$ in experimental microcosms constructed with confining-bed material and incubated with a H_2 headspace. Interestingly, there was no observed generation of methane in these microcosms. This implies that methanogenic bacteria were absent from confining beds, which is consistent with previous observations.

The reasons why fermenting bacteria, such as the acetogens, are able to inhabit confining beds whereas respirative bacteria such as Fe(III) reducers, sulphate reducers, and methanogens do not is not known. There are, however, several possibilities. The acetogens cultured from confining-bed sediments by Chapelle and Bradley were gram-positive. One ecological advantage of the gram-positive cell wall is that it helps micro-organisms deal with the lack of water. This is probably one reason why gram-positive bacteria are so abundant in the unsaturated zone. Water in highly compressed confining beds is held tightly at the surface of clay minerals. Thus, while confining bed sediments are water-saturated, this water is much less available to micro-organisms. It is possible, therefore, that the gram-positive cell walls of acetogenic bacteria provide them with a competitive advantage over gram-negative respiring bacteria in confining beds. This, in turn, may explain the predominance of acetogenic metabolism in confining-bed environments.

REGIONAL FLOW SYSTEMS

A regional flow system is defined as having its recharge area at a groundwater divide and its discharge area at the bottom of the regional basin. A regional flow system is characterised by long flowpaths and extremely slow rates of groundwater flow (Fig. 24.1). Because of the long residence times of groundwater in regional systems, the water is often highly mineralised and unsuitable for human consumption, animal consumption, or irrigation. Also, the great depth of many regional flow systems, often greater than 300 metres, makes water-well drilling prohibitively expensive. For these two reasons, regional flow systems are not often tapped for supplies of potable water.

The deep sedimentary basins in which regional flow systems develop are, however, very important economically because they are often sites of hydrocarbon accumulation. It was recognised very early on that bacterial processes such as sulphate reduction and methanogenesis had important impacts on the production of oil and gas from wells. In addition, it was soon recognised that bacterial processes may have been involved in petrogenesis, the processes that convert buried organic matter into petroleum. For these and other reasons, studies of bacterial processes in regional flow systems were undertaken early in the twentieth century and have continued to the present.

Early Observations from Petroleum Reservoirs

The single largest problem for a petroleum prospector is obviously how to find accumulations of oil. To this end, enormous effort was invested in understanding the geology, hydrology, and geochemistry of known petroleum reservoirs in order to identify leads for finding new reservoirs. One such lead was that brines associated with petroleum tended to be characterised by low concentrations of sulphate and high concentrations of sulphides. The practical importance of this was that the simple analysis of water for sulphate and sulphide was a possible indicator of the presence or absence of nearby petroleum accumulations. This observation by itself, however, was unsatisfactory until its cause could be identified. This was accomplished by Bastin when he showed that sulphate-reducing bacteria were present in oil field brines and that the activity of these micro-organisms could account for the observed phenomenon.

Distribution of Bacteria in Regional Flow Systems

Sampling sediments from regional flow systems presents extreme technological problems. While auger drilling rigs are easily adapted for aseptically sampling local flow systems, and mud-rotary rigs can be used to aseptically sample local and intermediate flow systems, aseptically sampling sediments from deep regional flow systems is much more difficult. One of the most obvious problems is cost. Drilling into regional flow systems is expensive under any circumstances and coring sediments is prohibitive. While cores are routinely taken in such environments, they are usually taken for stratigraphic control or for analysis of organic carbon content. In addition, procedures for obtaining cores that are not contaminated by drilling fluids are difficult to apply. For these reasons, micro-organisms in regional flow systems are most commonly recovered from groundwater.

Culture techniques

As in local and intermediate flow systems, various culture techniques have been used to investigate the presence and kinds of micro-organisms in regional flow systems. The work of Bastin and Davis is fairly typical of this approach. More recent examples of how the distribution of micro-organisms in regional flow systems were given by Dockins and Olson, who documented the presence of methanogenic and sulphate-reducing bacteria in water produced from the Madison aquifer system of Montana. The work of Dockins and Olsen is particularly notable in that the investigators showed that the bacteria grew at the temperatures of the aquifers (~70°C) but would not grow at room temperature (~20°C). This showed that the bacteria were acclimated to the specialised conditions of the aquifer and ruled out the possibility that the observed bacteria were contaminants from the near surface.

Because of the slow movement of fluids in regional aquifers and because of their isolation from the surface, there has been considerable interest in using them as repositories of hazardous and radioactive wastes. Due to the potential for micro-organisms to affect waste mobility, evaluations of such flow systems as waste repositories has led to microbiological studies. One such study was described by Pedersen and Ekendahl. These investigators showed that groundwater produced from a deep (860 metres) granitic aquifer contained on the order of 105 cells/ml total bacteria, of which between 10^2–10^3 were cultureable under anaerobic conditions. Sulphate-reducing and methanogenic bacteria were part of the ambient microflora as well as members of the genera *Shewanella* and *Pseudomonas*.

Pedersen and Ekendahl were microbiologists; they approached the problem of micro-organisms in the deep granitic terrain using classic microbiologic techniques. When confronted with the problem of determining which microbial processes were active *in situ*, however, it was clear that the mere presence of certain micro-organisms was equivocal. Consequently, these investigators, as had the earlier petroleum geologists, turned to evidence provided by groundwater chemistry. They noted that sulphate (and sulphide) was present in the groundwater and that sulphate reducers largely exclude the activity of methanogens. The conclusion, therefore, was that sulphate reduction was the most likely predominant process *in situ*.

Geochemical methods

There are obvious problems with relying solely on culture methods to study the distribution of micro-organisms in regional flow systems. In addition to the problems of media selectivity and determining which processes are most active *in situ*, there is the problem of how representative micro-organisms in groundwater are of the sediment-bound population. Davis recognised this problem explicitly, as did Pedersen and Ekendahl. In both cases, it was suggested that groundwater geochemistry gave important information as to *in situ* microbial processes.

Using groundwater chemistry data and a computer program for making equilibrium chemistry calculations, Thorstenson identified sulphate reduction as the predominant microbial process in the intermediate portion of the flow system, which graded to methanogenesis as the predominant process in the regional portion of the flow system. This showed that geochemical modelling was one of the best available methods for deducing the distribution of microbial processes in inaccessible regional flow systems.

Molecular methods

Recent advances in molecular methods for analysing microbial communities without culture methods have been applied to intermediate and regional flow systems. An example of this was given by Fry. This study used molecular methods to study the microflora of an intermediate flow system (~300 metres below land surface) and an underlying regional flow system (~1200 metres below land surface) of the Columbia River Basalt aquifer. Micro-organisms from the two aquifers were collected by filtering pumped groundwater. Nucleic acids from these samples were extracted, and the microbial community examined using oligonucleotide 16S RNA probes supplemented by analysis of rDNA analysis. The results of this study indicated that, while members of the domain Bacteria dominated the microflora (92–64 per cent), members of the Archaea (~2 per cent) and Eucarya (5–14 per cent) were also present. A substantial percentage of gram-positive bacteria (7–12 per cent) were present, and possibly indicate the presence of acetogenic micro-organisms.

The groundwater chemistry of the two sampling locations suggests that the shallower aquifer was dominated by methanogenic respiration (methane concentrations $\sim 10^{-4}$ M; sulphate concentrations <0.05 mg/l) and the deeper aquifer dominated by sulphate reduction (methane concentrations $\sim 10^{-6}$ M; sulphate concentrations ~ 140 mg/l). Interestingly, there was little apparent difference in the presence of methanogens (*Archaea*) or sulphate reducers (*Desulfovibrio*) between the two sampled horizons. This may reflect bias introduced into the analysis by considering free-living micro-organisms as opposed to rock-bound micro-organisms, or by bias inherent in the primers and oligonucleotide probes used. Alternatively, it may reflect a reality that both methanogens and sulphate reducers were equally abundant in each horizon, but that their activity differed. At present, there are too few studies using this approach to make reliable conclusions. Nevertheless, the molecular approach holds much promise for understanding the complex microbial ecology of regional aquifer systems.

To sum up in dealing with microbial processes in subsurface environments, it is first necessary to classify the kinds of environments that are most commonly encountered. In this chapter, a classification is given that divides subsurface environments on the basis of their hydrologic attributes and is independent of depth. The two broad environments recognised are: (i) the unsaturated zone characterised by the presence of both gas and water in interstices, and (ii) the saturated zone where water fills all available interstices. The saturated zone is further subdivided into local, intermediate, and regional flow systems. These divisions are made on the basis of quantitative considerations of groundwater flow in closed basins as described by Toth. Local systems are characterised by abundant local recharge, fast groundwater flow velocities (metres/day), and are generally aerobic. Intermediate systems are not directly affected by transient precipitation events, have slower flow velocities (metres/year), and are generally anaerobic. Regional systems are largely isolated from surface influences, often have very sluggish flow velocities (metres/century), and are ubiquitously anaerobic.

These different subsurface environments are characterised by different microbial processes. The unsaturated zone and local flow systems are dominated by aerobic heterotrophic bacteria that are present in numbers of between 10^6 and 10^8 cells/gram sediment. These systems may contain measurable numbers of eukaryotic micro-organisms as well. Intermediate flow systems often contain very diverse populations of anaerobic and facultatively anaerobic micro-organisms in numbers ranging from 10^4 to 10^6 cells/gram sediment. Regional flow systems often contain micro-organisms as reflected in their presence in groundwater. However, little information is available on the numbers or kinds of micro-organisms presence in the rocks or sediments of these deeply buried systems. Geochemical modelling is a useful tool for studying microbial processes in regional flow systems. However, the methods of molecular microbial ecology show promise for increasing our understanding of these inaccessible subsurface ecosystems.

Chapter 25

Microbiological Sampling of Subsurface Environments

INTRODUCTION

Subsurface environments, whether they are shallow local flow systems or deep regional flow systems, share a common characteristic. They are relatively inaccessible and therefore very difficult to sample. Simply obtaining materials from subsurface environments, which may consist of unconsolidated sediments or fully lithified rocks, is difficult under most circumstances. However, if the additional requirement is made that these samples must be free of chemical and biological contamination induced by the sampling process, this becomes an even more formidable undertaking.

Early work on the microbiology of deep subsurface environments, such as the isolation of sulphate-reducing bacteria by Bastin, relied on water produced from oil or water wells as a sample source. This is certainly the easiest and most direct method for sampling the deep subsurface. Unfortunately, there are many potential problems with such methods. The most serious is that the drilling process, for either water wells or oil wells, can potentially contaminate the environment under study. Drilling fluids and grouting materials contain substantial quantities of micro-organisms, and these can 'inoculate' subsurface environments during drilling and well construction. When sampling wells, therefore, it is difficult to judge whether the micro-organisms recovered were indigenous or introduced by the drilling process.

A second problem with using wells for sampling the subsurface is that the construction of a well introduces a new 'environment' that did not exist prior to drilling. This new environment includes the borehole, the gravel pack, the casing and screen, and the grout. Since all of this material is artificially introduced, it may substantially alter conditions in the vicinity of a well. This alteration of the physical conditions may significantly affect the microbial processes near the well. Perhaps the best example of this is the potential introduction of oxygen to the subsurface. Wellbores, being open to the surface, may transport oxygen into previously anaerobic environments. This, in turn, may support growth of aerobic bacteria that are not active under undisturbed conditions. Clearly, the possibility of such alteration must be carefully accessed when using wells for microbial sampling.

A third problem with using water from wells for microbiologic analysis is that many micro-organisms in the subsurface tend to be sediment-bound. Thus, the kinds and diversity of free-living micro-organisms may be significantly different from those attached to sediment surfaces. This possibility, together with the other problems associated with sampling well water, introduces considerable uncertainty into the data obtained.

For these reasons, modern studies of microbial processes in subsurface environments have tended to emphasise direct sampling of subsurface materials. Significant advances have been made since the early 1970s in techniques for obtaining uncompromised samples of unconsolidated sediments from subsurface environments for microbiological analysis. These sampling techniques have probably been more important in the investigation of microbial processes in subsurface environments than any other factor. Much less attention has been paid, however, to sampling consolidated rock materials. In this chapter, different methods of sampling subsurface environments for micro-organisms and microbial processes are outlined.

SAMPLING THE UNSATURATED ZONE

The unsaturated zone is characterised by a negative hydraulic pressure. That is, water in the unsaturated zone tends to be bound to sediments by capillary forces and does not flow readily to wells. This characteristic has a major impact on methods for sampling sediments from the unsaturated zone. The relative lack of water pore pressure means that the sediment framework is supported by grain-to-grain contacts rather than by hydrostatic pressure. This greatly increases the cohesiveness of the sediment and makes it much easier to sample.

A sampling system for subsurface sediments must simultaneously solve three problems. First, and most obviously, the system must be able to bore a hole from the surface to the horizon of interest. Second, the system must be capable of removing sediments from the hole as it is bored. This sediment retrieval may or may not be coupled to the sampling process. Third, the system must be able to maintain the sides of the borehole so that it does not collapse. The cohesive nature of sediments in the unsaturated zone allows each of these problems to be solved with relatively simple technology.

Hand Augering

By far the simplest method of obtaining sediment samples from the subsurface is by the use of hand augers. A typical hand auger is equipped with a simple blade system that enables the tool to cut into unconsolidated sediments. The head of the auger also is designed to capture and hold sediments so that they can be retrieved.

The sediments that are retrieved serve both to deepen the hole and as the sample material as well. Thus, for hand-augering systems, there is no distinction between hole construction and sample recovery. The stem of the hand auger is equipped with a handle that enables the head of the auger to be twisted into the ground. Once the head of the auger is twisted into the ground, the handle serves as a means of pulling the head out of the hole. The handle of the auger, and thus the length of the auger stem, can be extended by additional lengths of rod.

The usual procedure for obtaining sediment samples by means of a hand auger is as follows. First, the auger is twisted into the ground the approximate length of the auger head; this is usually about 6 or 8 inches. Second, the sample is retrieved by pulling the head of the auger back to land surface. Finally, the sediment recovered is subsampled by means appropriate to the problem at hand. If the sediments are to be used for microbiologic analyses, subsampling is accomplished using flame-sterilised instruments and the subsamples placed into appropriate sterilised containers.

It is generally possible to sample subsurface sediments to a depth of about 20 or 30 feet using hand augers. Beyond 20 feet or so, the weight of the augers as well as the difficulty in twisting them into the ground and retrieving quickly becomes limiting. It is often difficult to hand-auger below the water-table. The increased pore pressure of the water tends to separate the sediment grains, thus decreasing the sediment cohesiveness. Sand-sized sediments will usually collapse once the auger is removed so that it becomes impossible to deepen the hole. Even in clayey sediments that are more cohesive, pulling the auger out of the sediments tends to collapse the hole. In general, hand augers are not suitable for routine sampling of sediments below the water table.

The advantage of hand augering is the technical ease of obtained subsurface sediments and its applicability to a number of unsaturated subsurface environments. However, because of the limited depth obtainable with this method, the inability to penetrate hard or indurated sediments, and the difficulty of sampling saturated sediments, its applicability to many subsurface environments is greatly reduced.

Air Drilling and Coring

In air drilling and coring, a large compressor is used to force air down the drillpipe, out the bit, and up the outside of the borehole. As the core barrel cuts the subsurface materials, the air serves to blow the cuttings out of the hole as well as to cool the core barrel. Often, a small amount of water or surfactant is injected into the airstream to control dust and to aid with cooling the bit. However, use of these materials increases the possibility of contamination of the recovered sediments.

An example of the use of air coring for recovering unsaturated sediments for microbiologic analysis has been given by Colwell. In this case, air coring was considered to be the sampling method of choice because it minimised the possibility of sample contamination by circulating fluids. To reduce the possibility of airborne micro-organisms compromising the sampling procedure, the air being circulated into the borehole was first filtered though a 0.3 micron filter. Because water or surfactants were not used to help cool the core barrel, drilling had to proceed fairly slowly in order to avoid overheating the core barrel. Clearly, if the core barrel was allowed to become too hot, the microbiological integrity of the samples could be compromised. The main disadvantage of this method, therefore, is that sample recovery must proceed fairly slowly. The advantage is that the probability of sample contamination by the coring process is fairly low.

SAMPLING LOCAL FLOW SYSTEMS

Water-saturated sediments, which by definition occur below the water table, are much less cohesive than unsaturated sediments. When a hole is bored into saturated unconsolidated sediments, therefore, there is much more tendency for the hole to collapse. A major technical problem in sampling saturated sediments is to be able to hold a borehole open. There are numerous solutions to this technical problem with saturated sediments. By far the most common technique in relatively shallow water table aquifers (local flow systems) is the use of auger drilling combined with split spoon and shelby-tube core sampling.

Split-Spoon Sampling

A widely used method for sampling sediments from relatively shallow subsurface environments is by means of hollow-stem auger drilling combined with split-spoon coring. The split-spoon technique was developed for use by civil engineers in foundation studies for buildings, bridges, and roads. The dual purpose of this technique was to provide estimates of the *in situ* strength of the subsurface materials and to provide samples for laboratory testing. Because of the ease of using this technique in relatively shallow systems, however, it has been widely adapted for groundwater applications. In turn, this technique has been widely adapted for microbiologic sampling of subsurface environments.

A typical system for hollow-stem coring includes hollow auger flights, a plug for closing off the hole in the auger flights, a cutter bit on the auger flights, a pilot bit on the plug, and rods that are coupled to the plug and pilot bit. Both the pilot and auger cutter bits are designed to penetrate unconsolidated sediments.

The main advantage of the hollow-stem auger method of drilling is that the auger flights remain in place during sampling, thus preventing collapse of the borehole. Also, because the screw action of the auger flights is what removes cuttings from the borehole, circulating fluids are not required. This greatly decreases the potential for contamination of the sampled sediments.

The principal disadvantage of the split-spoon method is that the depth of sampling is greatly limited. Because of their great weight, the number of auger flights that can be manipulated by the drilling rig is limited. A typical drilling rig designed for hollow-stem augering has a depth limitation of about 100 feet. If more auger flights are added, the rig would be incapable of holding the added weight and the drill string would be lost. Another disadvantage of this method is that the diameter of recovered sediment samples is typically limited. A sample diameter of 1½ inches, the diameter of core recovered by a standard split spoon, is the minimum size for obtaining sediments for microbiological analysis. In clayey-textured sediments that are fairly cohesive, this sample size is not a significant problem. In some coarse sandy sediments, however, the lack of cohesion often makes it difficult to obtain or manipulate uncompromised sediments. In such sediments, or for applications where more sample material is needed, other drilling techniques must be used.

Push-Tube (Shelby Tube) Sampling Methods

For many applications, particularly involving aseptic technique for microbiologic analysis, the limited sample diameter provided by split-spoon sampling is insufficient. An alternative is provided by push-tube sampling methods. These methods are notable from a historical perspective because they were among the first used to obtain subsurface sediments that were uncontaminated by surface-sediment micro-organisms.

An important limitation of the push-tube methods for obtaining subsurface sediments is that the auger flights must be removed prior to sampling. This, in turn, requires the sediments to be sufficiently cohesive so as to prevent the borehole from caving in as the flights are withdrawn. This is possible in some, but not all, locations.

Direct Push Sampling

Most direct push methods, such as Geoprobe, are designed for sampling depths between 30 and 60 feet (9–18 metres). However, it is possible to sample as deep as 100 feet (90 metres). Depending on the kinds of tools used, it is possible to obtain sediment cores, soil-gas samples, and groundwater samples. For sampling soils or other unsaturated zone sediments, the open-tube method can often be used [Fig. 25.1(a)]. With this method, a core barrel equipped with a plastic liner is simply driven into the ground, filling the barrel with sediment so that it can be retrieved. For sampling saturated zone sediments that often collapse, a closed-piston method is used [Fig. 25.1(b)]. With this method, the core barrel is equipped with a retractable point that allows the closed barrel to be driven to the desired depth. Next the point is retracted, and the core barrel filled by driving it into the sediments as with the open-barrel method. In both cases, sediments obtained are subject to various kinds of contamination, which must be carefully considered when sampling for microbiologic purposes.

Aseptic Technique with Split-Spoon, Shelby Tube and Direct Push Sampling

The goal of sediment sampling for microbiologic analysis is the recovery of materials that are representative of the subsurface environment under study. In order to accomplish this, the most obvious requirement is that the samples not be contaminated by micro-organisms from other horizons. This is particularly important for subsurface sediments that contain fewer micro-organisms than surface soils or aquatic sediments. Because of this, the possibility for contamination needs to be carefully assessed in each individual application.

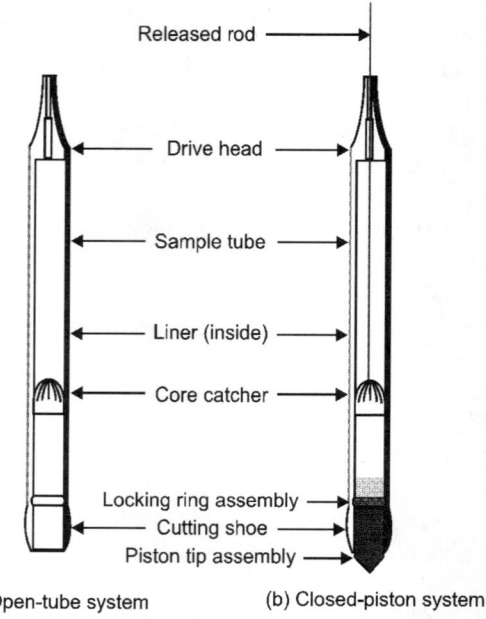

Released rod

Drive head

Sample tube

Liner (inside)

Core catcher

Locking ring assembly
Cutting shoe
Piston tip assembly

(a) Open-tube system   (b) Closed-piston system

Fig. 25.1. Diagram of a direct-push coring device: (a) open tube system useful for surface sediments, and (b) a closed-piston system used for capturing sediments below the water table.

Tool contamination

There are numerous possible sources of contamination when using any kind of soil or sediment sampling device. First, there is possible contamination from the sampling tool itself. This possibility has been emphasised by Dunlap and Wilson, who stress that the tools must be sterilised prior to sampling. Wilson favour flame sterilisation, in which the tool is rinsed with alcohol and then ignited until the alcohol is completely burned off. This is probably the most convenient method of sterilising sampling tools and is common in field applications.

Another method of sterilising sampling tools is with the use of compressed steam. It is common practice, when drilling sites characterised by chemical contamination, for drillers to steam-clean auger flights and sampling tools in order to prevent cross-contamination between holes. These steam cleaners operate at temperatures in excess of 100°C and are capable of sterilising the tools as well as cleansing them of chemical contamination. Because the steam is delivered at atmospheric pressure, there is some doubt that bacterial spores can be inactivated.

Down-hole contamination

Another possible mode of sample contamination is from down-hole contamination. This is to say that materials that originated from the surface or from the side of the borehole fall to the bottom of the hole and are recovered by the sampling tool. Such contamination is often readily apparent upon sample recovery by the presence of loose sediment at the top of the sample container.

SAMPLING INTERMEDIATE AND REGIONAL SYSTEMS

The technology involved with split-spoon, push-tube, or direct push sampling is relatively simple and straightforward. However, it is rarely feasible to use these methods for sampling sediments deeper than about 100 feet. It is possible, with special equipment, for auger rigs to penetrate as deep as 250 feet. However, this is rare and seldom done in practice. When dealing with these deeper systems, which generally include intermediate and regional groundwater systems, more sophisticated technology is required. This technology introduces several more problems related to aseptic technique that must be carefully evaluated.

Mud-Rotary Drilling

In mud-rotary drilling, the hole is bored by a rotating bit, as was the case with auger drilling. However, the twin problems of cuttings removal and hole stabilisation (that is, preventing the hole from collapsing) are solved by the use of drilling fluid. Drilling fluids are continually circulated in the hole to remove cuttings and to apply pressure to the borehole in order to keep it from collapsing. This technique has been widely used in microbiologic investigations of deep subsurface environments, and considerable effort has been invested in solving the sample-integrity problems encountered with this technique.

A mud-rotary rig has a mast (also called a boom), which can be raised for drilling and lowered for travelling by means of hydraulic cylinders. The hole is cut by a bit that has ports to allow drilling fluid to be pumped into the hole. The drilling fluid is pumped into the drilling rods through the water swivel (since the rods are rotating) with a specially designed mud pump. The fluid travels down through the rods, out of the bit, and back up the hole. As the fluid moves up the hole, cuttings are carried out of the hole. The cuttings are then separated from the fluid, and the fluid is reused.

Mud-rotary drilling is most commonly used to construct a borehole for later construction into a well. However, mud-rotary technology can also be used for coring applications. For subsurface microbiology applications, mud-rotary coring is the most common method for obtaining samples of relatively deep sediments (100–3000 feet). This method has been extensively used in studying the microbiology of coastal plain settlements.

Drilling Fluids

Drilling fluids are generally referred to as mud by drillers and geologists. This term, while being a fairly accurate description, gives little hint as to the technological sophistication involved. Drilling fluid is a marvel of applied chemistry and has undergone extensive research and development since its invention in the nineteenth century.

Drilling fluids are designed to accomplish four main functions:
1. To remove cuttings from the hole during drilling: After cuttings are generated by the drill bit, it is critical to immediately remove them from the hole. If the cuttings are allowed to settle around the drill bit, they will prevent the bit from turning freely. Ideally, cuttings are immediately entrained by the fluid stream and carried out of the hole.
2. To keep the borehole from collapsing: When sediments, particularly sandy sediments, are saturated by water, they have little cohesive strength. Under most conditions, a borehole drilled into sandy sediments will not remain open without some sort of support. Drilling fluid provides this support by virtue of the pressure it exerts on the borehole wall. If the pressure exerted by the drilling fluid is greater than the pressure in the formation, the borehole will not collapse. A borehole that is prevented from collapsing with drilling fluid is referred to as a stable hole.

3. To cool and lubricate the bit: Drilling bits, due to friction with the sediments being drilled, generate considerable amounts of heat. Drilling fluids prevent a buildup of heat in the drilling tools by lubricating the bit, thus minimising heat generation, and by transferring heat away from the bit.

4. To control fluid loss into the formations being penetrated: When drilling in highly permeable formations, drilling fluids will tend to migrate from the borehole into the formation. Drilling fluids are designed to control this fluid loss by generating a filter cake on the outside of the borehole. This filter cake is made primarily from low-permeability clays that inhibit migration of fluid out of the borehole.

Each of these functions is addressed by one or a combination of drilling fluid properties. The principal properties of water-based drilling fluids are listed in Table 25.1.

Table 25.1. Principal properties of water-based drilling fluids.

1. Density (weight)	4. Gel strength
2. Viscosity	5. Fluid-loss-control effectiveness
3. Yield point	6. Lubricity (lubrication capacity)

Density

The density of a drilling fluid is defined as the weight per unit volume. Thus, it is common practice to use the terms density and weight interchangeably. Density is usually expressed in terms of pounds per gallon (lbs/gal) or pounds per cubic foot (lbs/ft^3).

Viscosity

The viscosity of a drilling fluid refers to the sheer stress that it can exert on surfaces or particles. The ability of drilling fluids to lift cuttings from the hole are a function of fluid viscosity and fluid velocity. Curiously, viscosity and density are not functionally related to each other and are measured in different units.

In the field, the viscosity of drilling fluid is measured with a Marsh funnel. This is a funnel that carries a standard volume and has a standard-sized opening at the bottom. Viscosity is expressed as the number of seconds required for the fluid to drain from the funnel. A Marsh viscosity reading of 35–40 seconds is sufficient to remove most fine sands from the hole. Coarser sediments, such as coarse gravels, may require a Marsh viscosity of 75–85 seconds.

Yield point

Water is referred to as a Newtonian fluid because it deforms in direct proportion to an applied stress. Drilling fluids that contain clay minerals and other additives, however, tend to behave as plastics. This is to say that they do not deform until a significant amount of stress is applied. The yield point of a drilling fluid is the pressure at which the pump is able to move the drilling fluid, and the fluid begins to behave as a Newtonian fluid. Below the yield point, the viscosity of the drilling fluid changes with increasing stress. Above the yield point, the drilling fluid viscosity does not increase with increasing stress.

Gel strength

Gel strength is a measure of a drilling fluid's ability to support particles when the fluid is not moving. In clay-based drilling fluids, gel strength is provided by the electrical interaction of the clay particles. The flat surfaces of clay particles are typically positively charged whereas the edges are negatively charged. When the fluid is at rest, the attraction between clay particles provided by these charge differences provides its gel strength. Gel strength can also be enhanced by means of polymer additives such as guar gum.

The gel strength of a drilling fluid is maximised if the clay particles and polymers are well-dispersed, allowing them to align themselves according to their electrical and/or chemical properties. For this reason, a good driller takes care to thoroughly mix clays and polymers in the fluid prior to use.

Fluid-loss control

An important function of drilling fluid additives is to prevent fluid migration into the formations being drilled. Because fluid pressure in the borehole must exceed the hydrostatic pressure in the formation to prevent borehole collapse, the tendency is for drilling fluids to migrate into the formation. This is undesirable for two reasons. First of all, drilling fluids are expensive and loss during drilling is a significant economic drain on the driller's operation. Second, the presence of substantial quantities of drilling fluid in a water-producing formation can degrade water quality, rendering any well useless for water-chemistry studies or for water supplies. A combination of clay minerals and organic additives are generally used to control fluid loss during drilling operations.

Lubricity

One function of drilling fluid is to lubricate and cool the bit during drilling operations. The flat structure of clay minerals used in drilling fluids confers excellent lubricating properties. When physically pressed together, clay minerals tend to align themselves along their flat, positively charged surfaces. Because the positive charges repel each other, there is little tendency to stick together, and the clay surfaces slide easily past each other. The lubricating characteristics of drilling fluids maximises the efficiency of drilling and minimises down-hole heat buildup.

Mud-Rotary Coring

The technology that has been most extensively utilised for microbiological sampling of intermediate and regional flow systems is mud-rotary coring. This coring technology has developed parallel to mud-rotary drilling methods and utilises many of the same tools and drilling fluids. However, there are significant philosophical differences between drilling and coring technologies. The primary goal of drilling technology is the production of a borehole that can be used for constructing a well.

In coring, however, the primary goal is the recovery of sediment samples that have been disturbed as little as possible, and the borehole itself is generally of less importance. The net affect of these differences is that coring technology has become a narrowly focused speciality in the drilling trades that is practiced by relatively few individuals. It is important for the subsurface microbiologist to recognise that successful coring operations require a coring specialist and cannot be carried out by just any drilling contractor.

Equipment used for mud-rotary coring

The essential features present on a coring rig include a boom for raising and lowering the drill string, a rotary table for turning the drilling tools, and a mud pump for circulating drilling fluid. In all of these particulars, a coring rig is virtually identical to any mud-rotary rig. The primary differences are in the tools that are used.

Role of the driller

It is difficult for the casual observer to readily appreciate the skill involved in mud-rotary coring. When watching a coring operation, one sees the physical work of raising and lowering the inner barrel, adding rods to the drilling string, and manipulating the rig's controls. These physical activities are by far the easiest and least important aspects of the driller's job. Far more important are things that are not immediately apparent.

When coring a particular run, as a length of core barrel is called, the driller continuously adjusts three variables:

1. Down-hole pressure on the drilling bit.
2. Drilling fluid pressure at drill bit.
3. Rate of drill bit rotation.

It is important for the geoscientist and microbiologist working with cored sediments to be at least cursorily familiar with the drilling problems involved in core recovery. Many of the same factors that enter into successful core recovery also are involved with potential core contamination by drilling fluids. For example, many drillers tend to keep as much drilling fluid pressure at the drilling bit as possible. This is because if the fluid jets become clogged with sediments, the entire string of rods must be removed in order to clean them. For a drilling string 1000 feet long, this could easily become a one- or two-day job. However, high fluid pressures also increase the possibility that drilling fluids will penetrate into the core and contaminate them. When informed of the special requirements of microbiological sampling—that is, samples that are not contaminated with drilling fluids—most drillers will adjust their techniques to accommodate the more exacting standards. Communicating these special requirements, however, it is much more effective if the scientists understand the many technical problems facing the drillers.

Drilling fluid technology and coring

Proper application of drilling fluid technology is critical to successful coring operations. In simple borehole drilling, drilling fluids are mixed to maximise drilling rates and to minimise use of expensive drilling fluid additives. Thus, a borehole driller may choose to work with a less viscous drilling fluid and make up for cutting removal efficiency by increasing the rate of fluid circulation in the hole. As we have already seen, however, the coring driller often does not have this extra flexibility, as excessive fluid velocity inhibits core recovery.

DRILLING-FLUID CONTAMINATION OF CORED SEDIMENTS

The use of drilling fluids in coring operations introduces a significant source of possible chemical and biological contamination in recovered sediments. The possibility of such contamination must be carefully evaluated when using cored sediments for microbiological or pore-water chemistry investigations.

Down-hole Saturation Contamination

There are several ways in which drilling fluids can penetrate and contaminate cores during coring operations. First, if the pressure of the drilling fluid at the bit face is excessive relative to the permeability of the formation, the fluid will saturate sediments several inches below the bit. When this drilling-fluid-saturated sediment is subsequently recovered by the inner barrel, the entire sample is always contaminated. This type of contamination, which is referred to as down-hole saturation contamination, is very common when coring sandy or gravelly material.

There are several techniques that an experienced driller can use to minimise or eliminate down-hole saturation contamination. The most important technique, and one that a good coring driller practices as a matter of course, is to minimise the drilling fluid pressure at the bit. When coring silty or clayey sediments, this practice is often sufficient to prevent down-hole saturation.

Core Seepage Contamination

A second kind of contamination is called core seepage. When cored material is pushed into the inner barrel, some drilling fluid is present on the outside of the core and can seep into the centre of the core. This type of contamination can become particularly severe if the driller maintains an incorrect ratio between fluid pressure and drill stem pressure. When coring sandy material, excessive fluid pressure relative to drill stem pressure results in the core being slightly smaller than the diameter of the inner barrel. This, in turn, allows excessive quantities of drilling fluid in with the core and encourages drilling fluid seepage into the core.

Core-Fracture Contamination

Another common way for cored sediments to become contaminated by drilling fluids is by core fracturing. It is fairly common for cored sediments to come under considerable compressional stress as they are pushed into the inner barrel. If some portion of the sediment has less compressional strength than another portion, the sediment can fracture. Once the core is fractured, drilling fluid from the outside of the core can seep inward and contaminate the inner portion of the core. Core fracture is the most common way that clayey sediments can become contaminated.

Evaluating Drilling Fluid Contamination

Using cores recovered by mud-rotary techniques for biological or pore-water chemistry studies requires that the extent of drilling-fluid contamination be quantitatively evaluated. As all of the cores are contaminated to some extent, the problem is to determine the extent of contamination and then to judge whether the cores are usable for the purpose at hand.

There are at least four classes of tracers that have been used in subsurface microbiological investigations. These include:

1. The drilling fluid itself.
2. Chemical additives to drilling fluid.
3. Particulate tracers.
4. Biologic tracers.

SAMPLING GROUNDWATER FOR MICRO-ORGANISMS

Sampling groundwater for micro-organisms is fraught with numerous potential contamination problems. Drilling fluid used in constructing boreholes for wells contain substantial numbers of micro-organisms. The drilling process, therefore, effectively 'inoculates' any aquifer system being drilled. It may well be that the majority of bacteria introduced by drilling die off with time in their new environment; however, this has never been studied quantitatively. It must be assumed that some micro-organisms recovered from well water may represent introduced, rather than indigenous, species.

For some applications, however, it is just as important to consider introduced micro-organisms as well as indigenous ones. For example, micro-organisms introduced by drilling procedures into wells to be used for water supply are probably of more concern from a health standpoint than indigenous species. For some applications, therefore, sampling groundwater may be the procedure of choice.

The procedures used to sample groundwater for micro-organisms are generally fairly simple. Olson, in a study of methanogenic and sulphate-reducing bacteria in the Madison aquifer of Montana, allowed the completed well to purge for 8 hours prior to sampling. A sterile piece of rubber tubing was then inserted into the discharge pipe, and this tube used to fill the sterile sample bottles.

Hirsch and Rades-Rohkohl used a more sophisticated setup to sample groundwater from a shallow aquifer in Schleswig-Holstein, Germany. For sampling fully penetrating wells, these investigators sterilised a submersible pump and its attached tubing with 70 per cent ethanol alcohol and lowered it into the wells to be sampled. Three casing volumes were then pumped from the well and discarded in order to flush out traces of ethanol and to remove free standing water from the wellbore. Groundwater was then collected in a sterile 500 ml sampling flask equipped with a screw cap for subsequent plate count and isolation procedures. These procedures included spread-plating 0.1 ml of groundwater onto triplicate plates of solid media, followed by standard streak-plate isolation methods.

In addition to sampling fully-penetrating wells in the shallow aquifer, Hirsch and Rades-Rohkohl sampled a series of multilevel screened wells. As before, about three well-volumes of groundwater was removed from each well prior to sampling. Plate counts of this water that had stood in the wellbores showed higher numbers of bacteria than present in water pumped directly from the aquifer. Water was lifted from the wells by suction into a desiccator and a 500 ml sterile sampling flask was filled with water for microbiologic analysis.

Most strategies for sampling groundwater from wells for microbiologic analysis are similar to those employed by Olsen and Hirsch and Rades-Rohkohl. It is assumed that free-standing water in the borehole is not representative of aquifer water and so at least three well volumes, and often tens of well volumes, are evacuated prior to sampling. After well development, the groundwater samples are collected in appropriate pre-sterilised vessels. Similar techniques have been employed in more recent years by Stetzenbach in the Tucson Basin and by Pedersen and Ekendahl in a deep granitic aquifer in southeastern Sweden.

To sum up thus, subsurface sediments can be sampled for microbiological analysis using a number of methods. The unsaturated zone is most conveniently sampled using a hand auger to a depth of about 20 feet. For relatively thick unsaturated sediments, air drilling combines ease of sample recovery with relatively low risk of sample contamination. Local flow systems, which are usually relatively shallow, are most easily sampled using hollow-stem augering combined with split-spoon or push-tube methods. Hollow-stem augering is the method of choice for sample depths less than about 100 feet.

Intermediate and regional flow systems, being deeper, often cannot be sampled using auger rigs. In these cases, mud-rotary coring has been most often employed for sampling. Mud-rotary coring is technologically sophisticated and uses a number of highly specialised drilling tools and drilling fluid additives. While coring, the driller constantly adjusts drillstem pressure, fluid pressure, and drill bit rotation. The efficiency of core recovery depends in large measure on the skill of the drill and the degree to which these three variables are optimised. Coring efficiency also may be enhanced by the proper use of specialised drilling fluid additives.

The use of drilling fluid in mud-rotary coring often results in contamination of the core. There are three common ways for cored sediments to be contaminated by drilling fluids. One is down-hole saturation, where drilling fluid saturates the sediments ahead of the core barrel. Another is core seepage, where drilling fluid seeps into the core while it is in the inner barrel. Fracturing of the core once it has entered the inner barrel will also contaminate cored sediments, particularly relatively hard clays or indurated sediments.

Because of the possibility of core contamination by drilling fluids, it is necessary for each core used for microbiologic analysis to be screened for contamination. To this end, a number of different types of tracers can be used. Some components of drilling fluids, such as barium or sulphate from barite additives, are potential tracers. It is also possible to use micro-organisms that are present in the drilling fluid but not in subsurface sediments, such as coliform bacteria, as tracers. Often, however, the best indicators of contamination are chemical or particulate tracers added to drilling fluids. Rhodamine dye or bromide are examples of chemicals used as chemical tracers and florescent microspheres are examples of particulate tracers.

The microbiology of subsurface environments can also be sampled by producing groundwater from properly constructed wells. While it is fairly easy to preclude water contamination after it has left the well, it is difficult to rule out sample contamination by micro-organisms present in drilling fluids or micro-organisms growing gratuitously in the wellbore. Furthermore, the similarity between waterborne microbial communities and sediment-bound communities is problematic. Nevertheless, these techniques are suitable for some applications.

Chapter 26

Microbial Acclimation to Groundwater Contamination

INTRODUCTION

The presence of micro-organisms in what are called extreme environments plays an important role in microbial acclimation to groundwater contamination. A handy definition of an extreme environment is one that is either so cold or so hot or so saline or so oligotrophic or so chemically contaminated that it strains human credulity that life can exist at all. Micro-organisms have been found living in boiling water emanating from volcanic terrains, on the snowpack of Arctic and Antarctic wastelands, in brines three or four times more concentrated than seawater, in the bottom waters of deep ocean basins, and in sediments contaminated with the most noxious, poisonous, xenobiotic chemicals imaginable. As a general rule, most microbiologists have ceased expressing amazement at the ability of microbes to adapt themselves to difficult circumstances. Rather, understanding the biochemistry of microbial acclimation to stressed environments has become an active field of investigation.

Understanding the mechanisms of microbial acclimation to various stresses is interesting in its own right, but it also has very practical applications to pollution-control technology. Numerous studies of soils, sediments, and aquifer systems have shown that indigenous micro-organisms are capable of acclimating to a variety of chemical stresses imposed by human activity. In the process of acclimating themselves to these stresses, micro-organisms often accelerate the natural degradation of the chemicals involved. Thus, understanding the acclimation response is basic to the utilisation of micro-organisms in bioremediation technology.

In this chapter, aspects of microbial acclimation to chemical stresses in groundwater systems are considered. The first step in this discussion is to consider different manifestations of microbial acclimation to chemical stresses. The second step is to discuss the various mechanisms that lead to microbial acclimation. As we shall see, there is a wide variety of biochemical and genetic mechanisms that enable micro-organisms to acclimate themselves to various conditions. The final step is to give some examples of microbial acclimation in aquifer systems, and to consider how such acclimation may affect the efficiency of bioremediation efforts.

MICROBIAL RESPONSE TO ENVIRONMENTAL CHANGES

When micro-organisms collected from just about any natural environment are returned to the laboratory and placed in enrichment media, there is often a characteristic lag time before they begin to grow and reproduce actively. The generally accepted reason for this phenomenon is that the conditions encountered in the enrichment media are considerably different from those in the environment to which the micro-organisms are accustomed. Even if a microbiologist were careful to provide a carbon source, electron acceptor, trace nutrients, and a physical environment similar to that of the natural environment, there would inevitably be some differences. For example, the enrichment media may contain higher concentrations of carbon sources than are present in the natural environment. The temperature, ionic strength, pH, or one of a dozen other environmental factors may be different from what the micro-organisms are used to. Presented with these differences, micro-organisms typically require an adjustment period to acclimate themselves to the new conditions. This period of acclimation is the reason for the observed lag time before active growth commences.

The extent of the lag time is dependent upon a number of factors. If the conditions offered are reasonably similar to those in the environment, the lag time generally is fairly short. On the other hand, if there are significant differences between the media and the environment, the lag time generally is more extended. If the differences are too extreme, the micro-organisms will be unable to acclimate themselves and will become inactive or eventually die.

Once acclimation to laboratory conditions has been achieved, however, microbial growth becomes exponential and continues until nutrients are exhausted or a build-up of toxic by-products limits growth. When acclimated micro-organisms are reinoculated onto fresh media, growth continues at an exponential rate with no lag time. Thus, one indication that micro-organisms have acclimated themselves to particular conditions is the absence of a lag time.

In many ways, groundwater contamination is analogous to the case of micro-organisms acclimating themselves to a particular growth media under laboratory conditions. In both cases, conditions for growth are substantially changed and micro-organisms must adapt themselves to the new conditions. Also, many kinds of contamination offer potential substrates for micro-organisms to grow on, although these substrates are often much different from those encountered in pristine aquifers. If this analogy held, one would predict that micro-organisms subjected to contamination in subsurface environments would behave in at least two ways. First, when the contaminant was introduced, there should be a lag time as microbial populations adjust themselves to the new conditions. Second, one would expect that micro-organisms that were acclimated to the contamination would utilise the contaminants more efficiently than micro-organisms that are were not acclimated to the contamination.

This kind of behaviour is widely observed in studies of aquatic sediments and contaminated groundwater systems. An example of this observed behaviour was given by Madsen in a study of a shallow water table aquifer contaminated by coal tar derivatives. In sediments cored at the water table (WT) and in the shallow saturated (SS) zones of the pristine part of the aquifer, there was a pronounced lag of ten or twelve days before mineralisation of radiolabelled p-hydroxybenzoate was apparent (Fig. 26.1). In sediments cored from within the contaminated zone, on the other hand, mineralisation of p-hydroxy-benzoate occurred with no lag at all (Fig. 26.1).

This behaviour, which has been observed in a number of studies, has important implications for bioremediation of contamination in groundwater systems. First of all, it is evident that rates of biodegradation will be a function of the acclimation of indigenous micro-organisms to the contaminated conditions. In turn, this implies that biodegradation rates will change over time as acclimation proceeds. Finally, it is evident that acclimation has the potential for substantially increasing biodegradation rates. Thus, technologies for improving the acclimation of indigenous micro-organisms to specific compounds are potentially powerful tools for remediating contaminated environments. For these reasons, understanding the metabolic, physiologic, and genetic mechanisms involved in microbial acclimation is especially important.

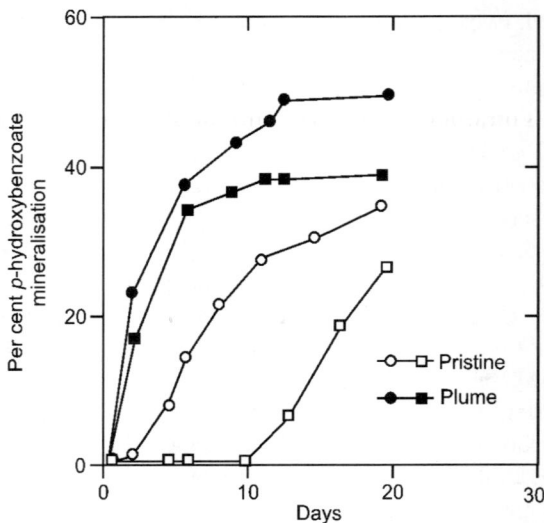

Fig. 26.1. Evidence for a lag time in the mineralisation PHB by subsurface bacteria in pristine sediments and the lack of a lag time in PHB-contaminated sediments.

MECHANISMS OF ACCLIMATION

Microbial cells are capable of adapting themselves to a wide variety of environmental conditions. For example, bacteria regulate cellular water content by adjusting the properties of their cell walls. Within limits, bacteria can reduce the osmotic flux of water into or out of the cell by adjusting the membrane properties of the cell wall. In similar fashion, bacterial cells can adjust to different pH and temperature conditions. In addition to adapting to various environmental conditions, micro-organisms are capable of adjusting their metabolism to new substrates. From the viewpoint of micro-organisms present in groundwater systems, contamination events often introduce previously unavailable substrates. It is generally recognised that microbial populations adapt to new substrates by means of three distinct mechanisms: (i) induction of specific enzymes not present (or present at low levels) before exposure; (ii) selection of new metabolic capabilities produced by genetic changes; and (iii) an increase in the number of organisms able to metabolise newly available substrates.

Induction

The process of induction has been discussed previously in the context of bacterial metabolism. Induction refers to the series of processes by which microbial cells produce enzymes specific to a particular substrate only when that substrate is actually present. One of the best-studied examples of induction in bacteria is that of the lactose operon in *E. coli*. Metabolism of lactose requires the presence of three enzymes: β-galactosidase, galactoside permease, and thiogalactoside transacetylase. β-galactosidase cleaves the disaccharide lactose into the monosaccharides galactose and glucose, and galactoside permease is required to transport lactose across the cell membrane. The role of thiogalactoside transacetylase is still unclear.

The induction of the enzymes needed to metabolise lactose involves the activation of specific regions of the bacterial genome (Fig. 26.2). The lactose operon consists of a promoter region (P), a regulatory gene (R) that codes for a repressor protein, and an operator region (O) that occurs between the promoter and the genes coding for the lactose enzymes. When lactose is not present in the cell [Fig. 26.2(a)], the regulatory gene is expressed and a repressor protein binds to the promoter region of the lactose operon. This effectively blocks transcription of the Lac genes that code for lactose-metabolising enzymes. However, when lactose is present [Fig. 26.2(b)], a derivative of lactose, allolactose, acts as an inducer by binding to the repressor protein. This inactivates the repressor and allows the synthesis of the three enzymes needed to metabolise lactose.

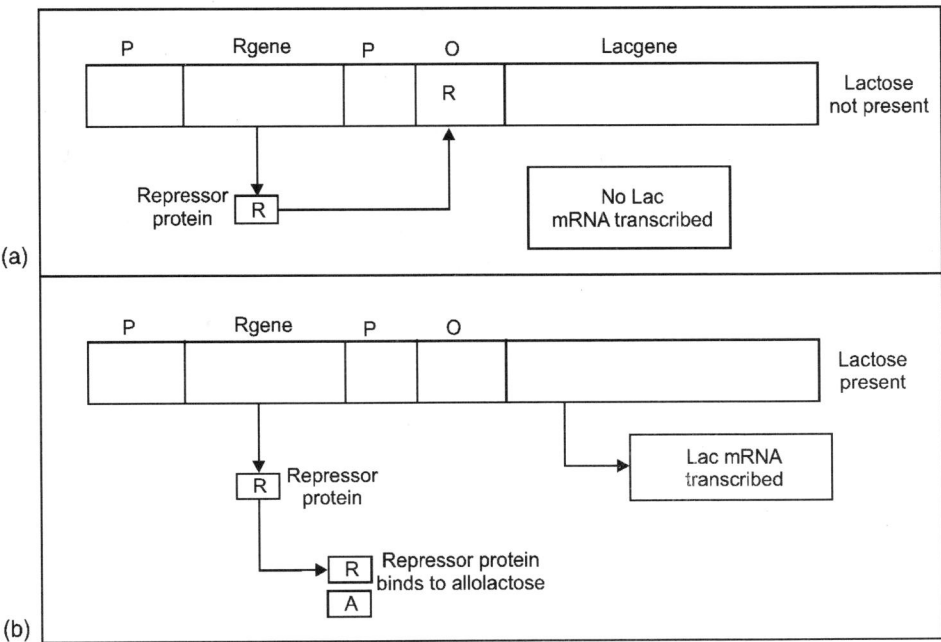

Fig. 26.2. (a) Repression of LAC genes in the absence of lactose, and (b) expression of LAC genes in the presence of lactose.

Hydrocarbon-degrading enzymes

The lactose operon is a particularly clear example of how induction can lead to the acclimation of micro-organisms to specific substrates. These concepts are important in the subsurface microbiology of contaminated sites, since many contaminant-degrading enzyme systems are inducible. There are numerous examples of this in the literature, but one particularly good example is the induction of aliphatic hydrocarbon-degrading enzyme systems.

The lactose operon is present on the chromosomal DNA of *E. coli*. However, it is commonly observed that enzyme systems for degrading xenobiotic compounds are found in extrachromosomal DNA (plasmids).

Well-known examples of this are the OCT and TOL plasmids of *Pseudomonas*, which code for the degradation of *n*-octane and toluene, respectively.

The OCT plasmid of *Pseudomonas putida* has been shown to be inducible by a wide range of straight-chained compounds. Induction promotes the synthesis of two enzymes, alkane hydroxylase and alcohol dehydrogenase, which are involved in alkane degradation. In the case of the lactose operon, only one compound (allolactose) served as the inducer. In contrast, alkane hydroxylase and alcohol dehydrogenase are induced by a wide variety of straight-chained compounds. Any straight-chained compound in the C_6-CI_O range will induce the synthesis of these enzymes. Synthesis occurs even if carbonyl, hydroxyl, or methyl groups are present on the terminal carbon of one side of the chain. Interestingly, these enzymes may be induced by compounds against which the enzymes have no activity. For example, dicyclopropyl ketone and dicyclopropyl methanol will induce alkane hydroxylase activity, even though they are not degraded by the enzyme. These compounds are referred to as gratuitous inducers. Conversely, some compounds, such as undecane, will not induce enzyme production but are degraded if the enzymes are produced gratuitously.

Catabolite Repression

Induction allows micro-organisms to synthesise degradative enzymes if a particular substrate becomes available. However, the mere availability of a substrate does not necessarily mean that its utilisation is in the micro-organism's best interest. For example, if two usable substrates, A and B, are present but more energy is available from substrate B, clearly the micro-organism is better off suppressing A-degrading enzymes and producing B-degrading enzymes. In fact, micro-organisms have evolved mechanisms for distinguishing between available substrates. One of these mechanisms is termed catabolite regression.

Jacques Monod discovered the phenomenon called diauxie, or the ability of certain micro-organisms to utilise one substrate present in growth media preferentially to another substrate. The classic example of this is the sequential utilisation of glucose and lactose. When a glucose-lactose growth medium is inoculated with *E. coli*, there is a brief lag time as the cells acclimate themselves to the new conditions. After growth commences, growth is supported only by glucose oxidation with lactose not being used at all. After the glucose is exhausted, there is another brief tag as the cells acclimate themselves to lactose utilisation, followed by growth at the expense of lactose. The phenomenon of diauxie typically produces a biphasic growth curve.

Catabolite repression of organic compound oxidation

There is some evidence to suggest that catabolite repression of organic contaminants may occur under some conditions. As is the case with glucose-lactose, micro-organisms can obtain more energy from glucose oxidation than from oxidation of petroleum hydrocarbons. It is, therefore, in their interest to oxidise hydrocarbons only in the absence of glucose as a substrate.

Van Eyk and Bartels showed that *n*-alkane oxidation in *Pseudomonas aeruginosa*, presumably conferred by the OCT plasmid, could be suppressed by addition of glucose or malate to the growth media. There is also some evidence that the mechanism of the catabolite repression is similar to that of glucose repression of lactose oxidation. Dalhoff and Rehm showed that the repression could be reversed by adding cyclic AMP plus an inducer substrate to the growth media.

Catabolite repression probably explains the observation that certain petroleum hydrocarbons, such as toluene, are observed to be preferentially utilised relative to compounds like benzene or MTBE in contaminated groundwater systems. Thus, this phenomenon is important to consider when considering bioremediation strategies.

Although it is important to stimulate growth of natural populations in order to facilitate degradation, it is equally important not to add nutrients or carbon sources that might repress degradation of organic contaminants.

Genetic Mutations

The dual mechanisms of induction and catabolic repression provide micro-organisms with powerful tools for acclimating their metabolism to different conditions. Clearly, however, micro-organisms have evolved these pathways because they were advantageous in helping them deal with the environmental conditions they encountered. Since enzymatic mechanisms for degrading particular substrates have developed in the past, it is reasonable to expect that micro-organisms might evolve new mechanisms for degrading xenobiotic compounds.

For example, Betz and Clarke showed that point mutations produced changes in substrate specificity of *Pseudomonas aeruginosa* amidase. In this case, these changes enabled the mutant strains to grow on amides that the wild-type strain was unable to use. Furthermore, sequential mutations were shown to convert what had been solely an acetamidase to a phenylacetamidase. These and other experiments have shown that the specificity of an enzyme can be changed by point mutation of the genome, and that these point mutations can sometimes improve the ability of an enzyme to degrade particular substances. In addition to point mutations, the metabolic capabilities of bacteria can be significantly altered by deletions and additions of DNA via transposable genetic sequences (transposons).

These are just a couple of examples of the tremendous capacity for change inherent in the bacterial genome. It is reasonable to assume that point mutations, transposon deletions, and transposon insertions occur as a matter of course in micro-organisms. Most of the time, such genetic changes result in a decrease in metabolic efficiency. However, if a micro-organism is subjected to novel substrates, there is a finite chance that a fortuitous genetic change will increase its efficiency. Thus, it is entirely possible for micro-organisms to develop the ability to degrade even the most xenobiotic compounds.

Acclimation to Available Electron Acceptors

Pristine water table aquifers are often aerobic systems. This is because, under pristine conditions, the input of dissolved oxygen from percolating recharge typically exceeds the input of organic carbon substrates. Thus, micro-organisms are more carbon-limited than oxygen-limited. With the advent of organic chemical contamination, however, the situation often rapidly changes. With the sudden influx of potential carbon substrates, carbon limitations disappear and oxygen is rapidly consumed. Where aerobic microbial metabolism might have dominated before contamination, anaerobic metabolism often dominates after contamination. Thus, in addition to acclimating to new carbon substrates, indigenous micro-organisms must acclimate to electron acceptors other than oxygen.

Production of Proteins in Response to Chemical Stresses

We have touched briefly on the genetic basis of microbial acclimation to changing environmental conditions such as available electron donors and electron acceptors. It is important to keep in mind, however, that the expression of genetic capabilities results ultimately in the production of proteins that enable micro-organisms to carry out useful functions. If the genome of a micro-organism is the sum of all of its genetic capabilities, the proteome can be defined as the sum of those proteins that have been actually produced to allow a micro-organism to adjust to certain environmental conditions. The study of proteins expressed by micro-organisms subjected to particular chemical and environmental stresses is one way to evaluate the acclimation response of microbial populations to particular stresses.

The technical problem associated with examining proteins present in micro-organisms is, of course, that there are so many of them. Even the simplest bacterial cell produces thousands and thousands of proteins, and so evaluating what proteins are present under certain conditions is not a trivial undertaking.

It is possible to separate proteins extracted from cells using gel electrophoresis. However, because of the large numbers of proteins present, and because many proteins have similar isoelectrical properties, this does not provide sufficient separation of the proteins. The breakthrough in proteomics came in 1975 when several researchers developed 2-D gels. The basis of this technique was to first separate proteins on electrophoresis gels on the basis of isoelectric point, and then to turn the gel 90° and further separate the protein on the basis of molecular weight. With this 2-D gel technology, it was possible to resolve thousands of individual proteins and it became feasible to identify proteins produced by particular physical and chemical stresses.

FACTORS AFFECTING MICROBIAL ACCLIMATION

The ability of micro-organisms to adapt to new conditions imposed by chemical contamination is influenced by a number of factors. Because induction and catabolite repression are important mechanisms in the acclimation response, it can be anticipated that the kinds of organic compounds available and their relative concentrations will be important factors. In addition, because the characteristic lag time associated with microbial acclimation varies widely with the kinds of organic compounds available, the time of exposure to contamination will likely influence the degree of adaptation attained. Finally, the similarity of introduced contaminants to substrates normally utilised by micro-organisms will influence microbial acclimation. If they are very similar, then acclimation may entail only the induction of appropriate degradative enzymes. On the other hand, if there is little similarity, than acclimation may involve serendipitous genetic mutations. This, in turn, probably would vastly increase the length of time involved in microbial acclimation. Evaluating the factors involved in microbial acclimation to chemical stress is an important component of designing bioremediation strategies.

Rates of Acclimation

Rates of microbial acclimation to xenobiotic compounds typically are extremely variable. In many cases, it is difficult to determine the precise reasons for the rate at which a particular microbial community adapts to a xenobiotic compound. Nevertheless, it is possible to make some generalisations.

The general pattern of acclimation response shown by Spain and van Veld has been observed by a number of investigators. This pattern includes a wide range of acclimation times to a particular xenobiotic that cannot be predicted *a priori*. Sediment samples from the different locations on the same river exhibited acclimation times that varied as much as fourfold. This variability is even more pronounced in sediments from groundwater systems.

Concentration Effects

Concentration of xenobiotic compounds has been shown to have important effects on achievement of acclimation and on the length of the lag time.

The mechanism of this minimum threshold effect was not systematically investigated by Linkfield and others. However, they suggest that a concentration-dependent receptor mechanism may explain the observed effect. According to this hypothesis, a xenobiotic compound must be present in sufficient concentration before binding to inducer proteins regulating gene expression. Alternatively, this effect may simply reflect competitive utilisation (i.e. diauxie) of alternative substrates in the sediment. If the xenobiotic compound is present below a particular threshold, then alternative carbon sources may be more attractive to indigenous micro-organisms and preclude induction of degradative enzyme systems.

The minimum threshold effect of acclimation has been observed in sediments cored from groundwater systems. In a study of acclimation of pristine aquifer sediments to xenobiotic compounds, Aelion showed that relatively low concentrations (14 ng/g sediment) of *p*-nitrophenol induced little or no acclimation and mineralisation. In contrast, sediments exposed to higher concentrations (529 ng/g sediment) exhibited rapid mineralisation of *p*-nitrophenol following a 40-day acclimation period.

Cross-Acclimation of Xenobiotic Compounds

If induction of enzymatic systems is a principal mechanism in the acclimation of microbial populations to xenobiotic compounds, it might be expected that acclimation to one particular xenobiotic compound would confer acclimation to related compounds. This would be the case if several xenobiotics were degraded via the same enzymatic pathway. This behaviour has, in fact, been reported in the case of reductive dehalogenation of halobenzoates. In a series of experiments, these investigators showed that acclimation to reductive dehalogenation of one halobenzoates reduced acclimation times to other halobenzoates. For example, the acclimation time of lake sediments to 3-iodobenzoate was on the order of 2–3 weeks. Once acclimation was achieved, however, these sediments were able to degrade other halobenzoates completely in less than one week. This effect is important to bioremediation technology since it implies that acclimation to one class of xenobiotic may confer enhanced degradation rates to a number of related compounds.

Chemical Structure of Xenobiotics

The chemical properties of organic compounds depend largely on their chemical structures. Because of this, it is reasonable to expect that the acclimation of microbial communities to xenobiotic compounds will be affected by chemical structure.

There is evidence that this is the case for microbial communities in aquatic sediments adapting to reductive dehalogenation of halobenzoates.

In this study, it was shown that acclimation periods depended largely on the kinds of halogens substituting on benzoate and on their position on the molecule.

ACCLIMATION TO XENOBIOTICS IN GROUNDWATER SYSTEMS

Much of the research concerning microbial acclimation to xenobiotic chemicals has been focused on aquatic surface sediments or soils. However, there is a small but significant literature concerning microbial acclimation in groundwater systems. Two general approaches have been used in these studies. One approach has been to study sites that exhibit contamination by particular classes of compounds. Differences of xenobiotic utilisation between contaminated and pristine zones of the same aquifer are then used to document the presence or absence of acclimation effects. Another approach, which is similar to studies of aquatic sediments, is to obtain sediments from pristine subsurface environments, amend them with xenobiotics, and observe the lag period before mineralisation commences.

Acclimation Response in a Contaminated Aquifer

In order to test this hypothesis, sediments were cored from within the contamination plume and in nearby pristine sediments. Two sites within the contaminated area and a site outside the contaminated area were examined for differences in mineralising various polyaromatic hydrocarbons (PAH) compounds.

The results showed that sediments from the contaminated site exhibited significantly more mineralisation activity than sediments from the pristine site. Even though sediments from one contaminated site contained no measurable PAH initially, amending these samples with PAH and oxygen showed rapid consumption relative to autoclaved controls. In contrast, sediments from the pristine site showed little or no consumption of PAH. These results are important because they document the importance of the acclimation process to degradation rates of pollutants in aquifer systems.

While these experiments provided documentation of acclimation in the contaminated sediments, it is apparent that, if *in situ* degradation rates were as high as those measured in the laboratory, the contamination would have been entirely degraded since operations at the creosote plant closed (approximately ten years). Since contamination was still evident, it was clear that the degradation experiments overestimated actual *in situ* rates.

Acclimation Response in Pristine Aquifer Sediments

The study of Wilson showed that acclimation occurred in response to chemical contamination of a shallow aquifer system. The next logical question to be addressed was at what rate does such acclimation occur in pristine sediments exposed to various kinds of contamination.

In real instances of chemical contamination, it is rare that one compound or even one class of compounds is involved; rather, a complex mixture of xenobiotics, as well as nonxenobiotic chemical compounds, typically are delivered simultaneously. The nature of the mixture would be expected to affect acclimation of indigenous micro-organisms. In addition, if it were possible to increase rates of the acclimation response by means of chemical addition, this would have important implications for bioremediation technology.

Acclimation of Eukaryotic Micro-organisms

Eukaryotic micro-organisms such as fungi and protozoa are present in low numbers in many groundwater systems. Fungi, which are usually aerobic heterotrophs, compete directly with bacteria in many environments. Thus, oligotrophic conditions that limit the population density of bacteria in groundwater systems also limit populations of fungi. The low population density of protozoa in pristine aquifer systems is also related to low populations of bacteria. Because protozoa prey upon bacteria, substrate limitations for bacterial populations translate to substrate limitations for protozoa.

It has been observed that one manifestation of microbial acclimation to polluted groundwater is an increase in the number of protozoa. This implies that microbial acclimation to xenobiotics is not limited to the bacteria but includes acclimation of higher eucaryotic micro-organisms as well.

Fungi, which may be limited by oxygen availability, are not evidently able to increase their population. Protozoa, on the other hand, which often feed upon bacterial cells, show a noticeable population increase. This, in turn, suggests that these micro-organisms are capable of acclimating to the chemical stress, as are the bacteria. The grazing effects of the protozoa are apparently the principal reason that total numbers of bacteria do not increase.

Acclimation in Bioremediation Technology

Available evidence clearly shows that microbial populations exposed to organic chemicals exhibit an acclimation response that, in general, results in increased rates of biodegradation. This is shown by field studies in which micro-organisms in sediments exposed to contamination show greater biodegradation than unexposed sediments and by laboratory studies showing that degradation rates increase over time.

The extent to which the acclimation response can be utilised in bioremediation technology is much less clear. Numerous practical considerations intervene. Available evidence shows that significant microbial acclimation occurs in time periods of less than a year. Considering that many instances of groundwater contamination are discovered tens of years after the fact, implementing measures to increase acclimation rates does not seem particularly advantageous. Furthermore, other factors, such as electron-acceptor availability, often are much more limiting to degradation rates than the extent of microbial acclimation. Thus, while the acclimation response certainly affects degradation of organic compounds in groundwater environments, its application to bioremediation technology is often impractical.

ACCLIMATION TO METAL TOXICITY

Metals play important roles in the activity of many enzymatic systems in prokaryotic and eukaryotic cells. Obvious examples include the utilisation of iron in electron transport mechanisms, the presence of molybdenum in nitrate reductase, or the role of zinc in carbonic anhydrase. However, when one particular metal interferes with enzyme function by competing with or replacing another particular metal, toxicity results. Metal toxicity in this context simply refers to the slowing or complete cessation of cellular growth as a direct result of the presence or absence of a particular metal in the cell's immediate environment.

It is reasonably common for micro-organisms, particularly those inhabiting soils or aquatic sediments, to encounter potentially toxic metals. For example, micro-organisms inhabiting soils developed on metal ore bodies would be exposed to fairly high metal concentrations. Micro-organisms inhabiting aquatic sediments or aquifer systems developed from erosion of ore bodies would suffer similar metal exposure with potentially toxic effects. As a result of such exposure, some micro-organisms have developed elaborate biochemical mechanisms for dealing with metal toxicity. Because many of these mechanisms involve induction of enzymatic systems, microbial communities having such biochemical capabilities often exhibit an acclimation response to metal exposure.

The acclimation of micro-organisms to metal toxicity and mechanisms of metal resistance have been widely studied. Much of this work has been motivated by the observation that microbial transformations of metals greatly affect metal mobility in the environment. For this reason, microbial acclimation to metal toxicity is relevant to the microbiology of groundwater systems as well.

Metal Detoxification Mechanisms

Micro-organisms employ several different biochemical mechanisms in detoxifying metals in their immediate surroundings. These include: (i) binding metals to the cell surface in order to prevent transport into the cell, (ii) biotransformation of the metal to a less toxic form, and (iii) depositing the metal in an insoluble form. Some of these mechanisms, such as metal deposition as sulphides, are gratuitous in that they result from other metabolic functions, such as sulphide production. Other mechanisms, such as biotransformations, are highly specific and are determined at the genetic level. The ability to transform metals to less toxic or more volatile forms often are coded on plasmids or transposons.

Metal binding

The outer layers of bacterial cells are characterised by numerous anionic sites that are available to bind positively charged metal ions of metabolic importance, such as Mg^{2+}, Fe^{3+}, Cu^{2+}, Na^+, and K^+. This binding, which often is the first step in active transport into the cell, results largely from interactions with anionic sites on the cell wall such as phosphodiesters, carboxyl groups of peptidoglycan, and hydroxyl groups of carbohydrates. Peptidoglycan is especially efficient at metal binding. Such heavy metals as Hg^{2+} and Pb^{2+} are also gratuitously bound by anionic sites on a cell wall. However, the cell wall typically shows less affinity for the heavy metals than for lighter metals such as Mg^{2+}. Once bound to the cell's surface, a potentially toxic metal is immobilised and is effectively prevented from entering the cell, where it might interfere with metabolic functions. Thus, metal binding confers some level of metal resistance to all micro-organisms.

Biotransformations

An important class of metal resistance mechanisms available to micro-organisms is that of biotransformation. These biotransformations include oxidation-reduction reactions and organic-inorganic conversions. In both cases, the mobility or volatility of the metal, or both, may be greatly affected. For example, reduction of Hg^{2+} to $HgO°$ greatly increases the volatility of mercury and may contribute to its transport away from the micro-organism's immediate environment. Similarly, the attachment of alkyl groups to metals may greatly change its solubility and volatility. For example, some methylated metals are highly volatile and virtually insoluble in aqueous solution.

Thus, they may more easily escape from an aquatic environment. There is some evidence that alkylation of metals serves as a primary metal-resistance mechanism. Conversely, alkyl substitutions greatly increase the affinity of the metal complex for lipids. This increased lipophilicity is an important factor in metal toxicity in micro-organisms and in higher organisms. It has been shown that some micro-organisms have the inducible ability to dealkylate metals. In the case of methyl mercury compounds, the first step in this process is demethylation by means of organomercurial lyases followed by the reduction of Hg^{2+} to $HgO°$. The genetic basis of mercuric reduction is one of the best-documented of all metal-resistance mechanisms and is discussed in detail in next section.

Metal deposition

The ability to precipitate metals in an insoluble form is an attractive mechanism for metal resistance. Most metal sulphide compounds are highly insoluble, and deposition of a metal in this form effectively prevents its transport into the cell. In some cases, as in *Desulfovibrio*, sulphide production and metal immobilisation is gratuitous and is simply a by-product of the micro-organism's normal metabolism. In other cases, as in the case of some *Clostridia*, sulphide production is inducible and is used as a metal resistance mechanism. The ability of some micro-organisms to precipitate metals has applications in biotechnology. Metal-precipitating micro-organisms have been used to remove metals from industrial waste solutions. Conversely, metal-solubilising microbial mechanisms have been used to leach and recover metals from ores.

Plasmid-Encoded Metal Resistance Mechanisms

Metal resistance mechanisms present in prokaryotic micro-organisms are often plasmid-encoded. This is to say that the genetic material coding for the relevant enzyme systems are present on extrachromosomal DNA. However, as many of these genetic sequences are transposable, they may also be found in chromosomal DNA.

Plasmid-encoded resistance mechanisms have been documented for a wide range of metals, including cobalt and zinc, cadmium, silver, and mercury. Of these, the mechanism for mercury resistance in *Staphylococcus aureus*, an opportunistic human pathogen, has been particularly well documented. As such, it serves as a convenient model for illustrating the genetic basis for metal resistance in other micro-organisms as well. A number of different plasmids and transposons, each exhibiting slightly different details, have been shown to confer mercury resistance in *Staphylococcus aureus*. These functions are located on plasmid DNA and are referred to collectively as the 'mer operon'.

The basic components of the mer operon and their functions in conferring mercury resistance are shown in Fig. 26.3. The regulatory gene mer R is inducible and codes for a protein that, when bound to the promoter-operator (OP) region, initiates synthesis of the mercury resistance enzyme system. This enzyme system is composed of several different functions.

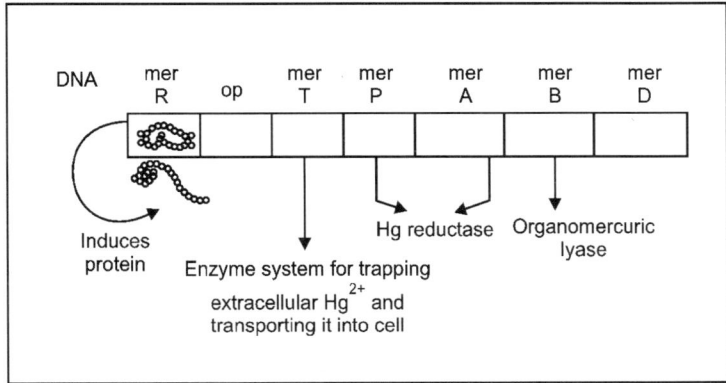

Fig. 26.3. Basic components of the mer operon.

The mer T and mer P regions code for a system that traps extracellular Hg^{2+} and transports it through the cell wall and cell membrane to the cytoplasm. The mer A region codes for mercuric reductase, which is present in the cytoplasm. Hg^{2+} delivered to the cytoplasm by the transport mechanism is then reduced to the volatile and less toxic $Hg°$ form. The mer B region codes for an organomercurial lyase system that serves to split Hg^{2+} from organic complexes so that it can be subsequently reduced by mercuric reductase.

Depending on the components of the mer operon actually present in plasmid DNA, a wide variety of responses to mercury are possible. For example, if the mer A region, which codes for mercuric reductase, is removed from the operon, the resulting cells have the ability to trap Hg^{2+} but not to detoxify it. Such cells are actually hypersensitive to mercury stress. Cells that have broad spectrum resistance to mercury contain both the mer A and mer B genes. Alternatively, cells containing mer A but not mer B exhibit resistance to Hg^{2+}, but less resistance to organomercuric compounds such as methylmercury.

While the mer operon is certainly the most studied of the mercury-resistance mechanisms, others have been documented. For example, *Clostridium cochlearium* has an entirely different plasmid-encoded mercury resistance mechanism. This mechanism does not include mercuric reduction but, rather, the precipitation of mercuric sulphide. Organomercuric compounds are split, presumably by means of a lyase, and sulphide is generated. Both of these functions appear to be plasmid mediated in the case of *C. cochlearium*. This serves to emphasise the wide range of resistance mechanisms present in micro-organisms and illustrates their flexibility in dealing with metal stress.

Acclimation to Mercury Toxicity

The fact that the mer operon, and other plasmid-mediated metal resistance mechanisms, are inducible enzymatic systems suggests that an acclimation period would be expected in response to mercuric stress. Furthermore, this suggests that microbial populations acclimated Hg^{2+} should be much more efficient at Hg^{2+} reduction than non-acclimated populations. It has been widely observed that aquatic sediments exposed to mercuric stress exhibit a higher incidence of mercury-resistant bacteria than nonexposed sediments. In one such study, Barkay and Olson showed that lake sediments exposed to high concentrations of mercury, up to 40 µg/g, showed higher counts of mercury resistant bacteria and a higher incidence of the mer operon than did sediments that had not been exposed. Subsequent research has demonstrated the inducible nature of mercury resistance in aquatic-sediment microbial communities. For example, Barkay showed that loss of Hg^{2+} from water samples of estuarine showed a distinct lag time. However, samples pre-exposed to Hg^{2+} showed no such lag time. Furthermore, the half-life of Hg^{2+} in pre-exposed samples was between two and fourfold shorter than in unexposed sample.

To sum up this chapter has discussed that micro-organisms employ a wide variety of mechanisms for acclimating themselves to changing conditions. In many cases, acclimation consists of inducing appropriate enzyme systems or, alternatively, repressing systems that are inappropriate. These changes are effected at the genetic level but do not involve fundamental changes of the micro-organism's genome. It is also possible that genetic changes, initiated by point mutations or wholesale deletions of genetic material, may result in advantageous metabolic capabilities. However, these changes are unpredictable and have an uncertain impact on xenobiotic degradation in natural systems.

The hallmark of the acclimation response is the presence of a characteristic lag time in the microbial growth curve that follows exposure to new substrates. These acclimation periods have been observed in microbial communities from numerous environments responding to a broad spectrum of contaminants, including xenobiotic organic compounds and metals. Microbial communities in shallow groundwater systems appear to acclimate to a variety of chemical stresses in time periods as short as several months. The acclimation time for particular compounds can be reduced somewhat by judicious addition of trace elements or vitamins. However, it is not always practical to apply this to bioremediation strategies.

Micro-organisms possess a number of different mechanism for acclimating to the toxic affects of metals. These include metal binding to the cell wall, biotransformations of metal species, and metal precipitation processes. Metal-resistance mechanisms are often plasmid encoded and are induced by metal exposure. Metals resistance has been most extensively studied in aquatic sediments. But studies indicate that groundwater systems contain similar metals-resistance mechanisms as well.

SECTION VII

Aeromicrobiology

27. **Aeromicrobiology: A Review**   437

28. **Bioaerosol Control and Biosafety in Laboratory**   456

29. **Beneficial and Pathogenic Microbes in Agriculture**   465

30. **Airborne *Mycobacterium* Spp.**   475

Chapter 27

Aeromicrobiology: A Review

INTRODUCTION

Aerobiology is the study of the aerosolisation, aerial transmission, and deposition of biological materials. It can also be defined as more specifically the study of diseases that may be transmitted via the respiratory route. Despite the variations in definition, this relatively new science is becoming increasingly important in many aspects of such diverse scientific fields as public health, environmental science, industrial engineering, agricultural engineering, biological warfare, and space exploration.

The first part of this chapter will introduce the basics of aerobiology, including the nature of aerosols, the fundamentals of the aeromicrobiological (AMB) pathway, common bioaerosol sampling methods, and aerobiological transport modelling. The remainder of the chapter will then focus on a subset of the science that we shall term aeromicrobiology. Aeromicrobiology, as defined for the purpose of this text, involves various aspects of intramural (indoor) and extramural (outdoor) aerobiology as they relate to the airborne transmission of environmentally relevant micro-organisms, including viruses, bacteria, fungi, yeasts, and protozoans.

IMPORTANT AIRBORNE PATHOGENS

Generally, as indicated by the previous definitions of aerobiology, one usually associates airborne micro-organisms with disease occurrence in humans, animals, or plants. Numerous plant pathogens are spread by the aeromicrobiological pathway (Table 27.1). Up to 70 per cent of all plant diseases are caused by fungi such as wheat rusts that can be spread by airborne transmission.

Aerial transmission is capable of transporting these phytopathogens many thousands of kilometres. The impact of airborne plant pathogens, especially fungi, on the economy of the agricultural industry is in the billions of dollars each year.

There are also numerous airborne pathogenic micro-organisms that infect animals (Table 27.2). Infection of pets and livestock by airborne micro-organisms also costs the public and livestock owners billions of dollars each year.

For example, foot-and-mouth disease virus is known to be transmitted by the aeromicrobiological pathway. An outbreak of foot-and-mouth disease in England that lasted only four months affected over 2300 farms and resulted in the loss of almost 4,50,000 animals. Finally, the airborne transmission of many pathogens (Table 27.3) such as *Legionella pneumophila*, *Mycobacterium tuberculosis*, and newly recognised pathogens such as the Sin Nombre virus (hantavirus) is associated with human infection and disease.

Table 27.1. Important airborne plant pathogens.

Plant diseases	Pathogens
Fungal diseases	
Dutch Elm disease	*Ceratocystis ulmi*
Apple rust	*Gymnosporangium* spp.
Potato late blight	*Phytophthora infestans*
Banana leaf spot	*Mycosphaerella musicola*
Blossom infection	*Sclerotinia laxa*
Cedar rust	*Gymnosporangium* spp.
Leaf rust	*Puccinia recondita*
Crown rust of oats	*Puccinia coronata*
Fusiform rust of southern pines	*Cronartium fusiforme*
Loose smut of wheat	*Ustilago tritici*
Beef downy mildew	*Perospora* spp.
Downy mildew	*Pseudoperonospora humuli*
Maize rust	*Puccinia sorghi*
Annosus root rot	*Fomes annosus*
Powdery mildew of barley	*Erysiphe graminis*
Southern corn leaf blight	*Helminthosporium maydis*
Stem rust of wheat and rye	*Puccinia graminis*
Tonbacco blue mould	*Peronospora tabacina*
Sigatoka disease of bananas	*Mycosphaerella musicola*
White pine blister rust	*Cronartium ribicola*

Table 27.2. Important airborne animal pathogens.

Diseases of animals	Pathogens
Bacterial diseases	
Tuberculosis	*Mycobacterium bovis*
Glanders	*Actinobacillus mallei*
Brucellosis	*Brucella* spp.
Salmonellosis	*Salmonella* spp.
Fungal diseases	
Aspergillosis	*Aspergillus* spp.
Cryptococcosis	*Cryptococcus* spp.
Coccidioidomycosis	*Coccidioides immitis*
Viral diseases	
Canine herpes	Herpesviridae
Eastern equine	Alphavirus
Encephalomyelitis	

(Contd ...)

Diseases of animals	Pathogens
Hog cholera	Pestivirus
Influenza	Influenza virus
Feline distemper	Morbillivirus
Fowl plaque	
Rabies	Rhabdoviridae
Canine distemper	Morbillivirus
Newcastle disease	
Infectious bronchitis	Influenza, others
Foot and mouth disease	Aphthovirus
Rhinderpeste	Morbillivirus
Ephemeral fever	
Infectious laryngotracheitis	

Table 27.3. Important airborne human pathogens.

Human diseases	Pathogens
Bacterial diseases	
Brucellosis	*Brucella melitensis*
Pulmonary tuberculosis	*Mycobacterium tuberculosis*
Glanders	*Actinobacillus mallei*
Pneumonia	*Clamydia psittaci*
Pneumonia	*Klebsiella pneumoniae*
Pulmonary anthrax	*Bacillus anthracis*
Staph. respiratory infection	*Staphylococcus aureus*
Strep. respiratory infection	*Streptococcus pyogenes*
Legionellosis	*Legionella* spp.
Meningococcal infection	*Neisseria meningitidis*
Pneumonic plague	*Yersinia pestis*
Typhoid fever	*Salmonella typhi*
Whooping cough	*Bordetella pertussis*
Tuleremia	*Francisella tularensis*
Diptheria	*Corynebacterium diptheriae*
Fungal diseases	
Aspergillosis	*Aspergillus fumigatus*
Blastomycosis	*Blastomyces dermatiridi*
Coccidioidomycosis	*Coccidioides immitis*
Cryptococcosis	*Cryptococcus neoformans*
Histoplasmosis	*Histoplasma capsulatum*
Nocardiosis	*Nocardia asteriodes*
Sporotrichosis	*Sporotrichum schenckii*

(Contd ...)

Human diseases	Pathogens
Viral diseases	
Influenza	Influenza virus
Hemorrhagic fever	Bunyavirus
Hantavirus pulmonary syndrome	Hantavirus
Hepatitis	Hepatitis virus
Chicken pox	Herpes virus
Common cold	Picornavirus
Yellow fever	Flavivirus
Dengue fever	Flavivirus
Lyssa fever	Lyssavirus
Pleurodynia	Coxsackievirus, Echovirus
Rift Valley fever	Phlebovirus
Rubella	Rubivirus
Measles	Morbillivirus
Protozoal diseases	
Pneumocystosis	*Pneumocystis carinii*

Many of these pathogens such as the rhinovirus, which causes the common cold, affect almost everyone during their lifetime. The purchase of medications and treatments for symptoms of the common cold keeps numerous companies, such as those that make tissue paper and cold medicines, economically viable. This indicates how important the AMB pathway is to public health, not to mention the economy.

IMPORTANT AIRBORNE TOXINS

Also to be considered by aeromicrobiologists are microbial toxins. One example is a toxin from *Clostridium botulinum* (botulinal A toxin) that is a potential biological warfare agent. Botulinal toxin is a neurotoxin that is normally associated with ingestion of contaminated food. However, the lethal dose is so small that aerosolisation can also be a means of dissemination. The lethal dose for botulinal toxin by inhalation is 0.3 µg of toxin, with death expected 12 hours after exposure. Symptoms are caused by inhibition of the production of acetylcholine at nerve endings. Death is a result of asphyxiation caused by the paralysis of respiratory muscles.

Another toxin produced by bacteria is staphylococcal enterotoxin. This toxic protein is highly resistant to inactivation in association with the AMB pathway. It is rarely associated with long-lasting effects but can incapacitate those exposed to it. On occasion this toxin can be fatal. The lethal dose is estimated to be 25 µg by inhalation. The symptoms include cramping, vomiting, and diarrhea, which occur within 1 hour of exposure by aerosolisation.

Other examples of airborne toxins are lipopolysaccharides (LPS). LPS is derived initially from the cells of gram-negative bacteria and, when released into the air, can cause various respiratory distress syndromes. LPS molecules, also referred to as endotoxins, are highly antigenic biological agents that, when associated with airborne particles such as dust, are often associated with acute respiratory symptoms such as chest tightness, coughing, shortness of breath, fever, and wheezing. There are many sources associated with the production of high levels of LPS, such as cotton mills, haystacks, sewage treatment plants, solid waste handling facilities, swine confinement buildings, poultry houses, and even homes and office buildings. LPS is liberated when gram-negative bacteria in these environments are lysed but can also be released when they are actively growing.

Lipopolysaccharides are one of the major constitutive elements of the outer membranes of gram-negative bacteria. Because gram-negative bacteria are ubiquitous in the environment, LPS is also ubiquitous and is considered by some to be the most important aerobiological allergen.

NATURE OF BIOAEROSOLS

To better understand bioaerosols and the AMB pathway, one must understand the nature of aerosols in general. Bioaerosols vary considerably in size, and composition depends on a variety of factors including the type of micro-organism or toxin, the types of particles they are associated with such as mist or dust, and the gases in which the bioaerosol is suspended. Bioaerosols in general range from 0.02 to 100 µm in diameter and are classified on the basis of their size. The smaller particles (<0.1 µm in diameter) are considered to be in the nuclei mode, those ranging from 0.1 to 2 µm are in the accumulation mode, and larger particles are considered to be in the coarse mode. As shown in Fig. 27.1, particles in nuclei or accumulation mode are considered to be fine particles and those in coarse mode are considered coarse particles.

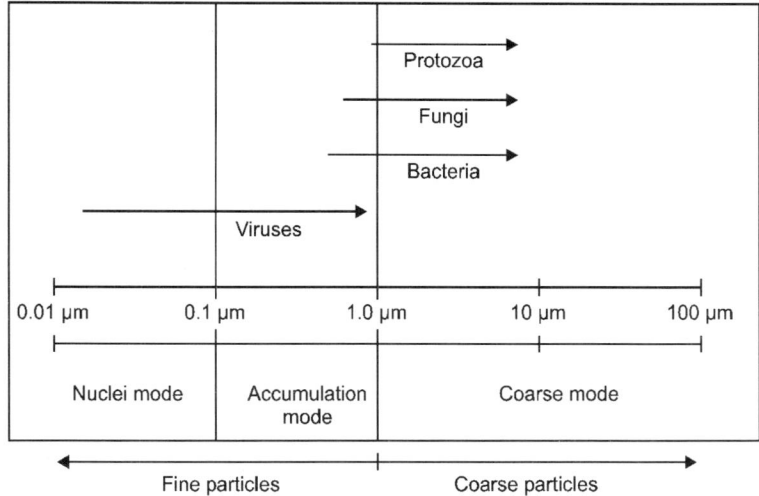

Fig. 27.1. Diagrammatic representation of the relative sizes of bioaerosols. The depictions of the various kinds of organisms are indicative of their potential sizes when associated with airborne particles (rafts). The terminologies used to describe the various sizes of the bioaerosols are also indicated.

The composition of bioaerosols can be liquid or solid or a mixture of the two and should be thought of as micro-organisms associated with airborne particles or as airborne particles containing micro-organisms. This is because it is rare to have micro-organisms (or toxins) that are not associated with other airborne particles such as dust or mist. This information is derived from particle size analysis experiments, which indicate that the average diameter of airborne bacterial particles is greater than 5 µm. By comparison, the average size of a soil borne bacterium, 0.3 to 1 µm, is less than one-fifth this size. Similar particle size analysis experiments show the same to be true for aerosolised micro-organisms other than bacteria, including viruses.

ATMOSPHERE

The AMB pathway by its nature involves the atmosphere. The layer of most interest and significance in aeromicrobiology is the boundary layer, which is the name given to the earth's atmosphere extending to a height of about 0.1 km from the earth's surface. It should be noted, however, that airborne transport of micro-organisms is by no means limited to this layer and it is not uncommon to have micro-organisms associated with layers of the troposphere above the turbulent boundary layer. However, it is the surface boundary layer that is largely responsible for the transport of particles over both short and long distances. The boundary layer consists of three parts: the laminar boundary layer, the turbulent boundary layer, and the local eddy layer.

The laminar boundary layer is a layer of still air associated with the earth and all projecting surfaces, whether solid or liquid. This layer can be anywhere from 1 µm to several metres thick, depending on weather conditions. Still conditions cause the thickness of this layer to increase, and windy conditions minimise it to a very close association with surfaces. The turbulent boundary layer is the layer that is considered to be always in motion and responsible for horizontal transport phenomena (wind dispersion), which occur whenever micro-organism-associated particles are launched either indoors or outdoors. In the lower levels of the turbulent layer, the linear flow of air is interrupted by surface projections and their associated laminar boundary layers.

This interaction results in the formation of friction against the airflow. This friction, which is apparent in the form of local areas of 'swirling' turbulence, gives the final layer its name. The local eddy layer is the actual zone of interaction between the still laminar boundary layer of surface projections and the turbulent boundary layer.

AEROMICROBIOLOGICAL PATHWAY

The aeromicrobiological pathway describes the launching of bioaerosols into the air, the subsequent transport via diffusion and dispersion of these particles, and finally their deposition. An example of this pathway is that of liquid aerosols containing the influenza virus launched into the air through a cough, sneeze, or even through talking. These virus-associated aerosols are dispersed by a cough or sneeze, transported through the air, inhaled, and deposited in the lungs of a nearby person, where they may begin a new infection. Traditionally, the deposition of viable micro-organisms and the resultant infection are given the most attention, but all three processes (launching, transport, and deposition) are of equal importance in understanding the aerobiological pathway.

Launching

The process whereby particles become suspended within the earth's atmosphere is termed launching. Because bioaerosols must be launched into the atmosphere to be transported, it is important to understand this process. The launching of bioaerosols is mainly from terrestrial and aquatic sources, with greater airborne concentrations or atmospheric loading being associated with terrestrial sources than with aquatic sources. Some researchers speculate that there may even be atmospheric sources of bioaerosols in addition to terrestrial and aquatic ones. This phenomenon is related to the limited potential for micro-organisms to reproduce while airborne. This, however, is an area of aeromicrobiology for which there is little available information.

Launching into the surface boundary layers can include, but is certainly not limited to, diverse mechanisms such as: air turbulence created by the movement of humans, animals, and machines; the generation, storage, treatment, and disposal of waste material; natural mechanical processes such as the action of water and wind on contaminated solid or liquid surfaces; and the release of fungal spores as a result of natural fungal life cycles.

Airborne particles can be launched from either point, linear, or area sources. A point source is an isolated and well-defined site of launching such as a pile of biosolid material before it is applied over the field. Point sources tend to display a general conical-type dispersion (Fig. 27.2). Point sources can be further defined on the basis of the type of launching phenomenon: (i) instantaneous point sources, for example, a single event such as a sneeze, and (ii) continuous point sources, from which launching occurs over extended periods of time, such as the biosolid pile.

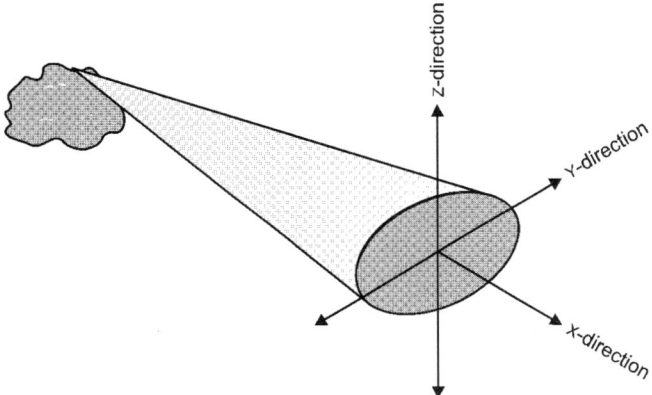

Fig. 27.2. Schematic representation of the type of bioaerosol distribution expected from a point source. This figure shows the three plains of diffusion: (i) x-direction is the mean direction in which the wind is blowing, (ii) y-direction is the lateral diffusion, and (iii) z-direction is the vertical diffusion.

In contrast to point sources, linear sources, and area sources, involve larger, less well-defined areas. When considered on the same size scale, linear and area sources display more particulate wave dispersion as opposed to the conical type of dispersion displayed by point sources. Linear and area sources can also be divided into instantaneous and continuous launching points of origin. For example, an instantaneous linear source might be a passing aircraft releasing a biological warfare agent. A continuous area source might be exemplified by release or bioaerosols from a large field that has received an application of biosolids.

Transport

Transport or dispersion is the process by which kinetic energy provided by the movement of air is transferred to airborne particles, with resultant movement from one point to another. This 'energy of motion' gained by airborne particles is considerable and can result in dissemination of airborne micro-organisms over long distances. Transport of bioaerosols can be defined in terms of time and distance. Submicroscale transport involves short periods of time, under 10 minutes, as well as relatively short distances, under 100 metres. This type of transport is common within buildings or other confined spaces. Microscale transport ranges from 10 minutes to 1 hour and from 100 metres to 1 km and is the most common type of transport phenomenon. Mesoscale transport refers to transport in terms of days and distances up to 100 km, and in macroscale transport, the time and distance are extended even further. Because most micro-organisms have limited ability to survive when suspended in the atmosphere, the most common scales considered are the submicroscale and microscale. It should be noted, however, that some viruses, spores, and spore-forming bacteria have been shown to enter into mesoscale and even macroscale transport. The hoof-and-mouth disease outbreak in England involved transport by wind of over 60 km (mesoscale). Other studies have shown that coliforms aerosolised from sewage treatment plants have been transported over 1.2 km (macroscale). Another interesting example is influenza pandemics. Pandemics are epidemics that occur over a wide geographic area, and influenza has been shown to spread from east to west around the world.

These pandemics have been positively correlated with the prevailing trade winds. This can be considered global transport and illustrates how important the aeromicrobiological pathway can be in microbiological dissemination, especially in relation to disease.

As bioaerosols travel through time and space, different forces act upon them such as diffusion, inactivation, and ultimately deposition. Diffusion is the scattering and/or dissipation of bioaerosols in response to a concentration gradient as well as gravity, and is generally aided by airflow and atmospheric turbulence. The amount of turbulence associated with airflow and, thus, the relative amount of diffusion that may occur in association with particulates such as bioaerosols can be estimated using the method of Osbert Reynolds. Reynolds found that factors associated with man wind velocity, the kinetic viscosity of the air, and the relative dimension of the interfering structures could provide an indication of the amount of turbulence associated with linear airflow. Without turbulence, airborne particles from a point source would travel in a concentrated stream directly downwind.

When dealing with particulate transport over time and distance, Tayler indicated that diffusion during horizontal transport could be viewed as an increase in the standard spatial deviation of particles from the source over time. What does this mean? For an instantaneous point source under the influence of a mean wind direction, spread would be a standard spatial deviation from a linear axis (x) extending from the source (origin) in the mean direction of wind flow, with diffusion caused by turbulence occurring in the lateral (y) and vertical (z) axes (Fig. 27.2). The standard deviation of particulate diffusion cannot be considered constant over a particular spatial orientation but is instead dependent on the time taken to reach the particular distance. Mathematical models that attempt to estimate the transport of airborne particles use this basic premise as a foundation for predictions. To picture this concept, imagine standing at the door of a room, where someone is holding a smoking candle. If there is no air current in the room the smoke will still eventually reach you at the door but it will be very diffuse as it is spreading in every other direction as well. However, if there is a fan behind the person holding the smoking candle and this fan is pointed at the door, then the smoke from the candle will be carried by this air current. It will travel the same distance as it did before, but it will travel faster, undergo less diffusion, and as a result be more concentrated when it reaches you. This is the principle of time-dependent diffusion as indicated by Tayler's theory.

Deposition

The last step in the AMB pathway is deposition. An airborne bioaerosol will eventually leave the turbulence of the suspending gas and will ultimately be deposited on a surface by one or a combination of interrelated mechanisms. These mechanisms are discussed in the following sections and include: gravitational settling, downward molecular diffusion, surface impaction, rain deposition, and electrostatic deposition, to name a few of the most important. These processes are linked in many ways, and even though viewed separately, they all combine to create a constant, if not steady, deposition of particles.

Gravitational settling

The main mechanism associated with deposition is the action of gravity on particles. The force of gravity acts upon all particles heavier than air, pulling them down and essentially providing spatial and temporal limitations to the spread of airborne particles.

Downward molecular diffusion

Downward molecular diffusion, as indicated by the name, can be described as a randomly occurring process caused by natural air currents and eddies that promote and enhance the downward movement of airborne particulates. These random movements exist even in relatively still air and tend to be in the downward direction because of gravitational effects. As a result, measured rates of gravitational deposition tend to be greater than those predicted by the Stokes equation. The increase in the rate of deposition is due to the added effects of downward molecular diffusion. Molecular diffusion is also influenced by the force of the wind. Molecular diffusion-enhanced deposition rates tend to increase with increasing wind speed and turbulence.

Surface impaction

Surface impaction is the process by which particles make contact with surfaces, such as leaves, trees, walls, and computers. With impaction there is an associated loss of kinetic energy. In nature, it is rare to find flat, smooth surfaces on which wind currents are unobstructed. Thus, surface impaction is a very critical factor influencing transport and deposition, especially for bioaerosols.

   Impaction potential is the relative likelihood that an airborne object will collide with another object in its path. Impaction does not necessarily result in permanent deposition, however. Once a particle collides with an object, it has the potential to bounce. Bouncing off a surface causes the particle to re-enter the air current at a lower rate, which can have one of two effects: (i) it can allow subsequent downward molecular diffusion and gravitational settling to occur, resulting in deposition on another nearby surface, and (ii) it can allow the particle to escape the surface and once again re-enter the air current. Studies have shown that impaction is influenced by the velocity and size of the particle as well as the size and shape of the surface it is approaching. To visualise the two extremes of this effect, consider a small airborne particle travelling slowly toward another small round object (Fig. 27.3). This small particle has much less chance of impaction with the small object than a larger particle travelling at a greater velocity toward a much larger flat object. Thus, the larger particle has a higher impaction potential.

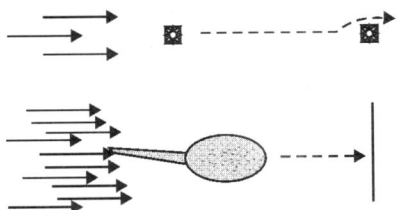

Fig. 27.3. Schematic representation of forces that influence impaction potential. In the top figure a small particle represented by a icosahedral virus is travelling slowly toward another small particle (target) also represented by an icosahedral virus. The small size and slow velocity of the red virus and the small size and round shape of the target result in a low impaction potential. On the other hand a large object with greater mass, represented as a bacterium, travelling at a high velocity toward a large flat surface has a very high impaction potential.

Rain and electrostatic deposition

Rainfall and electrostatic charge also can affect deposition. Rainfall deposition occurs as a condensation reaction between two particles (raindrop and bioaerosol), which combine and create a bioaerosol with a greater mass, which settles faster. The overall efficiency of rain deposition also depends on the spread area of the particle plume. Larger, more diffuse plumes undergo stronger impaction than smaller, more concentrated plumes. Rain deposition is also affected by the intensity of the rainfall. The heavier the rainfall, the greater the overall rates and numbers of the condensation reactions and the greater the subsequent increase in rain deposition.

Electrostatic deposition also condenses bioaerosols, but is based on electrovalent particle attraction. All particles tend to have some type of associated charge. Micro-organisms typically have an overall negative charge associated with their surfaces at neutral pH. These negatively charged particles can associate with other positively charged airborne particles, resulting in electrostatic condensation. The major phenomenon occurring may be a coagulation effect between particles (much like the condensation of the clostridial spore with the water droplet), which would increase the bioaerosol mass and enhance deposition. It might also be assumed that as an electromagnetically charged bioaerosol comes into close proximity with an electromagnetically charged surface, electroattractive or electrorepulsive influences may be present.

MATHEMATICAL MODELLING

When micro-organism-associated particles are aerosolised or launched into the atmosphere, the major question becomes, how far will they travel and in what concentration? Although the answer to this question is complex, mathematical equations have been developed to attempt to simulate this process. Initial models such as Tayler's time dependent diffusion model were simplistic in reasoning but provided a solid foundation which subsequent modellers have built upon to modify and develop more sophisticated models. Modelling the aerobiological pathway is intended for use in prediction of airborne bioaerosol concentrations in the vicinity of a contaminated source. Models are useful in that they can help determine effective sampling locations and in some cases provide valuable information when sampling and analysis are not possible.

Point Source Modelling

The classical model of plume spread from a point source was developed in 1961 by Pasquill, who described the transport of airborne particles in general. The model requires the input of four variables: (i) the mean wind speed, (ii) the atmospheric stability class (which influences diffusion), (iii) the downwind distance from the origin of the point of interest (sampler), and (iv) the source height.

Area Source Modelling

The previous model is used to describe plume spread from a point source. Another model, by Parker, describes continual launching from line or area sources such as a land application site immediately after biosolids have been applied.

Indoor Air Modelling

The point source and area source models and variations of these models can be used for a wide variety of outdoor situations. Indoor modelling of bioaerosols, however, is very different and more complex in many ways. This is because buildings are isolated environments made up of many individual compartments. In each of these compartments, distinctly different air mixing patterns can occur, which increases the complexity of the modelling efforts. Central heating and cooling systems can introduce additional sources of air turbulence and means for the dissemination of organisms. Heating and cooling systems can also be the actual origins of microbial contamination, as in the case of legionellosis.

SAMPLING DEVICES FOR THE COLLECTION OF BIOAEROSOLS

Many devices have been designed for the collection of bioaerosols. Choosing an appropriate sampling device is based on many factors, such as availability, cost, volume of air to be sampled, mobility, sampling efficiency (for the particular type of bioaerosol), and the environmental conditions under which sampling will be conducted. Another factor that must be taken into account, especially when sampling for micro-organisms, is the overall biological sampling efficiency of the device. This factor is related to the maintenance of microbial viability during and after sampling. In this section, several types of commonly used samplers are described on the basis of their sampling methods: impingement, impaction, centrifugation, filtration, and deposition. Impingement is the trapping of airborne particles in a liquid matrix; impaction is the forced deposition of airborne particles on a solid surface; centrifugation is the mechanically forced deposition of airborne particles using inertial forces of gravity; filtration is the trapping of airborne particles by size exclusion; and deposition is the collection of airborne particles using only naturally occurring deposition forces. The two most commonly used devices for microbial air sampling are the all glass AGI-30 impinger and the Anderson six-stage impaction sampler.

Impingement

The AGI-30 (Fig. 27.4) operates by drawing air through an inlet that is similar in shape to the human nasal passage. The air is transmitted through a liquid medium where the air particles become associated with the fluid and are subsequently trapped. The AGI-30 impinger is usually run at a flow rate of 12.5 l/min at a height of 1.5 metres, which is the average breathing height for humans. The AGI-30 is easy to use, inexpensive, portable, reliable, easily sterilised, and has high biological sampling efficiency in comparison with many other sampling devices. The AGI-30 tends to be very efficient for particles in the range of 0.8 to 15 μm. The usual volume of collection medium is 20 ml and the typical sampling duration is approximately 20 minutes, which prevents evaporation during the sampling of warm climates or freezing of the liquid medium when sampling at lower temperatures. Another feature of the impingement process is that the liquid and suspended micro-organisms can be concentrated or diluted, depending on the requirements for analysis. Liquid impingement media can also be divided into subsamples in order to test for a variety of micro-organisms by standard cultural and molecular methods. The impingement medium can also be optimised to increase the relative biological recovery efficiency. This is important, because during sampling the airborne micro-organisms, which are already in a stressed state because of various environmental pressures such as ultraviolet (UV) radiation and desiccation, can be further stressed if a suitable medium is not used for recovery. Sampling media range from simple to complex. A simple medium is 0.85 per cent NaCl, which is an osmotically balanced, sampling medium used to prevent osmotic shock of recovered organisms. A more complex medium is peptone (1 per cent) which is used as a resuscitation medium for stressed organisms. Finally, enrichment or defined growth media can be used to sample selectively for certain types of organisms. The major drawback when using the AGI-30 is that there is no particle size discrimination, which prevents accurate characterisation of the sizes of the airborne particles that are collected.

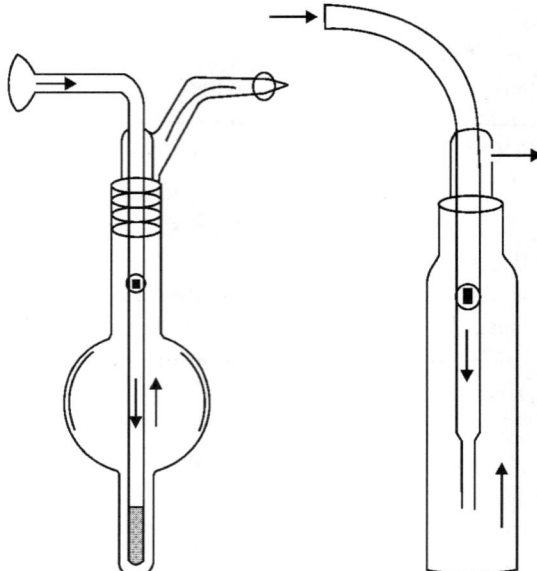

Fig. 27.4. This is a schematic representation of two all glass impingers (AGI). The impinger on the right is the classic AGI-30 impinger. Arrows indicate the direction of air flow. The air enters the impinger drawn by suction. As bioaerosols impinge into the liquid collection medium contained in the bottom of the impinger, the airborne particles are trapped within the liquid matrix.

Impaction

Unlike the AGI-30, the Anderson six-stage impaction sampler (Anderson 6-STG) provides accurate particle size discrimination. It is described as a multilevel, multiorifice, cascade impactor. The Anderson 6-STG (Fig. 27.5) was developed by Anderson in and operates at a input flow rate of 28.3 l/min. The general operating principle is that air is sucked through the sampling port and strikes agar plates. Larger particles are collected on the first layer, and each successive stage collects smaller and smaller particles by increasing the flow velocity and consequently the impaction potential.

The shape of the Anderson sampler does not conform to the shape of the human respiratory tract, but the particle size distribution can be directly related to the particle size distribution that occurs naturally in the lungs of animals. The lower stages correspond to the alveoli and the upper stages to the upper respiratory tract. The Anderson sampler is constructed of stainless steel with glass petri dishes, allowing sterilisation, ease of transport, and reliability. It is useful over the same particle size range as the AGI-30 (0.8 to over 10 µm), corresponding to the respirable range of particles. It is more expensive than the AGI-30, and the biological sampling efficiency is somewhat lower because of the method of collection, which is impaction on an agar surface. Analysis of viruses collected by impaction is also somewhat difficult, because after impaction, the viruses must be washed off the surface of the impaction medium and collected before assay. In contrast, bacteria or other micro-organisms can be grown directly on the agar surface. Alternatively, these microbes can be washed off the surface and assayed using other standard methodologies. The biggest single advantage of the Anderson 6-STG sampler is that particle size determinations can be obtained. Thus, the two reference samplers (AGI-30 and Anderson 6-STG) complement each other's deficiencies.

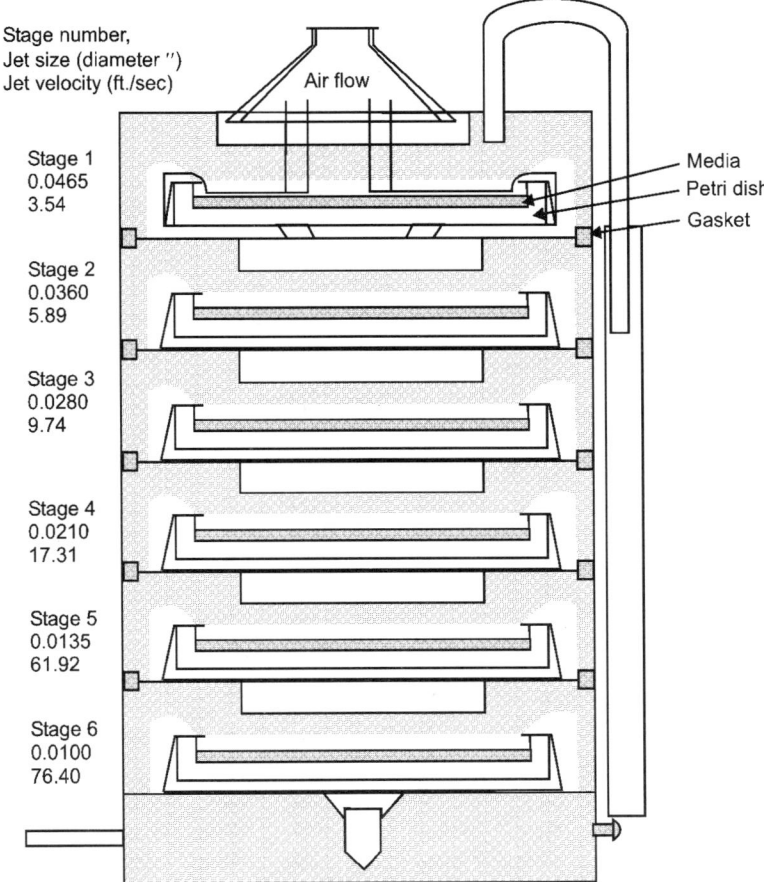

Stage number,
Jet size (diameter ")
Jet velocity (ft./sec)

Air flow

Stage 1
0.0465
3.54

Media
Petri dish
Gasket

Stage 2
0.0360
5.89

Stage 3
0.0280
9.74

Stage 4
0.0210
17.31

Stage 5
0.0135
61.92

Stage 6
0.0100
76.40

Fig. 27.5. This is a schematic representation of the Anderson six stage impaction air sampler. Air enters through the top of the sampler and larger particles are impacted upon the surface of the petri dish on stage 1. Smaller particles, which lack sufficient impaction potential follow the air stream to the subsequent levels. As the air stream passes through each stage the air velocity increases thus increasing the impaction potential so that particles are trapped on each level based upon their size. Therefore larger particles are trapped efficiently on stage 1 and slightly smaller particles on stage 2 and so on until even very small particles are trapped on stage 6. The Anderson six stage thus separates particles based upon their size.

Centrifugation

Centrifugal samplers use circular flow patterns to increase the gravitational pull within the sampling device in order to deposit particles. The Cyclone, a tangential inlet and return flow sampling device, is the most common type (Fig. 27.6). These samplers are able to sample a wide range of air volumes (1–400 l/min), depending on the size of the unit. The unit operates by applying suction to the outlet tube, which causes air to enter the upper chamber of the unit at an angle. The flow of air falls into a characteristic tangential flow pattern, which effectively circulates air around and down along the inner surface of the conical glass housing. As a result of the increased centrifugal forces imposed on particles in the airstream, the particles are sedimented out.

The conical shape of the upper chamber opens into a larger bottom chamber, where most of this particle deposition occurs. Although these units are able to capture some respirable-size particles, in order to trap micro-organisms efficiently, the device must be combined with some type of metered fluid flow that acts as a trapping medium. This unit, when used by someone proficient can be effective for microbiological air sampling. It is relatively inexpensive, easily sterilised, and portable, but it lacks high biological sampling efficiency and particle sizing capabilities. Analysis is performed by rinsing the sampler with an eluent medium, collection of the eluent, and subsequent assay by standard methodologies.

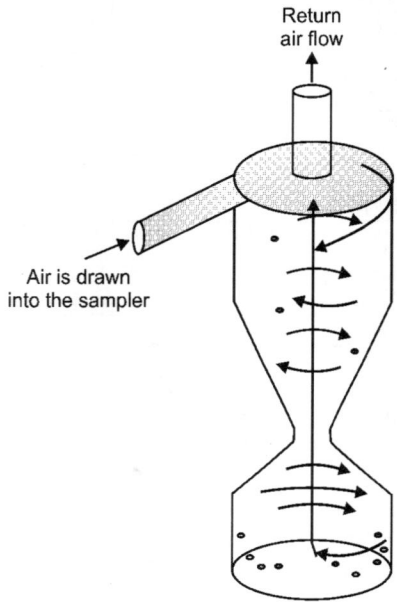

Return
air flow

Air is drawn
into the sampler

Fig. 27.6. This is a schematic representation of a tangential inlet and returned flow centrifugal air sampler. Air is drawn into the sampler at an angle (tangential) to the walls of the device so that it circulates around and down the walls. As it circulates the decrease in the diameter of the sampling body causes a dramatic increase in the velocity of the air and subsequently on particle's terminal velocity. This increase in gravitational settling potential causes the particles to be trapped in the lower collection chamber because their 'centrifugally increased' mass prevents them from exiting with the return air flow.

Filtration and Deposition

Filtration and deposition methods are both widely used for microbial sampling for cost and portability reasons. Filter sampling requires a vacuum source and involves passage of air through a filter, where the particles are trapped. Membrane filters can have variable pore sizes tend to restrict flow rates. After collection, the filter is washed to remove the organisms before analysis. Filtration sampling for micro-organisms is not highly recommended because it has a low overall sampling efficiency and it is not portable. However, in many cases the low cost makes it an attractive method.

One case in which filtration is routinely used is in sampling for airborne LPS. The sampling and analysis procedure for airborne LPS levels is slightly different from methods used for analysis of airborne micro-organisms. The most efficient means of sampling is usually filter collection using polyvinyl chloride or glass fibre membrane filters. Quantification analysis is usually done using a chromogenic *Limulus* amoebocyte lysate assay.

This system uses a *Limulus* amoebocyte lysate obtained from blood cells of horseshoe crabs. The lysate contains a enzyme-linked coagulation system, which is activated by the presence of LPS. With the addition of a substrate, the system is able to quantitate, by luminescence, the amount of environmental LPS by comparison with a standard curve.

Deposition sampling is by far the easiest and most cost-effective method of sampling. Deposition sampling can be accomplished merely by opening an agar plate and exposing it to the wind, which results in direct impaction, gravity settling, and other depositional forces. The problems with this method of sampling are: low overall sampling efficiency because it relies on natural deposition, no defined sampling rates or particle sizing, and an intrinsic difficulty in testing for multiple micro-organisms with varied growth conditions. Analysis of micro-organisms collected by depositional sampling is similar to impaction sample analysis.

MICROBIAL SURVIVAL IN THE AIR

The atmosphere is an inhospitable climate for micro-organisms mainly because of desiccation stress. This results in a limited time frame in which microbes can remain biologically active. Many micro-organisms, however, have specific mechanisms that allow them to be somewhat resistant to the various environmental factors that promote loss of biological activity. Spore-forming bacteria, moulds, fungi, and cyst-forming protozoa all have specific mechanisms that protect them from harsh gaseous environments, increasing their ability to survive aerosolisation. For organisms that have no such specific mechanisms the survival in aerosols can often be measured in seconds. In contrast, organisms with these mechanisms can survive indefinitely.

As a result, viability is highly dependent on the environment, the amount of time the organism spends in the environment, and the type of micro-organism. In addition, microbes may be viable but nonculturable, but for simplicity in this chapter we will use the term viable rather than the term culturable. Many environmental factors have been shown to influence the ability of micro-organisms to survive. The most important of these are relative humidity and temperature. Oxygen content, specific ions, UV radiation, various pollutants, and AOFs (air-associated factors) are also factors in the loss of biological activity. Each of these factors is discussed in the following sections.

Relative Humidity

The relative humidity or the relative water content of the air has been shown to be of major importance in the survival of airborne micro-organisms. Wells and Riley were among the first to show this phenomenon, indicating that as the relative humidity approaches 100 per cent, the death rate of *Escherichia coli* increases. In general, it has been reported that most gram-negative bacteria associated with aerosols tend to survive for longer periods at low relative humidities. The opposite tends to be true for gram-positive bacteria, which tend to remain viable longer in association with high relative humidities. Thus, the ability of a micro-organism to remain viable in a bioaerosol is related to the organism's surface biochemistry. One mechanism that explains loss of viability in association with very low relative humidity is a structural change in the lipid bilayers of the cell membrane. As water is lost from the cell, the cell membrane bilayer changes from the typical crystalline structure to a gel phase. This structural phase transition affects cell surface protein configurations and ultimately results in inactivation of the cell.

Early studies by Loosli showed that the influenza virus was also adversely affected by an increase in relative humidity. More recent work suggests that viruses possessing enveloped nucleocapsids (such as the influenza virus) have longer airborne survival when the relative humidity is below 50 per cent, whereas viruses with naked nucleocapsids (such as the enteric viruses) are more stable at a relative humidity above 50 per cent. It should be noted that viruses with enveloped nucleocapsids tend to have better survival in aerosols than those without. Some viruses are also stable in the AMB pathway over large ranges of relative humidity, which makes them very successful airborne pathogens.

Temperature

Temperature is a major factor in the inactivation of micro-organisms. In general, high temperatures promote inactivation, mainly associated with desiccation and protein denaturation, and lower temperatures promote longer survival times. When temperatures approach freezing, however, some organisms lose viability because of the formation of ice crystals on their surfaces. The effects of temperature are closely linked with many other environmental factors, including relative humidity.

Radiation

The main sources of radiation damage to micro-organisms including bacteria, viruses, fungi, and protozoa are the shorter UV wavelengths, and ionising radiation such as X-rays. The main target of UV irradiation damage is the nucleotides that make up DNA. Ionising radiation or X-rays cause several types of DNA damage, including single strand breaks, double strand breaks, and alterations in the structure of nucleic acid bases. UV radiation causes damage mainly in the form of intrastrand dimerisation, with the DNA helix becoming distorted as thymidines are pulled toward one another. This in turn causes inhibition of biological activity such as replication of the genome, transcription, and translation.

Several mechanisms have been shown to protect organisms from radiation damage. These include association of microbes with larger airborne particles, possession of pigments or carotenoids, high relative humidity, and cloud cover, all of which tend to absorb or shield bioaerosols from radiation. Many types of organisms also have mechanisms for repair of the DNA damage caused by UV radiation. An example of an organism that has a radiation resistance mechanism is *Dienococcus radiodurans*. *D. radiodurans* is a soil bacterium that is considered the most highly radiation-resistant organism that has been yet been isolated. An important component of its radiation resistance is the ability to enzymatically repair damage to chromosomal DNA. The repair mechanism used by these bacteria is so highly efficient that much of the metabolic energy of the cell is dedicated exclusively to this function.

Oxygen, OAF and Ions

Oxygen, open air factors (OAFs), and ions are environmental components of the atmosphere that are difficult to study at best. In general, it has been shown that these three factors combine to inactivate many species of airborne microbes. Oxygen toxicity is not related to the dimolecular form of oxygen (O_2), but is instead important in the inactivation of micro-organisms when O_2 is converted to more reactive forms. These include superoxide radicals, hydrogen peroxide, and hydroxide radicals. These radicals arise naturally in the environment from the action of lightning, UV radiation, pollution, etc. Such reactive forms of oxygen cause damage to DNA by producing mutations, which can accumulate over time. The repair mechanisms described in the previous section are responsible for control of the damaging effects of reactive forms of oxygen.

Similarly, the open air factor (OAF) is a term coined to describe an environmental effect that cannot be replicated in laboratory experimental settings. It is closely linked to oxygen toxicity and has come to be defined as a mixture of factors produced when ozone and hydrocarbons (generally related to ethylene) react. For example, high levels of hydrocarbons and ozone causing increased inactivation rates for many organisms, probably because of damaging effects on enzymes and nucleic acids. Therefore, OAFs have been strongly linked to microbial survival in the air.

The formation of other ions, such as those containing chlorine, nitrogen, or sulphur occurs naturally as the result of many processes. These include the action of lightning, shearing of water, and the action of various forms of radiation that displace electrons from gas molecules, creating a wide variety of anions and cations not related to the oxygen radicals. These ions have a wide range of biological activity. Krueger showed that positive ions cause only physical decay of micro-organisms, e.g. inactivation of cell surface proteins, whereas negative ions exhibit both physical and biological effects such as internal damage to DNA.

EXTRAMURAL AEROMICROBIOLOGY

Extramural aeromicrobiology is the study of micro-organisms associated with outdoor environments. In the extramural environment, the expanse of space and the presence of air turbulence are two controlling factors in the movement of bioaerosols. Environmental factors such as UV radiation, temperature, and relative humidity modify the effects of bioaerosols by limiting the amount of time aerosolised micro-organisms will remain viable. This section is an overview of extramural aeromicrobiology that includes several topics: the spread of agricultural pathogens; the spread of airborne pathogens associated with waste environments; and germ warfare. This section is not intended as a review of all aspects of extramural aeromicrobiology but instead attempts to show the wide diversity of the science.

Agriculture

Contamination of crops and animals via bioaerosols has a huge economic impact worldwide. The list of agricultural pathogens in Table 27.1 shows the diversity of airborne micro-organisms that infect plants and animals. As the earth's population increases, the need for a larger, more stable supply of food becomes increasingly important. Rice and wheat are the two major staple crops that are paramount to world food security. Major pathogens of such crops are the wheat rust fungi. These spore-forming fungi cause some of the most devastating of all diseases of wheat and other grains. In 1993, one type of wheat rust (leaf rust) was responsible for the loss of over 40 million bushels of wheat in Kansas and Nebraska alone. Even with selective breeding for resistance in wheat plants, leaf rust continues to have major economic impacts. The high concentration of wheat in areas ranging from northern Texas to Minnesota and up into the Dakotas makes this whole region highly susceptible to rust epidemics.

Spores of wheat rust are capable of spreading hundreds if not thousands of kilometres through the atmosphere. The airborne spread of rust disease has been shown to follow a predictable trend, which starts during the fall with the planting of winter wheat in the southern plains. Any rust-infected plant produces thousands of spores, which are released into the air by either natural atmospheric disturbance or mechanical disturbance during the harvesting process. Once airborne, these spores are capable of long-distance dispersal, which can cause downwind deposition onto other susceptible wheat plants. The generation time of new spores is measured in weeks, after which, new spores are again released from vegetative fungi into the AMB pathway. For example, during the harvest of winter wheat in Texas, the prevailing wind currents are from south to north, which can allow rust epidemics to spread into the maturing crops farther north in Kansas and up into the young crops in the Dakotas. This epidemic spread of wheat rust and the resulting economic destruction produced are indicative of the impact that airborne microbial pathogens can have on agriculture. In addition, this example indicates the extent and rate at which the AMB pathway can spread such contagion.

A factor that complicates the control of such diseases is that chemical treatment for the control of pathogens is viewed as undesirable. This is because many pesticides have extremely long half-lives and their residence in an ecosystem can be extremely harmful. Therefore, instead of using wheat rust fungicides, attempts are being made to breed strains of wheat that are more resistant to the fungi. Another method used for controlling phytopathogenic (plant pathogenic) fungi is spore monitoring as a disease control strategy. In this approach, the life cycle of the fungi especially the release of spores is monitored, and fungicide application is timed to coincide with spore release. This approach minimises use of harmful chemicals. Thus, efficient AMB pathway sampling, monitoring, detection, and modelling have the ability to aid in the control of airborne pathogens.

The airborne spread of pathogenic micro-organisms is also highly important in the animal husbandry industry. The occurrence of foot-and-mouth disease is an example of the importance of bioaerosols in the spread of airborne disease. It has long been thought that bioaerosol spread is linked primarily to respiratory pathogens, but there is growing evidence that gastrointestinal pathogens are also important in airborne transmission of disease among animals. One example of bioaerosol spread of a gastrointestinal pathogen is transmission of *Salmonella typhimurium* among calves that are housed individually in small pens. The potential for bioaerosol spread of this pathogen was recognised because the initial symptoms resembled those of pneumonia and appeared randomly within these animals, two factors that are not characteristic of oral transmission. Oral transmission generally occurs sequentially from one pen to the next, whereas aerial transmission can carry organisms past nearby pens, infecting calves randomly. Furthermore, Wathes, showed that *S. typhimurium* could survive for long periods in an airborne state, and calves and mice exposed to aerosolised *S. typhimurium* developed symptoms, proving that gastrointestinal pathogens could be spread by aerosolisation. Finally, Baskerville showed that aerosolised *Salmonella enteritidis* could infect laying hens. These hens showed clinical symptoms and were shedding the test strain of salmonellae in their feces within a few days. Thus, the AMB pathway can be important even in the spread of diseases for which pathogens not normally considered airborne.

Waste Disposal

Waste disposal is a multibillion dollar industry in the United States and other developing countries. However, there are many hazards inherent in the treatment and disposal of waste-water and biosolid material. Major hazards associated with waste effluents are pathogenic micro-organisms including bacteria, viruses, protozoa, and helminths. Waste-water treatment plants utilise activated sludge and trickling filter systems and all of these treatment processes potentially create relatively large amounts of aerosols, which have been shown to include pathogenic micro-organisms. Other aspects of the treatment process such as composting and land disposal are also associated with the generation of aerosols containing pathogenic micro-organisms.

One of the primary methods for the disposal of biosolids is agricultural land application. This type of disposal is steadily becoming one of the more widely used alternatives because of restrictions on ocean and landfill dumping. Dowd showed the presence of a wide variety of airborne pathogens and pathogen indicators that were collected with an AGI-30 sampler, including *Salmonella* sp., *Clostridium* sp., *E. coli*, and F+ coliphages. The major concern associated with the aerosolisation process in relation to waste disposal operations is the exposure of waste disposal workers to pathogenic micro-organisms, although nearby population centres are also potential exposure risks. At present, epidemiological studies are being conducted to determine the health risks associated with waste-water and waste disposal operation, but no clear-cut risk has yet been identified.

Germ Warfare

Biological warfare has been a combat weapon for centuries. As early as A.D. 1346, Tartars besieging the walled city of Kaffa used catapults to launch plague-infested bodies into the city. During the French and Indian War, the English offered blankets to Native Americans who had taken control of a military outpost. The English had purposely exposed the blankets to the smallpox virus knowing the lack of immunity of the native people.

The United States began field-scale experiments in the 1950s using inert substances (fluorescent dyes) to simulate biological warfare agents. These aerosols were released into the air circulation system of a subway system and into the air off the coast of San Francisco.

The studies indicated that 100 per cent of the associated populations inhaled what was considered lethal doses of the inert indicators. In Russia, an accident at a biological warfare research institute caused the widespread exposure of nearby populations to a genetically modified strain of *Bacillus anthracis*. In 1995, Tokyo police found large quantities of *Clostridium botulinum* toxin during one of several raids on a terrorist-controlled facility. In the late 1990s, Iraq was investigated of having production facilities for biological warfare agents. Thus, we see that biological warfare is a reality in today's society.

Detection of biological warfare agents is an area that requires sophisticated equipment and training. The war between the United States and Iraq known as Desert Storm is a recent example of these facts. The United States needed the ability to detect biological warfare agents in order to give early warning to troops in case biological warfare agents were released. Unfortunately, the technology available, including the level of sensitivity, and the specificity required for rapid detection of biological and chemical warfare agents was and still is inadequate. With this in mind, the United States is presently developing a more advanced antibiological warfare defense programme.

Chapter 28

Bioaerosol Control and Biosafety in Laboratory

INTRODUCTION

A bioaerosol is a biological aerosol. These particles are very small and range in size from less than one micrometer to one hundred micrometers. Bioaerosols react to air currents and move quickly or slowly depending on the environment. Bioaerosols are impacted by gravity but due to their size air density and air currents play a large role in their movement.

METHOD OF CONTROLLING BIOAEROSOLS

The control of airborne micro-organisms can be handled in a variety of ways. Launching, transport, and deposition are all points at which the airborne spread of pathogens can be controlled. The mechanisms used to control bioaerosols include ventilation, filtration, UV treatment, biocidal agents, and physical isolation. These are discussed below.

Ventilation

Ventilation is the method most commonly used to prevent the accumulation of airborne particles. This mechanism involves creating a flow of air through areas where airborne contamination occurs. This can mean simply opening a window and allowing outside air to circulate inward, or use of air-conditioning and heating units that pump outside air into a room. Ventilation is considered one of the least effective methods for controlling airborne pathogens, but is still very important. Ventilation relies on mixing of intramural air with extramural air to reduce the concentration of airborne particles. However, in some cases the addition of extramural air can actually increase airborne particles. For example, one study showed that hospitals in Delhi, India that relied on ventilation alone contained airborne fungal loads that were higher inside the hospital than those outside. This indicates that ventilation alone may not be sufficient to significantly reduce circulating bioaerosols. Thus, for most public buildings, especially hospitals, other forms of bioaerosol control need to be implemented.

Filtration

Unidirectional airflow filtration is a relatively simple and yet effective method for control of airborne contamination. Some filters, for example, high-efficiency particulate air (HEPA) filters, are reported to remove virtually all infectious particles. These types of filters are commonly used in biological safety hoods. However, because of their high cost, they are not often used in building filtration systems. Instead, other filtration systems that rely on baghouse filtration (a baghouse works on the same principle as a vacuum cleaner bag) are used. Typically, air filters (baghouse, HEPA, etc.) are rated using the dust-spot percentage, which is an index of the size of the particles efficiently removed by the filter, with higher percentages representing greater filtration efficiencies. The typical rating for the filters used in most buildings is 30 to 50 per cent. Studies have shown that a 97 per cent dust-spot rating is required to effectively remove virus particles from the air. Other factors that influence filtration efficiency are related to the type of circulation system and how well it mobilises air within the building, the type of baghouse system used, and the filter material chosen (nylon wound, spun fibreglass, etc.) as well as the filter's nominal porosity (1 μm, 5 μm, etc.). All these factors combine to influence the efficiency of the air filtration and removal of particles including bioaerosols. Inspite of the high level of efficiency that can be achieved with filtration, many systems still cannot stop the circulation of airborne micro-organisms, especially viruses, and added treatments may be required to ensure that air is safe to breathe.

Biocidal Control

Biocidal control represents an added treatment that can be used to eradicate all airborne micro-organisms, ensuring they are no longer viable and capable of causing infection. Many eradication methods are available, for example, superheating, superdehydration, ozonation, and UV irradiation. The most commonly used of these methods is UVGI or ultraviolet germicidal radiation. UVGI has been shown to be able to control many types of pathogens, although some microbes show various levels of resistance. The control of contagion using UV irradiation was tested in a tuberculosis (TB) ward of a hospital. Contaminated air was removed from the TB ward through a split ventilation duct and channelled into two animal holding pens that contained guinea pigs. One pen received air that had been treated with UV irradiation; the other received untreated-air. The guinea pigs in the untreated-air compartment developed TB, but none of the animals in the UV-treated compartment became infected. The American Hospital Association indicated that, properly utilised, UV radiation can kill nearly all infectious agents, although the effect is highly dependent on the UV intensity and exposure time. Thus, major factors that affect survival (temperature, relative humidity, UV radiation, ozone, etc.) in the extramural environment can be used to control the spread of contagion in the intramural environment.

Isolation

Isolation is the enclosure of an environment through the use of positive or negative pressurised air gradients and airtight seals. Negative pressure exists when cumulative airflow travels into the isolated region. Examples of this, as previously mentioned, are the isolation chambers of the tuberculosis wards in hospitals used to protect others outside the TB wards from the infectious agent generated within these negative-pressure areas. This type of system is designed to protect other people in the hospital from the pathogens (*Mycobacterium tuberculosis*) present inside the isolation area. Air from these rooms is exhausted into the atmosphere after passing through a HEPA filter and biocidal control chamber.

Positive-pressure isolation chambers work on the opposite principle by forcing air out of the room, thus protecting the occupants of the room from outside contamination. One can reason that the TB ward is a negative-pressure isolation room while the rest of the hospital, or at least the nearby anterooms, are under positive-pressure isolation. Other examples are the hospitals critical care wards for immunosuppressed patients such as organ transplant, human immunodeficiency virus (HIV)-infected, and chemotherapy patients. These areas are protected from exposure to any type of pathogen or opportunistic pathogens. The air circulating into these critical care wards is filtered using HEPA filters, generating purified air essentially free of infectious agents.

BIOSAFETY IN THE LABORATORY

Microbiological laboratories are rooms or buildings equipped for scientific experimentation or research on micro-organisms. Many microbiological laboratories work specifically with pathogenic micro-organisms, some of which are highly dangerous, especially in association with the AMB pathway. Also, many types of equipment, such as centrifuges and vortexes, that are commonly used in microbiological laboratories can promote the aerosolisation of micro-organisms. Thus, laboratories and specialised equipment used in these laboratories (e.g. biosafety cabinets) are designed to control the spread of airborne micro-organisms. There are essentially four levels of control designed into laboratories, depending on the type of research being conducted. These levels of control are termed biosafety levels 1–4, with 1 being the lowest level of control and 4 the highest level of control. Within these laboratories, biosafety cabinets are essentially isolation chambers that provide safe environments for the manipulation of pathogenic micro-organisms. In this section, we will discuss biosafety cabinets and biosafety suits, followed by a short discussion of the actual biosafety levels imposed to achieve specific levels of control.

Biological Safety Cabinets

Biological safety cabinets (BSCs) are among the most effective and commonly used biological containment devices in laboratories that work with infectious agents. There are two basic types of biosafety cabinets currently available (Class II, and Class III), each of which has specific characteristics and applications that dictate the type of micro-organisms it is equipped to contain. Properly maintained biosafety cabinets provide safe environments for working with micro-organisms. Class II biosafety cabinets are characterised by having considerable negative-pressure airflow that provides protection from infectious bioaerosols generated within the cabinet, and Class III biosafety cabinets are characterised by total containment. Class I cabinets are also in existence, but they are no longer produced and are being replaced by Class II cabinets for all applications.

Class II biosafety cabinets, of which there are several types, are suitable for most work with moderate-risk pathogens (Table 28.1). Class II biosafety cabinets operate by drawing airflow past the worker and down through the front grill. This air is then passed upward through conduits and downward to the work area after passing through a HEPA filter. Room air is also drawn into the cabinet through the top of the unit, where it joins the circulating air and passes through the HEPA filter and into the work area. About 70 per cent of the air circulating in the work area is then removed by passing it through the rear grill of the cabinet, where it is discharged into the exhaust system.

Table 28.1. Examples of classification of biological agents according to risk.

Class	Type of agent	Agent
Class I	Bacterial	All those which have been assessed for risk and do not belong in higher classes
	Fungal	Influenza virus reference strains
	Protozoal	Newcastle virus
	Viral	Parainfluenza virus 3, SF4 strain
Class II	Bacterial	*Actinobacillus* spp.
		Bordetella spp.
		Borrelia spp.

(Contd ...)

Class	Type of agent	Agent
		Campylobacter spp.
		Clostridium spp.
		E. coli spp.
		Klebsiella spp.
		Listeria spp.
		Mycobacteria spp.
		Shigella spp.
		Vibrio spp.
		Salmonella spp.
	Fungal	*Cladosporium* spp.
		Blastomyces spp.
		Penicillium spp.
		Cryptococcus spp.
		Microsporum spp.
	Protozoal	*Cryptosporidium* spp.
		Giardia spp.
		Encephalitozoon spp.
		Enterocytozoon spp.
		Babesia spp.
		Echinococcus spp.
		Entamoeba spp.
		Fasciola spp.
		Leishmania spp.
		Plasmodium spp.
		Schistosoma spp.
		Trypanosoma spp.
	Viral	Adenoviruses
		Corona viruses
		Cowpox virus
		Coxsackie A and B viruses
		Echoviruses
		Hepatitis viruses A, B, C, D and E
		Epstein-Barr virus
		Influenza viruses
		Vaccinia virus
		Rhinoviruses
Class III	Bacterial	*Bartonella* spp.
		Brucella spp.
		Codiella burnetti

(Contd ...)

Class	Type of agent	Agent
		Mycobacterium bovis
		Mycobacterium tuberculosis
		Pseudomonas mallei
		Rickettsia spp.
		Yersinia pestis
	Fungal	*Coccidioides immitis*
		Histoplasma capsulatum
	Protozoal	None
	Viral	Dengue virus
		Lymphocytic choriomeningitis virus
		Monkey pox virus
		Yellow fever virus
	Bacterial	None
Class IV	Fungal	None
	Protozoal	None
	Viral	Absettarov
		Hemmorrhagic fever agents
		Ebola fever virus
		Guanarito
		Hanzalova
		Lassa virus
		Marburg virus
		Tick borne encephalitis viruses
		Herpesvirus simiae

The remaining 30 per cent is passed through the front grill, essentially recirculating in the cabinet (Fig. 28.1). Laboratory personnel require special training in order to properly use Class II cabinets and to ensure proper containment of bioaerosols.

One of the major hazards associated with Class II cabinets is the potential for the disruption of the negative airflow. Many mechanical actions can disrupt the protective airflow, such as repeated insertion and withdrawal of arms, opening or closing of doors in the laboratory, or even someone walking past the cabinet while it is in use. Any of these actions can potentially allow the escape of bioaerosols from the cabinet. Practices that minimise the release of bioaerosols include the following:

1. Delaying manipulation of materials for 1 minute after inserting arms or hands inside the cabinet.
2. Keeping the arms raised slightly off the front grill of the cabinet to prevent room air from flowing directly into the work area.
3. Performing all operations at least 4 inches in front of the front grill.
4. Running the cabinet 24 hours a day to help to maintain the room-cabinet air balance (at a minimum, cabinets should be turned on at least 5 minutes before use).
5. Wiping work surface, interior walls, the surfaces of all introduced objects, and interior surface of the window with 70 per cent ethanol or other appropriate disinfectant prior to work with pathogens.

6. Placing plastic-backed absorbant towelling on the work surface to reduce splatter and aerosol formation.

7. Placing all materials toward the rear of the cabinet, with bulky items placed to one side and aerosol-generating objects such as centrifuges and vortexes as far to the back of the cabinet as possible.

8. Open flames should be avoided in biosafety cabinets, because they cause turbulence, which disrupts the pattern of air supplied to the work surface. Open flames are unnecessary in the nearly microbe-free environment created within the biosafety cabinet work area. Touch-plate burners equipped with pilot lights or small electric furnaces can be used to decontaminate bacteriological loops when necessary.

9. Finally, for decontamination of the hood, the surfaces of all materials should be wiped with 70 per cent ethanol or other appropriate disinfectant before removal, biohazardous waste materials should be contained within appropriate biohazard bags or containers and their surfaces wiped down with disinfectant before removal from the cabinet, all interior surfaces should then be wiped down including the interior of the window, and finally gloves and gowns should be removed and the investigator's hands washed.

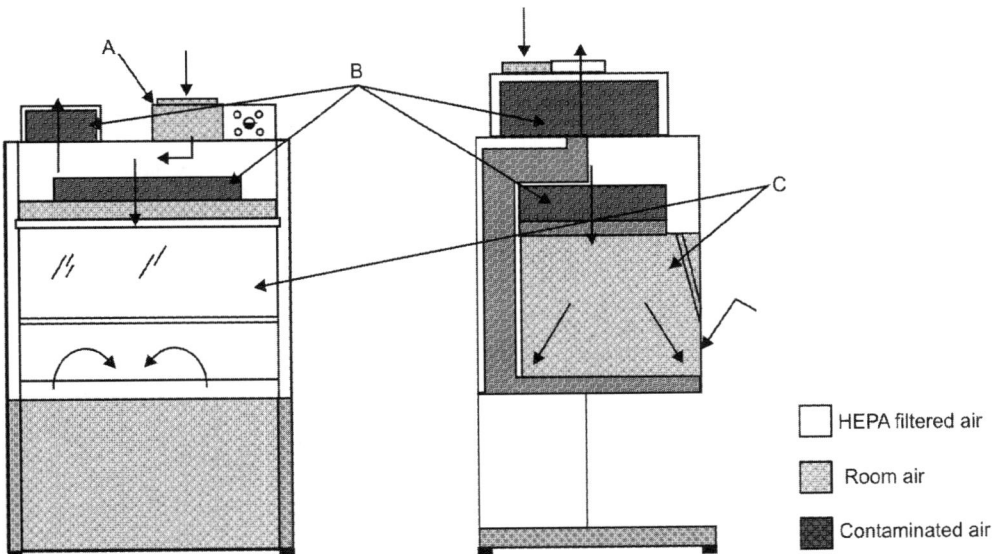

Fig. 28.1. This is a schematic representation of the airflow paths within a typical class II biosafety cabinet. Room air is drawn in from the top and from the front as indicated by arrows. The non-purified atmospheric air entering from the top of the cabinet is drawn in by an air pump (A), and then is purified by a HEPA filter (B) as it enters the workspace. Non-purified air from the room entering from the front of the cabinet passes into the front grill and passes up through the top of the cabinet where it also passes through a HEPA filter before entering the workspace (C). This prevents the experiments in the workspace from being contaminated by airborne room contaminants. As the purified room air is exposed to the work environment and becomes contaminated it is passed through yet another HEPA filter before being exhausted to the atmosphere. This pattern of airflow and purification ensures that the worker and the atmosphere are not exposed to the biohazards contained within the biosafety cabinet.

This list is not meant as an exhaustive account of the procedures required in the use of biosafety cabinets; rather, it serves as an indication of the extent of the containment procedures required to safely handle pathogens.

The Class III biosafety cabinet (Fig. 28.2) is a completely enclosed environment that offers the highest degree of personnel and environmental protection from bioaerosols. Class III cabinets are used for high-risk pathogens (Table 28.1). All operations in the work area of the cabinet are performed through attached rubber gloves. Class III cabinets use complete isolation to protect workers. All air entering the cabinet is filtered using a HEPA filter, and the air leaving the cabinet is filtered by two HEPA filters in series. The exhaust may also include biocidal treatment such as incineration following the HEPA filtration to further ensure complete biological inactivation. In addition to these safeguards, Class III cabinets are connected with airtight seals to all other laboratory equipment (such as incubators, refrigerators, and centrifuges) that is needed for working with the pathogens while using the cabinet. The Class III cabinet must also be connected to autoclaves and chemical dunk tanks used to sterilise or disinfect all materials entering or exiting the cabinet.

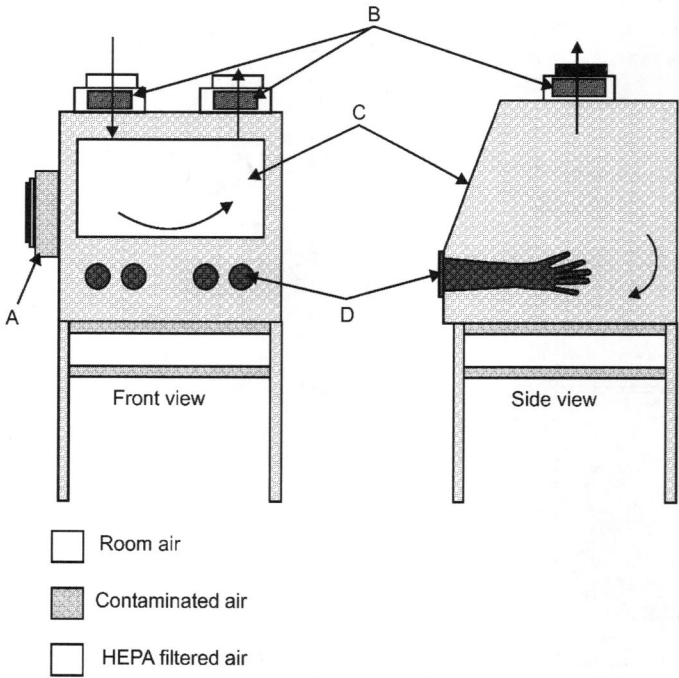

Fig. 28.2. This is a schematic representation of a Class III biological safety cabinet. This cabinet is completely sealed from the environment. Any materials entering or leaving the cabinet are passed through a chemical dunk tank or autoclave (A) in order to sterilise them and prevent environmental contamination. Air Entering or leaving these cabinets is passed through HEPA filters (B). Access to the workspace is by means of rubber gloves (D) and the workspace is visualised through a sealed window (C). These biosafety cabinets are utilised when working with highly pathogenic micro-organisms to protect workers and the environment. Class III cabinets can be used to work with all biohazardous agents except those specifically designated for level 4 containment.

Another type of containment that typically provides the same level of protection as a Class III biosafety hood is the biological safety suit. The biological suit, unlike biosafety cabinets, operates under positive pressure created by an external air supply, thus protecting the wearer. Like the biosafety cabinets, the biosafety suit isolates the laboratory worker wearing it from bioaerosols. Biosafety suits are typically used in airtight complete biocontainment areas, and are decontaminated by means of chemical showers upon exiting the biohazard area.

Some biosafety suits are portable and can be used in environments outside the laboratory such as 'hot zones' (epidemiological areas that are currently under the influence of epidemic cases of diseases caused by high-risk pathogens) so that microbiologists and physicians working in these areas can minimise their risk of exposure to pathogens. As in biosafety cabinets, the air entering and leaving the biosafety suit passes through two HEPA filters.

Biosafety Laboratories

Biosafety laboratories are carefully designed environments where infectious or potentially infectious agents are handled and/or contained for research or educational purposes. The purpose of a biosafety laboratory is to prevent the exposure of workers and the surrounding environment to biohazards. There are four levels of biohazard control which are designated as biosafety levels 1 through 4.

Biosafety level 1, as defined by the centres for disease control (US department of health and human services: CDC-NIH, 1993), indicates laboratories where well-characterised agents that are not associated with disease in healthy adult humans are handled. In general, no safety equipment is used other than sinks for hand washing and only general restrictions are placed on public access to these laboratories. Work with the micro-organisms can be done on bench tops using standard microbiological techniques. A good example of a biosafety 1 laboratory is a teaching laboratory used for undergraduate microbiology classes.

Biosafety 2 indicates an area where work is performed using agents that are of moderate hazard to humans and the environment. These laboratories differ from biosafety 1 laboratories in that the personnel have specialised training in the handling of pathogens and access to the work areas is limited. Many procedures that may cause aerosolisation of pathogenic micro-organisms are conducted in biological safety level II cabinets or other physical containment equipment to protect the laboratory workers.

Biosafety 3 indicates laboratories where agents that can cause serious or fatal disease as a result of AMB exposure are handled. As with biosafety 2, all personnel are specifically trained to handle pathogenic micro-organisms. All procedures involving these infectious agents are conducted in biological safety level II cabinets or other physical containment devices.

These facilities also have permanent locks to control access, negative airflow, and filtered ventilation in order to protect the public and the surrounding environments. With certain pathogens used in biosafety 3 laboratories, Class III safety hoods may also be used and clothes must be changed before leaving the premises.

Biosafety 4 is the highest level of control and is indicated for organisms that have high potential for life-threatening disease in association with aerosolisation. To work in these facilities, personnel must have specialised training beyond that required for biosafety levels 2 and 3. Biosafety level 4 laboratories are 100 per cent isolated from other areas of a building and may even be separated from other buildings altogether.

Work in these areas is confined exclusively to Class III biological safety cabinets unless one-piece positive-pressure ventilation suits are worn, in which case Class II biosafety cabinets may be used.

These laboratories are also specially designed to prevent micro-organisms from being disseminated into the environment. The laboratories have complete containment and require personnel to wear specialised clothing, which is removed and sterilised before leaving the containment areas. Personnel are also required to shower before leaving the facility. In general, all air into and out of these laboratories is sterilised by filtration and germicidal treatment. These facilities represent the ultimate in our ability to control the AMB pathway.

Chapter 29

Beneficial and Pathogenic Microbes in Agriculture

INTRODUCTION

Pathogenic microbes are microbes that cause infectious diseases. The organisms involved include pathogenic bacteria, causing diseases such as plague, tuberculosis and anthrax; protozoa, causing diseases such as malaria, sleeping sickness and toxoplasmosis; and also fungi causing diseases such as ringworm, candidiasis or histoplasmosis. However, other diseases such as influenza, yellow fever or AIDS are caused by pathogenic viruses, which are not living organisms and are not, therefore, micro-organisms. This chapter is dedicated to human pathogenic microbes.

SOIL–PLANT–MICRO-ORGANISM SYSTEM

As we have seen, most normal soils do not contain abundant microbial nutrients, because microbial communities utilise any nutrients that are available. In contrast, the rhizosphere is a unique soil environment found in close proximity to plant roots, where nutrients are more abundant because of the influence of the plant itself. Increased nutrient availability, in turn, results in enhanced microbial activity and numbers. Thus, the rhizosphere exists because of soil–plant–micro-organism interactions (Fig. 29.1). Ultimately, microbial gene expression in the rhizosphere is controlled by these interactions, which, in turn, are influenced by direct or indirect environmental factors. Overall, the microbial populations within the soil–plant–micro-organism system can affect plant growth in beneficial or detrimental ways (Fig. 29.2). In this chapter, we will examine the role of these microbes in the agricultural arena.

RHIZOSPHERE ENVIRONMENT

The term rhizosphere was coined by Hiltner in 1904 to describe the part of the soil that is influenced by plant roots. Originally, the rhizosphere was thought to extend 2 mm outward from the root surface. Now it is recognised that the rhizosphere can extend 5 mm or more as a series of gradients of organic substrate, pH, O_2, CO_2, and H_2O. Essentially two regions of the rhizosphere are now recognised: (i) the rhizosphere soil, and (ii) the soil in direct contact with the plant root, which is the rhizoplane.

Micro-organisms also inhabit the root itself and are known as endophytes. The portion of the root occupied by microbes was formerly known as the endorhizosphere but this term is no longer used by soil microbiologists. Finally, note that the rhizosphere effect occurs almost as soon as a seed is planted, with the area of increased microbial activity around a seed bring known as the spermosphere.

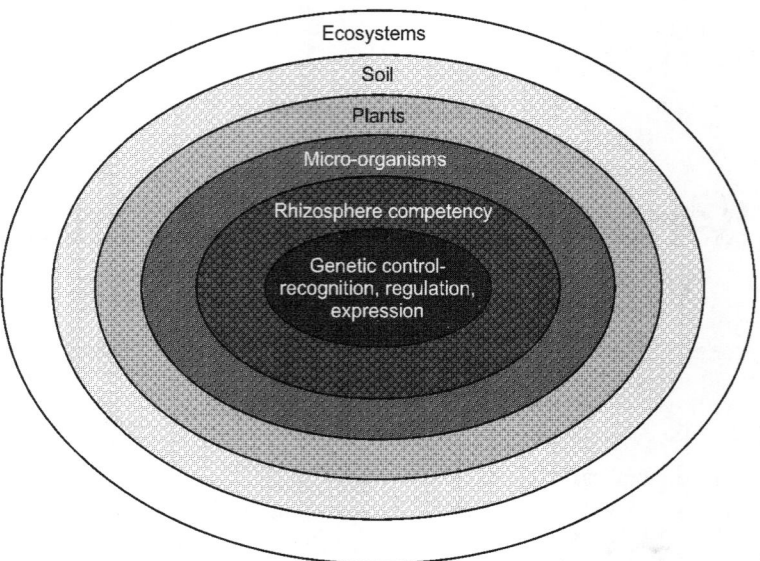

Fig. 29.1. The soil–plant–micro-organism system. Here, the influence of the environment on both soil and plant controls gene expression in the rhizosphere.

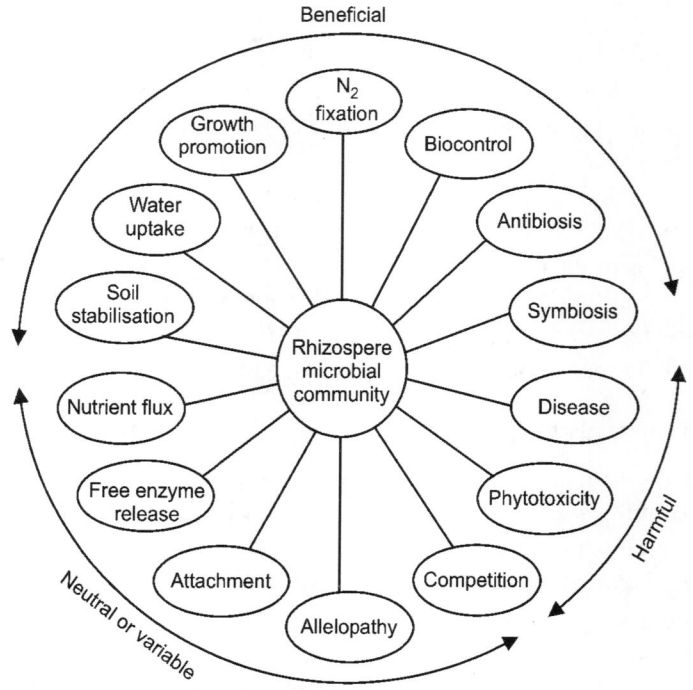

The beneficial, harmful, and neutral or variable effects of the rhizosphere microbial community on plant growth

Fig. 29.2. Potential influences of the rhizosphere microbial community on plant growth.

The rhizosphere effect is caused by the release of organic and inorganic compounds from the plant roots. In particular, the rhizosphere is influenced by living root border cells that are released by the root.

Because of these releases and because of the influence of the plant roots themselves, rhizosphere soil is thought to be quite different from non-rhizosphere or bulk soil. However, despite hundreds of different studies, very little can actually be said with certainty about rhizosphere soil. Part of the problem lies in the methods used to sample rhizosphere soil. Despite even the most sophisticated of analyses performed on 'rhizosphere soil', most studies are restricted due to the historically crude method of obtaining such a sample. Typically, this has involved extracting a plant from soil and shaking the roots until most of the soil particles fall off. To this day, this is still the method of choice. Other problems include the subtle interactions between specific plants and specific soils in specific environments. Thus, an infinite array of different 'rhizosphere environments' is possible. This is evidenced by a perusal of rhizosphere literature, which tends to be confusing and inconsistent. This is not an indictment of the scientists who have worked on the rhizosphere, rather it illustrates the difficulties of studying the complex rhizosphere ecosystem. As an example, rhizosphere soil can have pH values one unit higher or lower than those of the bulk soil, depending on nitrogen nutrition and other factors. It is, however, known that rhizosphere soil tends to be drier than bulk soil because of plant transpiration, and it also contains greater concentrations of organics because of plant-released compounds.

ORGANIC COMPOUNDS RELEASED BY PLANTS

In natural vegetation systems, plant roots are in intimate contact with soil particles. Soil exists as a discontinuous environment with a matrix of organic and inorganic constituents combined in diverse conditions and is, therefore, a unique environment for many micro-organisms. These organisms include viruses, bacteria, actinomycetes, fungi, algae, protozoa, and nematodes. Aerobic and anaerobic microsites exist in close proximity, allowing organic and inorganic substrates to be metabolised by organisms with different modes of nutrition. These conditions permit billions of organisms to coexist in soil. Populations vary with the soil, environment, and method of analysis, but reasonable values for normal soils are shown in Table 29.1.

Roots are, therefore, surrounded by organisms and exist as part of the soil-plant-micro-organism system, which can be termed the rhizosphere. The complexity of the rhizosphere is shown in Fig. 29.1, where the two inner circles depict, respectively, the early events necessary for colonisation. In addition, the figure shows subsequent factors that contribute to rhizosphere competence and the ability to metabolise and reproduce in the rhizosphere in the presence of other organisms. The components of this system (micro-organisms, plant, and soil) interact with each other, which distinguishes the rhizosphere from the bulk soil. The activity of root micro-organisms is affected by soil environmental factors or by environmental factors operating indirectly through the plant. Root micro-organisms can affect the plant and plant nutrient uptake, directly by colonising the root and modifying its structure or indirectly by modifying the soil environment around the root. The spokes of the wheel in Fig. 29.2 depict not only the significant processes that affect plant growth but also those that have potential for enhancement through improved cultural practices, genetic manipulation, and modelling.

Substrates released from roots have many origins and were originally classified by Rovira as:
1. Exudates: Compounds of low molecular weight that leak non-metabolically from intact plant cells.
2. Secretions: Compounds metabolically released from active plant cells.
3. Lysates: Compounds released by the autolysis of older cells.

4. Plant mucilages: Polysaccharides from the root cap, root cap cells, primary cell wall, and other cells.
5. Mucigel: Gelatinous material of plant and microbial origin.

Table 29.1. Numbers of micro-organisms in the rhizosphere (R) of wheat (*Triticum aestivum*) and Non-rhizosphere Soil (S) and their resultant *R/S* ratio.

Micro-organisms	Rhizosphere	Non-rhizosphere	R/S ratio
	CFU[a] g soil		
Bacteria	120×10^7	5×10^7	24.0
Fungi	12×10^5	1×10^5	12.0
Protozoa	2.4×10^3	1×10^3	2.4
Ammonifiers	500×10^6	4×10^6	125.0
Denitrifiers	1260×10^5	1×10^5	1260.0

[a] CFU, colony-forming units.

More recently, it has been demonstrated that living root border cells affect the rhizosphere ecology more than any other plant source of carbon substrate. The terms 'exudates' and 'exudation' were sometimes used collectively and perhaps incorrectly to include all of the organic compounds released from roots.

Exudates, Secretions, and Lysates

The release of soluble organic compounds (loosely known as root exudates) is also responsible for some of the rhizosphere effect. Loss of substrates from roots can change the pH, the structure of rhizosphere soil, the availability of inorganic nutrients, and can induce toxic or stimulatory effects on soil micro-organisms. The major mechanisms are leakage and secretion. Leakage involves simple diffusion of compounds because of the higher concentrations of compounds within the root as compared with the soil. Secretion can occur against concentration gradients but requires the expenditure of metabolic energy. Polysaccharides in particular are susceptible to secretion. Almost any plant metabolite has the potential to be exuded including carbohydrates, amino acids, organic acids and lipids, growth factors, enzymes, and miscellaneous compounds. Of these, the carbohydrates and the amino acids, which also represent a source of nitrogen, are particularly important as substrates.

Organic acids and lipids reduce the pH of the rhizosphere and also have a role in the chelation of metals. Growth factors, including vitamins, and enzymes stimulate microbial activity and low growth of organisms with complex heterotrophic requirements. Miscellaneous compounds including volatiles can physiologically stimulate or inhibit organisms. When viewed collectively, it is apparent that the rhizosphere is a unique ecosystem in soil that provides a constant supply of substrate and growth factors for organisms.

Root Border Cells and Mucigel

As the root cap extends through soil, viable root border cells and some nonviable material (sloughed cells) are released into the soil. The role of border cells in controlling the rhizosphere ecosystem is discussed later in this chapter. The amount of sloughed material can be considerable. In solution culture, peanut plants released 0.15 per cent of the plant's carbon, nitrogen, and hydrogen per week. One would predict that much more material would be lost in soil because of its abrasive nature. From the root tip to the root hair zone, the root is frequently covered with a layer composed of sloughed root border cells and polysaccharides of plant and microbial origin, which is termed mucigel. These plant products are excellent substrates for microbial growth, in particular soil bacteria, which are extremely competitive at metabolising simple sugars. Thus, mucigel is in intimate contact with bacteria that consume the material, as well as bacteria that contribute bacterial polysaccharides to the mucigel. The amount of mucigel on a particular root depends on the net production and consumption of the material, so that in some instances parts of the root may have no mucigel. Mucigel may protect the root rip from injury and desiccation as well as play a role in nutrient uptake through its pH-dependent cation-exchange capacity (COO⁻ groups).

Factors Affecting the Release of Compounds

Major factors affecting release of organic compounds include plant species and cultivar, age and stage of plant development, light intensity and temperature, soil factors, plant nutrient, plant injury, and soil micro-organisms. Because so many factors affect the release of compounds, generalisations are difficult, including the actual rate of exudation. However, it is known that plant genes control the release of root border cells.

RHIZOSPHERE POPULATIONS

Rhizosphere populations are influenced by many plant, soil, and environmental factors. Crop plant roots tend to have greater rhizosphere populations than tree roots. Different cultivars of the same plant species may have different rhizosphere populations.

Soils directly affect the growth and vigour of plants and therefore influence shoot growth, photosynthesis, and the amount of exudation into the rhizosphere. The concentration of oxygen in the rhizosphere is usually lower than in nonrhizosphere soil as a result of its utilisation by large rhizosphere populations. Hence, in heavy-textured soils oxygen may become limiting, resulting in reduced rhizosphere populations compared with coarser-textured soils. The physical environment around the plant and its roots also affects rhizosphere populations by affecting the amount of organic material released into the soil. Factors such as light, moisture, and temperature can all cause changes in plant metabolism and the rhizosphere effect. In summary, rhizosphere populations are dependent on many diverse interacting factors, and care must be taken when interpreting different studies.

Overall, a vast number of different kinds of micro-organisms are found in the rhizosphere, and their numbers generally decrease from the rhizoplane outward toward bulk soil. The rhizosphere effect is often evaluated in terms of R/S ratios, where R = the number of microbes in the rhizosphere and S = the number of similar microbes in bulk soil. Thus, the greater the R/S ratio, the more pronounced the rhizosphere effect (Table 29.1).

Microflora

Bacteria including actinomycetes are the most numerous inhabitants of the rhizosphere, and R/S ratios can typically be 20:1 (Table 29.1). Pseudomonads and other gram-negative bacteria are especially competitive in the rhizosphere. Typical actinomycete R/S ratios are 10:1. Overall R/S ratios are useful in delineating the rhizosphere effect, but they are only estimates and vary with different crop plants and different soil environments. The mechanisms that allow rhizosphere competence are discussed later.

Fungal plate counts are generally less than bacterial counts and in any case are often biased toward sporeforming species. However, fungal inhabitants of the rhizosphere are prevalent and can be extremely important because they can be beneficial, as in the case of mycorrhizal fungi, or harmful when they are pathogenic to plants.

Microfauna

Most research on the microfauna has centred on protozoa, nematodes, and the microarthropods. Soil protozoa are mostly rhizopods and flagellates, with smaller numbers of ciliates. Protozoan populations tend to mimic bacterial populations, because bacteria are their major food supply. Thus, the rhizosphere should contain large populations of protozoa. Rouatt reported R/S ratios for protozoa of 2:1 in wheat rhizospheres (Table 29.1). Darbyshire reported even higher protozoan populations in ryegrass rhizospheres. Many soil nematodes, including *Heterodera* and *Fylenchus*, are plant parasites that feed on underground roots. Little research has been conducted on nematodes in the rhizosphere, but populations have been reported to be higher in rhizosphere than in nonrhizosphere soil. The Acari (mites) and Collembola (springtails) are important members of soil microarthropods. Mites are predatory on nematodes, and the rhizosphere would be expected to be a favourable habitat for the Acari, but studies in the rhizosphere have been limited in scope. Springtails have been shown to be abundant in cotton rhizospheres with *R/S* ratios of 4:1 in a sandy loam soil, but the reasons for their attraction to roots are not clear.

BENEFICIAL ROOT-MICROBIAL INTERACTIONS

The fact that there are many beneficial root-microbial interactions can easily be demonstrated by growing plants in the laboratory in sterilised soil and comparing plant growth with that achieved in nonsterilised soil. Inevitably, growth in the nonsterilised system is superior to that in the sterile system. There are many ways that micro-organisms can beneficially influence plant growth, but two of the predominant mechanisms involve the macroelements nitrogen and phosphorus. The prokaryotic bacteria can enhance plant nitrogen uptake through the process of biological nitrogen fixation, whereas eukaryotic fungienhance plant phosphorus uptake through mycorrhizal associations.

Biological Dinitrogen Fixation

Nitrogen is critical for plant growth and is often applied to plants as organic or inorganic fertilisers. However, organic forms of nitrogen can be converted to nitrate by the microbial processes of ammonification and nitrification and subsequently denitrified to an inorganic form, nitrogen gas or nitrous oxide. Clearly, some microbial process that converts nitrogen gas back to ammonia must be present, or all nitrogen would ultimately end up as nitrogen gas. This process, biological dinitrogen fixation, is mediated only by prokaryotes, including bacteria, cyanobacteria, and the actinomycete *Frankia*. These nitrogen-fixing organisms can exist as independent free-living organisms or as part of complex interactions with other microbes, plants, and animals. Organisms that can utilise atmospheric nitrogen gas as their sole source of nitrogen for growth are known as diazotrophs.

In terms of benefits to agriculture, the following major systems can be delineated:

1. Symbiotic relationships.

Symbiont	*Host*
Rhizobia (bacterium)	Legumes
Frankia (actinomycete)	Nonlegume
Anabaena (cyanobacterium)	Azolla (fern)

2. Associative symbiotic relationships involving free-living diazotrophs.

Microbe	Benefitting crop
Acetobacter	Sugarcane
Azotobacter	Tropical grasses

The preceding list is by no means inclusive of all of the possible associations between microbes and plants, but in terms of importance to agriculture, these are the major players. They are also the systems that have been most easily manipulated by human activity, including that of environmental microbiologists. Also note that there are about 100 true diazotrophs that can exist free living and that can contribute fixed nitrogen into the rhizosphere and other environments. Each of these nitrogen-fixing associations will be examined, with emphasis on the rhizobia-legume symbiosis, because it is the best studied system and is critical to agricultural crop production. First, however, we will examine the enzymatic process of biological nitrogen fixation, because many characteristics are common to all nitrogen-fixing associations.

PATHOGENIC MICROBES IN AGRICULTURE

The major source of detrimental micro-organisms that affect plant growth is, of course, plant pathogens. The importance of these pathogens has led to the emergence of the discipline known as plant pathology. This can be defined as the study of the causes, mechanisms, environmental factors, and control of diseases of plant caused by micro-organisms, many of which are soilborne. Clearly, a massive amount of literature exists on this subject including many excellent reference books such as 'Plant Pathology'. Therefore, the intent here is to present an overview of the important plant pathogens, which include viruses, bacteria, fungi, protozoa, and nematodes. The focus will be on *Agrobacterium tumefaciens*, a bacterial pathogen that has been extensively studied at the molecular level and exhibits a unique interaction between a prokaryotic microbe and the eukaryotic higher plants. Finally, important aspects of biological control of pathogens will be presented.

Plant Disease Caused by Fungi

Most plant pathogenic fungi have a filamentous structure known as a mycelium with individual branches known as hyphae. Almost all plant pathogenic fungi spend some of their time on the host plant and the remainder of their lives in soil or in plant debris within the soil. Thus, the survival and effects of the pathogen are controlled mainly by soil environmental factors including biotic (microbial) and abiotic factors such as temperature and moisture. The scope and diversity of plant fungal pathogens are extensive, and these organisms are responsible for billions of dollars of crop damage worldwide in all countries where agriculture is practiced. Some examples of important plant fungal pathogens and their plant host are shown in Table 29.2. Almost all commercial crops are subject to plant fungal attacks, which can result in diseases of seeds, roots, stems, leaves, fruit, or grain kernels.

Table 29.2. Examples of important fungal plant pathogens.

Fungal pathogen	Plant host	Disease or symptom
Pythium	Almost all plants	Seed or root rot (damping off)
Phytophthora	Vegetables, fruit trees	Root rot
Plasmopara	Grapes	Downy mildew
Rhizopus	Fruits and vegetables	Soft rot of fruit or vegetables

(Contd ...)

Fungal pathogen	Plant host	Disease or symptom
Podosphaera	Fruit trees	Powdery mildewy growth
Alternaria	Vegetable	Leaf blight
Fusarium	Vegetable and field crop	Leaf wilting
Puccinia	Cereals and grains	Leaf and stem rusts
Ustilago	Cereals and grains	Corn smut
Rhizoctonia	Herbaceous plants	Root and stem rot
Armillaria	Fruit trees	Root rot

Diseases Caused by Bacteria

Most plant pathogenic bacteria are rod shaped, with the exception of *streptomyces*, which is a filamentous actinomycete. Almost all plant pathogenic bacteria occur within the host plant as parasites or on plant leaves as epiphytes. They also exist within plant debris or in soil as saprophytes. Some bacterial pathogens such as Erwinia predominate in the plant host, whereas others such as *Pseudomonas solanacearum* predominate in soil. However, most bacterial plant pathogens enter soil via the host tissue, and bacterial numbers often remain high only as long as the host tissue is still present within the soil. Important plant bacterial pathogens are shown in Table 29.3. Fruits and vegetables are particularly prone to pathogenic bacterial attack. Of all of the bacterial plant pathogens, *Agrobacterium tumefaciens* has perhaps been the best studied organism because of its mode of attack, which involves nucleic acids.

Table 29.3. Examples of important bacterial plant pathogens.

Bacterial pathogen	Plant host	Disease or symptom
Pseudomonas syringae	Tobacco, vegetables	Leaf spots
Pseudomonas fluorescens	Potatoes	Soft rot
Xanthomonas campestris	Cereals, fruits	Several leaf spots (blights)
Xanthomonas campestris	Crucifers, e.g. cabbage	Black rot
Erwinia tracheiphila	Cucumbers, melons	Vascular wilts
Erwinia carotovora	Fruits and vegetables	Soft rot
Agrobacterium tumefaciens	Fruit trees	Crown gall
Streptomyces scabies	Potatoes	Potato scab
Xylella fastidiosa	Grape	Pierce's disease

Crown gall disease–agrobacterium tumefaciens

Crown gall disease is caused by the soilborne pathogen *Agrobacterium tumefaciens*. The disease manifests itself in uncontrolled cell division in the host plant, which results in the formation of a tumor or gall typically around the crown of the root. The disease is induced in a variety of dicotyledonous plants, particularly stone fruits, roses, and grapes. The majority of the bacterial genes necessary to induce the disease are plasmid borne.

Plant Diseases Caused by Viruses

A variety of viruses can infect plants causing disease. Typically, plant viruses enter cells only through wounds made mechanically or perhaps via an infected pollen grain that is deposited in an ovule. Viruses which can contain RNA or DNA typically result in leaf lesions. Although viral attacks can result in catastrophic crop losses, most virus diseases occur on crops year after year and cause small to moderate losses. Typical viral plant pathogens are shown in Table 29.4. Because of the small size of viral pathogens, their presence has been indicated primarily by the symptoms exhibited by the plant host. More recently, viruses are being identified by new molecular techniques including polymerase chain reaction (PCR) and reverse transcriptase (RT)-PCR analyses.

Table 29.4. Examples of important viral plant pathogens.

Viral pathogen	Type of nucleic acid	Plant host	Disease or symptom
Tobamovirus (tobacco mosaic virus)	(ssRNA)[a]	Tobacco	Leaf chlorosis and distortion
Furovirus (wheat mosaic virus)	(ssRNA)	Wheat	Dwarf and mottled leaves
Potexvirus	(ssRNA)	Potato	Stunted plants
Potyvirus	(ssRNA)	Beans	Mottled, chlorotic leaves
Phytoreovirus	(dsRNA)	Rice	Galls or tumors
Caulimovirus	(dsDNA)	Cauliflower	General poor plant growth

[a] ss, single-stranded; ds, double-stranded.

Soil Biological Control of Plant Diseases

A relatively new approach to the control of plant pathogens is that of using micro-organisms instead of chemicals. This can be done by either introducing microbes into a particular soil or manipulating the indigenous microflora. In either case, the objective is to reduce the numbers and activity of specific pathogens. Biological control can occur within the plant root itself, within the rhizosphere, or in the bulk soil in the vicinity of the root.

Antagonists are biological agents that reduce numbers or activities of pathogens through antibiosis, competition, or hyperparasitism. Antibiosis occurs when the pathogen is inhibited or lethally affected by metabolic products of the antagonist such as enzymes, acidic agents, or antibiotics. In contrast, competition can be for nutrients, growth factors, oxygen, or occasionally space. Hyperparasitism is due to the invasion of the parasite by the secretion of lytic enzymes. All of these mechanisms can result in decreased activities of pathogens, but because biological control acts by altering the biological equilibrium of the soil community, such control may take longer to act than chemical methods, and the efficacy of such methods may be more difficult to predict. On the other hand, when successful, biological control can last longer than chemical control. Finally, note that biological control methods are often most successful when used with integrated pest management strategies.

Pathogen-suppressive soils are soils in which a particular pathogen does not establish itself or persist, or if it does establish itself, it causes no damage or the disease becomes less severe with time. The main purpose of biological control is to maximise soil suppressiveness.

Maintenance of suppressive soils

Biological control through the manipulation of resident antagonists can be controlled by crop or soil management practices.

Soil management practices

Soil management practices that enhance suppressiveness include crop rotations and soil tillage, both of which reduce potential pathogen populations by reducing the incidence of specific crop residues that may harbour pathogens. Other practices include incorporation of organic amendments into soil, which apparently enhances the population of antagonists in the soil relative to pathogen populations.

Introduced biological control agents

Biological control agents produce a number of chemical metabolites that can participate in controlling plant disease. Microbes that produce these metabolites can be inoculated into the rhizosphere as just discussed, or the metabolites themselves can be added. Perhaps the best known example of a metabolite used as a biological control agent is the crystal toxin produced by *Bacillus thuringiensis*, a naturally-occurring soil microbe. *B. thuringiensis* produces a paracrystalline body during its growth that is toxic to specific groups of insects.

Function of root border cells as biocontrol agents

Plant roots must move into new soil areas in order to obtain nutrients, and to do this, new root tissue must be generated by the root meristem. Evidence has suggested that the direction of root growth, and in fact the rhizosphere ecosystem, is controlled by root border cells. Under controlled conditions, border cells and their associated products can contribute up to 98 per cent of carbon released as root exudates.

Biological control of crown gall disease

One very successful example of biological control worthy of mention is the control of crown gall disease by *Agrobacterium tumefaciens* strain K84. This commercially important biological control method was developed in Australia. Inoculation of planting stock with the nonpathogenic strain K84 is often successful in preventing the disease. The mechanism of control involves the production of a bacteriocin that inhibits closely related bacteria — in this case other virulent strains of *A. tumefaciens*. The bacteriocin is known as agrocin 84 and is a fraudulent adenine nucleotide that inhibits DNA synthesis. It is taken up by *Agrobacterium* strains that synthesise nopaline or agrocinopine. Part of the agrocin molecule is similar to the structure of agrocinopine; thus strains that contain agrocinopine permease are capable of taking up agrocin 84. It is believed that bacterial colonisation of the root surface may also be involved in the control mechanism. However, virulent bacterial mutant strains that do not take up agrocinopine are not sensitive to the bacteriocin.

Chapter 30

Airborne *Mycobacterium* Spp.

INTRODUCTION

The only genus in the family Mycobacteriaceae is *Mycobacterium*. The mycobacteria are slightly curved or straight bacilli, 0.2 to 0.7 μm in diameter by 1.0 to 10 μm in length, and are sometimes branching. Mycobacteria have cell walls with a high lipid content that includes waxes having characteristic mycolic acids with long, branched chains. The minimal standards for including a species in the genus *Mycobacterium* are (i) acid-alcohol fastness, (ii) presence of mycolic acids containing 60 to 90 carbon atoms which are cleaved to C_{22} to C_{26} fatty acid methyl esters by pyrolysis, and (iii) a guanineplus-cytosine content of the DNA of 61 to 71 mol per cent.

The genus *Mycobacterium* includes obligate parasites, saprophytes, and opportunistic pathogens. Most species are free living in soil and water. However, the more publicised mycobacteria, *Mycobacterium tuberculosis* complex and *M. leprae*, amplify in the tissues of humans and other warmblooded animals.

Mycobacteria can be divided into two main groups based on growth rate. The slowly growing species require more than 7 days to form visible colonies on solid medium, while rapidly growing species require less than 7 days. In general, the slowly growing species are often pathogenic for humans or animals, while rapidly growing species are usually considered nonpathogenic for humans, although important exceptions exist. Mycobacterial species, complexes, and groups are listed in Table 30.1.

In general, environmental specimens are collected from the air, soil, and water. Clinical specimens include human sputum, urine, blood, fecs, cerebrospinal fluid, tissue biopsy specimens (e.g. liver, bonemarrow, and lymph nodes), or any suspected source of infection. In addition, specimens may be isolated from contaminated equipment, dressings, or other materials. Clinical specimens may be aerosolised, thus becoming airborne environmental specimens. Whether samples are taken from an environmental or clinical matrix, a mixed bacterial, and possibly fungal, flora should be expected.

BACTERIOLOGY, EPIDEMIOLOGY AND PATHOGENESIS OF MYCOBACTERIAL INFECTIONS

M. tuberculosis Complex

The *M. tuberculosis* complex includes the species *M. bovis*, *M. microti*, *M. africanum*, and *M. tuberculosis*. *M. bovis* causes tuberculosis in cattle, humans, and other primates. *M. bovis* bacille Calmette-Guerin, usually referred to as *M. bovis* BCG or BCG, is an attenuated strain of *M. bovis*. This strain was attenuated by serial passage. This vaccine strain was found to be a avirulent to guinea pigs, rabbits, and horses. It was first given to a human in 1921. BCG vaccines are the most extensively used human vaccines worldwide, with more than 3 billion doses administered. Because of the history of conflicting results from clinical trials, the efficacy of BCG remain unknown. *M. microti* causes naturally acquired generalised tuberculosis in warm-blooded animals. *M. africanum* is a cause of human tuberculosis in tropical Africa.

Table 30.1. Key biochemical tests for identification of mycobacteria.

Descriptive term	Species	Complex or group
Tuberculosis	M. tuberculosis	M. tuberculosis complex
	M. bovis	
	M. africanum	
	M. microti	
Hansen's disease	M. leprae	
Nonchromogens	M. avium	M. avium complex
	M. intracellulare	
	M. xenopi	
	M. haemophilum	
	M. malmoense	
	M. simoidei	
	M. genavense	
	M. celatum	
	M. ulcerans	
	M. terrae	M. terrae complex
	M. triviale	
	M. nonchromogenicum	
	M. gastri	
Photochromogens	M. kansaii	
	M. marinum	
	M. simiae	
	M. asiaticum	
Scotochromogens	M. gordonae	
	M. scrofulaceum	
	M. szulgai	
	M. flavescens	
Rapid growers	M. fortuitum	M. fortuitum group
	M. chelonae	M. chelonae group
	M. smegmatis	
	M. phlei	
	M. vaccae	

Tuberculosis was present in paleolithic times but remained an unimportant disease for humans until the necessary environmental and social changes occurred in feudal Europe, thus releasing the 'great white plague'. Koch first described the tubercle bacillus, *M. tuberculosis*, and demonstrated it to be the cause of tuberculosis.

Tuberculosis is spread almost exclusively by airborne residues of tiny droplets produced by infectious individuals while coughing, speaking, laughing, and sneezing. Using high-speed photography, Jennison estimated more than 20,000 droplets released during a sneeze and only a few dozen to a few hundred expelled while coughing. More recently, Loudon and Roberts collected aerosolised particles from subjects coughing and talking. Subjects expelled an average of 470 particles per cough (50 per cent were 5 to 10 µm or smaller) and 1800 particles while counting from 0 to 100 (50 per cent were 30 µm or smaller).

Nottrebart described a situation in which a number of hospital employees were awarded compensation after developing active disease. Exposure to autopsy material and machines used to suction secretions from tuberculosis patients, the adjusting of tuberculosis patient's beds, and the application of an ointment to a tuberculosis ulcer were implicated. An additional incident of nosocomial transmission of tuberculosis occurred after exposure to a hospitalised patient who underwent surgical incision, drainage, and syringe irrigation of an abscess with debridement of the surrounding necrotic tissue. The patient spent 13 days in a positive pressure patient room and 4 days in the intensive care unit. Four of five (80 per cent) surgical suite employees, 28 of 33 (85 per cent) general medical floor employees and 6 of 20 (35 per cent) intensive care unit employees had skin test conversions. More recently, transmission of *M. tuberculosis* was transmitted from an infectious crew member to other crew members on an aircraft. The risk of infection of frequent flyers also increased with increasing hours of exposure to the index case.

Despite the efforts of numerous health organisations worldwide, the eradication of tuberculosis has never been imminent. Fox estimates that nearly half of the world's population is infected with *M. tuberculosis*, with approximately 8 million new cases and 3 million deaths attributable to tuberculosis yearly. After decades of decline, tuberculosis has risen in recent years, even in the Unites States, where up to 10 million individuals believed to be infected.

Mycobacteria other than *M. tuberculosis*

Modern leprosy (Hansen's disease) is caused by *M. leprae*, first identified by Gerhard Armauer Hansen in Norway in 1874. There are two distinct types of Hansen's disease, lepromatous (skin) and tuberculoid (nerve). Estimates by the World Health Organisation put the global prevalence of Hansen's disease at 10 to 12 million. An individual with untreated lepromatous Hansen's disease may discharge up to 8×10^8 acid-fast bacilli in a single nose blow. Naturally occurring *M. leprae* infections of the nine-banded armadillo have been documented in Texas and Louisiana. The mode of transmission has never been validated; some researchers suspect that it is transmitted through air, while others say that only direct contact will result in infection.

M. avium complex an *M. kansaii* are the most common pulmonary pathogens. *M. avium* complex organisms are ubiquitous and have been isolated from water, soil, and air. One of the most common types of *M. avium* complex isolated from humans and animals is serotype 8. In general, these bacteria are of low pathogenicity and frequently infect immune-competent humans without causing disease. This lack of disease complicates the interpretation of air sampling results. In an aquatic study, airborne culturable mycobacteria in natural aerosols from the James River were collected by using six-stage Andersen samplers. *M. avium*, *M. intracellulare*, and *M. scrofulaceum* were detected. In general, the collected mycobacterial aerosols were associated with particles larger than 5 µm. In a study of acid, brown-water swamps of the southeastern United States, *M. avium* complex organism were found in higher concentrations than inland rivers. Kirschner concluded that swamp soils, waters, and aerosols may have played a role in the epidemiology of infections in the swamp area.

M. kansaii and *M. xenopi* have been found in hot and cold water systems. Several other species of mycobacteria are known to be pathogenic; however, their routes of transmission are either unknown or not suspected to be via aerosols.

SAMPLING CONSIDERATIONS

Bioaerosol sampling may be performed to verify airborne transmission of mycobacteria during epidemiological investigations and research studies to evaluate engineering controls. If air sampling is deemed appropriate, the user must keep in mind that false-negative results are quite possible and should interpret all negative findings with caution. False-positive results due to poor sampling and/or analytical techniques are also possible.

Investigators should use appropriate personal protective equipment and practice good personal hygiene when conducting indoor environmental quality, infectious disease outbreak, and agricultural health investigations that have resulted in medically diagnosed symptoms. Such equipment may include respiratory protection gear to prevent inhalation of contaminants and micro-organism-resistant clothing to prevent the transmission to investigators.

SAMPLING AND ANALYTICAL METHODS

Sampling for Airborne Mycobacteria

General guidelines for aerobiological sampling are discussed. Collection methods for airborne mycobacteria include culturable bioaerosol sampling and nonculturable and nonviable bioaerosol sampling. Because free bacteria (singel cells) can be mycobacterial aerosols of interest in some environmental investigations, the sampling method must collect these droplet nuclei (5 µm). Often, however, the bioaerosols will be clumps of micro-organisms or micro-organisms attached to another particle such as a skin scale or sputum droplet. When one is sampling a clean room or other environment with extremely low levels of culturable bioaerosols, the lower limit of 30 CFU for statistical comparisons may not be achievable. In such a situation a qualitative representation must be used without being able to accommodate statistical validity. When sampling for nonviable micro-organisms or when culturability is not of concern, collection efficiency is the overriding concern.

Analytical Methods

Equally important, the analytical method chosen must be compatible with the sampling method and capable of detecting the organism of interest. The physiology of mycobacteria has been reviewed by Ratledge and by Barksdale and Kim.

Microscopic assay

Simple and differential staining may be performed; however, members of the genus *Mycobacterium* do not stain readily. The Ziehl-Neelsen (acid-alcohol) staining procedure was first proposed in 1883 and, with minor modifications, is used today to classify organisms as acid fast. Fluorescence microscopy for the direct count of micro-organisms has been described in a number of studies. In the fluorescence technique, the organisms are stained with auramine or some other fluorescent dye. The preparation is then examined within 2 hours if not protected from light. In comparisons of clinical techniques, fluorescence microscopy techniques for *M. tuberculosis* gave results similar to or slightly poorer than those obtained by using PCR.

Culture

The ideal culture medium for mycobacteria should promote rapid growth of small numbers and be simple to prepare from readily available ingredients. Many media have been recommended for the cultivation of *M. tuberculosis*; however, there is no general agreement as to which of them is best. Table 30.2 lists some suggested media for the cultivation of mycobacteria, and several references are available. *M. leprae* has not been shown to grow in solid or liquid media. However, *M. leprae* grows naturally in armadillos and humans. Large quantities of *M. leprae* (10^6 organisms) may be cultivated in the footpads of BALB/c (nude) mice.

Table 30.2. Suggested media for the cultivation of mycobacteria.

Solid
Agar based
Middlebrook 7H10
Middlebrook 7H11
Mitchison's selective 7H11
Dubos oleic acid albumin
Egg based
Wallenstein
Lowenstein-Jensen with RNA
Lowenstein-Jensen with pyruvic acid
Liquid
BACTEC 12B medium
Middlebrook 7H9 broth
Dubos albumin

Inoculated agar plates are incubated at the appropriate temperature for times ranging from days for rapidly growing mycobacteria to weeks for multidrug-resistant *M. tuberculosis*. Sometimes colonies of rapidly growing mycobacteria may take weeks to appear on initial isolation and reveal their shorter generation times only on subculture. This delay may occur as a result of bacterial injury (structural or metabolic) or the adaptation process to permit the organism to multiply on artificial medium. Petri dishes containing agar media may be conveniently incubated in gas-permeable bags that maintain moisture during lengthy incubation. An atmosphere enriched with 5 to 10 per cent CO_2 enhances the growth of many mycobacteria. However CO_2 has not been shown conclusively to enhance the growth of mycobacteria on egg media. Most mycobacteria multiply at 35 to 37°C. A small group of mycobacteria (e.g. *M. marinum* and *M. ulcerans*), optimally grow at 30°C. A distinctive property of culturable mycobacteria includes their growth rates at 24, 31, 37, and 45°C. The incubator should also be humidified to minimise desiccation of the medium. If a liquid medium is used, the tubes should be well aerated but desiccation of the medium must be prevented. Tubes less than half filled have sufficient oxygen to support adequate growth. Laboratory medium banks and field medium blanks to detect accidental contamination must be handled in the same manner as samples.

When sampling for mycobacteria with a modified six-stage sample (stages 1 and 6 only) and Middlebrook 7H10, Macher lost 33 of 51 samples as a result of fungal overgrowth within the first three weeks after collection. Rapidly growing mycobacteria were recovered on three occasions at one location.

If impinger samplers are used, the liquid collection medium may be filtered, directly inoculated on solid medium, or incubated directly if it will support the growth of the target mycobacterium. By using direct filtration techniques and the direct inoculation technique, the airborne concentrations may be estimated.

Distinctive properties of culturable mycobacteria include accumulation of nicacin, susceptibility to thiopene-2-carboxylate hydrazide, nitrate reduction, production of catalase (37° and 68°C), Tween 80 hydrolyse, tellurite reduction, tolerance to 5 per cent NaCl, iron uptake arylsulphatase breakdown of phenolphthalein disulphate, growth on MacConkey agar, and ability to process pyrazinamidase; however, these individual assays are time-consuming. The BACTEC system has reduced the isolation time for mycobacteria to approximately 10 days.

This system is based on the principle that mycobacteria multiply in Middlebrook 7H12 broth and metabolise ^{14}C-containing palmitic acid. $^{14}CO_2$ is released into the headspace of the vial. The instrument samples the headspace three times weekly for 6 weeks and measures the radioactivity. *M. tuberculosis*, *M. bovis*, and mycobacteria other than *M. tuberculosis* can be diffenentiated by using *p*-nitro-α-acetylamino-β-hydroxypropiophenone or thiophene-2-carboxylate hydrazide.

Mycolic acids (cellular fatty acids) of mycobacteria are structural in nature, occurring in the cell membrane or cell wall of all bacteria. When the bacteria are grown under standardised growth condition, the mycolic acid profiles are reproducible within a genus, down to the subspecies or strain level in some micro-organisms. The Microbial Identification System provides a chromatographic technique and software libraries capable of identifying various micro-organisms on the basis of mycolic acid composition. This chromatographic technique is also known and gas chromatography fatty acid methyl ester analysis, and a database contain the analysis libraries for culturable mycobacteria is available.

Immunoassay

Many immunoassays are now readily available from commercial sources, permitting laboratories to rapidly develop in-house immunochemical analytical capability without lengthy antibody preparation. Some of the more widely used formats are radioimmunoassays, fluorescence immunoassays, and enzyme immunoassays. Witerbsky and Conville summarised recent developments in diagnostic mycobacteriology, including radioimmunoassays, with emphasis on laboratory capabilities relevant to the more rapid and accurate detection and identification of mycobacterial pathogens. Enzyme-linked immunosorbent assays (ELISAs) are now highly automated, and efforts are underway to commercially develop well-standardised kits containing appropriates controls and materials. ELISAs for *M. tuberculosis* and *M. leprae* have been used in laboratories for identification.

Molecular techniques

Diagnostic mycobacteriology is rapidly adapting molecular biology techniques in addition to classical identification methods to identify organism. Genus-specific and species-specific DNA probes have been developed for mycobacteria. This approach has been successfully used to detect various organism, including *M. tuberculosis*. Mycobacteria other than *M. tuberculosis* have been successfully identified, and Richter developed a three-primer PCR to detect *M. leprae*. Membrane filters may be used to remove micro-organisms from liquids and analysed by PCR. If air samples are collected with a liquid impinger, then the collection fluid may be filtered and analysed by PCR for a specific micro-organism. Air samples were collected by Sawyer on filters and analysed by PCR for varicella-zoster virus.

Finally, Schafer proposed sampling for airborne *M. tuberculosis* by using a method similar to that used by Sawyer. Restriction fragment length polymorphism analysis is widely used to distinguish genetic changes within a species. Such analysis has been used in epidemiological studies to track the transmission of *M. tuberculosis* in New York City. *M. leprae* isolates from different sources were found to have identical sequence of the spacer region between two RNA genes.

To sum-up Riley estimated the indoor M tuberculosis droplet nuclei concentration in the air to be as low as 1 infectious unit per 310 m^3 (1 infectious unit per 12,000 ft^3) and stated that the infectious dose is less than 10. Because indoor airborne mycobacterial concentrations are thought to be very low, bioaerosol monitoring of *Mycobacterium* spp. is problematic. Available methods include the measurement of viable (culturable and nonculturable) and nonviable micro-organisms in both indoor (e.g. industrial, office, or residential) and outdoor (e.g. agricultural and general air quality) environments. Investigators must ensure that the purpose of sampling is clear, the organism of interest is identified, and the appropriate analytical technique is chosen prior to air sampling. Air sampling may be appropriate to include in epidemiological investigations, in research studies, or in situations indicated by a physician.

SECTION VIII

Biotransformation and Biodegradation

31. **Biodegradation: A Review**   **485**

32. **Biodegradation of Phenol**   **491**

33. **Biodegradation and Bioremediation of Halogenated Organic
    Compounds**   **503**

34. **Microbial Fertilisers**   **511**

35. **Microbial Production of Biosurfactants**   **535**

Chapter 31

Biodegradation: A Review

INTRODUCTION

Biodegradation is the breakdown of organic contaminants that occurs due to microbial activity. As such, these contaminants can be considered as the microbial food source or substrate. Biodegradation of any organic compound can be thought of as a series of biological degradation steps or a pathway that ultimately results in the oxidation of the parent compound. Often, the degradation of these compounds results in the generation of energy.

COMPLETE BIODEGRADATION

Complete biodegradation or mineralisation involves oxidation of the parent compound to form carbon dioxide and water, a process that provides both carbon and energy for growth and reproduction of cells. Figure 31.1 illustrates the mineralisation of an organic compound under either aerobic or anaerobic conditions. The series of degradation steps constituting mineralisation is similar whether the carbon source is a simple sugar such as glucose, a plant polymer such as cellulose, or a pollutant molecule. Each degradation step in the pathway is catalysed by a specific enzyme made by the degrading cell. Enzymes are most often found within a cell but are also made and released from the cell to help initiate degradation reactions. Enzymes found external to the cell are known as extracellular enzymes. Extracellular enzymes are important in the degradation of macromolecules such as the plant polymer cellulose. Macromolecules must be broken down into smaller subunits outside the cell to allow transport of the smaller subunits into the cell. Degradation by either internal or extracellular enzymes will stop at any step if the appropriate enzyme is not present (Fig. 31.2). Lack of appropriate biodegrading enzymes is one common reason for persistence of organic contaminants, particularly those with unusual chemical structures that existing enzymes do not recognise. Thus, contaminant compounds that have structures similar to those of natural substrates are normally easily degraded. Those that are quite dissimilar to natural substrates are often degraded slowly or not at all.

Partial Biodegradation

Some organic contaminants are degraded partially but not completely. This can result from absence of the appropriate degrading enzyme as mentioned earlier. A second type of incomplete degradation is cometabolism, in which a partial oxidation of the substrate occurs but the energy derived from the oxidation is not used to support microbial growth. The process occurs when organisms coincidentally possess enzymes that can degrade a particular contaminant. Thus, such enzymes are nonspecific.

Cometabolism can occur during periods of active growth or can result from interaction of resting (nongrowing) cells with an organic compound.

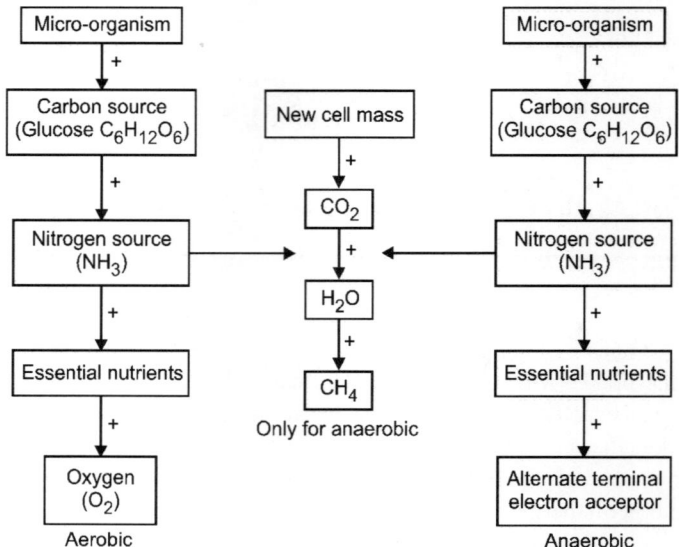

Fig. 31.1. Aerobic or anaerobic mineralisation of an organic compound.

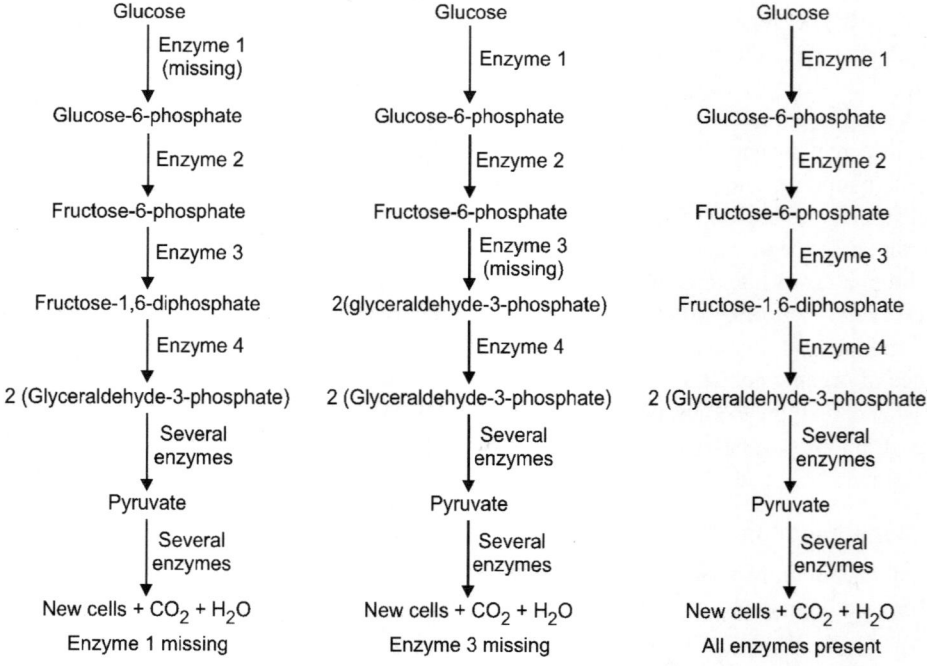

Fig. 31.2. Stepwise degradation of organic compounds. A different enzyme catalyses each step of the biodegradation pathway.

Cometabolism is difficult to measure in the environment but has been demonstrated for some environmental contaminants. For example, the industrial solvent trichloroethylene (TCE) can be oxidised cometabolically by methanotrophic bacteria that grow on methane as a sole carbon source. TCE is currently of great interest for several reasons. It is one of the most frequently reported contaminants at hazardous waste sites, it is a suspected carcinogen, and it is generally resistant to biodegradation. As shown in Fig. 31.3, the first step in the oxidation of methane by methanotrophic bacteria is catalysed by the enzyme methane monooxygenase. This enzyme is so nonspecific that it can also cometabolically catalyse the first step in the oxidation of TCE when both methane and TCE are present. The bacteria receive no energy benefit from this cometabolic degradation step. The subsequent degradation steps shown in Fig. 31.3 may be catalysed spontaneously, by other bacteria, or in some cases by the methanotroph. This is an example of a cometabolic reaction that may have great significance in remediation. Research is currently investigating the application of these methanotrophs to TCE-contaminated sites. Other cometabolising micro-organisms that grow on toluene, propane, and even ammonia are also being evaluated for use in TCE bioremediation.

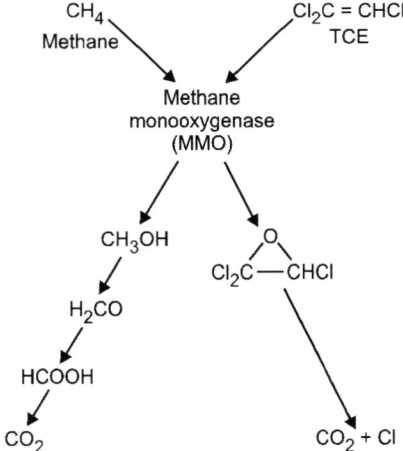

Fig. 31.3. The oxidation of methane by methanotrophic bacteria is catalysed by the enzyme methane monooxygenase. Subsequent degradation steps may be catalysed spontaneously, by other bacteria, or in some cases by the methanotroph.

Partial or incomplete degradation can also result in polymerisation or synthesis of compounds more complex and stable than the parent compound. This occurs when initial degradation steps, often catalysed by extracellular enzymes, create reactive intermediate compounds. These highly reactive intermediate compounds can then combine with each other or with other organic matter present in the environment.

ENVIRONMENTAL FACTORS AFFECTING BIODEGRADATION

A number of parameters influence the survival and activity of micro-organisms in any environment. One factor that has great influence on microbial activity is organic matter, the primary source of carbon for heterotrophic micro-organisms in most environments. Surface soils have a relatively high and variable organic matter content and therefore are characterised by high microbial numbers and diverse metabolic activity. In contrast, the subsurface unsaturated (vadose) zone and saturated zone usually have a much lower content and diversity of organic matter, resulting in lower microbial numbers and activity.

Exceptions to this rule are some areas of the saturated zone that have high flow or recharge rates, which can lead to numbers and activities of micro-organisms similar to those found in surface soils.

Occurrence and abundance of micro-organisms in an environment are determined not only by available carbon but also by various physical and chemical factors. These include oxygen availability, nutrient availability, temperature, pH, salinity, and water activity. Inhibition of biodegradation can be caused by a limitation imposed by anyone of these factors, but the cause of the persistence of a contaminant is sometimes difficult to determine. Perhaps the most important factors controlling contaminant biodegradation in the environment are oxygen availability, organic matter content, nitrogen availability, and contaminant bioavailability. Interestingly, the first three of these factors can change considerably depending on the location of the contaminant.

Oxygen

Oxygen is very important in determining the extent and rate of contaminant biodegradation. In general, aerobic biodegradation is much faster than anaerobic biodegradation. For example, petroleum-based hydrocarbons entering the aerobic zones of freshwater lakes and rivers are generally susceptible to microbial degradation, but oil accumulated in anaerobic sediments can be highly persistent. It follows that as oxygen is depleted, a reduction in the rate of hydrocarbon degradation can be expected. Oxygen is especially important for degradation of highly reduced hydrocarbons such as the alkane hexadecane ($C_{16}H_{34}$). Biodegradation of hexadecane was found to occur only in the presence of oxygen, although an oxygen tension as low as 1 per cent of full oxygen saturation was enough to allow degradation to occur.

Benzene, oxygenated aromatics such as benzoate or phenols, and alkylated aromatics such as toluene have been shown to be biodegraded under anaerobic conditions when nitrate, iron, and sulphate are available for use as terminal electron acceptors. Nevertheless, it is still accepted that the biodegradation of such compounds is much slower under anaerobic conditions and requires much longer adaptation periods than degradation in an aerobic environment. Some studies have shown that anaerobic bio-degradation by specifically adapted consortia can achieve rapid loss of contaminants, indicating that anaerobic degradation may become more important as a bioremediation tool as more is understood about this process.

Organic Matter Content

Surface soils have large numbers of micro-organisms. Bacterial numbers are generally 10^6 to 10^9 organisms per gram of soil. Fungal numbers are somewhat lower, 10^4 to 10^6 per gram of soil. In contrast, microbial populations in deeper regions such as the deep vadose zone and groundwater region are often lower by two orders of magnitude or more. This large decrease in microbial numbers with depth is due primarily to differences in organic matter content. Both the vadose zone and the groundwater region have low amounts of organic matter. One result of low total numbers of micro-organisms is that a low population of contaminant degraders may be present initially. Thus, biodegradation of a particular contaminant may be slow until a sufficient biodegrading population has been built up. A second reason for slow biodegradation in the vadose zone and groundwater region is that because a low amount of organic matter is present, the organisms in this region are often dormant. This can cause their response to an added carbon source to be slow, especially if the carbon source is a contaminant molecule that has low bioavailability or to which the organisms have not had prior exposure.

Because of these trends in oxygen availability and organic matter content, several generalisations can be made with respect to surface soils, the vadose zone, and the groundwater region (Fig. 31.4):

1. Biodegradation in surface soils is primarily aerobic and rapid.
2. Biodegradation in the vadose zone is also primarily aerobic, but significant acclimation times may be necessary for significant biodegrading populations to build up.
3. Biodegradation in the deep groundwater region is also initially slow because of low numbers, and can rapidly become anaerobic because of lack of available oxygen. Biodegradation in shallow groundwater regions is initially more rapid because of higher microbial numbers but is similarly slowed by low oxygen availability.

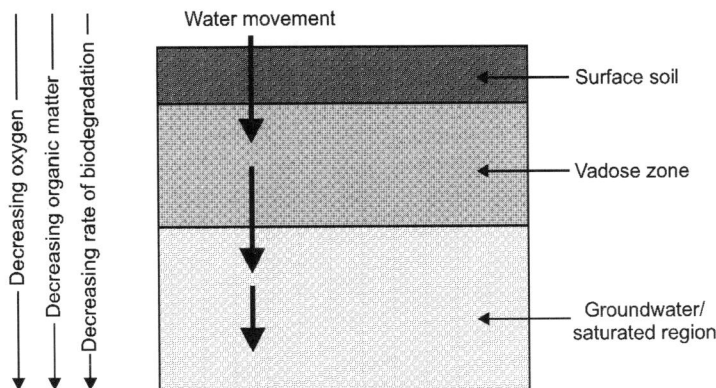

Fig. 31.4. There are three major locations where contamination can occur in terrestrial ecosystems: surface soils, the vadose zone, and the saturated zone. The availability of both oxygen and organic matter varies considerably in these zones. As indicated, oxygen and organic matter both decrease with depth, resulting in a decrease in biodegradation activity with depth.

Nitrogen

Microbial utilisation of organic contaminants, particularly hydrocarbons composed primarily of carbon and hydrogen, creates a demand for essential nutrients such as nitrogen and phosphorus. Ward and Brock monitored seasonal variations in hydrocarbon degradation in a temperate lake and found that variations in the available forms of nitrogen and phosphorus limited degradation. Maximum rates of hydrocarbon degradation were evident in early spring, when available nitrogen and phosphorus levels were high, but rapid consumption of these nutrients reduced the rate of hydrocarbon degradation during the summer months. Thus, biodegradation can often be improved simply by the addition of nitrogen fertilisers. This is particularly true in the case of biodegradation of petroleum oil spills, in which nitrogen shortages can be acute. In general, microbes have an average C:N ratio within their biomass of about 5:1 to 10:1 depending on the type of micro-organism. Therefore, a ratio of approximately 100:10:1 (C:N:P) is often used in such sites. However, in some instances, quite different ratios have been used. For example, Wang and Bartha found that effective remediation of hydrocarbons in soil required the addition of nitrogen and phosphorus additions to maintain a C:N ratio of 200:1 and a C:P ratio of 1000:1. Why were the C:N and C:P ratios maintained at levels so much higher than the cell C:N and C:P ratios? It is because much of the hydrocarbon carbon that is metabolised is released as carbon dioxide so that much of the carbon is lost from the system. In contrast, almost all of the nitrogen and phosphorus metabolised is incorporated into microbial biomass and thus is conserved in the system.

Other Environmental Factors

Temperature

Hydrocarbon degradation has been reported to occur at a range of temperatures between close to freezing and more than 30°C. Bacteria can adapt to temperature extremes in order to maintain metabolic activity; however, seasonal temperature fluctuations in the natural environment have been shown to affect the rate at which degradation occurs. For example, the degradation rates of hexadecane and naphthalene in a river sediment were reduced approximately 4.5-fold and 40-fold, respectively, in winter (0°–4°C) compared with summer (8°–21°C) samples.

pH

In soils, the rate of hydrocarbon degradation is often higher in alkaline conditions than in acidic conditions. In acidic soils, fungi are more competitive than bacteria, which prefer a neutral environment. Therefore, at lower soil pH, fungi become more important in hydrocarbon degradation. Acidic soils favour the growth of fungi, which degrade hydrocarbons but usually at a slower rate than soil bacteria, which prefer a more neutral to alkaline environment.

Smith examined the effect of pH on hydrocarbon degradation in a salt marsh sediment. The pH of the sediment ranged from 6.5 to 8.0, but when incubated at different pH levels, lower rates of hydrocarbon degradation were evident at pH 5.0 and 6.5 than at pH 8.0.

Salinity

Hydrocarbon degradation has been shown to occur in saline environments. Samples of freshwater sediment incubated under saline conditions showed a reduced rate of hydrocarbon degradation. In contrast, hydrocarbon degradation in estuarine sediments incubated with increasing levels of salinity were little affected, although hypersaline conditions were reported to reduce the rate of hydrocarbon degradation in sediments sampled from a saline lake.

Water activity

Optimal conditions for activity of aerobic soil micro-organisms occur between 38 and 81 per cent of the soil pore space (also referred to as per cent saturation) because in this range of water contents, water and oxygen availability are maximised. At higher water contents, the slow rate of oxygen diffusion through water limits oxygen replenishment, thereby limiting aerobic activity. At lower water contents, water availability becomes limiting. Why is the optimal per cent saturation range so broad? It is because optimal activity really depends upon a combination of factors including water content and available pore space. Available pore space is measured as bulk density which is defined as the mass of soil per unit volume (g/cm^3). This means that in any given soil, increasing bulk density indicates increasing compaction of the soil. In a soil that is loosely compacted (lower bulk density), a water saturation of 70 per cent represents more water (more filled small pores and pore throats) than in a highly compacted soil. Thus, for soils of low bulk density, oxygen diffusion constraints become important at lower water saturation than for highly compacted soils. For example, Neilson and Pepper showed that respiration in a clay loam soil was maximal at different per cent saturation depending on bulk density. At a bulk density of 1.1 g/cm^3, respiration was optimal between 38 and 45 per cent water saturation. At a bulk density of 1.6 g/m^3, respiration was optimal at a higher water saturation, 81 per cent.

Chapter 32

Biodegradation of Phenol

INTRODUCTION

Two white-rot basidiomycetes, *Phanerochaete chrysosporium*, well-known to degrade lignin and also to degrade many toxic or organohalides into non-toxic or noncarcinogenic compounds, and *Pleurotus sajor-caju*, with low ligninolytic activities (lignin peroxidase – LiP, and manganese peroxidase – MnP), were used to study degradation of phenol. Both the test fungi were able to tolerate phenol concentration up to 800 ppm. Phenol appeared to have some stimulatory effect at a concentration of 100 ppm on the growth of *P. sajor-caju*. However, the hyphal growth of *P. chrysosporium* was affected at all the concentrations and mycelial biomass decreased with increase in concentration of phenol. Nitrogen- or carbon-limited culture conditions were not necessary for the degradation of phenol for both the test fungi except that degradation was adversely affected at very low or very high concentrations of nitrogen in the cultures. Degradation of phenol was observed during the exponential as well as during the stationary and declining phases of growth of both of the test fungi. In general, *P. sajor-caju* was able to degrade phenol in a shorter time than *P. chrysosporium*. Although, literature indicates that high LiP or MnP activities, associated with nitrogen- and carbon-limited conditions and secondary metabolism, are correlated for the degradation of lignin and certain environmental pollutants. However, study showed that it was not so in the case of both of these test fungi. The data obtained in this study indicated that there may be some enzymes other than the known ligninolytic enzymes that are responsible for the degradation of pollutants like phenol.

Phenol has been classified as a priority pollutant by the US Environment Protection Agency. It is a common soil and water contaminant at phenolic resins and caprolactam fabrication facilities, where the commercial formulations of the product are used. Several existing chemical or physical treatment processes can be used to remove phenol from contaminated media, but most of them are either too expensive or introduce other compounds in the environment that are at least as toxic as phenol itself.

Much effort has recently been directed toward affecting biodegradation of toxic compounds in the environment. Some higher fungi responsible for white-rot of wood are known to produce extracellular enzymes, ligninases, that catalyse depolymerisation of lignin, a constitutive material in plant and the most abundant of natural aromatic substances next to cellulose. Because of their properties to depolymerise lignin, the white-rot fungi have also been tested for degradation of recalcitrant pollutants. The strong ligninolytic activities (lignin peroxidase LiP, and manganese peroxidase-MnP) discovered in *Phanerochaete chrysosporium*, a white-rot basidiomycete, have been found to be the two-electron oxidants directly involved in pollutants degradation. The ligninolytic activities are commonly known to occur under carbon-nitrogen limiting conditions. *Pleurotus sajor-caju* is another white-rot basidiomycete, but it produces very low ligninolytic activities and a change of dye colour in various hues by it was correlated to weak ligninolytic activities. However, in another study it was found to be good in depolymerisation of lignin into oligolignols of low molecular weight. There is some evidence, as yet unexplained, that no direct involvement of ligninolytic enzymes (LiP and MnP) has been shown for degradation of lignin.

The present study was undertaken to determine the ability of *P. sajor-caju*, a basidiomycete with weak ligninolytic activity, to degrade phenol whether the phenol degradation by this fungus occurs under conditions favouring ligninolytic activities. We have included *Phanerochaete chrysosporium* to our study because it is the most intensively studied fungus for production of ligninolytic enzymes and for bioremediation of environmental pollutants.

MATERIALS AND METHODS

Micro-organisms

Pleurotus sajor-caju NRRL 18757 and *Phanerochaete chrysosporium* ATCC # 24725, white-rot fungi, were used in this study. They were maintained on Mandels and Weber agar medium containing 2 per cent delignified wheat straw and stored at 4°C. The growth temperature was 30°C for *P. sajor-caju* and 35°C for *P. chrysosporium.* The culture medium (Mandels medium) contained per litre of distilled water: glucose 10 gms, KH_2PO_4 2 gms, $(NH_4)_2$ SO_4 1.4 gms, urea 0.3 gm, $MgSO_4 \cdot 7H_2O$ 0.3 gm, $CaCl_2$ 0.3 gm, $FeSO_4 \cdot 7H_2O$ 5 mg, $ZnSO_4 \cdot 7H_2O$ 1.4 mg, $CoCl_2 \cdot 6H_2O$ 3.67 mg, and yeast extract 0.5 gm. The glucose was replaced with 2 per cent delignified wheat straw for maintaining the cultures on solid agar medium in Petri plates.

Nitrogen Concentration

The Mandels medium described above is a nitrogen-sufficient medium. The effect of nitrogen-sufficient, nitrogen-limited, nitrogen-excess and nitrogen deficient media on degradation of phenol was examined for both of the micro-organisms. The desired nitrogen concentration was obtained by either increasing or reducing the amount of urea and $(NH_4)_2SO_4$. Cultures were grown (in triplicate) in 100 ml of nitrogen-sufficient (N), nitrogen-limited (N/2), nitrogen-deficient (N/10) and nitrogen-excess (N × 10) media, containing 200 ppm of phenol and were monitored for decrease in phenol concentration for 12 days.

Sensitivity to Phenol

These experiments were carried out in order to find out the maximum phenol concentration the test fungi could tolerate on both solid and liquid Mandels media. The tolerance to phenol was determined by the hyphal extension (in solid medium) and by the dry weight of mycelial biomass (in liquid medium). The hyphal extension was determined by placing 0.5 cm disks of actively grown cultures in the centre of the Petri plates with phenol concentrations ranging from 0 to 800 ppm and measuring the average daily increase in colony diameter in two perpendicular directions. The daily increase in mycelial biomass in different phenol concentrations was measured by filtering the samples on tared filter paper and drying them in an oven at 80°C for 24 hours.

Adaptation of the Test Fungi to Phenol

For adaptation of the test fungi, the phenol concentration in the Mandels agar medium with glucose was increased weekly by one half of the previous concentration. Maximum concentration tolerated, at which the test organism could grow, was determined by measuring the hyphal extension daily. The hyphal extension rate was recorded as the average daily increase in colony diameter measured in two perpendicular directions.

For determination of the maximum concentration at which the test fungi could grow, the cultures that showed no growth after four consecutive days were considered to be either killed or inhibited by the corresponding phenol concentration.

Age of Cultures

For determining the effect of age of the culture, the test fungi were grown for 0, 3, 7, 14, and 21 days prior to phenol addition to the culture media. All cultures were run in triplicate with biotic (inoculated) and abiotic (uninoculated) controls. The test pollutant was phenol, 99 per cent pure, purchased from Fischer Scientific Company, New Jersey.

Analytical Methods

Phenol was analysed with a waters 510 high-pressure liquid chromatography apparatus coupled with a Waters 481 Lambda-Max LC spectrophotometer and a Professional 350 Digital computer. The column used was a reverse-phase NovaPak C_{18} (3.9 × 150 nm). Samples (0.66 ml) were mixed with acetonitrile (0.33 ml) containing 0.33 per cent acetic acid, centrifuged for 5 minutes at 5500 gms in a microcentrifuge and injected onto the column.

The mobile phase was composed of solvent A (water:acetonitrile:acetic acid in the proportions 70:30:0.1) and solvent B (same compounds in the proportion 50:50:0.1, respectively). A linear gradient of solvent A to solvent B in 5 minutes with a flow rate of 2 ml/min was used to separate phenol. The detection wavelength of 220 nm was used.

Phenol Biodegradation

The test fungi were grown in 250 ml Erlenmeyer flasks on a rotary shaker at 200 rpm, at 30°C for *P. sajor-caju* and 35°C for *P. chrysosporium.* The flasks were plugged with polyether foam stoppers. Flasks containing 100 ml of liquid media were aseptically inoculated (in triplicate) by transferring a 5 mm disc cut from the outer edge of an actively growing culture on an agar plate for 3 to 4 days. The mycelial growth obtained by this method was called pre-inoculum. The pre-inoculum was macerated at high speed in a Waring blender for 30 seconds.

The macerated mycelium was used to inoculate 100 ml of fresh media (10 per cent, v/v), and incubated for 2 days. The mycelial growth obtained by this method was called inoculum. Experimental flasks were inoculated with the macerated inoculum at 10 per cent v/v. Liquid cultures were incubated for 3 to 4 days prior to addition of phenol for *P. chrysosporium* and 4 to 6 hours with *P. sajor-caju* except where indicated otherwise.

The concentration of the mycelium in the inoculum was 0.5 gm (dry)/l. Reported phenol concentrations are averages (≤10 per cent coefficient of variation) of three separate samples. The depletion of phenol is correlated to its degradation.

RESULTS AND DISCUSSION

Sensitivity to Phenol

It has been reported that tolerance of fungi to pentachlorophenol varies to a great extent. In order to find out how much phenol concentration the test fungi could tolerate, their sensitivity to phenol was evaluated by measuring the hyphal extension on Mandels agar medium with glucose (1 per cent) and phenol concentrations ranging from 0 to 800 ppm and the dry weight of fungal mycelium on Mandels liquid medium. The hyphal extension and the mycelial biomass production of *P. sajor-caju* was stimulated on both solid and liquid media by phenol concentration of 100 ppm (Figs 32.1 and 32.3). On the other hand the hyphal extension of *P. chrysosporium* was affected at all the concentrations and mycelial biomass production decreased with the increase in concentration of phenol (Figs 32.2 and 32.4). Nevertheless, both of the test fungi were able to tolerate phenol concentrations up to 800 ppm. The tolerance to high concentrations of the contaminants by an organism is one of the important advantages for bioremediation of highly contaminated sites.

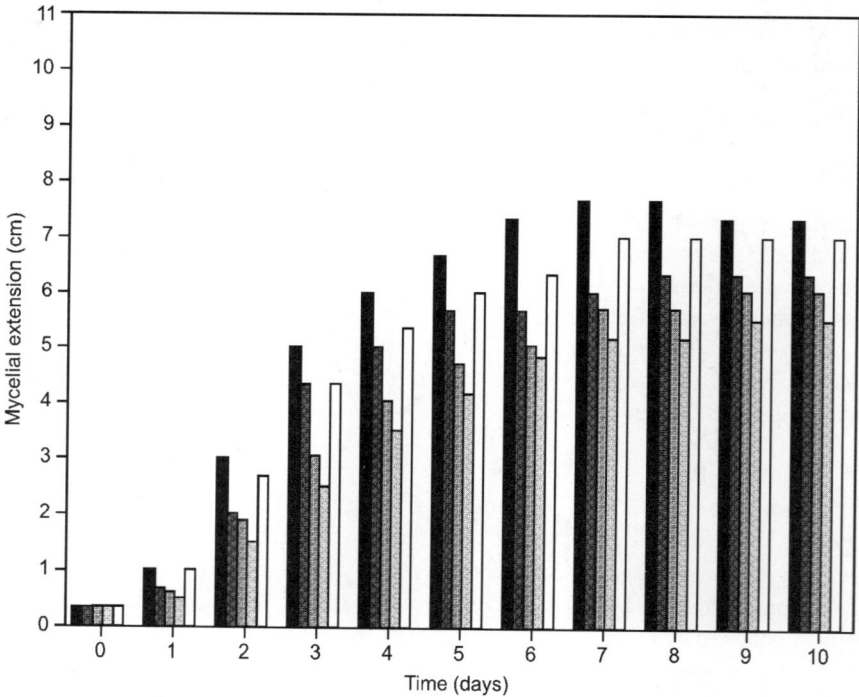

Fig. 32.1. Hyphal extension of *P. sajor-caju* with different phenol concentrations. From darkest to lightest, the bars represent: 100 ppm, 200 ppm, 600 ppm, 800 ppm and 0 ppm, respectively.

Effect of Adaptation

Mileski had observed with other toxins that the higher the pollutant concentration, the lower the degradative capacity of *P. chrysosporium*. Therefore, it was thought that adaptation of test fungi to high concentration of phenol may help to achieve more efficient degradation of phenol. For this experiment the inocula of *P. sajor-caju* and *P. chrysosporium* were adapted on 800 ppm and 600 ppm, respectively, by the serial transfers on higher and higher concentration of phenol. In spite of the adaptation to high phenol concentrations, both of the test fungi failed to show any significant increase in their ability for phenol degradation, however, *P. sajor-caju* was able to degrade phenol completely and faster than *P. chrysosporium* (Fig. 32.5).

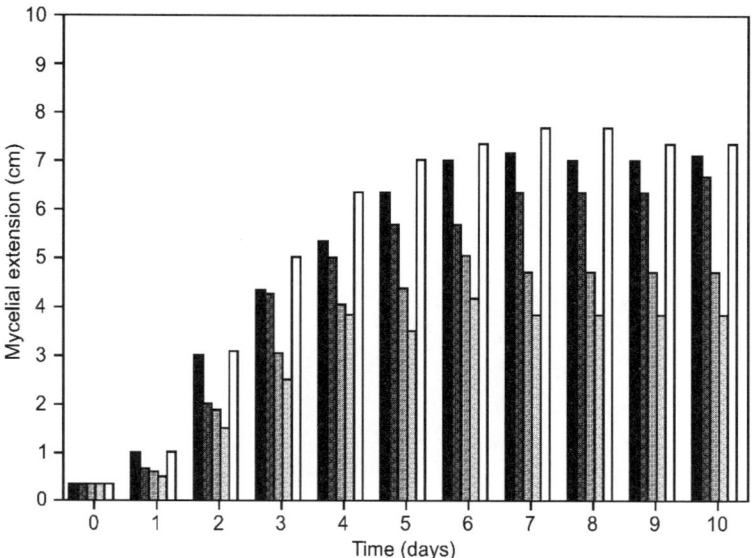

Fig. 32.2. Hyphal extension of *P. chrysosporium* with different phenol concentrations. From darkest to highest, the bars represent: 100 ppm, 200 ppm, 600 ppm, 800 ppm and 0 ppm, respectively.

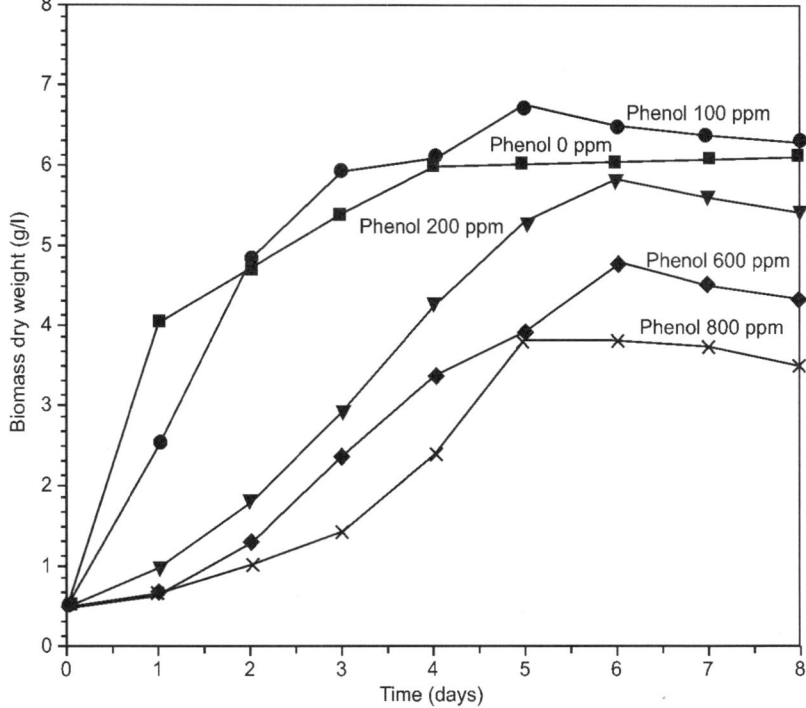

Fig. 32.3. Sensitivity to phenol of *P. sajor-caju* on liquid medium with different phenol concentrations. (■) 0.0 ppm, (●) 100 ppm, (▼) 200 ppm, (♦) 600 ppm and (×) 800 ppm.

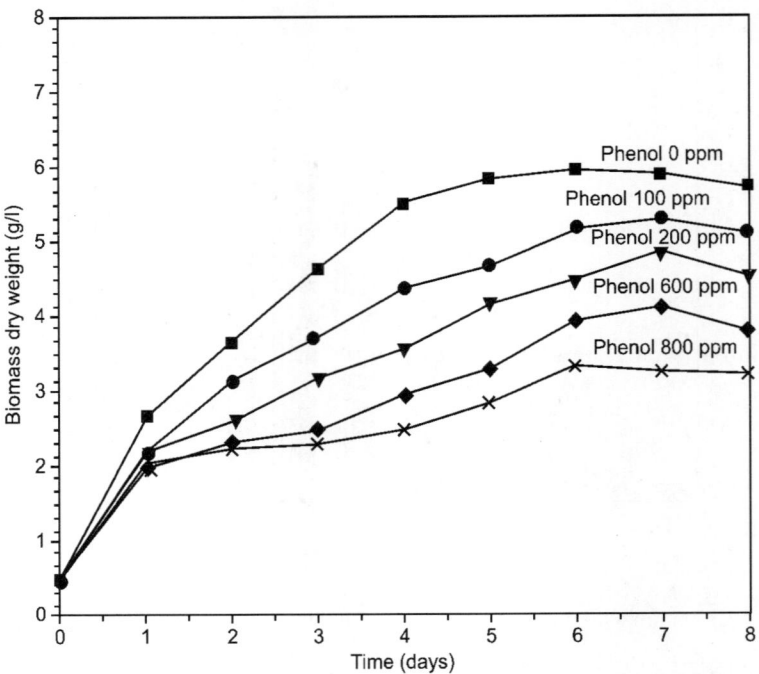

Fig. 32.4. Sensitivity to phenol of *P. chrysosporium* on liquid medium with different phenol concentrations. (■) 0.0 ppm, (●) 100 ppm, (▼) 200 ppm, (♦) 600 ppm and (×) 800 ppm.

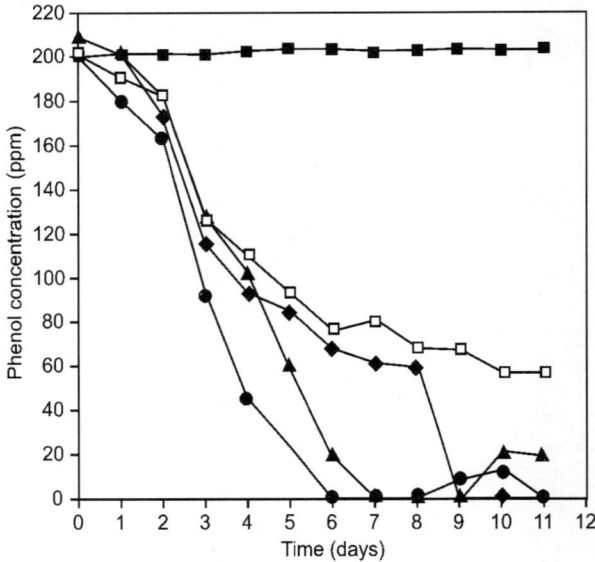

Fig. 32.5. Effect of adaptation to phenol on its degradation by *P. chrysosporium* and *P. sajor-caju*. (▲) *P. sajor-caju* (wild strain), (♦) *P. sajor-caju* (adapted strain, (□) *P. chrysosporium* (wild strain), (●) *P. chrysosporium* (adapted strain) and (■) abiotic control.

Effect of Nitrogen Concentration on Phenol Degradation

It has been known for several years that ligninolytic activities of *P. chrysosporium* appear as a secondary metabolic event and in response to nitrogen- and carbon-limitation, and lignin degradation is also carried out under similar cultural conditions. Hyphae in their primary growth are known to be non-ligninolytic.

It was observed that neither the amount nor the time necessary for complete phenol degradation by *P. sajor-caju* was affected by nitrogen-sufficient, nitrogen-limited, and nitrogen-deficient media (Fig. 32.6). Complete phenol degradation was observed at 7–9 days under these conditions. On the other hand, *P. chrysosporium* took 8–11 days to degrade the phenol under similar nitrogen conditions in the media (Fig. 32.7). However, when cultivations were carried out with a ten-fold higher concentration of nitrogen, phenol degradation was affected to a great extent with both the test fungi. On the other hand, when the nitrogen concentration was too low (N/10), there was little growth of fungi, consequently, there was some delay in degradation of phenol in the case of *P. sajor-caju* and some reduction in degradation of phenol in the case of *P. chrysosporium*. It appears that a certain balance of C:N is required for optimum phenol degradation.

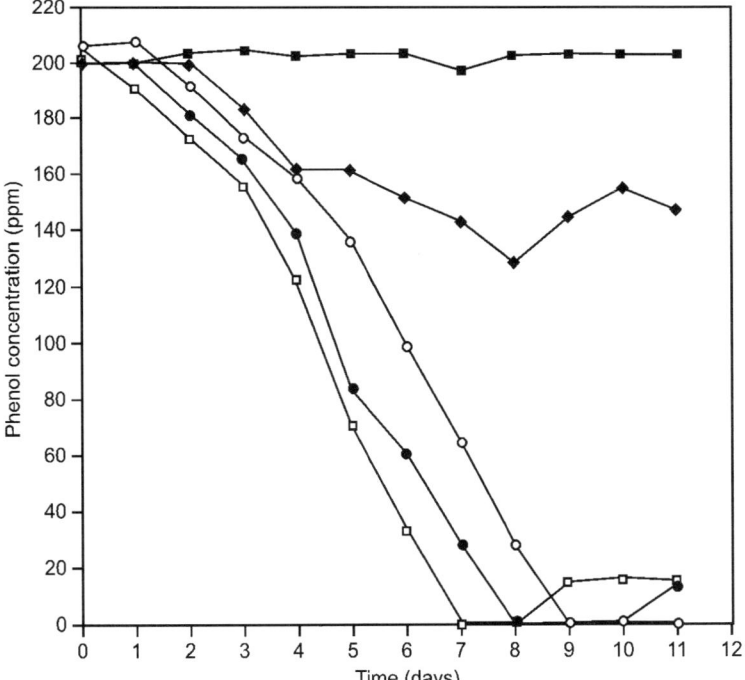

Fig. 32.6. Effect of nitrogen concentration on phenol degradation by *P. sajor-caju*, (□) nitrogen-sufficient, (◆) nitrogen-limiting, (○) nitrogen-deficient, (●) nitrogen-excess, and (■) abiotic control.

It could be deduced that both of the test fungi can degrade phenol under nitrogen-sufficient, nitrogen-limited, and nitrogen-deficient conditions in the media. However, *P. sajor-caju* even having weak ligninolytic activities, was faster in phenol degradation than *P. chrysosporium*, which has strong ligninolytic activities. Therefore, it is possible that there may be some enzymes other than the known ligninases (LiP and MnP) that are produced by these fungi, that are responsible for phenol degradation.

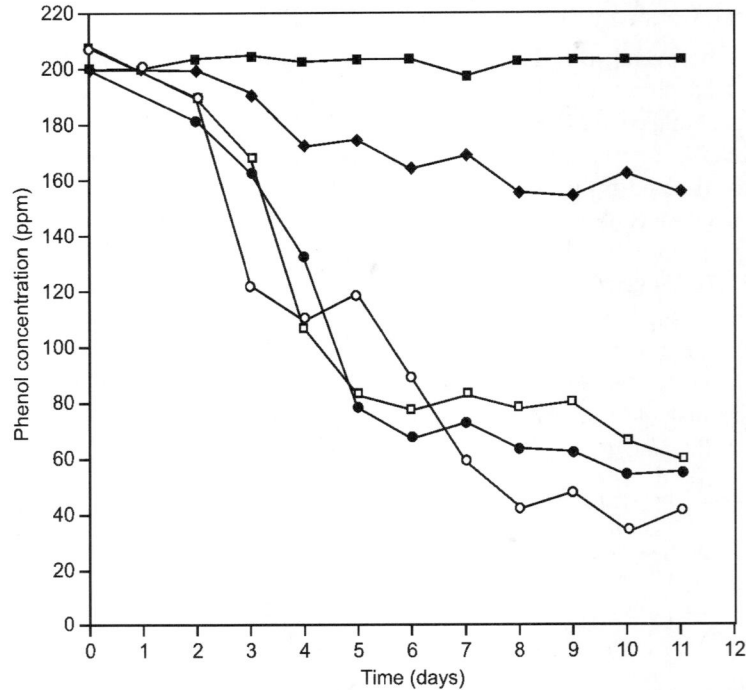

Fig. 32.7. Effect of nitrogen concentration on phenol degradation by *P. chrysosporium*, (□) nitrogen-sufficient, (♦) nitrogen-limiting, (○) nitrogen-deficient, (●) nitrogen-excess, and (■) abiotic control.

Effect of Culture Age

It has been reported that the time of the production, the concentration and the type of peroxidases of *P. chrysosporium* are a function of the age of the culture. The trend of phenol degradation by *P. chrysosporium* was almost the same with 0-, 7-, and 14-day old cultures (Fig. 32.8). On the other hand, *P. sajor-caju* degraded more phenol when the phenol was added to the culture medium at the time of inoculation, i.e. 0-day old culture, and its ability decreased with the increase in the age of the culture (Fig. 32.9). However, both micro-organisms were unable to degrade phenol when the cultures were 21 days old, probably it had become too old to produce the required enzymes and could not grow further because of exhaustion of nutrients from the medium. The data indicated that it may be necessary to use cultures of *P. chrysosporium* pre-grown for one to two weeks in order to maximise phenol degradation, but the cultures of *P. sajor-caju* were efficient in phenol degradation even at its early stages of growth. Mileski had also reported that degradation of pentachlorophenol was markedly better in pre-grown cultures of *P. chrysosporium*.

Effect of Glucose Concentration

The quantity of phenol degraded by *P. chrysosporium* was almost the same with glucose concentration of 5 g/l and 10 g/l (Fig. 32.10), but it was retarded to a great extent when glucose concentration was 20 g/l. On the other hand, phenol degradation was not affected by *P. sajor-caju* with glucose concentration from 5–20 g/l (Fig. 32.11). Degradation of pentachlorophenol was reported to be adversely affected when the glucose was ≥10 g/l in the cultures.

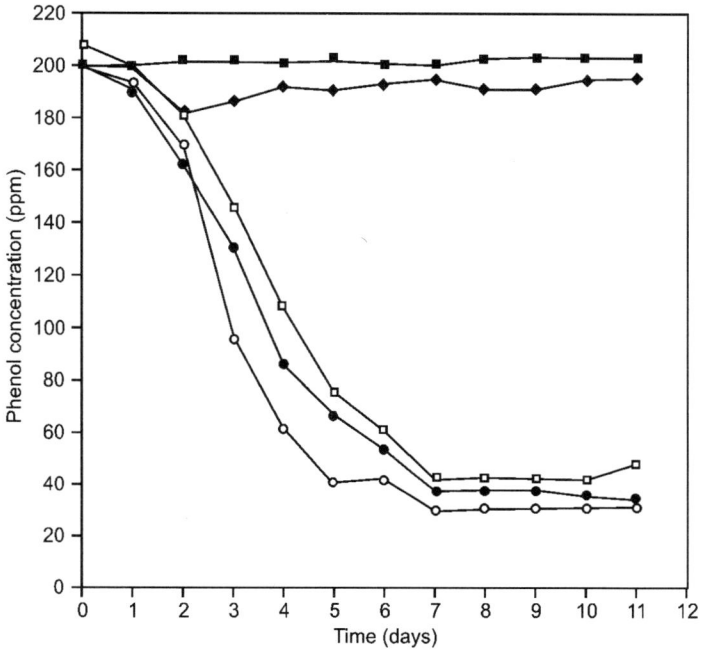

Fig. 32.8. Effect of the culture age on phenol degradation by *P. chrysosporium*. (□) 0 day, (◆) 7 days, (○) 14 days, (●) 21 days, and (■) abiotic control.

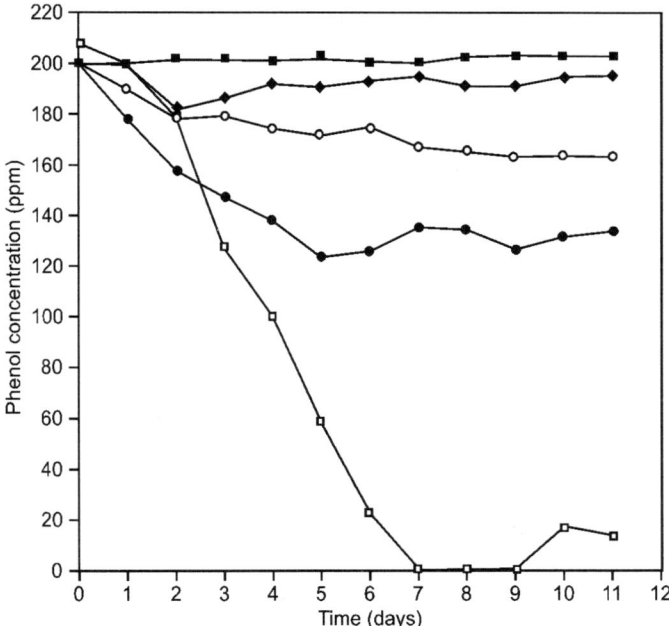

Fig. 32.9. Effect of the culture age on phenol degradation by *P. sajor-caju*. (□) 0 day, (◆) 7 days, (○) 14 days, (●) 21 days, and (■) abiotic control.

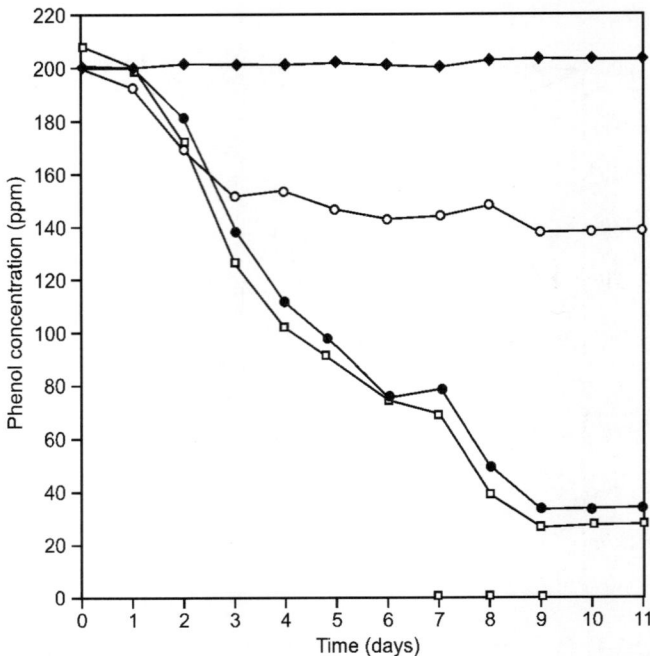

Fig. 32.10. Effect of glucose concentration on phenol degradation by *P. chrysosporium*. (□) 5 g/l, (●) 10 g/l, (○) 20 g/l, and (♦) abiotic control.

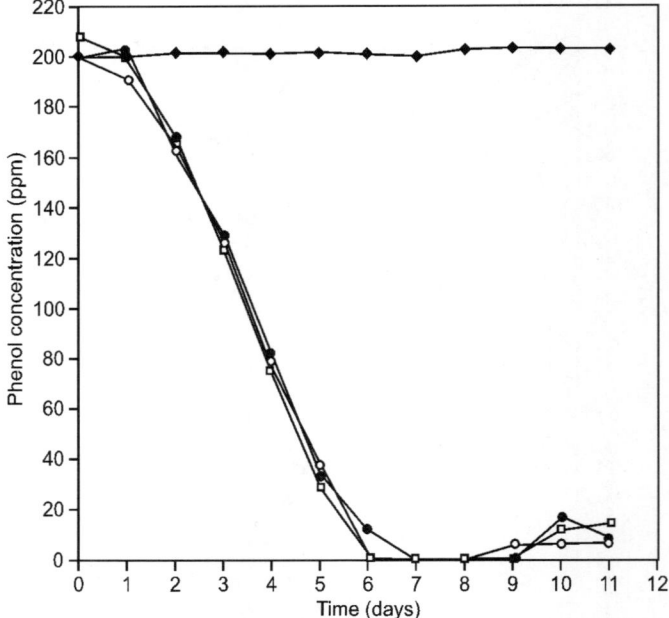

Fig. 32.11. Effect of glucose concentration on phenol degradation by *P. sajor-caju*. (□) 5 g/l, (●) 10 g/l, (○) 20 g/l, and (♦) abiotic control.

Effect of Carbon Source

Different sources of carbon: glucose, cellulose, delignified wheat straw, and sucrose, were examined to determine their effect on phenol degradation (Table 32.1).

Table 32.1. Effect of the carbon source on the quantity of phenol degradation and mycelial biomass production by *P. sajor-caju* and *P. chrysosporium* during exposure to 200 ppm of phenol over 7 days.

Species	Carbon source (10 g/l)	Phenol concentration (ppm)		Biomass (g/l)	
		Initial	Final	Initial	Final
Pleurotus	Glucose	201.33	16.25	0.50	5.14
sajor-caju	Cellulose	200.22	11.25	0.53	5.64
	Delignified wheat straw	200.88	0.00	0.51	5.81
	Sucrose	202.25	81.34	0.55	3.85
Phanerochaete	Glucose	199.45	44.25	0.54	5.54
chrysosporium	Cellulose	203.24	38.54	0.52	5.85
	Delignified wheat straw	200.58	24.00	0.55	5.88
	Sucrose	204.55	125.25	0.50	4.58

All sources of carbon supported good growth except the sucrose. However, the nature of carbon source had tremendous effect on degradation of phenol.

The best results of degradation of phenol were obtained with both micro-organisms on delignified wheat straw. It might be due to the fact that wheat straw was very similar of the natural habitat, wood, of both the test fungi.

However, *P. sajor-caju* was able to degrade phenol completely when delignified wheat straw was used as a carbon source.

In general, *P. sajor-caju* proved to be more efficient in degradation of phenol than *P. chrysosporium* on all the carbon sources tested.

Relationship between Growth and Degradation of Phenol

In this experiment glucose at 10 g/l was selected as a carbon source. Because when delignified wheat straw was used it was difficult to find out the actual mycelial biomass produced due to some difficulties in determination of unutilised wheat straw which was mixed with the mycelial biomass.

In the case of *P. sajor-caju,* the biomass production stopped on the 3rd day of cultivation and thereafter it started to decline, however, the degradation of phenol was going on right from the beginning and was complete at the 6th day of incubation (Fig. 32.12).

In case of *P. chrysosporium,* the growth continued until the 5th day, and thereafter the growth remained stationary, however, degradation of phenol continued up to the 10th day of incubation.

The growth rate and phenol degradation of *P. chrysosporium* was lower than that of *P. sajor-caju.* Nevertheless, it appears that both of the test fungi were degrading phenol during the exponential as well as during the stationary and declining phases. Bourbonnais, also reported that lignin degradation by *P. sajor-caju* occurred during its growth.

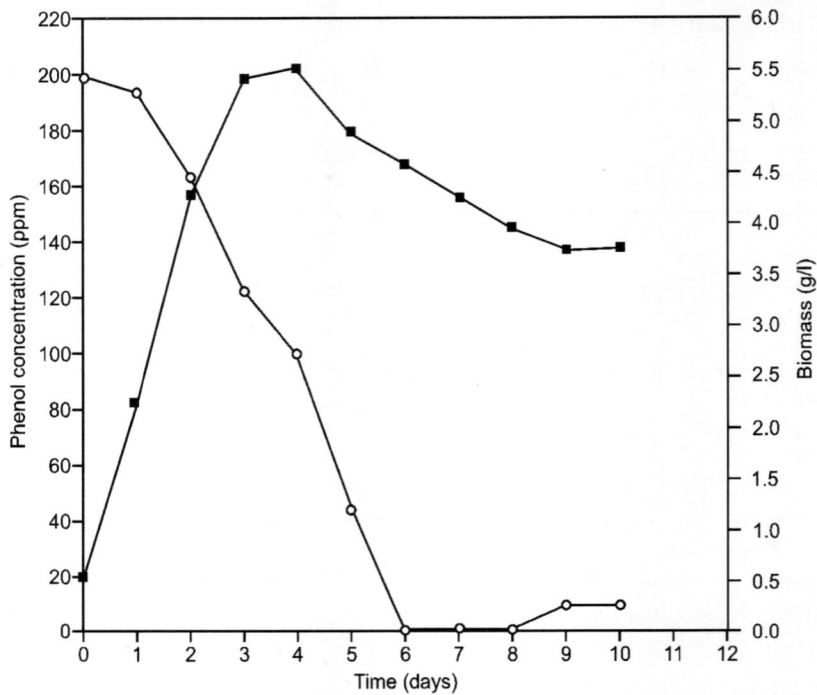

Fig. 32.12. Relationship between growth and phenol degradation by *P. sajor-caju*, (○) Phenol concentration, and (■) biomass.

Chapter 33

Biodegradation and Bioremediation of Halogenated Organic Compounds

INTRODUCTION

Halogenated organic compounds include some of the most useful and economically important chemicals available to industry and agriculture. Alkyl halides — aliphatic organic compounds that are substituted with chlorine, bromine, iodine or fluorine — are used in many applications. Chlorinated alkanes and alkenes, such as trichloroethane and trichloroethene, are used as dry-cleaning fluids, refrigerants, degreasing agents, solvents, and in the production of caffeine-free coffee. Brominated alkyl compounds, such as ethylene dibromide (EDB), are used as pesticides. Aryl halides — aromatic compounds such as chlorobenzenes, chlorophenols, and chlorinated biphenyls — are extensively used as herbicides, insecticides, fungicides, heat transfer media, insulators, and lubricants.

Some of the uses of these compounds — pesticides, for example — involve intentional release to the environment. Pesticides used in agriculture may be mobilised from the soil zone and carried to shallow groundwater aquifers by percolating recharge. Many uses for these compounds — solvents or refrigerants, for example — do not involve intentional release to the environment. Nevertheless, improper disposal techniques (seepage pits) and inadvertent chemical spills have led to extensive release of these compounds to groundwater systems. In either case, the potential for subsurface micro-organisms to degrade these compounds influences their mobility in groundwater systems. This chapter considers the various microbially mediated processes that affect the fate and transport of halogenated organic compounds in groundwater systems.

Chlorinated organic compounds are widely used for industrial and agricultural purposes. Alkyl halides such as trichloroethylene (TCE) are used as solvents and cleaning agents. Aryl halides such as dichlorobenzene are used for household and industrial deodourising. Other chlorinated organic compounds are used as insecticides, herbicides, antifungal fumigants, dielectric agents, lubricants, and a host of other applications. It is fairly safe to assert that modern industrial society could not function without the use of these chemicals. As a result of their wide usage, however, these compounds have contaminated a large number of groundwater systems and represent a significant health hazard in some cases.

Micro-organisms are capable of transforming chlorinated organic compounds under a wide range of environmental conditions. However, the mechanisms, pathways, and rates of degradation vary widely and depend largely on the structure of the chemical. Alkyl halides such as TCE are not significantly degraded by aerobic heterotrophic bacteria. Consequently, alkyl halides tend to persist in aerobic groundwater systems. These compounds are, however, degraded by the methane monooxygenase enzyme present in methanotrophic bacteria. Methanotrophs cannot obtain energy for growth by TCE oxidation, and methane competitively inhibits this reaction. Nevertheless, stimulation of methanotrophic activity is presently being evaluated as a potential *in situ* and *ex situ* bioremediation strategy. Under anaerobic conditions, reductive dehalogenation reactions are the most important alkyl halide-degrading processes. In electron-donor-poor groundwater systems, however, reductive dehalogenations tend to be incomplete, resulting in the accumulation of vinyl chloride.

Reductive dehalogenation reactions may also be important microbial processes involved in degrading aryl halides. Unlike alkyl halides, however, aryl halides are capable of supporting the growth of heterotrophic aerobic bacteria. Consequently, aryl halides tend to be rapidly degraded in aerobic groundwater systems.

While reductive dehalogenation reactions have been shown in laboratory studies to degrade halogenated benzoates, the importance of reductive dehalogenation of aryl halides in anaerobic groundwater systems is not clear. Studies have reported that dichlorobenzenes persist for long periods of time in some anaerobic groundwater systems.

Chlorinated pesticides vary widely in their susceptibility to microbial attack in groundwater systems. Some insecticides such as DDT are recalcitrant and degrade only partially. As a result of this recalcitrance, DDT has been virtually banned from agricultural practice in the United States. Herbicides such as atrazine are widely used in agricultural practice and are biodegradable in the soil zone. However, degradation rates of atrazine in nutrient-poor aquifer systems is relatively slow and appear to be limited to deethylation reactions. Thus, if rates of atrazine delivery to shallow groundwater systems exceeds rates of degradation, atrazine can accumulate in solution.

Contamination of groundwater systems by phenolic compounds is fairly common and often accompanies wood-treatment operations. Creosote and, later, pentachlorophenol (PCP) were widely used as wood preservatives, and wastes from these operations often reach shallow groundwater systems. Phenolic compounds are rapidly degraded in groundwater systems under either aerobic or anaerobic conditions. However, the most common chlorinated phenol, pentachlorophenol, appears to be much more recalcitrant. Nevertheless, field evidence suggests that degradation does occur in some systems.

Available microbiologic and biochemical evidence is sufficient to explain many patterns of chlorinated hydrocarbon degradation observed in groundwater systems. However, because field conditions typically involve multiple contaminants, the presence of inhibiting chemicals or competing substrates, and other factors that make microbial processes difficult to evaluate, it is clear that much more field-oriented research is needed before a consistent picture of degradation in groundwater systems can emerge.

MICROBIAL DEGRADATION OF HALOGENATED ORGANIC COMPOUNDS

Because environmental contamination by halogenated organic compounds is so common, considerable efforts have been made to understand the microbial processes that affect their fate and transport in aquatic systems.

In general, the kinds of operative microbial degradation processes depend upon the chemical characteristics of the individual compounds and upon the aerobic or anaerobic nature of the environment. It is useful, therefore, to consider the microbial degradation of each compound separately.

Chlorinated Ethenes

Chlorinated ethenes are subject to a variety of microbial degradation processes that include reductive dechlorination, aerobic oxidation, anaerobic oxidation, and aerobic cometabolism. Because of this variety, assessing ongoing degradation processes in groundwater systems is often complex. However, because the conditions under which these processes occur have now been well-documented, it is often possible to identify the presence or absence of particular degradation processes at a given site.

Reductive dechlorination

Chloroethenes contain chlorine atoms in their molecular structure, and this makes them relatively oxidised compounds. Because of this, chloroethenes can act as oxidants (that is, electron acceptors) in microbial metabolism. In reductive dechlorination, molecular hydrogen reacts to replace a chlorine on a chlorinated ethene molecule.

Because this causes a net reduction of the molecule by removing a chlorine, it is called reductive dechlorination. In addition, because it is now known that these reduction processes can be used by some micro-organisms to support respiration, they are also referred to as chlororespiration.

Aerobic oxidation

While the tendency of chloroethenes to undergo reductive dechlorination decreases as the number of chlorines present in the molecule decreases, the tendency to undergo oxidation increases. As the least chlorinated of the chloroethenes, vinyl chloride (VC) has the greatest tendency to undergo oxidation. Rapid microbial degradation of vinyl chloride, including mineralisation, has been observed in laboratory cultures and aquifer samples under aerobic conditions.

Moreover, under aerobic conditions, VC can be used as sole carbon source for growth and metabolism. Dichloroethene (DCE), on the other hand, has been shown to oxidise under aerobic conditions in liquid culture, but this oxidation apparently does not support microbial growth.

Because the production of DCE and VC generally occurs by reductive dechlorination under anaerobic conditions, the aerobic oxidation of these compounds is often limited in groundwater systems. However, where anaerobic conditions that produce DCE and VC grade to more oxic conditions, which often happens on the fringes of contaminant plumes, aerobic oxidation of these compounds can be significant. Furthermore, the aerobic degradation of VC and DCE can be stimulated in groundwater systems by providing a source of oxygen.

Anaerobic oxidation

The anaerobic oxidation of DCE and VC is a process that has only been recognised recently. Given the early emphasis on reductive dechlorination under anaerobic conditions as a principal degradation pathway, the fact that these compounds also oxidise anaerobically was unexpected. Nevertheless, in a series of papers, it was shown that VC, and to a lesser extent DCE, could oxidise to carbon dioxide under Fe(III)-reducing conditions.

At first, the fact that Fe(III) reduction could support the anaerobic oxidation of VC was attributed to the fact that Fe(III) is a fairly strong oxidant. However, further investigation showed that a portion of both ^{14}C-VC and ^{14}C-DCE was mineralised to ^{14}CO$_2$ not just under Fe(III)-reducing conditions but under sulphate-reducing conditions and even methanogenic conditions. In addition, it was shown that humic acids could serve as an electron acceptor for VC and DCE oxidation. Clearly, the simple explanation that a strong oxidant such as Fe(III) was solely responsible for anaerobic oxidation could not account for observed oxidation under methanogenic conditions.

While it has been shown that DCE is also subject to a net oxidation under anaerobic conditions, it is presently unclear whether DCE can be transformed directly, or whether DCE transformation is initiated by reductive dechlorination to VC.

The fact that the kinetics of DCE mineralisation are different from those for VC, however, suggests that the processes differ, and this may indicate the reductive dechlorination to VC is a necessary step in the complete biodegradation of DCE.

Aerobic cometabolism

Another process that leads to microbial degradation of chloroethenes in groundwater systems is cometabolic oxidation. Wilson reported that methanotrophic (methane-oxidising) bacteria were capable of oxidising TCE to CO_2 under aerobic conditions. Since then, a wide variety of aerobic micro-organisms, which are able to oxidise TCE, DCE, and VC to CO_2 without accumulation of toxic intermediates, have been identified.

These include methane oxidisers, propane oxidisers, ethene oxidisers, aromatic compound oxidisers, ammonium oxidisers, isoprene oxidisers, and vinyl chloride oxidisers. Cometabolic oxidation of chloroethenes does not supply energy for microbial growth or metabolism. Rather, for each of these processes, the responsible micro-organisms contain nonspecific oxygenases that fortuitously oxidise chloroethenes to CO_2.

Consequently, aerobic cometabolism of chloroethenes requires the presence of oxygen and a primary substrate to initiate the production of a suitable oxygenase.

Because it is rare for significant concentrations of methane (or other cosubstrates) to coexist with dissolved oxygen, cometabolic oxidation is probably rare in groundwater systems. However, cometabolic oxidation of chloroethenes by subsurface micro-organisms has been successfully exploited for engineered remediation of chloroethene-contaminants in groundwater. TCE contamination in aerobic aquifers can be biodegraded by methanotrophic micro-organisms when methane is supplied to the subsurface in sufficient quantity to stimulate and support methanotrophic activity. Because methane and oxygen do not typically co-occur in groundwater systems, however, methanotrophic oxidation of chloroethenes is unlikely under non-engineered circumstances.

A number of *ex situ* investigations have demonstrated that toluene oxidisers can cometabolise chloroethenes under aerobic conditions. Moreover, many groundwater contaminant plumes are the result of mixed waste releases and, as a consequence, contain chloroethenes and aromatic compounds like toluene. Unfortunately, contaminant plumes that contain hydrocarbon compounds in sufficient concentration to support cometabolic oxidation of chloroethenes are characteristically anaerobic because of oxygen depletion by aerobic respiration.

Consequently, significant aerobic cometabolic oxidation of chloroethenes in such systems would be, at most, a transient phenomenon. However, it is possible to artificially create conditions favourable to cometabolic oxidation by injecting methane (or some other cometabolic substrate) together with oxygen into groundwater systems.

Chlorinated Benzenes

Chlorinated benzenes, like halogenated aliphatic compounds, are subject to biodegradation under both aerobic and anaerobic conditions. Outwardly, there is a good deal of similarity between biodegradation of chlorinated benzenes and ethenes. Under aerobic conditions, lightly chlorinated benzenes and ethenes can both serve as electron donors to support microbial growth. Similarly, under anaerobic conditions more heavily chlorinated ethenes and benzenes are both subject to reductive dechlorination. However, the biochemistry of these processes is different, and this leads to differences in the efficiency of chlorinated ethene and chlorinated benzene biodegradation.

Aerobic degradation

A number of micro-organisms have been described that are capable of utilising chlorinated benzenes as primary substrates. For example, Reineke and Knackmuss described a bacterium capable of utilising chlorobenzene as a sole carbon and energy source. Studies with this chlorobenzene-utilising bacterium led to the elucidation of the operative degradation pathway.

The general pathway of chlorobenzene degradation is shown in Fig. 33.1. The general strategy is the formation of chlorocatechol followed by *ortho* cleavage of the ring. The substituted muconic acid is then dechlorinated and the nonchlorinated intermediate products metabolised to 3-oxoadipate, which then enters the cell's tricarboxylic acid cycle. Note that this is virtually identical to the pathway of aerobic benzene degradation. The only substantial difference is the necessary elimination of the chloride from the organic compound at some point in the pathway.

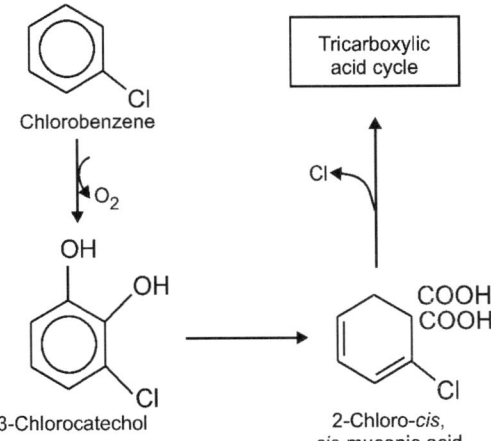

Fig. 33.1. Aerobic biodegradation of chlorobenzene.

Anaerobic degradation

In contrast to aerobic biodegradation, processes affecting chlorinated benzenes in groundwater systems is less well understood. Qualitatively, it seems reasonable that highly chlorinated benzenes such as di- and trichlorobenzene would be subject to reductive dechlorination under anaerobic conditions.

Polychlorinated Biphenyls

Polychlorinated biphenyls (PCBs) are used extensively as transformer fluids. The resistance of PCBs to microbial degradation is one property that made these compounds so useful in industrial applications. Nevertheless, they are subject to degradation under certain conditions. PCBs with relatively few chlorines per molecule are subject to oxidation reactions under aerobic conditions. For example, Clark described degradation of PCBs by mixed microbial cultures taken from soils and aquatic sediments. Furukawa described an *Acinetobacter* species capable of degrading 2,4,4′-trichlorobiphenyl. The pathway of degradation implied by these studies involves the familiar pattern of hydroxylation followed by *meta* cleavage of the alicyclic ring. These studies have documented the relative recalcitrance of heavily chlorinated PCBs to aerobic microbial oxidation. This may reflect interference of the hydroxylation and ring cleavage steps by chlorine molecules. Alternatively, this may simply reflect the fact that heavily chlorinated PCBs are more oxidised compounds than lightly chlorinated PCBs, and that this makes oxidation less energetically favourable.

Under anaerobic conditions, the predominant microbially mediated degradation process is reductive dechlorination. It was observed that reductive dehalogenation is analogous to the reductive processes documented for halogenated benzoic acids.

The picture that emerges from consideration of microbial attack on PCBs under aerobic and anaerobic conditions is that highly chlorinated molecules are resistant to aerobic oxidation but are subject to reductive dehalogenation. Conversely, lightly chlorinated PCBs are resistant to reductive dehalogenation but are subject to aerobic oxidation. This pattern has not been lost on researchers investigating possible bioremediation strategies.

Specifically, it has been suggested that a sequence of anaerobic degradation followed by aerobic degradation can result in complete mineralisation of PCBs. Unfortunately, the sequence normally followed in natural aquatic sediments—initial shallow burial and aerobic degradation, followed by deeper anaerobic degradation—is just the reverse of the pattern that would be most effective. However, it is evident that microbial processes have the potential to completely degrade these compounds if the proper sequence of redox conditions can be achieved.

Organochlorine Insecticides

Organochlorine insecticides such as DDT, lindane, and chlordane were extensively used in agriculture. However, because of their environmental persistence and because of their tendency to accumulate in the lipids of aquatic organisms, their use has been largely discontinued in the United States. They have been replaced by such insecticides as the pyrethroids, permethrin, and fenvalerate, which require much lower application rates. Nevertheless, because of DDT's environmental persistence it remains an issue in some areas.

Although DDT is very persistent in the environment, degradation does occur. The first step that must occur in order for degradation to proceed is the reductive dechlorination of DDT to DDD (dichloro-diphenyldichloroethane). This process occurs only under highly reducing conditions and does not appear to require microbial mediation. However, this reaction may be facilitated by reduced iron porphyrins present in mixed populations of anaerobic bacteria. In addition, DDT may be reduced to DDE (dichlorodiphenyldichlorethene). This appears to be a dead-end pathways, as further degradation of DDE is not generally observed.

Like DDT, other organochlorine insecticides are rather recalcitrant to degradation. The cyclodiene insecticides aldrin, chlordane, endrine, and heptachlor are generally considered to be persistent in most environments. Heptachlor may be oxidised to heptachlor epoxide by soil micro-organisms, and endrin is metabolised to form ketones and aldehydes that retain the attached chlorines. Lindane, the most soluble of the chlorinated insecticides, is also one of the most degradable, forming chlorinated benzenes with the mediation of fermentative soil bacteria such as *Bacillus* and *Clostridium.*

Chlorinated Herbicides

While PCBs and organochlorine insecticides are relatively toxic and resistant to microbial degradation, their mobility in groundwater systems is limited by their tendency to partition onto sediments and organic matter. Chlorinated herbicides such as atrazine and 2,4-D are much more soluble and tend to be transported more easily in the unsaturated and saturated zones of groundwater systems. Because of this relative mobility, relative rates of microbial degradation are important to their fate and transport.

Atrazine belongs to the general class of herbicides known as the s-triazines, a name that refers to the central ring structure that contains three nitrogen groups. Attached to the ring structure of atrazine is a chlorine atom, an isopropylamino group, and an ethylamino group. Atrazine is degraded by a number of biotic and abiotic mechanisms that involve removal of one or more of the functional groups as well as ring cleavage. The most important mechanisms in atrazine degradation are hydroxylation (H) and dealkylation (D) reactions. The dechlorination and hydroxylation of atrazine to produce hydroxyatrazine occurs both biotically and abiotically. When occurring abiotically, this process has been shown to be pH-dependent, with maximal rates occurring at a pH of about 8.0. The formation of hydroxyatrazine results in detoxification of atrazine.

Dealkylation of both the isopropyl and ethyl groups has been reported. Behki showed that particular strains of *Pseudomonas* were capable of utilising atrazine as a sole carbon source and that atrazine was metabolised via dealkylation reactions. This study also suggested that formation of deisopropylatrazine was favoured over deethylatrazine and that particular species were capable of dechlorinating either of these metabolites when another carbon source was provided. It was also observed that dechlorination of atrazine tended to proceed after dealkylation.

Chlorinated Phenols

Chlorinated phenols released into surface and subsurface sediments from wood-treatment facilities or as degradation products of chlorinated herbicides are subject to both aerobic and anaerobic biodegradation. As with other chlorinated organic compounds, however, the mechanisms and pathways involved differ substantially under aerobic and anaerobic conditions.

Aerobic degradation

Under aerobic conditions, two distinct mechanisms for the metabolism of chlorophenols have been identified. The first of these mechanisms is the methylation of one or more hydroxyl groups associated with the phenolic compound. Neilson showed that a species of *Arthrobacter* carries out this process. The chloroanisoles that are produced by this process are volatile and tend to escape to the atmosphere at land surface. As chlorophenols are typically present in the wood shavings used for litter in poultry-raising operations, the production of volatile chloroanisoles may result in contamination of chicken eggs or meat. The second mechanism for the metabolism of chlorophenols that has been extensively studied involves the formation of chlorocatechols from chlorophenols. A number of *Pseudomonas*, *Alcaligenes*, and *Nocardia* species have been described that carry out this and similar biochemical transformations.

For example, Hellwig showed that a *Pseudomonas* species could grow using 4-chlorophenol as a sole carbon and energy source with the complete degradation of the 4-chlorophenol. The initial step in this metabolic pathway was the conversion of 4-chlorophenol to 4-chlorocatechol with subsequent cleavage of the alicylic ring via either the *ortho* or the *meta* pathway.

Whether these or other biochemical mechanisms are involved in the aerobic degradation of chlorophenols in natural systems is difficult to say. However, the aerobic degradation of chlorinated phenols has been observed in soils, and this process has been suggested as a means for the bioreclamation of chlorophenol contaminated soils.

Anaerobic degradation

The anaerobic degradation of chlorinated phenols has been extensively studied because of the delivery of these compounds to sewage treatment plants and their subsequent degradation by activated-sludge micro-organisms. Boyd showed that chlorinated phenols were completely degraded by a combination of reductive dehalogenation and anaerobic metabolism.

The reductive dechlorination of chlorophenols was studied by Bryant, who showed that the order of chlorine removal from pentachlorophenol was predictable. In the case of one adapted sediment community, the chlorine in the *para* position tended to be removed preferentially to that in the ortho position and that the chlorine in the *meta* position tended to be removed last. These or similar reductive processes are probably the principal mechanisms involved in chlorophenol degradation in anaerobic aquifer systems.

Chapter 34

Microbial Fertilisers

INTRODUCTION

The term bacterial (microbial) fertilisers refers to preparations containing primarily active strains of the micro-organisms mainly bacteria in sufficient numbers. They are used either to fix atmospheric nitrogen or to solubilise plant nutrients like phosphates or to otherwise stimulate plant growth through synthesis of growth promoting substances. In addition to bacteria, blue-green algae have also been shown to fix atmospheric nitrogen, if they are inoculated into the soil and established in paddy fields. Preparations containing these can also be considered as fertilisers under a broad term microbial fertilisers. Of the different bacterial (microbial) fertilisers nitragin, a peat, lignite or oil based preparation containing effective strains of *Rhizobium* (root nodule bacteria) specific for different leguminous crops in adequate numbers is the most important. The other two bacterial fertilisers are *Azotobacterin*, a preparation containing cells of *Azotobacter chroococcum* grown on agar and phosphobacterin, a kaolin based preparation containing cells of *Bacillus megaterium* var. *phosphaticum*. Besides these a concentrated preparation of dried cells of blue-green algae are also useful for enhancing nitrogen content of paddy field soils.

The development of acetylene reduction technique and its use for measurement of N_2 fixed *in situ* and that of callus—*Rhizobium* studies enabled extensive and intensive investigations on biological nitrogen fixation in order to evaluate its ecological and agricultural significance. Using acetylene reduction technique, studies with soyabean and peanut crops were made by Hardy and others. They found that activity was initiated to 20 to 30 days in soyabeans, and 40 to 50 days in peanuts, and continued until senescence in soyabeans to 125 days and in peanuts to 145 days. Activity was very low during the initial stage mainly of vegetative growth; increased rapidly during fruit formation and maturation in apparent response to the needs of the plants and finally declined precipitously at senescence. In both cases less than 10 per cent of the total nitrogen fixation occurred prior to flowering, while greater than 90 per cent occurred during fruit formation and maturation. About 250 mg or less than 25 per cent of the nitrogen of each mature plant was supplied by fixation. About 100 kg nitrogen per hectare was fixed per season by soyabeans.

Indian data showed that through legume seed inoculation, an additional nitrogen to the extent of 11 kg/ha was taken up by gram grain over the uninoculated control and that in the case of soyabean the additional amount of nitrogen removed by beans in the inoculated plot was found to vary from 40 to 120 kg/ha depending upon the crop variety, soil and climatic condition.

Rhizobium survey conducted in the country showed that in about half the area the nodulation was poor for one reason or the other, though the legumes are grown in the area for a long time, and this might be one of the reasons for the poor yields. If atmospheric nitrogen present to the extent of 88,000 tonnes over an hectare of land can be effectively tapped by selection and use of efficient strains of *Rhizobium* specific for the variety of crop and suited to the local environmental conditions coupled with proper soil amendments, nitrogenous fertilisers can be conserved and a greater part be diverted for the use of new varieties of cereal crops where these are required in large doses.

While lack of quality control of cultures was mainly responsible for the absence of consistent responses in the earlier trials, the use of peat-based cultures of *Rhizobium* produced with adequate quality control resulted in significantly increasing the yields of leguminous crops. In this chapter, the techniques of production of peat-based cultures on a mass scale and quality control have been described. Quality control includes ensuring through serological technique that the culture used is true to type; the broth produced is free from contamination and the final product has the requisite number of *Rhizobium* cells/g of the material.

High temperatures reduce rapidly the count of *Rhizobium* cells. So, the commercially prepared cultures need be stored preferably at 5°C.

For the production of animal protein of quality, leguminous fodder is essential and for this purpose as well as for building up of soil structure, growing of legumes and grasses together is important. When new areas are reclaimed and new leguminous fodder crops are introduced, legume inoculation is a must. Pulse proteins are a rich source of essential amino acids and these as well as animal protein of high quality can be made available to the population through the growing of different food and fodder legumes, using efficient strains of *Rhizobium* and tapping atmospheric nitrogen cheaply through biological fixation.

Studies using acetylene reduction technique showed that the utilisation of added sucrose for nitrogen fixation by a soil, enriched in nitrogen-fixing bacteria (*Azotobacter vinelandii* or *Clostridium pasteurianum*), is extremely unsatisfactory. 200 kg of added sucrose caused 1 kg of additional nitrogen to be fixed and demonstrated the extreme limitation of free living nitrogen-fixing bacteria as a source of agricultural nitrogen. However, when associated with photosynthetic organisms the position is different. Blue-green algae and photosynthetic bacteria can carry out their activities in the presence of light. In the photosynthetic organisms, the bond energy is provided by photochemical reactions that convert radiant energy into chemical bond energy which is then used to drive the machinery of the cell.

In a system where we have blue-green algae, *Azotobacter* and photosynthetic bacteria *Rhodopseudomonas*, the photosynthetic evolution of oxygen by the blue-green algae, can provide an aerobic environment to *Azotobacter*. The utilisation of oxygen by *Azotobacter* may result in a partial or complete anaerobiosis in the immediate vicinity, where the anaerobic *Rhodopseudomonas* can thrive well. *Rhodopseudomonas* can use lower fatty acids and carbohydrates. In this course the bacteria can fix nitrogen and assimilate carbon dioxide. In the heterotrophic course, the heterotrophic bacteria *Azotobacter* can use the carbohydrates and excrete lower fatty acids. The carbon dioxide released in the heterotrophic dark course can be assimilated by the photosynthetic bacteria and blue-green algae. The evidence so far obtained stresses the need to harness the beneficial associative effects of blue-green algae, photosynthetic bacteria and *Azotobacter* for increasing nitrogen fixation in tropical countries like India.

It was found that 'Azotobacterin' increased biological activity in terms of total nitrogen fixed in the mixed culture of *Azotobacter chroococcum*, *Rhodopseudomonas capsulatus*, and *Cylindrospermum muscicola* than in pure cultures of each organism, the Indian data further showed that significant increases in paddy crop yields were obtained through algal inoculation. With high-yielding varieties of paddy where heavy doses of nitrogenous fertilisers were used, blue-green algal inoculation increased the yield at different levels of fertilisers, probably due to the synthesis of growth-promoting substances, like vitamins, amino acids and gibberellin-like substances. The yield per unit nitrogen applied was more when this was accompanied by algal inoculation than without it. But since experiments conducted to elucidate this were not many, there is need to examine this further through systematically conducted field trials at different places.

Field experiments conducted with *Azotobacter* showed that in the majority of experiments significant increases in crop yields were obtained. In the Soviet Union, 50 to 70 per cent of the field crops were found to increase in yield by about 20 per cent. Most of the estimates of non-symbiotic fixation under field conditions are in the range of 0 to 60 kg N/ha annually. Although the value of such inoculation might be considered due to the result of an enhancement of nitrogen fixation with subsequent release of fixed nitrogen in forms available to plants, the beneficial effects appear to be more due to production of growth-promoting substances like gibberellins and vitamins, since the response to *Azotobacter* inoculation was more in fertile soils, or in the case of vegetable crops, where large doses of organic manures and fertilisers were applied.

Owing to the low cost, however, the treatment would be profitable even if the increases in yield were relatively small. Accordingly, a careful examination of the effects of inoculation is desirable on a wider scale than hitherto employed. In one experiment where a combination of *Azotobacter* and *Rhodopseudomonas* was used it was observed that the rhizosphere of wheat crop supported very high counts of *Azotobacter* and resulted in significant increase in crop yield and nitrogen uptake. This effect can be attributed to the associative beneficial effects of *Azotobacter* and the photosynthetic organism *Rhodopseudomonas* discussed earlier. This suggests the possibility of tapping atmospheric nitrogen in tropical countries cheaply for the non-legumes, and needs more intensive study.

The Indian work with phosphate solubilising organisms showed that under specific conditions soil phosphate and/or fertiliser phosphate can be solubilised by the organisms *Bacillus megaterium* or *B. circulans* and crops were able to utilise this phosphorus for their nutrition recording significant increases in some field trials. Details are discussed in the section on 'Phosphobacterin'. However, more experiments need be conducted to define the exact conditions under which this can happen. Search for more useful organisms and methods of their manufacture and application is desirable in view of the paucity of phosphatic fertilisers and high-grade phosphate rock. The limited number of experiments conducted with radioactive phosphorus showed the possibility of utilising low-grade phosphate rocks more effectively through the use of phosphate solubilising bacteria. Here again the beneficial effect was considered to be partly due to synthesis of growth-promoting substances.

While there is need for more laboratory and field work at the research stations for elucidating the usefulness and limitations of inoculation with phosphate-dissolving organisms or *Azotobacter*, a stage has been reached where the development of legume inoculation is of immediate importance. Because of the obvious need for inoculation, the marked crop responses resulting from the use of suitable strains and the short time required to prepare, evaluate and distribute such bacterial preparations, work in this field is of highest priority.

Micronutrient deficiency like that of molybdenum, susceptibility of *Rhizobium* to high temperature and low moisture regimes, and suppression by pesticides are among the problems that are to be tackled in *Rhizobium* programme. Though these have been solved to a considerable extent through pelleting techniques and use of fungicides compatible with the survival of *Rhizobium*, there is further scope for research in these and allied fields. Most of the Indian soils are deficient in nitrogen, and as such, its presence or absence in adequate amount determines to a great extent the level of the yield. Nitrogen is present in the atmosphere to the extent of 88,000 tonnes over an hectare of land. Therefore, in the economy of nitrogen in the soils, due attention needs to be paid towards fixing the nitrogen to the maximum extent. In India, 17 per cent of the total cultivated land is under pulse crops which amounts to 24 million hectares. The amount of nitrogen fixed by root nodule bacteria associated with each legume is determined by soil and local conditions.

Recent work with soyabean indicated that peat-based culture gave increase in yield comparable to that obtained with imported nitrogen and additional amount of nitrogen removed by the beans in the inoculated plots was found to vary from 40 to 120 kg/ha depending upon the nature of the crop variety, soil and climatic conditions.

Most of the nitrogen varying from 50 to 80 per cent is concentrated in the plant tops of fodder legumes. So, for the soil to make appreciable gain in nitrogen, the plant tops have to be returned to it directly or indirectly through the grazing animal. Nutman observed that the effectively nodulated legumes growing vigorously can provide itself all the nitrogen it needs, even when none is available from the soil. There is also evidence that under certain ill-defined conditions the living legume root may excrete appreciable amounts of nitrogen.

Legumes have a special role to play in areas poor in protein because legumes are an excellent source of high quality protein. *Rhizobia* are widely distributed in soils of the tropics. Even in newly developed lands, not previously supporting economic legumes, these bacteria are to be found. However, many indigenous bacteria are wholly or largely incapable of bringing about appreciable fixation on roots of the appropriate host. Selection and use of efficient strains of *Rhizobium* together with other agronomic practices would lead to increased supply of quality protein to the population and maintain soil fertility at a high level.

Among the asymbiotic nitrogen fixers, *Azotobacter, Clostridium, Beijerinckia* and the photosynthetic nitrogen-fixing blue-green algae are important. In waterlogged paddy soils blue-green algae are important sources of nitrogen. The role of photosynthetic bacteria in tropical soils was shown by recent researches in this country to be highly significant specially in association with other asymbiotic nitrogen fixers like *Azotobacter.* Some of these organisms are known to synthesise growth-promoting substances and antibiotics and as such are important in enhancing plant growth and in combating plant diseases, though the amount of nitrogen fixed may not be large.

Phosphorus differs fundamentally from the carbon, nitrogen and sulphur cycles in that no natural channel exists for the return of the annual net loss. As such, it is a critical limiting factor for the continued functioning of the biosphere. High-yielding crops may remove as much as 10 per cent of all the phosphorus in the upper 15 cm of a fertile soil. But many of our tropical soils lock up large amounts of phosphorus that are made available after very lengthy periods. Micro-organisms play an important role in the mobilisation and immobilisation of phosphorus.

Solublisation of insoluble inorganic compounds is brought about by some of the species of *Bacillus, Pseudomonas, Mycobacterium, Micrococcus, Flavobacterium, Penicillium, Sclerotium* and *Aspergillus.* These bacteria and fungi bring about solubilisation by the production of organic acids. The chemo-autotrophs, ammonium and sulphur oxidising bacteria produce nitric and sulphuric acids, which convert the tricalcium phosphate to di- and monobasic phosphates leading to increased availability of phosphorus to the plants. The amount of phosphate brought into solution by heterotrophs varies with the carbohydrates oxidised and the transformation proceeds only if the carbonaceous substrate is converted into organic acids. The organic acids specially hydroxy acids promote solution of the mineral-liberate hydrogen sulphide which reacts with ferric phosphate to yield ferrous sulphide liberating the phosphate. Phosphate is released from iron and aluminium phosphate when the environment becomes oxygen deficient.

There is ample evidence for the biological alterations in the availability, solubility or oxidation state of potassium, manganese, selenium, tellurium, arsenic, zinc, copper, calcium, magnesium, aluminium and molybdenum. Since micro-organisms need a variety of anions and cations for their own development, immobilisation of the nutrients also takes place.

As long as there is decay and microbial cell synthesis, both mineralisation and immobilisation take place. For example, phosphorus content governs not the absence of one or the other transformations but rather the greater rate of uptake or release of the nutrient. If the substrate contains more phosphorus than the critical level (0.2 per cent), some is released. If the material has less than the critical level, less than needed by the microflora, phosphorus disappears from the environment as the net effects due to immobilisation.

Similarly, for the mobilisation of other nutrients critical values have been worked out, the value for nitrogen being 2 per cent.

NITROGEN: PREPARATION AND USAGE

Among the bacterial fertilisers, the most important are those containing the legume root nodule bacteria known as *Rhizobium* which functions in association with the host plant and helps in fixing atmospheric nitrogen. Nitragin is a peat, lignite or soil based preparation containing effective strains of *Rhizobium* (root nodule bacteria) specific for different leguminous crops, in adequate numbers. Its preparation, distribution and usage are discussed below, as it involves some of the principles involved in the production in general of bacterial fertilisers on a mass scale. Very briefly, the process of nodule formation begins with the infection of a root-hair or epidermal cell by the rhizobia with the appearance of a hypha-like 'infection thread' containing the bacteria.

The nodule itself is initiated when the penetrating infection thread approaches a preformed tetraploid cell in the cortex. This cell and neighbouring diploid cells rapidly divide giving rise to the nodule with four distinct zones; an outer cortical layer surrounded by parenchymatous tissue, meristematic area, vascular tissue and bacteroid zone or zone of nitrogen fixation.

Two types of nodules are found among leguminous plants, one the 'effective' and the other 'ineffective'. Effective nodules are fewer, larger in size, with a pink central zone which is the area of active nitrogen fixation. The host plant in all these cases is healthy in appearance. They vary in shape and size according to host and bacterial strain. They may be simple or compound, round, elongated or club shaped, single or in clusters, and are generally located about the main root and first formed lateral roots. They range in size from 2 to 6 mm for *Lotus* species, clover and vetch; 3 to 10 mm for medics; and 5 to 25 mm for lupins. The nodules of the serradella, cowpea and soyabean are generally round and vary in size from 3 to 10 mm.

Ineffective nodules are small, white, very numerous and scattered over the entire root system. The plants are generally yellow in colour, except subterranean clover which tends to be reddish, and are stunted in growth.

Six host or 'cross-inoculation' groups are recognised on the basis of ability of the rhizobia to form nodules with legume plants, as given in the Table 34.1.

Table 34.1. Host groups which form nodules with legume plants.

Host group	Rhizobium species	Representative hosts in groups
1.	*Rhizobium meliloti*	*Medics* and *Melilotus* spp.
2.	*Rhizobium trifolii*	Clovers
3.	*Rhizobium leguminosarum*	Pea and vetches
4.	*Rhizobium phaseoli*	Beans
5.	*Rhizobium lupini*	Lupins and serradella
6.	*Rhizobium japoricum*	Soyabeans, cowpea, peanut, *Glycine* spp., *Stylosanthes* spp., *Pueraria* spp., *Phaseolus atropurpureus*, *Desmodium* spp., *Lotus* spp., *Lotononis* spp., and other species

While this is a convenient and useful grouping for general purposes, there are certain exceptions, particularly in the group of hosts listed for *R. japonicum*. Some of these hosts, as *Lotononis*, have very specific rhizobial requirements and do not readily cross-inoculate with the other hosts listed.

The six host groups listed can be further sub-divided based on the effectiveness of the association formed with the infecting rhizobia. This aspect and the specific cases associated with the *R. japonicum* group are particularly important in legume seed inoculation. Establishment failure may result from: (i) the failure of the plant to nodulate, as the strain 'TA1' of *R. trifolii* does not nodulate under field conditions with the Woogenellup cultivar of subterranean clover; and (ii) ineffectively nodulated plants, as strain 'WU290' of *R. trifolii* is suitable for most cultivars of subterranean clover, but is ineffective in nitrogen fixation with white clover. Such host and strain specificities must be recognised if seed inoculation is to be successful.

For prompt nodulation, the inoculum rhizobia must multiply in the rhizosphere to a population sufficient for infection of a majority of the first available foci of infection. These foci are available for a limited time only and change position with the growing root tip. Thus for subsequent infections, the rhizobia must move down the rhizosphere with the root tip. The soil environment is not always suitable for rhizobial multiplication so that considerable delays can be incurred before enough sites are infected to satisfy the nodule requirement of the plant. Only a certain proportion of infections develop into nodules. This proportion can vary greatly with each host species and rhizobial strain. In the field where the establishing plant is dependent on an early nodulation for its nitrogen supply, the delay in the formation of nodules will reduce the chances of the plant becoming well established.

Therefore, emphasis is to be placed on the number of rhizobia in the inoculum and on the survival of the inoculum on the seed. For this the bacterial fertilisers 'Nitragin' should contain a high count of 10^9 efficient *Rhizobium* cells per gram of the material.

The important points to be noted are: (i) the longevity of the culture before use, (ii) survival of the organism on the seed before and after sowing, (iii) establishment of the organism near the rhizosphere of the plant, and (iv) the infection of the root area and subsequent proliferation and invasion in the root, forming effective nodules leading to fixation of atmospheric nitrogen.

Some of the *Rhizobium* strains can tolerate wide variation in pH or temperature and moisture regimes and are robust, while others are very sensitive to environmental conditions. This naturally leads to selection of strains for the specific purpose.

After obtaining a number of strains from the nodules of healthy and vigorously growing leguminous crop, the first step is to test the different strains in agar slants for its infectibility and formation of nodules on the host plant. Later, some of these selected on the basis of good nodulation and the maximum plant weight are tested in pot culture experiments: (i) using sand with nutrient solution excluding nitrogen, (ii) in sterilised soil, and (iii) in unsterilised soil. Based on their performance in the pot culture a few strains are tested in statistically laid field experiments, then multiplied on a mass scale and supplied to cultivators in proper carrier, which may be either agar slant or soil culture or peat-based or lignite-based culture.

Manufacture of *Rhizobium* Inoculants

Having selected a suitable bacterial strain that can form effective nodules in association with selected legumes over suitable range of environmental conditions, it is necessary to lyophilise them to maintain their efficiency, as subculturing at monthly intervals on artificial medium ultimately leads to deterioration of the activity of the organisms. Whenever it is required for use, the tube containing the lyophilised material is cut open and the lyophilised cells are reconditioned. These are grown on sufficient number of agar slants to be used for mass multiplication during the season.

It is to be recognised that the rhizobia are sufficiently mutable to require their regular continued evaluation with the relevant hosts. Such a programme should be combined with the systematic long-stored, untouched culture in lyophilised condition as mentioned above. This is necessary to assure continued symbiotic efficiency of the cultures issued as seed inoculum.

At the beginning of each season an ampoule or paraffin culture of each strain is opened and a series of 12 to 15 new paraffin cultures prepared. Of these, 12 serve as 'mother' cultures for the preparation of manufacturer mother cultures at monthly intervals.

These are supplied to manufacturers at monthly intervals so that manufacturers do not have the responsibility of ensuring that strains remain true to type. In this way, continuity of strains within and between years can be assured. The remaining paraffin cultures are held in different collections as reference material.

While under laboratory conditions, it is practically easy to produce a broth having a titre value varying from 10^3 to 10^6 cells/ml, when the scaling up operation to produce large quantities of broth is taken up, certain special precautions have to be observed. No contamination should be allowed because even a single contaminant which has a mean generation time of a few minutes reduces the *Rhizobium* count considerably during the period of 5 days in the case of slow-growing species of *Rhizobium* like that of *Rhizobium japonicum* specific for soyabean. In practice, therefore the growth of a broth culture of *Rhizobium* requires strict aseptic control in a vigorously aerated culture. Other practical considerations are the economics of essential substrates, their method of sterilisation and the question of sufficient aeration.

The manufacturer inoculates the broth in the fermentor (Figs 34.1 and 34.2) with the growth washed from the surface of a large agar slope which amounts to 0.2 per cent inoculum. The side-arm flask containing the agar is connected prior to sterilisation by means of its side-arm and rubber tubing with a second side-arm flask that contains the requisite amount of water, which is used for washing the growth from the agar slope in the first suction flask.

A rigorous system has been developed to guard against contamination at the time the vessel is inoculated. The brass entry tap of the fermentor is thoroughly heated, cooled with sterile water to prevent jamming and the rubber tube connecting the two flasks is cut and attached to the brass tap with great care while the fermentation vessel is under positive pressure. Inoculation is done when the broth is under 30°C and growth allowed to take place at 27°C.

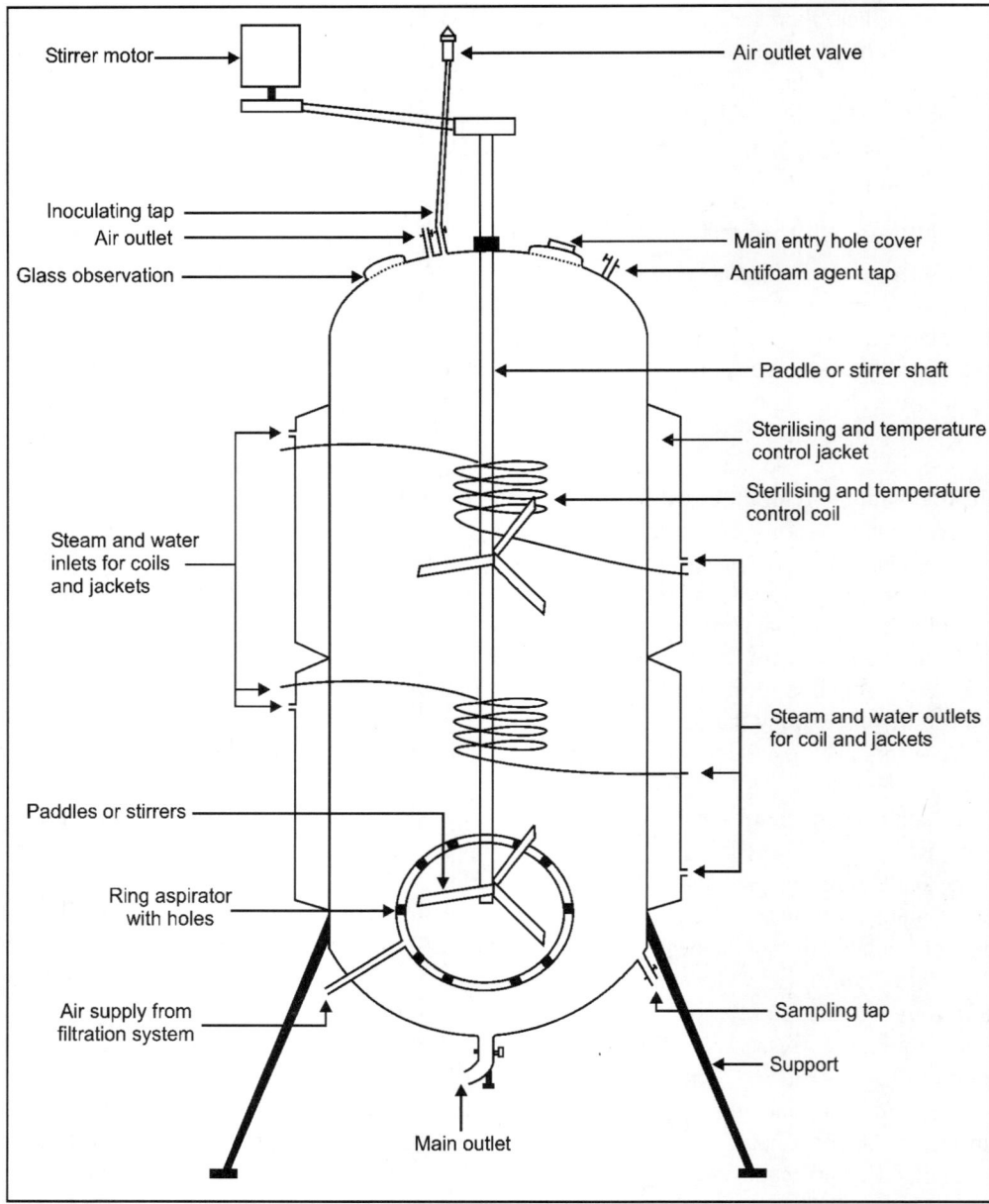

Fig. 34.1. Fermentor vessels of 1000 litres capacity.

When the broth is produced in fermentors, the sterilised medium is allowed to circulate from the bottom to the top while the stirrers provided with plates, one situated at the bottom, and another at the centre stir up air throughout the liquid. Aeration is regulated by forcing sterile air through spargers at the bottom of the fermentor. An oxygen partial pressure of 0.15 atmosphere is optimum for respiration. Growth is not increased by violent aeration or agitation. Approximately 5 litres of air/hr/l will be adequate.

Maximum growth is reached in about 24 to 72 hours, depending upon whether the organism is a slow grower or fast one. A good quality broth should have at least 500×10^6/ml viable rhizobia of the specified type and should be free of contaminants.

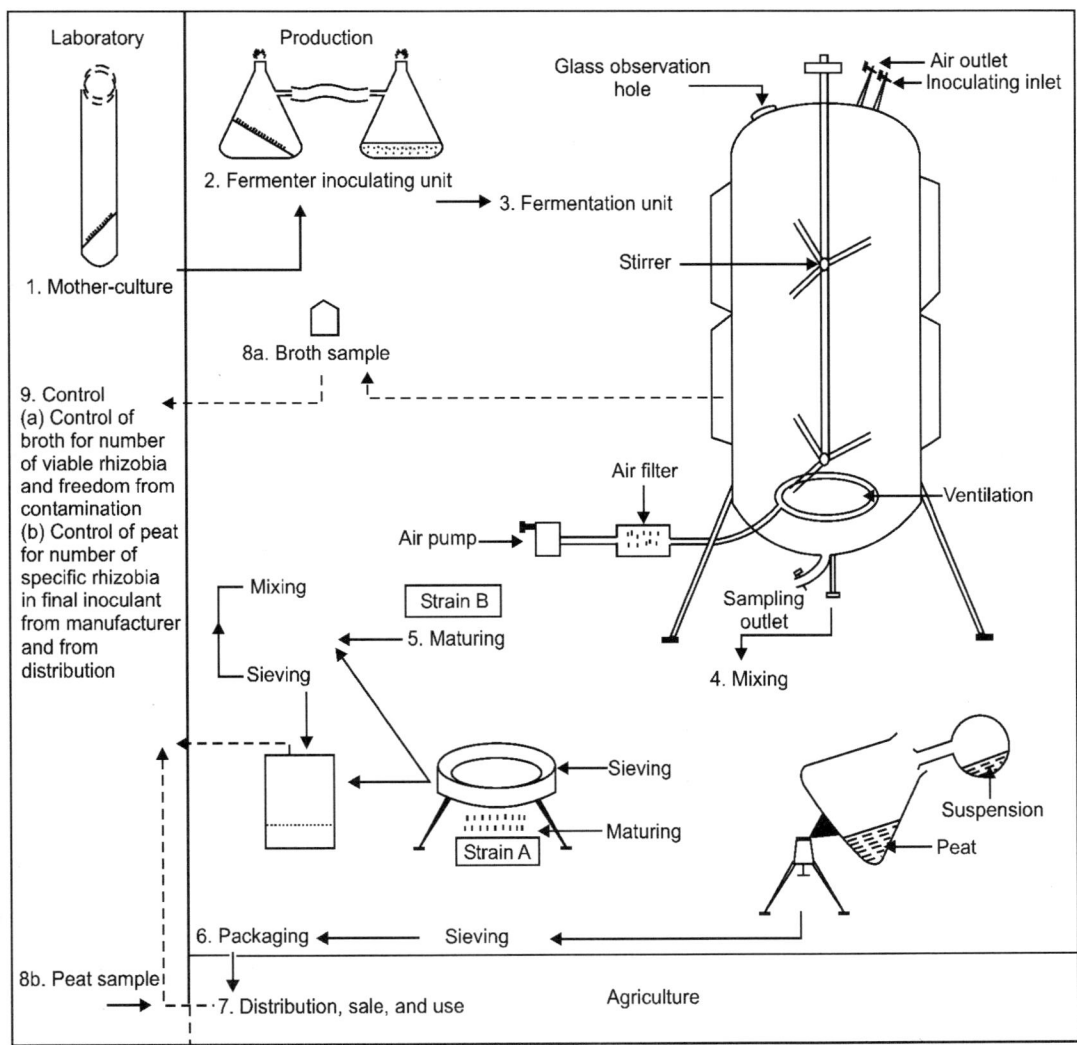

Fig. 34.2. Flow-sheet—Commercial production of legume inoculants using a peat carrier-base.

This broth is added aseptically to finely ground peat, or lignite sieved through at least a 300 mesh sieve. Peat is sterilised at 120°C for 2 hours on two consecutive days in thin layers in trays. This peat should normally contain about 90 per cent organic matter. Peat containing lower percentage can be used, but the keeping quality is considerably less.

The acid peat is to be neutralised with lime prior to sterilisation. However, frequently this method does not completely sterilise the peat, but the contaminants that are left over are usually not found to inhibit the growth of *Rhizobium*. Depending upon the water-holding capacity of the peat usually the ratio of peat to broth added is 2:1 to 3:1. After thorough mixing the peat blended with broth is spread in thin layers in covered trays and incubated for 48 to 72 hours at 25°C. During this period heat generated by the wetting of the dry powder is dissipated.

The aggregates are broken up by intimate mixing or milling and finally a pulverised homogeneous product is obtained. This should pass through 250 mesh sieve. This is finally packed in polythene film (0.038–0.051 mm gauge) bags.

Ultimately, the peat-based culture should contain at least 10^9 cells/g after incubating it for 3 to 4 weeks, as per the standards revised by the Australian Inoculants Research and Control Service, for pure culture inoculants and 0.5×10^9/g for clover, lucerne and vetch which are based on unsterilised peat. When this peat culture is added to the seed, it should be possible to establish 10^3 to 10^4 cells per seed and under adverse conditions, it is essential that the count of *Rhizobium* established per seed should be of the order of 10^5 or even 10^6 by raising the quantity of peat culture applied. When testing the quality of peat cultures, the plate counts give fairly reliable results, in high count peat cultures, but with older cultures difficulties in detecting *Rhizobium* in the presence of other organisms necessitate the use of a 'most probable number' method based on the ability of aliquots of successive dilutions to produce nodules on plant growing aseptically from surface sterilised seeds. For small seeds the agar tube method and for large ones the bottle-jar method is used.

Problems like seed-coat toxicity are known to adversely affect the survival of *Rhizobium* on the seed. It was found that soaking the seed in water for a few hours removes the water-soluble toxic substance and later coating the seed with the peat-based culture usually results in establishing the requisite count of *Rhizobium* on the seed.

Pelleting with finely-divided lime (calcium carbonate) provides a fair protection against acidity in soil and fertiliser at the time of sowing. The pelleting material may also provide a small physical barrier between the rhizobia and the seed-coat and thus protect against the toxic effects of the latter. Lignite may be used in place of peat.

Soil + Farmyard Manure-based Cultures

Where peat or lignite is not available the following procedure may be followed:
1. Obtain good fertile garden soil, sieve through 70 mesh and autoclave at 15–20 lbs for 2 hours.
2. Obtain some well-rotted farmyard manure, sieve through 70 mesh and autoclave at 15–20 lbs for 2 hours.
3. Sterilise mineral medium containing 10 per cent sugar (molasses or *gur*) and autoclave.
4. Mix equal amounts of (1) and (2) above and add 100 ml of medium-c per kg of the mixture.
5. Add the standardised culture of the rhizobia to (4) above. The amount is to be determined on the basis of standard curve of turbidity and viable counts. The final product should be such that 400 grams wet weight should provide 10^6 viable rhizobia per seed and sufficient for the number of seeds required for one hectare. This will have to be predetermined by actual weighing of 100 seeds and taking average and calculating on the basis of seed rate for each crop. About 400 grams of this can be added to each polythene bag and sealed.

Quality Control of Legume Inoculants

The following are the tests on samples of manufacturer broth culture.

Qualitative

1. At a pH: At a pH below 6 or greater than 8 a broth is suspected as being contaminated.
2. Agglutination with specific antiserum: A broth should give a definite agglutination with 1:100 specific antiserum. Partial clearing can indicate gross contamination.
3. Absence of growth on glucose-peptone agar: (Media no. 1—Annexure I). A loopful of broth is streaked on to this medium and incubated at 30°C overnight. Rhizobia grow only poorly, if at all, under these conditions so that any appreciable growth indicates gross contamination. Contaminant growth is frequently associated with a pH change which can be observed by the inclusion of an indicator such as bromo-cresolpurple in the medium.
4. Smear and gram stain: A smear prepared from undiluted broth should be free of Gram-positive cells. A few Gram-positive cells in occasional fields, which may be due to dead cells in the medium, can be disregarded.
5. Streak on yeast-mannitol agar: (Media no. 2—Annexure I). A loopful of broth culture is streaked onto a plate of this medium as a check on the characteristic colony growth of the strain concerned. It also serves as a means of detection of contaminants that can be missed by the Gram-stained smear, e.g. a Gram-negative rod of similar morphology to *Rhizobium*, or by the glucose-peptone agar, e.g. an organism not growing on this medium.

Quantitative

1. Total count: This is done by using a 1:100 dilution of the broth sample in sterile water. A sample of this material is examined on a Petroff-Hausser bacterial counting chamber using phase contrast illumination. This provides an immediate first assessment of the count and can also reveal gross contamination such as cocooid forms or 'chains' of cells of various bacilli or actinomycetes.
   The total count will include non-viable cells and therefore can be used as a guide. A broth that fails to provide 1000×10^6 rhizobial cells per ml in the total count is likely to provide less than 500×10^6 viable cells per ml in the plate count.
2. Viable or plate count: Dilutions are prepared in the usual way in sterile tap water and 1 ml aliquot plated at 10^{-6}, 10^{-7}, 10^{-8}, in yeast-mannitol agar, glass beads, sufficient to cover the bottom of the first dilution bottle, help in the dispersion of cultures. Typical colonies of the fast growing rhizobia are counted after 4 to 6 days at 25°C or after 7 to 10 days for the slow growing forms. Plates are examined also for the presence of contaminants.
3. Acceptance or rejection of broths: Only broths that have at least 500×10^6 viable rhizobia per ml of the specified type and that are free of contamination are accepted for the manufacture of peat culture. The peat cultures are prepared by mixing one part of broth to two parts of dried finely ground peat.

This series of tests demands that broth cultures be produced as pure cultures and thus imposes a difficult task since the rhizobia grow slowly as compared to most other organisms. They take 3 to 10 days to reach maximum viable numbers compared to 1 to 2 days with many other industrially used organisms. In addition the rhizobia do not have the advantage of growing in a selective medium and are weakly competitive for growth in the presence of more vigorously growing contaminants.

The following tests to be adopted on peat cultures: Plate counts give reliable estimates of rhizobia for high count peat cultures. Where more accuracy is required and in older cultures where a high proportion of other organisms make rhizobial colony recognition difficult, a plant infection technique must be used. The plant-infection technique is based on the ability of aliquots from successive dilutions to produce modules on aseptically growing plants.

However, a shortened process may be used for routine evaluation. This consists of preparing a 10-fold dilution series of the sample by weighing 30 grams peat to 300 ml water to provide a 10^{-1} dilution followed by mechanical shaking for 15 minutes. Further dilution is made by adding 1 ml of the 10^{-1} dilution to 99 ml of water to give a 10^{-3} dilution, and thus to provide 10^{-5}, 10^{-6}, 10^{-7}, 10^{-8} dilutions. Aliquots of 0.2 ml of appropriate dilutions; i.e. 10^{-6} and 10^{-7} are spread on the surface of a yeast-mannitol-agar medium containing 1 per cent w/v of an 1/400 aqueous solution of congo red. Aliquots of 0.2 ml are also taken to single tubes containing the host plant growing aseptically in a nitrogen-free plant nutrient medium at dilutions 10^{-6}, 10^{-7} and 10^{-8}. This permits a culture to be defined as below standard, 10×10^6 rhizobia/g; doubtful, 10 to $100 \times 100 \times 10^6$/g; satisfactory, 100 to 1000×10^6 g; and very satisfactory, 1000×10^6/g.

The below standard and doubtful material is discarded. Under exceptional circumstances doubtful material can be used with a 2-month expiry period. The satisfactory and very satisfactory groups have an expiry of 6 months. These expiry periods were determined on the survival of the rhizobia in peat culture so that the number of rhizobia in a peat culture graded as doubtful would be expected to have the same number of rhizobia at the end of 2 months, as a satisfactory culture at 6 months. No extra expiry is given for very satisfactory cultures.

Legume inoculant manufacturers are given within 24 hours of a preliminary result for broths based on the pH, agglutination, glucose-peptone agar, Gram-stained smear and total count tests as described above. With a satisfactory preliminary result, broth is mixed with peat and held as single strain material. The final broth result based on a satisfactory viable count permits manufacturer to blend single strain peats, where more than one strain is recommended, and to dispense into packets. This material is withheld from distribution until the viable plate counts on representative peat samples of the batch have been obtained. The expiry times of two or six months are based on this test and confirmed by the plant nodulation results.

Quality control of legume inoculants in Australia, by the U-DALS group, completed its tenth year of operation. In this period, inoculant quality has increased from useless and poor quality cultures to a situation where consistently high-quality cultures are available to farmers. The first and perhaps most important phase of control is that a collection of suitable cultures be assembled and maintained in a form that minimises the chances of variation and contamination. Each culture in the collection should be checked regularly for viability and nitrogen-fixing ability so that if a recommended strain is found faulty it can be replaced immediately by a strain of known performance.

The second phase of control is that of quality of cultures. The quality of broth and peat cultures, as judged by the procedures above, showed a very marked improvement during the first three years and has been maintained at this level. The information in Table 34.1 illustrates the improvement that has been made. It should be noted that a relatively constant number of rejections due to contamination were recorded and serve to emphasise the need for continued control. By the fourth and fifth seasons of control rejections based on low counts of rhizobia, fell from 20 to 25 per cent to a low level of 5 to 8 per cent. A corresponding increase in percentage broths approved was obtained as problems were resolved. During the last 7 years the percentage of broths approved has remained between 80 and 90 per cent. However, the critical part of the quality control programme consists of evaluating the quality of the peat cultures prepared from approved broths. The results of tests for quality of peat at the time of preparation show that quality continued to improve throughout the 10-year period.

The proportion of cultures graded as 'failure' declined rapidly throughout the period and 'doubtful' cultures declined significantly in the second half of the period. The percentage of very satisfactory cultures rose steadily during the last five years so that approximately 50 per cent of cultures reaching the market in 1965/66 were of this category. This improvement was not fully reflected in the results tests on material sampled on the retail market and indicated that high quality of cultures was not maintained during distribution.

The contribution made by research to inoculant quality may be considered as a third phase of control. The information gained from studies on growth and peat cultures has contributed substantially to the development of techniques and procedures which have resulted in improved quality.

The need for quality control of legume inoculants and to establish control groups has been recognised in Uruguay also. Locally prepared inoculants were not good and imported materials gave variable results. Within a two-year period a manufacturing unit, capable of meeting the entire needs of Uruguay's rapidly developing pasture improvement programme was established along with the selection and testing of the necessary rhizobial strains, and quality control facilities.

Large-seeded legumes

The following bottle-jar method is used: A 20-oz bottle, that has had its bottom cut-off so as to provide a flat finally-ground finish, is inverted in a quartz preserving jar in such a way that it sits firmly on the rim of the jar and has the lip of the bottle clear of the jar. A slotted and perforated screw cap holds a wick so that it makes ample contact with the sand in the bottle and projects down to the bottom of the jar. A little cotton-wool on the screw cap prevents the contents of the bottle falling into the preserving jar. Well-washed course sand is used to fill the bottle to 5.18 cm of the top and is moistened with a dilute solution of KNO_3 (0.001 per cent). The lower compartment (jar) is filled to within 2.54 cm of its top with the seedling solution of medium 3, diluted to one-fifth. The open end of the bottle is capped with a 10 cm petri dish lid (or beaker) and the whole is covered with a waxed paper-bag for sterilisation (4 hours at 121.6°C). The portion that covers the junction between bottle and jar is left in place when the upper part is opened at sowing.

Seeds are sterilised and germinated as under Method 1, and planted aseptically just below the surface of the moist sand. When the young seedlings are established, they are inoculated with 1 ml aliquots from the successive dilutions and the surface of the sand carefully covered with a 1–2 cm layer of small, dried and sterilised gravel to prevent aerial contamination. From this stage the petri dish or beaker can be left off to permit full plant development. No difficulty is experienced in keeping controls free of nodulation even under fairly dusty conditions in the glass-house. Uninoculated and nitrate controls and units inoculated with counted suspensions of the specific *Rhizobium* need also to be provided.

Plants should be lifted and the roots washed free of sand at about 8 weeks or when the units known to have had rhizobia added are well nodulated. Again, if the effectiveness of fixation is also to be assessed the nodule count will have to be delayed until there is sufficient differentiation between N-free and nitrate controls (seedling solution containing 0.05 per cent KNO_3) (Annexure I).

Methods of Application of Nitragin (*Rhizobium* Inoculants)

Proper use of the inoculum, prompt sowing and soil improvement are essential to avoid undue loss of viability between inoculation and sowing and permit rhizobia to multiply to a level that saturates the root zone. Peat based or soil-based cultures are suspended in 12.5 per cent sugar or gur (jaggery) solution in water. The sugar or gur solution is boiled and cooled before the culture is sprinkled on the seeds and the seeds are thoroughly mixed so as to have uniform coating. A count of 1000 viable cells per seed is to be attained at the time of treating the seed and the quantity of culture used to be accordingly adjusted.

The manufacturer's instructions contain details of quantities of materials to be used for different sizes of seeds, to reach the required standard of inoculation.

Usually, 400 grams of peat or lignite based culture or 900 grams of soil based culture would be sufficient for the quantity of seed required per hectare. Under adverse conditions, it is essential that the count of rhizobium established per seed should be of the order of 10^5 or even 10^6. This is accomplished by pelleting the seed using adhesives like gum arabic. A 40 per cent solution of gum arabic (high grade colourless one with particle size 8 mesh is preferred) is made, the peat culture is added to it and thoroughly mixed. The seeds are evenly coated with the above suspension. Finely-ground lime passing through 300 mesh sieve is added and the seeds are mixed, so as to have even coating of lime. The pellets so formed should appear dry without loose lime left on the surface. Pellets should be firm enough to drop on the floor without damage. When dry, the pellets should withstand light rolling between the fingers.

Rhizobium inoculants

Experiments conducted by various scientists showed that berseem rotated with cowpea and wheat resulted in significant increase of soil nitrogen and organic carbon when berseem crop was adequately manured with phosphates. They found vigorous microbial activity and higher microbial population in plots treated with phosphates. Ammonification, nitrification and nitrogen fixation were also increased in these plots.

The cultivation of inoculated legumes with the application of phosphates helps in the reclamation of saline or alkaline lands. In trials at Mettumaradur (Tamil Nadu) a crop of *dhaincha*, raised in the lands after the preliminary correction of soil alkalinity using gypsum as an amendment, proved effective in improving soil fertility and increasing paddy yields up to 3363 kg/ha in the treated area where no crop grew before.

It has been observed that the protein content of cereal crops is increased when grown in association with legumes in suitable proportions.

To derive maximum benefit by growing legumes for increasing yields and high nitrogen content and building soil fertility and reclamation: (i) the seeds of the legumes should be inoculated with the relevant strains of *Rhizobium* suited to the locality, and (ii) lime or phosphorus or minor elements like boron, molybdenum and copper should be applied, as and when required for correcting soil acidity or as fertilisers.

The right kind of bacteria must be present in the soil for each leguminous crop. When a legume is grown for a number of years, the right type of nodule bacteria usually gets established in the soil, unless it is destroyed by bacteriophage or parasitic fungi. This can be avoided by inoculating the legume seed at the planting time with nitrogen-fixing bacteria, as it ensures that the seedlings are exposed to the right kind of nitrogen-fixing bacteria early in the growing season.

Where the soils are very poor with respect to nitrogen there should be addition of adequate amount of nitrogen to meet the crop requirements till nodules are well established (i.e. usually up to six weeks from the date of germination).

The use of legume inoculants has given consistent increase in yields in different soils all over the world. When grown for the first time, 25 to 60 per cent increase in the yield of berseem and other legumes was obtained and in some cases the increases were even 200 per cent.

Pelleting

Inoculation followed by pelleting either with rock-phosphate or with lime considerably increased the dry-weight of shoots of soyabean plants.

The results show that pelleting the seeds with lime and peat culture of *Rhizobium japonicum* proved beneficial to nodulation, especially when the moisture in the soil was conserved by mulching with wheat straw or covering the top of the soil with sand. The dry-weight of the plants in the wheat straw mulched treatment was superior to others. These results reflect that pelleting of seeds coupled with mulching of soil may ameliorate the soil conditions unfavourable to nodulation, like cracking heavily in the absence of moisture in some of the soils.

As for the practice of *Rhizobium* inoculation in the USA, Burton observed that *Rhizobium* inoculants are prepared for alfalfa, clover, pea, bean, lupine, soyabean, cowpea and lotus groups mainly. Discussing the field results he observed: 'Legume inoculation has been a tremendous asset to agriculture in the United States. Effective cultures of rhizobia properly applied to legume seeds bring about large increases in yield and improvement of quality in leguminous crops. Culture of such plants as alfalfa (*Medicago sativa*), soyabean (*Glycine max*), lupine (*Lupinus* spp.), and birdsfoot-trefoil (*Lotus corniculatus*) would not have been practical without the aid of nodule bacteria.'

Nonetheless, substantial increases in yield are often obtained from inoculating even in fields which have grown the particular leguminous crop for several years. Recent tests with soyabeans in Illinois showed yield gains of 14 to 25 per cent, despite the fact that soyabeans had been grown in this particular field for many years, and that no visible differences in growth or colour were apparent between the inoculated and non-inoculated plants at any time during the growing season. The nodulation was superior on the inoculated plants.

Under field conditions, effective rhizobia produce clusters of large nodules on the main tap root and nodules which are red or pink on the inside. In contrast, the soil-borne rhizobia produce small, anaemic nodules which are scattered throughout the root system. These lack the leghaemoglobin associated with effective nitrogen fixation.

In the light of the remarkable gains in crop production, which are possible through the use of effective cultures of the rhizobia, one wonders why all legume seeds are not inoculated. Yet, it is known that in some areas the soils do harbour an abundant supply of effective root nodule bacteria for the crops which are grown. Approximately 42 per cent of the legume seeds, planted in USA each year, are inoculated. The percentage of inoculated seeds planted varies with different legumes. It is estimated that 80 per cent of all lupine (*Lupinus angustifolius*) seed is inoculated whereas with beans (*Phaseolus vulgaris*) only 6 per cent is inoculated.

Selection of rhizobial strains

Burton stressed the need to screen rhizobial strains not only for nitrogen-fixing ability on specific host, but for ability to survive on seeds, proliferate in the soil, and induce nodulation under natural soil and climatic conditions where the host is adapted, depending on the rhizobia's ability to compete with natural flora of ineffective rhizobia capable of nodulating the test plant. The ability to compete is influenced by the inoculum level and method of application to the seed.

Problems in Legume Establishment

Legume establishment may fail for one or several reasons. The more common reasons for failure are:

Failure of seed to germinate

This is due to the use of poor quality seed and lack of moisture sufficient for seed germination.

Failure of seedlings to become nodulated

In this case the majority of young seedlings would show nitrogen deficiency symptoms while a few plants may appear healthy and be nodulated. An insufficiency of rhizobia is generally the cause of this condition and may be due to several factors:

Poor-quality inoculant

This may be the result of an initially low number of rhizobia in the culture but occurs most frequently where inoculants have been used beyond their recommended expiry date or where inoculants have been badly handled. A common error is frequently made by leaving inoculants stand in hot milk cans, cars, trucks; sheds and shop windows where temperatures very often reach levels detrimental to the survival of the root-nodule bacteria.

Adverse conditions

Such conditions for the multiplication and survival of the rhizobia at sowing are caused by contact with acid soils and fertilisers. The limiting values of soil acidity for rhizobial survival and nodule-formation are lower for clovers, e.g. pH 4.7 to 4.8, than for medics, e.g. 5.8 to 5.9. Neutral to slightly acid conditions, or pH 6.5 to 7.0 are generally considered optimal for most rhizobia.

Soils which are more acid than the optimum for nodule formation are frequently the cause of failure where there are long delays between sowing and seed germination. In soils where the pH is more acid than 4.5 to 5.0 the rhizobia frequently do not survive even when seed germination takes place immediately.

Similarly, rhizobia are killed when inoculated seed comes into contact with acid fertilisers at sowing. Superphosphate with a pH of approximately 2 is the most toxic of the commonly used pasture fertilisers. These hazards can be greatly reduced by adopting such precautions as lime-pelleting, sowing with neutralised fertiliser, and band seeding so as to avoid direct contact between the rhizobia and the acid conditions.

Inoculation with the wrong inoculant

The use of clover inoculant, for example to inoculate lucerne seed will result in plants without nodules. The major host groups that are previously described should be used as a guide. However, there are some exceptions. For example, the use of general clover inoculant (strain TA1) with Woogenellup subterranean clover results in only a few effectively nodulated plants. In another case the host species *Lotononis bainesii* requires a specific strain of *Rhizobium* for nodulation. The farmer should be guided by the recommendations on the packets and, if any doubt exists, then it is a wise precaution to seek advice.

Strains of rhizobia that survive and compete only poorly even under relatively favourable conditions

The failure of *Dolichos lablab* to become well nodulated under sod-seeded conditions in Australia has been attributed to an inherent failure of the inoculum strain to colonise the immediate root zone and soil. Even at high inoculum levels only 50 to 60 per cent of plants were nodulated.

Pesticide treatment of seed with concurrent inoculation

The majority of the commonly used fungicides, insecticides, nematicides and herbicides are toxic to the root-nodule bacteria. Dieldrin when used at rates recommended for insect control is relatively safe. Thiram can be used safely. Compounds of copper and mercury and the systemic insecticide, Rogor, are toxic. Lime-pelleting reduces the severity of the effect of these materials on the inoculum but not sufficient to recommend concurrent inoculation and pesticide treatment.

Ineffective nodulation

In this case the majority of plants are nodulated but show nitrogen deficiency symptoms. This situation may result from:

1. The presence of a large volunteer soil population of rhizobia which competes with the rhizobia applied on the seed for nodule formation and which is ineffective in fixing nitrogen with the introduced legume. This does not imply that the volunteer strains are more competitive. The advantage may well be quantitative where the volunteer population outnumbers the rhizobia applied to the seed, although seed applied rhizobia are considered to have a positional advantage being concentrated near the emerging root tip at seed germination. Strain competitiveness is relative and can be judged only for each situation. In one such instance, Ireland and Vincent found that an inoculum level per seed, for subterranean clover, at least equivalent to the per gram of soil population of volunteer white clover rhizobia was required for successful nodulation by the inoculum strain. Where soil numbers are relatively high, e.g. greater than 1,00,000 per gram, equivalent seed numbers cannot always be achieved unless high quality inoculants are used.
2. The use of inoculants containing ineffective strains of rhizobia: This situation is likely to arise where no previous information is available for the strain requirements of the host being used and where inoculants are produced without any control being exercised in respect of strains used.
3. The use of the wrong inoculant: For example, *Rhizobium trifolii*, (strain WU 290), which is used specifically for Woogenellup subterranean clover is ineffective with white clover.

Nutritional deficiencies

These may be divided into two groups: (i) those responsible for acute nitrogen deficiency symptoms, such as soil acidity, calcium, molybdenum, and (ii) those indirectly influencing the fixation of nitrogen, such as phosphorus, potassium, magnesium, manganese, sulphur, boron, cobalt, copper and zinc.

High levels of iron and aluminium associated with pH levels of less than 5.5 to 6.0 result in phosphorus fixation with an overall reduced availability of phosphorus. Increasing acidity below pH 6.0 reduces molybdenum availability and induces aluminium and manganese toxicities. Correction for acidity is important in such cases.

Phosphorus deficiency is important because of its role in protein synthesis. An effectively nodulated plant under phosphorus deficient conditions will make no better growth than a nitrogen-deficient plant. Calcium is required for nodule formation as a component in cell wall development. It also helps in counteracting soil acidity.

Molybdenum is important because of its role in the nutrition of *Rhizobium* and in the nitrogen-fixing process. Nodules rich in molybdenum fix at least two or three times as much nitrogen as nodules deficient in molybdenum.

Plant competition

Failure of the legume plant to become promptly nodulated and established is frequently due to competition for such factors as moisture, nutrients, light and space. Competition from grass is an important factor in sod-seeding of both temperate and tropical legumes.

In summary then, many of the difficulties associated with the failure of legumes to become established can be attributed to an insufficiency of rhizobia. Therefore, any treatment that will counteract adverse conditions such as acid soils and fertilisers or dehydration on the seed surface or that will stimulate multiplication of the rhizobia will be advantageous in legume establishment.

It is essential that through the production of *Rhizobium* cultures of right quality in adequate amounts and their use, atmospheric nitrogen should be tapped for the nitrogen nutrition of legumes (pulses, groundnut; and leguminous fodder crops) which cover an area of 30 million hectares. This will enable the use of artificial nitrogenous fertilisers which are costly and in short supply for the newly introduced high-yielding varieties of cereal crops.

'One gains the impression that inoculation is not a general practice, even when the microbiologist in his laboratory, remote from the agricultural situation is convinced that the practice is worthwhile. What is lacking is information, dissemination, a propaganda effort directed at the farmer and extension and advisory worker. It is criminal to see remarkable results of inoculation trials in experimental stations and then find that there is almost no national programme for the use of inoculants.' However, the present situation is somewhat different. Some efforts are being made to produce the inoculants and also carry on propaganda when useful results are achieved in the field. But a systematic organised effort to produce the inoculants on a sufficiently large scale with adequate quality control is still to be undertaken both at the state and national level.

AZOTOBACTERIN

This is a preparation containing cells of *Azotobacter chroococcum* grown on agar. These organisms fix atmospheric nitrogen deriving energy through the decomposition of organic matter present in the soil.

Preparation and Usage

Azotobacterin is prepared in half-litre bottles by growing *Azotobacter chroococcum* on agar slants. In every half litre, there will be 40 to 50 millions *Azotobacter chroococcum* cells. The colonies are also grown on agar spread in the entire inner surface of the bottle. The young cells are not pigmented while the older ones are pigmented.

The chief defect of the agar medium is that the cells of *Azotobacter* slowly die in the medium. One can use the agar slants for two months. After that the counts rapidly fall. The second defect is that in India, during severe summer months, it is impossible to use agar cultures stored in bottles for long time as the agar dries up very soon. If dry powder of *Azotobacter* can be prepared it is the best. But since it is a nonspore organism, it is difficult to preserve without taking special steps. In Kiev (USSR), they produce liquid azotobacterin using Winogradsky's medium by deep cultivation with aeration. But this method also suffers from shortcomings. After 30 hours, they have in 1 ml of liquid azotobacterin, 150×10^6 cells of *Azotobacter*. After one month, in the same liquid, they have about 800×10^6 and after 2 months 200×10^5 to 300×10^6 *Azotobacter* cells. The counts are made by plating technique.

The Leningrad School does not encourage liquid cultures of *Azotobacter* because bottles break due to freezing in winter. They produce first azotobacterin and then with the help of special separators, the cells are separated from the liquid. The cells after separation appear like a paste. This paste is dried by freezing (Lyophilisation). But the results are not quite good, as the cells die fast. Immediately after preparation their count was 15×10^9 per gram. After 10 days the count fell to 200×10^6 per gram.

Method of application

Azotobacterin growth on agar slants in 500 ml bottles is scraped by adding water and shaking the bottle, washed into a basin and sprinkled on the seeds spread in thin layer. The seeds are mixed so that the material is evenly distributed, and allowed to dry in the shade. The seeds are now to be sown.

Field Experiments (Usage)

It was found that paddy crop yields increased significantly in the treatment with sulphate of ammonia + superphosphate + *Azotobacter* inoculation as compared to the treatment, sulphate of ammonia + superphosphate. This brings out the role of *Azotobacter* as a synthesiser of plant growth-promoting substances, since the amount of N fixed by *Azotobacter in situ* is very small compare to the dose of ammonium sulphate added and *Azotobacter* inoculation alone did not increase the paddy crop yields significantly over the uninoculated.

Azotobacter is known to synthesise vitamin B complex components. Table 34.2 shows the relative amounts of these components synthesised by *Azotobacter* and beer yeast.

Table 34.2. Vitamins synthesised by *Azotobacter* and yeast.

	Thiamin	Riboflavin	Nicotinic	Biotin	Pantothenic acid
	(μg per gram of dry cells)				
Azotobacter	96	304	480	4.23	184
Beer yeast	40	50	560	2.00	50

Azotobacter is also known to synthesise gibberellin-like compounds. Further the fungistatic compounds synthesised by this organism against pathogens like *Alternaria* are also responsible for increasing crop yields. Where the soils were heavily manured and fertility was built up as in the case of vegetable crops the beneficial effects of azotobacterin were pronounced probably because of the synthesis of growth-promoting substances by the organism *Azotobacter*.

Table 34.3 shows the amount of vitamin B_{12} like substances synthesised by *Azotobacter* and *Rhizobium* strains. Iswaran showed that nitrogen fixation and nitrification processes were accelerated by vitamin B_{12}. The effect of bacterisation thus appears to be qualitatively real but inconsistent and unpredictable. It is mostly relatively small, probably not related to nitrogen fixation but to synthesis of growth-promoting substances. But considering the cheapness of the process, the small increments obtained may be of advantage when fertilisers are in scarce supply and there is need to conserve them by maximising their utilisation.

Table 34.3. Vitamin B-12 synthesis by *Rhizobium* and *Azotobacter* strains.

Organism	$m\mu g/ml$
Rhizobium of berseem	19.9
Rhizobium of soyabean	18.9
Rhizobium of pea	14.4
Rhizobium of cowpea	13.0
Azotobacter R	7.3
Azotobacter C	7.0

There is controversy in the literature as to survival and multiplication of *Azotobacter* in the rhizosphere of crops and also the practical utility of *Azotobacter* in increasing crop yields. Larger populations of *Azotobacter* according to some investigators are found near and on the root than at a distance away. The rhizosphere of beet, tobacco, mustard and several legumes seems to exert a favourable effect on *Azotobacter* whereas the bacteria are uncommon under onions, wheat and corn. The results of many of the studies indicate either that the aerobic non-symbiotic N_2-fixers are unaffected by proximities to the root system or that the increase is too slight to be of consequence for crop production. Climate or soil differences may account for some of the conflicting views. There is little doubt that for certain crops in a number of areas, the root exerts a favourable influence upon *Azotobacter* spp. Whether this stimulation is economically important, however, is still subject to question. Venkataraman showed that *Azotobacter* population was greater in the rhizosphere of wheat crop than in the control both in the manured and in the unmanured soil at different stages of plant growth.

After detailed examination the following conditions were found necessary for effective functioning of *Azotobacter*:

1. *Azotobacter* should be used in soils rich in organic matter. In soils poor in organic matter, organic manure should be applied. These may be farmyard manure, compost or green manure. *Azotobacter* is better for vegetable crops because large amount of organic matter is added to these crops and the inoculated *Azotobacter* in association with other organisms like cellulose dissolving bacteria utilises this as energy source and fixes more nitrogen.

2. Since *Azotobacter* is not effective in acid soils, these should be limed.

3. Adequate moisture should be present in the soil. If the soils are too dry the count of *Azotobacter* rapidly falls.

4. Bacterial fertilisers like azotobacterin are not intended to replace mineral or organic fertilisers. On the contrary best results are obtained in the cases where bacterial fertilisers are used in good proportion along with organic and mineral fertilisers. Added manures and fertilisers are better utilised when bacterial fertilisers are added.

Inoculation with *Azotobacter* was found to increase vitamin C content in tomatoes. Venkataraman observed increased biological activity in terms of the total nitrogen fixed in the mixed culture of *Azotobacter chroococcum*, *Rhodopseudomonas capsulatus* and *Cylindrospermum muscicola* than in pure cultures of each organism. Joshi, showed that high counts of *Azotobacter* were established in the rhizosphere of wheat when a mixed culture of *Azotobacter chroococcum* and *Rhodopseudomonas capsulatus* was used.

Since the mutual association of *Azotobacter* and the photosynthetic bacteria *Rhodopseudomonas capsulatus* results in providing the energy source, it will be of interest to further explore the practical value of the inoculation of seed with a combination of these two micro-organisms in a tropical country like India.

A course of events may be visualised as occurring in an ecosystem involving three organisms of blue-green algae, *Azotobacter* and photosynthetic bacteria. Both the blue-green algae and photosynthetic bacteria can carry out their metabolic activities in presence of light. In the special case of the photosynthetic organism, the bond energy is provided by photochemical reactions that convert radiant energy into chemical bond energy, which is then used to drive the machinery of the cell. The photosynthetic evolution of oxygen by the blue-green algae can provide an aerobic environment to *Azotobacter*. The utilisation of oxygen by *Azotobacter* may result in a partial or complete anaerobiosis in the immediate vicinity, where the anaerobic *Rhodopseudomonas* can thrive well. *Rhodopseudomonas* can use lower fatty acids and excrete carbohydrate. In this course, the bacteria can fix nitrogen and assimilate carbon dioxide. On the other hand, in the heterotrophic course, heterotrophic bacteria can use the carbohydrates and can excete lower fatty acids. The carbon dioxide released in the heterotrophic dark course can be assimilated by the photosynthetic bacteria and blue-green algae. Though certain micro-algae are capable of utilising organic acids, there is yet no information that the nitrogen fixing blue-green algae can also use them.

Balasundaram and others showed that the treatment of paddy seedlings with *Beijerinckia* along with soil application of 40 N kg/ha in the form of urea resulted in increasing yield significantly over control while urea at 40 kg/ha did not give significant increase in yield over control. They observed that the yield with urea at 40 kg N/ha + *Beijerinckia* did not significantly differ from that obtained with urea at 80 kg/ha.

BLUE-GREEN ALGAE

Preparation of Blue-green Algae Inoculants

The tank culture unit consists of an inner rectangular culture chamber, an outer thermostatic water chamber, an axial stirrer and provisions for gas inlets and illumination.

Coupled with production, the preservation of these algae is equally important, since they tend to putrefy, if kept moist for long. Watanabe's method of growing them on volcanic gravels is noted for its ingenuity and simplicity. Venkataraman developed a 'dry sand culture', in which the algae suspension was mixed with washed quartz sand and slowly sun dried. The dried algae were found to retain their viability unimpaired for well over two years. This method is economical and facilitates easy handling of bulk material for distribution.

Notwithstanding these advantages, the quartz particles on inoculation into the paddy soils, have a tendency to sink down into the mud along with the algae, thus impeding the quick development of the algae. An ideal carrier material must, therefore, be porous, harmless to the algae and crop plant, and have a larger surface area.

Synthetic sponge pieces have been found to meet all these requirements. The sponge is cut into rectangular pieces, boiled in water, soaked in nitrogen-free mineral medium and sterilised at 121.6°C for 30 minutes. These pieces are spread on metallic boats, covered with thin plastic or vinyl sheets which are supported by a number of arch-shaped metallic wires attached to the sides of the boats. The algae suspension is sprayed over the sponge pieces and incubated under weak illumination. Within three to four days, the sponge pieces are entirely covered by the algal growth, which can be subsequently stored in plastic bags or glass bottles.

On inoculation into the paddy soils, these sponge pieces float on the surface of the water and the algae adhering to them begin to grow and spread over the water surface. With adequate care, the sponge pieces can be retrieved just before draining the fields and used for further inoculations. Seiji Inagaki of Tokyo, Japan, suggested a 'drifting fused magnesium-calcium phosphate fertiliser' as the carrier material for the blue-green algae, in which foam pieces (1 cm^3) were soaked in liquid binders (liquid synthetic paste or water glass) to which powdered fused magnesium phosphate (FMP) was added. The excess liquid was squeezed out and the foam pieces were air dried. The FMP provides sufficient nutrients for the algae to grow.

Method of Application of Blue-green Algae

Small algal beds may be carved out in the farmers' fields and enclosed by a small earth embankment. Useful strains of blue-green algae may be inoculated into these beds just before the rainy season, or where water is available, the beds may be kept waterlogged. Prior to seeding, the beds may be sprinkled with little lime to promote the growth of the inoculum. With the water accumulating into these beds, the inoculated algae will grow to form a scum, which may be scooped out, dried, and broadcast over the paddy field together with lime. It may also be possible to allow the inflow of irrigation water itself to flow over these beds, so that the algal material may get distributed over the entire field. This practice will require only a little material for the initial inoculation of the algal beds. Lime sprinkling will facilitate in suppressing other algae and also in lowering the acidity of the water to a favourable level.

For the promotion of the indigenous algae (local algal populations) and also for the introduced algae (inoculated algae) in the fields, the regimen recommended may be followed, which consists of the application of 1000 kg lime, 100 kg superphosphate, and 0.25 kg of sodium molybdate.

Field experiments (usage)

It was found that different varieties respond to algal inoculation even in the presence of high level of nitrogen fertiliser. Field experiments with high-yielding varieties at different places showed that algae stimulated crop growth even in the presence of nitrogen at 60 to 150 kg/ha.

When high-yielding short-duration varieties of paddy were grown, and heavy doses of fertilisers were applied, the returns in yield and cash were more in the presence of blue-green algae than without at different levels of nitrogenous fertilisers (30, 60 and 120 kg/ha). When fertilisers are in short supply, the blue-green algal mixture increases crop yields moderately and gives good return for the investment.

There are great differences in the amount of nitrogen fixed by various genera and sometimes by the same species from different localities. The influence of genetic constitution has also been found to be an important factor which determines the relative capacity of the blue-green algae to fix nitrogen. A mixture of blue-green algae is recommended to be applied.

If the soil is deficient in one strain of useful blue-green algae and if that is present in the mixture applied, it will multiply and enhance crop yields.

Nature and Production of Extra-cellular Substances

The important factors in regulating the extent and accomplishment of increased soil nitrogen by the algal activity are: (i) the relative efficiency of their nitrogen-fixing capacity, (ii) their carbohydrate status, (iii) rate of excretion of soluble nitrogenous substances and their availability to the crop plants, and (iv) the soil and root environment. About 50 per cent of the total nitrogen fixed by these organisms appears in the medium in soluble form. The deficiencies of potassium and iron in the medium have been found to markedly increase the liberation of extra-cellular nitrogen.

The extra-cellular products may be ecologically important in many ways; those liberated are directly utilised by other organisms, thus short circuiting the nutrient cycle, or some of the extra-cellular enzymes may play a part in the breakdown of organic substances. They may also be responsible in determining the existence and abundance of a particular form in a given environment. Some of the extra-cellular products of algae have chelating properties. The filtrates of an endophytic *Nostoc* strain have given indication for the presence of some thermolabile growth-promoting substances. While the extra-cellular nitrogenous substances are composed of complex polypeptides in some cases, as in *Caldothrix brevissima* and in a *Nostoc* strain, considerable amounts of free amino acids, like aspartic acid, glutamic acid and alanine have been observed. Recently, Henriksson reported that polysaccharides and vitamins belonging to B complex (Thiamine, Riboflavin, Nicotinic acid, Pantothenic acid and Biotin) are excreted by *Nostoc* associated with the lichen *Collema*.

Venkataraman showed that the beneficial effect of blue-green algae on rice plants was more due to vitamins and auxins synthesised and supplied by these algae to the crop than to the nitrogen alone fixed by them. This accounts for the additional effect of algae even in presence of high level of nitrogenous fertilisers.

Role of Blue-green Algae in Rice Fields

The conditions of rice fields may be clearly distinguished into three periods: (i) the waterlogged period from transplantation to harvest time, (ii) the dry period after harvest, and (iii) the desiccation period during which the soil temperature frequently exceeds 50°C. During the latter periods the algae remain dormant and their activity is almost nil. The available evidence at present shows that under waterlogged rice field conditions the nitrogen fixation is essentially an algal process.

The appearance of an abundant growth of algae in waterlogged rice fields is a very common phenomenon in the tropics, and the fertility of the rice fields in the tropical countries may be at least partly ascribable to the activity of these algae.

PHOSPHOBACTERIN

Preparation of Inoculants of Phosphate Solubilising Bacteria

The technology of the production of this bacterial fertiliser has been perfected. The duration of the use of phosphobacterin is longer than that of azotobacterin. A packet of phosphobacterin can be used for 12 months while agar slants of *Azotobacter* must be used within two months. The cells of *Bacillus megaterium* var. *phosphaticum* are cultivated by submerged cultivation technique in fermenters, separated by centrifuging and then lyophilised (dried by freezing under vacuum). This is later mixed with kaolin. The quantity of phosphobacterin produced in the USSR can cover an area of 14 million hectares while that of azotobacterin is sufficient to cover 3 million hectares.

Good quality phosphobacterin should contain 8×10^9 cells/g. Five grams of phosphobacterin are suspended in one litre of water and allowed to stand for two hours to allow the spores to germinate. Later, the material is sprinkled on the seeds to be sown in one hectare. The seeds are well mixed and allowed to dry in the shade before sowing. In the case of corn or potatoes, 15 grams of phosphobacterin should be used.

When agar slant of *B. megaterium* are used, the quantities are so adjusted as to give similar count of *B. megaterium* per seed as those obtained with the above quantities of solid phosphobacterin. This is usually obtained by scraping the growth on agar slant in 500 ml capacity bottle for small-size seeds and growth from slants of three bottles for large-size seeds. Usually, a count of 1000 viable cells per seed would be adequate.

Field Experiments (Usage)

It was found that the phosphate uptake by cowpea crop and percentage utilisation of added phosphate by wheat crop from superphosphate or apatite with or without phosphobacterin (PB) or an Indian strain of *Bacillus circulans* (I_2) in pot culture experiments. The results show that significantly greater amount of phosphate was taken up by cowpea crop with either of the organisms from superphosphate or apatite treatments when a basal dressing of farmyard manure was applied, as compared to the corresponding uninoculated. In the case of wheat crop, phosphobacterin was effective in increasing the percentage utilisation of added superphosphate by wheat, when farmyard manure and ammonium sulphate were added as basal dressing. *Bacillus circulans* (I_2) was effective in enhancing percentage utilisation of added superphosphate or apatite by wheat crop when ammonium sulphate was used as basal dressing, showing thereby that this organism may be more suited in Indian soils poor in organic matter in mobilising added phosphate.

Bacillus megaterium var. *phosphaticum* was shown to synthesise vitamin B_{12}, paraminobenzoim acid, indole acetic acid, biotin, nicotinic acid, pantothenic acid and vitamin B_1. The increases in crop yields due to inoculation with this organism may be due to the growth promoting effect of these substances, in addition to release of phosphate in available form under specific conditions.

ANNEXURE I

Media Used for Rhizobium Culture

Media No. 1: Glucose-peptone-agar

Glucose 5.0 grams; peptone 10.0 grams; bromocresol purple 1.6 per cent; alcoholic 10.0 ml; water distilled 1 litre; agar 15.0 grams. Melt for 5 minutes—then add bromocresol purple tube and sterilise at 115°C for 10 minutes.

Media No. 2: Yeast-mannitol-agar

Mannitol 10.0 grams; K_2HPO_4 0.5 gram; $MgSO_4$, $7H_2O$ 0.2 gram; NaCl 0.1 gram; yeast water (pH 6.8) 100 ml @; water distilled 900 ml; agar 15.0 grams.

@For preparing yeast water: Prepare a water extract of starch-free baker's yeast, 10 per cent (wet basis). Mix the yeast with cold water and let it stand at room temperature to 1–2 hours. Autoclave for 40–60 minutes, allow to settle and use the clear supernatant.

Alternatively, 0.5 gram of Oxoid yeast extract or 'vegemite' can be substituted and water made up to 1 litre.

Media No. 3: Seedling solution and agar

$CaHPO_4$ 1.0 gram; K_2HPO_4 0.2 gram; $Mg SO_4$ $7H_2O$ 0.2 gram; NaCl 0.2 gram; $FeCl_3$ 0.1 gram; water distilled 1.0 litre; agar 8.0 grams; pH adjusted to 6.5.

Do not filter; add 10 ml of N/10 NaOH per litre of medium to adjust pH: after first autoclaving.

Chapter 35

Microbial Production of Biosurfactants

INTRODUCTION

Biosurfactants are those chemicals which are produced by micro-organisms but which have both clearly defined hydrophilic and hydrophobic groups. They occur in nature in bacteria, yeasts, and fungi, and in particular in bacteria which grow on a water-immiscible substrate, using it as a food source. By evolution these bacteria have adapted themselves to feeding on these substrates by manufacturing and using a surface active product that helps the bacteria to adsorb, emulsify, wet, or disperse or solubilise the water-immiscible material. The four main types of biosurfactant are: (i) glycolipids, (ii) phospholipids, (iii) lipoproteins or lipopeptides, and (iv) polymeric. By growth of the bacterium *Pseudomonas fluorescens* (NCIMB 11712) on virgin olive oil, production of a glycolipid in the form of a rhamnolipid is thought to have taken place. This biosurfactant group is based on the rhamnose structure, which is a methyl pentose monosaccharide. However, rhamnolipids can also be based on the disaccharide by condensing two moles of rhamnose together. The link to the hydrophobic group is by way of an acetal group; however, the 'lipid' part of the molecule contains ester and carboxyl groups. Biosurfactants over the years have found a great many uses in industry, for example, (i) oil recovery, (ii) oil spill clean-up, (iii) textiles, (iv) pharmaceuticals, and (v) cosmetics. By this programme of research, it is proposed to produce a biosurfactant for use in the detergents/cleaning materials industries with the specific aim being to produce a biosurfactant which could be used in place of chemical surfactants in a detergent formulation for use in the household.

In recent years detergent phosphates have been blamed for eutrophication in certain inland lakes and ponds — leading to the depletion of aquatic life in these areas. Oil spillages have destroyed marine life and the synthetic detergents used to clean up these spillages have often led to more destruction of the environment. From an environmental viewpoint it is important that all substances released into the environment are biodegradable, firstly to assess their potential for causing environmental damage and secondly to safeguard against the possibility of future harm due to build-up in the environment. Micro-organisms, because of their large surface-to-volume ratio and diverse synthetic capabilities are promising candidates for widening the present range of surfactants. Biosurfactants are those chemicals which are produced by micro-organisms but which have both clearly defined hydrophobic and hydrophilic groups. They are produced by bacteria, yeasts and fungi (Table 35.1), and particularly in bacteria which are in a state of growth on a water-immiscible substrate which is a source of food, for example, crude oil spillage treated with selected micro-organisms. By evolution, the bacteria have adapted themselves to feeding on water-immiscible materials by manufacturing and using a surface active product that helps the bacteria which are in the aqueous phase to adsorb, emulsify, wet, and disperse or solubilise the water-immiscible material.

Table 35.1. Microbial biosurfactants and bioemulsifiers.

Micro-organism	Biosurfactant/bioemulsifier
Torulopsis species	Sophorolipids
Pseudomonas species	Rhamnolipids
Rhodococcus erythropolis	Trehalose lipids
	Trehalose mycolates
	Sucrose and fructose lipids
Rhodococcus species H13-A	Trehalose lipids
Candida species	Mannosyl erythritol lipid
Candida bogoriensis	Sophorolipid
Acinetobacter species	Fatty acid, glycerides, emulsan
Corynebacterium lepus	Corynemycolic acids
Candida petrophilum	Peptidolipid
Bacillus subtilis	Cyclic lipopeptide
Bacillus licheniformis	Cyclic lipopeptide
Candida tropicalis	Mannan-fatty acid complex
Corynebacterium hydrocarboclastus	Proteo-lipid-carbohydrate complex

TYPES OF BIOSURFACTANTS

The main types of biosurfactant are:
1. Glycolipids.
2. Phospholipids.
3. Lipopeptides and lipoproteins.
4. Polymeric.

Glycolipids

These are compounds of a carbohydrate and a lipid; the linkage is by way of either an ether or an ester group (Fig. 35.1). The main glycolipids which are found to occur and are most often investigated are:
1. Rhamnolipids.
2. Mycolates of mono, di- and tri-saccharides
3. Sophorolipids.

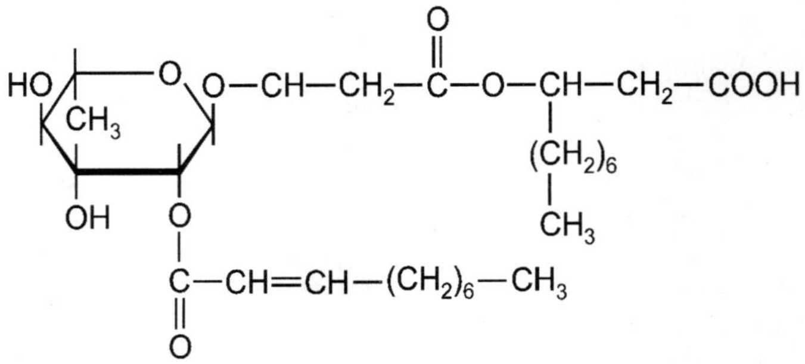

Fig. 35.1. A glycolipid produced by a *Pseudomonas* strain.

Phospholipids

These are the esters formed between the alcohol groups on a lipid and a phosphate.

Lipopeptides and Lipoproteins

These consist of a lipid attached to a polypeptide chain (Fig. 35.2).

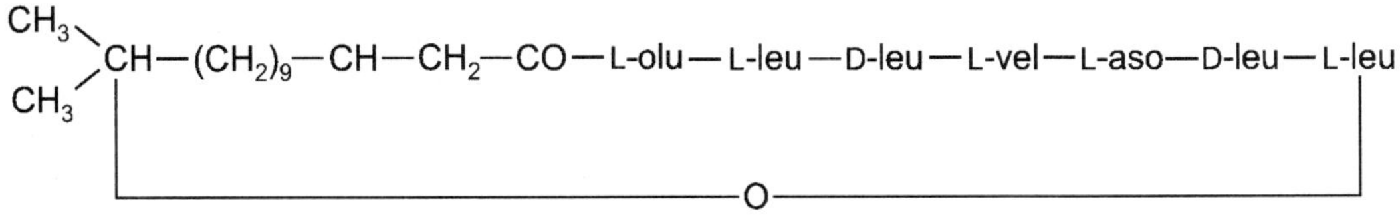

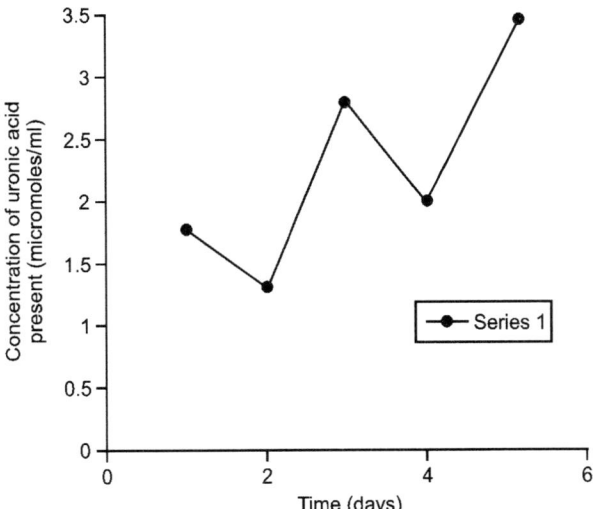

CH$_3$
   \
    CH—(CH$_2$)$_9$—CH—CH$_2$—CO—L-olu—L-leu—D-leu—L-vel—L-aso—D-leu—L-leu
   /
CH$_3$
                                        |
                                        O

Fig. 35.2. A lipopeptide structure (surfactin) produced by *Bacillus subtilis*.

Polymeric

These are products again formed between saccharide units and fatty acid residues; but they are polymeric in nature. Of the four types listed, it is the glycolipid and lipopeptide groups which are of particular interest, since the bacterial species used in this study, i.e. *Pseudomonas fluorescens* (NCIMB 11712) is known to produce both types, especially the former (Fig. 35.3).

Fig. 35.3. Variation in the concentration of uronic acid present with time [NaNO$_3$ conc. = 2.5 g/l].

EXPERIMENTAL PROCEDURE

Production of Biosurfactants

Culture conditions and subsequent analysis and olive oil was chosen as one of the substrates due to the waste produced during its processing. Initially microbial growth and biosurfactant production experiments were carried out in 500 ml shake flasks containing 250 ml of medium, i.e. 225 ml of sodium nitrate solution of the following concentrations: 1, 2, 5, 4 and 6 g/l (using distilled water for dilution), and

2.5 g/l using tap water and 25 ml of olive oil. A cell suspension from 24 hours nutrient agar cultures was used as innoculum and flasks were incubated at 30°C in a New Brunswick G100 gyratory-incubator shaker at 200 rpm. Cells were separated from the culture broth by centrifugation at 8000 × gm for 15 minutes.

Chemical analysis of biosurfactants

Isolation of surface active compounds from culture supernatant was performed using conventional methods. The wet cell paste containing about 40–50 mg of cells (dry weight) was diluted to 1 ml with water and to the suspension 3.75 ml of methanol/chloroform (2:1 v/v) was added. The mixture was shaken and left at room temperature for several hours with intermittent shaking.

After centrifugation, the supernatant extract was decanted and to the residue 4.75 ml of methanol/chloroform/water (2:1:0.8) was added; the mixture was then shaken and centrifuged. To the combined supernatant extracts, 2.5 ml of both chloroform and water were added and the mixture centrifuged. The lower chloroform phase was withdrawn, diluted with benzene (to aid removal of traces of water) and brought to dryness in a rotary evaporator (30°–35°C). The lipid residue was immediately dissolved in chloroform/methanol (1:1) and the solution centrifuged and made to a known volume with chloroform.

Rhamnolipid concentration

This was estimated by colourimetric determination of rhamnose concentration by the method of Chandrasekaran and Bemiller.

Uronic acid

Samples were heated in concentrated sulphuric acid containing borate and then reacted with carbazole.

Reagents

1. Borate-sulphuric acid: 3.82 gms of sodium borate was dissolved in 10 ml of hot water and 390 ml of well-cooled concentrated sulphuric acid was added by stirring with a glass rod. This reagent was stored in a refrigerator in a glass bottle.
2. 0.2 per cent carbazole: 100 mg of carbazole was dissolved in 50 ml of absolute ethanol and stored in a refrigerator in a brown glass bottle.
3. Standard glucuronolactone: 17.6 mg of D-glucuronolactone was dissolved in 100 ml of deionised water (1 μmol/ml) just before use.

Procedure

Aliquots of the standard solution of glucuronolactone 0 to 0.2 ml (by 0.02 ml increments) were transferred by pipette to test tubes. After bringing the volume in each tube to 0.5 ml with water, 3 ml of cold borate-sulphuric reagent was added with immediate mixing. The tubes were heated in a boiling water bath for 20 minutes. After cooling to 0°C, 0.1 ml of 0.2 per cent, carbazole solution were added. The solution was shaken well and heated again in the boiling water bath for 10 minutes. After cooling at room temperature for 15 minutes the colour was read at 530 nm against the blank. Three different aliquots of the sample were subjected to this reaction to determine the amount of uronic acid present.

Hexosamines

A sample containing 0.3–0.5 μmol of uronic acid in 0.2 ml of water was placed in a small ampoule (2 ml capacity), mixed with 0.2 ml of concentrated hydrochloric acid, sealed and heated in a thermoblock at 100°C for 4 hours. After hydrolysis, the ampoule was opened and placed in a vacuum desiccator containing sodium hydroxide pellets in a beaker. After complete drying, the residue in the ampoule was dissolved in 0.3 ml of water (Fig. 35.4).

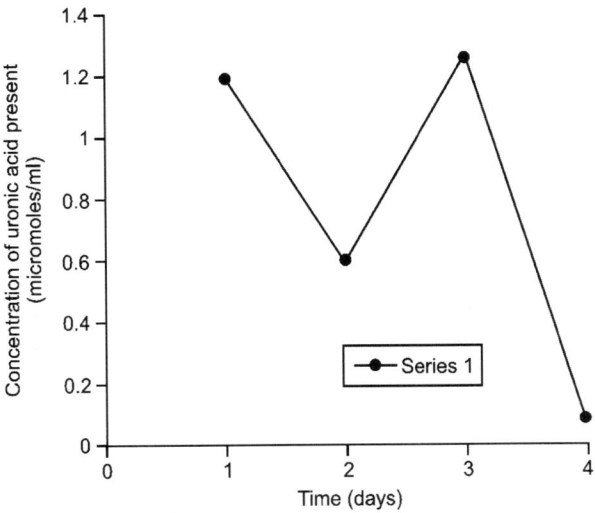

Fig. 35.4. Variation in the concentration of uronic acid present with time [NaNO$_3$ conc. = 2.5 g/l].

Procedure

Hexosamines were reacted with an alkaline solution of 2,4-pentanedione to form chromogens, which yield colour with *N,N*-dimethyl-*p*-aminobenzaldehyde in acid. Both glucosamine and galactosamine form chromogens when this reaction is done at 100°C, whereas only galactosamine forms chromogens when the reaction is done at 0°C. Hence, the method can be used to give the glucosamine/galactosamine ratio.

Reagents

1. Alkaline 2,4-pentanedione: 0.15 ml of 2,4-pentanedione was diluted to 5 ml with 0.7 M sodium carbonate. This reagent was prepared just before use.
2. Ehrlich reagent: 100 mg of *N,N*-dimethyl-*p*-aminobenzaldehyde is added to 3.2 ml of 60 per cent perchloric acid and then made up to 10 ml with 95 per cent ethanol. This reagent must be freshly prepared before use.
3. 90 per cent ethanol.
4. Hexosamine hydrochloride solution, 2 μmol/ml.

Total hexosamine

Aliquots of the standard solution of hexosamine hydrochloride (0 to 0.1 ml by 0.01 ml increments) were transferred by pipette to screw top tubes. After making up the solutions to 0.1 ml with water, 0.025 ml of 1.5 M hydrochloric acid was added with thorough mixing. The tubes were capped and heated in a boiling water bath for 20 minutes, then cooled to approximately 20°C by immersion in cold water. Then 2 ml of 90 per cents, ethanol and 0.5 ml of Ehrlich reagent were added with thorough mixing, and the tubes allowed to stand at 20°–25°C for 1 hour. The colour was read at 535 nm against the blank (Fig. 35.5).

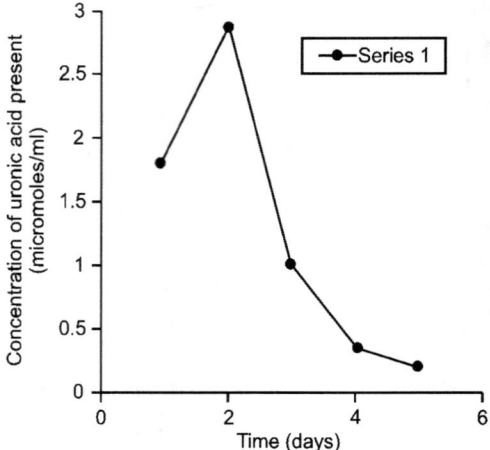

Fig. 35.5. Variation in the concentration of uronic acid present with time for a control run [not inoculated, NaNO$_3$ conc. = 2.5 g/l].

Galactosamine

Standards of galactosamine were prepared as for total hexosamine; several blanks containing 0.2 µmol or more of standard glucosamine were also prepared. The tubes were placed in an ice/salt bath at 0°C and to these were added. First, 0.25 ml of 1.5 M hydrochloric acid were added (0°C) and then 0.25 ml of pentanedione reagent (0°C). The tubes were capped, the contents were mixed thoroughly, and samples kept at 0°C for 18 hours. After incubation, 2 ml of 90 per cent ethanol and 0.5 ml of Ehrlich reagent were added; the contents were mixed, and the samples heated in a water bath at 70°C for 1 hour. These were then kept at 20°–25°C for 1 hour and the absorbance of each recorded at 525 nm.

It must be noted that although all the tests included in this chapter were used, only that for uronic acid proved positive and it was decided at this point to use only uronic acid determination as a marker for the production of a rhamnolipid by the bacteria under the conditions employed for each particular fermentation. The variation in the concentration of uronic acid present with time is shown in Fig. 35.6.

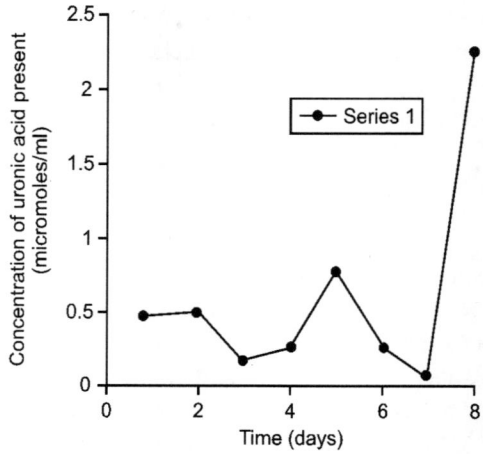

Fig. 35.6. Variation in the concentration of uronic acid present with time [NaNO$_3$ conc. = 2.5 g/l].

RAPID METHOD FOR MONITORING MAXIMUM BIOSURFACTANT PRODUCTION OBTAINED BY ACETONE PRECIPITATION

Isolation of the Biosurfactant

The organisms were cultivated as before in a medium of olive oil and a sodium nitrate solution, the culture broth was centrifuged at 10000 rpm for 30 minutes at 4°C, and the clear supernatant recentrifuged at 10000 rpm for 30 minutes at 4°C, after which the cell free supernatant was treated with 3 volumes of chilled acetone. The precipitate was collected by centrifugation at 5000 rpm for 10 minutes and dried under a current of air. Variation in the concentration of uronic acid present with time is shown in Fig. 35.7.

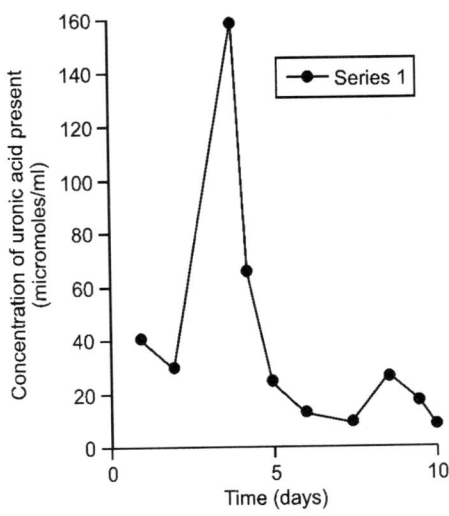

Fig. 35.7. Variation in the concentration of uronic acid present with time [$NaNO_3$ conc. = 2.5 g/l; oil/aqueous phase ratio, 1:3 v/v].

Estimation of Biomass

Discarded cells from the isolation of the biosurfactant process were centrifuged at 12000 rpm for 15 minutes and extracted with a mixture of acetone/hexane (3:1) to remove adhering hydrocarbon. This was followed by centrifugation with hexane and drying overnight to obtain dry biomass.

CHEMICAL ANALYSIS

Protein Estimation

Protein in the isolated biosurfactant was estimated by the Pierce BCA method; this involves the use of a sensitive reagent, bicinchonic acid (BCA) for the pectrophotometric determination of protein concentration in solutions. It combines the reaction of protein with Cu^{2+} in an alkaline medium (yielding Cu^+) with BCA. A purple reaction product results whenever two molecules of BCA interact with one of Cu^+. This complex is water soluble and exhibits a strong absorbance at 562 nm.

Three protocols may be used:

1. Standard protocol: 37°C for 30 minutes.
2. Room temperature protocol: room temperature for 2 hours.
3. Enhanced protocol: 60°C for 30 minutes.

Working reagent

One part reagent B was added to 50 parts reagent A with thorough mixing, this is stable for 1 day at room temperature. Reagent A = sodium carbonate

Reagent A:

Sodium carbonate

Sodium bicarbonate

BCA detection reagent

Sodium tartrate in 0.2 N NaOH

Reagent B:

4 per cent copper sulphate solution

Method

First, 0.1 ml of each standard or unknown protein sample were pipetted into appropriately labelled test tubes, water being used as a blank. To each tube 2 ml of working reagent was added and mixed well, the tubes were incubated at the selected protocol after which they were all allowed to cool to room temperature, and the absorbance was read at 562 nm, that for the blank solution being subtracted from the values determined (Fig. 35.8).

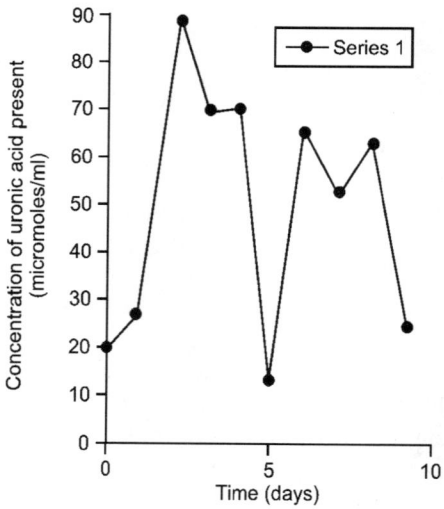

Fig. 35.8. Variation in the concentration of uronic acid present with time [$NaNO_3$ conc. = 2.5 g/l; oil/aqueous phase ratio, 1:3 v/v].

Standards consisted of: 0, 50, 100, 150, 200 and 250 µg/ml of B.S.A. protein standard as provided.

Carbohydrate estimation

Anthrone reagent

Sulphuric acid (66 per cent): 340 ml distilled water and 660 ml of concentrated sulphuric acid (specific gravity 1.84). First, 500 mg of recrystallised anthrone and 10 gms thiourea were added to 1.1 of 66 per cent sulphuric acid and the mixture warmed to 80–90°C, shaking the flask occasionally. This was stable for 2 weeks in a refrigerator (Fig. 35.9).

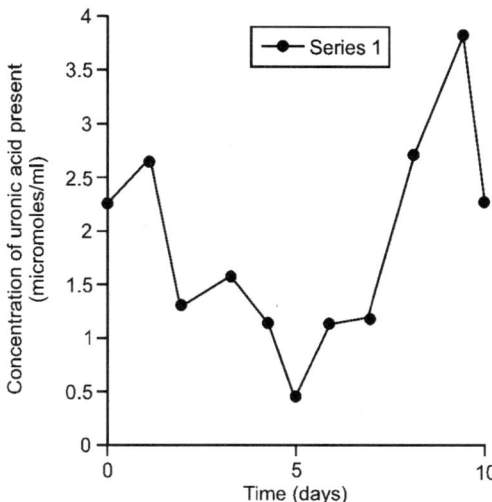

Fig. 35.9. Variation in the concentration of uronic acid present with time [NaNO$_3$ conc. = 2.5 g/l; oil/aqueous phase ratio, 1:1 v/v].

Standard glucose solution

Stock solution: glucose powder was dried in an oven at 60°–70°C, after which 100 mg were dissolved in 100 ml of saturated benzoic acid solution. Working standard: 10 ml of stock solution were transferred to a 100 ml volumetric flask and the volume made up with saturated benzoic acid solution. Then, 1 ml of this solution containing 0.1 mg of glucose was used as a standard.

Deproteinising reagents: 5 per cent trichloroacetic acid.

Method

The sample was deproteinised with 5 per cent trichloroacetic acid, dilution being 1:10. To three appropriately labelled test tubes were added the following: 1 ml of sample, 1 ml of glucose standard, and 1 ml of distilled water. To each, 10 ml of anthrone reagent were added with thorough mixing the tubes were stoppered and placed in water at room temperature for 3–5 minutes, after which they were placed in a boiling water bath for 15 minutes with light being excluded. The tubes were then removed to water at room temperature again and allowed to stand for a further 20–30 minutes. The absorbance of the samples were then measured at 620 nm is variation in the concentration of uronic acid present with time shown in Fig. 35.10.

The following equation was used to calculate the amount of carbohydrate present:

DU/DS × 0.1 × dilution of sample × 100 = mg glucose/100 ml,

where,

DU: optical density of sample
DS: optical density of standard
0.1: mg glucose/ml standard solution.

Lipid estimation

To determine the lipid content, 0.5 gm of the isolated material was extracted with 25 ml of diethyl ether on a water bath at 50°C for 30 minutes. The ether extract was dried using anhydrous Na$_2$SO$_4$ and the lipid content determined by gravimetric estimation.

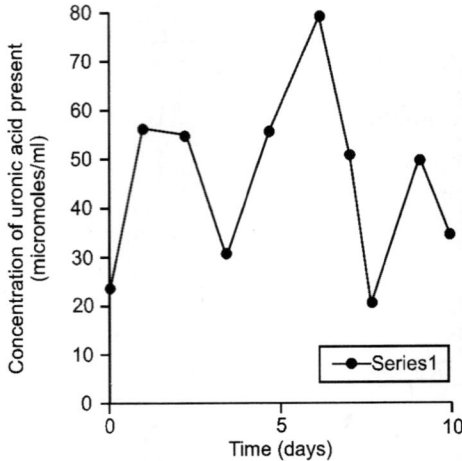

Fig. 35.10. Variation in the concentration of uronic acid present with time [NaNO$_3$ conc. = 2.5 g/l; oil/aqueous phase ratio, 1:1 v/v].

Measurement of emulsification activity

To estimate the emulsification activity, 6 ml of *n*-dodecane was added to 4 ml of the culture broth in a graduated tube and vortexed at high speed for 2 minutes. The emulsion stability was determined after 24 hours. The emulsification index, E$_{24}$ (per cent) was calculated by dividing the height of the emulsion layer by the total height of the mixture and multiplying it by 100. For measurement of the emulsification activity of the acetone precipitated biosurfactant 4 ml of the biosurfactant solution (1 mg/ml) in water was taken. To this, 6 ml of *n*-dodecane was added and the emulsification activity was determined as above variation in the concentration of uronic acid present and the pH with time for a control run is shown in Fig. 35.11.

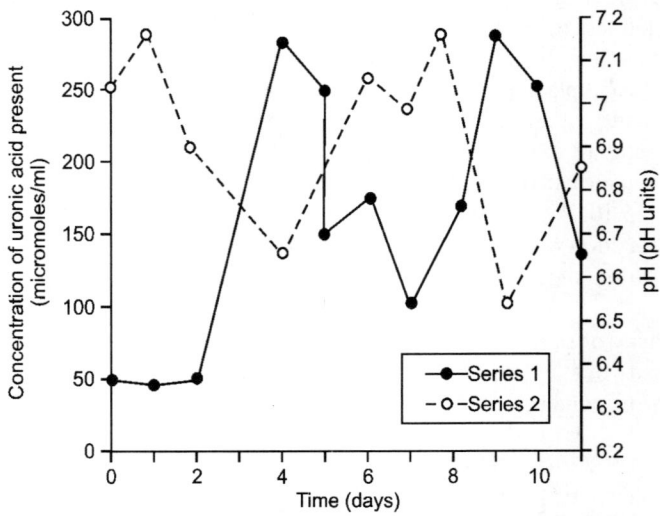

Fig. 35.11. Variation in the concentration of uronic acid present and the pH with time for a control run [run inoculated, NaNO$_3$ conc. = 2.5 g/l; oil/aqueous phase ratio, 1:1 v/v].

Surface tension measurement

This was measured by a 'White' surface and interfacial tension torsion balance, model OS.

DISCUSSION AND CONCLUSIONS

It is thought that the bacteria in the fermentation medium metabolise the oil to produce glucose, the terminal CH_2OH group on the glucose molecule is then oxidised to a carboxyl group, producing a uronic acid — glucuronic acid. Thus, uronic acid determination was used as a marker for biosurfactant production and as such was also used to determine the optimum concentration of sodium nitrate solution in the fermentation medium, thus the levels of uronic acid for each concentration of sodium nitrate solution used, as can be seen from the results the maximum concentration of uronic acid present was for a 2.5 g/l* (where * indicates that the solution was made with tap water and not distilled water) solution. It was decided then to use this sodium nitrate solution in the fermentation medium, and also to use a solution of the same concentration but one which was made with distilled water to run in parallel with these fermentations as a comparison. Initially analyses of all incubations involved determination of uronic acid as an indicator of biosurfactant production by *Pseudomonas fluorescens*. The latter is well known as both a biosurfactant producer and a secretor of specific lipases for the hydrolysis of fats. As oils (in this case virgin olive oil) consist virtually of fat only, then either or both production of a biosurfactant or lipases would be possible on incubation with *Pseudomonas fluorescens*.

As can be seen from Figs 35.3 to 35.13 depicting uronic acid levels in the incubation media, the levels in some cases are very high, then it can be fairly sure that a biosurfactant is being produced. No attempt was made to assess for lipase production in this work since it is the primary desire to achieve maximum biosurfactant production.

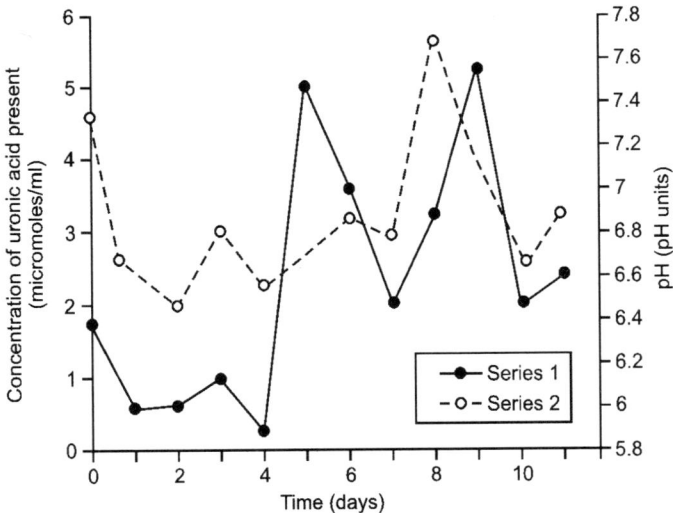

Fig. 35.12. Variation in the concentration of uronic acid present and the pH with time for a run using chip-pan oil [NaNO$_3$ conc. = 2.5 g/l; oil/aqueous phase ratio, 1:1 v/v].

However, evidence at a later stage emerged that protein was present. Hence, it is now assumed that a rhamnolipid and a proteinaceous material are both present but at different stages.

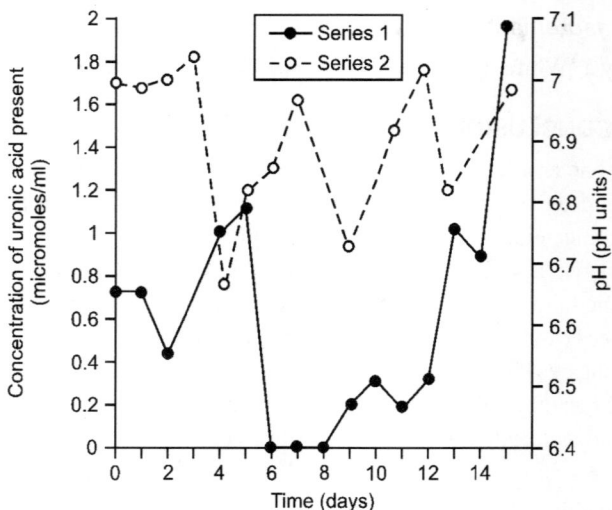

Fig. 35.13. Variation in the concentration of uronic acid present and the pH with time [NaNO₃ conc. = 2.5 g/l*; oil aqueous phase ratio, 1.3 v/v].

With the levels of uronic acid, as detected in the samples chosen, varying from high to low and then returning to high again, with the trend being repeated, it is thought that for biosurfactant production to continue to help achieve maximum emulsification then a rhamnolipid is produced followed by either a second, lipoprotein biosurfactant or enzyme (lipase) activity. The latter has yet to be proven.

This work on virgin olive oil has been used to, in the first instance, induce *Pseudomonas fluorescens* to produce a biosurfactant of the rhamnolipid class, and the presence of uronic acid is indicative of this. The presence of the proteinaceous material and its exact role has yet to be clarified. The work is now to be extended to waste chip-pan oil using a microprocessor-controlled bioreactor with the parameters (pH, dissolved oxygen, temperature and agitation rate) to be optimised.

Analysis of samples taken from a fermentation using olive oil and a 2.5 g/l* solution of sodium nitrate as the fermentation medium with an oil to aqueous phase ratio of 1:3 yielded the following results (Fig. 35.13):

Protein concentration: 0.908 g/l

Carbohydrate concentration: 0.022 mg glucose/l

Lipid concentration: approximately. 0.1 g/l

Emulsification activity: E_{24} (per cent) precipitate = 6.15 per cent

CHN analysis:

C:28.15 per cent

H:4.10 per cent

N:8.83 per cent

Surface tension: 0.033 Nm³

Biomass: 0.8 g/l

Analysis of samples taken from a fermentation using olive oil and a 2.5 g/l solution of sodium nitrate as the fermentation medium with an oil to aqueous phase ratio of 1:1 yielded the following results (see also Fig. 35.10):

Protein concentration: 0.011 g/l

Carbohydrate concentration: 0.00008 mg glucose/l

Lipid concentration: approximately 0.3 g/l

Emulsification activity:

E_{24} (per cent) broth = 0.083 per cent

E_{24} (per cent) precipitate = 1.67 per cent

CHN analysis:

C : 1.17 per cent

H : 2.69 per cent

N : 11.84 per cent

Surface tension measurement: 0.032 Nm^2

Biomass: 0.1 g/l

SECTION IX

Special Topics

36. **Environmental Friendly Products and Processes for the 21st Century**   551

37. **Microbial Production of Energy from Industrial Waste**   560

38. **Patents and Secret Processes**   567

Environmental Friendly Products and Processes for the 21st Century

INTRODUCTION

Due to life and social style modernisation, many kinds of organic synthetic high polymers have been frequently used as powerful and economical agents for a wide range of applications. However, some of these chemicals are harmful to the environment and constitute dangerous sources of pollution that can affect future generations. Thus, safe biodegradable alternatives, that is, biopolymers, which will minimise environmental and health risks, are both urgently needed and are attracting wide research interest as a biologically compatible way to protect the environment and also to prevent pollution. For this purpose, strategies for a 'design concept' employed at each step in the production of environmentally friendly products, should be important to reduce environmental impact by changing the type of materials used in the product. Bioflocculants and bioabsorbants, i.e. new biomaterials, will be discussed.

Rhodococcus erythropolis produced a kind of microbial flocculant. This bioflocculant, which is composed of protein, could efficiently flocculate all inorganic and organic suspended solids. *Alcaligenes latus* produces a new super bioabsorbent which can absorb up 1000 times its own weight of water. This bioabsorbent is a polysaccharide composed of four different sugars such as glucose and glucuronic acid. These biopolymers are environmentally safe for almost all applications, including desertification prevention.

Scientists have succeeded in creating a strain of bacteria that normally lives only in an aqueous solution, allowing the bacteria to survive in an organic solvent such as petroleum.

Environmental impacts occur at all stages of a product life cycle, such as material extraction, material processing, manufacturing product use, and waste management. Bioprevention technology has great potential for preserving our environmental quality in coping with these problems.

The concept of 'design' can be employed to reduce these threatening impacts by changing the amount and type of materials used in the product, by creating more efficient manufacturing operations and processes, by reducing the energy and materials consumed during use, and by improving the recovery of energy and materials during waste management.

Environmental friendly products and processes—'green products' and 'green processes'—can be expected to grow in scale and to encourage, promote and stimulate further development of bioprevention technology to preserve our earth's environmental quality.

In a world where population growth and economic growth put increasing pressures on natural resources and ecosystems, the dominant paradigm upon which environmental policies are based can be expected to evolve from environmental protection involving resource management and eco-development.

How should one view the significance of green design as a competitive and environmental strategy? As a competitive strategy, green design can help manufacturers generate less waste and reduce production costs at the same time. As waste disposal costs and regulatory compliance costs increase, the environmental attributes of products will necessarily become more important to consumers and investors. Some OECD member countries are already moving aggressively to integrate 'clean' products into their industrial strategies for future competitiveness, and international trade will increasingly be influenced by environmental concerns. All of these trends suggest that having an environmental dimension to one's design capabilities will be an important competitive asset in the future.

Many organic synthetic high polymers have been frequently used as powerful and economical agents. However, among them, some of these high polymer agents are harmful to the environment and are a dangerous source of pollution that can adversely affect future generations. Thus, a safe biodegradable high polymer, that is, a biopolymer, that will minimise environmental and health risks is both urgently needed and is attracting wide research interest.

As an environmental strategy, green product design offers a new way of addressing environmental problems. For example, organic synthetic high polymer flocculating agents have been used for a wide range of applications, which include waste-water treatment, dredging and industrial processes. In particular, polyacrylamide derivatives have frequently been used because they are economical and effective flocculating agents. However, studies indicate that the monomer of acrylamide is both neurotoxic and a strong carcinogen in the human body.

Another example is a water absorbent. Due to life style modernisation, the use of various kinds of sanitary products and disposable baby diapers is increasing annually, with most of these water absorbent articles being currently made of synthetic high polymer materials such as polyacrylate and polyacrylamide derivatives. These organic synthetic high polymer absorbents are both cost-effective and high-activity agents. However, many of these products do not easily biodegrade and are disposed of, discharged, and remain in the environment; they pose a significant environmental hazard. It has been reported that the synthetic high polymer agent monomer, acrylic acid, is a strong irritant to the human body.

Another area of ecological global concern is the accelerated expansion of the world's deserts, which is a serious environmental protection problem because their area is reportedly expanding by about 60,000 km^2 each year. In some countries, soil aridity prevention and desert greening experiments are being done using synthetic high polymer absorbent materials. One major disadvantage of this technique is apprehension that these nonbiodegradable synthetic high polymer materials will remain in the soil for a long period or a fine powder of the polymer and residual monomers will move from the soil to the rivers, thus contributing to environmental pollution. After plants begin to grow, the bioabsorbent will be expected to cause no damage to the soil, because it is composed of natural materials, as a polysaccharide, and is expected to be more easily decomposed.

Biocatalysts such as microbes are normally hydrophilic and cannot live in an organic solvent. Most chemical substances are not soluble in water but can be dissolved readily in an organic solvent. Therefore, most chemical industries utilise organic solvent processes, and with a few rare exceptions, biocatalysts such as microbes are rarely used as catalysts in chemical industries.

To solve these ecological problems, the development of environmentally friendly products and processes, i.e. 'green product and green process' which are more compatible with the environment, are urgently needed. For example, these new biopolymers may be expected to overcome the problems associated with conventional synthetic high polymer agents because they are protein bioflocculants and polysaccharide bioabsorbents produced by a micro-organism and therefore do not pose a secondary pollution environmental hazard.

ENVIRONMENTAL FRIENDLY PRODUCTS (GREEN PRODUCTS)

New Bioflocculant Produced by *Rhodococcus erythropolis*

Screening for and characteristics of microbial flocculants

Fourteen strains of the genera *Rhodococcus*, *Nocardia* and *Corynebacterium* were found to produce substances which flocculate kaolin clay.

Rhodococcus erythropolis produce a kind of microbial flocculant which had very interesting and unique flocculating characteristics. It could efficiently flocculate all suspended solids in the aqueous solutions tested. Among those effectively tested were micro-organisms such as *E. coli* and alcohol yeast, activated sludge, *Microcystis aeruginosa* (AOKO), kaolin clay, muddy water, river dredging muddy water, river bottom sediment (HEDORO), ash from a steam-power station, and charcoal. These results indicated that the flocculant from *R. erythropolis* had a wide flocculating activity against both organic and inorganic materials. Cations such as Ca^{2+} also significantly increased the flocculating activity.

Application to the waste-water from the livestock industry: The flocculant produced by *R. erythropolis* S-1 was used for the primary treatment of pig urine and excrement waste-water. As shown in Table 36.1, pig urine and excrement did sediment effectively with the culture broth of *R. erythropolis*. These results show the feasibility of the use of the flocculant of *R. erythropolis* in waste-water treatment for domestic stockbreeding.

Table 36.1. Flocculation of pig urine and excrement waste-water with culture broth of *R. erythropolis*.

	Precipitate volume (%)	Supernatant after 10 minutes		
		TOC (ppm)	*TN* (mg/l)	*Clarification* (OD660)
Waste-water	0	1420	420	8.60
Waste-water + Ca^{2+}	0	1420	420	8.60
Waste-water + Ca^{2+} + culture broth				
1.0 ml	3	1250	310	6.90
5.0 ml	11	425	215	>0.02*
10.0 ml	13	504	213	>0.02**

* 10 ml of this culture broth contained about 0.1 mg of crude flocculant.
** This supernatant was colourless and clear.

Culture conditions for production of microbial flocculant by Rhodococcus erythropolis

1. General conditions: Addition of yeast extract and casamino acids stimulated the production of a flocculant by *R. erythropolis*. The favourable substrates for flocculant formation were fructose (or glucose), urea, and yeast extract. Optimum culture temperature and optimum culture pH for the production of the flocculant were 30°C and pH 8.5–9.5 respectively. Excess aeration decreased the production of the flocculant. The production of the flocculant paralleled the growth curve.

2. Low cost production techniques for microbial flocculant (NOC-1) by *Rhodococcus erythropolis*:
Both the flocculant production and cell growth were greatly affected by the addition of yeast extract and casamino acids as the organic nitrogen source. The most suitable concentration of the yeast extract

was 0.2 per cent. However, it is too expensive for commercialisation of the microbial flocculant, because it represents about 70–80 per cent of the total cost of the medium.

Among economical organic nitrogen sources tested, seafood processing waste-water was effective for flocculant formation and cell growth. This waste-water added medium can yield about a 70–80 per cent cost reduction of the production medium, compared with the conventional medium containing yeast extracts. Especially, the addition of this nonsterilised waste-water as an organic nitrogen source can greatly increase the flocculating activity in the culture broth and also substantially reduce culture time for the microbial flocculant production (Table 36.2). Bovine blood medium also appeared favourable for both flocculant production and cell growth. The bovine blood medium used alone can greatly reduce the total medium cost by about 80–90 per cent.

Table 36.2. Maximum flocculation activity of each medium.

Medium	pH	Experiment 1		Experiment 2	
		Maximum F.A.	Time (hour)	Max. F.A.	Time (hour)
Conventional	7	2.6	142	3.3	188
(Sterilised)	8	–	–	3.6	188
Waste-water	7	4.0	41	5.8	20
(Non-sterilised)	8	–	–	3.5	20
Waste-water	7	1.1	75	1.1	137
(Sterilised)	8	–	–	5.7	137

Ethanol as a carbon source was favourable for flocculant production, and it produced about a 30 per cent reduction of the medium cost.

Correlation between flocculant production and morphological changes in Rhodococcus erythropolis S–1

Morphological changes were observed in the bacterium during cultivation when *R. erythropolis* S-1 was cultured for flocculant production in a medium containing 0.5 per cent glucose and 0.5 per cent fructose as carbon sources. Elongation of the cells was observed during the early and middle log phases (2–6 d), and the flocculant was simultaneously produced. In the stationary phase (6–10 d), the elongated cells started to divide and no additional flocculant was produced. This elongation seems to occur simultaneously with flocculant production when the carbon source was easily assimilated. Morphological change in the bacterial cells is believed to be the result of metabolic changes. These phenomena may offer some insight into the relationship between the secretion of microbial products and morphological changes as well as reveal the physiological roles of the products.

A. protein bioflocculant produced by Rhodococcus erythropolis

The micro-organism was inoculated in 100 ml of the following culture medium and then incubated for 4–5 days at 30°C. The culture medium contained 10 grams of glucose, 0.5 gram of yeast extract, 0.5 gram of urea, 5 grams of K_2HPO_4, 2 grams of KH_2PO_4, 0.1 gram of NaCl, and 0.2 gram of $MgSO_4 \cdot 7H_2O$ in eleven of distilled water. After cultivation, cells were removed by centrifugation (10000 × gram for 5 minutes) and cell-free culture broth was obtained.

A kaolin clay suspension was used to test for flocculating activity. Because multivalent cations were indispensable for flocculation of kaolin by NOC-1, kaolin was suspended in 1.2 mM $Al_2(SO_4)_3$ solution at a concentration of 5 g/l. The sample was added to 1 ml of this suspension, and the assay mixture was then vortexed. The aggregate that formed was sedimented by weak centrifugation at 40 × gram for 15 seconds, and the optical density (OD) of the aqueous phase was measured at 660 nm (OD sample). The OD of the sample-free kaolin suspension was also measured (OD blank). Flocculating activity was calculated by the following equation:

$$\text{Flocculating activity} = \frac{\text{OD blank} - \text{OD sample}}{\text{OD blank}}$$

Forty grams of $(NH_4)_2SO_4$ and 100 ml of waste-saturated *n*-butanol were added to 100 ml of the culture broth, followed by centrifugation. Three phases appeared after centrifugation, i.e. a lower water phase, an *n*-butanol upper phase, and an insoluble layer boundary phase. The dry weight of the insoluble layer was 20–50 mg, while the amount of both butanol and water phases was less than 2 mg. It must be emphasised here that only the insoluble layer had flocculating activity. The active layer was further washed with acetone, suspended in distilled water and then dialysed; the dialysate was designated the flocculant preparation.

The flocculant preparation thus obtained lost its activity upon filtration through a membrane filter (0.22 µm), showing that it formed micelles. When a flocculation test was done at different concentrations of the flocculant preparation, sufficient activity was observed in the range of 10–20 mg/l. The activity decreased when an excess of the preparation (more than 20 mg/l) was supplied.

Elemental analysis showed that the nitrogen content was 11.0 per cent. In an infrared spectrum, two significant absorption bands (1500–1550 and 1650–1700 cm^{-1}) that were considered to correspond to mono-substituted amide can be observed. Three mg of the flocculant preparation was suspended in 10 ml of 0.04 M potassium phosphate buffer (pH 7.5) containing 6 µg/ml of Pronase E (protease, type XXV, Sigma) and in 10 ml of 0.01 N HCl containing 6 µg/ml of pepsin. After incubation at 37°C for 18 hours the mixtures were dialysed and a flocculation test was done. Flocculating activity was decreased to less than 10 per cent of the initial activity, suggesting that the protein portion is indispensable for flocculation.

When the flocculant preparation was analysed by filter paper electrophoresis, a single protein band was detected.

Several polymers that were produced by micro-organisms have been reported to be flocculants. Two of them had protein portions, but their activities were not lost on Pronase treatment. In contrast to that, the activity of NOC-1 was lost on protein-hydrolysis enzymatic digestion, suggesting that the protein portion is important.

New Super Polysaccharide Bioabsorbent from *Alcaligenes latus*

A. latus B-16 bioabsorbent production

After 4–6 days of cultivation in the basal production mediums, 20–25 grams of crude bioabsorbent was produced per litre of culture broth with either fructose and sucrose the most effective for bioabsorbent formation. Both urea, added as an inorganic nitrogen source, and yeast extract (0.05 per cent), as an organic nitrogen source, appeared to enhance bioabsorbent production.

Water absorption capacity

The bioabsorbent's water absorption capacity was measured using two different purified bioabsorbent (SP and FP) samples. Six control samples ware also tested, i.e. pulp, silica gel, ion exchange resin, a high-grade water-absorbing synthetic high polymer, poly(vinyl alcohol) (PVA), and an anionic synthetic high polymer absorbent.

Table 36.3 clearly shows that the presented bioabsorbent absorbs the most water (up to 1000 times its own weight), even more than the control group's synthetic high polymer water absorbents. It should be noted that this bioabsorbent water absorption capacity is more than 3–5 times greater than that of the currently used synthetic high polymer water absorbents.

Table 36.3. Water absorption capacities of various absorbents.

	Sample description	*Water absorption capacity per gram of dried sample (g)*
Test group	SP[a]	1349.0
	FP[a]	1295.4
Control group	Pulp	3.8
	Silica gel	1.4
	Ion-exchange resin	2.5
	PVA	4.6
	High-grade synthetic high-polymer absorbent[b]	249.4
	Anionic synthetic high-polymer absorbent[c]	363.6

[a] Bioabsorbent samples were produced using different culture conditions, i.e. by changing the culture medium's carbon sources. SP: Sucrose, FP: Fructose.

[b] High grade synthetic high-polymer absorbent: polyacrylate/PVA derivative (copolymer of acrylate and vinyl alcohol).

[c] Anionic synthetic high polymer absorbent: polyacrylamide derivative, MW 350×10^4.

Moisture absorption capacity

The moisture absorption capacities of the SP bioabsorbent and a mixture of SP and FP (1:1) bioabsorbent(MIX) were measured and compared with several control samples, with the results showing a high moisture absorption capacity.

Moisture retention capacity

The moisture retention capacities of the SP bioabsorbent and a mixture of SP and FP bioabsorbents (MIX) were measured and compared with the same control samples using three different desiccators containing a saturated solution of sodium nitrate/phosphorus pentoxide (data not shown) and a saturated solution of magnesium chloride (Fig. 36.1) at the respective relative humidities (20°C) of 64.8, 34 and 33 per cent.

Figure 36.1 clearly indicates that the new bioabsorbent has a high moisture retention capacity, i.e. in a dry environment (34 per cent and 33 per cent relative humidity) the SP bioabsorbent retained 82 per cent of its moisture after 24 hours.

Water absorption capacity in the presence of NaCl

The new bioabsorbent was placed in saline solution of varying NaCl concentrations to confirm its water absorption effectiveness, with the results being shown in Fig. 36.2.

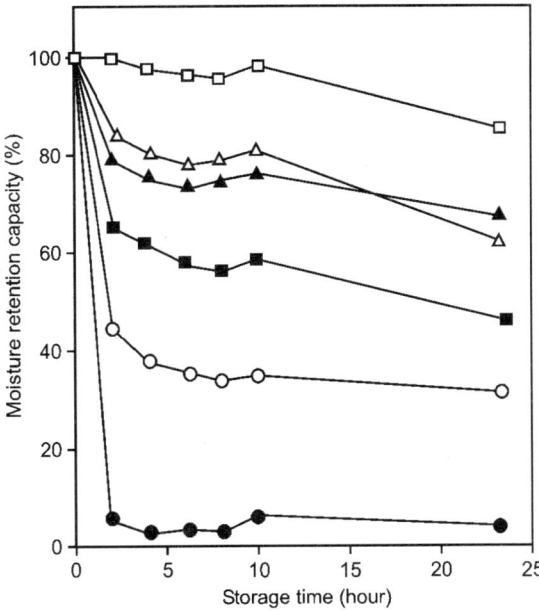

Fig. 36.1. Moisture retention capacities of various absorbents. Conditions: 33 per cent relative humidity, 20°C, in desiccator containing a saturated solution of magnesium chloride.□, biosorbent (SP sample); •, urea; △, glycerine; ▲, PVP; ■, PEG 200; ○, anionic polymer.

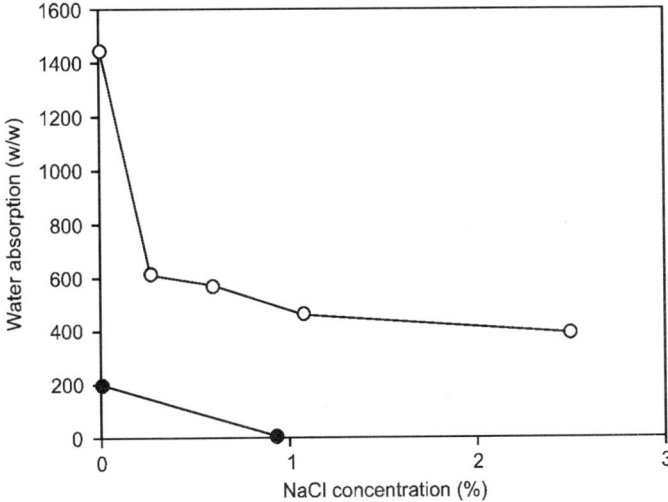

Fig. 36.2. Bioabsorbent and synthetic high polymer absorbent water absorption capacities in the presence of various NaCl concentrations. ○, bioabsorbent (SP sample); •, high-grade water absorbing synthetic high polymer (Sumika Gel S-50).

The water absorption capacity decreased in the presence of NaCl, however, this decrease was significantly less when compared to that of the synthetic high polymer absorbents (Sumika Gel S-50).

This water absorption capacity in the presence of 0.9 per cent NaCl enabled the bioabsorbent to absorb 450–550 times its own weight, which is 20 times greater than that of a currently used synthetic high polymer absorbent (22 times, our test data).

As shown in Fig. 36.2, the bioabsorbent was capable of absorbing water at 450 and 360 times its own weight at respective NaCl concentrations of 1 per cent and 2.5 per cent. It should be noted that these values are still slightly higher than those of conventional synthetic super water-absorbent high polymers for absorption of pure distilled water.

Retention of supplied water in sand with the bioabsorbent

A huge amount of water is needed for irrigation of the greening desert, and most irrigated water immediately disappears into the earth, causing injury from salt, and also disappears into the air at high temperatures.

Table 36.4 clearly shows that sand with the bioabsorbent has high water-retention capacity, meaning that sand with the bioabsorbent retains much of the supplied water even at a high temperature (70°C) like that of a hot desert surface in daytime.

Table 36.4. Keeping capacity of supplied water in the sand with bioabsorbent.

	Keeping capacity (%)		
	5 minutes	*1 hour*	*24 hours*
Bioabsorbent	100	86	42
Sodium alginate	13	7	8
Xanthan gum	8	0	0
Synthetic acrylamide high polymer	8	0	0
Synthetic high polymer absorbent	8	7	7

Culture broth for production of bioabsorbent

Glucose and sucrose were favourable carbon sources for the production of bioabsorbent from *A. latus* B-16. Yeast extract at a concentration of 0.5 per cent was also effective for growth. Biopolymers from *A. latus* were purified and separated into two types of biopolymers; a high-molecular weight component (acid-biopolymer) and a low-molecular weight component (neutral-biopolymer). These two types of biopolymers were shown to be homogeneous on electrophoresis and GPC (gel permeation chromatography on HPLC). The high-molecular weight component was a bioabsorbent and was able to absorb water at more than 1000 times its own weight. The total yield of acid and neutral-biopolymers from *A. latus* B-16 was 20–25 grams (dry weight) per litre of culture broth, and they were produced at a ratio of 20:1 (acid: neutral).

Constituent sugars of bioabsorbent

The constituent sugars of the polysaccharide bioabsorbent from *A. latus* B-16, which can absorb water at more than 1000-fold (maximum 2000-fold) its own weight, were identified by four methods such as thin-layer chromatography and high-performance liquid chromatography. This polysaccharide bioabsorbent is composed of glucose, rhamnose, fucose and glucuronic acid.

Environmental Friendly Process (Green Process)

Creation of bacterium living in petroleum

Biocatalysts such as microbes are normally hydrophilic and cannot live in an organic solvent. Most chemical substances are not soluble in water but can be readily dissolved in an organic solvent. Therefore, most chemical industries involve organic solvent processes, and with a few rare exceptions, biocatalysts such as microbes are rarely used as catalysts in chemical industries.

However, experiments were conducted on 400 million types of bacteria by mutation such as ultraviolet irradiation, then adding an organic solvent to cause mutation, from which three strains of bacteria capable of resisting organic solvents were created. Bacteria capable of living in organic solvents will be free of the limitations of utilising microbes in aqueous solutions and enable application to efficiency improvement of various industrial processes. The research team observed that the technology is applicable to various microbes and is engaged intensively in further research for early commercialisation.

To create organic solvent-resisting bacteria, a bacterium can be used that decomposes a sulphur compound known as dibenzothiophene (DBT). First, the bacterium culture was spread on a specified agar plate, then irradiated with ultraviolet rays for about 20 seconds, followed by the addition of an organic solvent such as heptanol and then culturing. These processes were repeated 20 times while gradually increasing the concentration of heptanol.

Heptanol is highly toxic and ordinary bacteria will not survive in it, but it was possible to detect three strains of organic solvent-resisting bacteria from among the roughly 400 million which were cultured. They survived even in a highly concentrated solution consisting of almost 100 per cent heptanol. The details of the mechanism of evolution are as yet unknown, but a big change has been confirmed in the fatty acid structure of the cell membrane, where the saturated fatty acids were changed to unsaturated fatty acids and the fluidity increased, which are the factors allowing resistance to organic solvents.

The capacity to decompose DBT was about 80 per cent compared with that before the evolution, but the technology developed by the joint research team has a broad range of applications, such as using microbes to produce diverse chemical products.

Scientists have succeeded in creating a strain of bacteria that normally lives only in an aqueous solution, allowing the bacteria to survive in an organic solvent such as petroleum.

To sum up environmental impacts occur at all stages of a product's life cycle. Design can be employed to reduce these impacts by changing the amount and type of materials used in the product, by creating more efficiently designed products, by reducing materials consumed during use, and by improving recovery of materials during waste management.

Biopolymer produced by micro-organism is composed of polysaccharide (or protein) and is completely biodegradable. Therefore, it is one of the safer, more ideal and promising 'green products' to preserve environmental quality for the next generation. Organic solvent-resistant micro-organisms have great potential for application to the chemical industry, i.e. 'green process'. For biotechnology R&D concentrated on 'bioprevention', the 'green design concept' will play an important role.

Chapter 37

Microbial Production of Energy from Industrial Waste

INTRODUCTION

Methane, a major gaseous fuel can be produced in abundant quantities from industrial wastes. This would substitute either partial or total energy needs of an industry along with stabilisation/disposal of effluent. Several process improvement approaches such as innovative fermentation modes, application of novel reactor designs, treatment of feeds, use of biostimulants may be used in order to optimise methane production.

Methane, a major gaseous fuel is obtained by microbial transformation of substrates in a variety of natural or man-made anaerobic environments. The conditions necessary for microbial production of methane include:

1. The presence of degradable organic substrates.
2. The presence of suitable nutrients.
3. The presence of sufficient water to submerge or dissolve organic substances and to dilute soluble compounds to below toxic levels.
4. The presence of suitable type of bacteria.
5. Optimum pH 6.6–7.6 and temperature (mesophilic, 30° to 38°C and thermophilic, 49° to 57°C).

This process can accept pure as well as heterogeneous feeds including almost all types of organic wastes. The bioconversion of complex organic compounds to methane is achieved by anaerobic digestion, which is carried out in airtight reactors/digesters.

Due to rapid industrial growth, the problem of effluent disposal of a wide variety of waste-waters has been attracting global attention and it has been proved that it is possible to recover biogas (methane) from the waste which in turn can be stabilised. It is, therefore, possible to generate clean fuel, methane with simultaneous alleviation of difficult environmental problems. This chapter will focus on the advances made in improving the understanding of biomethane production and its application in energy production from industrial wastes.

ANAEROBIC DIGESTION OF COMPLEX SUBSTRATES

The biological conversion of the organic matter to methane and carbon dioxide can be divided in three main stages (Fig. 37.1):

1. The first step involves hydrolysis of high molecular weight compounds such as proteins, fats and carbohydrates into compounds suitable for use as a source of energy and cell carbon.

2. The second step involves the bacterial conversion of compounds resulting from the first step into identifiable, lower molecular weight intermediate compounds, e.g. organic acids, H_2, CO_2.
3. The third step involves the bacterial conversion of intermediate compounds into simpler end products principally methane and carbon dioxide.

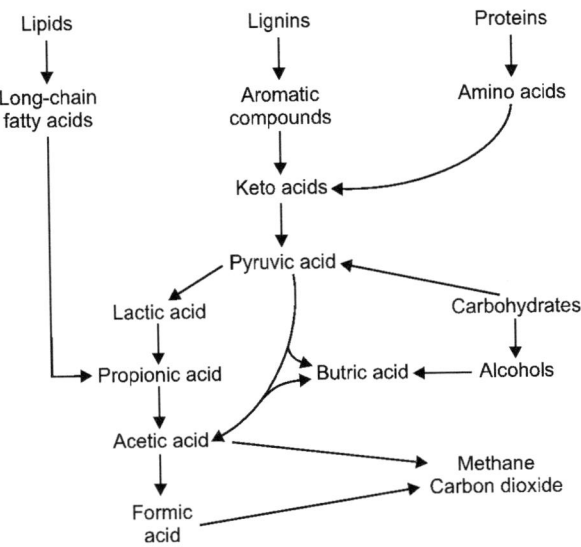

Fig. 37.1. Biochemical pathways for methane formation from complex organic compounds.

The first two steps described above are carried out by non-methanogenic organisms consisting of facultative and obligate anaerobic bacteria, also terms as 'acid formers'. On the other hand, methanogenic organisms or methane formers are strict anaerobes and converts the organic acids formed by the first groups to methane gas and carbon dioxide. A strict balance is required between non-methanogenic and methanogenic organisms for efficient conversion of complex substrates to methane.

PERFORMANCE OF ANAEROBIC DIGESTERS

To maintain a dynamic equilibrium between the two groups of microflora, the digester content should be void of dissolved oxygen and free from inhibitory concentrations of such constituents as heavy metals and sulphides. The pH of the aqueous environment should range from 6.6 to 7.6 and the temperature maintained in the range of 30° to 38°C and 49° to 57°C for mesophilic and thermophilic digestion respectively. A sufficient amount of nutrients, such as nitrogen and phosphorus, must also be available to ensure proper microbial growth.

It is considered that rate limiting step in methane fermentation is phase transfer of methane and carbon dioxide out of culture and faster fermentation is promoted by inducing faster gas transfer from liquid to gas phase by vigorous agitation, low gas pressure, and/or elevated temperature. Hashimoto showed that at a reduced gas pressure of 0.96 atm, methane production was only 5 per cent higher than that at atmospheric pressure at a 4-day hydraulic retention time (HRT), but about the same at a 6-day HRT. Andrews and Graef suggested that maintenance of a large head space above the digesting culture and recirculation of CO_2 scrubbed digester gas could prevent pH drops and enhance gas production.

CONVENTIONAL PROCESS CONFIGURATIONS

The following process configurations are practised commercially as shown in Fig. 37.2:

1. Lower or standard rate digestion with unmixed and heated digesters maintained at 30°–35°C and operated at hydraulic retention time (HRT) of 30 to 60 days at loading rates of 0.5 to 1.6 kg vs/m³d.

2. High rate digestion with 'completely mixed' heated digesters maintained at 30–35°C and operated at HRT's of 14 to 20 days and at loading rates of 1.6 to 3.2 kg vs/m³d.

3. Two stage digestion utilising a high rate and a low rate (unmixed) digester in series. The primary function of the second stage is to separate the digested solids from the supernatant liquor for increased gas production.

4. Anaerobic contact process. In this process high rate digester is operated in series with an anaerobic settler and settled sludge containing dense microbial population is recycled to high rate tank.

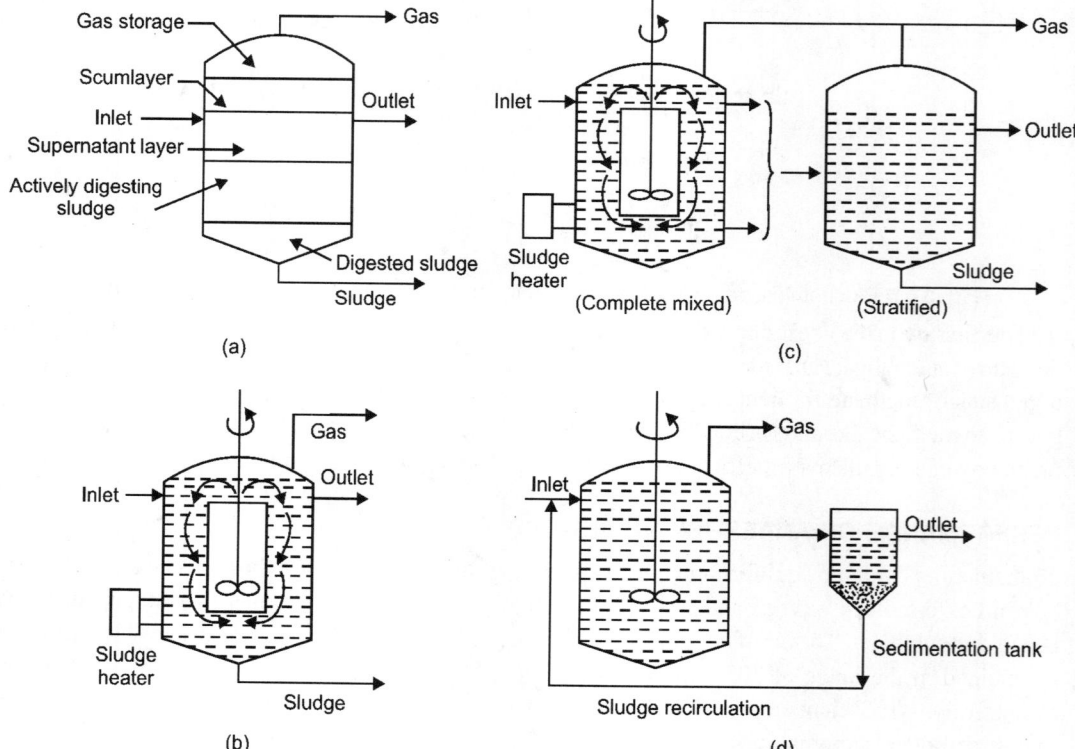

Fig. 37.2. Conventional digesters for biomethanation of organic waste: (a) standard rate digesters; (b) heat rate digester; (c) two-stage process; and (d) anaerobic contact process.

The conventional single-stage digestion process was designed to maximise sludge stabilisation for pollution control and not to optimise methane production. It was operated at long HRT and low loading rates to ensure volatile solids destruction and process stability. These conditions would result in high capital and operating costs and low net energy productions. The operation strategy to maximise net energy production efficiency would follow in the coming section.

Limitations of Conventional Digestion and Process Improvement Approaches

For economical biomethane production it is necessary to maximise net methane production rate, which may be achieved by:

1. Digester operation with the highest feasible feed concentration.
2. Operation at the highest feasible substrate and microbial solid residence times and lowest possible HRT.
3. Enhancement of feed biodegradability achievable by feed pre-treatment, residue post-treatment and recycling or the addition of metabolic stimulants.

The first two improvement strategies cannot be instituted in conventional single-stage complete mixed digesters which become unstable at high feed concentrations and short HRT's and where the SRT is the same as the HRT.

Process improvement approaches

The following approaches could be used to overcome the limitations of conventional digestion, to enhance net energy production:

1. Apply novel reactor designs to permit low HRT and high SRT operations.
2. Apply an innovative digestion mode (two-phase digestion) to optimise liquefaction-acidogenic and methanogenic fermentations.
3. Development of cost-effective feed pre-treatment and residue post-treatment and re-feeding techniques to increase feed biodegradability.
4. Develop biostimulants to enhance substrate metabolism.
5. Develop genetically superior organisms that can degrade substrates faster and are tolerant to process inhibitors.

Novel Bioreactor Designs

In the past one decade considerable advances have taken place in anaerobic biotechnology. Basically two approaches have been used for improved digester performance:

1. Maintaining high microbial density in the reactor, i.e. high SRT and low HRT to ensure reduced reactor volume.
2. Optimal operating conditions for improved microbial activity.

Some of the novel bioreactors shown in Fig. 37.3 are described below.

Immobilised packed bed reactors

The microbial biomass may be immobilised on various non-toxic, non-biodegradable support materials, e.g. clay, PVC, polyurethane foam, etc. Both upflow (also known as UAE, upflow anaerobic filter) and downflow (also known as DSFF, downflow stationary fixed film reactors) modes of operation have been employed and each has specific features.

The upflow anaerobic filter consists of a random packed medium such as stones, and rasching rings. The influent is introduced at the bottom and the effluent overflows at the top. The majority of the cells are retained in the packing interstices as unattached clumps, less than half of the cells are immobilised on the support matrix. Some of the problems encountered were in selection of packing, its plugging and difficulties in using concentrated wastes and wastes containing large amount of solids. However, plugging can be avoided by using coarse packing with considerable void space as well as substantial void space under the packing. Plugging problems occur because of suspended growth in the bottom one-third of the reactor. The development of DSFF reactor started in 1976 to circumvent the problems encountered in operation of upflow anaerobic filter (UAF) reactor. It consisted of packing with an oriented geometry that forms vertical channels that run the length of the packing (Fig. 37.4). The downflow mode of operation in combination with the oriented support medium configuration allows counter-current interaction of liquid and evolved digester gas to enhance digester mixing.

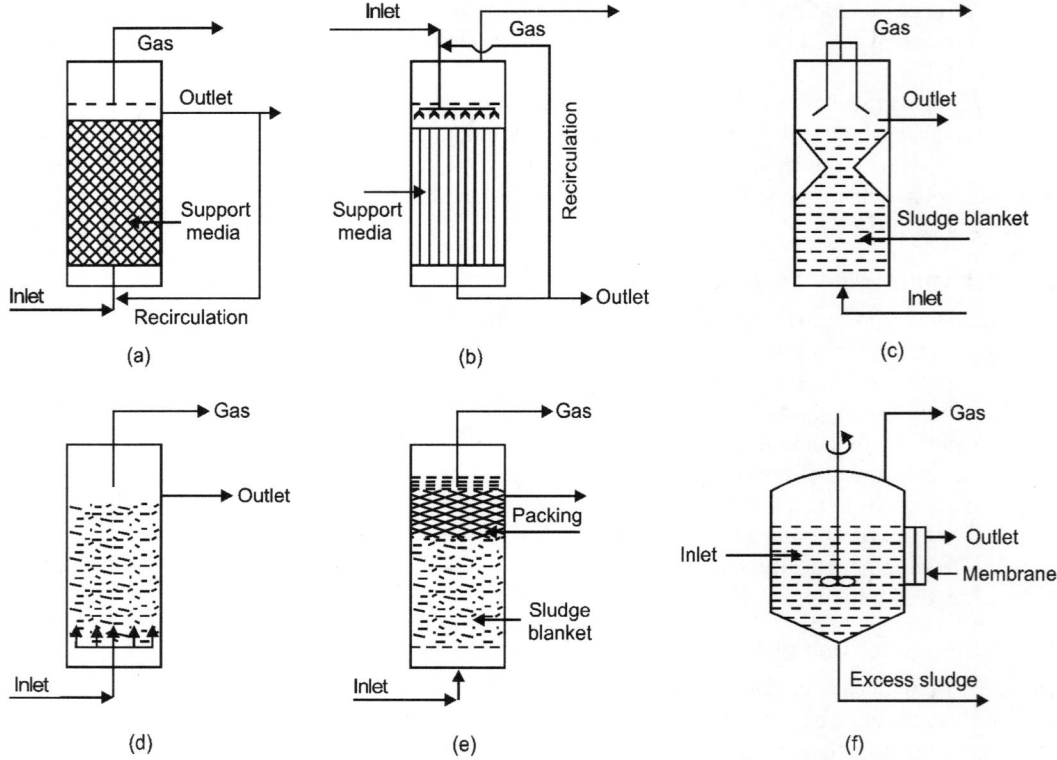

Fig. 37.3. New bioreactor designs: (a) upflow packed bed; (b) downflow stationary fixed film; (c) upflow sludge blanket; (d) fluidised bed; (e) upflow sludge blanket filter; and (f) membrane reactor.

The microbial cells are attached to the packing and the excess biomass is removed from the support surface by fluid shear thus preventing blockage of the channels. The important features of DSFF are:

1. Hydraulic overloading affects DSFF reactors less than suspended growth reactors.
2. Higher organic shock loads are not detrimental to DSFF reactors.
3. Waste with large amount of suspended material may be used.

The performance of DSFF reactors using different type of wastes is shown in Table 37.1.

Table 37.1. Performance of DSFF reactors.

Type of waste	Process details	Rate of methane production m^3 (STD)/M^3d	COD removal %
Dairy waste COD, 4.0 kg/m³	Needle punched polyster support	2.5	77
Sugar waste COD, 10.0 kg/m³	Needle punched polyster support	3.0	91
Chemical industry waste, COD, 10.0 kg/m³	Clay support medium	2.7	82
Barley stillage waste, COD, 53 kg/m³	Clay support medium	2.7	82

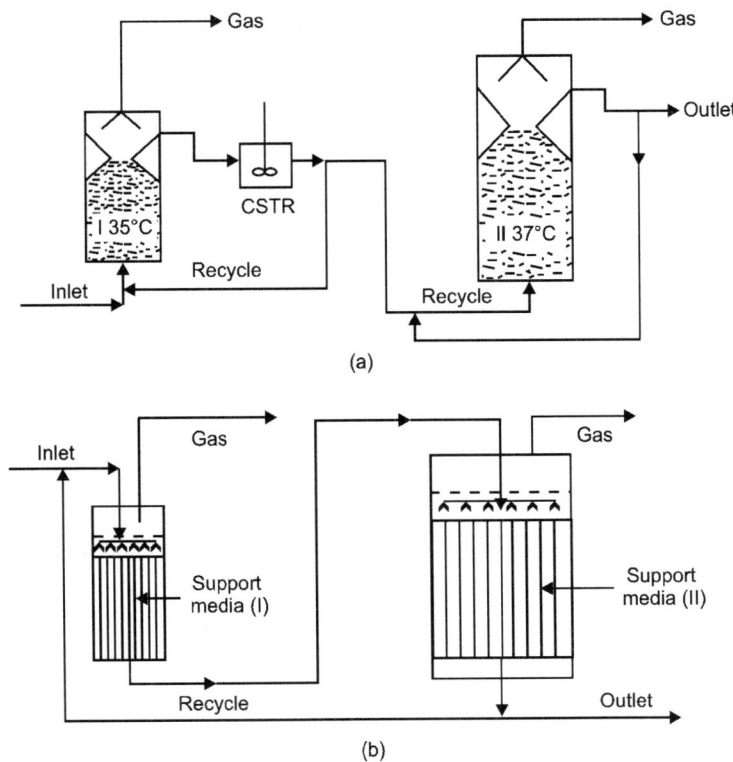

Fig. 37.4. Two phase biomethanation of industrial wastes: (a) two stage UASB reactor; and (b) two stage DSFF reactor.

Expanded bed/fluidised bed reactors

The expanded/fluidised bed reactors consist of inert sand particles in a column which expand with the upward flow of the waste through the column. Low density carriers such as PVC beads, ion-exchange resins, porous alumina may be used to reduce fluidisation energy requirement. In expanded bed reactors low upflow liquid velocity is employed to give a bed expansion of 10–20 per cent compared to 50–75 per cent in fluidised bed processes. Large biomass concentration is achieved on the large surface area provided by the small solid particles. These reactors have the advantage of minimum diffusional limitations and elimination of clogging problems.

Upflow anaerobic sludge blanket (UASB) reactor

It consists of a dense bed of biomass, either as flocs or granular sludge at the bottom of the reactor. The biomass concentration upto 80/1 is obtainable, and the process does not require any mixing. The rising gas maintains biomass granules and flocs in a more or less fluidised stage, and the resulting turbulence aids in detaching the gas bubbles from flocs in the upper part of the reactor. Since larger amounts of active biomass can be retained in a UASB reactors, these are potentially capable of higher loading rates and sometimes high conversion than DSFF reactors. The baffled reactor is another version of UASB. Multiple baffling of reactor provides staging and enhances cell retention.

Two Phase Digestion—An Innovative Fermentation Mode

Anaerobic digestion is a two-step process carried out by two groups of microflora, e.g. acidogents and methanogens. These organisms have different environmental and growth requirements. Therefore, optimisation of these dissimilar microbial phases in separate digesters has many advantages, e.g. enhanced net energy production, increased digester size and plant capital and operating costs, and production of a higher methane content gas.

Phase separation can be achieved by inhibition of methane formers, dialysis separation of kinetic control. Various two-phase reactor configurations have been proposed such as CSTR followed by UASB, two-phase UASB reactors and two-phase immobilised downflow reactors (Fig. 37.4).

Digestion of Preheated Feed and Use of Biostimulants

Most complex feed components are not degraded at high conversion efficiencies. However, this can be enhanced by special chemical and enzymatic hydrolysis. In addition certain biochemical additives stimulate microbial growth and metabolism, as they catalyse the reactions conductive to vigorous acidogenic and methanogenic fermentations, e.g. addition of thermophilic whey fermentation product to the digester enhanced methane yield and production rate.

Biomethanation of Industrial Wastes

Recently many reports have appeared on methane production from industrial wastes and simultaneous waste stabilisation. Some of the data have been summarised in Table 37.2.

Table 37.2. Biomethanation of industrial wastes.

Type of waste	Process details	Biogas yield/production rate
Distillery effluent	Two-phase immobilised down flow reactor 100 l, UASB reactor	0.69 m³/kg COD removed (75% methane 91/1.d (75% methane)
Sugarcane press mud	10 m³ pilot plant digester, feed: 100 kg press mud/day at 30 days RT (loading rate, 2.1 kg vs/m³d)	5.5 m³ biogas/d (65–80% methane)
Confectionary waste-water	Anaerobic expanded bed reactor, feed COD, 2937 mg/l (loading rate, 5.0 kg TOC/m³d)	4.9 m³/m³d (70% methane)
Coffee liquid waste	Two-stage upflow polyurethane foam reactors (loading rate, 15.2 kg vs/m³d)	2.66 m³/m³d (81% methane)

Thus, it can be expected that biomethane production from industrial waste would substitute either partial or total energy needs of an industry draining out its effluent with high organic load. Several improved process designs may be adopted in near future to optimise the net methane income production rate.

Chapter 38

Patents and Secret Processes

INTRODUCTION

The economic and competitive position of a fermentation process depends on several factors, some of which are recoverable yields, research costs, size of the market, profit potential, and patent or secret process position of the fermentation process or product. The latter consideration is particularly important, because patents and secret processes provide a degree of protection to the competitive position of a commercial fermentation. But which approach should be taken by an industrial concern? Should a fermentation process or product be patented, assuming that they are patentable, to provide 17 years of monopoly, or should the fermentation process and know-how be kept a secret so as to possibly allow many additional years of protection for the process? A yet more basic question, however, concerns what can and cannot be patented. The answers to these and other questions relating to the patenting of inventions will become more clear if the basic concepts behind the granting of patents by the patent office of the government are examined.

Patents are granted to inventers in return for a public disclosure of their inventions. These disclosures add to the knowledge of the respective art and help to advance the state of that art. The patent, in turn, gives the inventer the right to exclude others from making, using, or selling his particular invention as disclosed in the 'claims' of the patent. Most inventers will settle for the 17 years of protection rendered to them by a patent and will apply for a patent on their invention instead of attempting to maintain secrecy during the use of their invention. Obviously, secrecy about a fermentation process is difficult to maintain. Special contracts must be negotiated with trusted employees so that they will not reveal their knowledge of the process and, in addition, other problems arise which will not be discussed here. Regardless of these considerations, however, in a few instances, secrecy concerning a fermentation process has allowed the maintenance of a good competitive position for a commercial fermentation for many years in excess of the 17 years monopoly of a patent. A notable example is the commercial production of citric acid by *Aspergillus niger* as the fermentation process is conducted by the Charles Pfizer Company, Incorporated.

A fermentation process or product protected by a patent, however, also may have secrets associated with it. Special fermentation and recovery techniques associated with obtaining high yields and good product recoveries are often secret information known only to the inventer and his associates and not disclosed in the patent itself, since such data are generally developed after the application for the patent has been filed. Also, the patent describes a workable process and a product without necessarily describing minor variations in process technique and product recovery, considerations that may be, however, of extreme importance in maintaining the competitive position of the fermentation. There is some protection available from state or Central courts for this type of information. Thus, an individual or individuals can be prosecuted for stealing microbial cultures and technical data such as secret process information.

The individual working in an industrial research laboratory or any laboratory in which fermentation processes of potential economic value are under study should know-how to read a patent in order to be able to determine the points of the invention which are actually protected by the patent. This knowledge should prevent his infringing on the right, of other inventers. He should also understand the types of information that are required for filing a patent application so that research can be directed towards obtaining this information. He should be able to decide the extent of process variations or ranges in variations in chemical structure of a product which should be claimed in a patent application. As we shall see, claiming too little or too much about the process or product can be disastrous. Guidance in these problem areas frequently can be obtained by consulting a qualified patent attorney.

It will be easier to understand patents if we first consider how the concept of patents has developed during the past few hundred years. Patents in one form or another have been in existence at least since 1332, when Bartolomeo Verde of Venice received a revocable 12-year patent on a Windmill. A patent statute enacted in Venice in 1474 allowed the granting of exclusive rights for 10 years to 'inventers of new arts and machines'. In 1501, Aldus in Venice obtained a patent of *italic* type. Galileo, in 1594, received a patent for a machine to be used for raising water and irrigating land. This patent was for a 20-year term, and within one year of granting of the patent, he had to construct his new form of machine. This patent also provided for the punishment of infringers, and it was based on the assumption that the machine has not previously been invented or thought of by others, and that it had never been the subject of a previous patent or grant. Thus, Galileo's patent included several principles found in modern patent law. It provided a reward to the inventer and to the first inventer only, it provided a requirement for compulsory working of the invention, and the right to exclude others from practicing the invention.

Monopolies of a different sort were commonplace in England during the reign of Queen Elizabeth. These monopolies were granted by the Crown on a favouritism basis to various individuals, although this practice was illegal, based on English common law. Such products as salt, starch, glass and paper were included in these monopolies. The power to enforce the monopoly was also granted to the holders of the monopolies and, obviously, this led to high prices and poor quality. Regardless of these monopolies, however, a patent was granted by the Crown in 1565 for a furnace, and this probably was the first example in England of a reward to an inventer. In contrast to monopolies, the granting of a patent such as this was legal under English common law. This question of the legality of patents versus monopolies was decided when the Statute of Monopolies was enacted in 1623 during the reign of King James, since statute protected inventions by the granting to inventers of monopolies (patents) by the Crown, but is prohibited the Crown from granting other forms of monopolies. These early English patents did not include a written description of the invention, nor were drawings included. In fact, it was not until the middle of the eighteenth century that these became a part of a patent.

The early American colonists were well aware of the injustice that had occurred in England through the granting of monopolies. Therefore, a Massachusetts statute of 1641 stated that monopolies should not be granted or allowed, although such new inventions that might be profitable to the colony might receive a monopoly for a short period of time. In succeeding years, patents were granted by the various states and, in fact, an inventer had to obtain a separate patent from each state. The constitution of the United States, however, provided for patents, and the first Federal patent act enacted in 1790 allowed the patenting of 'any useful art, manufacture, engine, machine, or device or improvement therein not before known or used'. There have been many changes in the patent laws of the United States, since this first Federal patent act, and further changes are under consideration at the present time.

COMPOSITION OF A PATENT

A patent consists of three parts; the grant, specifications and claims. The grant is filed at the patent office and is not published. It is a signed document and is the agreement that grants patent rights to the inventer. The specifications and claims are published as a single document which is available to the public at a minimal charge from the Patent Office. Thus, this is the part of a patent that one normally sees. The specifications section is a narrative description of the subject matter of the invention and of how the invention is carried out. Therefore, anyone skilled in the particular branch of learning relating to the patent should be able to reproduce the invention on reading the specifications section. The claims section specifically defines the scope of the invention to be protected by the patent. That which others may not practice is defined here and, if the patent relates to industrial fermentations, it should be clearly understood from the claims just how the invention differs from known products and processes. The inventer may claim a part of all of that which is described in the specifications. In fact, the exact wording of the claims is important, because it states exactly what is to be protected and what is not. Thus, a patent stands or falls depending on the statements included in the claims section. It is obvious, then, that the inventer must decide as best he can that which he should and should not claim. If he does not claim certain variations in the process or product, it may be possible for others utilising these variations to carry out the process or make the product without infringing on the rights of the inventer. In contrast, if in the claims the inventer attempts to protect all possible variations in the process or product without experimentally establishing the validity of each of these variables, he may find that the patent is in jeopardy because of non-workability of some of the claims. Thus, the inventer should claim those variables of which he is sure and, if possible, he should experimentally test all other variables which, if not claimed, might allow others to circumvent the patent.

SUBJECT MATTER AND CHARACTERISTICS OF A PATENT

Inventions are divided into various categories or classes of subject matter. Several of these classes do not apply to microbial processes or products, however, and therefore are not considered in the present discussion. Microbial processes or products usually fall under one of two classes: an 'art or process' or a 'composition of matter'. The second class pertains to new chemical compounds or novel compositions or mixtures, and a micro-organism or its enzymes may be used to aid the accomplishment of a chemical synthesis or in producing such compositions. The art or process class of patents pertains to methods, including microbial fermentations, for bringing about useful chemical or physical results.

Implicit in the patentability of a microbial process is the concept that man has adjusted the environment of the micro-organism to such an extent that the organism will carry out, in the laboratory or commercial production plant, a process that it could not carry out, to any extent, under the conditions occurring in nature. Thus, *Clostridium acetobutylicum*, in nature, possibly might produce small amounts of acetic and butyric acid, but it is considered to be highly unlikely that natural environmental and nutritional conditions would be favourable for the further formation of acetone and butanol. Likewise, many investigations designed to demonstrate antibiotic production by micro-organisms in soil, for the most part, have yielded negative results. If this phenomenon does occur in non-sterilised soil, it is probably on a microscale in the immediate vicinity of the individual micro-organism or on the surface of a particle of readily decomposable organic matter.

The micro-organism utilised in a microbial process cannot be patented, and this statement also is true for mutants. However, the utilisation of a micro-organism not previously described or of a newly derived mutant does lend 'novelty' or newness to an invention. This consideration has caused considerable difficulty for the classification of industrial micro-orgainsms, because a large number of supposedly new micro-organisms have been given species names in patent applications, although these organisms often differed only slightly from recognised species. In fact, extensive studies of certain groups of these organisms are underway to establish their taxonomic relationships so as to reduce the numbers of species described in the patent literature to those that really differ taxonomically.

It is possible that the inventer of a microbial process may have the alternatives of patenting either the process or the product, or of patenting both. Obviously, if the product is a compound already well known (for example, the amino acid L-lysine), a product patent cannot be obtained, and the patent application must cover only the process for manufacture. In contrast, newly discovered product not previously described, such as certain new antibiotics, may allow application for a product patent. This product patent has distinct advantages over a process patent in that there may be more than one microbiological route or even a chemical route to formation of the product. A specific example is the Pfizer product patent, on the antibiotic tetracycline, since there are at least two processes involving micro-organisms yielding this antibiotic. Thus, tetracycline can be produced by a direct fermentation, or chlorotetracycline can be first accumulated through fermentation followed by its chemical conversion to tetracycline.

The United States has an 'examination' patent system in contrast to the 'registration' system in use in some other countries. The examination system requires that a patent application be studied for 'novelty' in the light of 'prior art' and for usefulness or 'utility'. The prior art consists of all printed material, including patents, that was available previous to the time of patent application. This material may be from any country and in any language. The invention must be new in respect to use in this country, so that any unpatented process, even if secret, already in use cannot be patented. For example, Pfizer's secret citric acid process cannot now be patented. Prior experimental use, an abandoned experiment, or lost art does not affect novelty. Based on this, a research worker should not abandon experiments in his research book, and he should not state in writing that he has given up on the experimental approach. A research worker also must be careful that his own scientific publications do not constitute a bar to obtaining a patent. Thus, the Patent Office considers that a patent for which the application was filed within one year after a pertinent publications or public use by the inventer still may be granted, but that longer time periods forfeit novelty. Publications or prior uses by others than the inventer can be antedated.

Too many patent applications are presented to the US Patent Office for a rapid determination of novelty. This backlog results in a prolonged period between the time of patent application and patent issue — as much as three years or more.

The requirement for usefulness or utility means that the invention must perform some beneficial function. Thus, it is considered that utility is absent for inventions that are inoperative, frivolous or injurious to the health, morals, and good order of society. No particular degree of utility is required, however, and there need not be a presently existing practical usefulness. For example, a compound may be useful under the patent statutes if it can be employed for research purposes, as in the case of intermediates that may be used in the synthesis of other compounds of a useful class.

The basic criteria for patentability are set forth in the patent statutes. For example, Section 101 of Title 35 USC states:

'Whoever invents or discovers any new and useful process, machine, manufacture, or composition of matter, or any new and useful improvement thereof, may obtain a patent therefore, subject to the conditions and requirements of this title.'

Thus, apart from utility, the basic requirements for patentability are new and non-obvious subject matter and adequacy of disclosure. In regard to the latter point, in microbiological applications, the drafting of a proper disclosure is quite complex and involves problems that require competent professional assistance. Also, where a new organism is involved, in addition to a proper description in the specification, a culture should be deposited with a recognised depository culture collection prior to filling, although public access to the culture may be restricted until the patent issues.

Based on novelty and utility, the Patent Office decides whether a patent should be granted for a particular invention. If this decision should be unfavourable as it relates to a particular patent application, an appeal can be made to a Board of Appeals within the Patent Office, and one can even appeal to various federal courts. Also, amendments can be made to the original patent application in order to make the invention more acceptable in the eyes of the Patent Office.

There is no requirement that a patent granted in the United States must be put to actual use. In fact, there are many 'defensive' patents and patents on small improvements that are never sold or used commercially. Despite such nonuse, these patents seldom impede economic development, since they provide valuable information and encourage invention of modifications or alternatives. Also, in many cases, a patent of this type may be licensed to 'unblock' another invention. A study by the Patent, Trademark, and Copyright Foundation of George Washington University (Washington, D.C.) showed that 55 to 65 per cent of assigned United States patents and 40 to 50 per cent of unassigned United States patents were actually produced for sale or used for making commercial articles at some time during the life of the patent, which indicates a rather high rate of commercial utilisation of patents.

The patent concepts discussed thus far apply not only to 'basic' patents but also to 'improvement' patents that follow and are closely related to the basic invention. Thus, if an inventer obtains a basic patent for an invention, but later finds an improved way of carrying out the invention, he may obtain an improvement patent. The situation is more complex, however, if an inventer other than the original inventer obtains the improvement patent, since it is likely that the second inventer cannot utilise the improvement without permission or license from the holder of the basic patent. Thus, the solution usually calls for negotiation and sale of rights so as to allow utilisation of the improvement patent. A specific example of basic and improvement patents is associated with a fermentation process for the manufacture of L-lysine. The basic patent (US 2771, 396) pertains to a fermentation process for utilising and *Escherichia coli* mutant to produce 2,6-diaminopimilic acid (DAP), with this compound then being decarboxylated to L-lysine by a wild-type strain of *Aerobacter aerogenes*. This basic patent was followed two years later by an improvement patent (US 2841, 532), which was secured to protect the decarboxylation of (DAP) by *Escherichia coli* back mutants that accumulated during inoculum growth and production, a situation in whch the DAP decarboxylase of the *Aerobacter aerogenes* was not needed. In this instance, negotiations were not required, since both the basic and improvement patenst were assigned to the same fermentation company.

WHO IS THE INVENTER: WHO OWNS AN INVENTION

In the United States, a valid patent is granted only to the first inventer, regardless of filling dates for patent applications. It is not always clear, however, just who really is the first inventer and, therefore, the patent laws provide for 'interference proceedings' to determine priority of invention. These interference proceedings may be initiated by the Patent Office before issue of the patent, or they may be initiated by another applicant who has a copending application or who files within one year after the patent issues. The inventer is considered to be that individual who first conceived the idea of the invention, regardless of whether the idea has at that time been reduced to practice. But, to be considered as the inventer, one also must have demonstrated reasonable diligence in carrying out the invention (reducing it to practice), or be able to prove reduction to practice prior to the other party's conception. As can be seen, it is of utmost importance for the inventer to establish the actual date on which he conceived the idea of the invention. To establish this date, the inventer should immediately get the incentive idea down in writing, sign and date the document who can verify the date and content of the description of the invention. This recording of the invention is often done in a research book, and it should include what is considered to be the results and means for obtaining the results of the invention. In addition to recording the invention, every page in the research book on which further experimental work has been recorded should also be signed and dated by the inventer and by one or two witnesses who understand the invention. These witnesses possibly may be called on at a later date to identify the writing in the research book and to establish its date and content. The procedures outlined above establish both the date of conception of the invention and the progress in its reduction to practice but, to prove reduction to practice, the inventer must be able to produce corroboration that the invention was actually carried out, generally through witnesses who actually observed the experimental work. Alternatively, however, a patent application itself can serve as a 'constructive' reduction to practice without further proof being required, but by so doing, the invention is placed in a less favourable position as regards considerations for patentability.

Implicit in the concept of granting of patents is the premise that a 'spark of genius' or 'inventive ingenuity' is involved. In other words, individuals other than the inventer, even with the knowledge at hand which was available to the inventer, could not have conceived the invention. At times however, questions arise as to who among several individuals working in a single laboratory or for an industrial concern is the actual inventer. In other words, who actually possessed the inventive ingenuity? If a supervisor presents an idea and the experimental approach for reducing it to practice to laboratory technician, then the invention is the property of the supervisor and not of the technician who merely carried out instructions to reduce it to practice.

However, the supervisor who presents the idea without a solution may possibly not be the sole inventer. At times, the actual inventer may be difficult to decide, and the names of more than one inventer may appear on the patent application as joint inventers. Usually, however, invention is thought to be conceived by a single individual.

Aside from determining who the inventer is, there is the question of who owns the patent. This is a particularly pertinent question, because patents can be of great economic value in addition to the monopoly that they afford, since they can be sold, licensed for a return of royalties, or assigned to an industrial concern or to the federal goverment. Also, patents are heritable property and can be part of an estate. Research workers employed by an industrial concern often sign contracts stating that they will assign their inventions to the concern so that the resulting patents are actually owned by the concern. In fact, it is often a condition of employment that such agreements be signed. Problems arise at times when an individual with such a contract invents something 'in his basement' which may be directly related to the interests of the company, or when the individual utilises company equipment and facilities to make an invention, whether or not the invention is directly related to the interests of the company. Although there are some ground rules applying to these situations, it still may be difficult to make a fair decision as to actual ownership of such a patent.

A published patent states at the top of its page the type of ownership that is to be associated with the patent. There are three categories. The first category, in which most patents occur, is that in which commercial rights are retained by the inventer or the concern to which the patent has been assigned, although these rights may be further licensed, assigned, or sold. In the second category are those patents that are assigned to the federal government, for instance, those assigned to the US Department of Agriculture. The assigning of patents to the federal government allows it some control in the licensing of the use of these patents. The last category includes those patents 'dedicated to the public'. Such patents do not provide a monopoly for any individual or any industrial concern but make the inventions available for all. In regard to this category, the absence of exclusivity for these patents may not provide adequate incentives to invest in the development or marketing of products.

PROTECTION OF THE RIGHTS OF THE INVENTER: INFRINGEMENT

A patent grant to an inventer confers on that inventer the right to exclude others from making, using or selling his invention for a period of 17 years from the date of patent issue. In case of 'infringement', that is, a violation of these patent rights, the inventer can call on the federal courts for help. In the courts, the inventer may sue the infringer for up to triple damages plus court costs, and the damages include profits that the infringer has accumulated in the practice of the invention. Thus, the infringer has a great deal to lose when he violates the patent rights of others. However, the potential infringer may feel that he has not infringed the subject matter specifically stated in the claims of the patent. Also, he may be able to show that certain of the claims, or all of the claims, are seriously defective so that the patent grant is not valid. If, as a result of these proceedings, certain of the patent claims are declared invalid by the courts, the patent may still be preserved by filing a 'disclaimer' with the Patent Office for the faulty claims. Also, a 're-issue' patent may be granted to correct errors in the patent as originally granted. In certain instances, there are apparent infringements of inventers rights in which United States court generally do not enjoin or penalise the infringer. This occurs when someone uses information from the patent claims on an experimental basis, or for private, non-commerical, use. For example, a patent covering a new process for the manufacture of wine might be employed by an individual in his basement to make a small batch of wine for home consumption. While this is technically an infringement, it is difficult to conceive of expensive infringement proceedings being brought against such an individual.

Drugs, pharmaceuticals, and other fermentation products manufactured in a foreign country and imported for use or sale in the United States present special problems. If such products are protected in the United States by a product patent, then importation and use or sale of the product constitute infringement. A specific example can be cited for the antibiotic tetracycline as produced by fermentation in Italy where patent coverage cannot be obtained for pharmaceutical products or processes of manufacture. In this instance, importation and use or sale of tetracycline by American concerns not licensed by the holder of the tetracycline product patent are considered to constitute infringement, although the Italian companies themselves cannot be sued.

COST OF A PATENT

The costs for patent application and issuance have recently been greatly increased. Also, there are additional fees for drawings and appeals, and patent attorney fees may be considerably more. In regard to this point, in rare instances patent attorneys will accept a percentage of royalties from the monies accruing from use of the patent instead of a specific fee. Research personnel for an industrial concern do not pay fees since these are absorbed by the employer as a part of the agreement for assignment of patents. Obviously, the real expenses associated with a patent occur if an infringer is challenged. Neverthless, usually the high potential monetary returns from a useful patent outweigh all these considerations.

POSSIBLE CHANGES IN UNITED STATES PATENT LAW

As previously stated, over a period of years, various changes have occurred in United States patent law, and further changes are still being considered. For instance, under consideraion is a change to grant the patent to the first-to-file rather than to the actual first inventer, a procedure that would totally eliminate patent interferences. Another change would permit the filing of one or more preliminary patent applications prior to the filing of a complete application within one year. In addition, all applications would automatically be published 18 to 24 months from the earliest filing date. Finally, the life of the patent would be for 20 years from the date of filing of the patent application, rather than for the present 17 years from the date of granting of the patent.

PATENTS IN OTHER COUNTRIES

To protect his invention, an inventer may well wish to apply for patents in countries other than the United States. However, this is more complicated than it would seem, because the procedures for application, the structures of patents, and the protection afforded by a patent vary markedly from one country to another, so that separate and differing patent applications must be prepared for each country. Several countries use a 'registration' system for filing patent applications instead of the 'examination' system employed in the United States. In the registration system, the proper filing of the patent for granting of a patent. This system is operative in France, Switzerland, Italy, Belgium, Luxemburg, and some other countries. Also, the small and less industrially developed countries utilise this system, because it is simpler and less costly. Some of the countries utilising the registration system do not require a definite listing of claims, but merely a short resume. Other countries require that the claims be stated, but this is only to help in deciding whether more than one invention is involved. With the registration system, the scope of the patent is not limited by the exact wording of the claims and, if infringement litigation ensues, the entire disclosure is studied in light of prior art. Thus, the registration system for patents makes no decision on scope and validity before granting of the patent, so that these considerations are left to the decision of the courts.

Many countries, excluding the United States and Canada, consider the inventer to be the individual who first files a patent application, and not necessarily the one who first conceives the invention and reduces it to practice. Also, in many countries, the determination of novelty is prior to the date of filing of a patent application. This is in contrast to the one-year grace period allowed to an inventer in the United States between the time of published description or use of an invention and the filing of a patent application.

At present, pharmaceuticals and the microbiological processes for making them cannot be patented in Italy. In contrast, Germany and the Netherlands provide patent grants for processes for making chemicals and pharmaceuticals, but the products themselves cannot be patented.

Most countries other than the United States require the payment of periodic maintenance fees to keep a patent in force. Also, many countries require that a patent actually be used within the country granting the patent if the patent is to stay in force.

In fact, if the government feels that the public would benefit from the exploitation of a patent not being used, it may request the granting of a compulsory license to others who are willing to use the invention. In a few countries, a patent may be revoked if not used.

The many variations in patents and patent application procedures make it difficult for an inventer to know how to proceed in obtaining patents in countries other than the United States.

However, a summary guide for the inventer may be found in a special report entitled 'Common patents for the common market' published in chemical and engineering news. In this guide, the following patent considerations are compared for 22 countries: (i) what is considered as prior art, (ii) what cannot be patented, (iii) language patent written in, (iv) type of examination, (v) opposition, (vi) life of patent, (vii) maintenance fees, and (viii) working requirements.

Glossary

Abiogenesis	:	Abiogenesis is the theory of spontaneous generation of living cells from nonliving material.
ABPA	:	Allergic bronchopulmonary aspergillosis.
Abscissa	:	X-axis.
Absolute risk	:	Absolute risk is the incidence of a disease in a population.
Abstriction	:	Abstriction is the formation of spores by the growth of septa thereby cutting off successive sections of a conidiogenous cell.
Acanthamoeba	:	Protozoan; freel-living amoebae commonly found in natural waters.
Accession number	:	Sequential numbering of specimens according to the order in which they are obtained; also used for nucleic acid sequences in a database such as GenBank.
Accidental fecal release (AFR)	:	AFR inadvertent discharge of human fecal material into recreational waters such as beaches, swimming pools, and jacuzzi tubs.
Acellular	:	Without a cell structure.
Acerose	:	Stiff and needle-like.
Acetobacter	:	Acetobacter is a bacterial genus; acid tolerant, aerobic, gram-negative bacilli, an acetic acid bacterium that ferments ethanol to acetic acid but also has the needed enzymes to complete the cycle by oxidising acetic acid to CO_2 and water; commonly used in the commercial production of vinegar.
Acetogenesis	:	Acetogenesis is the formation of acetate from CO_2 as the result of the metabolic processes of certain bacteria, in contrast to methanogenesis; three recognised processes are the acetyl-CoA pathway, the glycine synthase-dependent pathway, and the reductive citric acid cycle.
Acetogenic bacteria	:	Acetogenic bacteria are approximately 40 recognised species isolated from diverse anaerobic habitats that conduct acetogenesis using the acetyl-CoA pathway for the conservation of energy and growth; a variety of bacillus and coccoid-shaped gram-negative and gram-positive bacteria that are mostly mesophilic, although some psychrophilic and thermophilic species have been isolated.
Acetyl-CoA pathway	:	It is a autotrophic CO_2 fixation conducted by obligate anaerobes including methanogens and sulphate-reducing bacteria.
Achlorophyllous	:	Achlorophyllous is nonphotosynthetic; lacking chlorophyll.
Acholeplasma laidlawii	:	Microbial species, a member of the Mollicutes but these organisms do not require sterol or serum; initially isolated from sewage, compost, and soil.
Acidophile	:	Acidophile is a micro-organism that preferentially grows at an acidic pH; cells that have an affinity for acidic dye.
Acid rock drainage (ARD)	:	The bacterial mediated leachate resulting from the oxidation of sulphide minerals exposed to air and water, and the products of the interaction of alkaline rock and water with acidic metal-containing solutions.

Acinetobacter	:	Bacterial genus; member of the Neisseriaceae; gram-negative, oxidase negative coccobacillus that is saprophytic; isolated in soil, water, sewage, and clinical samples, currently 7 recognised species.
Acinetobacter baumannii	:	These are bacterial species; isolated from air samples associated with waste-water treatment operations.
AcNPV	:	The designation of a nucleopolyhedron baculovirus that infects the alfalfa looper (*Autographa californica*).
Acropetal	:	Acropetal is the characteristic of fungal conidia arranged in chains in which the youngest is located at the tip of the chain; in contrast to basipetal.
Actinomycetes	:	Bacteria in which species are characterised by the formation of branching and/or true filaments.
Acuminate	:	Acuminate gradually tapering at the tip.
Acute bronchitis	:	Disease that presents with cough productive of sputum that is often secondary to a viral or bacterial infection and the inhalation route of exposure to high concentrations of bioaerosols in agriculture facilities.
Acute reaction	:	It is characterised by a relatively short time period between exposure and effect; generally involves a dramatic onset of an adverse health effect.
N-acyl homoserine lactone (AHL)	:	AHL is a small, diffusible signal molecules used by gram-negative bacteria for intercellular communication.
Adenosine triphos- phate (ATP)	:	ATP is a carrier composed of adenosine and three phosphate groups that transports phosphate and energy in biological systems; uniformly present in all micro-organisms and the concentration is relative to cell carbon but dependent on the physiological state of the cell and lost rapidly with cell death; used as a measurement of microbial biomass; detected by the luciferin-luciferase assay and high performance liquid chromatography.
Adjuvant	:	A substance such as aluminium sulphate that increases the efficiency of a vaccine.
Aerial hyphae	:	Hypha structures that project upward from the surface of a fungal colony.
Aerobiology	:	The study of bioaerosols including airborne bacteria, fungal spores, fragments of micro-organisms, and by-products of microbial metabolism.
Agar	:	A derivative of marine seaweed used as a solidifying agent in many microbiological media.
Agglutinins	:	Antibodies that participate in agglutination reactions.
Antibody	:	A highly specific protein molecule produced by plasma cells in the immune system, antibodies function in humoural immunity.
Antigent	:	Any chemical substance that elicits a response by the body's immune system.
Autograft	:	Tissue taken from one part of the body and grafted to another.
Bacillus cereus	:	Bacterial species; associated with both diarrheal food poisoning and emetic food poisoning; endospores survive normal cooking procedures and germinate during improper storage after cooking.
Bacillus licheniformis	:	Bacterial species; opaque colony with a dull to rough surface and hair-like outgrowths; colony is firmly attached to the agar surface and aged cultures may become brown in colour; bacilli are often in chains; endospores occur in soil and may survive extreme heat; produces bacitracin.

Bacitracin	:	Bacitracin are peptides produced by *Bacillus licheniformis* that inhibit synthesis in gram-positive bacteria.
Bacteria	:	These are ubiquitous, single celled prokaryotic micro-organisms; the domain level of taxonomy of micro-organisms that are distinctly different from prokaryotes in the domain Archaea and the eukaryotes in the domain Eukarya.
Bacterial lawn	:	Confluent, uniform distribution of bacterial cells across the surface of an agar medium.
Bacteriocins	:	A group of bacterial proteins toxic to other bacteria.
Bacteriology	:	The study of bacteria and members of the Archaea.
Bioaccumulation	:	General term used for a process by which chemicals are taken up by organisms from ambient medium directly or through consumption of food containing the chemicals, i.e. via. food chain.
Bioassay	:	Bioassay is a biological-based analysis.
Bioconcentration	:	A process by which there is a net accumulation of a chemical directly from the ambient medium into organisms resulting from simultaneous uptake and elimination.
Bioconversion	:	A biocatalyst-mediated conversion of one substance to another.
Biodegradation	:	The process of chemical breakdown of a substance to smaller products caused by micro-organisms or their enzymes.
Biodeterioration	:	The chemical or physical alteration of a product that decreases the usefulness of that product for its intended purpose, caused by micro-organisms or their enzymes.
Biogeochemical	:	Biogeochemical involving biological, geological, and chemical aspects.
Biohazard	:	Biological material that is likely to cause a risk to human health.
Bioluminescence	:	Bioluminescence is the generation of visible light by micro-organisms.
Biomass	:	The amount of living material present; a measurement of the quantity of energy being stored in a segment of the biological community expressed in units of weight.
Biopreservation	:	The use of micro-organisms or microbial by-products to prevent spoilage and extend the shelf-life of foods.
Biotransformation	:	Enzyme-catalysed conversion of one chemical, other than the normal body constituents of live organisms, into another. Normal metabolism refers to such conversions restricted to carbohydrates, fats, proteins, etc. taking place inside the body.
Caespitose	:	Caespitose in dense groups forming tufts.
Calicivirus	:	Viral genus; in the family Caliciviridae containing many species, including the Norwalk virus, which is responsible for many water and foodborne outbreaks of gastroenteritis.
Calvin cycle	:	Autotrophic CO_2 fixation.
Campylobacter	:	Bacterial genus; gram-negative, microaerophilic, chemoorganotrophic, motile with a corkscrew motion, spiral-shaped curved bacteria; many species are pathogenic to humans and animals; isolated from the intestinal tract, oral cavity and reproductive organs of humans and animals.
Capsomere	:	It is a protein subunit in a viral capsid; the cluster of viral proteins or protomers.
Carboxydotrophic bacteria	:	Bacteria that aerobically utilise carbon monoxide as both a carbon source and as an energy source, for example, *Pseudomonas carboxidoflava* and *Pseudomonas carboxidohydrogena*.

Carboxysomes	:	Enzymatic polyhedral shaped inclusion bodies necessary for the Calvin cycle; first purified and studied in *Halothiobacillus neapolitanus*, formerly termed *Thiobacillus neapolitanus*.
Carcinoma	:	A malignant tumour that arises in epithelial cells, infiltrates the surrounding tissue, and spreads to distant sites in the body.
Carnitine	:	An amino acid ($C_7H_{15}NO_3$) that is used as an osmolyte.
Carrier	:	An individual that transmits an infectious agent to others, but does not show any symptoms of the disease.
CBOD	:	Carbonaceous biological oxygen demand.
CCD camera	:	Charge-coupled device camera.
CCL	:	Contaminant candidate list.
CDC	:	Centres for Disease Control and Prevention.
CDFF	:	Constant-depth film fermenter.
cDNA	:	Complementary DNA.
CEGLs	:	Continuous exposure guidance levels.
Chemotaxis	:	It is a movement of a micro-organism in response to a chemical gradient.
Chimera	:	A plasmid engineered to contain a fragment of foreign DNA.
Chlamydias	:	Gram-negative bacteria that are obligate intracellular parasites causing respiratory and urinary-genital tract diseases in humans.
Chronic	:	Continuous, over an extended period of time.
Chronic health effect	:	It is a long-term adverse reaction.
Citrobacter	:	Member of the family Enterobacteriaceae; human pathogen with potential environmental exposure via aerosols generated during waste-water treatment practices; member of the total coliform group.
Clones	:	It is a group of recombinant DNA molecules, genetically identical cells, or genetically identical organisms that are derived from a single ancestral molecule, cell, or organism.
Coliforms	:	Gram-negative, lactose-fermenting, enteric rods as *E. coli*.
Complement	:	A group of proteins that functions in a cascading series of reactions during the response by the body to certain antigens, the complement cascade is stimulated by antigen-antibody activity.
Cosmid	:	Phage-plasmid artificial hybrids.
Dark repair	:	Mechanism of DNA repair following exposure to ultraviolet light that occurs in the absence of light; also termed excision repair; in contrast to photo repair.
Dauxie	:	The utilisation of one substance at a given rate before the utilisation of another substance at a different rate.
DBP	:	Disinfection by-product.
DBT	:	Dibenzothiophene.
DDBJ	:	DNA Data Bank of Japan.
Death phase	:	Final period of a growth curve of a microbial culture in which the number of organisms decreases over time.
Decussate	:	It is arranged in pairs at right angles to the subsequent pair.
Deep agar	:	Test tube with medium solidified without a slant that is generally inoculated with a needle in a single stab.
Denaturation	:	The separation of double-stranded DNA into two single strands by manipulation of the ionic conditions of the solution; in contrast to melting for separation; in

contrast of hybridisation for the construction; also refers to the breaking of hydrogen bonds to alter tertiary structure of proteins.

Deoxynevalenol	:	Mycotoxin; also called vomitoxin; may cause laboratory changes in immuno-globulins with exposure via handling of contaminated foodstuffs.
Depauperate	:	Appearing starved or underdeveloped.
Depth of field	:	The distance between the closest and farthest objects in focus within a field of view of a lens.
DFA	:	Direct fluorescent antibody staining method.
Diazotroph	:	An organism that is capable of nitrogen fixation.
DIC	:	Differential interference contrast microscopy.
Diphasic fungi	:	Characterisation of some fungal genera in which there is a filamentous phase and a yeast phase.
Diphtheroid	:	Shaped like a diphtheria bacillus.
Disinfect	:	Disinfect is to treat a surface or a liquid with a substance that will kill micro-organisms.
Disinfectant	:	A chemical used on surfaces or in water to kill micro-organisms; may cause harm to host tissue.
Dispersion	:	The spread of a substance from the source; in statistics it is the measurement of how closely the data cluster around a typical value or central tendency.
DNA microarray	:	Molecular method for detection of specific micro-organisms in which hundreds to thousands of oligonucleotide probes are fixed onto a support matrix, labelled sample DNA is added, hybridisation occurs, and the identity of the complimentary sequences is determined.
DNA polymerase	:	An enzyme that directs the synthesis of DNA by the addition of nucleotides in the order prescribed by a template.
Dose	:	The amount of a substance, generally that ingested or applied at one time.
Droplets	:	Airborne particles of mucus and sputum from the respiratory tract that contain disease organisms.
Eaton agent	:	An alternative name for *Mycoplasma pneumaniae*.
Eccentric	:	Asymmetrical in growth.
E. coli	:	*Escherichia coli*.
Ecology	:	The study of the relationships between organisms and the environment; derived from the Greek *oikos* (dwelling) and *logos* (law).
Ecosystem	:	A community of organisms in their natural environment.
Ectoparasite	:	Organism that lives on or within the skin.
Edema	:	A swelling of the tissues brought about by an accumulation of fluid.
Effluent	:	The liquid discharge from sewage treatment and industrial plants.
EFM	:	Epifluorescent microscopy.
EHEC	:	Enterohemorrhagic *Escherichia coli*.
EHS	:	Extremely hazardous substances.
ELISA	:	Enzyme-linked immunosorbent assay.
ELSD	:	Evaporative light scattering detection.
Eluate	:	The liquid suspension resulting from elution.
Emigration	:	The movement of a population out of an environment; in contrast to immigration.
EMS	:	Ethyl methane sulphonate.
Encephalitis	:	An inflammation of the brain.

Encrusted	:	Covered with a layer of mineral.
Endogenous	:	Developed or living within an organism.
Endolithic	:	Micro-organisms that live within the rock matrix; in contrast to lithobiotic.
Endotoxin	:	A metabolic poison produced chiefly by Gram-negative bacteria, endotoxins are part of the bacterial cell wall and consequently, are released on cell disintegration, they are composed of lipid polysaccharide-peptide complexes.
Enterotoxin	:	A toxin that is active in the gastrointestinal tract of the host.
Enterovirus	:	A virus that infects intestinal cells.
Episome	:	A plasmid attached to the chromosome of a bacterium.
Erythema	:	A zone of redness in the skin due to accumulation of blood.
Exotoxin	:	A metabolic poison produced chiefly by Gram-positive bacteria, exotoxins are released to the environment; they are composed of protein and affect various organs and systems of the body.
FACS	:	Fluorescent-activated cell sorting.
Facultative	:	Capable of growth in the presence or absence of an environmental factor.
Faculative parasite	:	An organism that can exist either as a parasite or as a saprophyte.
False negative	:	Result that incorrectly identifies the absence of the analyte of interest when it is actually present in the sample; in contrast to a false positive.
False positive	:	Result that incorrectly identifies the presence of the analyte of interest when it is absent from the sample; in contrast to a true positive.
FAME analysis	:	Fatty acid methyl ester analysis.
Fastidious	:	Having very specific nutritional and environmental requirements.
FCM	:	Flow cytometry measurement.
Fermenter	:	A micro-organism that uses organic compounds as both primary electron donor and the ultimate electron acceptor.
Fibrobacter	:	Bacterial genus; many species are present in the complex of micro-organisms isolated from the rumen.
Fimbriae	:	Short, hairlike structures used by bacteria for attachment, sometimes used as an alternative expression for pili.
Fixed factor	:	Statistical term to denote that a condition either exists or it does not; in contrast to a random factor.
Flocculation	:	A process that causes flocs to form.
Fomites	:	Inanimate objects such as clothing or utensils that carry disease organisms.
Fungemia	:	Dissemination of fungi through the circulatory system.
Gall	:	A tumorous growth on infected plants; symptom of plant disease caused by *Agrobacterium* spp.
Gastroenteritis	:	An inflammation of the lining of the stomach and intestines resulting in symptoms such as nausea, vomiting, diarrhea, fever; caused by a variety of enteric micro-organisms such as Norwalk viruses, *Salmonella* spp., and rotavirus.
Gastromycetes	:	A grouping a Basidiomycetes that includes puffballs, earth stars, stinkhorns, and bird's nest fungi; spores are not actively discharged.
Gel	:	An inert polymer, generally composed of agarose or polyacrylamide.
GEMs	:	Genetically-engineered micro-organisms.
Gene	:	The basic unit of heredity, consists of one or more segments of DNA required to produce a single polypeptide.

Gene library	:	Collection of cloned DNA sequences that comprises the genetic information from a single organism.
Genetically engineered micro-organisms (GEMs)	:	The micro-organisms with *in vitro* altered gene sequences to enhance a particular function.
Genetic engineering	:	The *in vitro* manipulation of gene sequences.
Geniculate	:	Descriptive of a fungal conidiophore that is bent, resembling a knee joint.
Genome	:	The nucleic acid core of the virus.
Germ	:	Informal term used to describe a micro-organism that is capable of causing disease.
Giardiasis	:	Foul-smelling diarrhea and cramping caused by infection with *Giardia lamblia*.
GM	:	Geometric mean.
GMOs	:	Genetically-modified organisms; genetically engineered micro-organisms.
Good laboratory practice (GLP)	:	A system of quality assurance for processes, procedures, and conditions in environmental and pharmaceutical laboratories.
Gonococcus	:	A colloquial expression for *Neisseria gonorrhoeae*.
Gradostat	:	A multistage chemostat that provides spatial gradients for the study of microbial populations in liquid culture.
Grazing	:	A means of ingestion of masses of micro-organisms that are colonised on submerged surfaces that is used by a variety of aquatic organisms such as snails and urchins with scraping mouth organs; in contrast to filter feeding.
Green bacteria	:	Anoxygenic phototrophic bacteria that conduct photophosphorylation using Bchl *c* and Bchl *d* chlorophyll pigments that are located within chlorosomes, in contrast to the purple bacteria.
Green fluorescent protein (GFP)	:	Protein produced by some bacteria due to the introduced *gfp* gene that is used as a biosensor for the discrimination of bacteria from background matrix.
Green sulphur bacteria	:	Autotrophic phototrophic bacteria that oxidise reduced sulphur compounds under anaerobic conditions resulting in the formation of elemental sulphur that is stored outside of the bacterial cell.
Growth	:	An increase in the number of micro-organisms.
Growth curve	:	Depiction of the cycle of a microbial population in culture in which the population increases, stabilises, and decreases; defined for each microbial population with a lag phase, an exponential phase, a stationary phase, and a death phase.
Halobacteria	:	Generalised term to describe members of the Archaea that are extreme halophiles.
Halophiles	:	Organisms requiring NaCl for growth; extreme halophiles grow in concentrated brines.
Halorespiring	:	The coupling of reductive dehalogenation and electron transport-coupled phosphorylation catalysed by specific enzymes in some bacterial species that results in degradation of haloorganic compounds in anoxic polluted soils, aquifers, and sediments; in contrast to use of metal-containing cofactor for reductive dehalogenation by many anaerobic bacteria.
Hansen's disease	:	An alternative name for leprosy.
Harvest	:	In cell culture, describes the collecting of cells or the growth medium in which cells exposed to a sample were grown, so that the cells or medium can be analysed or the cells can be used as a seed for further propagation of the cell culture.

Helical	:	In the shape of a coil or spiral; many viruses have capsids that exhibit helical symmetry in which the capsomeres are packed around the nucleic acid in the shape of a spiral.
Hemagglutination	:	The agglutination of red blood cells.
Hemagglutinin	:	An enzyme on the surface spikes of certain influenza viruses that allows the virus to bind to red blood cells.
Hemolysins	:	Enzymes that dissolve red blood cells, produced by streptococci, staphylococci, gas gangrene bacilli, and other micro-organisms.
Hepatotoxin	:	A compound that is toxic to the liver.
Heterocyst	:	A specialised cyanobacterial cell that conducts nitrogen fixation.
Hetrotroph/ Heterotrophic	:	A micro-organism that utilises organic carbon as a nutrient source for growth.
Heterotrophic potential	:	The ability of microbial populations to utilise an organic substrate.
Heterotrophic succession	:	Shift in microbial populations due to the insufficient supply of organic material for the survival of heterotrophic micro-organisms.
Hierarchical stepwise regression	:	Statistical analysis in which a single variable or cluster of variables are examined by stepwise regression but the order of the introduced variables is predetermined.
Hirsute	:	Hairy in appearance.
Homothallic	:	Spontaneous sexual reproduction of a single fungal isolate by identical nuclei undergoing fusion; in contrast to heterothallic.
Horizontal gene transfer	:	The transfer of genetic information between organisms in the same niche; also termed lateral gene transfer; in contrast to vertical gene transfer to progeny.
HP	:	Hypersensitivity pneumonitis.
HPC	:	Heterotrophic plate count.
HPLC	:	High performance liquid chromatography.
HSP	:	Heat-shock protein.
HTS	:	High throughput screening.
Humoural immunity	:	Immunity arising from the activity of antibodies directed against antigens in the blood stream.
Humus	:	The organic portion of the soil remaining after microbial decomposition.
Hybridoma	:	A mass of cells produced by the fusion of myeloma cells with antigen-stimulated plasma cells, produces monoclonal antibodies.
Hydrophobic	:	Molecule that does not interact readily with water, e.g., lipid; contrast with hydrophilic.
Hypha	:	Singular term of hyphae.
Hyphae	:	Septate or nonseptate filaments of fungi.
Icosahedron	:	A symmetrical figure composed of 20 triangular faces and 12 points, one of the major shapes taken by the virus.
IDHLs	:	Immediate danger to life or health exposure levels.
Imidazoles	:	A group of antifungal drugs that interfere with sterol synthesis in fungal cell membranes, includes micronazoles and ketoconazoles.
Immobilised enzyme	:	An enzyme bound to a solid support.
Immunoglobulin	:	An alternative term for antibody.
Immunosuppressed	:	Having a lower resistance to disease than the typical individual, may be caused by exposure to chemotherapeutic drugs.
Immunotoxic	:	Negatively affects the immune system.

Imperfect state	:	The asexual form of a fungus.
Implementing procedure (IP)	:	Protocol
IMS	:	Immunomagnetic separation.
Incidence density	:	The number of occurrences within a defined area.
Incidence rate	:	The number of new occurrences of an event (e.g. illness) per population at risk within a defined period of time.
Incubator	:	An enclosed, temperature-regulated chamber for the growth of micro-organisms in culture.
Independent variable	:	That condition of which is being changed or manipulated during the conduct of an experiment.
Inducer	:	A substances that may activate the operon of the cell by combining with and negating the repressor protein.
Infect	:	To enter another organism and multiply within it.
Infectious	:	Capable of invading a host cell and multiplying thereby causing an infection.
Infections dose	:	The number of organisms required to produce an infection in an exposed individual.
Infrared spectrometry (IR)	:	An analytical technique used for the characterisation of surfaces and interfaces between surfaces.
Inhibition	:	The prevention of a function; this condition may be temporary.
Inoculating loop	:	A flat, open circle device used to apply micro-organisms to a surface or liquid.
Inoculum	:	The micro-organisms, cells, or other biological material that are added to growth medium to start a culture.
In situ	:	Latin term for the original place; used to denote experiments conducted at the site or on location.
Integration	:	The incorporation of DNA into another genome.
Interaction	:	A collaborative action between two or more entities.
Interferon	:	An antiviral protein produced by body cells on exposure to viruses; interferon triggers production of a second protein that binds to mRNA coded by the virus and thereby inhibits viral replication.
Interleukins	:	Lymphokines produced by white blood cells that act on other white blood cell, important in cellular immunity.
Interquartile range	:	The middle 50 per cent of data; also termed mid-spread.
Interstitial	:	Situated in the space between things.
Intramatrical	:	Within a substrate; in contrast to extramatrical.
In vitro	:	Latin term 'in glass' used to denote experiments conducted outside of a cell.
In vivo	:	Latin term 'in cell' used to denote experiments conducted in living cell.
IP	:	Implementing procedure.
Isolation	:	The separation of two or more entities.
J	:	Symbol of equitability.
Japan collection of micro-organisms (JCM)	:	Culture collection entity located in Wako, Japan that catalogs and sells standard strains of micro-organisms.
Jaundice	:	A yellowing of the skin, mucous membranes, and eyes due to rapid destruction of blood cells or abnormalities of the liver; a symptom of hepatitis.

JCM	:	Japan Collection of Micro-organisms.
Job's syndrome	:	An immune disorder characterised by defective chemotaxis between phagocyte and micro-organism.
Joist	:	Horizontal beam as a structural element of a building.
Joule (J)	:	Unit of energy equal to 10^7 ergs.
kb	:	Kilobase.
Kelly's medium	:	Culture medium for the isolation of *Borrelia* spp.
Kilobase pair	:	A segment containing 1000 base pairs.
Kingdom	:	The second highest taxonomic ranking; the taxonomic ranking below domain.
Klebsiella	:	Bacterial genus, member of the family *Enterobacteriaceae*; nonmotile, gram-negative bacilli some of which are human pathogens while other are commensals in humans and animals, or phytopathogens.
Kogure technique	:	The use of nalidixic acid and yeast extract with acridine orange and light microscopy to enumerate swollen or elongated cells as viable.
Korarchaeota	:	A kingdom of hyperthermophilic Archaea.
K strategists	:	Micro-organisms that depend on physiological adaptations to environmental resources or the carrying capacity of the environment for continued survival within a community so they reproduce slowly and are successful in environments that are limited in nutritional resources.
Kurung-Yegian agar	:	A fungal culture medium used for the conversion of *Histoplasma capsulatum* and *Blastomyces dermatitidis* to the yeast phase.
Laccases	:	Multi-copper blue oxidase enzymes implicated in fungal conidia production, lignin degradation, pathogenicity, and fruiting-body production.
Lactose broth	:	A liquid nutrient medium containing lactose sugar that is used for the detection of total coliform bacteria or fecal coliform bacteria.
Lag phase	:	Initial period of time in the growth curve of micro-organisms in which growth does not occur immediately; the period to time prior to exponential growth.
LAL assay	:	*Limulus* amebocyte lysate assay.
Latent virus	:	A virus whose genome is integrated into the host's genome, but is not expressed; upon activation (e.g. by stress or exposure to ultraviolet irradiation), infective virus particles are produced and symptoms of infection appear.
LDL	:	Lower detection limit/lower limit of detection.
Leaching	:	The removal of metal from ore by chemical or microbial activity; the transport of dissolved materials from upper soil layers deeper into the subsurface.
Legionellaceae	:	Bacterial family; characterised as intracellular parasites or endosymbionts of free-livng parasites.
Lentic	:	Stagnant, nonflowing water.
Leukemia	:	A cancer of the white blood cells.
Leukocidin	:	An enzyme that destroys phagocytes thereby preventing phagocytosis of the parasite.
Lyme disease	:	An acute inflammatory disease with a rash that may result in arthritis and heart and nervous system problems; caused by an infection with *Borrelia burgdorferi* resulting from a tick bite; first reported in Old Lyme, Connecticut.
Lymphoblast	:	The young cell to which the T-lymphocyte reverts, lymphoblasts secrete lymphokines.
Lymphocyte	:	A type of leukocyte that functions in the immune system.

Lymphokines	:	Proteins that increase the efficiency of phagocytosis at the antigen sites in cellular immunity.
Lymphopoietic cells	:	Primitive cells that arise from stem cells and are modified to form B-lymphocytes or T-lymphocytes.
Lyophilisation	:	Process in which cold temperature and air evacuation are used for preservation of micro-organisms; also termed freeze-drying.
Lysogeny	:	The phenomenon in which a virus remains in the cell cytoplasm as a fragment of DNA or attaches to the chromosome, but fails to replicate in or destroy the cell.
μ	:	Greek letter mu used as the symbol for microbial growth, formerly used as the symbol of micron which was replaced by micrometer (μm)
MA	:	Muramic acid.
Mab	:	Monoclonal antibody.
MAC	:	MacConkey's agar; *Mycobacterium avium* complex.
Macroconidia	:	Plural of macroconidium.
Macroconidium	:	A single, large, multicellular conidium.
Macromonas	:	Bacterial genus; cylindrical to bean-shaped cells that oxidise sulphur and sulphur compounds; may accumulate calcium carbonate with sulphur globules; found in seawater.
Macrophage	:	Long-lived phagocytic cell in mammals that can engulf bacteria and parasites, and may become activated in response to foreign materials resulting in the release of substances that stimulate other cells of the immune system.
Magnetospirillum	:	Bacterial genus; magnetotactic organisms that participate in biomineralisation of magnetosomes.
Magnetotaxis	:	Motility of micro-organisms directed by a geomagnetic field.
Magnification	:	Observed increase in size of an object when viewed through a microscope or lens.
Manifold	:	A chamber with multiple channels or openings for the flow of liquid or air.
Mannan	:	A cell wall constituent in yeast; a mannose-containing polysaccharide storage material in plants.
Masking	:	Interference that results in the inability to detect the analyte.
Maxiprep	:	The colloquial term use to describe the purification of plasmids from bacterial suspensions of 100–500 ml in volume; in contrast to megaprep, midiprep, and miniprep.
MCYSTs	:	Microcystins.
Mean deviation (MD)	:	The average of the absolute deviation of data.
Median	:	The statistical point where half of the individual data points are greater and half of the data points are lower in value; in contrast to arithmetic mean.
Meningitis	:	An inflammation of the membranes surrounding the brain and spinal cord.
Meningoencephalitis	:	An inflammation of the brain and the membranes surrounding the brain and spinal cord.
Mesotrophic	:	Moderately productive lake or surface freshwater resulting from an intermediate amount of nutrient.
Metabolism	:	The collection of biochemical reactions within a cell that generate or require energy; generally classified as either anabolic (synthetic) or catabolic (breakdown) reactions.

Methylotroph	:	Micro-organism that utilises single carbon compounds such as methane, methanol formate, and carbon monoxide as their carbon source.
Metulae	:	Plural of metula.
Microbial induced corrosion (MIC)	:	The accelerated corrosion of a metal or an alloy by the activity of a consortium of micro-organisms.
Microbial pest control agent (MPCA)	:	A nonpathogenic micro-organism applied to agricultural crops to minimise the colonisation of a microbial phytopathogen.
Microconidia	:	Plural of microconidium.
Microconidium	:	A small, generally unicellular, asexual cell; may function in sexual reproduction in some genera.
Microcosm	:	A small-scale experimental model that is designed to reproduce the environmental conditions of interest as closely as possible; used in laboratory experiments to define environmental conditions and test biological populations.
Microenvironment	:	The physical and chemical conditions in the area immediately surrounding an organism.
Micrometer (μm)	:	Unit of measure describing the length of 10^{-6} metre.
Micro-organism	:	A microscopic organism that exists as a single cell or in an aggregate of cells, or as an acellular entity (i.e. virus).
Microscope slide	:	Glass surface used for the mounting of specimens for analysis by microscopy.
Microtubule	:	A structural entity of eukaryotic flagella.
Miscible	:	Liquids that are soluble in all proportions.
mmHg	:	Millimetres of mercury, used to express atmospheric pressure.
Moniliaceous	:	In a chain or a series of segments.
MPA	:	Microscopic particulate analysis.
MPCA	:	Microbial pest control agent.
mRNA	:	Messenger RNA.
Mu	:	A temperate virus that is also a transposable phage and a mutator phage used in genetic engineering.
Mucedinoid	:	Musty or mouldy odour.
Multiplicity of infection (MOI)	:	The average number of virus particles per cell.
Muriform	:	Structures that are transversely and longitudinally septate.
Mycobiont	:	The consumer population in a mutualistic relationship.
Mycorrhizal fungi	:	The fungi involved in mycorrhizae.
NA	:	Numerical aperture.
Naked virus	:	A virus without an envelope surrounding the capsid.
Nanobacteria	:	Smallest cell-walled bacteria currently recognised.
Nanometer	:	A unit of measurement equivalent to one billionth of a metre, the unit is designated as nm and is often used in measuring viruses and the wavelength of energy.
Naumanniella	:	Bacterial genus; bacilli that oxidise iron and are encapsulated with iron oxides; widely distributed in iron-rich waters.
NBOD	:	Nitrogenous biochemical oxygen demand.
Necrosis	:	Death of cells or tissue; also called rot in plant pathology.
Neoplasm	:	New or abnormal growth, such as a tumour.
Nested variables	:	Statistical condition in which each variable occurs at only one level.
Neural	:	Relating to the nervous system.

Neurotoxin	:	A toxin that is active in the nervous system of the host.
Neurotrophic	:	Having an affinity for nerve tissue.
Neutralisation	:	A type of antigen-antibody reaction in which the activity taking place between reactants is not visible.
Neutralism	:	A relationship of microbial populations in which there is no interaction between the populations; occurs when populations of organisms have different metabolic capabilities, populations are spatially distant from each other, when environmental conditions are unfavourable for active growth, or when organisms are in a resting state.
Night soil	:	Human feces sometimes used as an agricultural fertiliser.
Nitrifying bacteria	:	A group of bacteria that oxidises ammonia to nitrite or nitrite to nitrate.
Nitrobacteriaceae	:	Bacterial family; characterised as nitrifying bacteria; commonly isolated from soil, freshwater, and marine environments.
Node	:	A joint, point of origin of fungal hyphae, or an enlarged area on a fungal hypha.
Nod genes	:	Genetic sequences in nitrogenic fixing bacteria that direct specific steps in the formation of a root nodule.
Normal flora	:	Micro-organisms that are routinely present on or in another organism.
OA	:	Occupational asthma.
OAF	:	Open air factor.
Obclavate	:	An inverted club shape with the larger portion located at the base.
Obligate aerobe	:	A micro-organism that requires the presence of oxygen.
Obligate anaerobe	:	A micro-organism that cannot grow in the presence of oxygen.
Ochrobium	:	Bacterial genus; iron oxidising ellipsoidal to rod-shaped cells that are surrounded by a capsule containing iron oxides; distributed in iron-bearing freshwater.
OFAGE	:	Orthogonal field alternation gel electrophoresis.
Olfactory	:	Related to the sense of smell.
Oligonucleotide	:	A short (generally, up to 20) sequence of nucleotides.
Oligotrophic	:	Low carbon or low nutrient content; in contrast to copiotrophic.
Operon	:	A sequence of DNA that contains one or more structural genes plus regulatory genetic elements.
Outlier	:	A value that is inconsistent with the other data obtained during statistical analysis.
Oxylabile	:	Killed or inactivated by the presence of oxygen.
P-A	:	Presence-absence coliform test.
PAB	:	Propionic acid bacterium.
PADs	:	Phenolic acid decarboxylases.
Paecilomyces variotii	:	Fungal species; yellow-brown colonies; thermotolerant fungus that produces a sweet aromatic odour; $Aw = 0.84$; deteriorates jute fibre and paper and causes soft rot of timber.
PAGE	:	Polyacrylamide gel electrophoresis.
PAHs	:	Polyaromatic hydrocarbons.
Palindrome	:	A sequence of DNA that is linked to the same sequence in reverse order.
Papilloma	:	A tumour of the skin tissue.
Papovavirus	:	Any virus in the family Papovaviridae.
Parasite	:	An organism that lives on or in a host.
Parasitism	:	Relationship of microbial populations in which one population derives its nutritional requirements at the harm of the other population.

Partial F test	:	Statistical analysis used to determine the significance of contribution of a single variable after all the other variables are in the equation.
Passive sampling	:	Collection of material without the use of a mechanical device; gravitational sampling.
Pathogen/Pathogenic	:	An organism that infects a host and is capable of causing disease.
PBBs	:	Polybrominated biphenyls.
PCA	:	Principal component analysis.
PCBs	:	Polychlorinated biphenyls.
PCP	:	Pentachlorophenol.
PCR	:	Polymerase chain reaction.
Perfect state	:	Sexual state of a fungus.
Perithecia	:	Plural of perithecium.
Permissive host cell	:	Virus replication is accomplished within the host cell; in contrast to nonpermissive host cells.
PFA	:	Polyunsaturated fatty acid.
PFGE	:	Pulsed field gel electrophoresis.
PFU	:	Plaque-forming unit.
PGA	:	Poly(γ-D-glutamic acid).
Phage	:	A short form of the word bacteriophage.
Phagocyte	:	A cell that practices phagocytosis.
Phagocytosis	:	A process in which solid particles are taken into the cell, important in nutritional processes and in defence against disease.
Phagosome	:	A vesicle that contains particles of phagocytised material.
Picornavirus	:	A small virus containing RNA in its genome.
Pili	:	Physical appendages on bacteria, similar in structure to fimbriae but are longer; generally only one or two are visualised on the bacterial cell when they serve as specific receptors of certain types of virus particles.
Pinocytosis	:	A type of phagocytosis in which materials dissolved in fluid are taken into the cell.
Pip	:	A membrane protein that serves as a receptor protein for bacteriophage.
Precipitins	:	Antibodies that participate in precipitation reactions.
Prions	:	Infectious particles of protein, possibly involved in human disease of brain.
Properdin	:	A protein that functions in the alternative pathway of complement activation.
Prophage	:	The DNA segment of a temperate phage.
R	:	Multiple correlation coefficient.
R^2	:	Coefficient of determination.
R2A agar	:	A low nutrient culture medium used for the isolation of oligotrophic bacteria, primarily for isolation of bacteria in water samples.
RAB	:	Rotating annular bioreactor.
Racemes/Racemic	:	A mixture of substances that does not rotate plane polarised light.
Racquet hyphae	:	Fungal hyphae that are composed of cells that are inflated at one end, resembling the shape of a tennis racquet or a snowshoe.
Radioimmunoassay (RIA)	:	Immunoassay method that uses radioisotopes to increase the sensitivity of antibody detection.
Recalcitrant	:	Totally resistant to biodegradation.
Recombinant DNA	:	A segment of DNA that contains nucleic acid from two or more sources.

Redox reaction	:	The oxidation of one compound paired with the reduction of another compound.
Reducing agent	:	Substance that combines with oxygen or loses electrons in a reaction.
Repressor protein	:	A protein that inhibits the activity of certain genes, lysogeny is established when repressor protein is produced under direction of a virus.
Retrovirus	:	Any virus in the family Retroviridae; viruses are icosahedral in shape, surrounded by a lipid envelope, contain single-stranded RNA; human pathogens in this family cause different types of cancer, and the human immunodeficiency viruses (causative agent of AIDS) are members of the lentivirus genus in this family.
Reverse transcriptase	:	An enzyme that synthesises a DNA molecule from the code supplied by a RNA molecule.
Rhicadhesin	:	A calcium binding adhesion protein on the surface of *Rhizobium* and *Bradyrhizobium*.
Rhinovirus	:	A genus of viruses in the family Picornaviridae; the most commonly isolated viruses in persons with the common cold.
Rhizobia	:	Common name for members of the Rhizobiaceae.
Rhizobiaceae	:	Bacterial family; characterised by their ability to fix atmospheric nitrogen.
Rhizoid	:	Root-like structure of branched filaments used by some fungi to obtain nutrients.
RIFA	:	Radioimmunofocus assay.
Risk factor	:	A feature or characteristic of an individual that is associated with an increased probability of acquiring a particular disease.
RMP	:	Risk management programme.
Rot	:	Plant pathology term used to describe the symptom of some plant diseases that result in death of cells or tissue.
Rugose	:	Coarsely wrinkled in appearance.
Sabin vaccine	:	A type of polio vaccine prepared with attenuated viruses, the vaccine is taken orally.
Sabouraud dextrose agar (SDA)	:	A fungal culture medium containing 40 grams of dextrose per litre; generally used for the isolation of medically important organisms.
Saccharomyces	:	Fungal genus; round to oval, multilateral budding yeast cells with short pseudohyphae.
Saccharomyces carlsbergensis	:	Fungal species; a brewery bottom yeast that is used in the production of lager beer.
Saccharomyces cerevisiae	:	Fungal species; a brewery top-fermenting yeast that is used in the production of ale.
SAED	:	Selected-area electron diffraction.
Safranin	:	Dye used as the counterstain in the Gram stain procedure to impart a pink to red colouration to bacteria.
Sake	:	A type of rice beer produced primarily in the orient.
Salmonella enterica	:	Bacterial species; foodborne pathogen transmitted via the ingestion route of exposure through the consumption of raw tomatoes.
Salmonella paratyphi	:	Bacterial species; waterborne and foodborne pathogen transmitted via the ingestion route of exposure.
Salmonellosis	:	Self-limiting gastrointestinal disease caused by the ingestion of *Salmonella*; onset of symptoms occurs after several days due to the amplification of the bacteria in the intestine.
Sampling stress	:	Physical conditions that occur during collection of a sample that result in loss of viability of the micro-organisms in the sample.

Sarcoma	:	A tumour of the connective tissues.
Scotochromogenesis	:	Formation of pigment only when the micro-organism is cultured in dark; used in the classification of some *Mycobacterium* spp.; in contrast of photochromogenesis.
Selection	:	The growth of a particular microbial population due to the presence or absence of a nutrient or physical condition.
Selection bias	:	The introduction of a systematic error due to the manner in which the test and control populations are selected in contrast to surveillance bias and misclassfication bias.
SE_M	:	Standard error of the mean.
Septicemia	:	A generalised bacterial infection of the blood stream due to any of several organisms including streptococci and staphylococci, once known as blood poisoning.
Serology	:	The branch of immunology that studies serological reactions.
Serratia marcescens	:	Bacterial species; most strains produce colonies in culture with an orange-red pigment, prodigiosin, but some are nonpigmented; has been used as tracer to demonstrate atmospheric dispersal of bioaerosols.
Short-term emergency	:	Former term for short-term public emergency guidance levels.
Siderophore	:	A natural iron chelator that binds iron present at low concentrations.
Sinuous	:	Wavy or serpentine in appearance.
S-layer	:	An outer wall layer of protein or glycoprotein that is found in many prokaryotic organisms.
Solfatara	:	A hot, generally acidic, sulphur-rich environment inhabited by thermophilic Archaea.
SOP	:	Standard operating procedure.
Sordelli's method	:	A drying technique for the preservation of micro-organisms that uses an inner tube containing a microbial suspension in horse serum and an outer tube containing P_2O_5 that is sealed with a vacuum.
sp.	:	Abbreviation for species that is placed after a genus name; used to denote a single species; in contrast to spp.
SP	:	A single-stranded RNA, male-specific bacteriophage in the Leviviridae family.
Specificity	:	The ability to distinguish the analyte of interest from other materials.
SPEGLs	:	Short-term emergency guidance levels.
SRB	:	Sulphate reducing bacteria.
SSCP	:	Single-strand-conformation polymorphism.
ssDNA	:	Single-stranded DNA.
SSR	:	Simple sequence repeat.
ssRNA	:	Single-stranded RNA.
STA	:	Slit-to-agar sampler.
Stationary phase	:	Period of time in the growth curve of a microbial population in which there is no net increase or decrease in the number of organisms.
Streptococcus faecalis	:	*Enterococcus faecalis.*
Streptococcus suis	:	Bacterial species; pathogen of pigs; associated with severe illness in humans.
Swab sampling	:	Method to assess microbial contamination on smooth or textured surfaces using a sterile cotton swab with analysis by culture or polymerase chain reaction amplification.
Taenioella	:	Fungal genus; produces dark brown conidia that may be collected in spore trap air samples, transparent tape samples, and from bulk material in indoor

environments, but is not cultured on laboratory media; present on leaves and wood in nature, on plants and lumber in the indoor environment.

Target cells	:	Cells to which IgG fixes during type II hypersensitivity.
T-DNA	:	A segment of the Ti plasmid of *Agrobacterium* that is transferred to plant cells.
Teliospore	:	Thick walled, binucleate resting spore; dry, powdery airborne stage of smuts; difficult to distinguish microscopically from myxomycetes.
TEM	:	Transmission electron microscopy.
Thallic	:	The transformation of a segment of fungal hyphae into conidia by the formation of cross-walls.
Thallus	:	The vegetative and reproductive structures that make up a fungal colony.
Thermal death point	:	The temperature required to kill an organism in a given length of time.
Thermophiles	:	Organisms that grow at a high temperature ranges of 40° to 90°C
Thiospira	:	Bacterial genus; spiral-shaped cells with pointed ends that oxidise sulphur compounds; found in water overlaying sulphur-rich mud.
Tissue typing	:	An immunological procedure used to locate compatible tissue types for transplantation.
Titer	:	The most dilute concentration of antibody that will yield a positive reaction with specific antigen, a method of expressing the amount of antibody in a sample of serum.
T-lymphocyte	:	A lymphocyte that is modified in the thymus gland and is associated with the system of cellular immunity, also called a T-cell.
Toxin	:	A poisonous substance produced by a species of micro-organism, bacterial toxins are classified as exotoxins or endotoxins.
Toxoid	:	An immunising agent produced from an exotoxin that elicites antitoxin production by the body.
Transcription	:	The synthesis of RNA using a DNA template.
Transduction	:	Genetic recombination process mediated by virus; DNA is incorporated from one cell to another with the assistance of virus by generalised transduction or specialised transduction.
Transgenic micro-organism	:	A micro-organism with a cloned DNA sequence from another organism.
Transposon	:	A segment of DNA that moves from one site on a DNA molecule to another site, transposons carry information for protein synthesis, also known as jumping genes.
Trihalomethanes (THMs)	:	Any one of a group of chemicals that result from the chlorination of waters containing organic materials; many of these compounds are known or suspected carcinogens.
tRNA	:	Transfer ribonucleic acid.
Tween	:	A surfactant used as an amendment to buffers and aqueous solutions to assist in the suspension of particulate material.
Type strain	:	Archived specimen of a micro-organism used for taxonomic comparison.
Ultrapasteurisation	:	A pasteurisation process in which milk is heated at 82°C for 3 seconds.
Umber	:	Dark brown or reddish brown in colour.
Umbo	:	A raised knob or mounded in the centre.
Undulant	:	Wavy in appearance.
Undulate	:	Having a wavy-like motion.
Unicellular	:	Single cell.

Upper detection limit (UDL)/Upper limit of detection	:	Largest quantity that can be accurately determined.
Upper respiratory tract	:	The nose, mouth, sinuses, and pharynx.
Urocystis	:	Fungal genus; member of the Basidiomycotina; causative agent of smut of onion.
Vaccine	:	Any antigenic preparation administered to stimulate the recipient's immune defense mechanisms with respect to given pathogens or toxic agents.
Vaccinia	:	The alternative name for cow-pox.
Vacuole	:	A large, fluid-filled sac located in the cytoplasm.
Vaginitis	:	A general term for disease of the vagina.
Valine ($C_5H_{11}NO_2$)	:	An essential amino acid that cannot be synthesised by the human body and therefore must be ingested in foods.
Vampirococcus	:	Name given to predatory bacteria that use the cytoplasmic constituents of other bacteria as a nutrient source.
Variable number of tandem repeats (VNTR)	:	Sequences of repetitive DNA; used in DNA fingerprinting.
Varicella	:	An alternative name for chicken pox, means little vessel, a reference to small chicken pox lesions.
Varicella-Zoster immune globulin (VZIG)	:	A preparation of purified antibodies from blood donors that give some protection of chicken pox.
VB broth	:	Vogel-Bonner broth.
VBNC	:	Viable (but) non-culturable.
Vector	:	A living organism that transmits the agents of disease.
Ventricose	:	Swollen or enlarged in the middle.
Vermiform	:	Worm-shaped.
Virion	:	A completely assembled virus outside its host cell.
Viroids	:	Tiny fragments of nucleic acid associated with certain plant diseases, possibly associated with animal disease.
VFF	:	Vortex flow filtration.
VLPs	:	Virus-like particles.
VNA	:	Viral nucleic acid.
VNC	:	Viable nonculturable.
VNTR	:	Variable number of tandem repeats.
Wallemia sebi	:	Fungal species; slow growing fan-like or stellate, powdery colonies that are orange-brown to blackish brown in colour with cylindrical, pale brown, smooth-walled conidiophores; osmophilic airborne fungus isolated from forest soil, textiles, timber, paper, pecan nuts, salted fish and beans, and stored bay and may cause ulcerative abscesses in humans; cultured on high osmotic medium.
Wandering cells	:	Cells of the reticulo-endothelial system that move about actively within the tissues.
Warm vent	:	A hydrothermal vent that emits temperatures of 6°–23°C.
Waterborne transmission	:	Transfer of microbial contaminants via water; generally an ingestion route of exposure that primarily results in intestinal disease although dermal contact of parasites in water can result in disease.
Water holding capacity (WHC)	:	The maximum amount of moisture that can be retained in a substance, usually used in studies with soil.

Water moulds	:	Aquatic fungi that are members of the Oomycetes.
Waterwashed disease	:	Illness caused by organisms that originate in feces and are transmitted through contact because of inadequate sanitation.
WHC	:	Water holding capacity.
Whey	:	A waste liquid of the dairy industry containing lactose and minerals that is used in industrial processes as supplemental carbon.
Whorl	:	A group of branches that radiates from a common point.
Windrow	:	Long, generally rectangular, pile of material in a static pile composting facility.
Wood-rotting fungi	:	Referring to basidiomycetes; commonly classified as brown rot and white rot fungi.
x	:	Symbol used to denote an unknown quantity.
X	:	Used to denote the value for one data point.
$\bar{X}$	:	Symbol for arithmetic mean.
Xanthomonadins	:	Yellow, membrane-bound, halogenated aryl polyene pigments that are produced by *Xanthomonas* spp. and may provide some protection against photodamage.
Xanthomonas	:	Bacterial genus; member of the family Pseudomonaceae; chemoorganotrophic, gram-negative, straight, obligate aerobic, bacillus that is motile by a single polar flagellum; many species are phytopathogens.
Xanthomonas campestris	:	Bacterial species; phytopathogen, causative agent of black rot of crucifers.
Xanthomonas oryzae	:	Bacterial species; phytopathogen, causative agent of blight of rice.
Xanthomonas vascularum	:	Bacterial species; phytopathogen, causative agent of gumming of sugar cane.
XPS	:	X-ray photoemission spectroscopy.
Yeast	:	Unicellular fungus that reproduces by budding; most are members of the Ascomycetes; some have a filamentous phase; for example, *Saccharomyces cerevisiae* are used extensively in food production for leavening of bread and in beer and wine fermentation.
Yersinia	:	Bacterial genus; member of the family Enterobacteriaceae; facultative, gram-negative, nonsporulating bacilli with 10 established species, most with simple nutritional requirements; previously classified as the genus *Pasteurella*.
Yersinia enterocolitica	:	Bacterial species; causative agent of an intestinal infection in humans and warm-blooded wild and domestic animals; also isolated from soil, surface water, reptiles, fish and shellfish; ingestion route of exposure with contaminated food and water.
Yo Pro-1	:	A nucleic acid-standing fluorescent dye used with epifluorescent microscopy that has been shown to provide a brighter fluorescence than 4′, 6′-diamidino-2-phenylindole (DAPI) staining.
Zearalenone	:	A patented growth stimulant for animals and an anabolic steroid; produced by *Fusarium culmorum*.
Zeta potential	:	A measure of the electrostatic potential, determined by the presence of charged groups on the surface of a cell or molecule.
Zonate	:	Arranged in zones or rings radiating from the centre.
Zone of inhibition	:	The area in which an antimicrobial substance prevents growth.
Zooglea	:	Referring to a mass of bacteria and/or lower algae that are held together by a slime matrix and water.
Zooglea ramigera	:	Bacterial species; produces extracellular polysaccharide slime matrix during sewage treatment.

Zoophilic	:	Prefers animals rather than humans or soil.
Zygomycetes	:	Fungal class; rapid growing non-septate fungi with sporangiospores produced in sporangia and some species have rhizoids and stolons; sexual reproduction produces a dark thick-walled zygospore.
Zygomycosis	:	Diseases caused by members of the Zygomycetes.
Zygospore	:	The sexual spore produced by members of the Zygomycotina that results from the fusion of two gametania.
Zygosporium	:	Fungal genus; conidia are colourless without a distinctive morphology; isolated from damp interior walls, dead leaves and soil; no information available on human health effects, toxicity, or allergenicity.
Zymocide	:	A factor present in some yeast cells that is toxic to other yeasts.
Zymomonas	:	Bacterial genus; tolerant of low pH and ethanol concentrations up to 10 per cent; large, gram-negative bacillus that ferments sugars to ethanol; active in fermentation of plant sap for industrial production (e.g. fermentation of agave in Mexico to produce tequila and palm sap in tropical areas); responsible for spoilage of fruit juices and production of an odour of rotten apples in spoiled beer.

References

Alvarez, G.M., *Biohazards,* Academic Press, London.

Arceivala, K.J., *Waste-water Treatment and Disposal,* Marcel Dekker Inc., New York.

Benaim Pinto, C., *Sampling and Analysis of Airborne Micro-organisms,* Prentice-Hall, London.

Brown, M.H., *Environmental Microbiology,* Cambridge University Press, Cambridge.

Chapelle, J. and Crossland, J., *Water Pollution,* Harcourt Brace Jovanovich, New York.

Coolingwood, R.W., *Microbial Ecology,* John Wiley & Sons, New York.

Downe, S.A., *Identifying Future of Water Contaminations,* John Wiley & Sons, New York.

Dugan, P.R., *Biochemical Ecology of Water Pollution,* Plenum Publishing Corporation, London.

Edward, C.F., *Biology of Freshwater Pollution,* Longman Group Ltd., London.

Goldman, M., *Bioaerosols,* Gordon and Breach, Science Publishers, New York.

Harding, G., *Metals and Micro-organisms,* Prentice-Hall, London.

Hiddy, A.J., *Microbial Transformation and Degradation,* John Wiley & Sons, New York.

James, A. and Evison, L., *Biological Indicators of Water Quality,* John Wiley & Sons, New York.

Jarvis, B., *Waste-water Microbiology,* John Wiley & Sons, New York.

Kim, C.K., *Water and Waste-water Disinfection,* Marcel Dekker, New York.

Lechevallier, M.W., *Biological Conversion of Waste from Pulp and Paper Mill,* Academic Press, London.

Mitchell, R., *Microbiology,* Pergamon Press, Oxford, London.

Morris, R.D., *Sampling and Analysis of Biological Organic Compounds,* McGraw-Hill, Tokyo.

Phillips, D.J.H., *Quantitative Aquatic Biological Indicators,* Applied Science Publishers, London.

Reid, G.L. and Herman, S.G., *Environmental Toxicology of Pesticides,* Academy Press, New York.

Rheinheimer, M., *Aquatic Microbiology,* Applied Science Publishers, London.

Rochelle, P.A., *Ecology of Inland Waters and Estuaries,* Reinhold Publishing Corporation, New York.

Rudd, R.L., *Pesticides and the Living Landscape,* University of Wisconsin Press, Madison, Wisconsin.

Smith, H.S., *Waste-water Microbiology,* Chilton Book Company, Radnor, Pennsylvania.

Stephens, W. and Stumm-Zollineger, E., *Water Pollution Microbiology,* Wiley Interscience, New York.

Stephens, W.D., *Encyclopedia of Bioprocess Technology,* John Wiley & Sons, New York.

Stumbo, C.R., *Airborne Viruses,* Pergamon Press, New York.

Taylor, K.C. and Borrel, N., *Sampling and Analysis of Air Borne Analysis,* McGraw-Hill, New York.

Vollenweider, R., *Environmental Engineering Handbook, Vol. 2. Air Pollution,* Chilton Book Co., USA.

Wetzel, J.C., *Introduction to Environmental Microbiology,* Academic Press, London.

Index

A

16S rRNA-based approaches, 292
A. protein bioflocculant produced by *Rhodococcus erythropolis*, 554
A. latus B-16 bioabsorbent production, 555
Acclimation in bioremediation technology, 430
Acclimation of eukaryotic micro-organisms, 430
Acclimation response in a contaminated aquifer, 429
Acclimation response in pristine aquifer sediments, 430
Acclimation to available electron acceptors, 427
Acclimation to mercury toxicity, 433
Acclimation to metal toxicity, 431
Acclimation to xenobiotics in groundwater systems, 429
Activity measurements in microcosms, 378
Adaptation of the test fungi to phenol, 492
Adenine incorporation into DNA and RNA, 286
Adenylate energy charge (AEC), 189
Advanced oxidation process, 159
Aerobic cometabolism, 506
Aerobic degradation, 507, 509
Aerobic oxidation, 505
Aeromicrobiological pathway, 442
Aeromonas spp., 100
Aeromonas, 107
Affinity chromatography, 207
Age of cultures, 493
Air drilling and coring, 412
Air-water interface, 255
Algae, 18
Alkaline phosphate as an indicator of phosphorus limitation, 358
Anaerobic degradation, 507, 509
Anaerobic digestion of complex substrates, 560
Anaerobic oxidation, 505
Analytical methods, 478, 493
Antagonism, parasitism and predation, 388
Antibody affinity, 194
Antibody specificity, 194
Antibody, 193
Antiglobulins, 194
Antimetabolite, 10
Aquatic environments, 243
Aquatic microbes: food for the future, 259
Archaea, 15

Area source modelling, 446
Aseptic technique with split-spoon, shelby tube and direct push sampling, 413
Aspects of behavioural diversity among prokaryotes, 64
Assessing populations, 328
Assessing protistan community structure, 299
Autotrophic and heterotrophic bacteria, 7
Azotobacterin, 528

B

Bacteria identification, 11
Bacteria, 3
Bacterial and its growth, 6
Bacterial respiration, 8
Bacteriology, 11
Bacteriology, epidemiology and pathogenesis of mycobacterial infections, 475
Bacteriophages, 98, 108
Bacterivory, 313
Barriers to the application of molecular methods for detection of pathogens in shellfish, 143
Beneficial root-microbial interactions, 470
Benthic habitat, 237
Benthic versus pelagic ecosystems, 309
Biocidal control, 457
Biofilms, 240
Biogeochemical cycles, 321
Biological control of crown gall disease, 474
Biological dinitrogen fixation, 470
Biological safety cabinets, 458
Biology and ecology, 113
Biomass measurements in soil microbiology, 398
Biomethanation of industrial wastes, 566
Biosafety in the laboratory, 458
Biosafety laboratories, 463
Biotransformations, 432
Blue-green algae, 531
Brackish water, 249

C

Calculating the risk estimation, 175
Carbohydrate estimation, 542
Carbon cycle, 324
Carbon dioxide evolution, 188
Carbon fixation and energy flow, 325
Carbon reservoirs, 324

Carbon respiration, 327
Cartridge filtration apparatus, 209
Catabolite repression of organic compound oxidation, 426
Catabolite repression, 426
Caveat emptor principle, 177
Cell culture-based detection methods for viruses, 186
Chemical analysis of biosurfactants, 538
Chemical structure of xenobiotics, 429
Chemistry of ozone, 159
Chlorinated benzenes, 506
Chlorinated ethenes, 504
Chlorinated herbicides, 508
Chlorinated phenols, 509
Choosing water sources and watershed protection, 145
Classification of subsurface environments, 393
Classification of water-related diseases, 77
Closed-vessel systems, 166
CO_2 uptake by gas analysis, 264
Commensalism, 386
Commonly used indicators, 96
Community-level physiological profile (CLPP), 48
Competitive ELISA, 196
Complete biodegradation, 485
Composition of a patent, 569
Constituent sugars of bioabsorbent, 558
Control of gas chlorination, 152
Conventional methods for the detection of bacteria in
    shellfish, 136
Conventional methods for the detection of human
    protozoan parasites in shellfish, 137
Conventional process configurations, 562
Core seepage contamination, 419
Core-fracture contamination, 419
Correlation between flocculant production and
    morphological changes in rhodococcus
    erythropolis S–1, 554
Cost of a patent, 573
Creation of bacterium living in petroleum, 559
Cross-acclimation of xenobiotic compounds, 429
Crown gall disease–agrobacterium tumefaciens, 472
Cultivation of microeukaryotes, 46
Cultivation, enrichment and isolation, 350
Cultivation, enrichment, enumeration and isolation, 343
Cultural media for bacteria, 182
Cultural methods for algae and cyanobacteria, 185
Cultural methods for enumeration of bacteria, 179
Cultural methods for fungi, 185
Culture broth for production of bioabsorbent, 558
Culture conditions for production of microbial flocculant
    by Rhodococcus erythropolis, 553
Culture media, 8

Culture techniques, 407
Culture, 479
Culture-dependent methods, 47
Culturing, 290
Current regulations regarding drinking and
    surface waters, 102

D

Deep unsaturated zone, 402
Defining the exposure level, 170
Defining the main variables incorporated in risk estimates
    for the level of infectious disease associated with
    ingestion of water, 170
Defining the probability of death, 173
Defining the probability of illness, 173
Defining the probability of infection, 170
Depth and seasonal distributions, 310
Describing protistan community structure, 301
Detection and quantitation, 114
Detection of infectious agents in water, 92
Detection of viruses, 212
Digestion of pre-heated feed and use of biostimulants, 566
Digestive enzyme activity approach, 317
Direct gas feed, 156
Direct physico-chemical flocculation and phase
    separation, 207
Direct probing, 229
Direct push sampling, 413
Directed adsorption, 206
Directed adsorption, elution and reconcentration of
    viruses, 207
Diseases caused by bacteria, 472
Disinfection with chlorine dioxide, 157
Disinfection with ozone, 159
Disinfection with ultraviolet light, 162
Dissimilatory nitrate reduction and denitrification, 335
Dissolved substances, 165
Distribution of bacteria in intermediate flow systems, 405
Distribution of bacteria in local flow systems, 404
Distribution of bacteria in regional flow systems, 407
Distribution of bacteria in the unsaturated zone, 401
Diversity of habitats exploited for growth of
    prokaryotes, 65
Diversity: insights from macromolecular analyses, 62
DNA-DNA hybridisation, 291
Down-hole saturation contamination, 418
Downward molecular diffusion, 445
Drilling fluid 415
Drilling fluid technology and coring, 418
Drilling-fluid contamination of cored sediments, 415, 418
Drinking waters, 111

E

Early observations from petroleum reservoirs, 407
Ecological and biogeochemical significance, 262
Effect of carbon source, 501
Effect of glucose concentration, 498
Effect of nitrogen concentration on phenol degradation, 497
Electrochemical techniques, 273
Electron acceptor, 187
Emerging protozoans of potential health significance
    in water, 128
Emerging waterborne pathogens, 85
Endemic waterborne disease, 80
Enlarging perspectives, 66
Enrichment and isolation, 222
Enteric and aquatic bacteria, 81, 86
Enteric bacteria, 132
Enteric protozoa, 84
Enteric viruses transmitted by bivalve molluscs, 132
Enteric viruses, 83, 85
Enterococci and fecal streptococci, 106
Enteropathogenic protozoa, 134
Enumeration and isolation techniques, 179
Environmental determinants that govern
    extreme environments, 255
Environmental factors affecting biodegradation, 487
Environmental microbiology, 20
Environmental water, drinking water and sewage
    effluent, 213
Environmental friendly process (green process), 559
Environmental friendly products (green products), 553
Enzyme assays, 189
Enzyme-linked immunosorbent assays, 196, 231
Enzymes of bacteria, 7
Enzyme-specific tests, 108
Epidemic waterborne disease, 80
Equipment configurations, 166
Equipment used for mud-rotary coring, 417
Escherichia coli, 97
Estimation of biomass, 541
Etiology of mollusc-associated outbreaks of illness, 131
Eucarya, 13
Evaluating drilling fluid contamination, 419
Expanded bed/fluidised bed reactors, 565
Experiments to determine empirical conversion factors for
    the TdR and Leu methods, 286
Extramural aeromicrobiology, 453
Exudates, secretions and lysates, 468

F

Factors affecting bacteria, 9
Factors affecting microbial acclimation, 428

Factors affecting the release of compounds, 469
Failure of seed to germinate, 525
Failure of seedlings to become nodulated, 526
FC/FS ratio, 101
Fecal streptococci and enterococci, 97
Feeding sodium hypochlorite, 155
Field experiments (usage), 533
Filtration and deposition, 450
Fluorescent antibodies (FA) techniques, 231
Fluorescent immunolabelling, 195
Format of the equations, 175
Free-living protozoa, 114
Freshwater environments, 243
Freshwater versus marine ecosystems, 309
Function of root border cells as biocontrol agents, 474
Fungi, 19

G

Gaia hypothesis, 322
Gene probes and probing, 200
General media used for culturing bacteria, 182
General steps in the isolation and detection of pathogens
    from shellfish, 134
Generation of chlorine dioxide, 157
Genetic mutations, 427
Geochemical methods, 381, 408
Germ warfare, 455
Germicides and bacteriostatic agents, 10
Giant marine bacteria, 258
Glycolipids, 536
Groundwater, 146
Group B rotavirus, 91

H

H_2S producers, 100
Hand augering, 411
HAV, 133
Hepatitis E virus, 90
Hexosamines, 538
High solute, 256
HPC, 101
Human caliciviruses, 133
Human diseases associated with water, 69
Hunters and the hunted, 31
Hydrocarbon-degrading enzymes, 425

I

Immobilised packed bed reactors, 563
Immunoaffinity chromatography assays, 198
Immunoassays, 194
Immunocytochemical assays, 198

Immunological approaches, 290
Immunological methods, 192
Immunomagnetic separation assays, 197
Immunoprecipitation assays, 198
Impaction, 448
Impingement, 447
Important airborne pathogens, 437
Important airborne toxins, 440
Incorporation of radio-labelled tracers into cellular
    macromolecules, 189
Indoor air modelling, 446
Influence of dispersion, 165
Influence of water quality, 165
Inner cell mass, 188
Intermediate flow systems as a microbial habitat, 405
Intermediate flow systems, 404
Intermediate unsaturated zone, 401
Introduced biological control agents, 474
Involvement and homeostasis are evolutionary
    consequences, 28
Isolation of the biosurfactant, 541
Isotopic versus nonisotopic techniques, 228

K

Kinetics of disinfection, 150

L

Lakes, 244
Large-seeded legumes, 523
Leucine Incorporation into protein, 285
Limitations in comprehension of prokaryote diversity, 52
Limitations of conventional digestion and process
    improvement approaches, 563
Lipid analysis, 291
Lipid estimation, 543
Lipopeptides and lipoproteins, 537
Local flow systems as a microbial habitat, 402
Local flow systems, 402
Low pH, 257
Low-molecular-weight RNA profiles, 291

M

M. tuberculosis complex, 475
Maintenance of suppressive soils, 473
Managing combined chlorine (chloramine) residuals, 156
Manipulation methods, 314
Manufacture of Rhizobium inoculants, 517
Marine water, 250
Mathematical modelling, 446
Measurement of diversity, 383
Measurement of emulsification activity, 544

Measurement of phosphorus uptake and incorporation by
    using ^{32}P and differential filtration, 354
Measurement of respiration gases (CO_2 and O_2), 188
Measurements of microbiological water quality, 92
Mechanisms involved in viral concentration methods, 206
Mechanisms of acclimation, 424
Media used for rhizobium culture, 534
Membrane filter method, 13
Metabolism, 7
Metal binding, 431
Metal deposition, 432
Metal detoxification mechanisms, 431
Method of application of blue-green algae, 531
Method of controlling bioaerosols, 456
Methods for measuring primary production in aquatic
    habitats, 264
Methods for the detection of pathogens in shellfish, 134
Methods in subsurface microbial ecology, 376
Methods of application of nitragin (Rhizobium
    inoculants), 523
Methods of disinfection commonly used in water
    treatment, 149
Microautoradiography, 277
Microbial activity measurements in environmental
    samples, 188
Microbial activity measurements in pure culture, 187
Microbial agents associated with waterborne disease, 81
Microbial degradation of halogenated organic
    compounds, 504
Microbial diversity and niches in aquifer systems, 383
Microbial habitats in the aquatic environment, 235
Microbial indicator organisms, 93
Microbial mats, 239
Microbial nitrogen cycling, 328
Microbial processes in confining beds, 405
Microbial response to environmental changes, 422
Microbial survival in the air, 451
Microbiological, chemical and molecular tools, 49
Microfauna, 470
Microflora, 469
Micro-organisms, 492
Microscopic assay, 478
Microscopic examination, 123
Microscopy, 46, 289
Moisture absorption capacity, 556
Moisture and gas content, 397
Moisture retention capacity, 556
Molecular approaches to studying protistan community
    structure, 311
Molecular biological methods, 227
Molecular methods for the detection of pathogens in
    shellfish, 137

Molecular methods, 110, 408
Molecular microbial ecology, 381
Molecular probes and detection techniques, 228
Molecular techniques for density and diversity measurement, 348
Molecular techniques, 480
Morphologies and sizes of prokaryotic cells, 55
Morphology, 6
Most probable number (MPN), 12
Most probable number technique, 181
MPN (multiple-tube technique), 103
Mud-rotary coring, 417
Multiple fermentation tube method, 12
Multiple-barrier approach, 93
Mycobacteria other than M. tuberculosis, 477
Mycobacteria, 99

N

N₂ fixation, 328
Nature and production of extra-cellular substances, 532
Nature of bioaerosols, 441
Nature of the hazards associated with pathogenic micro-organisms in aquatic environments, 169
Neighbourhood, cliques and cliches, 24
Neutralism, 385
New bioflocculant produced by Rhodococcus erythropolis, 553
Niches and sources of microbial diversity, 384
Nitrification, 332
Nitrogen concentration, 492
Nitrogen: preparation and usage, 515
Novel bioreactor designs, 563
Nucleic acid sequence analysis, 48
Nucleic acid-based methods of analysis, 199
Nucleic acid-based methods, 200
Nutritional deficiencies, 527

O

Organic compounds released by plants, 467
Organic sulphur oxidation, 352
Organochlorine insecticides, 508
Other bacterial pathogens, 226
Other commonly used indicators, 99
Other DNA-based approaches, 292
Other enteric viruses, 134
Other trophic pathways in microbial food webs, 320
Oxygen source, 162
Oxygen, OAF and ions, 452
Ozone decay and ozone demand, 159

P

Parasitic protozoa, 115
Particulate matter, 165
Partial biodegradation, 485
Passive adsorption, 206
Patents in other countries, 574
Pathogenic bacteria transmitted by bivalve molluscs, 132
Pathogenic microbes in agriculture, 471
PCR conditions, 228
PCR methods for the detection of enteric viruses in shellfish, 138
PCR methods for the detection of human protozoan parasites in shellfish, 143
PCR methods for the detection of vibrio species in shellfish, 138
Performance of anaerobic digesters, 561
Phenol biodegradation, 493
Phosphobacterin, 533
Phospholipid fatty acid analysis, 378
Phospholipids, 537
Phosphorus cycling, 353
Photopigment radiolabelling, 276
Phototrophic processes, 352
Physiological diversity among the prokaryotes, 57
Physiological methods, 187
Pigment-based estimates of primary productivity, 275
Planktonic environment, 235
Plant disease caused by fungi, 471
Plant diseases caused by viruses, 472
Plasmid-encoded metal resistance mechanisms, 432
Plating methods, 180
Point source modelling, 446
Polychlorinated biphenyls, 507
Polyclonal and monoclonal antibodies, 194
Polymerase chain reaction, 201
Population interactions, 385
Possible changes in United States patent law, 574
Postgate's media, 343
Preparation of blue-green algae inoculants, 531
Preparation of inoculants of phosphate solubilising bacteria, 533
Presence-absence (P-A), 107
Prevention and control of waterborne disease, 92
Prey addition methods, 316
Primary production, 236
Primer selection, 227
Probe-microautoradiography combination, 297
Problem with interpretation of biochemical mass, 48
Problems in legume establishment, 525

Procedures for estimating BBP from incorporation rates of TdR, 284
Process improvement approaches, 563
Process of primary production and relevant micro-organisms, 261
Production of biosurfactants, 537
Production of proteins in response to chemical stresses, 427
Protection of the rights of the inventer: infringement, 573
Protein estimation, 541
Protistan abundance and biomass, 301
Protistan herbivory, 318
Protistan species diversity, 299
Protozoa, 19
Protozoa, fungi and algae, 88
Pseudomonas spp., 99
Pure culture approaches, 46
Pure culture, 47
Push-tube (Shelby tube) sampling methods, 413

Q

Quality control of legume inoculants, 520
Quantification, 229

R

r and K strategies in microbial ecosystems, 389
r and K strategies in the aquifer environment, 390
Radiotracer studies and isotope exchange, 350
Rain and electrostatic deposition, 446
Rapid method for monitoring maximum biosurfactant production obtained by acetone precipitation, 541
Rapid tests, 109
Rates of acclimation, 428
Rationale, advantages and disadvantages, 281, 285, 286
Raw waste-water, 217
Reactivation, 164
Recently recognised and suspected waterborne pathogens, 85
Recently recognised waterborne pathogens in developing countries, 90
Recognition of prokaryote diversity: background, 52
Recombinant DNA techniques, 202
Recovery of injured bacteria, 110
Reductive dechlorination, 505
Regional flow systems, 407
Relationship between growth and degradation of phenol, 501
Relative humidity, 451
Reproduction and growth of bacteria, 8
Research needs for the detection of pathogens in shellfish, 144

Residual control system, 152
Retention of supplied water in sand with the bioabsorbent, 558
Rhamnolipid concentration, 538
Rhizobium inoculants, 524
Rhizosphere environment, 465
Rhizosphere populations, 469
Risk assessment approaches, 94
Rivers and streams, 243
Role of blue-green algae in rice fields, 533
Role of the driller, 418
Root border cells and mucigel, 468
Routes by which human acquire waterborne infections, 70
Routes of pathogen acquisition, 170

S

Salmonella detection, 222
Salmonella isolation media, 223
Sample filtration and elution, 119
Sample preparation, 230
Sampling and analytical methods, 478
Sampling apparatuses and processing equipment, 209
Sampling considerations, 478
Sampling devices for the collection of bioaerosols, 447
Sampling for airborne mycobacteria, 478
Sampling groundwater for micro-organisms, 419
Sampling intermediate and regional systems, 415
    mud-rotary drilling, 415
Sampling local flow systems, 412
Sampling point, 153
Sampling requirements, transportation and preservation, 221
Sampling the unsaturated zone, 411
Scope of subsurface microbial ecology, 372
Screening for and characteristics of microbial flocculants, 553
Secondary production, 237
Selection of rhizobial strains, 525
Sensitivity to phenol, 492, 493
Significance of indicators to public health, 102
Soil + farmyard manure-based cultures, 520
Soil biological control of plant diseases, 473
Soil management practices, 473
Soil–plant–micro-organism system, 465
Sources of ultraviolet light, 164
Split-spoon sampling, 412
Springs, 243
Staphylococci, 98
Statistical techniques, 49
Staying at home versus venturing to new places, 34
Sterilisation requirements, 211

Sterols and other chemical indicators of sewage contamination, 101
Storage and feeding of ammonium sulphate, 156
Storage and feeding of anhydrous ammonia, 156
Stress and microbial diversity, 384
Studies of waterborne disease, 80
Subject matter and characteristics of a patent, 569
Substrate disappearance, 187
Subsurface microbiology and contaminant biodegradation, 368
Subsurface microbiology and microbial ecology, 366
Subsurface microbiology and the geosciences, 366
Subterranean water, 255
Sulphite-reducing clostridia, 99
Sulphur cycling, 339
Sulphur oxidation, 349
Surface freshwaters, 111
Surface impaction, 445
Surface tension measurement, 545
Surface water sources, 145
Synergism and symbiosis, 386

T

Techniques for field sterilisation of sampling equipment, 211
Temporal and spatial changes in community structure, 309
Thermotolerant coliforms (fecal coliforms), 96, 106
Thymidine (TdR) incorporation into DNA, 281
Total coliforms, 96, 106
Traditional methods for the detection and enumeration of bacterial pathogens, 221
Treatment and disinfection, 148
Trophic interactions among aquatic microbes, 312

Two phase-digestion— an innovative fermentation mode, 566
Types of biosurfactants, 536

U

Ultrafiltration, 206
Ultraviolet light, 163
Understanding ecology, 41
Understanding the nature of community life, 27
Unsaturated zone as a microbial habitat, 397
Unsaturated zone, 395
Upflow anaerobic sludge blanket (UASB) reactor, 565
Uptake of ^{14}C-labelled CO_2, 265
Uronic acid, 538

V

Validation of the risk estimation technique, 176
Ventilation, 456
Vibrios, 132
Viral assay techniques, 219
Viral concentration techniques, 213
Viral ecology, 17
Viruses, 16, 47

W

Waste disposal, 454
Waste-water sludge, 218
Water absorption capacity, 556
Water absorption capacity in the presence of NaCl, 556
Waterborne disease of unknown etiology, 91
Watershed protection, 146
Western immunoblotting assays, 197
Widdel and Pfennig medium, 344
Winkler method, 271